OIL IN THE SEA
INPUTS, FATES, AND EFFECTS

Steering Committee for the Petroleum in the Marine Environment Update
Board on Ocean Science and Policy
Ocean Sciences Board
Commission on Physical Sciences, Mathematics, and Resources
National Research Council

NATIONAL ACADEMY PRESS
Washington, D.C. 1985

NATIONAL ACADEMY PRESS   2101 Constitution Avenue, NW   Washington, DC 20418

Library of Congress Catalog Card Number 85-60541

International Standard Book Number 0-309-03479-5

Printed in the United States of America

Cover Photograph: Courtesy, Exxon Corporation and American Petroleum Institute Photographic and Film Services.

Ocean Sciences Board

Board on Ocean Science and Policy

Commission on Physical Sciences, Mathematics, and Resources

Study Contributors

Steering Committee for the Petroleum in the Marine Environment Update

GORDON A. RILEY, Bedford Institute of Oceanography, Canada (Retired), *Cochairman*
WILLIAM M. SACKETT, University of South Florida, *Cochairman*
RITA R. COLWELL, University of Maryland
JOHN W. FARRINGTON, Woods Hole Oceanographic Institution
C. BRUCE KOONS, Exxon Production Research Company
JOHN H. VANDERMEULEN, Bedford Institute of Oceanography, Canada

Workshop and Section Area Chairmen

Inputs, C. BRUCE KOONS
Methods, RITA R. COLWELL and JOHN W. FARRINGTON
Microbial Studies, G. D. FLOODGATE
Fates, WILLIAM M. SACKETT and GORDON A. RILEY
Biological Processes, RICHARD F. LEE
Effects, JOHN H. VANDERMEULEN
Processes, CARL SINDERMAN
Food Chain, DONALD W. WESTLAKE
Ecosystems, Seeps and Spill Case Histories, Plankton, GORDON A. RILEY and
    JOHN W. FARRINGTON
Benthos, DONALD F. BOESCH

Workshop Participants

Thomas Albert, The North Slope Borrough
R. C. Allred, Conoco, Inc.
*Jack W. Anderson, Battelle Pacific
  Northwest
Karl Banse, University of Washington
*Richard Bartha, Rutgers University
*Wayne Bell, Hamilton College
*William L. Berry, Shell Oil Company
Norman J. Blake, University of South Florida
*Paul Boehm, Energy Resources Company

*Donald F. Boesch, Louisiana Universities
  Marine Consortium
*James M. Brooks, Texas A&M University
R. G. B. Brown, Canadian Wildlife Service
*James N. Butler, Harvard University
John Calder, NOAA Office of Marine
  Pollution

———————————

*Contributed background paper.

vi

G. P. Canevari, Exxon Research and Engineering Company

Judith M. Capuzzo, Woods Hole Oceanographic Institution

D. W. Chamberlain, Atlantic Richfield Company

*R. B. Clark, The University Newcastle-upon-Tyne, England

*Robert C. Clark, Jr., NOAA National Marine Fisheries Service

Rita R. Colwell, University of Maryland

Bruce C. Coull, University of South Carolina

Elmer Danenberger, U.S. Geological Survey

*Robert A. Duce, University of Rhode Island

*F. Rainer Englehardt, Northern Environmental Protection, Canada

John W. Farrington, Woods Hole Oceanographic Institution

G. D. Floodgate, University College of North Wales

H. I. Fuller, Rookery Farm, United Kingdom

*Robert B. Gagosian, Woods Hole Oceanographic Institution

William D. Garrett, Naval Research Laboratory

Richard A. Geyer, Texas A&M University

*Edward S. Gilfillan, Bowdoin College

Jack R. Gould, American Petroleum Institute

Otto Grahl-Nielson, University of Bergen, Norway

Judith Grassle, Woods Hole Oceanographic Institution

William O. Gray, Exxon Corporation

George Harvey, NOAA, Atlantic Marine Laboratory

*Carl Hershner, Virginia Institute of Marine Science

*Robert W. Howarth, The Ecosystems Center

Thomas Hruby, Massachusetts Audubon Society

Walter Japp, Florida Department of Natural Resources

James Kittredge, University of Southern California

C. Bruce Koons, Exxon Production Research Company

*Keith Kvenvolden, U.S. Geological Survey

Paul LaRock, Florida State University

*John Laseter, University of New Orleans

*Richard F. Lee, Skidaway Institute of Oceanography

Olof Linden, IVL, Studsvik, Sweden

*Arlene Crosby Longwell, NOAA National Marine Fisheries Service

Yossi Loya, Tel Aviv University, Israel

*Donald MacKay, University of Toronto

*Donald C. Malins, NOAA National Marine Fisheries Service

James P. Marum, Mobil Oil Corporation

Leo T. McCarthy, U.S. Environmental Protection Agency

Richard Y. Morita, Oregon State University

*Jerry M. Neff, Battelle New England

*Bori Olla, NOAA National Marine Fisheries Service

*Candace Oviatt, University of Rhode Island

*E. H. Owens, Woodward-Clyde Consultants

*James R. Payne, Science Applications, Inc.

David Peakall, Canadian Wildlife Service

*Erman A. Pearson, University of California at Berkeley (Retired)

William Pequegnat, College Station, Texas

*Jon A. Percy, Arctic Biological Station

Fred Piltz, U.S. Department of the Interior

*James G. Quinn, University of Rhode Island

*James P. Ray, Shell Oil Company

*Stanley Rice, NOAA National Marine Fisheries Service

Gordon A. Riley, Bedford Institute of Oceanography, Canada (Retired)

*Cal W. Ross, Mobil Oil Canada, Ltd.

William M. Sackett, University of South Florida

Alain Saliot, Université Pierre et Marie Curie, France

*Y. S. Sasamura, International Maritime Organization, England

Ted Sauer, Exxon Production Research Company

---

*Contributed background paper.

vii

*David G. Shaw, University of Alaska

Carl Sinderman, NOAA National Marine Fisheries Service

June Lindstedt-Siva, Atlantic Richfield Company

*Woollcott K. Smith, Woods Hole Oceanographic Institution

Norman B. Snow, Petro Canada

*Robert B. Spies, University of California

John H. Steele, Woods Hole Oceanographic Institution

*John J. Stegeman, Woods Hole Oceanographic Institution

Dale Straughan, University of Southern California

*John M. Teal, Woods Hole Oceanographic Institution

Howard Teas, University of Miami

Sunniva Lonning Vader, University of Tromso, Norway

*Chase Van Baalen, University of Texas

John H. Vandermeulen, Bedford Institute of Oceanography, Canada

Edward S. Van Vleet, University of South Florida

*Gabriel Vargo, University of South Florida

*Sandra Vargo, Florida Institute of Oceanography

Richard C. Vetter, National Research Council

Jan Vorhees, University of South Florida

*John D. Walker, U.S. Environmental Protection Agency

Neill Weaver, American Petroleum Institute

*Peter G. Wells, University of Toronto, Canada

*D. W. S. Westlake, University of Alberta, Canada

R. B. Wheeler, Exxon Production Research Company

Mary Wolff, City University of New York

*Oliver Zafiriou, Woods Hole Oceanographic Institution

---

*Contributed background paper.

Preface

The 1975 National Research Council (NRC) report, *Petroleum in the Marine Environment,* has proven to be an extremely important document. It has been used as a primary source by individuals and groups ranging from scientific investigators to concerned laymen. However, in mid-1980 it became clear that an update of the 1975 report was necessary. Much of the published material used as a basis for the earlier report predates a workshop held in 1973 that provided most of the background for the 1975 report. Since then, significant new data and information have been published. Thus, the U.S. Coast Guard requested that the Ocean Sciences Board (OSB) (now the Board on Ocean Science and Policy) undertake a new examination of this subject. The OSB appointed a steering committee consisting of cochairmen Gordon A. Riley, Halifax, Nova Scotia, and William M. Sackett, University of South Florida, along with Rita R. Colwell, University of Maryland; John W. Farrington, Woods Hole Oceanographic Institution; C. Bruce Koons, Exxon Production Research Company; and John H. Vandermeulen, Bedford Institute of Oceanography. Later, the National Oceanic and Atmospheric Administration, the Environmental Protection Agency, the Bureau of Land Management (now the Minerals Management Service), Mobil, Exxon, and the Andrew W. Mellon Foundation joined the U.S. Coast Guard in providing financial support for the project.

The steering committee took the following major steps.

1. A public meeting was held on November 13, 1980, at which representatives from oil industry, university, government, and environmental groups were invited to make presentations on important topics for consideration by the steering committee.

2. In February 1981, 46 expert contributors were invited to prepare summary papers on all aspects of petroleum in the oceans. These were reviewed and commented on by other experts selected by the steering committee.

3. An international workshop was held November 9-13, 1981, where contributors, reviewers, and other invited scientists came together to discuss the main issues brought out from the previous two steps and to make recommendations concerning future research needs. Approximately 90 of the participants came from U.S. university, governmental, and industrial organizations. Another 22 came from Canada, the United Kingdom, France, Germany, Norway, Israel, and Sweden, providing a strong expert background and a wide range of institutional and foreign governmental expertise to this new report.

4. In February 1982 the steering committee began the task of preparing the new report, based on the input, ideas, and comments obtained by the previous steps. The writing process involved several review steps. Drafts from these iterations were carefully reviewed at several meetings of the entire steering committee. The review process was completed in November 1984.

Contents

List of Figures

List of Tables

OIL IN THE SEA
INPUTS, FATES, AND EFFECTS

Introduction

This report follows up and expands on the earlier National Research
Council (NRC) report <u>Petroleum in the Marine Environment</u> published in
1975. As noted in the preface to the present publication, much of the
earlier report was based on studies and findings predating a 1973
workshop that formed a basis for the earlier study. Since then, much
research has provided further insight into the problems of dealing with
petroleum in the world's oceans. The new interpretations and data
bases that have developed in the intervening years are summarized in
the present document. The report is the result of an intensive and
multidisciplinary effort and, to the best of our ability, represents a
fair assessment of the problems relating to petroleum-derived hydrocarbons in the marine environment.

This report is an extension of the earlier report in several ways.
It follows the same basic format in that the major chapters deal with
inputs, analytical methods, fates, and effects. However, all chapters
are larger than those in the 1975 report, the significantly expanded
chapter on effects reflecting the extensive scientific effort in recent
years that has gone into determining effects of petroleum on marine
organisms. To this purpose, extensive lists of references are provided
at the ends of the chapters for readers who wish more detailed information on various subjects. We have also included discussions of
petroleum contamination in such geographic areas as the polar environment and the mangrove and coral reef systems of tropical waters--areas
not considered in the earlier report. In many instances we have gone
into considerable detail, especially in those disciplines where new
work appears promising or new ideas have added to our understanding of
the hazards of petroleum contamination. Finally, where necessary, we
have taken extra space for discussions of controversial subjects such
as the potential hazards to human health, the impact of oil-related
activities in the northern Gulf of Mexico, and the potential impact of
petroleum on fisheries.

To avoid the space-consuming repetition of scientific and organiza-
tional names, abbreviations are used extensively throughout the report.
To aid the reader, a list of abbreviated terms is given in Appendix B.

Petroleum is a naturally occurring substance composed of a highly
complex and variable mixture of hydrocarbons with minor amounts of
compounds containing nitrogen, sulfur, and oxygen atoms in their mole-

1

cules. For the purposes of this report, the term "petroleum" will be used generically; i.e., it will include, in large part, the commercially produced materials ranging from methane-rich natural gases to extremely viscous heavy oils. It will also include refined products such as gasoline, kerosene, and lubricating oil, but not commercially transformed products such as halogenated hydrocarbons, ethylene glycol, formaldehyde, and phthalic acid.

Because many of the compounds found in petroleum can also originate from sources other than petroleum--for example, hydrocarbons from forest and grass fires, and from combustion of fossil fuels (e.g., coal, fuel oil, and gas)--we caution that it was often difficult to separate these types of sources of hydrocarbons in the marine environment. We have attempted to address this issue explicitly in appropriate sections of this report.

It would have been very satisfying if we could have found answers to all the questions relating to petroleum in the marine environment. In fact, progress has been made in many areas, and the result is better understanding and more credible evaluations of potential hazards. However, uncertainties about various physical, chemical, and biological processes in the ocean and the complexity and variety of petroleum types and products contribute to doubts that remain regarding the various inputs, fates, and effects of petroleum in the marine environment.

Special mention should be made of two relatively recent parallel studies that culminated in published reports while this report was being prepared. One was prepared by the Marine Pollution Subcommittee of the British National Committee on Oceanic Research for the Royal Society in 1980 and entitled The Effects of Oil Pollution: Some Research Needs. The other is Oil Pollution of the Sea, prepared for the Royal Commission on Environmental Pollution in 1981. These reports appeared to be designed primarily to help assess the relative importance of oil pollution research within the context of other environmental pollution and research priorities as applied to the United Kingdom, although information of generic use was assembled in these reports. The focus of the present report is to provide an independent assessment of the knowledge and state-of-the-art research concerned with petroleum in the marine environment. We gained this assessment by convening a public hearing, commissioning review papers by 46 experts, seeking extensive peer review of those papers, and convening a large workshop, followed by repeated distillations, summations, and evaluations of this large body of information by the six-member steering committee and additional peer review at key stages in the final document development.

Executive Summary

OVERVIEW

The study of petroleum pollution in the ocean deals with two potentially
opposing aspects of man's activities: on the one hand is pollution
arising from activities undertaken to meet man's needs--the extraction,
transport, and use of petroleum for energy and chemical feedstocks--and
on the other hand is the strong desire to preserve living marine
resources both for current uses and for a legacy for future generations.

In this examination of marine petroleum pollution we recognized
this duality and attempted to examine it from a broad and, at times,
somewhat distant perspective, without bias whenever possible, so as to
avoid pitfalls of misinterpretation. Petroleum is a naturally occurring
substance, derived from organic materials once living but since transformed into a complex mixture of chemicals, consisting mainly of
hydrocarbons and small amounts of other organic compounds. A small
amount of petroleum has seeped into the world's oceans for at least
centuries and probably millions of years, and portions of the oceans
have accommodated long-term influx of some petroleum into their
communities and ecosystems.

The modern influx of petroleum into the marine environment is on a
different scale, occurring more rapidly and over a wider area, and
probably is of a different kind. The product entering the oceans
today, both from chronic effluent release and runoff and from sudden
catastrophic spills, represents a sudden and significant input of
contaminants when viewed against the much longer, but much lower,
continuous presence of seepage petroleum. Also, the chemical composition of this modern petroleum input often differs from that of the
seepage oil, the latter being altered by the degradation processes,
both physical/chemical and microbial, occurring in the marine sediments
and crustal layers. In this context it must be noted that inputs of
petroleum are not the only sources for many of the compounds of
concern. For example, combustion of coal yields several polynuclear
aromatic hydrocarbons similar to or also found in petroleum. Finally,
modern petroleum input to the oceans is no longer restricted to the
seep locations but now includes many waters formerly held unpolluted
and pristine. Even those areas themselves free of oil exploration and

production activity are nonetheless subject to potential pollution resulting from petroleum tanker traffic.

Our mandate from the Ocean Sciences Board (now the Board on Ocean Science and Policy) was to review the accomplishments since the 1975 NRC Report, <u>Petroleum in the Marine Environment</u>, to arrive at conclusions on the basis of our newer understanding of the behavior and fate of petroleum in the marine environment, and in the end to make recommendations concerning possible further research. We recognize potential environmental problems requiring further study as well as areas where much less concern is required, either as a result of new findings or because investigations are essentially completed.

Inevitably, the potential impact of petroleum as part of, or together with, other contaminants in the marine environment was considered. While in some instances petroleum itself is readily seen as the identifiable pollutant, as for example, in tanker spills or in known cases of chronic petroleum pollution, there are many regions where petroleum hydrocarbons are thought to form part of a more general pollution threat to the health of those environments. Waters near or receiving the effluent of urban and industrial regions serve here as primary examples.

GENERAL ADVANCES: 1973-1983

Progress in oil-pollution-related research during this past decade has been impressive. Knowledge and understanding of its problems have come about in each of the areas identified in the 1975 NRC report. Most significant of the advances in these areas are the reduction in the uncertainties regarding the rates of input and amounts of marine petroleum pollution, the increasing sophistication of the analytical methodology applied to chemical and biological studies, clearer identification of the various processes acting on petroleum in the oceans, and the clear identification of problem areas in the effects of petroleum on biota.

Inputs

There is now a better understanding of the data base with respect to input of petroleum into the world's oceans, especially for urban runoff, which is a major source. Progress has been made also in the design and implementation of procedures for measurement of atmospheric inputs of petroleum hydrocarbons to the marine environment, although more data are needed. In addition, over the past 8 years there has been a better definition of the various sources of input, which has led to the recognition and elimination of the problem of double bookkeeping, i.e., including a particular source in more than one category in previous estimates. These advances came as a result of the discussion and deliberations of the 1973 workshop, the recommendations from which were carried forward to the present effort.

Methods

Underlying the progress made in the area of inputs are the improvements in chemical methods. Several new sampling methods and devices, including specialized water samplers that avoid contamination from surface film, have been developed and have led to greater confidence in subsequent data. The analysis of petroleum hydrocarbons, pyrogenic hydrocarbons derived from the atmosphere and resulting from incomplete combustion of oil, coal, wood and gas, products of metabolic alteration of petroleum by organisms, and products of chemical and biochemical transformation after oil enters the sea (e.g., metabolites and photo-chemical oxidation products), has made great strides, especially with the parallel development of glass capillary gas chromatography, high-pressure liquid chromatography, and glass capillary gas chromatography/mass spectrometry computer systems. As a result, there exists today an improved capability to analyze a wide range of petroleum compounds, including volatile components, in water, sediments, and organisms, that was not possible or was achieved only with difficulty at the time of the first workshop (1973). The intercomparability of data from different laboratories is more reliable as a result of several quality control exercises and workshops for intercalibration.

Fate

Describing the fate of petroleum in the marine environment is possible in a way that was not conceivable in 1973. Methods of modeling are now approaching the level of sophistication where they are of potential use in spill impact prediction. Much more is now known of the various environmental processes acting on petroleum, and much has been added to our understanding of the properties and factors relevant to the fate and effects of oil. For example, whereas the process of photochemical oxidation was only recognized at the time of the first report, today it is being actively researched.

A major advance since the last report is the interactive biological/chemical approach to studying petroleum in the marine environment, i.e., recognition of the need of these two disciplines to work intimately on this problem. Indeed it is rare to find studies involving one without the other. As a direct result, the role of biological alteration of petroleum is much better understood. Such processes as microbial degradation are now recognized as significant to the fate of petroleum in the marine environment, and work on establishing rates of microbial degradation for different environments is now being done. The broad outlines of metabolic pathways of petroleum degradation are being developed for bacteria, phytoplankton, and higher animals, and con-siderable work is being carried out on specialized detoxification mechanisms where these are known. It has also been shown that animals exposed to sublethal dosages take up hydrocarbons, but in most cases are able gradually to alter the various petroleum components, in some cases by conversion to compounds that are more soluble and more readily excreted, although a few of these are themselves highly toxic. Hence, accumulation by itself, under conditions of low concentrations, may not be severe. Indeed there is very little evidence of increased accumulation in the higher predatory members of the food web. In terms of human health the available data do not indicate that consumption of oil-polluted seafood is a widespread problem, although we caution against complacency in the application of broad generalizations to individual local situations.

Effects

Studies on the biological effects of petroleum have benefited greatly from improvements in experimental study design, from the application of modeling made available by computer technology, and from the development of specialized experimental equipment. A major step forward has been the move away from lethal studies to sublethal studies, using specialized flow-through systems and the concomitant measurement of petroleum composition and concentration.

In terms of the measurement and evaluation of the impact of petroleum, advances have been made along several lines of study--cellular, organismic, population, and ecosystem. Much of this advancement is due to a better understanding of the interaction of petroleum or of its components with water, sediment, and tissues and to the specialized analytical procedures developed over the last 8 years. These in turn have led to a better predictive ability to anticipate and evaluate the potential impact of oiling, whether acute (as from spills) or chronic.

Extensive studies on a variety of marine organisms have been carried out on the toxic effects of petroleum and on selected individual components. These include laboratory experiments and observations in the field made after oil spills. Significant differences in the tolerance of individual species and of different life stages in a given species have also been recorded, so that a better understanding of possible impact is available.

FINDINGS AND RECOMMENDATIONS

Major Findings

The authors of this report conclude, based upon the evidence available, that there has been no evident irrevocable damage to marine resources on a broad oceanic scale, by either chronic inputs or occasional major oil spills. However, specific information that would enable unequivocal assessment of the impact of oil on the environment does not yet exist, particularly, with regard to certain specific environments and conditions. We, therefore, recommend further research in a number of areas as discussed below.

Despite the considerable advances made in the past 10 years, as detailed in the preceding pages, a number of significant questions remain unanswered. Frequently, the evidence on which conclusions regarding petroleum impact are based is either circumstantial or insufficient. The former may occur, as in the situation where a conclusion regarding the fate or effect of a given hydrocarbon compound has to be based on data obtained for a similar or related compound, because the relevant information is not available. In some cases, impact on one marine species must be inferred from observations made on a different but perhaps related species. The evidence is often insufficient, as when data are available for only a single life-cycle stage in an organism that normally has several such stages, each with potentially differing susceptibilities. On the other hand, assembling information on each and every organism or life-cycle stage would require an extreme effort that might not be justifiable. It therefore appears to us that more data on a number of select species would be preferable to a large amount of scattered and possibly nonintegrated data on several species.

In reviewing the past 10 years, it is probably fair to say that these years have yielded a much improved understanding of the general impact of petroleum on the marine environment. Furthermore, we conclude on the evidence available that there has been no evident irrevocable damage to marine resources on a broad oceanic scale, by either chronic inputs or occasional major oil spills. Lacking, however, is the specific understanding relating to details that would enable unequivocal assessment of such impact. That understanding does not yet exist. We therefore recommend further research in a number of areas as discussed below. This recommendation stems from the many findings that very low levels of petroleum, below 0.1 mg/L, can affect such delicate biological entities as fish larvae.

Continued research will result in a much improved capability to predict and assess the impact of both sudden and chronic petroleum pollution, especially in those coastal regions where oil exploration and production activities coincide with harvestable marine resources.

Such further research should also lead to a better understanding of the apparent link between the polycyclic aromatic hydrocarbons, some of which may be coming from pyrolytic sources, and the more general problem of what seem to be pollution-related diseases found in commercially important fish stocks in waters receiving a mixture of contaminants.

Input of Petroleum

Major Finding

The estimated range for the total input of petroleum from all sources is between 1.7 and 8.8 million metric tons per annum (mta), with a best estimate of 3.2 mta.

We believe that this range, rather than a single-number estimate, is a more accurate summary of the state of knowledge, and reflects the uncertainties that exist in all source data. Calculations of the total input of petroleum are complex, for there are many sources, and in many cases those data that are available are minimal for the purpose. There are also wide geographical gaps in information on sources, particularly in the southern hemisphere and tropics.

This range of 1.7-8.8 mta includes the single-number estimate of 6.1 mta made in the 1975 NRC report. The difference (decrease) between that earlier estimate and the current best estimate of 3.2 mta does not necessarily represent a decline in annual input of petroleum hydrocarbons into the marine environment during this period but indicates a better estimating of individual inputs.

Assessment of the available data on inputs and sources of petroleum entering the marine environment confirms the earlier conclusions of the 1975 NRC report that a considerable portion of marine petroleum pollution is due to non-gas- and oil-related activities, and originates from other human activities (see Table 2.22). These include river and terrestrial runoff from municipal, urban, and industrial sources as well as from seeps and through atmospheric transport. A significant source of petroleum pollution originates from bilge cleaning.

If we are concerned over the continuing health of our oceans and if we are to develop a better understanding of the cycling and transport processes of contaminants in the marine environment, then we suggest the following steps to narrow these existing uncertainties in the inputs of oil.

1. Improved documentation of continental margins to determine the extent of submarine seepages, and to more accurately gauge their flow rates.
2. Continued monitoring of all facilities discharging petroleum hydrocarbon containing effluent.
3. Better methods for distinguishing petroleum hydrocarbons from the "oil and grease" and naturally generated hydrocarbons currently applied to municipal and industrial effluents.
4. A better accounting of petroleum inputs for the southern hemisphere, where currently there is great uncertainty as to amounts and rates of input to the oceans.

Major Recommendation

It is recommended that atmospheric transport of petroleum hydrocarbons, particularly by rainwater, be given priority status for research.

Rain scavenging of atmospheric particles is commonly thought to be the major pathway for petroleum into the oceans from the atmosphere. Unfortunately, today there exists little or no scientific evidence or information on this potentially significant pathway. Concurrent with this is the need to determine the various processes and reactions affecting these compounds as they are transported from sources through the atmosphere into the oceans.

Study Methods: Chemical

Major Finding

Marked progress has been made in the application and development of analytical methods to quantify petroleum components in air, water, sediments, and biota of the sea and to distinguish these components from other sources of hydrocarbons.

While the purpose and main thrust of this report are an examination of petroleum in the seas, reaction products and other sources of hydrocarbons also require attention in order to provide more accurate and meaningful measurements in support of research and monitoring on inputs, fates, and effects of petroleum. These products and other sources include (1) products of metabolic alteration of petroleum by marine organisms; (2) products of chemical and photochemical alteration after oil enters the oceans; and (3) pyrogenic hydrocarbons produced during combustion of oil, coal, and other carbonaceous materials. Some of these compounds are toxic or biologically active in ways that may be deleterious to natural resource populations or to man.

At the time of the 1975 report, existing analytical methods were inadequate to deal effectively with this problem, but marked progress has been made in the intervening years. Thus it is the general consensus that at the present time, chemical methods are available to make useful measurements on metabolites, reaction products, and nonhydrocarbon constituents of petroleum.

Study Methods: Biological

Major Finding

Techniques for experimental exposure of organisms to petroleum have advanced significantly. The use of controlled environment systems (mesocosms) in particular has been a significant step toward understanding the impact of petroleum on communities.

More experiments need to be conducted in the field to coordinate laboratory and field components and to validate laboratory findings. Biological interactions of different species are complex, and the presence of degradation products of petroleum in the natural environment adds to the complexity, creating a gap between field-work observations and laboratory results.

Physical/Chemical Fate of Petroleum

Major Findings

Considerable progress has been made toward rigorously defining the various processes affecting the movement of spilled oil, and its ultimate fate. It is now estimated that evaporation can account for the loss to the atmosphere of from one- to two-thirds of the oil spilled onto the sea surface.

Present knowledge suggests that almost all of the evaporated oil becomes photochemically oxidized in the atmosphere. Photochemical oxidation has also been identified as a significant process acting on oil at the sea surface.

Modeling of the drift of spilled oil remains a complex and difficult problem, in part because of the wide spectrum of oil types and the changing environmental conditions occurring during a spill. The best estimate at the present time for drift velocity still is between 3 and 4% of the wind speed. It appears now that dissolution into the water column is considerably less important than evaporation in determining the ultimate fate of spilled oil, because of the low aqueous solubility of most components.

Considerable work has been devoted in the past decade to developing a better understanding of the interaction of spilled oil with the water column, including "mousse" formation, oil droplet entrainment into the water column, and sedimentation of spilled oil through eventual sorption onto particulate matter in the water column.

Major Finding

For all these processes, the relationship between chemical composition and the formation and stability of oil-water emulsions and of the sorption characteristics between oil and organic particulate matter are only poorly known.

If there is to be a better understanding of the physical and chemical behavior and fate of petroleum hydrocarbons in the marine water column, then more basic research and experimentation in this area are needed. This will yield useful information for quantitative models of greater ability for assessing the fate of petroleum inputs in various areas of the world's oceans.

Biological Fate of Petroleum

Major Findings

Uptake of petroleum by animals, and apparently by plants, from food and/or water is universal. However, animals and, apparently, plants are able to clear their tissues by releasing the accumulated petroleum back into the water after the removal of the pollution source(s).

There appear to exist in those living systems examined various enzymatic mechanisms capable of metabolically transforming a range of petroleum hydrocarbon compounds. One notable exception appears to be the bivalves.

In many cases, as in low level contamination, metabolic degradation and/or the clearing of petroleum from the tissues back to the water can balance uptake, without significant bioaccumulation of these compounds in the tissues.

Microbial degradation is a major clearing mechanism for removal of petroleum pollutants from marine environments. Environmental parameters affecting the rate of biodegradation are now being defined, and some progress has been made in measuring rates of biodegradation in the oceans. Refinement and standardization of methodology are required before rate projections will be sufficiently reliable.

Despite marked progress in the study of biological fates, some important aspects remain to be clarified. Thus much less is known about the fate of petroleum in marine plants, including macro-algae and phytoplankton, than in animals.

As well, little is known of either the distribution, fate, or turnover of the metabolic products of petroleum hydrocarbon, after their formation within the tissues of marine organisms. It has been demonstrated that some of these derivatives may be either toxic or mutagenic.

Major Recommendation

It is recommended that further studies be encouraged to examine the formation and fate of metabolic derivatives of petroleum hydro- carbons taken up by marine organisms.

Amounts of Hydrocarbons in the Oceans

Several U.S. agencies, including the Bureau of Land Management, the National Oceanic and Atmospheric Administration, and the Environmental Protection Agency, as well as agencies of other countries, have actively supported studies on the amount of petroleum found in the water column, sediments, and organisms in the oceans. These studies yield the follow- ing major findings.

Major Findings

Petroleum hydrocarbon concentrations in the water column can vary by several orders of magnitude and are generally related in their amount to the proximity of petroleum sources, e.g., offshore and shore-based coastal production and refining activities, and to transportation routes and accidents.

Tar balls and other floating oil residues also vary by orders of magnitude, with highest concentrations associated with tanker shipping lanes and some mid-ocean gyres such as the Sargasso Sea. Significant decreases in these concentrations have not been observed.

Petroleum hydrocarbon contamination (PHC) of marine sediments parallels that for the water column, with PHC concentrations directly related to the proximity of sewage and industrial outfalls, dumping sites, and accidental discharges.

Relatively little information is available on PHC concentrations in pelagic organisms, mainly because of analytical problems, that is, in differentiating between PHC and hydrocarbons produced by organisms in nature and micro tar balls caught in plankton net tows made in heavily traversed oceanic areas. PHCs are usually detected in samples of benthic organisms collected from polluted areas, but not from areas free of spills or other sources of input.

Although the more immediate concerns are in the coastal areas that receive the major amounts of petroleum inputs, the paucity of data for the larger open ocean areas requires attention because it is a handicap to our understanding of the long-term fate of petroleum in the marine environment on a global scale.

Effects

Prior to the 1975 report, much of the work on effects was focused on establishing toxic and lethal thresholds and on the assessment of hydrocarbon concentrations in environmental samples.

Major Finding

Research since 1975 has resulted in considerable advances in the understanding of the toxicities of various petroleum components, of the effects on organisms and their life-cycle stages, and on the relative vulnerability of various marine ecosystems.

This information has come both from experimental studies in the laboratory and from the examination of spill situations. Much has been learned, for example, regarding the impact of petroleum on intertidal and coastal ecosystems and about the effects of petroleum on various metabolic and physiological processes, especially in fish and in invertebrates such as mollusks and crustacea.

Major Finding

Little is known about the impact of petroleum on pelagic organisms and populations. There are also obvious gaps in our understanding of oil impact on macro-algae, on larval fish, and on polar and tropical organisms.

In addition, research activity has broadened to include work on the site of action of petroleum compounds, regarding their effects as dynamic processes affecting the living organism at its various levels-- enzymatic, metabolic, ultrastructural, and molecular. This has become evident from the breadth of studies brought together in this report, a range of effects extending from the ecosystem level down to the chromosomal.

Effects-related research in recent years can be characterized according to two significant advances: a shift toward gaining an understanding of the sublethal toxicities of petroleum, and recognition that both species and life-cycle stages vary widely in sensitivity and response. An increasing scientific awareness of a need to obtain and interlock chemical, physiological, and ecological data has provided additional momentum. As a result, the trend is toward studies of petroleum pollution on a solid basis of chemical and biological data.

However, <u>increased depth of understanding of the effects of petroleum in the marine environment can be gained through work in certain areas, as outlined below.</u>

<u>Mutagenicity/Tumorigenicity</u>. Only a small amount of information is available on the occurrence, kinds, and detection of mutagenic and tumorigenic problems in marine invertebrates and vertebrates and in marine plants.

<u>Alteration of Behavior</u>. Perturbation of normal behavior at very low concentrations of petroleum (less than 0.1 mg/L) is of particular concern. Change in or cessation of feeding is an early indication of oil's toxic effects in many test animals. Yet most available data are largely anecdotal, and at least in higher organisms, effects on behavior are only poorly understood.

<u>Mechanisms of Toxicity</u>. A focus of research on perturbations of physio- logical processes is recommended. While respiration, photosynthesis, AIP production, carbon assimilation, lipid formation, and related processes are known to be affected by individual hydrocarbons such as naphthalene, the ultimate site, or sites, of toxic action of these compounds have yet to be determined.

<u>Polar and Tropical Environments</u>. The polar environment poses a special problem because of almost year-round ice cover and relative inacces- sibility, compounded by large gaps in the basic data base on polar biology. While some information exists on oil and the polar environ- ment, the potential impact of a major oil spill on polar ecosystems cannot be estimated with confidence at this time.

With respect to tropical regions, rather limited data are available on the effects of oil on tropical ecosystems including mangroves, coral reefs, and their associated biota. Nonetheless, these ecosystems represent a large part of the tropical coastline and are often very near heavy tanker traffic or petroleum activity. They are also bio-logically highly productive and are of great economic and cultural importance. The preliminary data that do exist indicate these tropical systems are as sensitive to oil as, if not more sensitive than, temperate coastlines.

Synergistic Toxicity. The interaction of petroleum with other non-petroleum contaminants is not well understood. Further work is needed, especially as chronic pollution of inshore waters often involves more than one contaminant.

Ecosystem Effects. Population changes caused by an oil spill or by chronic pollution inevitably result in additional effects by altering food web relationship and interspecific competition in the ecosystem as a whole. Each situation is different, and the effects of any given spill or input can be quite unexpected. Underlying these concerns are the various findings that low concentrations (less than 1 mg/L) of petroleum hydrocarbons can apparently interfere with the normal behavior of marine organisms, especially the more fragile components such as the larval and juvenile forms of the marine foodchain. Continued study of a few key examples of recovery after oil spills or from well-defined chronic input sources, with adequate controls, is essential if eco-system effects and their economic significance are to be well-defined.

Major Recommendations

Study the effect of low concentrations of petroleum hydrocarbons on the behavior of marine organisms, particularly larval and juvenile forms.

Conduct studies to examine the apparent coincidence between elevated concentrations of mutagenic/carcinogenic petroleum hydrocarbons (PAHs) and pollution-related diseases in certain fish from waters receiving a mix of contaminants.

Conduct research into impacts of petroleum on polar and tropical environments.

Finally, a major problem throughout this report has been the difficulty of transferring information from laboratory studies to field conditions, i.e., the difficulty in predicting impact in the field from experimental data. Little is known of the effects of petroleum on zooplankton and ichthyoplankton and of the potential impact on larval fish stocks. Nonetheless, the potential exists that under certain conditions of prolonged exposure, such impact could become significant. Also, one of the most difficult aspects of this problem has been to assess the potential impact of spills on commercially important stocks of fish and shellfish.

Major Recommendation

We recommend that research into the effect of petroleum on fish stocks, including larval and juvenile stocks, be extended so as to enable sensible assessment of the impact on these marine resources.

The major uncertainty preventing such assessments, as with most ecosystem effects, is a limited understanding of natural fluctuations in populations and ecosystems. We think that major progress in determining the effects of petroleum, by itself and in concert with other contaminants, on populations and ecosystems will largely depend upon an increased understanding of short-term (years) and long-term (decades) natural fluctuations in populations and ecosystems.

SUMMARY AND CONCLUSIONS

Where oil has had an effect, subsequent monitoring has shown biological recovery taking place. Hydrocarbons from seeps and pyrolytic sources are part of the long-term evolution of the oceans, and results of observations made to date indicate that most living organisms can co-exist with hydrocarbons when concentrations are very low (less than 0.1 mg/L) and when the oil is weathered.

It is also revealing that, of the petroleum hydrocarbons entering the marine environment, an estimated 39% derives directly from oil and gas production and transportation. However, more than 45% originates from other shipping activities and from industrial, municipal, urban, and river runoff.

There is no clear indication so far that commercially important fish stocks have been severely disrupted by either chronic or catastrophic oiling of their environment. However, present census techniques remain too crude to provide clear knowledge of standing fish stocks, while natural variabilities in the stocks probably mask such impact from petroleum as may exist. The fragmented evidence on the effects of petroleum on some larval fish and fish eggs from a few laboratory and field studies indicates that such impact is possible, although it has not been rigorously examined. This inability to transfer information obtained from laboratory studies to field conditions has been an intractable problem throughout this report.

Petroleum can have a seriously adverse affect on local environments, persisting, in some cases unaltered, for decades. Moreover, some petroleum compounds are carcinogens and/or mutagens and can bind to nucleic acids. Metabolic products of petroleum degradation also can be potentially hazardous. However, the data are not available to indicate that such a hazard has occurred in populations in affected environments.

The greatest impact due to oiling clearly occurs in coastal areas, especially those with shallow water, and in areas where local current systems tend to contain or entrain the contaminant. Of special concern are situations of local chronic oiling where there is low level (less than 1 mg/L) but continuous exposure, as in waters near industrialized or heavily populated coastal regions. There is a clear need to continue research on these local situations, not only because of the intrinsic toxicity of petroleum, but also because of its poorly understood but suspected synergistic impact with other contaminants.

Particular concern is expressed about the potential impact of oil on tropical coastal environments--mangrove systems and coral reefs. These represent a major part of the coastline in tropical and subtropical regions and are highly significant in terms of fisheries and other resources. They have unique physical and biological characteristics that make them highly vulnerable to the effect of oiling. Unfortunately, the research effort on these ecosystems has been confined to comparatively few studies.

1
Chemical Composition of Petroleum Hydrocarbon Sources

INTRODUCTION

Sources of hydrocarbons entering the marine environment are numerous, and the number of individual hydrocarbon components are quite large. Thus, the analytical chemist faces a challenge in deriving detailed compositional data on a given environmental sample as does the biogeochemist in associating a given complex hydrocarbon assemblage with one or more sources. The interpretive dilemma becomes further aggravated by chemical and microbial alterations that occur after introduction of a particular set of hydrocarbon compounds to the marine environment, a set originally attributable to a source but subsequently modified. This section highlights the main compositional characteristics of the hydrocarbon sources to the marine environment and distinguishes features of each source.

CRUDE OILS

Petroleum formation and composition have been discussed in detail recently by Tissot and Welte (1979) and Hunt (1980), and unless otherwise noted, those texts are the sources for the following information. The chemical composition of crude oils from different producing regions, and even from within a particular formation, can vary tremendously. Crude oils contain thousands of different chemical compounds owing to processes during petroleum formation resulting in "molecular scrambling." Hydrocarbons are the most abundant compounds in crude oils, accounting for 50-98% of the total composition (R.C. Clark and Brown, 1977), although the majority of crude oils contain the higher relative amounts of hydrocarbons. While carbon (80-87%) and hydrogen (10-15%) are the main elements in petroleum, sulfur (0-10%), nitrogen (0-1%), and oxygen (0-5%) are important minor constituents present as elemental sulfur or as heterocyclic constituents and functional groups. Compounds containing N, S, O as constituents are often collectively referred to as NSO compounds. Crude oils contain widely varying concentrations of trace metals such as V, Ni, Fe, Al, Na, Ca, Cu, and U (Posthuma, 1977).

Table 1-1 presents examples of the composition of crude oils and fuel oils. Petroleum hydrocarbons (Figure 1-1) consist of alkanes, cycloalkanes, and aromatic compounds containing at least one benzene ring. The alkanes, or aliphatic hydrocarbons, consist of the fully saturated normal alkanes (also called paraffins) and branched alkanes of the general molecular formula (C_nH_{2n+2}), with n ranging from 1 to usually around 40, although compounds with 60 carbons have been reported. Above C_{13}, the most important group of branched compounds is the isoprenoid hydrocarbons consisting of isoprene building blocks. Pristane (C_{19}) and phytane (C_{20}) are usually the most abundant isoprenoids, and while the C_{10}-C_{20} isoprenoids are often major petroleum constituents, extended series of isoprenoids (C_{20}-C_{40}) have been reported (Albaiges, 1980).

Many of the cycloalkanes or saturated ring structures, also called cycloparaffins or naphthenes, consist of important minor constituents that, like the isoprenoids, have specific animal or plant precursors (e.g., steranes, diterpanes, triterpanes) and that serve as important molecular markers in oil spill and geochemical studies (Albaiges and Albrecht, 1979; Dastillung and Albrecht, 1976).

Aromatic hydrocarbons, usually less abundant than the saturated hydrocarbons, contain one or more aromatic (benzene) rings connected as fused rings (e.g., naphthalene) or lined rings (e.g., biphenyl). Petroleum contains many homologous series of aromatic hydrocarbons consisting of unsubstituted or parent aromatic structures (e.g., phenanthrene) and like structures with alkyl side chains that replace hydrogen atoms. Alkyl substitution is most prevalent in 1-, 2-, and 3-ringed aromatics, although the higher polynuclear aromatic compounds (>3 rings) do contain alkylated (1-3 carbons) side groups. The polycyclic aromatics with more than 3 rings consist mainly of pyrene, chrysene, benzanthracene, benzopyrene, benzofluorene, benzofluoranthene, and perylene structures. The naphthenoaromatic compounds consist of mixed structures of aromatic and saturated cyclic rings. This series increases in importance in the higher boiling fractions along with the saturated naphthenic series. The naphthenoaromatics appear related to resins, kerogen, and sterols. Petroleum generation usually involves the formation of some naphthenoaromatic structures.

The nonhydrocarbon petroleum constituents, such as examples in Figure 1-1c, can be grouped into six classes according to Posthuma (1977): sulfur compounds, nitrogen compounds, porphyrins, oxygen compounds, asphaltenes, and trace metals. Sulfur compounds comprise the most important group of nonhydrocarbon constituents. Most sulfur present is organically bound, e.g., heterocyclic. The organosulfur compounds consist of thiols, disulfides, sulfides, cyclic sulfides (e.g., thiacyclohexanes), and thiophenes. The benzothiophenes and dibenzothiophenes are important constituents of the higher-molecular-weight aromatic fractions of environmental samples, with the tetramethyl dibenzothiophenes apparently having the highest molecular weight of the sulfur heterocyclics (Jewell, 1980).

Nitrogen is present in all crude oils in compounds such as pyridines, quinolines, benzoquinolines, acridines, pyrroles, indoles, carbazoles, and benzcarbazoles (R.C. Clark and Brown, 1977; Posthuma,

TABLE 1-1a Physical Characteristics and Chemical Properties of Several
Crude Oils

Characteristic or Component	Crude Oil		
	Prudhoe Bay[a]	South Louisiana[b]	Kuwait[b]
API gravity (20°C) (°API)*	27.8	34.5	31.4
Sulfur (wt %)	0.94	0.25	2.44
Nitrogen (wt %)	0.23	0.69	0.14
Nickel (ppm)	10	2.2	7.7
Vanadium (ppm)	20	1.9	28.
Naphtha fraction[c] (wt %)	23.2	18.6	22.7
Paraffins	12.5	8.8	16.2
Naphthenes	7.4	7.7	4.1
Aromatics	3.2	2.1	2.4
Benzenes	0.3[d]	0.2	0.1
Toluene	0.6	0.4	0.4
C_8 aromatics	0.5	0.7	0.8
C_9 aromatics	0.06	0.5	0.6
C_{10} aromatics	--	0.2	0.3
C_{11} aromatics	--	0.1	0.1
Indans	--	--	0.1
High-boiling fraction[e] (wt %)	76.8	81.4	77.3
Saturates	14.4[f]	56.3	34.0
n-paraffins	5.8[g]	5.2	4.7
C_{11}	0.12	0.06	0.12
C_{12}	0.25	0.24	0.28
C_{13}	0.42	0.41	0.38
C_{14}	0.50	0.56	0.44
C_{15}	0.44	0.54	0.43
C_{16}	0.50	0.58	0.45
C_{17}	0.51	0.59	0.41
C_{18}	0.47	0.40	0.35
C_{19}	0.43	0.38	0.33
C_{20}	0.37	0.28	0.25
C_{21}	0.32	0.20	0.20
C_{22}	0.24	0.15	0.17
C_{23}	0.21	0.16	0.15
C_{24}	0.20	0.13	0.12
C_{25}	0.17	0.12	0.10
C_{26}	0.15	0.09	0.09
C_{27}	0.10	0.06	0.06
C_{28}	0.09	0.05	0.06
C_{29}	0.08	0.05	0.05
C_{30}	0.08	0.04	0.07
C_{31}	0.08	0.04	0.06
C_{32} plus	0.07	0	0.06
Isoparaffins	--	14.0	13.2
1-ring cycloparaffins	9.9	12.4	6.2
2-ring cycloparaffins	7.7	9.4	4.5

TABLE 1-1a (continued)

Characteristic or Component	Crude Oil		
	Prudhoe Bay[a]	South Louisiana[b]	Kuwait[b]
3-ring cycloparaffins	5.5	6.8	3.3
4-ring cycloparaffins	5.4	4.8	1.8
5-ring cycloparaffins	--	3.2	0.4
6-ring cycloparaffins	--	1.1	--
Aromatics (wt %)	25.0	16.5	21.9
Benzenes	7.0	3.9	4.8
Indans and tetralins	--	2.4	2.2
Dinaphthenobenzenes	--	2.9	2.0
Naphthalenes	9.9	1.3	0.7
Acenaphthenes	--	1.4	0.9
Phenanthrenes	3.1	0.9	0.3
Acenaphthalenes	--	2.8	1.5
Pyrenes	1.5	--	--
Chrysenes	--	--	0.2
Benzothiophenes	1.7	0.5	5.4
Dibenzothiophenes	1.3	0.4	3.3
Indanothiophenes	--	--	0.6
Polar materials[h] (wt %)	2.9	8.4	17.9
Insolubles[i]	1.2	0.2	3.5

NOTE: These analyses represent values for one typical crude oil from each of the geographical regions; variations in composition can be expected for oils produced from different formations or fields within each region.

[a]Adapted from Thompson et al. (23) and Coleman et al. (24).
[b]From Pancirov (25).
[c]Fraction boiling from 20° to 205°C.
[d]Reported for fraction boiling from 20° to 150°C.
[e]Fraction boiling above 205°C.
[f]Reported for fraction boiling above 220°C.
[g]Prudhoe Bay crude oil weathered 2 weeks to duplicate fractional
  distillation equivalent to approximately 205°C n-paraffin percentages from
  gas chromatography over the range C_{11}-C_{32} plus for the Prudhoe Bay
  crude oil sample only (R. C. Clark, Jr., unpublished manuscript, 1966).
[h]Polar material: clay-gel separation according to ASTM method D-2007 (10;
  part 24) using pentane on unweathered sample.
[i]Insolubles: pentane-insoluble materials according to ASTM method D-893
  (10; part 23).

*API gravity = 141.5/ (specific gravity at 60°F or 16°C) - 131.5.

SOURCE: R.C. Clark and Brown (1977). Numbers in parentheses in footnotes above are reference numbers in R.C. Clark and Brown (1977).

TABLE 1-1b Physical Characteristics and Chemical Properties of Two Refined Products

Characteristic or Component	No. 2 Fuel Oil[a]	Bunker C Fuel Oil
API gravity (20°C) (°API)*	31.6	7.3
Sulfur (wt %)	0.32	1.46
Nitrogen (wt %)	0.024	0.94
Nickel (ppm)	0.5	89
Vanadium (ppm)	1.5	73
Saturates (wt %)	61.8	21.1
n-paraffins	8.07	1.73
$C_{10} + C_{11}$	1.26	0
C_{12}	0.84	0
C_{13}	0.96	0.07
C_{14}	1.03	0.11
C_{15}	1.13	0.12
C_{16}	1.05	0.14
C_{17}	0.65	0.15
C_{18}	0.55	0.12
C_{19}	0.33	0.14
C_{20}	0.18	0.12
C_{21}	0.09	0.11
C_{22}	0	0.10
C_{23}	0	0.09
C_{24}	0	0.08
C_{25}	0	0.07
C_{26}	0	0.05
C_{27}	0	0.04
C_{28}	0	0.05
C_{29}	0	0.04
C_{30}	0	0.04
C_{31}	0	0.04
C_{32} plus	0	0.05
Isoparaffins	22.3	5.0
1-ring cycloparaffins	17.5	3.9
2-ring cycloparaffins	9.4	3.4
3-ring cycloparaffins	4.5	2.9
4-ring cycloparaffins	0	2.7
5-ring cycloparaffins	0	1.9
6-ring cycloparaffins	0	0.4
Aromatics (wt %)	38.2	34.2
Benzenes	10.3	1.9
Indans and tetralins	7.3	2.1
Dinaphthenobenzenes	4.6	2.0
Naphthalenes	0.2[b]	
Methylnaphthalenes	2.1[b]	2.6
Dimethylnaphthalenes	3.2[b]	

Table 1-1b (continued)

Characteristic or Component	No. 2 Fuel Oil[a]	Bunker C Fuel Oil
Other naphthalenes	0.4	
Acenaphthenes	3.8	3.1
Acenaphthalenes	5.4	7.0
Phenanthrenes	0	11.6
Pyrenes	0	1.7
Chrysenes	0	0
Benzothiophenes	0.9	1.5
Dibenzothiophenes	0	0.7
Polar materials[c] (wt %)	0	30.3
Insolubles (pentane)[c] (wt %)	0	14.4

NOTE: These analyses represent typical values for two different refined products; variations in composition can be expected for similar materials from different crude oil stocks and different refineries. From Pancirov (25).

[a]This is a high aromatic material; a typical No. 2 fuel oil would have an aromatic content closer to 20-25%. From Vaughan (26).
[b]From Vaughan (26).
[c]See footnotes h and i for Table 1-2.

*API gravity = 141.5/(specific gravity at 60°F or 16°C) - 131.5.

SOURCE: R.C. Clark and Brown (1977). Numbers in parentheses in footnotes above are reference numbers in R.C. Clark and Brown (1977).

1977; Hunt, 1979; Tissot and Welte, 1978). The porphyrins are nitrogen-containing compounds derived from chlorophyll and consisting of four linked pyrrole rings. Porphyrins occur as organometallic complexes of vanadium and nickel.

Oxygen compounds in crude oils (0-2%) are found primarily in distillation fractions above 400°C and consist of phenols, carboxylic acids, ketones, esters, lactones, and ethers.

Petroleum contains a significant fraction (0-20%) of material of higher molecular weight (1,000-10,000), consisting of both hydrocarbon and NSO compounds called asphaltenes. These compounds, consisting of 10-20 fused rings with aliphatic and naphthenic side chains, contribute significantly to the properties of petroleum in geochemical formations and in spill situations in relation to emulsification behavior.

TABLE 1-1c Examples of Individual Polynuclear Aromatic Hydrocarbon Concentrations in Petroleum (10^{-6} g/g Petroleum)

	South Louisiana Crude	Kuwait Crude	No. 2 Fuel Oil	Bunker C
Pyrene	4.3	4.5	41	23
Fluoranthene	6.2	2.9	37	240
Benzanthracene	3.1	2.3	1.2	90
Chrysene	23	6.9	2.2	196
Triphenylene	13	2.8	1.4	31
Benzo[a]pyrene	1.2	2.8	0.6	44
Benzo[e]pyrene	3.3	0.5	<0.1	10

SOURCE: Pancirov et al. (1980).

Vanadium and nickel are the most abundant metallic constituents of crude petroleum, sometimes reaching thousands of parts per million. They are primarily present in porphyrin complexes and other organic compounds (R.C. Clark and Brown, 1977; Yen, 1975).

The stable isotope ratio of ^{13}C to ^{12}C in whole crude oils, in oil fractions, and in the total and lipid fractions of sediments and organisms is being used to identify sources of carbon and to characterize or "fingerprint" various types of petroleum. Values in the literature are generally expressed in terms of $\delta^{13}C$, where $\delta^{13}C$ (in o/oo) = $[(^{13}C/^{12}C)sample/(^{13}C/^{13}C)standard - 1]1,000$ and the standard is the Chicago PDB material (Craig, 1953). The $\delta^{13}C$ compositions of natural crude oils range from about -18 to -35 o/oo (Silverman and Epstein, 1958). Relative to the composition of a given whole crude, C, CH_4 values are as negative as -40 o/oo and C values can be as positive as +2 o/oo for up to C_{15} hydrocarbons (Silverman, 1963).

REFINED PRODUCTS

Refined petroleum products introduced to the marine environment include gasoline, kerosene, jet fuels, fuel oils (No. 2, No. 4, No. 5, No. 6) or Bunker fuel oils, and lubricating oils. Figure 1-2 illustrates the types of common products obtained from crude oil distillation and cracking. As refining processes and terminologies differ worldwide, comparisons of compositions of refined products vary widely. For example, distillation, catalytic and thermal cracking, polymerization, and reforming yield products that are blended together to achieve desired chemical properties. They contain all of the hydrocarbon classes previously mentioned, but with narrower boiling ranges than

n-ALKANES

CH_4 $CH_3 - CH_3$ $CH_3 - (CH_2)_n - CH_3$ (n = 1-58)

METHANE   ETHANE

ISOALKANES

$CH_3 - CH - CH_2 - CH_3$
          |
          CH_3

ISOBUTANE

$CH_3 - \overset{\displaystyle CH_3}{\underset{\displaystyle CH_3}{C}} - CH_2 - \overset{}{\underset{\displaystyle CH_3}{CH}} - CH_3$ ISOOCTANE

$CH_3 - \overset{\displaystyle CH_3}{\underset{\displaystyle CH_3}{C}} - (CH_2)_n - \overset{\displaystyle CH_3}{C} - (CH_2)_n - \overset{\displaystyle CH_3}{C} - (CH_2)_n - \overset{\displaystyle CH_3}{C} - CH_3$ (n = 3)

PRISTANE
(an isoprenoid hydrocarbon)

CYCLOALKANES

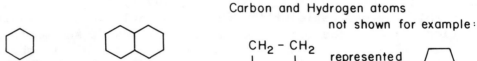

CYCLOHEXANE   DECALIN

Carbon and Hydrogen atoms
          not shown for example:

$CH_2 - CH_2$
  | |
CH_2 CH_2
     \   /
      CH_2

represented
by

CYCLOPENTANE

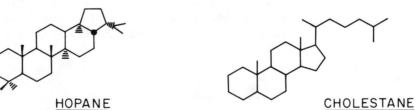

HOPANE
(general class of similar
structures are triterpanes.)

CHOLESTANE
(general class of similar
structures are steranes.)

FIGURE 1-1a Chemical structure of petroleum hydrocarbons. Isooctane
was formed in cracking process for gasoline production.

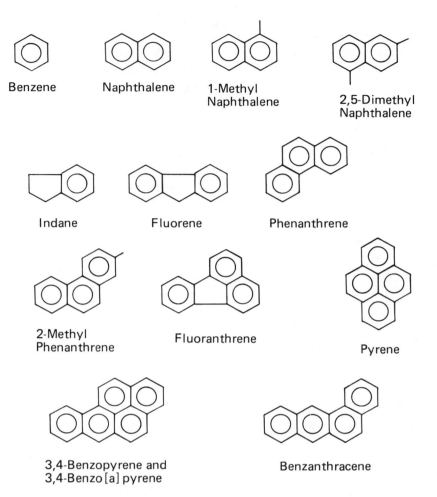

Benzene   Naphthalene   1-Methyl Naphthalene   2,5-Dimethyl Naphthalene

Indane   Fluorene   Phenanthrene

2-Methyl Phenanthrene   Fluoranthrene   Pyrene

3,4-Benzopyrene and 3,4-Benzo[a]pyrene   Benzanthracene

FIGURE 1-1b Chemical structure of petroleum hydrocarbons.

corresponding crude oils. In addition, cracking operations generate olefins (alkenes and cycloalkenes), which occur in concentrations as high as 30% in gasoline and about 1% in jet fuel. Olefins are not present in crude petroleum and are present only in minor amounts in other refined products. Alkylation processes yield many branched compounds such as isooctane (Figure 1-1a). An excellent discussion of the chemical properties of refined products pertinent to fate and effects in the environment is found in R.C. Clark and Brown (1977).

OIL SEEPS AND ANCIENT SEDIMENTS

The composition of seep oil is similar in many respects to crude oil pumped from wells, but can be influenced by a variety of physical, chemical, and biological processes to be discussed in a later section.

SULFUR COMPOUNDS

CH₃-CH₂-SH   CH₃-CH₂-S-S-CH₂-CH₃

Ethanethiol   3,4 Dithiahexane

Thiacyclohexane   Thiophene   Dibenzothiophene

NITROGEN COMPOUNDS

Pyridine   Quinoline   Indoline   Carbazole

OXYGEN COMPOUNDS

Fluorenone   Phenol   Dibenzofuran

FIGURE 1-1c Nonhydrocarbon petroleum constituents: NSO compounds.

Hydrocarbons and other compounds associated with ancient sediments range in composition from that of many crude oils to that of biogenic and early diagenetic compounds found in recent nonpolluted sediments.

BIOGENIC HYDROCARBONS

Hydrocarbons are synthesized by most marine plants and animals, including microbiota (Han and Calvin, 1969; J.B. Davis, 1968), phytoplankton (Blumer et al., 1971; Clark and Blumer, 1967), zooplankton (Blumer et al., 1969; Blumer and Thomas, 1965a,b; Avignan and Blumer, 1968), benthic algae (Youngblood et al., 1971; Youngblood and Blumer, 1973; Clark and Blumer, 1967), and fishes (Blumer et al., 1969; Blumer and Thomas, 1965b). Organisms can both produce their own hydrocarbons and acquire them from food sources.

Species of marine organisms synthesize limited numbers of hydrocarbon constituents over relatively narrow boiling ranges. For example, odd-numbered carbon chains predominate in marine biotic systems (C_{15}-C_{21} normal alkanes in phytoplankton), although the biogenic production of even-numbered carbon chains has been observed (R.C. Clark, unpublished manuscript, 1966). Pristane (C_{19}) is a major component of calanoid copepods and, consequently, of some fishes (Blumer et al., 1963; Blumer,

FRACTIONAL DISTILLATION DISTRIBUTION

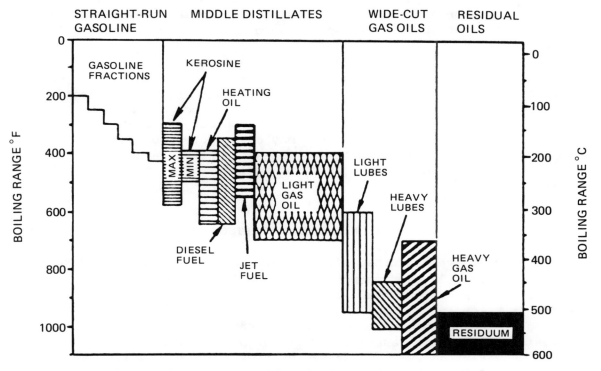

FIGURE 1-2 Boiling point range of fractions of crude petroleum.

NOTE: Bunker oil (not shown) is an oil of high viscosity used as a fuel oil. A given Bunker oil may be a mixture of two or more of the distillate cuts shown in the figure or it may be a residual oil from a distillation run.

SOURCE: Adapted from Bureau of Naval Personnel by R.C. Clark and Brown (1977).

1967). Although normal and branched alkanes are biosynthesized, alkenes are the most abundant biosynthetic compounds in all trophic levels.

Terrestrial plants (and sargassum) produce C_{21}-C_{33} odd-chain n-alkanes, with the C_{21}-C_{29} compounds dominating in marsh grasses and the C_{27}-C_{33} alkanes associated with the waxy coatings of grasses and leaves. These are major hydrocarbon components of most "clean" coastal sediments (Wakeham and Farrington, 1980; Simoneit, 1978).

Other compounds that have been detected in marine organisms include certain of the triterprenoid (hopane) hydrocarbons in marine bacteria (Ourisson et al., 1979) and naphthenes containing 1-3 rings in land herbs and plants (Blumer, 1969). Although there have been reports of the synthesis of polycyclic aromatic hydrocarbons by algae and higher plants (e.g., Borneff et al., 1968a,b), these contentions are disputed (Grimmer and Düval, 1970; Hase and Hites, 1976), and at present the issue remains unresolved.

DIAGENETIC SOURCES

Biogenic precursor molecules (e.g., terpenes, sterols, carotenoid pigments) may be altered after deposition in sediment by microbially mediated and chemical processes to yield a variety of chemical compounds.

Diagenetic hydrocarbon constituents include aliphatic hydrocarbons, cycloalkanes, sterenes, polynuclear aromatic hydrocarbons, and pentacyclic triterpanes. Among the most significant sets of diagenetic products are the PAH compounds, including some compounds that are also found in petroleum and other hydrocarbon sources (Wakeham et al., 1981). These diagenetic compounds may constitute important components of recent sediment hydrocarbon assemblages. Perylene and retene are among those compounds formed in reducing sediment (Hites et al., 1980; Aizenshtat, 1973).

COMBUSTION SOURCES AND COMPARISON TO PETROLEUM

Particulate matter in urban air contains saturated (Hauser and Pattison, 1972) and aromatic hydrocarbons formed during the high-temperature incomplete combustion or pyrolysis of fossil fuels (coal, oil, and wood) (M.L. Lee et al., 1977). Polynuclear aromatic hydrocarbons (PAH) formed during combustion processes are transported seaward via direct deposition on the sea surface or rainout over land followed by stormwater runoff. PAH compounds are, therefore, ubiquitous chemical components of marine systems throughout the world (Laflamme and Hites, 1978; Pancirov and Brown, 1977; Youngblood and Blumer, 1975; Windsor and Hites, 1979; R.A. Brown and Weiss, 1978).

Aromatic hydrocarbons from combustion sources are characterized by a lesser degree of alkylation than aromatics from petroleum. The degree of alkylation within a homologous series of aromatics (e.g., phenanthrenes) in a given PAH assemblage is dependent on the temperature of formation of the PAH; high temperature processes (incomplete combustion or pyrolysis) favor less alkylation, while relatively low temperature geological processes (petroleum maturation) favor higher degrees of alkylation. Figure 1-3 illustrates the principle that allows for the differentiation of combustion-related inputs from fresh and weathered petroleum by considerations of alkyl homolog distributions (AHD) (Blumer, 1976; M.L. Lee et al., 1977; Hites and Bieman, 1975; Youngblood and Blumer, 1975). Combustion sources contain relatively low quantities of 2-ringed aromatic families (e.g., naphthalenes). Therefore, in some cases the relative inputs of petroleum and combustion sources can be discerned from AHD plots of 2- to 5-ringed aromatics. Figure 1-4a illustrates a case where substantial quantities of petroleum- (2-3 rings) and combustion-related (3-5 rings) hydrocarbons are present, while Figure 1-4b gives results for a sample comprised mainly of 3- to 5-ringed aromatics of pyrolytic origin.

Saturated hydrocarbons derived from combustion and weathered petroleum may have similar composition and thus may not be diagnostic for interpreting source material. However, there has been less research on saturated hydrocarbons from combustion sources than on PAH.

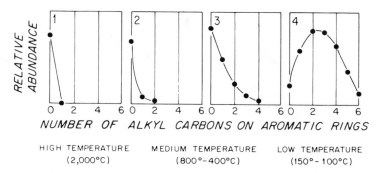

EXAMPLE: TWO ALKYL CARBONS
ON PHENANTHRENE

CH₃
CH₃

(2,3-DIMETHYLPHENANTHRENE)

RELATIVE ABUNDANCE

NUMBER OF ALKYL CARBONS ON AROMATIC RINGS

HIGH TEMPERATURE
(2,000°C)

MEDIUM TEMPERATURE
(800°-400°C)

LOW TEMPERATURE
(150°-100°C)

FIGURE 1-3 Relative abundance of parent aromatic compound and alkyl
substituents as influenced by temperature of formation.

OTHER SOURCES

Anthropogenic hydrocarbons may be introduced through a variety of
sources (dredge spoil, sewage sludge, fly ash, industrial wastes)
containing mixed inputs of hydrocarbon material (petroleum plus combustion material). In addition, the direct introduction of coal may be
significant in certain areas. Saturated and aromatic hydrocarbons in
coal (Tripp et al., 1981) are very similar to those in petroleum, both
being formed through low temperature processes, although careful evaluation of AHD plots may differentiate oil and coal (Hites et al., 1980).

DIAGNOSING SOURCES OF HYDROCARBONS

The task of determining the sources of hydrocarbons in environmental
samples is often difficult due to the multiplicity of sources that might
be present and due to the postintroduction or postdeposition environmental modification of source materials. However, the proper compositional evaluation is often as important as determining the absolute
values of hydrocarbon components.

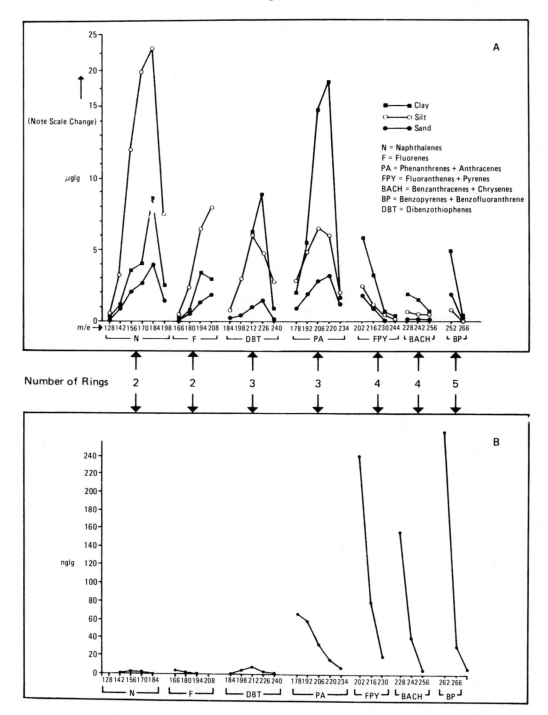

PAH Alkyl Homologue Distributions showing Sewage Sludge—Mainly Petroleum Derived (A) and Dredge Spoil—Mainly Combustion Derived (B); within each Homologous Series

FIGURE 1-4   PAH alkyl homolog distributions for sewage sludge and dredge spoils.

SOURCE: Boehm (1983).

Many analytical techniques can aid in compositional evaluation as well as determination of absolute amounts of individual compounds or groups of compounds. However, some of the methodology has been developed in only one or two laboratories for certain specific purposes and has yet to be rigorously tested in a variety of situations in several laboratories. Thus, while working guidelines can be presented for differentiating between various sources of input, many of these guidelines need further testing and may not provide sufficient sensitivity when applied to samples containing low concentrations of compounds (1-ppm level or less) from several different sources.

Biogenic Sources

Characteristics of biogenic hydrocarbons are controlled by biosynthetic/ metabolic pathways that usually result in mixtures of compounds of limited complexity due to structural specificity relations to specific biological functions. Thus:

1. Biogenic hydrocarbons of recent origin show a high predominance of odd-carbon $\underline{n}$-alkanes. Possible exceptions are the $\underline{n}$-C_{22} to $\underline{n}$-C_{30} alkanes present at low concentrations in phytoplankton and bacteria.

2. Terrigenous plant detrital inputs are characterized by odd-numbered $\underline{n}$-alkanes in the C_{23}-C_{33} region; marine biogenic inputs are often marked by the presence of odd-chain $\underline{n}$-alkanes $\underline{n}$-C_{15}, $\underline{n}$-C_{17}, and $\underline{n}$-C_{19}.

3. Biogenic inputs are often noted by the predominance of a single isoprenoid, usually pristane. The ratio of pristane to phytane is usually much greater than one in biogenic sources. Phytane is rarely found as a biolipid except in some bacteria.

4. One or several aliphatic olefins or cyclic alkenes may occur in narrow ranges of molecular weight.

5. Biogenic compounds often include polyolefins such as heneicosahexaene in algae and squalene in higher animals. Other polyolefins are abundant in many organisms and comprise most of the material in "aromatic/saturate" silica gel adsorption column eluates.

6. Biogenic aromatics do not occur often or in high concentrations.

7. Stable carbon isotope ratios of biogenic material are often much different than isotopically heavier petroleum (Degens, 1969).

Characteristics of Undegraded Petroleum

1. Petroleum contains a much more complex mixture of hydrocarbons over wider boiling ranges than biogenic inputs.

2. Crude oil contains no olefins.

3. The ratio of odd to even carbon number $\underline{n}$-alkanes in various molecular weight ranges expressed as either the odd-even preference (OEP; Scalan and Smith, 1970) or carbon preference index (CPI; Farrington and Tripp, 1977) is near unity. The alkane $\underline{n}$-C_{16} is rarely found in biolipids (Thompson and Eglinton, 1978).

4. Petroleum contains several homologous series of compounds (e.g., normal alkanes; branched alkanes; cycloalkanes; isoprenoid alkanes, including branched cyclohexanes, steranes, and triterpanes).

5. Petroleum contains homologous series of alkylated aromatics (e.g., mono-, di-, tri-, and tetra-methyl benzenes; naphthalenes; fluorenes; dibenzothiophenes; phenanthrenes).

6. Petroleum contains numerous naphthenic and naphthenoaromatic compounds.

7. Petroleum contains numerous heterocyclic compounds containing S, N, and O.

8. Petroleum contains trace metals, with Ni and V often present in μg/g quantities.

9. Hydrocarbons of a petroleum origin should have little ^{14}C activity.

10. Stable carbon isotope ratios are isotopically heavier than biogenic inputs.

All characteristics are attributable to petroleum and refined products, although the composition of distillate cuts is narrower in boiling range than the corresponding crude oil. Light distillate cuts may contain olefinic material and little, if any, trace metal content unless added.

However, one caveat pertains to carbon-number ratios. Smooth distributions of alkanes (CPI or OEP = 1) within the crude oil nonvolatile molecular weight range have been reported for marine bacteria (Han and Calvin, 1969) and have been detected in marine fish (Whittle et al., 1977a,b; Boehm, 1980). Thus, paraffinic tar and biogenic alkanes may be very similar in the C_{20}-C_{30} range. Furthermore, smooth n-alkane distributions have been noted in urban air (Hauser and Pattison, 1972) and in laboratory dust (Gelpi et al., 1970) samples. Thus, n-alkane distributions alone, in environmental samples and especially in marine fish, cannot be attributable to oil pollution without corroboration by other petroleum compositional features.

Characteristics of Petroleum Altered by Physical, Chemical, and Biological Processes

Most often, except for recent spill studies, environmental samples contain altered, rather than "fresh," petroleum. Thus, some diagnostic features associated with petroleum may be reduced in importance, and other or new diagnostic parameters become more important. The composition can be altered on time scales varying from days to years past the point where oil can be easily attributed to a particular source. Certain chemical marker compounds survive longer than others, and the time course and degree of change in composition vary with each spill or source of input, sample type, and environmental conditions. An example of a time series for fresh petroleum and samples subjected to alteration by natural processes is shown in Figure 1-5. The following changes in composition occur:

1. Loss of low boiling ($<C_{20}$) aromatic and saturated hydrocarbons through evaporation.

2. Loss of low boiling ($<C_{15}$) aromatic hydrocarbons through dissolution.

3. An increased relative importance of unresolved naphthenic and naphthenoaromatic compounds (i.e., the unresolved complex mixture or UCM).

4. An increased importance of highly branched aliphatic hydrocarbons (i.e., isoprenoids) relative to straight chain and singly methyl-branched molecules due to selective depletion of n-alkanes by biodegradation.

5. An increased importance of alkylated (dimethyl to tetramethyl) phenanthrene and dibenzothiophene compounds relative to other aromatics through combined weathering processes.

6. An increased importance of polycyclic aliphatic (e.g., pentacyclic triterpanes) compounds relative to all saturated components.

Weathered petroleum and the marine environment adjacent to weathering petroleum residues are affected by the oxidation products of both photochemical and microbial oxidation.

Photooxidation, photodecomposition, and polymerization of spilled oil components (Overton et al., 1979, 1980; Larson et al., 1977; Burwood and Speers, 1974; Freegarde et al., 1971; Parker et al., 1971) may result from sunlight-induced reactions. A variety of reaction products may result from free radical reactions, including aliphatic and aromatic ketones, aldehydes, carboxylic acids, fatty acids, esters, epoxides, sulfoxides, sulfones, phenols, anhydrides, quinones, and aliphatic and aromatic alcohols. In addition, polymerization reactions occurring during photolysis affect the chemical and physical properties of the oil, perhaps increasing its emulsification rate. Burwood and Speers (1974) observed an increase in UCM content of samples as photo-oxidation proceeded, the material presumably resulting from the formation of a variety of high-molecular-weight sulfoxides. As photolysis produces a variety of chemical reaction products, it also decomposes petroleum, thus affecting the residual hydrocarbon assemblage. The relative extent of photooxidation of spilled oil and the production of reaction products versus other weathering-induced changes in the oil's chemistry are largely unknown, although photolysis reactions are believed to play major roles in altering the oil's chemical composition after evaporation ceases to be the major process acting on oil.

The relative "competition" of photolysis and microbial oxidation/ degradation remains poorly understood. Microbial oxidation can result in the formation of the same class of reaction products as does photolysis (Overton et al., 1980). However, microbial processes are more selective due to enzymatic preferences for specific isomers, while photooxidative processes tend to be less selective. As a demonstration of this, Overton et al. (1980) compared the composition of samples of benzoic acid and its C_1-C_3 alkylbenzoic acid homologs from photochemical and microbial oxidation experiments and determined that the C_3+ homologs differed significantly between the experiments, the photochemical reaction products being more structurally diverse.

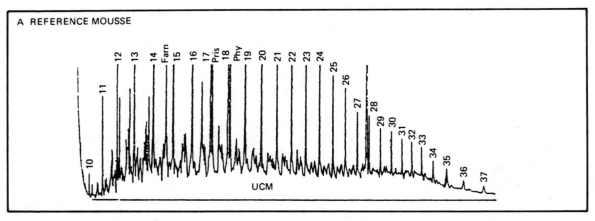

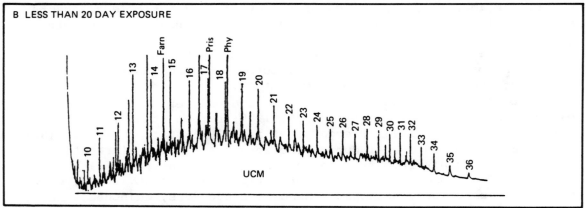

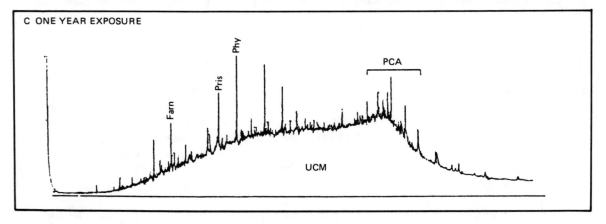

FIGURE 1-5   Glass capillary gas chromatograms of time series for fresh petroleum and petroleum subjected to alteration by natural processes in sediments.

To date, the various transformation products have not been used as markers for particular sources of inputs, although useful indicator compounds may emerge with further research.

Characteristics that have proven useful are the following:

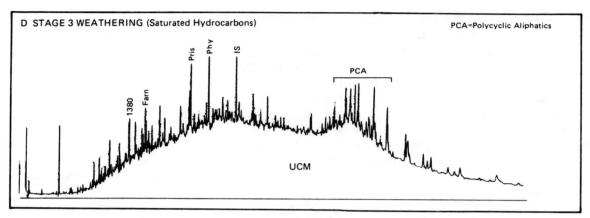

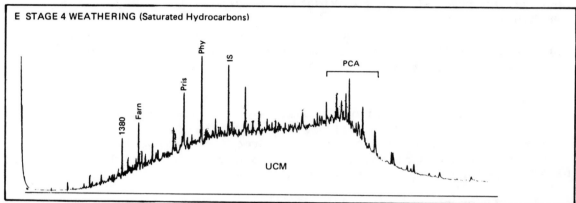

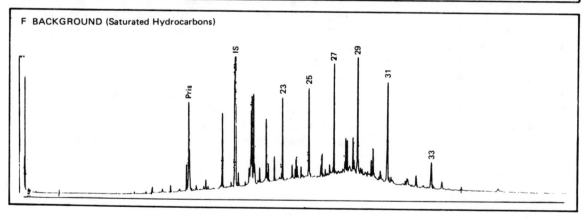

FIGURE 1-5 (continued).

1. An unresolved complex mixture is characteristic of weathered oils (see Chapter 3, Gas Chromatography section).

2. The normal alkane-to-isoprenoid ratio, which in fresh oils is much greater than 1, decreases as biodegradation proceeds (Boehm et al., 1981; Atlas et al., 1981). That is, pristane, phytane, and farnesane become dominant saturated hydrocarbon components of weathered oils until they, too, are degraded.

3. Extended series of isoprenoids (C_{20}-C_{40}) become useful indicators of petroleum in weathered samples (Albaiges, 1980).

4. C_{27+} pentacyclic triterpanes (e.g., hopanes; Figure 1-6), being relatively resistant to degradation, become prominent marker compounds (Dastillung and Albrecht, 1976; Albaiges and Albrecht, 1979; Boehm et al., 1981) as the presence of paired peaks (22R and S diastereomers) of the C_{31}, C_{32}, and C_{33} 17H,21H-hopanes is unique to petroleum (see Figure 1-6).

5. Alkylated phenanthrenes and alkylated dibenzothiophenes sometimes are the prominent aromatic components of weathered petroleum (Teal et al., 1978; Berthou et al., 1981; Boehm et al., 1981; Overton et al., 1981).

6. The relative amount of polar (N, S, O) material increases as degradation proceeds due to oxidation reactions (J.R. Payne et al., 1980a,b).

7. Stable isotope ratios of carbon, hydrogen, and sulfur do not vary greatly with weathering, and thus may be useful to identify weathered oils (Sweeney et al., 1980; Sweeney and Kaplan, 1978).

Characteristics of Combustion-Related Hydrocarbons

Little is known about the saturated hydrocarbon composition of the combustion products of fossil fuels. Most compositional information is based on polynuclear aromatic hydrocarbons.

1. PAH compounds generally occur in the 2- to 6-ring range.

2. Fluoranthrene and pyrene are often the most abundant PAH in pyrolysis-related samples together with phenanthrene, benzanthracene, chrysene, and the benzopyrenes.

3. Unsubstituted (nonalkylated parent) compounds are much more abundant than alkylated members of any homologous series. This important difference from weathered petroleum is often most striking for the phenanthrene series. Alkylated phenanthrene members are often the most abundant aromatic constituents of weathered oils.

4. Dibenzothiophene is relatively far less abundant than in oils.

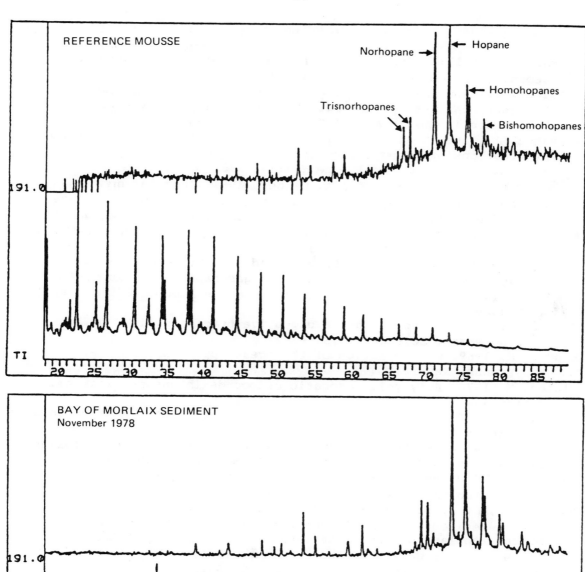

FIGURE 1-6   Gas chromatographic mass spectrometry selected ion searches for pentacyclic triterpanes (hopanes) in <u>Amoco</u> <u>Cadiz</u> reference oil and November 1978 weathered oil in sediments.

REFERENCES

Aizenshtat, A. 1973. Perylene and its geochemical significance. Geochim. Cosmochim. Acta 37:559-567.

Albaiges, J. 1980. Fingerprinting petroleum pollutants in the Mediterranean Sea, pp. 69-81. In J. Albarges, ed. Analytical Techniques in Environmental Chemistry. Pergamon, New York.

Albaiges, J., and P. Albrecht. 1979. Fingerprinting marine pollutant hydrocarbons by computerized gas chromatography-mass spectrometry. Intern. J. Environ. Anal. Chem. 6:171-190.

Atlas, R.M., P.D. Boehm, and J.A. Calder, 1981. Chemical and biological weathering of oil from the Amoco Cadiz oil spillage, within the littoral zone. Estuarine Coastal Mar. Sci. 12:589-608.

Avignan, J., and M. Blumer. 1968. On the origin of pristane in marine organisms. J. Lipid Res. 9:350-352.

Berthou, F., Y. Gourmelun, Y. Dreano, and M.P. Friocourt. 1981. Application of gas chromatography on gas capillary columns to the analysis of hydrocarbon pollutants from the Amoco Cadiz oil spill. J. Chromatogr. 203:279-292.

Blumer, M. 1967. Hydrocarbons in digestive tract and liver of a basking shark. Science 156:390-391.

Blumer, M. 1969. Oil pollution of the ocean, pp. 5-13. In D.P. Hoult ed. Plenum, New York.

Blumer, M. 1976. Polycyclic aromatic compounds in nature. Sci. Am. 234:34-45.

Blumer, M., and D.W. Thomas. 1965a. Phytodienes in zooplankton. Science 147:1148-1149.

Blumer, M., and D.W. Thomas. 1965b. Zamene, isomeric C_{19} monoolefins from marine zooplankton, fishes, and mammals. Science 148:370-371.

Blumer, M., M.M. Mullin, and D.W. Thomas. 1963. Pristane in zooplankton. Science 140:974.

Blumer, M., J. Gordon, J.C. Robertson, and J. Sass. 1969. Phytol-derived C_{19} and di- and tri-olefinic hydrocarbons in marine zooplankton and fishes. Biochemistry 8:4067-4074.

Blumer, M., R.R.L. Guillard, and T. Chase. 1971. Hydrocarbons of marine phytoplankton. Mar. Biol. 8:183-189.

Boehm, P.D. 1980. Gulf and Atlantic survey--Cape Hatteras to Gulf of Maine survey for selected organic pollutants in finfish and benthic animals. Final Report. NOAA Contract NA-80-FA-C-00046. NOAA/NMFS Northeast Fisheries Center, Sandy Hook, N.J.

Boehm, P.D. 1983. Coupling of organic pollutants between the estuary and continental shelf and the sediments and water column in the New York Bight region. Can. J. Fish. Aquat. Sci. 40(Suppl. 2):262-276, Fig. 6.

Boehm, P.D., D.L. Fiest, and A. Elskus. 1981. Comparative weathering patterns of hydrocarbons from the Amoco Cadiz oil spill observed at a variety of coastal environments, pp. 159-173. In Proceedings, Amoco Cadiz: Fate and Effects of the Oil Spill, November 19-22, 1979. Centre Nationale pour l'Exploitation des Oceans, COB, Brest, France.

Borneff, J., F. Selenka, H. Kunte, and A. Maximos. 1968a. Die synthese von 3,4-benzpyren and anderen plyzyklischen, aromatischem kohnlenwasserstoffen in pflanzen. Arch. Hyg. Bakt. 152:279-282.

Borneff, J., F. Selenka, H. Kunte, and A. Maximos. 1968b. Experimental studies on the formaton of polycyclic aromatic hydrocarbons in plants. Environ. Res. 2:22-29.

Brown, R.A., and F.T. Weiss. 1978. Fate and effects of polynuclear aromatic hydrocarbons in the aquatic environment. Publication 4297. American Petroleum Institute, Environmental Affairs Department, Washington, D.C.

Burwood, R., and G.C. Speers. 1974. Photooxidation as a factor in the environmental dispersal of crude oil. Estuarine Coastal Mar. Sci. 2:117-135.

Clark, R.C., Jr., and M. Blumer. 1967. Distribution of n-paraffins in marine organisms and sediment. Limnol. Oceanogr. 12:79-87.

Clark, R.C., Jr., and D.W. Brown. 1977. Petroleum: properties and analyses in biotic and abiotic systems, pp. 1-89. In D.C. Malins, ed. Effects of Petroleum on Arctic and Subarctic Marine Environments and Organisms. Vol. 1. Nature and Fate of Petroleum. Academic Press, New York.

Craig, H. 1953. The geochemistry of stable carbon isotopes. Geochim. Cosmochim. Acta 3:53-92.

Dastillung, M., and P. Albrecht. 1976. Molecular test for oil pollution in surface sediments. Mar. Pollut. Bull. 7:13-15.

Davis, J.B. 1968. Paraffinic hydrocarbons in the sulfate-reducing bacterium Desulfovibrio desulfuricans. Chem. Geol. 3:155-160.

Degens, E.T. 1969. Biogeochemistry of stable carbon isotopes, pp. 304-329. In G. Eglinton and M.T.J. Murphy, eds. Organic Geochemistry. Springer-Verlag, New York.

Farrington, J.W., and B.W. Tripp. 1977. Hydrocarbons in western North Atlantic surface sediments. Geochim. Cosmochim. Acta 41:1627-1641.

Freegarde, M., C.G. Hatchard, and C.A. Parker. 1971. Oil spilt at sea: its identification, determination and ultimate fate. Lab. Practice 20(1):35-40.

Gelpi, E., D.W. Nooner, and J. Oro. 1970. The ubiquity of hydrocarbons in nature: aliphatic hydrocarbons in dust samples. Geochim. Cosmochim. Acta 34:421-425.

Grimmer, G., and D. Duval. 1970. Investigations of biosynthetic formation of polycyclic hydrocarbons in higher plants. Naturforsch. 25b:1171-1175.

Han, J., and M. Calvin. 1969. Hydrocarbon distribution of algae and bacteria and microbiological activity in sediments, pp. 436-443. In Proceedings, National Academy of Sciences. Vol. 64. Washington, D.C.

Hase, A., and R.A. Hites. 1976. On the origin of polycyclic aromatic hydrocarbons in recent sediments: biosynthesis by anaerobic bacteria. Geochim. Cosmochim. Acta 40:549-555.

Hauser, T.R., and J.N. Pattison. 1972. Analysis of aliphatic fraction of air particulate matter. Environ. Sci. Technol. 6:549-555.

Hites, R.A., and W.G. Bieman. 1975. Identification of specific organic compounds in a highly anoxic sediment by gas chromatographic-mass spectrometry and high resolution mass spectrometry, pp. 188-201. In

R.P. Gibbs, ed. Analytical Methods in Oceanography. Advances in Chemistry Series 147. American Chemical Society, Washington, D.C.

Hites, R.A., R.E. Laflamme, and J.G. Windsor, Jr. 1980. Polycyclic aromatic hydrocarbons in marine/aquatic sediments, pp. 289-311. In L. Petrakis and F.T. Wiess, eds. Petroleum in the Marine Environment. Advances in Chemistry Series 185. American Chemical Society, Washington, D.C.

Hunt, J.M. 1979. Petroleum Geochemistry and Geology. W.H. Freeman and Co., San Francisco. 617 pp.

Jewell, D.M. 1980. The role of nonhydrocarbons in the analysis of virgin and biodegraded petroleum, pp. 219-235. In L. Petrakis and F.T. Weiss, eds. Petroleum in the Marine Environment. Advances in Chemistry Series 185. American Chemical Society, Washington, D.C.

Laflamme, R.E., and R.A. Hites. 1978. The global distribution of polycyclic aromatic hydrocarbons in recent sediments. Geochim. Cosmochim. Acta 42:289-304.

Larson, R.A., L.L. Hunt, and D.W. Blankenship. 1977. Formation of toxic products from a No. 2 fuel oil by photooxidation. Environ. Sci. Technol. 11(5):492-496.

Lee, M.L., G.P. Prado, J.B. Howard, and R.A. Hites. 1977. Source identification of urban airborne polycyclic aromatic hydrocarbons by gas chromatographics mass spectrometry and high resolution mass spectrometry. Biomed. Mass. Spec. 4:182-186.

Ourisson, G., P. Albrecht, and M. Rohmer. 1979. The hopanoids: paleo-chemistry and biochemistry of a group of natural products. Pure Appl. Chem. 51:709-729.

Overton, E.B., J.R. Patel, and J.L. Laseter. 1979. Chemical characterization of mousse and selected environmental samples from the Amoco Cadiz oil spill, pp. 169-214. In Proceedings, 1979 Oil Spill Conference (Prevention, Behavior, Control, Cleanup). American Petroleum Institute, Washington, D.C.

Overton, E.B., J.L. Laseter, W. Mascarella, C. Raschke, I. Nuiry, and J.W. Farrington. 1980. Photochemical oxidation of Ixtoc I oil, pp. 41-386. In Proceedings of the Conference on the Preliminary Scientific Results from the Researcher/Pierce Cruise to the Ixtoc I Blowout. NOAA, Office of Marine Pollution Assessment, Rockville, Md.

Overton, E.B., J. McFall, S.W. Mascarella, C.F. Steele, S.A. Antoine, I.R. Politzer, and J.L. Laseter. 1981. Petroleum residue sources identification after a fire and oil spill, pp. 541-546. In Proceedings, 1981 Oil Spill Conference (Prevention, Behavior, Control, Cleanup). American Petroleum Institute, Washington, D.C.

Pancirov, R.J., and R.A. Brown. 1977. Polynuclear aromatic hydrocarbons in marine tissues. Environ. Sci. Tech. 11:989-992.

Pancirov, R.J., T.D. Seal, and R.A. Brown. 1980. Methods of analysis for polynuclear aromatic hydrocarbons in environmental samples, pp. 123-142. In L. Petrakis and F.T. Weiss, eds. Petroleum in the Marine Environment. Advances in Chemistry Series 185. American Chemical Society, Washington, D.C.

Parker, C.A., M. Freegarde, and C.G. Hatchard. 1971. The effect of some chemical and biological factors on the degradation of crude oil at sea, pp. 237-244. In P. Hepple, ed. Water Pollution by Oil. Institute of Petroleum, London.

Payne, J.R., G. Smith, P.J. Mankiewicz, R.F. Shokes, N.W. Flynn, V. Moreno, and J. Altamirano. 1980a. Horizontal and vertical transport of dissolved hydrocarbons from the Ixtoc I blowout, pp. 239-266. In Proceedings of Symposium on the Preliminary Results from the Researcher/Pierce Cruise to the Ixtoc I Blowout. NOAA, Office of Marine Pollution Assessment, Rockville, Md.

Payne, J.R., N.W. Flynn, P.J. Mankiewicz, and G.S. Smith. 1980b. Surface evaporation/dissolution partitioning of lower-molecular-weight aromatic hydrocarbons in a down-plume transect from the Ixtoc I wellhead. In Proceedings of the Conference on the Preliminary Scientific Results from the Researcher/Pierce Cruise to the Ixtoc I Blowout. NOAA, Office of Marine Pollution Assessment, Rockville, Md.

Posthuma, J. 1977. The composition of petroleum. Rapp. P.-v. Reun. Cons. Int. Explor. Mer 171:7-16.

Scalan, R.S., and J.E. Smith. 1970. An improved measure of the odd-even predominance in the normal alkanes of sediment extracts and petroleum. Geochim. Cosmochim. Acta 34:611-620.

Silverman, S.R. 1963. Investigations of petroleum origin and evolution mechanisms by carbon isotope studies, pp. 92-102. In H. Craig, S.L. Miller, and G.J. Wasserburg, eds. Isotopic and Cosmic Chemistry. North-Holland, Amsterdam.

Silverman, S.R., and S. Epstein. 1958. Carbon isotopic compositions of petroleum and other sedimentary organic materials. Bull. Am. Assoc. Petrol. Geol. 42:998-1012.

Simoneit, B.R.T. 1978. Organic chemistry of marine sediments, pp. 233-311. In R. Chester and J.P. Riley, eds. Chemical Oceanography. Vol. 7. Academic Press, New York.

Sweeney, R.E., and I.R. Kaplan. 1978. Characterization of oils and seeps by stable isotope ratio. Proc. Energy/Environ. Calif. SPIB:281-293.

Sweeney, R.E., R.I. Haddad, and I.R. Kaplan. 1980. Tracing the dispersal of the Ixtoc I oil using C, H, S and N stable isotope ratios, pp. 89-118. In Proceedings, Symposium on the Preliminary Results from the Researcher/Pierce Cruise to the Ixtoc I Blowout. NOAA, Office of Marine Pollution Assessment, Rockville, Md.

Teal, J.M., K. Burns, and J. Farrington. 1978. Analyses of aromatic hydrocarbons in intertidal sediments resulting from two spills of No. 2 fuel oil in Buzzards Bay, Mass. J. Fish. Res. Board Can. 35:510-520.

Thompson, S., and G. Eglinton. 1978. Composition and sources of pollutant hydrocarbons in the Severn Estuary. Mar. Pollut. Bull. 9:133-136.

Tissot, B.P., and D.H. Welte. 1978. Petroleum Formation and Occurrence. Springer-Verlag, New York. 538 pp.

Tripp, B.W., J.W. Farrington, and J.M. Teal. 1981. Unburned coal as a source of hydrocarbons in surface sediments. Mar. Pollut. Bull. 12:122-126.

Wakeham, S.G., and J.W. Farrington. 1980. Hydrocarbons in contemporary aquatic sediments, pp. 3-32. In R.A. Baker, ed. Contaminants and Sediments. Vol. 1. Ann Arbor Science Publishers, Ann Arbor, Mich.

Wakeham, S.G., C. Schaffner, and W. Giger. 1981. Diagenic polycyclic aromatic hydrocarbons in recent sediments: structural information obtained by high performance liquid chromatography, pp. 353-363. In J. Maxwell and A. Douglas, eds. Advances in Organic Geochemistry. Macmillan, New York.

Whittle, K.J., J. Murray, P.R. Mackie, R. Hardy, and J. Farmer. 1977a. Fate of hydrocarbons in fish. Rapp. P.-v. Reun. Cons. Int. Explor. Mer 171:139-142.

Whittle, K.J., P.R. Mackie, R. Hardy, A.D. McIntyre, and R.A.A. Blackman. 1977b. The alkanes of marine organisms from the United Kingdom and surrounding waters. Rapp. P.-v. Reun. Cons. Int. Explor. Mer 171:72-78.

Windsor, J.G., Jr., and R.A. Hites. 1979. Polycyclic aromatic hydrocarbons in Gulf of Maine sediments and Nova Scotia soils. Geochim. Cosmochim. Acta 43:27-33.

Yen, T.F. 1975. The Role of Trace Metals in Petroleum. Ann Arbor Scientific Publishers, Ann Arbor, Mich.

Youngblood, W.W., and M. Blumer. 1973. Alkanes and alkenes in marine benthic algae. Mar. Biol. 21:163-172.

Youngblood, W.W., and M. Blumer. 1975. Polycyclic aromatic hydrocarbons in the environment: homologous series in soils and recent marine sediments. Geochim. Cosmochim. Acta 39:1303-1314.

Youngblood, W.W., M. Blumer, R.L. Guillard, and F. Fiore. 1971. Saturated and unsaturated hydrocarbons in marine benthic algae. Mar. Biol. 8:190-201.

2
Inputs

INTRODUCTION

Petroleum hydrocarbons (PHC) enter the marine environment from many sources. Estimates of these PHC inputs remain uncertain because the sources are interrelated and available data are minimal.

Figure 2-1 shows the international flow of petroleum. The width of that flow is representative of the amount of petroleum being transported along these routes. This pattern of flow may change significantly in future years, particularly in arctic areas where petroleum production is increasing.

A major fraction of the world's petroleum continues to be produced and transported from countries different from those in which the petroleum is refined and consumed. During the past decade the quantity of petroleum transported by sea, as well as the number and tonnage of ships in operation, has increased significantly (British Petroleum Company, Ltd., 1980; Lloyd's Register of Shipping, 1980). This increase is shown in Table 2-1.

Sources of PHC into the marine environment considered in this report include natural sources; offshore oil production; marine transportion (operational discharges, drydocking, marine terminals, bunker operations, bilge and fuel oil transfer, and accidental spillages); the atmosphere; coastal, municipal, and industrial wastes and runoff; and ocean dumping. Each source type will be addressed in the following sections.

NATURAL SOURCES

The direct input of PHC from natural sources is estimated to be 0.025-2.5 million metric tons per annum (mta), the best estimate being 0.25 mta. Natural seeps contribute the major fraction of this total. A minor contribution is estimated to come from erosional processes. These consensus estimates, developed at the 1981 workshop, are based on geological and geochemical principles, many of which were described by Wilson et al. (1973).

In this report on natural sources, hydrocarbons of a petroleum origin are the only ones considered. Biogenically produced hydrocar-

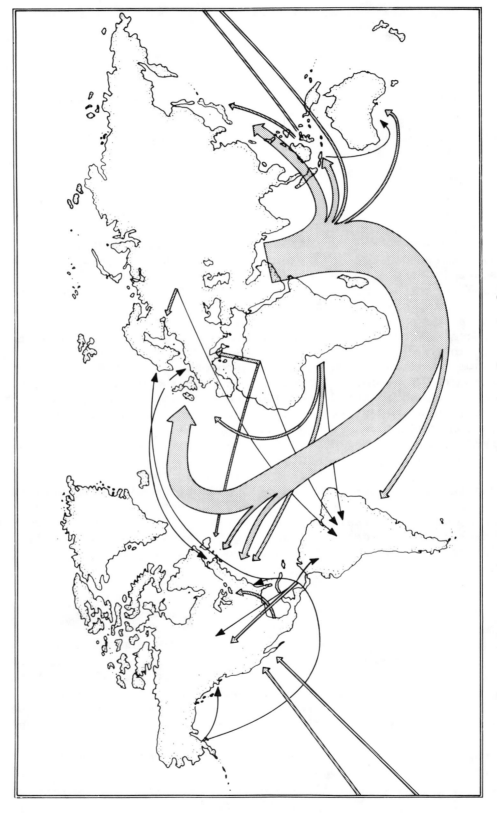

FIGURE 2-1 International marine transportation routes for petroleum.

SOURCE: Adapted from British Petroleum Co., Ltd. (1980).

TABLE 2-1 Petroleum Transport at Sea

	1971	1980	Ratio 1980/1971
Oil movement at sea (mta)			
Crude oil	1,100	1,319	1.19
Product oil	255	269	1.05
Total	1,355	1,588	1.16
World's merchant fleet			
Number of ships	55,041	73,832	1.34
Gross tonnage	247,200,000	419,911,000	1.70
World's tanker fleet			
Number of ships	6,292	7,112	1.13
Total deadweight tons	169,355,000	339,802,000	2.01
Average deadweight tons	26,900	47,800	1.78

bons, some of which have the same chemical structure as some PHC (e.g., n-alkanes and isoprenoid alkanes), are synthesized by marine organisms (see Chapter 3, Chemical Methods section).

Petroleum hydrocarbons, considered here as liquid petroleum and tar (hydrocarbons and other organic compounds with five or more carbon atoms), enter the marine environment naturally by means of two main processes--submarine seepage and erosion of sedimentary rocks. Estimating the contribution of each of these is a formidable problem for the following reasons:

1. Direct observation of submarine seeps is limited because the seeps are not normally visible. This invisibility leads to inaccurate estimates of seepage rates.
2. Submarine seeps flow intermittently, thus complicating both detection and estimation of seepage rates. The estimate is an average over geologic time, and in any particular year seepage events can exceed this estimate by orders of magnitude.
3. The potential area of continental margins where submarine seeps can occur is vast, whereas the areas of individual seepages are usually small, making an adequate inventory impossible with current technology and available monetary resources. In addition, the products of seepage cannot always be distinguished from petroleum pollution.
4. There are no direct measurements of the amount of petroleum entering the oceans by means of erosional processes, thus limiting the accuracy of any estimate.

Natural Seeps

Wilson et al. (1973) combined seepage rates on land with information on reported marine seeps, then extrapolated the data to the continental margins, which they classified into areas of potentially high, medium, and low seepage. They incorporated tectonic history, earthquake activity, and sediment thickness in their appraisal. Five basic assumptions were used in their estimates:

1. More seeps exist in offshore basins than have been observed.

2. Factors that determine the total seepage in an area (number of seeps per unit area and the daily rate for each seep) are related to the general geologic structure of the area and to the stage of sedimentary basin evolution.

3. Within each structural type, the number of seeps and, to a lesser extent, rate per seep are thought to depend primarily on the area of exposed rock and not on rock volume. This assumption presumes that there is sufficient sediment volume and organic matter for maturation and generation of petroleum.

4. Most marine seeps are clustered within the continental margins where the thickness of sedimentary rocks, which provides the needed source rocks for the seepage, exceeds a certain minimum.

5. Seepage rates are lognormally distributed.

Although the geologic relationships developed by Wilson et al. (1973) that affect seepage rates are reasonable and seem to agree with observations, the statistical arguments of the last assumption may be questionable. On purely abstract grounds, an exponential distribution of seepage rates is more likely than a lognormal distribution. While oil field volumes are generally lognormally distributed, the actual volumes of all oil accumulations (most of which are perhaps too small to be produced and thereby cannot be classified as fields) are likely to have an exponential distribution (Harbaugh and Ducastaing, 1981). The volumes of natural seepages are probably statistically distributed in a manner similar to the volumes of oil accumulations in general, because seeps do not necessarily need sources as large as oil fields. Considering the difficulties encompassed in the other assumptions, however, the form of the frequency distribution may be a minor matter.

Since Wilson et al. (1973, 1974) made their estimate, little new information has become available that would alter their worldwide estimates of marine seepage rates. Their compilations of 190 reported submarine seeps were derived mostly from Johnson (1971) and Landes (1973) and can be augmented by four newly identified seep areas (Scott Inlet, Canada; Buchan Gulf, Canada; Australian North Coast; and Laguna de Tamiahua, Mexico). All identified submarine seep areas are shown in Figure 2-2; 54 individual submarine seeps are represented by one dot off the California coast, and another 28 are so represented in the Gulf of Alaska. Of the four recent reports (Levy, 1978; Levy and Ehrhardt, 1981; McKirdy and Horvath, 1976; and Geyer and Giammona, 1980), none estimates rates of seepage.

The estimates available to Wilson et al. (1973) for Coal Oil Point (Santa Barbara Channel) and Santa Monica Bay ranged from 0.0007 to 0.05 mta. The more recent estimate of Fischer (1978) for the entire Santa Barbara Channel ranges from 0.002 to 0.03 mta, a span of values not greatly different from earlier estimates.

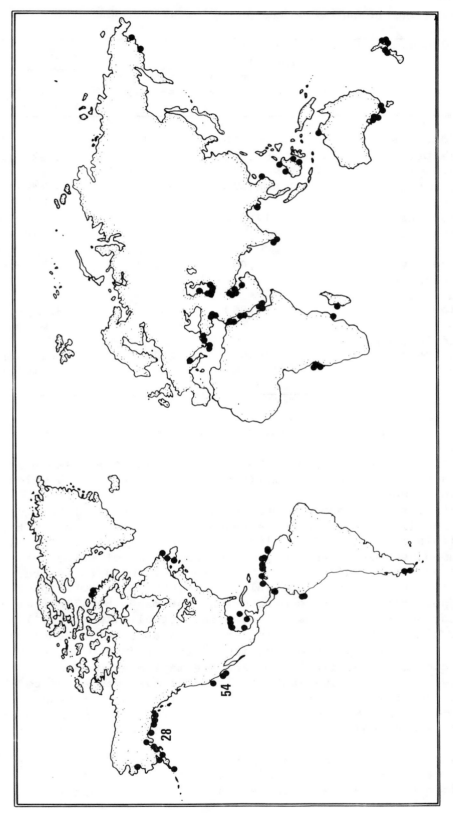

FIGURE 2-2 Location of reported marine seeps.

SOURCES: Wilson et al. (1973, 1974), Levy (1978), Levy and Ehrhardt (1981), McKirdy and Horvath (1976), Geyer and Giammona (1980), and Harvey et al. (1979).

TABLE 2-2 Petroleum Resource Estimates

| Source | Amount | | Reference |
	Millions of Metric Tons	Billions of Barrels	
Offshore oil "tar" sands	30,000	200	Weeks (1965)
Total offshore petroleum resources	350,000	2,500	Weeks (1965)
Total petroleum resources in place	1,000,000	7,200	Hunt (1972)
Proven oil reserves and potential resources offshore	14,000	100	Albers et al. (1973)
Total proven reserves and potential resources onshore and offshore	86,000	630	Albers et al. (1973)
World exploitable oil (discovered 163,000 mt; undiscovered 141,000 mt)	304,000	2,200	Halbouty and Moody (1980)
Large "tar" deposits	320,000	2,100	Demaison (1977)

Geological Implications of Seepage Rates

A comparison of estimated seepage rates with the amount available for seepage can be used to assess the maximum geologic time during which seepages could be sustained.

Table 2-2 lists petroleum resource estimates for several categories of petroleum and illustrates the wide range of resource estimates that have been calculated. Table 2-3 illustrates the comparison. The wide range of assumed seepage rates includes the estimates of Wilson et al. (1973, 1974) but extends downward to 0.02 mta and upward to 10 mta. At the low end, 10,000 mt is near the value of 14,000 mt of total proven reserves and potential resources offshore as estimated by Albers et al. (1973). At the other end of the scale, Wilson et al. (1973, 1974) accepted an estimate of 300,000 mt in place. Because this amount may represent only 1% of the petroleum mobilized from source beds, they assumed that the amount available for seepage may be as much as 30,000,000 mt. This scale available for seepage has been augmented to 100,000,000 mt to attempt to include unknowns with regard to the amount of petroleum that would have been available for seepage during geologic time and will become available in the future during the lifetimes of the seepage.

Table 2-3 shows that to maintain petroleum seepage for a span of geologic time of at least 50 million years (most of the Tertiary period) requires that seepage rates be equal to or less than 2 mta, and

TABLE 2-3   Maximum Lifetimes (Million Years) of World Oil Deposits

Oil Available for Seepage (mt)	Assumed Seepage Rates (mta)							
	0.02	0.04	0.20	0.60	1.0	2.0	6.0	10
10,000	0.5	0.25	.05	0.02	0.02	0.005	0.002	0.001
30,000	1.5	0.75	.15	0.05	0.03	0.015	0.005	0.003
100,000	5	2.5	0.5	0.2	0.1	0.05	0.02	0.01
300,000	15	7.5	1.5	0.5	0.3	0.15	0.05	0.03
1,000,000	50	25	5	2	1	0.5	0.2	0.1
3,000,000	150	75	15	5	3	1.5	0.5	0.3
10,000,000	500	250	50	20	10	5	2	1
30,000,000	1,500	750	150	50	30	15	5	3
100,000,000	5,000	2,500	500	200	100	50	20	10

at the same time the amount available for seepage must be equal to or greater than 1,000,000 mt. If seepage is maintained for 500 million years (most of the Phanerozoic), then seepage rates must be equal to or less than 0.02 mta and the amount available for seepage must be equal to or greater than 10,000,000 mt. The petroleum seepage rate that best seems to accommodate the requirements of reasonable geologic time and reasonable assumptions concerning availability for seepage is 0.2 mta with an uncertainty both upward and downward of an order of magnitude. Thus, the conclusion is reached that the average rate of petroleum seepage over time ranges from 0.02 to 2.0 mta, with a best estimate of 0.2 mta. This value is lower by a factor of 3 than the best estimate of Wilson et al. (1973, 1974) of 0.6 mta.

Erosional Inputs of Petroleum

The amount of petroleum that enters the marine environment by erosional processes has not been estimated before. Previous work by Wilson et al. (1973, 1974) considered only the marine input from natural seeps. Any estimation of erosional input of petroleum into the oceans can only be approximate.

There are at least three places where the erosional input of petroleum into the oceans could be studied in detail. Weaver (1969) showed examples of petroleum seeps at the beach and in the sea cliffs facing the Santa Barbara Channel, where erosion presently is taking place. Giammona (1980) described the Laguna de Tamiahua area where there are onshore and offshore seeps. The Marine Pollution Subcommittee of the British National Committee on Oceanic Research (1980) identified the Dorset coast of southern England as another place where petroleum source rocks as well as petroleum-containing reservoir rocks are exposed. They suggested this area for the study of natural seeps and erosional processes affecting the distribution of petroleum in the marine environment.

Because no direct information is available on erosional inputs of petroleum into the oceans, an indirect approach must be taken. This approach assumes that a portion of the organic carbon transported by all rivers is petroleum. Estimates of the organic carbon input into the oceans by rivers vary by nearly 2 orders of magnitude, from 30 to 1,000 mta, as summarized by Schlesinger and Melack (1981). They concluded, however, on the basis of two approaches, that the amount of organic carbon transported by rivers is 370-410 mta. Independently, Meybeck (1981) reached a similar estimate of 400 mta.

In estimations of the organic carbon content of rivers, no distinction has been made between carbon from modern biological sources, carbon from pollution, and ancient carbon indigenous to the eroded sediment being carried by the rivers. This latter class of carbon is of interest in estimating the erosional input of petroleum into the oceans.

The total organic matter content of rivers is divided about equally between dissolved organic carbon and particulate organic carbon (Meybeck, 1981). Meybeck further estimated that of the approximately 179 mta of particulate organic carbon that is transported by rivers, about 88 mta is ancient organic carbon. This ancient organic carbon is finely dispersed in clastic and carbonate rock particles, eroded from sedimentary rock formations on the continents (Ronov, 1976).

In ancient sedimentary rocks the amount of extractable organic matter constitutes, on the average, about 6% of the total organic matter (Hunt, 1979). If the extractable fraction in sediment particles in rivers is the same as that of source rocks, the amount in particulates in rivers would be 10.6 mta.

Most of the extractable organic material is dispersed in sedimentary rocks, but 0.5% of this material is petroleum (Hunt, 1972). If this factor is applied to the extractable organic matter of sediment particles in rivers, then the amount of petroleum transported from eroding outcrops by rivers to oceans is about 0.05 mta. This estimate may be high, because loss of organic carbon by oxidation during river transport and by sedimentation in estuaries leading to the oceans was not considered because of lack of data on these processes. Because of the numerous assumptions used to obtain this estimate, the uncertainty is at least an order of magnitude.

In estimating rates of seepage of petroleum into the marine environment, these rates have been compared with the amount assumed to be available for seepage over geologic time (Table 2-3). This same petroleum would be available for erosional processes over geologic time. The amount available is sufficient to sustain the estimated rates of natural seepage as well as rates of erosion of petroleum for an amount of time equivalent to the Tertiary period and probably longer.

OFFSHORE PRODUCTION

The amount of petroleum entering the marine environment from offshore petroleum production is estimated to be from 0.04 to 0.07 mta. Of these totals, major spills (>7 metric tons) from platforms contribute 0.03-0.05 mta, minor spills (<7 metric tons) 0.003-0.004 mta, and operational discharges 0.007-0.011 mta.

TABLE 2-4 Offshore Petroleum Production, 1979

Country	Production Rate x 10^9 bbl/year	x 10^6 mta
Saudi Arabia	1.03	147
United Kingdom	0.57	81
United States	0.39	56
Venezuela	0.38	54
Other countries	2.24	320
TOTAL	4.61	658

These estimates for the release of petroleum into the marine environment are lower, by about 30-50%, than the estimates generated earlier (NRC, 1975). Better data are available for operations, and major spill incidents in the United States have been more comprehensively documented since the earlier estimates were made. The available international data suggest that reductions have also been experienced outside the United States.

As reported by Burnet (1980), worldwide offshore petroleum production totaled approximately 658 mta in 1979. Over 50% of the production came from four countries: Saudi Arabia, the United Kingdom, the United States, and Venezuela. These data, which are the latest available published information, are summarized in Table 2-4. They are the basis for all subsequent calculations of the current petroleum input to the oceans from offshore petroleum production operations.

Operational (Produced Water) Discharges

In the United States, offshore produced water is normally discharged into the ocean after being processed to minimize the entrained petroleum content. Actual rates of discharge for produced water are not currently available. However, until 1976, the U.S Geological Survey (USGS) maintained records on these discharges from outer continental shelf operations in the Gulf of Mexico. At that time, 0.8 barrels of water were produced with every barrel of crude oil. This ratio is assumed to be still valid, and the same ratio is assumed to apply to all U.S. offshore production. This estimate is believed to be conservative, because recent Bureau of Land Management (BLM) environmental impact statements for outer continental shelf (OCS) lease sales assume a 0.6 water-to-crude-oil production ratio. The Department of Environment, U.K. (1976) report concerning discharges from offshore operations in U.K. waters stated: "The proportion of production water in crude oil will initially be less than 1 percent but will increase to some 30 percent as the reservoir becomes depleted, particularly when water injection is used on an increasing scale." This study assumed a proportion of 10% averaged over the life of the U.K. fields. For other countries an assumption of 30% was used. It should be noted that no water is produced off the Saudi Arabian shore.

Produced water regulations in the United States require that the daily maximum petroleum content not exceed 72 mg/L and that the monthly average be less than 48 mg/L. The Environmental Protection Agency (EPA, 1976) Development Document on which these guidelines are based also includes the results of an in-depth statistical analysis of all available data, which indicates that facilities meeting the above limitations will achieve a long term average petroleum content of 25 mg/L or less. These figures, however, do not include the C_6-C_{14} "volatile liquid" hydrocarbons, which are not determined by the solvent extraction technique used for "oil and grease" analysis. Therefore, a somewhat higher estimate of 35 mg/L hydrocarbons entering the oceans from U.S. produced water discharges was used. Because this regime does not include (1) upset and bypass situations in which higher discharge levels are probably experienced and (2) the fact that state-of-the-art equipment is not installed at all locations, a high estimated average is believed to be twice this level, 70 mg/L. A reasonable best estimate is 50:20 mg/L "volatile liquid" hydrocarbons and 30 mg/L higher-molecular-weight hydrocarbons ($>C_{14}$).

Similar arguments for the U.K. offshore operations and those in other countries (Table 2-5) lead to a range of 50-70 mg/L for the estimated hydrocarbon content of produced waters.

Based on these assumptions, the volume of petroleum entering the world's oceans from offshore produced water discharges is calculated to be between 0.0075 and 0.0115 mta, with a best estimate of 0.0095 mta (Table 2-6).

Specific estimates were not made for deck drainage, drilling fluid discharges, and other minor sources of petroleum (Schreiner, 1980). These sources are probably accounted for within the limits of confidence of the above numbers.

Minor Spills

Since 1971 the USGS has maintained a computerized OCS events file for Gulf of Mexico oil and gas operations (Danenberger, 1976). Included are data on all crude oil spills. The USGS classifies spills as minor (<7 metric tons or 50 barrels) or major (>7 metric tons or 50 barrels). Table 2-7 summarizes the minor spills in the Gulf of Mexico OCS for the 8-year period 1971-1978. The average spillage rate for this period was 0.00024% of total crude oil produced. The record for minor spills in offshore Alaska is better. The Lower Cook Inlet spillage rate for all spills from 1971 to 1980 is 0.0001% of total crude oil produced (Wondzell, 1981).

Similar data for operations in other U.S. areas and outside the United States are not readily available. Offshore operations are moving into more severe environments, such as the arctic regions.

TABLE 2-5 Offshore Produced Water Effluent Limitations

Country	Oil and Grease Content Limit (mg/L)	
	Average	Maximum
Abu Dhabi	–	15
Australia	30	50
Denmark	40	–
Egypt	–	60 (Mediterranean)
		15 (Red Sea)
France	–	20
Indonesia	30	–
Malaysia	100 (offshore)	–
	10 (coastal)	–
Netherlands	–	40
Nigeria	75	100
Norway	25-30	–
Spain	40 (Atlantic)	–
	60 (Mediterranean)	100
Trinidad	50	–
United Kingdom	40 (large facilities)	no more than 4%
	50 (small facilities)	greater than 100 mg/L
United States	48 (monthly)	72
	25 (long term)	
Venezuela	35	–

NOTE: Limitations shown here are from various sources. They are either existing government regulations, proposed government regulations (which could change), or limitations imposed by authorities for installations in operation in countries without regulations.

However, to balance this effect, there have been significant technological advances (such as warning systems and improved blowout preventers) that are reducing the occurrence of spills of all sizes. Average experience for all U.S. offshore operations probably is comparable to the Gulf of Mexico average, so a range of 0.00021-0.00030% is used for the United States. Another assumption is made that the worldwide percentage is about twice that of the United States, or 0.00042-0.00060 (Table 2-8). Clearly, there is uncertainty associated with this assumption.

With these data a range of 0.0027-0.0038 mta has been calculated as the estimate of petroleum entering the marine environment from minor spillage from offshore drilling and production activities worldwide. The best estimate is 0.003 mta, which is lower than the earlier NRC (1975) estimate of 0.01 mta.

TABLE 2-6 Oil to the Marine Environment From Offshore Produced Water Discharges

Country	1979 Offshore Oil Production (1,000 bbl)	Water Production (BW/BO)[a]	Water Production (1,000 bbl)	Low Oil Content (ppm)	Low Oil Discharged Annually (bbl)	Best Estimate Oil Content (ppm)	Best Estimate Oil Discharged Annually (bbl)	High Oil Content (ppm)	High Oil Discharged Annually (bbl)
United States	389,100	0.8	311,300	35	10,900	50	15,600	70	21,800
United Kingdom	573,600	0.1	57,400	35	2,000	60	3,400	70	4,000
Other	2,621,300	0.3	786,400	50	39,300	60	47,200	70	55,000
TOTAL					52,200 (0.0075 mta)		66,200 (0.0095 mta)		80,800 (0.0115 mta)

[a]BW is barrels of water, and BO is barrels of oil.

TABLE 2-7   Minor Oil Spills, Gulf of Mexico Outer Continental Shelf, 1971-1978

Year	Oil Production (1,000 bbl)	Number of Spills	Volume Spilled (bbl)	Percentage of Production Spilled
1971	386,400	1,245	1,500	0.00039
1972	391,000	1,159	1,000	0.00026
1973	375,300	1,171	900	0.00024
1974	343,900	1,129	700	0.00020
1975	316,000	1,126	700	0.00022
1976	303,100	948	500	0.00016
1977	293,000	864	600	0.00020
1978	282,500	873	600	0.00021
TOTAL	2,691,200 (~384 mta)	8,515	6,500 (~0.0093 mta)	0.00024 (avg.)

NOTE: Minor oil spills are defined as <7 metric tons or 50 barrels.

Major Spills

As was mentioned previously, the USGS has maintained a computerized OCS events file for Gulf of Mexico oil and gas operations (Danenberger, 1976).   The history of major oil spills (<7 metric tons or 50 barrels) from U.S. Gulf of Mexico operations for the period 1971-1978 is summarized in Table 2-9.   Because of the statistical distribution of spills, one large incident in a particular year greatly influences the annual figure.   However, the Gulf of Mexico 8-year average is fairly representative of any current year, i.e., the oil spillage rate from major incidents is 0.002% of the oil produced.   A similar average may apply nationwide.

Accurate worldwide information on major spills is often difficult to obtain.   Since 1979 the Oil Spill Intelligence Report (1979, 1980) has attempted to provide annual summaries of all incidents involving more than 20,000 gallons (or 68 metric tons) of oil.   However, these surveys are admittedly incomplete.   A single catastrophic incident is usually a major contributor to the annual total, but the probability of such an occurrence on an annual basis is very low and its amount is unpredictable.   Currently the world record spill resulted from the Petroleos Mexicanos (Pemex) Ixtoc I well blowout on 3 June 1979.   Until it was capped on 23 March 1980, a total of 0.44-1.4 mt of crude oil was released.   The uncertainty in the amount of crude oil spilled is related to the problems associated with estimating flow from the open hole.   Estimates of the amount burned vary from 30% (Ross et al., 1979) to as much as 58% (Program a Coordinada de Estadios Ecologicos en la

TABLE 2-8   Oil to the Marine Environment From Minor Spills

Country	1979 Oil Production (1,000 bbl)	Percentage of Production Spilled		Volume Spilled (bbl)	
		Low	High	Low	High
United States	389,000	0.00021	0.00030	820	1,170
Other	4,227,000	0.00042	0.00060	17,800	25,400
TOTAL				18,620 (~0.0027 mta)	26,570 (~0.003 mta)

NOTE: Minor spills are defined as <7 metric tons or 50 barrels.

Sonda de Campeche, 1980). No such massive incidents occurred in 1978. Only one major spill from offshore operations was reported that year, a spill of 0.003 mt in Indonesia (Oil Spill Intelligence Report, 1979). Major oil spills occur sporadically. In order to calculate a meaningful annual input, several years of experience have to be averaged.

A recent U.K. report on spills from offshore operations indicated that the average total oil spillage rate in U.K. waters for 1975 through 1979 was 0.00068% of production (Royal Commission on Environmental Pollution, 1981). This spillage rate is lower than that for the United States. We feel, however, that the worldwide spillage rate, excluding the United Kingdom, is probably higher than that of the United States because of less restrictive regulation of blowout prevention. As with minor spills, uncertainty is associated with this assumption.

Therefore, for purposes of this study, the estimated range of major crude oil spillage outside the United States (and the United Kingdom) is from 2 to 4 times the U.S. rate. The best estimate is 3 times the U.S. rate. The estimate of oil input to the oceans from major accidents during offshore oil and gas operations ranges from 0.025 to 0.05 mta with a best estimate of 0.04 mta. The calculations to obtain these figures are given in Table 2-10.

MARINE TRANSPORTATION

The estimated range in the amount of PHC discharged into the oceans due to maritime transportation activities is from 1.0 to 2.6 mta, with a best estimate of 1.45 mta. Just under half (about 0.7 mta) of this total is estimated to come from tanker operational discharges. The remainder is distributed among terminals (0.02 mta), dry-docking (0.03 mta), bilges and fuel oil from all ships (0.3 mta), and accidental spillages from tankers and other ships (0.4 mta). The earlier NRC

TABLE 2-9   Major Spills, Gulf of Mexico Outer Continental Shelf, 1971-1978

Year	Oil Production (1,000 bbl)	Number of Spills	Volume Spilled (bbl)	Percentage of Production Spilled
1971	386,400	11	1,300	0.0007
1972	391,000	2	200	0.0003
1973	375,300	4	22,200	0.0062
1974	343,900	8	22,700	0.0068
1975	316,000	2	300	0.0003
1976	303,100	3	4,700	0.0017
1977	293,000	4	700	0.0004
1978	282,500	3	1,100	0.0006
TOTAL	2,691,200 (~386 mt)	37	53,200 (~0.0076 mt)	0.0020 (avg.)

NOTE: Major oil spills are defined as >7 metric tons or 50 barrels.

(1975) estimated range for total marine transportation losses was 1.5-2.8 mta with a best estimate of 2.1 mta.

The quantity of oil discharged from ships depends on how effectively the standards developed for the control of oil pollution from ships are implemented. In 1973 the applicable rules and standards were the International Convention for the Prevention of Pollution of the Sea by Oil 1954, as amended in 1962 (OILPOL 1954/1962). The 1969 amendments to OILPOL 1954/1962 were adopted by the International Maritime Organization (IMO), formerly IMCO, assembly in 1973 but did not enter into force until February 1978. As of 9 November 1981, OILPOL 1954/1969 had been in force for more than 3 years by 66 nations representing approximately 95% of the world's merchant fleet.

The requirements of OILPOL 1954/1969 have been considerably strengthened by the adoption of the International Convention for the Prevention of Pollution from Ships, 1973, as modified by the Protocol of 1978 relating thereto (MARPOL 1973/1978). In particular, the worldwide implementation of the mandatory provision of segregated ballast tanks (SBT), dedicated clean ballast tanks (CBT), and crude oil washing systems (COW) for new and existing oil tankers would result in a significant reduction of the quantity of oil discharged into the oceans. The United States implemented these regulations 1 June 1981 with respect to U.S. flagships and foreign ships visiting U.S. ports.

MARPOL 1973/1978 has been ratified by the requisite number of nations and will enter into force on 2 October 1983. The majority of the requirements of MARPOL 1973/1978 pertaining to oil (Annex I) will become effective when the convention enters into force. The remaining

TABLE 2-10 Oil to the Marine Environment From Major Spills

Country	1979 Oil Production (1,000 bbl)	Percentage of Production Spilled		Volume Spilled (bbl)	
		Low	High	Low	High
United States	389,000	0.0020	0.0020	7,800	7,800
Other	4,227,000	0.0040	0.0080	169,100	338,200
TOTAL			18,620	176,900 (~0.025 mta)	346,000 (~0.05 mta)

NOTE: Major spills are defined as >7 metric tons or 50 barrels.

provisions will become effective no later than 2 October 1986. The present situation should therefore be regarded as a transitional period until MARPOL 1973/1978 is fully implemented.

The present estimates are based on a report prepared by a group of experts, consisting of representatives from maritime administrations and oil and shipping industries, under the auspices of the Marine Environment Protection Committee of IMO. This IMO workshop was convened prior to the November 1981 NRC workshop.

OPERATIONAL DISCHARGES

Crude Oil

During normal operations, oil tankers discharge into the sea a certain amount of oil contained in the ballast and tank washing water. OILPOL 1954/1969 stipulates that instantaneous rates of discharge from cargo tank areas of oil tankers must not exceed 60 L/mi, and the total quantity of oil discharged during any one ballast voyage must not exceed 1/15,000 of the total cargo carrying capacity (Tc). MARPOL 1973/1978 sets the same discharge standards outside special low pollution areas, but the maximum quantity of oil permitted to be discharged for new oil tankers has been reduced from 1/15,000 to 1/30,000 Tc.

In order to comply with the requirements of OILPOL 1954/1969, oil tankers should operate with load-on-top (LOT) procedures. At the time of the 1973 NRC study, 80% of the tanker fleet was assumed to be operating with LOT. Presently all crude oil tankers engaged on long haul voyages (exceeding 71 hours or 1,200 nautical miles) should operate with LOT, but tankers engaged on short haul voyages may not be able to do so. The International Association of Independent Tanker Owners (INTERTANKO) estimates that long and short haul voyages constitute 85 and 15%, respectively, of the world's crude oil movements.

During a ballast voyage, discharges of oil into the ocean may occur during two types of operation: discharge of departure (dirty) ballast without adequate separation and discharge of decanted water from slop tanks. After settlement, departure ballast separates into three parts, the largest part of which consists of water with an oil content on the order of 15-50 mg/L. An oil-water interface contains a relatively high oil content (in the order of 100-2,000 mg/L), and the oil layer on the surface is essentially pure oil.

By careful operation, the discharge of ballast water can be stopped as soon as the interface level is reached. In adverse sea conditions, the interface may be diffuse, and discharge is stopped when the oil content in ballast water rises above about 50 mg/L. Under less careful operation, the ballast may continue to be discharged for some time after the interface level has been reached. A similar situation might arise with discharge from slop tanks, which would have higher oil content but could be more easily controlled because of the slower pumping rate. Cargo lines that are not thoroughly flushed with water to slop tanks before being flushed to the sea may also cause oil discharges. In the worst cases, the LOT procedure may be completely ignored and the total oil-water mixture will be discharged into the sea.

The LOT operations emphasize "retention on board" procedures where dirty ballast water, tank washings, and oily residues are held in slop tanks for discharge at terminals. Before LOT procedures were initiated, these materials were discharged routinely into the sea.

Four oil companies (Socal, Mobil, Exxon, and Texaco) measured the quantities of retained slops for the period 1972-1977 (Gray, 1978). These data indicate a steady increase in the quantities of retained slops for the period 1972-1975. From 1975 to 1977 the quantities of retained slops leveled off or began to decline.

This decline of retained slops for company-owned tankers is probably attributable to the improved efficiency of pumping out cargo oil as well as the increasing use of COW systems, which, when fully implemented, reduces retained slops to a lower level.

Thus, it is not currently considered appropriate to use slop recovery data as a basis for estimating quantities of oil wastes discharged into the oceans. Smaller quantities of retained slops may not necessarily be an indication of the discharge into the oceans of larger quantities of oil wastes.

Tests have been carried out in various countries to evaluate the efficiency of LOT operations. The results of some of these tests are shown in Table 2-11.

The above data and experience by major oil companies indicate that perhaps two-thirds of crude oil tankers on long haul voyages already meet the OILPOL 1954/1969 discharge criteria of 1/15,000 Tc or better. Because of the dearth of data for oil tankers engaged on trades in which major oil companies are not involved, particularly oil trades on spot market, the assumption is made that half the long-haul-voyage crude oil tanker fleet meet the 1/15,000 Tc standard. As long haul tankers carry 85% of the 1,319.3 mta total, discharges from this source would be 0.037 mta (1,319.3 x 0.85 x 0.5 x 0.000067).

TABLE 2-11 Results of LOT Operation Efficiency Tests

Country or Oil Company	Oil Tanker	Vessel Size (thousand ton deadweight)	Deadweight/ Discharged Oil	Reference
Japan	Alriyadh	237	53,000	IMCO (1981)
Norway	Berge Princess	280	36,000	Overaas and Solum (1974)
Mobil	Three tankers	50-212	15,000	Desel (1972)
Exxon	Five tankers	52-254	11,000-200,000[a]	Gray (1978)

[a]Seven out of nine discharged less than 1/30,000 Tc.

Recent data (IMCO, 1981) show that of 650 tankers inspected during the years 1979-1980, approximately 2% arrived at their loading terminals (situated all over the world) with no slops at all on board, for unexplained reasons. The data from the four oil companies (Gray, 1978) show similar trends. These vessels probably represent performances better than the worldwide average, so the assumption is made that in the worst case, 5% of the long-haul-voyage crude oil tankers would discharge all their ballast, a total quantity of oil wastes equal to 0.4% of Tc. The discharge from this source would therefore be 0.224 mta (1,319.3 x 0.85 x 0.05 x 0.004).

The oil discharge for the remaining 45% of the tankers is estimated as follows: 30% would discharge 1/7,500 Tc (i.e., twice the OILPOL 1954/1969 criteria) producing, 0.045 mta, and 15% would discharge oil equal to 0.1% of Tc, producing 0.168 mta. The total discharged by long haul tankers would therefore be 0.474 mta (0.037 + 0.224 + 0.045 + 0.168).

Crude oil tankers engaged in short haul voyages may not be able to perform LOT; some of these tankers are provided with SBT or similar arrangements to avoid contaminated ballast. These tankers would contribute negligible pollution. Other short haul tankers are engaged in dedicated trades that include arrangements to transfer contaminated ballast to shore reception facilities or to long haul, very large crude carriers (VLCC) from which they take oil cargo. The remaining short haul tankers (estimated to be 50%) discharge into the sea oil amounting to 0.2% of Tc (which corresponds to the total oil content in dirty ballast tanks), or 0.198 mta (1,319.3 x 0.15 x 0.5 x 0.002).

The total annual discharge of crude oil into the sea resulting from the normal operation of crude oil tankers is estimated to be 0.672 mta (0.474 + 0.198).

Although there is a degree of uncertainty associated with the numbers, there does seem to be an improvement since the earlier NRC (1975) estimate was made (0.67 versus 1.08 mta), particularly when an increase in the amount of crude oil transported by sea from 1971 to

1981 is taken into account. Possible reasons for this improvement are as follows:

1. In 1973 tankers were allowed to discharge unlimited quantities of oily wastes outside the prohibited zone (which was normally 50 miles from land), whereas such discharge is illegal under OILPOL 1954/1969;

2. There has been significant improvement in the awareness of the master and crew, shipowners and operators, of the existence of and need to observe international rules for oil pollution prevention;

3. Surveillance and control of illegal discharges have been considerably tightened in many countries;

4. Dramatic increases in the price of oil in recent years have resulted in more careful handling of cargo oil at discharge ports, with less oil remaining on board after discharge;

5. Increased use of COW systems has enabled a higher proportion of tankers to comply with the 1/15,000 Tc standard; and

6. Inclusion of "cleaner seas" provisions in charter party agreements has alleviated the economic disadvantages for operators to retain oil residues on board.

At the same time, certain adverse factors must be borne in mind, such as the aging of the existing tanker fleet, the lack of well-equipped new tankers, the shift of the control of tankers from experienced to less experienced operators, and the increase in spot market oil trades.

Product Oil

Of 269 mta of product oil carried by tankers, one-fourth (67 mta) is estimated to be persistent (lubricating oil, fuel oil) and three-fourths (202 mta) nonpersistent (gasoline, kerosene). The discharge of persistent oil is subject to OILPOL 1954/1969, whereas the discharge of nonpersistent oil is not presently controlled. The NRC (1975) report did not provide specific figures for the operational discharge of product oil.

There are no measured data on the quantities of oil residues for persistent oil trades. The operation of tankers carrying persistent product oil is assumed to be similar to that of crude oil tankers on short voyages; namely, 50% are engaged in dedicated trades that include arrangements to transfer contaminated ballast to shore reception facilities or are provided with arrangements to avoid contaminated ballast. These ships will contribute negligibly to pollution.

The LOT performance for the remaining 50% of these tankers, compared with crude oil tankers, may be affected by the following: (1) the relatively higher viscosities of persistent product oil may result in higher clingage and (2) the relatively higher density of oil may provide some empty cargo tanks in which ballast water without contamination may be carried on a subsequent ballast voyage.

Considering the above factors, the remaining 50% of these tankers are likely on average to discharge persistent oil equal to 1/2,000 Tc, or 0.017 mta (67 x 0.5 x 0.0005).

The discharge of nonpersistent oil is permitted at present under OILPOL 1954/1969, but the quantity of such product discharged into the sea might be less than expected, because of less clingage, easy decanting, and higher rate of evaporation.

It is assumed that 50% of tankers carrying nonpersistent oil have means to avoid the discharge of contaminated water; hence, the discharge from this source should be negligible. The remaining 50% discharge, on average, oil equal to 1/5,000 Tc, and, hence, the discharge of nonpersistent oil from this source would be approximately 0.020 mta (202 x 0.5 x 0.0002).

Therefore, the annual operational discharge of product oil is estimated as 0.037 mta (persistent 0.017 mta, nonpersistent 0.020 mta).

The sum of crude oil plus product oil discharges is thus estimated at 0.71 mta (crude oil 0.67, product oil 0.04).

DRY-DOCKING

The NRC (1975) report estimated that half the tankers would arrive for dry-docking at average intervals of 18 months without tank washing residue. Since then the situation has changed considerably, including longer dry-docking intervals (2 years on average), the increased availability of reception facilities in repair ports, the reduction of sludge or slop due to more efficient stripping and COW systems, the increased degree of enforcement of OILPOL 1954/1969, and the increase in the value of crude oil.

On the basis of the above factors, it is estimated that of the world tanker fleet of 340 million deadweight tons (dwt), 5% of tankers discharge into the sea sludge or slop amounting to 0.4% of dwt prior to dry-docking at intervals of 2 years. The annual estimated loss to the sea then becomes 0.034 mta (340 x 0.5 x 0.05 x 0.004).

MARINE TERMINALS INCLUDING BUNKER OPERATIONS

The NRC (1975) study estimated discharges of 0.003 mta during terminal operations. This result was attributed to spillages that resulted from human error, such as overfilling tanks and disconnecting hoses without adequate drainage. There are other causes of spillages, including line or hose failures, submarine pipeline ruptures, or storage tank ruptures. Discharges under this category include spillages occurring during bunkering operations (filling the ship's fuel compartments) either at a terminal or from a bunkering barge.

The U.S. Coast Guard (1976, 1977, 1979, 1980) keeps statistics on marine terminal spillages from all types of incidents (such as hose breaks, tanker overfilling, line fractures, shore tank ruptures). The average marine terminal spillage for 1976, 1977, and 1979 was 0.0025 mta (the 1978 figure was not included in the average because it included a large tank rupture that was not a marine spill). As approximately 25% of the world oil movement by sea is around the U.S. coast and the rate of spillage is not greatly different elsewhere, an estimate can be made for marine terminal discharges of 0.010 mta (0.0025 x 4).

Major accidental spills in marine terminals, such as submarine pipeline and storage tank ruptures, although occurring rarely, may be the major causes of oil losses under this category (sometimes over 0.010 mt). However, no worldwide statistical data are available. For the present estimate, an average 0.010 mta of oil is assumed to be spilled into the oceans due to such accidents.

Thus, total spillage (discharges plus spills) from marine terminals is estimated at 0.020 mta (0.010 + 0.010).

BILGE AND FUEL OIL

Discharges under this category can be divided into three types: machinery space bilges, fuel oil sludge, and oily ballast from fuel tanks.

Machine Space Bilges

Steam tankers generate approximately 5 gal of bilge oil per day, while motor tankers generate about 15 gal per day. As there are about equal numbers of steam and motor tankers and tankers may operate some 300 days per year, the average quantity of bilge oil generated in a tanker per year is about 10 gal/42 x 7 x 300, or 10.2 metric tons.

The majority of the 7,100 world tankers retain such bilge oil in slop tanks for cargo oil or discharge it to shore reception facilities. Assuming that 10% of the total bilge oil generated in machinery spaces of tankers may be discharged into the sea, the annual discharge of bilge oil from tankers is estimated to be 10.2 tons x 7,100 x 0.1 = 7,242 tons, or 0.007 mta.

With a similar approach, the average quantity of bilge produced in cargo ships can be also estimated. The average size of tankers is approximately 25,000 gross tonnage (GT), with an average size of propulsion machinery of 20,000 horsepower (HP); in comparison, the average size of nontankers is approximately 3,700 GT, with an average size of propulsion machinery of 4,000 HP. Almost all nontankers are motor ships.

Therefore, the slop oil generated in each nontanker would on average be 3.0 gal per day (15 gal x 4,000/20,000), or 3.1 metric tons per year (3 gal/42 x 7 x 300). For the world's nontanker fleet of 66,700 the amount of total bilge oil produced would be 3.1 tons x 66,700 per year, or 0.207 mta.

The quantity of this bilge oil discharged into the ocean would depend on whether the ships are fitted with oily-water separators and on the availability and use of shore reception facilities. About half these ships are fitted with separators.

Assuming that ships with separators would discharge into the sea 10% of bilge oil and ships without separators two-thirds of bilge oil, the quantity of oil from bilge discharged into the sea per year is estimated to be 0.079 mta (with separators: 0.207 x 0.5 x 0.1 = 0.01 mta; without separators: 0.207 x 0.5 x 0.67 = 0.069 mta).

Thus, the rate of PHC input into the oceans from bilge discharges is 0.086 mta (tankers 0.007 mta, nontankers 0.079 mta).

Fuel Oil Sludge

Worldwide annual use of heavy residual bunker fuel for marine application is 108 mt. Tankers use 44 mt of bunker fuel, while nontankers use 64 mt of bunker fuel as well as 18 mt of gas oil.

Before being used in diesel engines, residual bunker fuel is purified to remove such impurities as sludge and water. The average sludge content in heavy fuel for marine application is 0.5%. For the present estimate, 0.3% of the quantity of heavy fuel oil used for diesel engines is assumed to be disposed of.

In the case of steam tankers, the usual practice is to retain sludge in cofferdams or slop tanks for cargo oil for eventual disposal to shore facilities. In the case of nontankers, the capacity of the sludge holding tank on board may not be sufficient to retain the sludge until the ship arrives at port. There would then be no alternative but to dispose of it into the sea.

If 20% of the sludge for motor tankers and 90% of the sludge for nontankers is discharged into the ocean, then the annual quantity of sludge discharged is estimated to be 0.186 mta (tankers 44 x 0.5 x 0.003 x 0.2 = 0.013 mta, nontankers 64 x 0.003 x 0.9 = 0.173 mta).

Oily Ballast From Fuel Oil Tanks

Water ballast for tankers is carried in cargo oil tanks or SBT, and no contamination of water ballast with fuel oil should occur. However, nontankers, which have to carry large quantities of water ballast for safety reasons, particularly fishing vessels, may have to carry water ballast in fuel oil tanks.

It is estimated that 2% of nontankers carry ballast water in fuel oil tanks, with an average clingage of 0.8% (including heavy and light fuel), a quarter of which will be discharged into the sea. Nontankers use some 64 mt of residual bunker fuel plus 18 mt of gas oil. Thus, the annual quantity of such oily ballast discharges is estimated to be 0.003 mta (82 x 0.02 x 0.008 x 0.25).

The total for the bilge and fuel oil inputs is 0.28 mta (bilge 0.086 mta, fuel oil sludge 0.186 mta, oily ballast 0.003 mta).

ACCIDENTAL SPILLAGES

Tanker Accidents

Various sources of data on tanker accidents producing oil pollution
have been available, including data from the International Tanker
Owners Pollution Federation Ltd. (ITOPF, 1981) and the French Institute
of Petroleum (IFP) (Bertrand, 1979), as shown in Table 2-12.

Nontanker Accidents

IFP (1981) average of annual oil spillages from nontanker accidents
over the years 1974-1979 is 0.017 mta. Therefore, the total quantity
of oil discharges due to maritime accidents is estimated to be 0.41 mta
(tanker 0.39, nontanker 0.02).

The estimated range for quantity of oil discharged annually into
the sea from transportation activities is 1.0-2.6 mta. This compares
with the earlier NRC (1975) range of 1.5-2.8 mta. Table 2-13 shows the
estimated range and best estimate of PHC discharged into the sea from
each category of transportation losses. In general a ±100% range was
considered realistic, but each category has been reviewed and slight
adjustments have been made. Tanker accident data were considered the
most reliable and were therefore assigned a ±10% range.

Although not addressed in this report in detail, spills caused
through acts of war should be considered, where appropriate, in future
discussions of inputs, particularly in areas such as the Persian Gulf.

TABLE 2-12 Annual Quantity of Oil Spills Due to Tanker Accidents

| Year | Quantity (mt) | |
	ITOPF	IFP
1975	0.368	0.362
1976	0.456	0.364
1977	0.316	0.297
1978	0.388	0.487
1979	0.760	0.649
1980	0.187	
TOTAL	2.731 (1974-1980)	2.374 (1974-1979)
Average	0.390 mta (1974-1980)	0.396 mta (1974-1979)

NOTE: The annual figure varies considerably, influenced primarily by a
few catastrophic incidents. For the purpose of the present estimate,
the average figure of 0.39 mta is appropriate.

TABLE 2-13 Summary of Transportation Losses (mta)

Type of Loss	Range	Best Estimate
Tanker operations	0.44-1.45	0.71
Dry-docking	0.02-0.05	0.03
Marine terminals	0.01-0.03	0.02
Bilge and fuel oil	0.16-0.60	0.28
Tanker accidents	0.35-0.43	0.39
Nontanker accidents	0.02-0.04	0.02
TOTAL	1.00-2.60	1.45

ATMOSPHERE

The estimated range of atmospheric input of PHC into the marine environment is 0.05-0.5 mta. The workshop panel working on atmospheric input agreed that they could not provide a "best" estimate because of the great uncertainty associated with their estimate. The primary pathway for this input appears to be removal by rain of particulate material. Secondary pathways involve dry deposition of atmospheric particulate matter, precipitation scavenging of trace gases, and direct gas exchange with the ocean.

Less is known about the global sources, distribution, and fluxes of organic matter than any other major class of chemical substances in the atmosphere. Aside from methane and certain halocarbons in the vapor phase, very few measurements of gaseous or particulate organic matter are available outside urban areas. Recent reviews by Duce (1978) and Simoneit and Mazurek (1981) have attempted to synthesize the available data and summarize our knowledge. The situation is complicated, of course, by the fact that there are probably thousands of different organic compounds emitted to the atmosphere from natural and pollution sources, and many other compounds are produced from atmospheric, particularly photochemically induced, reactions. Each of these substances has its own characteristic chemical and physical properties and associated atmospheric sources, residence times, and sinks.

In terms of these pathways, petroleum entering the sea via the atmosphere must first evaporate or be emitted into the atmosphere. National emission inventories identify vehicle exhaust and evaporation losses as the greatest source, followed by industrial losses through evaporation, particularly oil industry operations.

Rough estimates of the input of PHC to the ocean from the atmosphere have been made in the past. The Study of Critical Environmental Properties (SCEP, 1970) estimated that 9 mta of PHC entered the ocean from the atmosphere at that time and suggested this number could double by 1980. NRC (1975) estimated that the atmospheric input of PHC was much lower, about 0.6 mta. This latter estimate was not based on any measurements over the ocean, but simply on the total quantity of PHC injected into the atmosphere, its assumed reactivity in the atmosphere, and the general distribution of particles and the patterns of rainfall over the sea and the land.

The actual atmospheric input of petroleum to the ocean surface is very difficult to ascertain for several reasons. Petroleum is a complex mixture of many classes of compounds whose components have different reactivities and solubilities. For example, low-molecular-weight polynuclear aromatic hydrocarbon (PAH) and n-alkane reactivities with the OH radical span 5 orders of magnitude (Darnell et al., 1976). Chameides and Cicerone (1978) suggested that the photochemical lifetime of atmospheric ethane is about 25 days, while that of propane, butane, and pentane may be about 5 days. Zimmerman et al. (1978) and Hanst et al. (1980) have pointed out the potential importance of the photo-oxidation of nonmethane hydrocarbons as a source for atmospheric CO. Gas to particle conversion also occurs for organic material (Simoneit and Mazurek, 1981; Duce, 1978). During transport from continents to the sea via the atmosphere, particle fractionation may occur. Hence, if the organic composition is different for various particle size classes, the overall atmospheric particulate organic composition will change as a function of transport distance and time. Thus the organic chemical composition of petroleum-derived substances in the remote marine atmosphere may bear very little resemblance to what was emitted into continental air masses several thousand kilometers away. Many individual compounds in petroleum are also produced from other natural sources, such as $n-C_{15}$ and $n-C_{17}$ alkanes from marine phytoplankton, pristane from zooplankton, and $n-C_{27}$ and $n-C_{29}$ alkanes from land plants (see Chapter 3, Chemical Methods section). In several areas of the ocean, such as upwelling zones or downwind of major forests, these compounds may make up a significant portion of the hydrocarbons in the atmosphere. Finally, there is a paucity of data on petroleum organic compounds in rain, vapor, and particulate samples from open ocean areas, thus requiring a large number of simplifying assumptions to be made in any estimate of air to sea transport.

Taking into consideration the problems discussed above, the workshop panel on atmospheric input decided to concentrate on the n-alkane components of petroleum. The alkanes constitute approximately 30% of petroleum and some data are available, albeit very limited, to undertake estimates of their atmospheric input to the open ocean (Ketseridis and Eichmann, 1978; Eichmann et al., 1979, 1980; Hahn, 1981; Gagosian et al., 1981a; Gagosian et al., 1982; Atlas and Giam, 1981). Data for PAH are fewer and not sufficient to estimate their atmospheric input. Aerosol samples that had an oceanic origin were collected from coastal Norway by Bjorseth et al. (1979). The total PAH concentration averaged 1.6 ng/m^3 (nanogram per cubic meter). Hahn (1980) found PAH to be 80% of the n-alkane concentration for aerosol samples from the southern North Atlantic. His PAH values averaged 11 ng/m^3. However, no single PAH was detected in greater than 5-pg/m^3 (picogram per cubic meter) air concentration for particles at Enewetak Atoll, Marshall Islands, in the central North Pacific (Gagosian et al., 1981a). No PAH vapor or rain data have been reported from the open ocean.

As stated earlier, many of the n-alkanes produced by marine plankton and land plants are the same as those in petroleum. It is difficult to subtract out this "biogenic" component from many of the data sets available without introducing a preconceived bias. Rather than do this for selected samples, we did not do it for any of them. Atmospheric inputs to the ocean are thus derived for n-alkanes as a class of organic substances. The fraction of these n-alkanes that are of petroleum origin is uncertain. Thus the fluxes obtained represent maximum values relative to petroleum n-alkane input into the ocean from the atmosphere.

There are few data for n-alkanes over the ocean, limited basically to those of three research groups: the Ketseridis-Eichmann-Hahn group, the Gagosian-Duce-Zafiriou group, and the Atlas-Giam group. Data from over the open North Atlantic Ocean, from the Irish coast, and from Cape Grim, Tasmania, in the Indian Ocean are from the Ketseridis-Eichmann-Hahn group (Ketseridis and Eichmann, 1978; Eichmann et al., 1979, 1980; Hahn, 1981). Data from Enewetak Atoll, Marshall Islands, in the tropical North Pacific, have been obtained by the Gagosian-Duce-Zafiriou and Atlas-Giam groups (Gagosian et al., 1981, 1982; Zafiriou et al., 1982; Atlas and Giam, 1981).

The observed concentrations of particulate and vapor phase alkanes in the marine atmosphere are presented in Tables 2-14 and 2-15. Data from three North Atlantic sites are presented. Loop Head is on a peninsula on the west coast of Ireland at about 52°30'N, 9°50'W. Samples were collected from a cliff about 70 m above sea level. Samples collected only when the wind was from the ocean are reported here. Samples were also collected from a ship at the Joint Air/Sea Interaction (JASIN) site, located between Iceland and Scotland (60°N, 13°W). The tropical North Atlantic samples were also collected from a ship, in this case operating in the North Atlantic trade wind regime at approximately 15°N between Africa and the Caribbean Sea. The German data from Cape Grim, Tasmania, were obtained from the Australian Baseline Atmospheric Monitoring Station located on the northwest tip of Tasmania (40°41'S, 144°40'E). Samples were collected on a cliff 90 m above sea level. The samples obtained by the Gagosian-Duce-Zafiriou and Atlas-Giam groups were collected from a 20 m tower located on the windward coast of Bokandretok Island, Enewetak Atoll, Marshall Islands (11°20'N, 162°20'E). Sample collection on Enewetak was controlled automatically by wind speed and direction as well as atmospheric particle counts to avoid local contamination. Efforts were made in all studies to avoid local contamination.

Concerning the analytical methodology of measuring hydrocarbons in atmospheric samples, a recent review by Simoneit and Mazurek (1981) and reports by Ketseridis et al. (1976) and Gagosian et al. (1981a) discuss the necessity of ultraclean samplers and sampling conditions. The need to separate the hydrocarbon classes from other organic compound classes (usually by liquid chromatography) before gas chromatography (GC) and GC/mass spectrometry for quantitative analyses and structural determination was stressed by Gagosian et al. (1981a). The use of high resolution glass capillary GC for analysis is needed. These hydrocarbon measurements must be made in conjunction with micrometeorological studies of the sampling site and long range transport studies to

TABLE 2-14 Particulate n-Alkane Concentrations in the Marine
Atmosphere (ng/m^3 STP)

n-Alkane	Loop Head, Republic of Ireland[a]	JASIN Site[b]	Cape Grim[c]	Tropical North Atlantic[d]	Enewetak[e]
n-C_{15}	0.06	0.10	0.12	4.1	
n-C_{16}	0.13	0.12	0.27	3.2	
n-C_{17}	0.08	0.26	0.32	4.2	
n-C_{18}	0.19	0.17	0.13	2.5	
n-C_{19}	0.16	0.20	0.13	8.2	
n-C_{20}	0.33	0.24	0.17	1.3	
n-C_{21}	0.22	0.26	0.35	1.1	0.0017
n-C_{22}	0.26	0.28	0.11	3.0	0.0020
n-C_{23}	0.31	0.29	0.15	2.0	0.0023
n-C_{24}	0.45	0.21	0.18	0.4	0.0021
n-C_{25}	0.37	0.33	0.18	0.6	0.0030
n-C_{26}	0.27	0.28	0.20	0.5	0.0022
n-C_{27}	0.23	0.37	0.52	0.6	0.0067
n-C_{28}	0.22	0.19	0.40	0.3	0.0037
n-C_{29}					0.0170
n-C_{30}					0.0033

[a]Eichmann et al. (1979) and Hahn (1981).
[b]Hahn (1981).
[c]Hahn (1981) and Eichmann et al. (1980).
[d]Ketseridis and Eichmann (1978).
[e]Gagosian et al. (1981a, 1982); average of six samples.

ascertain the sources and transport pathways involved. Clearly, data
are also needed on other anthropogenic compounds, such as chlorinated
hydrocarbons, phthalate esters, and trace metals, along with source
marker information such as ^{210}Pb and δ^{13}C to interpret the
hydrocarbon data more fully.

Table 2-14 lists the particulate n-alkane data from C_{15}-C_{30},
and Table 2-15 presents the vapor phase n-alkane data for C_{10}-C_{30}.
Particulate n-alkane data for C_{10}-C_{14} were not presented, since
these compounds cannot be quantitatively recovered during the extraction
of the filter with organic solvent, the solvent evaporation, and the
liquid chromatography steps in the analytical scheme (Mackay and
Wolkoff, 1973).

As might be expected for such different oceanic regions, the
measured concentrations of particulate and vapor phase n-alkanes were
quite different at several of these locations. Data for particulate
n-alkanes (Table 2-14) from the Ireland, JASIN, and Cape Grim sites are
all rather similar, generally within a few tenths of a ng/m^3. All

TABLE 2-15 Vapor Phase n-Alkanes in the Marine Atmosphere (ng/m^3 STP)

n-Alkane	Loop Head, Republic of Ireland[a]	JASIN Site[b]	Cape Grim[c]	Enewetak[d] Florosil	Enewetak[d] Polyurethane Plugs	Enewetak[e] Polyurethane Plugs
n-C$_{10}$	12	15	21			
n-C$_{11}$	14	9	20			
n-C$_{12}$	11	5	8			
n-C$_{13}$	9	4	8	0.23		
n-C$_{14}$	9	3	6	0.19		
n-C$_{15}$	14	6	11	0.66		
n-C$_{16}$	8	4	8	0.13		
n-C$_{17}$	10	5	9	0.55		
n-C$_{18}$	12	5	8	0.07		
n-C$_{19}$	10	5	10	0.07		
n-C$_{20}$	18	6	9	0.07		
n-C$_{21}$	14	6	11	0.07		
n-C$_{22}$	20	5	4	0.07		
n-C$_{23}$	32	5	6	0.08	0.11	
n-C$_{24}$	22	3	6	0.09	0.14	0.032
n-C$_{25}$	16	3	6	0.10	0.14	0.095
n-C$_{26}$	9	2	5	0.08	0.10	0.088
n-C$_{27}$	7	2	3	0.06	0.08	0.055
n-C$_{28}$	6	1	2		0.06	0.024
n-C$_{29}$					0.006	0.019
n-C$_{30}$						0.013

[a]Eichmann et al. (1979) and Hahn (1981).
[b]Hahn (1981).
[c]Eichmann et al. (1980) and Hahn (1981).
[d]Atlas and Giam (1981).
[e]Zafiriou et al. (1982).

are much lower, however, than the data from the tropical North Atlantic. The reason for this difference is unclear. The latter sampling area is in the region of the Sahara dust plume, which carries large quantities of sand and soil-derived materials to the tropical North Atlantic in the northeast trade winds. However, the west coast of Africa would not be expected to be a significant source of petroleum-derived atmospheric n-alkanes, even though there is extensive tanker traffic along that coast. Viewed in the context of the other data presented in Table 2-14, it is tentatively concluded that the tropical North Atlantic data reported are not representative of that region. Clearly additional measurements to evaluate these data are needed. The JASIN, Ireland, and Cape Grim data are probably most representative of concentrations over the North Atlantic and in the coastal regions of the other oceans.

Particulate n-alkane concentrations from Enewetak Atoll are considerably lower than those over the North Atlantic or at Cape Grim. These data are probably more nearly representative of mid-North Pacific

TABLE 2-16   Summary of Atmospheric Inputs of n-Alkanes Into the Ocean (mta)

Mechanism	Case A	Case B
Rain scavenging of particles	0.023-0.23	0.0013-0.013
Rain scavenging of gases	1×10^{-7}-0.03	1×10^{-7}-0.002
Dry deposition of particles	0.0048-0.048	0.00022-0.0022
Direct gas exchange	0-0.02	0-0.0004
TOTAL	0.28-0.32	0.0015-0.018
GRAND TOTAL	0.03-0.3	0.03-0.3

Ocean, South Pacific Ocean, South Atlantic Ocean, and Indian Ocean regions far from continental influences.

Vapor phase n-alkane concentrations are presented in Table 2-15. Again, the Ireland, JASIN, and Cape Grim data are quite similar, but they are considerably higher than the Enewetak data, generally by a factor of 50-100. Note that there is, in general, good agreement between the Gagosian-Duce-Zafiriou and Atlas-Giam data at Enewetak for $n-C_{24}$ to $n-C_{29}$ alkanes. There are no vapor phase n-alkane data available from the tropical North Atlantic region. Again, the Ireland, JASIN, and Cape Grim data appear to be most representative of concentrations over the North Atlantic Ocean and in coastal regions, while the Enewetak data may be more representative of concentrations over the Indian, South Atlantic, South Pacific, and mid-North Pacific Oceans.

As can be seen from the data presented in Tables 2-14 and 2-15, the geographical coverage for atmospheric n-alkanes is very sparse. However, from this limited data, mean atmospheric particulate and vapor phase n-alkane concentrations over the world ocean have been derived, and input of this class of hydrocarbons into the oceans estimated. Details of the approach used are given by Duce and Gagosian (1982).

Table 2-16 presents a summary of calculations of the input of n-alkanes into the ocean. The total estimated input of atmospheric n-alkanes is 0.03-0.3 mta. The primary input mechanism clearly is via rain scavenging of n-alkanes on particles. However, better solubility data for $n-C_{20}$ to $n-C_{30}$ alkanes are needed before the importance of rain scavenging of gases and direct gas exchange in the deposition of n-alkanes to the sea surface can be fully assessed.

The estimates of the input of n-alkanes into the ocean via rain could be evaluated relatively easily by making measurements of the n-alkane concentrations in rain from samples collected, for example, in open North Atlantic and North Pacific regions--the regions in which most of the atmospheric petroleum hydrocarbons are apparently entering the oceans. Such rain measurements are strongly recommended.

As stated earlier, n-alkanes constitute approximately 30% of the organic components of petroleum. Cycloalkanes, PAH, and heteroatomic (nitrogen, sulfur, and oxygen) organic compounds make up the remainder. No data are available for the cycloalkane and heteroatomic compounds. Only a few numbers are available for PAH. Measurement of these other organic constituents of petroleum in vapor, aerosol, and rain samples are also strongly recommended.

The approach taken in using n-alkanes to estimate the input of petroleum into the ocean via the atmosphere is problematic. On one hand, using n-alkanes may give a maximum value of petroleum hydrocarbon atmospheric input, because many natural marine and terrestrially derived n-alkanes are included in the overall n-alkane deposition value. On the other hand, many organic components of petroleum, such as branched alkanes, cycloalkanes, and alkylated aromatics--the latter of which react very fast with OH radical to produce oxygenated species that fall to the ocean surface--are not included in the approach presented here. This suggests that using n-alkanes as an atmospheric input marker for the petroleum underestimates the input. On the basis of these facts, the estimate of PHC input is increased about two-thirds over that of the total n-alkane input. The range estimate for PHC input is thus about 0.05-0.5 mta.

More precise estimates of the atmospheric input of petroleum to the ocean will have to await information on the inputs of the various components of petroleum into the sea surface and further understanding of the reaction products, pathways, and rates of transformation of these compounds in the atmosphere.

COASTAL, MUNICIPAL, AND INDUSTRIAL WASTES AND RUNOFF

The estimated range of the input of PHC into the marine environment from municipal and industrial wastewaters, urban and river runoff, and ocean dumping is from 0.6 to 3.1 mta, with a best estimate of 1.2 mta (Table 2-17). Municipal wastewater appears as the largest contributor, followed by industrial discharges and urban runoff.

The earlier NRC (1975) study did not estimate a range of inputs of PHC from these sources, but made only a best estimate of 2.7 total mta. Many more data on these inputs have been accumulated over the past 7 years, so the lower estimates may be due in major part to better predictive capability and not necessarily to lower actual inputs.

Municipal Wastewaters

In 1979, Eganhouse and Kaplan (1981, 1982) analyzed 38 samples of treated municipal wastewater from five major wastewater pollution control plants in Southern California as reported by the Southern California Coastal Water Research Project (SCCWRP, 1980). The workshop panel decided to use four of these discharges in making their estimates for facilities serving approximately 9.8 million people in 1979.

TABLE 2-17 NRC Estimates of Hydrocarbons to World Ocean From Municipal and Industrial Wastes and Runoff (mta)

Source	NRC (1975)	1981 NRC Workshop	
		Most Probable	Likely Range
Municipal wastewater	0.3	0.75	0.4-1.5
Industrial			
Nonrefinery	0.3	0.2	0.1-0.3
Refinery	0.2	0.1	0.06-0.6
Urban runoff	0.3	0.12	0.1-0.2
River discharges	1.6	0.04	0.01-0.45
Ocean dumping	a	0.014	0.005-0.02
TOTAL	2.7	1.2	0.6-3.1

aNot estimated.

The wastewater samples were analyzed for total extractable organics and for total hydrocarbons (THC). The results of these analyses were compared with reported concentrations of oil and grease from the routine monitoring done by the wastewater management agencies as reported by SCCWRP (1980). Regression analysis indicates that THC accounts for approximately 38% of the oil and grease discharged from these treatment plants.

The total mass emission from the four discharges is estimated to be approximately 43 mta in 1979, resulting in an overall contribution of oil and grease of about 12 grams per capita per day (g/cap/d). These results can be used to calculate that the total per capita contribution of THC from the Southern California outfalls in 1979 was 38% of 12 g/d or 4.5 g/d.

The type and level of treatment given to the wastewater will affect the amounts of THC discharged. Based on general sanitary engineering experience with the removal of oil and grease in municipal wastewater, it is reasonable to assume an average removal of about one-third of the PHC in primary treatment and about 40% in secondary treatment. These removals can vary widely from plant to plant, depending on the plant design and operation. As most of the effluents in the Southern California wastewaters had been given primary treatment, the THC load in the untreated wastewater would be about 6.8 g/cap/d from municipal wastewaters (4.5 divided by 0.67).

In 1978, 120 million people lived within 50 miles of the coasts of the United States (U.S. Census Bureau, 1978). About 30% of this population lived on the West Coast of the continental United States, and 70% on the Gulf and East coasts. Assuming that the bulk of the wastewaters on the West Coast are given primary treatment, and those in the remainder of the country receive secondary treatment, the THC discharged to the U.S. coastal waters would be

$$[120 \times 10^6 \times 0.3 \times 6.8\ (10-0.33) \times 10^{-6} \times 365]$$
$$+\ [120 \times 10^6 \times 0.7 \times 6.8\ (1.00-0.40) \times 10^6 \times 365]$$
$$=\ 185{,}000\ \text{ta, or } 0.19\ \text{mta}$$

This calculation assumes that the oil and grease values reported by Eganhouse and Kaplan (1981, 1982) for the Los Angeles area are representative of discharges throughout the United States.

Evidence that this estimate of 0.19 mta for the entire United States is not too far out of line comes from Connell (1983). He reported that about 0.012 mta of petroleum hydrocarbons are going into the Hudson-Raritan Estuary from sewage discharge. This represents about 6-7% of the overall U.S. estimate for sewage discharges.

The calculated per capita THC discharge rate of 6.8 g/d cannot be used for other areas of the world because of the widely varying usage of petroleum products. In 1980, the United States used 18.3 million barrels per day (bbl/d) of petroleum products (International Petroleum Encyclopedia, 1980), and the estimated discharge of THC to the coastal waters if the wastewaters were untreated would have been about 298,000 ta ($120 \times 10^6 \times 6.8 \times 10^{-6} \times 365$), or about 16.1 ta for each 1,000 bbl/d consumed. This factor and an estimate of the extent of wastewater treatment in various areas of the world are used to estimate a global discharge, as shown in Table 2-18.

The estimated global discharge of 0.75 mta is based on a series of assumptions that are supported by few data. Note that one of these assumptions is the equivalency of THC with PHC. There is no doubt that PHC makes up a major fraction of THC, but the exact percentage is not known. However, the calculations do provide a rationale for the estimation procedure and show the areas in which measurements and data are required.

Nonrefinery Industrial Wastes

A sizable fraction of nonrefinery industrial waste discharges into municipal wastewater systems and its PHC have been accounted for in the previous section. However, there is a quantity of PHC that goes more or less directly into the marine environment through coastal nonrefinery effluent discharges. Extremely limited quantitation of this source has been made, and even less information is published for reasons of confidentiality. Previous estimates have been made by the NRC (1975) of 0.3 mta, and the Royal Commission on Environmental Pollution (1981) of 0.150 mta. Therefore, the estimate of this input of PHC is put at 0.2 mta, with full realization that the confidence in this number is quite limited.

TABLE 2-18 Global Discharge of Hydrocarbons Into Municipal Wastewaters

Area	1980 Petroleum Consumption[a] (millions of bbl/d)	Estimated Untreated THC Load[b] (mta)	Percent THC Removed by Treatment	Residual THC (= PHC) Discharged (mta)
North America				
United States	18.3	0.30	38[c]	0.19
Canada	1.8	0.03	38[c]	0.02
Latin America	4.2	0.07	0	0.07
Asia and Pacific	9.1	0.15	0	0.15
China	1.7	0.03	0	0.03
Middle East	2.0	0.03	0	0.03
USSR and Eastern				
Europe	10.5	0.17	30[d]	0.12
Western Europe	10.5	0.17	30[d]	0.12
Africa	1.2	0.02	0	0.02
TOTAL	63.1	0.97		0.75

[a]Source: International Petroleum Encyclopedia (1980).
[b]Assuming 16.1 ta of THC per 1,000 bbl/d consumed.
[c]1 - 185,000/298,000 = 0.38.
[d]Assumed.

Industrial Wastes From Refineries

This category of refinery discharges includes only those refineries that discharge PHC from their own wastewater facilities. Other refineries that do not have their own facilities are assumed to discharge their wastewater into municipal wastewater facilities.

Recently, estimates were made of the amount of PHC discharged with refinery industry effluents (National Petroleum Council, 1981). The National Petroleum Council (NPC) determined during 1977-1979 that for those refineries that treat and discharge their own wastewater, 0.002-0.004 mta of PHC were discharged. These values were based entirely on oil and grease analyses (one can assume that volatile liquid hydrocarbons were not analyzed).

The NPC related PHC discharge rates to total operating capacity of U.S. discharge refineries. It is estimated that 0.0025-0.005 kg of PHC is discharged annually for each 10^3 kg/yr of operating capacity. This value can be compared to a 1977 European value of 0.04 kg/10^3 kg (Royal Commission on Environmental Pollution, 1981). It also can be compared to the 1967 U.S. refinery survey value of 0.075 kg/10^3 kg (National Petroleum Council, 1981).

Selection of PHC discharge rates for the world is difficult. The United States seems to be unique in its rate estimate. Either its rate of 0.005 kg/10^3 kg (upper value of U.S. range) or the European rate of 0.04 kg/10^3 kg could be applied to the Canadian refinery rate. For these calculations, the higher rate was used for Canada. For the rest of the world, the PHC discharge rate is assumed to be no better than that which was measured in the 1967 U.S. refinery survey.

Estimating the fraction of PHC that reaches the ocean from all worldwide refinery sources is also difficult. In the United States, the Environmental Protection Agency (EPA) has determined the amount of refinery wastewater discharged directly into receiving bodies and indirectly into publicly owned treatment plants (Environmental Protection Agency, 1978, 1979). The following percentages are based on the processing capacity of refineries, not numbers of refineries.

Approximately 81% of all refineries discharge wastewater directly into receiving bodies; another 14% discharge indirectly to publicly owned treatment plants; the remaining 5% have no wastewater discharges. In California, 64% of all refineries discharge directly into receiving bodies; the rest into publicly owned treatment plants. In the continental United States, approximately 50% of all refineries discharging directly into receiving bodies are near the coast (an additional 7% occurs outside the continental United States). Of all the refineries discharging into public treatment plants, 63% are near the coast.

For the United States, the fraction of PHC directly reaching the world oceans from refinery sources is estimated to be 0.5. The fraction of PHC reaching the world oceans from other locations of the world is a rough estimate based on limited data.

The new values of 0.005 kg PHC/10^3 kg production for the United States, 0.04 for Canada and Europe, and 0.075 for the rest of the world have been used to obtain a refinery PHC global discharge estimate of 0.10 mta (Table 2-19). The Royal Commission on Environmental Pollution (1981) estimates a total global discharge from refineries to the sea of 0.06 mta.

Urban Runoff

The global input of petroleum hydrocarbons to coastal waters from urban runoff was estimated by NRC (1975) to be 0.3 mta. The value was based in part on the assumption that urban runoff contributed about half the amount of PHC contributed by municipal and nonrefinery wastewaters. The crudeness of this estimate was unavoidable because of the lack of measurements of PHC in urban runoff. The situation 8 years later is only slightly better because most of the studies undertaken in the intervening years have focused on analytical methods of characterizing the PHC fractions rather than on mass contributions of PHC. Part of this dilemma may be due to the difficulty of representative sampling of the runoff. Other problems are the determination of mean PHC concentrations and the volume of runoff, which permit accurate estimation of

TABLE 2-19 Estimated Global Discharge of Petroleum Hydrocarbons in Refinery Wastewaters

Geographic Area	Crude Oil Refinery Capacity 10^6 bbl/d	mta	Estimated Hydrocarbon Loss (kg/10^3 kg)	Total PHC Discharge in Refinery Wastewaters (mta)	Estimated Fraction to World Ocean	PHC Input Into Ocean (mta)
North America						
United States	18.4	960	0.005	0.0048	1/2	0.002
Canada	2.2	115	0.04	0.0046	1/5	0.001
Latin America	8.1	420	0.075	0.0315	4/5	0.025
Asia-Pacific	10.6	550	0.075	0.0413	1/2	0.021
China	1.8	90	0.075	0.0068	1/4	0.002
Middle East	3.7	190	0.075	0.0143	1/3	0.005
USSR and Eastern Europe	14.2	740	0.075	0.0555	1/3	0.018
Western Europe	20.22	1,050	0.04	0.0420	1/3	0.014
Africa	1.7	90	0.075	0.0068	1/2	0.003
TOTAL	80.9	4,205		0.1946		0.10 (best estimate)

NOTE: Conversion is 1 mta = 19,000 barrels per calendar day capacity.

mass PHC contributions. Estimates of PHC in runoff should be based on
factors such as runoff area, watershed characteristics, PHC usage, and
population density. Recent papers that have reported on the
characterization of PHC in urban runoff appear in Table 2-20.

Most of the studies in Table 2-20, except that of Hoffman et al.
(1982), do not provide sufficient data for the estimation of PHC
contributions from urban runoff based on watershed characteristics.
There have not been enough studies reporting watershed characteristics
to permit rational estimation. Recognizing the difficulties of
quantifying the mass of PHC contributed and considering hydrological,
physical, and land use variations in urban areas (as well as the
definition of urban), we conclude that the best estimate of urban PHC
runoff must be based on estimates of per capita contributions.
Population is one of the principal generating factors of urban PHC
runoff for a given petroleum consumption level. Table 2-21 shows data
for per capita estimates of PHC contributions from several locations.

Despite the gross variation in per capita PHC contribution, it is
believed to be the most accurate basis for current estimation of urban
PHC runoff. A per capita PHC contribution of 1.0 g/cap/d is probably
the most reliable estimate that can be made from present information.

Employing the unit per capita contribution of 1.0 g/cap/d per day
and a coastal population of about 120 million, one can estimate the
urban runoff contribution of the United States to be about 0.04 mta.
Assuming the United States uses about 0.3 of the world's hydrocarbons,
one can estimate the world urban runoff PHC contribution to the world
ocean to be about 0.12 mta, which is about one-third of the contribution
estimated by NRC (1975).

TABLE 2-20 Selected Urban Runoff Studies

Location	Drainage Area (ha)	Number of Storms Studied	Reference
Seattle	-	-	Wakeham (1977)
North Philadelphia	616		Hunter et al. (1979)
North Philadelphia	616	5	Hunter et al. (1979)
North Philadelphia	616		Whipple and Hunter (1979)
Trenton A	8	2	Whipple and Hunter (1979)
Trenton B	82	3	Whipple and Hunter (1979)
Los Angeles	210,000	1	Eganhouse and Kaplan (1981, 1982)
Leon County, Fla.	357	1	Byrne et al. (1980)
Narragansett Bay, R.I.	167,000	21	Hoffman et al. (1982)

River Discharges

Reexamination of the global input of hydrocarbons to the oceans indicates that the inclusion of a separate category for river discharges may be improper because of double accounting of hydrocarbon input. The major sources of hydrocarbons in rivers are the untreated and treated wastewater discharges, runoff (both urban and rural), and spills. All these sources are quantified and reported separately for coastal areas. If an additional 110 million people discharge PHC into the interior rivers of the United States (at a rate of 6.8 g/cap/d) and if 5% of these PHCs eventually reach the oceans, then this yields an annual flux of PHC from rivers to U.S. coastal waters of 0.013 mta. Assuming this amount is one-third of the world total, the river discharge of PHC to the ocean would be 0.04 mta.

OCEAN DUMPING

Some hydrocarbons are discharged into U.S. and world coastal regions in association with municipal wastewater treatment plant sludge/underflow. The sludge is generally discharged from dumping by barge or by discharges through pipelines. In the United States, this sludge is discharged by dumping in the New York Bight and by pipeline on the West Coast. In the New York Bight, approximately 7×10^6 wet tons of sludge are discharged per year. This material contains approximately 2,000 ppm of oil and grease, of which about 40% are hydrocarbons. This

TABLE 2-21 Per Capita Estimates of PHC Contributions in Urban Runoff

Location	Unit PHC Contribution (g/cap/d)	Reference
Philadelphia and Trenton	0.03	Whipple and Hunter (1979)
Narragansett Bay	2.7	Hoffman et al. (1982)
Los Angeles[a]	1.9	Eganhouse and Kaplan (1981)
Seattle	0.3	Wakeham (1977)
Sweden[b]	0.3	NRC (1975)

[a]Single storm extrapolated to annual runoff by author.
[b]Typical urban area (0.2 parking, 0.3 multifamily, and 0.6 single family).

amounts to 0.006 mta of THC. In addition, the Los Angeles pipeline discharges about 2,450 tons of oil and grease annually through the 7 mile sludge outfall (Eganhouse and Kaplan, 1981). This is estimated to be 0.001 mta of THC.

The annual worldwide discharge of wastewater sludge into the oceans is approximately 16 million tons. Thus, applying a similar ratio to that used for the United States, the total amount of hydrocarbons discharged worldwide by ocean dumping is about 0.02 mta.

Hydrocarbons are also released to the oceans from the dumping of dredge spoils. Dredge spoils are river and channel sediments that have been relocated by dredging and dumping operations. The hydrocarbons that accompany these spoils are accounted for in other sections of this report and are not included in the ocean dumping category.

GEOGRAPHICAL DISTRIBUTION OF INPUTS

The input of petroleum hydrocarbons into the ocean is certainly not distributed evenly. The geographical distribution of the inputs from each source is discussed below.

• Marine transportation (1.5 mta). The input of PHC from this source is concentrated along the principal transportation routes and in harbors and ports where oil tankers or other vessels are loaded or unloaded. About half the transportation total is derived from tanker operations (0.7 mta). Most of this loss is probably at sea along the prominent tanker routes from the Middle East to Europe, the American continents, or the Far East. Another major source in this category is tanker accidents (0.4 mta). These also tend to occur along the tanker routes, but in more congested areas near ports or in narrow straits. The third major source, that of bilge and fuel oils (0.3 mta), probably follows a similar distribution pattern to that of the tanker operations.

• Offshore oil production (0.05 mta). This relatively minor input occurs at offshore oil production facilities, and these tend to be near coastlines. The largest offshore producing areas are the Arabian Gulf, the North Sea, the Gulf of Mexico, offshore California, offshore Malaysia and Indonesia, and the west coast of Africa.

• Refineries (0.1 mta). This input of PHC into the sea is concentrated near the coasts of countries that do most of the refining of petroleum (e.g., the United States, Great Britain, Germany, France, Japan, Canada, Mexico, Kuwait, and Saudi Arabia).

• Nonrefinery wastes (0.2 mta). This input into the sea is concentrated near the coasts of the more industrialized nations in the world, such as the United States, the northern European countries, and Japan.

• Municipal wastes (0.75 mta). This input of PHC is distributed in much the same way as the nonrefinery industrial wastes. It would be concentrated near the coasts of the more highly industrialized and heavily populated nations. Best examples would again be the United States, the northern European countries, and Japan.

• Urban runoff (0.12 mta). This input of PHC closely follows the input from municipal wastes. The input would be primarily into coastal areas of countries with high industrialization and large populations.

• River runoff (0. 04 mta). This input is in coastal areas near the mouths of large rivers, such as the Mississippi, the Rhine, the Danube, the Saint Lawrence, and the Elbe.

• Natural sources (0.3 mta). Submarine seeps, at least those identified thus far, seem to be associated with tectonically active regions of the world and are usually near the coasts of continents. Such areas are offshore California and Alaska, the Arabian Gulf and the Red Sea, the northeast coast of South America, and the South China Sea.

• Atmosphere (0.05-0.5 mta). This input of PHC into the seas would be primarily downwind of heavily industrialized areas. Again, the inputs are greatest near the coastlines, with concentrations decreasing away from the coasts. The northwest Atlantic, the North Sea, and the northwest Pacific (near Japan) would probably have typically large atmospheric inputs of PHC.

Data are not available to estimate total PHC input by region except in an extremely qualitative manner. If one looks at information on the geographical distribution of each input, then one can say, qualitatively, that coastal areas off the United States, Europe, and Japan and the Arabian Gulf would probably have greater inputs.

SUMMARY AND RECOMMENDATIONS

The estimated range for total input of petroleum from all sources is 1.7-8.8 million mta. The best single-number estimate of total input is 3.2 mta. We believe that the range is a more accurate summary of the state of knowledge than a single-number best estimate. Uncertainties are particularly evident with certain sources, i.e., natural inputs (seeps and erosion), transportation, municipal/industrial runoff, and atmospheric inputs. There are also wide geographical gaps in information on sources, especially in the southern hemisphere. Table 2-22 presents sources, probable ranges, and best estimates for sources. The spread in probable range about the best estimate is a qualitative measure of the faith in the best estimate. For example, the tanker accident probable range is narrow (0.3-0.4 mta), so the best estimate is probably good. On the other hand, the marine seep probable range is wide (0.02-2.0 mta), indicating small reliability in the best estimate.

The 1975 NRC report gave only a single-number estimate of total input of petroleum, namely, 6.1 mta. No range was given. This number falls within the current estimated range of 1.7-8.8 mta. The difference in the two single-number estimates, 6.1 mta in 1975 and the current 3.2 mta, does not necessarily reflect a significant decline in input but indicates better estimation of individual inputs.

Although the amount of petroleum and petroleum products transported by sea, as well as crude oil produced offshore, has increased during the past 8 years, PHC input into the marine environment estimated at the 1981 NRC workshop does not appear to have followed this trend. This may be for the following reasons: (1) the individual input estimates are more accurate due to improved analytical data on PHC concentrations in effluent streams, (2) positive steps have been taken to reduce operational and accidental release of petroleum into the sea, and (3) double accounting of PHC inputs from sources has been reduced. Double accounting arises when it becomes difficult to distinguish PHC inputs from closely related sources (e.g., urban runoff, river runoff, industrial and municipal wastes). Thus, there may be the tendency to count the same PHC inputs twice or more times under different sources.

One source of PHC into the marine environment that was not estimated was PHC released from pleasure craft, primarily in near-coastal marine waters. Pleasure craft are primarily small inboard or outboard motorboats. While inputs from pleasure craft may be locally significant, we believe that the total amount of this input would not be on the same scale with the other inputs considered.

Major problems still remain in the estimation of PHC inputs into the marine environment. Certainly, significant improvements have been made in recent years in obtaining better analytical data on concentrations of PHC entering the marine environment from varied sources. However, additional work is still needed, particularly in the acquisition of improved data on PHC inputs from the atmosphere, from municipal and industrial waste sources, and from natural sources such as marine seeps and erosion of terrestrial sediments.

Following is a list of recommended research programs or projects that would address these problems:

TABLE 2-22 Input of Petroleum Hydrocarbons Into the Marine Environment (mta)

Source	Probable Range	Best Estimate[a]
Natural sources		
Marine seeps	0.02-2.0	0.2
Sediment erosion	0.005-0.5	0.05
(Total natural sources)	(0.025)-(2.5)	(0.25)
Offshore production	0.04-0.06	0.05
Transportation		
Tanker operations	0.4-1.5	0.7
Dry-docking	0.02-0.05	0.03
Marine terminals	0.01-0.03	0.02
Bilge and fuel oils	0.2-0.6	0.3
Tanker accidents	0.3-0.4	0.4
Nontanker accidents	0.02-0.04	0.02
(Total transportation)	(0.95)-(2.62)	(1.47)
Atmosphere	0.05-0.5	0.3
Municipal and industrial wastes and runoff		
Municipal wastes	0.4-1.5	0.7
Refineries	0.06-0.6	0.1
Nonrefining industrial wastes	0.1-0.3	0.2
Urban runoff	0.01-0.2	0.12
River runoff	0.01-0.5	0.04
Ocean dumping	0.005-0.02	0.02
(Total wastes and runoff)	(0.585)-(3.12)	(1.18)
TOTAL	1.7-8.8	3.2

[a]The total best estimate, 3.2 mta, is a sum of the individual best estimates. A value of 0.3 was used for the atmospheric inputs to obtain the total, although we well realize that this best estimate is only a center point between the range limits and cannot be supported rigorously by the data and calculations used for estimation of this input.

1. Improved methods should be developed for large scale, areal documentation of the continental margins to determine the extent of submarine seepages of petroleum. A program should be undertaken to gauge accurately flow rates for seeps of significantly different sizes, including probable microseeps.

2. There should be continued monitoring of all facilities discharging low levels of petroleum hydrocarbons dispersed or dispersed in

aqueous effluents (e.g., offshore platforms, refineries, and other industrial plants and transportation units such as tankers and terminals).

3. Rain samples collected from several locations on the ocean and near sea coasts should be analyzed for PHC content. This work is important since rain scavenging of atmospheric particles is believed to be the major pathway for petroleum into the ocean from the atmosphere. It is also necessary to determine reactions of, and changes occurring in, various petroleum components as they are transported from sources through the atmosphere across and into the oceans.

4. More applied investigations, including accurate measurements of PHC, are needed to better define municipal, industrial, and runoff inputs to the oceans. This is particularly needed in southern hemisphere countries. These investigations may lead to quantitative methods for distinguishing petroleum hydrocarbons from oil and grease and natural hydrocarbons found in municipal and industrial waste as well as samples of runoff.

5. Data should be collected on the C_2-C_{10} aliphatic hydrocarbons in vapor, particulate, and rain samples from over the oceans, to relate these to the distributions of other classes of organic compounds present in petroleum.

6. Better solubility data are needed for n-alkanes and polynuclear aromatic hydrocarbons to better ascertain the importance of rain scavenging of gases and air-sea gas exchange processes to the contribution of the flux of atmospheric petroleum hydrocarbons to the ocean.

7. There is a need to determine the reactions and organic compound class distributional changes that occur for the various organic compounds in petroleum, as this material is transported from its source through the atmosphere across the oceans.

8. Better solubility data are needed for n-alkanes, polynuclear aromatic hydrocarbons, etc., to better ascertain the importance of rain scavenging of gases and air-sea gas exchange processes to the contribution of the flux of atmospheric petroleum hydrocarbons to the ocean.

REFERENCES

Albers, J.P., M.D. Carter, A.L. Clark, O.B. Coury, and S.P. Schweinfurth. 1973. Summary of petroleum and selected mineral statistics for 120 countries, including offshore areas. Professional Paper 817. U.S. Geological Survey, Washington, D.C. 149 pp.

Atlas, E., and C.S. Giam. 1981. n-Alkane atmospheric input into the tropic North Pacific Ocean. Unpublished manuscript. Texas A&M University.

Bertrand, A.R.V. 1979. Les principaux accidents de diversements petroliers en mer et la banque de donnees de l'Institute Francais du Petrole sur les accidents de navires (1955-1979). Rev. Inst. Francais Petrole 34:3-7.

Bjorseth, A., G. Lunde, and A. Lindskog. 1979. Long range transport of polycyclic aromatic hydrocarbons. Atmos. Environ. 13:34-53.

British Petroleum Co., Ltd. 1980. BP statistical review of the world oil industry. London.

Burnet, B. 1980. Worldwide drilling and production. Offshore 40:62-70.

Byrne, C., C.R. Donahue, and W.C. Burnett. 1980. The effect of urban stormwater runoff on the water quality of Lake Jackson. Unpublished manuscript. Florida State University, Tallahassee.

Chameides, W.L., and R.J. Cicerone. 1978. Effects of non-methane hydrocarbons in the atmosphere. J. Geophys. Res. 83:947-952.

Connell, D.W. 1983. Sources and fates of petroleum hydrocarbons in the Hudson-Raritan Estuary. Coastal Ocean Pollut. Assessment News 2(4):39-39.

Danenberger, E.P. 1976. Oil spills, 1971-1975, Gulf of Mexico outer continental shelf. Circular 741. U.S. Geological Survey, Washington, D.C. 47 pp.

Darnell, K.R., A.C. Lloyd, A.M. Winer, and J.N. Pitts, Jr. 1976. Reactivity scale for atmospheric hydrocarbons based on reaction with hydroxyl radical. Environ. Sci. Technol. 10:692-696.

Demaison, G.J. 1977. Tar sands and super giant oilfields. Am. Assoc. Petrol. Geol. Bull. 61:1950-1961.

Department of the Environment. 1976. The Separation of Oil From Water for North Sea Oil Operations. Her Majesty's Stationery Office, London. 29 pp.

Desel, R.F. 1972. Improving the ocean environment through tanker operating techniques, pp. 81-91. In Proceedings of the Seventeenth API Tanker Conference. American Petroleum Institute, Washington, D.C.

Duce, R.A. 1978. Speculations on the budget of particulate and vapor phase non-methane organic carbon in the global troposphere. Pure Appl. Geophys. 116:244-273.

Duce, R.A., and R.B. Gagosian. 1982. The input of atmospheric $\underline{n}$-C_{10} to $\underline{n}$-C_{30} alkanes to the ocean. J. Geophys. Res., in press.

Eganhouse, R.P., and I.R. Kaplan. 1981. Extractable organic matter in urban stormwater runoff. I. Transport dynamics and mass emission rates to the ocean. Environ. Sci. Technol. 16:180-186.

Eganhouse, R.P., and I.R. Kaplan. 1982. Extractable organic matter in municipal wastewaters. 1. Petroleum hydrocarbons: temporal variations and mass emission rates to the ocean. Environ. Sci. Technol. 16:180-186.

Eichmann, R., P. Neuling, G. Ketseridis, J. Hahn, R. Jaenicke, and C. Junge. 1979. $\underline{n}$-Alkane studies in the troposphere. I. Gas and particulate concentrations in north Atlantic air. Atmos. Environ. 13:587-599.

Eichmann, R., G. Ketseridis, G. Schebeske, R. Jaenicke, J. Hahn, P. Warneck, and C. Junge. 1980. $\underline{n}$-Alkane studies in the troposphere II. Gas and particulate concentration in the Indian Ocean air. Atmos. Environ. 14:695-703.

Fischer, P.J. 1978. Natural gas and oil seeps, Santa Barbara Basin, pp. 1-62. In The State Lands Commission 1977, California Gas, Oil, and Tar Seeps. Sacramento, Calif.

Gagosian, R.B., E.T. Peltzer, and O.C. Zafiriou. 1981a. Atmospheric transport of continentally derived lipids to the tropical North Pacific. Nature 290:312-314.

Gagosian, R.B., E.T. Peltzer, and O.C. Zafiriou. 1981b. SEAREX News 4(2):31-35.

Gagosian, R.B., O.C. Zafiriou, E.T. Peltzer, and J.B. Alford. 1982. Lipids in aerosols from the tropical North Pacific: temporal variability. J. Geophys. Res., in press.

Geyer, R.A., and C.P. Giammona. 1980. Naturally occurring hydrocarbons in the Gulf of Mexico and Caribbean Sea, pp. 37-106. In R.A. Geyer, ed. Marine Environmental Pollution. Elsevier, Amsterdam.

Giammona, C.P. 1980. Biota near natural marine hydrocarbon seep in the western Gulf of Mexico, pp. 207-228. In R.A. Geyer, ed. Marine Environmental Pollution. Elsevier, Amsterdam.

Gray, W.O. 1978. Oil Tanker Pollution: Proceedings of Hearings Before a Subcommittee of the Committee on Government Operations, House of Representatives, Ninety-fifth Congress, Second Session, July 18-20. Washington, D.C. Pp. 94-111.

Hahn, J. 1980. Organic constituents in natural aerosols, pp. 359-376. In T.J. Kneip and P.J. Lioy, eds. Aerosols: Anthropogenic and Natural Sources and Transport. Annals N.Y. Acad. Sci. 338.

Hahn, J. 1981. $\underline{n}$-Alkane atmospheric input into the open North Atlantic Ocean, near the Irish coast, and Indian Ocean. Unpublished manuscript. Max-Planck Institute.

Halbouty, M.T., and J.D. Moody. 1980. World ultimate reserves of crude oil, pp. 291-302. In Proceedings of the Tenth World Petroleum Congress. Vol. II, Exploration Supply and Demand. Heyden and Son Ltd., London.

Hanst, P.L., J.W. Spence, and E.O. Edney. 1980. Carbon monoxide production in photooxidation of organic molecules in the air. Atmos. Environ. 14:1077-1088.

Harbaugh, J.W., and M. Ducastaing. 1981. Historical changes in oilfield populations as a method of forecasting field sites of undiscovered populations: a comparison of Kansas, Wyoming, and California. Subsurface Geology Series 5. Kansas Geological Survey, Lawrence, Kansas. 56 pp.

Harvey, G.R., A.G. Requejo, P.A. McGillivary, and J.M. Tokar. 1979. Observation of a subsurface oil-rich layer in the open ocean. Science 295:999-1001.

Hoffman, E.J., J.S. Latimer, G.L. Mills, and J.G. Quinn. 1982. Petroleum hydrocarbons in urban runoff from a commercial land use area. J. Water Pollut. Control Fed., in press.

Hunt, J.M. 1972. Distribution of carbon in the crust of the earth. Am. Assoc. Petrol. Geol. Bull. 56:2273-2277.

Hunt, J.M. 1979. Petroleum Geochemistry and Geology. W.H. Freeman and Sons, San Francisco, Calif. 167 pp.

Hunter, J.V., T. Sabatino, R. Gomperts, and M.J. MacKenzie. 1979. Contribution of urban runoff to hydrocarbon pollution. J. Water Pollut. Control Fed. 51:2129-2138.

Inter-Governmental Maritime Consultative Organization. 1981. Estimates on inputs of petroleum hydrocarbons into the oceans due to maritime transportation activities. Special Report from Meeting of Experts, convened by IMCO on 26-29 May 1981. London. 19 pp.

International Petroleum Encyclopedia. 1980. Volume 13. Penn Well Publishing, Tulsa, Okla. 464 pp.

International Tanker Owners Pollution Federation. 1981. Computer Data Base on Tanker Accidents Involving Oil Pollution. London.

Johnson, T.C. 1971. Natural oil seeps in or near the marine environment: a literature survey. Report Project No. 714141/002. U.S. Coast Guard, Washington, D.C. 30 pp.

Ketseridis, G., J. Hahn, R. Jaenicke, and C. Junge. 1976. The organic constituents of atmospheric particulate matter. Atmos. Environ. 10:603-610.

Ketseridis, G., and R. Eichmann. 1978. Organic compounds in aerosol samples. Pure Appl. Geophys. 116:274-282.

Landes, K.K. 1973. Mother nature as an oil polluter. Am. Assoc. Petrol. Geol. Bull. 57:637-641.

Levy, E.M. 1978. Visual and chemical evidence for a natural seep at Scott Inlet, Baffin Island District of Franklin, pp. 21-25. Current Research Part B. Paper 78-1B. Geological Survey of Canada.

Levy, E.M., and M. Ehrhardt. 1981. Natural seepage of petroleum at Buchan Gulf, Baffin Island. Mar. Chem. 10:355-364.

Lloyd's Register of Shipping. 1980. Statistical Tables. London.

Mackay, D., and A.W. Wolkoff. 1973. Rate of evaporation of low solubility contaminants from water bodies to atmosphere. Environ. Sci. Technol. 7:611-614.

Marine Pollution Subcommittee of the British National Committee on Oceanic Research. 1980. The Effects of Oil Pollution: Some Research Needs. London. 108 pp.

McKirdy, D.M., and Z. Horvath. 1976. Geochemistry and significance of coastal bitumen from southern and northern Australia. APEA J. 16:123-135.

Meybeck, M. 1981. River transport of organic carbon to the oceans, pp. 219-269. Carbon Dioxide Effects Research and Assessment Program Report 016-8009140. U.S. Department of Energy, Office of Energy Research, Washington, D.C.

National Oceanic and Atmospheric Administration. 1980. Proceedings of a Symposium on Preliminary Results from the September 1979 Researcher/Pierce Ixtoc I Cruise. U.S. Department of Commerce, Office of Marine Pollution Assessment, Washington, D.C.

National Petroleum Council. 1981. Environmental Conservation in the Oil and Gas Industry. National Petroleum Council, Washington, D.C. 80 pp.

National Research Council. 1975. Petroleum in the Marine Environment. National Academy of Sciences, Washington, D.C. 107 pp.

Oil Spill Intelligence Report. 1979. Oil spills in 1978--an international summary and review 2(12):1-20.

Oil Spill Intelligence Report. 1980. Oil spills in 1979--an international summary and review 3(21):1-32.

Overaas, S., and E. Solum. 1974. The load-on-top system for crude oil tanker--experience and possible design improvements, pp. 415-426. In Transactions of the 82nd Annual Meeting. Society of Naval Architects and Marine Engineers, New York.

Program a Coordinado de Estudios Ecologicos en la Sonda de Campeche. 1980. Report on work performed to control the _Ixtoc_ I well, to combat the oil spill, and to determine its effects on the marine environment. Petroleos Mexicanos, Mexico City.

Ronov, A.B. 1976. Global carbon geochemistry, volcanism, carbonate accumulation, and life. Geochem. Int. 13:172-195.

Ross, S.L., C.W. Ross, F. Lepine, and E.K. Langtry. 1979. _Ixtoc_ I oil blowout. Spill Technol. Newsl. 4:245-256.

Royal Commission on Environmental Pollution. 1981. Oil Pollution of the Sea. London. 307 pp.

Schlesinger, W.H., and J.M. Melack. 1981. Transport of organic carbon in the world's rivers. Tellus 33:172-187.

Schreiner, O. 1980. Discharge of oil-bearing waste water from the production of petroleum on the Norwegian continental shelf, pp. 137-153. In C.S. Johnston and R.J. Morris, eds. Oil Water Discharges. Applied Science Publishers, London.

Simoneit, B.R.T., and M.A. Mazurek. 1981. Air pollution: the organic components. Crit. Rev. Environ. Control, in press.

Southern California Coastal Water Research Project. 1980. Biennial Report 1979-1980. Long Beach.

Study of Critical and Environmental Properties. 1970. Man's Impact on the Global Environment--Assessment and Recommendations for Actions. MIT Press, Cambridge, Mass. 319 pp.

U.S. Census Bureau. 1978. Current population reports.

U.S. Coast Guard, Department of Transportation. 1976. Polluting incidents in and around U.S. waters--Calendar year 1975. Washington, D.C. 24 pp.

U.S. Coast Guard, Department of Transportation. 1977. Polluting incidents in and around U.S. waters--Calendar year 1976. Washington, D.C. 31 pp.

U.S. Coast Guard, Department of Transportation. 1979. Polluting incidents in and around U.S. waters--Calendar years 1977 and 1978. Washington, D.C. 44 pp.

U.S. Coast Guard, Department of Transportation. 1980. Polluting incidents in and around U.S. waters--Calendar years 1978 and 1979. Washington, D.C. 44 pp.

United States Environmental Protection Agency. 1976. Development Document for Interim Final Effluent Limitations Guidelines and Proposed New Source Performance Standards for the Oil and Gas Extraction Point Source Category, pp. 126-133. Washington, D.C.

U.S. Environmental Protection Agency. 1978. Draft Development Document Including the Data Base for the Review of Effluent Limitation Guidelines (BATEA). New Source Performance Standards, and Pretreatment Standards for the Petroleum Refinery Point Source Category. Washington, D.C.

U.S. Environmental Protection Agency. 1979. Economic Analysis of Proposed Revised Effluents Standards and Limitations for the Petroleum Refinery Industry. EPA-400/2-79-027. Washington, D.C.

Wakeham, S.G. 1977. A characterization of the sources of petroleum hydrocarbons in Lake Washington. J. Water Pollut. Control Fed. 49:1680-1687.

Weaver, D.W. 1969. Geology of the northern Channel Islands. Pacific Sections AAPG and SEPM Special Publication. Thousand Oaks, Calif. 200 pp.

Weeks, L.G. 1965. World offshore petroleum resources. Am. Assoc. Petrol. Geol. Bull. 49:1680-1693.

Whipple, W., and J.V. Hunter. 1979. Petroleum hydrocarbons in urban runoff. Water Resources Bull. 15:1096-1100.

Wilson, R.D., P.H. Monaghan, A. Osanik, L.C. Price, and M.A. Rogers. 1973. Estimate of annual input of petroleum to the marine environment from natural marine seepage. Trans. Gulf Coast Assoc. Geol. Soc. 23:182-193.

Wilson, R.D., P.H. Monaghan, A. Osanik, L.C. Price, and M.A. Rogers. 1974. Natural marine oil seepage. Science 184:857-865.

Wondzell, B.E. 1981. Crude oil production and oilspills in Cook Inlet offshore. Letter dated October 22. State of Alaska Oil and Gas Conservation Commission, Fairbanks.

Zafiriou, O.C., R.B. Gagosian, E.T. Peltzer, and J.B. Alford. 1982. Atmospheric transformations and fluxes of lipids at Enewetak Atoll. Unpublished manuscript.

Zimmerman, P.R., R.B. Chatfield, J. Fishman, P.J. Crutzen, and P.L. Hanst. 1978. Estimates on the production of CO and H_2 from the oxidation of hydrocarbon emissions from vegetation. Geophys. Res. Lett. 5:679-681.

3
Chemical and Biological Methods

INTRODUCTION

The purpose of this section is to set forth general principles in study design and in chemical and biological methodology. The specific examples cited should aid the reader in critically evaluating oil pollution research literature and present a general guide to the present state of the art in this area of research.

We strongly advocate that scientists be properly educated and trained in the fundamentals of the various disciplines of biology, chemistry, or marine sciences before engaging in the difficult research problems associated with petroleum pollution in the marine environment. Thus, it is not the intent of this section to provide a detailed account or "cookbook" approach to particular measurements.

General Strategies of Study Design

A discussion of statistical techniques used in studies of the fate and effects of oil in the environment would need to cover most of the areas in modern statistics. The more important general problems that have arisen in oil spill research will be briefly discussed and areas of study identified where recent developments in statistics may contribute to their solution. For the most part, discussions of the role of sampling and statistical design in oil spill studies have been confined to the use of classical statistical methods (see, for instance, G.V. Cox et al., 1977). In the past 10 years a number of new statistical methods have been developed for a wide range of environmental problems. These methods, used in water quality research, air pollution, and environmental sampling, have obvious application to the study of oil in the marine environment.

In comparison with the esoteric techniques of organic chemistry, fluid dynamics, and faunal analysis employed in studies of the effect of oil in the environment, the basic concepts of statistical design and analysis are relatively straightforward. Yet, statistical problems have often been cited as the cause of many of the deficiencies in oil pollution studies. The main issue is to design an experiment, set of experiments, or sampling program to answer a specific set of questions

89

within a specified set of cost, manpower, and time constraints. It is assumed that investigators can agree on a general model for a given system. When the questions, constraints, and model have been specified, the next step is to allocate samples and experimental treatments optimally so that the scientific and regulatory questions being asked can be answered clearly. If a problem is less sharply defined, the quality and efficiency of the statistical design decrease. Studies of oil pollution in the marine environment are particularly subject to this effect. In the past, but less frequently now, fortunately, several large environmental research programs were constructed by committees in which scientific questions, cost constraints, model processes, and statistical design were decided in a more or less haphazard and illogical order. Thus, rather than being the cause of the problem, improper statistical design and data analysis were more often symptoms of fundamental difficulties within the research plan.

The nature of pollution studies is such that some of the work must be done in areas where pollution has occurred, in what might be called an unplanned experiment. The objective, in most cases, is to assign a causal relationship between a particular event and some effect in the area. There is also an interest in drawing conclusions that could be applied in other related cases. In a classical randomized experiment, a set of areas is selected and then oil treatments are randomly assigned to half of the areas. In most field studies of oil in the environment, the treatment, for example, of the concentration of oil in the sediment, has not been assigned by the investigator as part of the experiment design, but rather is a consequence of human error and natural forces that dispersed the oil unevenly in different areas. This does not preclude making reasonable inferences about the causal nature of the event observed; however, one must make stronger assumptions about the underlying nature of the data and use rather different analyses. There is a large literature in statistics on the problems associated with observational studies. Recently McKinlay (1975) has reviewed the literature on observation studies, particularly those influencing human populations. Cochran (1951, 1968) was an early contributor in this area and still offers the definitive work. R.H. Green (1979) gives a relatively clear analysis of this problem in an environmental setting.

A simple example serves to illustrate the problem of observational data. The following is a fictitious data set used to illustrate "Simpson's paradox" (Lindley and Norick, 1981) in an environmental setting. Table 3-1 describes the results of a large, well-designed, random survey of an oil spill area. The 200 samples were classified as to whether they fell above or below the median oil concentration and whether the biomass of the benthic community fell above or below the median. Table 3-1 shows clearly that there is a positive correlation between high oil concentration and high biomass concentration. Suppose, however, we again divide the samples into two subareas, those below 5 m depth and those above 5 m depth, thus splitting the data into two separate tables. We see that the correlation in Table 3-1 is reversed when we look at the two subpopulations separately. The explanation for this is obvious: although the sampling program was random, the distribution of oil in the sediment (the treatment) was not; heavy

TABLE 3-1 Matrix of Oil Concentrations in Sediments and Correlations With Benthic Organism Biomass

| | Oil Concentration Groups (%) | |
| | Below Median Concentration | Above Median Concentration |
Biomass Concentration Groups		
Sediment at All Water Depths in Study Area		
Below median concentration	70	30
Above median concentration	30	70
Sediment Above and Below 5 m Water Depth		
Above 5-m water depth		
Below median concentration	70	30
Above median concentration	10	0
Below 5-m water depth		
Below median concentration	0	10
Above median concentration	20	70

NOTE: Hypothetical case (see text). Percentages are of total samples in entire data set.

concentrations of oil appear in the shallower areas exactly where the highest biomass is. Inference under these conditions has been studied by Lindley and Novick (1981) and by Blyth (1972). The reader is referred to these articles for various proposed solutions to this problem. Needless to say, this situation can arise in more subtle ways than that presented above; for instance, chronic oil pollution is usually associated with other industrial and urban activity which could also have an impact on the system.

Sampling Procedures and Equipment

The requirements of specialized sampling procedures and equipment are discussed in the following sections on chemical and biological methods.

Statistical Design of Analytic Procedures in the Laboratory

Often the complexities of environmental studies lead one to overlook the importance of careful design of laboratory analytical procedures. Proper design can prevent confounding of laboratory effects with natural processes that we wish to study. For example, where the same analysis is undertaken by two laboratories, samples should be assigned so that possible differences in analytic techniques are not confounded with natural variation. For instance, assigning all samples from area A to laboratory X and all samples from area B to laboratory Y will certainly confound the effect of analytic bias between laboratories with true variations in concentration between the two areas. There is a lengthy and very useful literature on statistical design and quality control in laboratory situations (Tietjen and Beckman, 1974).

Presenting Results From Complex Oil Pollution Studies

One of the more difficult problems in environmental oil pollution studies is to present the data and analysis in a comprehensible way. In most cases the data from a study consist of many different kinds of measurements made on samples distributed in both space and time. The data analysis problem is to decipher relationships among samples and variables and then to present these relationships in a compact way. Often, classical statistical analysis (analysis of variance, regression, etc.) used in reports and papers describing oil spill events provides little insight into the character of the processes and the data under investigation. Many of the techniques of exploratory data analysis can be used to investigate these relationships. J.W. Tukey and P. Tukey (personal communication, 1981) have described different techniques for displaying multivariate data. Cleveland and Kleiner (1975) and R.H. Green (1979) have given a number of methods for graphically presenting multivariate environmental data. This is one area where a great deal more work needs to be done.

Designing Ecological Experiments to Study the Effects of Pollutants

The discussion of problems one can encounter in observational studies should make clear the reasons for going to planned experiments to study the impact of oil or other marine pollutants. W. Smith et al. (1981) provide a general discussion of the design of ecosystem experiments. Large experiments, such as the No. 2 fuel oil experiment conducted at the University of Rhode Island Marine Ecosystems Research Laboratory (see Biological Methods section), present certain problems not encountered in usual experimental design situations. First, since such a system is partially open to the environment, all conditions within the system cannot be precisely controlled. The second problem is that since each experimental unit replicated is costly to maintain and measure, the number of experimental units replicated is relatively small.

These kinds of experiments occur over time, and thus, part of the experimental design is to determine the treatment time course. Careful design of time series experiments can help to overcome the problems mentioned above. There is a growing literature in this area (Mulholland and Gowdy, 1977; Glass et al., 1975; Box and Tiao, 1975; Huitema, 1979; Brockway et al., 1979).

An Oil Spill Survey Design

Environmental sampling problems have been well studied from a statistical design perspective. Sampling spatially and time-varying processes has been discussed by W. Smith (1979) and others. Moss and Tasker (1979) give a review of these methods in water quality research. In addition, composite sampling and other techniques for efficient sampling and analysis of environmental data have been investigated (Elder et al., 1981; R.H. Green, 1979; Smeach and Jernigan, 1977).

W. Smith (1979) developed a specific plan for the oil spill survey problem involving a two stage sampling scheme where, in the first stage, a large number of samples are collected in a systematic survey (grid or modified grid). Easy to record data, such as sediment type or if oil is present, are recorded in the first stage.

In the second stage of the sampling program, the data from the first stage are used to stratify the samples. The original samples can then be subsampled for the more costly chemical and biological analysis. In oil spill work this procedure has the great advantage that the first stage of the program can be implemented on short notice. The difficult sampling question can be postponed to the second stage of the program when the legal engineering and scientific questions that the oil spill raises have been better defined.

Summary

The interweaving of the various facets of observational and experimental studies of inputs, fate, and effects of petroleum in the marine environment demands careful attention to the combined requirements of many different types of sampling and measurement methods. It is not appropriate in a report of this type to review critically all aspects of environmental and engineering measurements relevant to oil pollution studies. For example, meteorological, physical oceanographic, and sedimentological measurements are often of importance to these studies. The reader is referred to texts and reviews in these areas of study for description and evaluation of methodology. Physical parameters such as specific gravity, viscosity, and pour point of crude oils and petroleum products are not extensively discussed because they have been well defined in American Society for Testing and Materials (ASTM) protocols (R.C. Clark and Brown, 1977) and have not been the subject of controversy in oil pollution research and monitoring activities.

Chemical methods for measuring petroleum and petroleum compounds and biological methods for measuring various levels of effects of petroleum in the marine environment have been subjects of continuing controversy and debate since an increase in studies of petroleum pollution in the late 1960s. For this reason chemical and biological methodology is described and evaluated in the following sections.

PART A
Chemical Methods

INTRODUCTION

There have been many significant advances in the application of chemical analyses to all aspects of petroleum pollution in the marine environment since the last National Research Council (1975) publication. However, no one method of analysis can measure all components of petroleum or answer all requirements for research and monitoring. Many techniques have been applied to oil spill studies, monitoring of long term sources of input such as sewage effluents and production platforms, and experimental studies of the fate and effects of petroleum in the marine environment. Concurrently, new equipment has been developed for the variety of sampling problems that have been encountered, and instrumental techniques for real-time monitoring of petroleum components in water near oil spills have been successfully tested.

The analytical methods applied to oil spill studies usually combine low resolution but relatively easily applied techniques, such as ultraviolet (UV) fluorescence spectrometry, with high resolution but more costly and time-consuming techniques, such as glass capillary/gas chromatography/mass spectrometry (GC^2/MS) computer systems. This also applies to monitoring of chronic inputs and analytical chemistry in support of experimental studies of fate and effects.

Two important issues that have sometimes been overlooked are (a) choosing the method(s) that will satisfactorily solve the analytical problem at hand; for example, gross levels of hydrocarbons in tissues as determined by nonspecific measurements such as ultraviolet fluorescence have minimal use when the problem is to distinguish between chronic background hydrocarbon pollution of combustion origin, chronic petroleum pollution, biogenic hydrocarbon inputs, and additions of petroleum hydrocarbons from a recent oil spill; and (b) quality control within and between laboratories. This latter point has been emphasized repeatedly as a priority item, but funding practices by federal agencies generally paid scant attention to this problem until a few years ago. The 1975 NRC report called attention to this, and recently the American Chemical Society (ACS) has issued guidelines for data acquisition and data quality evaluation in environmental chemistry (Keith et al., 1983), which address this fundamental issue for all types of environmental analytical chemistry.

The current status of quality control and laboratory intercomparison is not yet adequate to accomplish detailed comparisons of data sets from different laboratories or to be sure which specific chemicals in various petroleum fractions are responsible for observed effects. Generally, only comparisons of qualitative trends or large differences of factors of 10 or more are valid within quality control or inter-laboratory comparison experiences prior to 1979-1980. Although progress has been made, much more needs to be accomplished.

Two major developments in our knowledge of inputs, fates, and effects of petroleum in the marine environment since the 1973 literature review for the 1975 NRC report have an important bearing on analytical chemistry considerations. First, studies of polycyclic aromatic hydrocarbon sources and fates over the past 10 years have increased markedly and have revealed the global significance of chronic low level polynuclear aromatic hydrocarbons (PAH) inputs related to the incomplete high temperature combustion of fossil fuels. In many cases, analytical methods must try to distinguish between petroleum PAH inputs and pyrogenic PAH inputs. Second, present evidence substantiates a concern expressed in the 1975 NRC report that petroleum hydrocarbons readily undergo structural alterations by photochemical and biochemical metabolic oxidation. Postspill analytical programs based only on hydrocarbon measurements in seawater, sediments, and tissues cannot measure an important set of transformation products.

Acute needs have developed (1) to manage data and make them accessible (needs that have only occasionally been addressed in specific programs), (2) to evaluate existent data much more thoroughly to enable future efforts to be more targeted, and (3) to link divergent analytical developments. This latter concern arises from the fact that varieties of analytical techniques are being separately developed for petroleum chemistry research, marine chemistry research, forensic applications (e.g., U.S. Coast Guard techniques), general environmental chemistry research, and environmental regulatory or surveillance (e.g., U.S. Environmental Protection Agency (EPA) priority pollutant) methodologies. An overview of the literature confirms this and raises concerns that, in our efforts to monitor the environment, the methods being developed for and information derived from the various programs are diverging. This is apparent in the groups of marine chemistry and other environmental chemistry literature, citations from one omitting relevant literature from the other. Regulatory definitions of petroleum hydrocarbons must be more firmly based in current knowledge of the composition of petroleum inputs, fates, and effects in the environment.

The review of analytical techniques, methods, and strategies that follows has drawn from marine and nonmarine analytical chemistry and organic biogeochemical studies alike. Due to the great pool of recent literature, attempts have been made to include mainly post-1975 literature unless only pre-1975 information is available. As no single literature reference comprehensively covers many of the topics discussed, a number of references are cited in many cases. Several recent reviews have aided in this preparation and should be consulted for additional details: Petrakis and Weiss (1980), R.C. Clark and Brown (1977), Farrington et al. (1976a, 1980), R.A. Brown and Weiss (1978), Pancirov and Brown (1981), and Malins et al. (1980).

SAMPLING AND SAMPLE PRESERVATION

The nature and the quality of information derived from marine environmental samples are dependent on the quality of sampling methods used and the care taken in utilizing these methods. Of primary concern in both petroleum hydrocarbon baseline and oil spill samplings is the avoidance of sample contamination and cross contamination. R.C. Clark and Brown (1977) presented details of quality assurance aspects of collection techniques, which included attention to the cleanliness of sampling devices, subsampling implements, and storage containers and the exclusion of field (shipboard) contaminants from the samples. Details of collection methods of seawater, sediments, biota, and waterborne oil samples were presented in R.C. Clark and Brown (1977), D.R. Green (1978), and ASTM Method D 3325-78. Sampling strategies have been developed for each spill scenario, and usually provide for pre-impact (baseline) and postimpact samplings, reference samplings (unimpacted sites), and a postspill time series to examine details of recovery (e.g., Boehm et al., 1981b; Atlas et al., 1981; Teal et al., 1978; Burns and Teal, 1979; Keizer et al., 1978).

Sediments

Recent laboratory and field studies have revealed new, important subtleties related to sediment sampling, both in spill and nonspill situations. Gearing et al. (1980), in controlled experiments, and Boehm et al. (1981b), in a field assessment, pointed to the importance of sampling newly deposited hydrocarbon-bearing sediment (i.e., floc) in oil spill studies. Thompson and Eglinton (1978b) showed that different particle sizes and types within a given sediment have differing hydrocarbon compositions. Determination of petroleum and PAH chemical composition associated with different types and sizes of sediment particles may yield important information on availability of certain compounds for biological uptake in benthic communities.

A variety of sediment samplers has been used to obtain "surface sediment." These include box corers, which are most useful in soft bottoms and acquire a relatively undisturbed core of sediment; grab samplers (e.g., Smith-MacIntyre and Van Veen), which are useful in all sediment types, but may be subject to sample washout in gravelly or shelly sediment; gravity corers, which utilize a core liner (polycarbonate) to obtain a cylindrical core of sediment which may be subdivided for analysis; hydrostatically damped corers in multiple arrays (Pamatmat, 1971; Wakeham and Carpenter, 1976), which have damped rates of sediment penetration; sediment boundary layer suspension (floc) collectors (Bryant et al., 1980); diver and other manual collectors (Atlas et al., 1981; D'Ozouville et al., 1979). The selection of the sampling device is dictated by the sediment type being sampled and the informational needs of the particular program.

Sediment Traps

The design of sediment traps to ensure efficient collection and postcollection preservation of sedimenting material is an area of intense research and debate (e.g., Wakeham et al., 1980; W.D. Gardner, 1980; Jannasch et al., 1980). Traps have been utilized to examine the fluxes of suspended organics, including hydrocarbons, to open ocean and coastal sediments. The deployment of unsophisticated sediment traps in spill situations has provided critical information on the fate of

spilled oil (Boehm et al., 1981b; Johanssen et al., 1980; Boehm and
Fiest, 1980a).

Marine Organisms

A variety of sampling devices exists for the collection of pelagic and
benthic marine organisms (R.C. Clark and Brown, 1977; Grice et al.,
1972). These include plankton nets, trawls, and dredges of varied
design (see Biological Methods section). Extreme care must be taken to
avoid sample contamination from the sampling device, from the ship and
ship's discharges, from the sample containers, and from oil in the
water column. For example, collection of uncontaminated pelagic biota
samples from a ship during a spill event is very difficult, and it is
difficult to distinguish ingested from external oil (American Petroleum
Institute, 1977). Diver collections are preferable in these cases.
Again, the choice of sampling device and the sampling design depend on
the nature of the organism and the program's statistical design. For
example, in order to examine the relation of oil in the sediment to its
bioaccumulation in benthic organisms, animal samples should ideally be
obtained in close proximity to the sediment sample, with either divers
in subtidal areas or manually in intertidal areas, and from the same
sampling device (e.g., box corer).

A "sample" of marine organisms for analyses is defined by both
analytical and statistical considerations. An estimated 1-10 g dry
weight (100 g wet) are usually needed for prespill analysis and for
spill-impacted samples to achieve analytical detection limits. However,
the optimum sample size (i.e., number of organisms per sample) is dictated by several considerations, including whether information on a
population at a certain station is required, or knowledge of individual-
to-individual variation is desired (Boehm, 1978).

Seawater

Sampling of seawater to obtain information on hydrocarbon levels, in
both baseline and spill-related samples, is the most difficult of
samplings due to (1) the potential for contamination from the surface
film (Gordon and Keizer, 1974), (2) the potential for contamination
from the sampling device (Boehm and Fiest, 1978; Zsolnay, 1978a) or
from associated rigging and the sampling ship or platform, and (3)
possible problems with comparing data from samples obtained with
different sampling devices (Levy, 1979a; Boehm, 1980a).

Use of the various available devices for obtaining seawater samples
for petroleum hydrocarbon determinations has been reviewed recently by
D.R. Green (1978). Examples of the problems encountered are contamination by certain plastics and "O" rings. In addition, accumulator
systems (e.g., octadecylsilicic reversed phase adsorbents [May et al.,
1975; Eisenbeiss et al., 1978; Saner et al., 1979], XAD-2 macroreticular
resins [Ehrhardt, 1978], and polyurethane foam [e.g., deLappe et al.,
1980]) have been used with varying results to concentrate hydrocarbons

on solid phases. Alternatively, large volume water samples (10-90 L), which pass through the surface in a closed position, must be used to achieve analytical detection limits which allow sub-part-per-billion (μg/L) levels of hydrocarbons to be detected (e.g., deLappe et al., 1980; Boehm, 1980a; Farrington et al., 1976a). Chesler et al. (1976) and Keizer et al. (1977) utilized simple devices to obtain 4-10 L of sample using glass bottles which manually open below the surface. Recently, pumping systems have been applied successfully to the subsurface measurement of petroleum in the water column below surface oil slicks (Fiest and Boehm, 1981; Boehm and Fiest, 1980b; McAuliffe et al., 1980; J.C. Johnson et al., 1978).

The use of discrete versus continuous sampling systems is dictated by the sampling scenario. Continuous pumping systems can be used for separation of dissolved and particulate water column samples (Ehrhardt, 1978; Goutz and Saliot, 1980; deLappe et al., 1980; Boehm and Fiest, 1980b), although water from discrete samplers can be pressure filtered through glass fiber filters (Boehm, 1980a; J.R. Payne et al., 1980a). Ehrhardt (1976, 1978) and deLappe et al. (1980) describe continuous seawater pumping systems which pass large volumes of water through in-line glass fiber filters upstream of XAD-2 resin and polyurethane foam. Dissolved and particulate size fractionations are important in discerning the fate and pathways of biological uptake of spilled oil (Zurcher and Thuer, 1978; Boehm and Fiest, 1980b) and distribution of petroleum hydrocarbons in seawater in nonspill studies (Goutz and Saliot, 1980; Boehm, 1980a). However, the terms "dissolved" and "particulate" are operational in nature due to possibilities of passage of colloidal-sized particles through the filter and the likelihood of changing the pore size of the filter as filtration proceeds.

Sampling for Low-Molecular-Weight Hydrocarbons

Samples (sediment, seawater, biota) to be analyzed for low-molecular-weight hydrocarbons require special handling. After collection, water samples should be treated to avoid agitation or inclusion of air bubbles in storage bottles. Water samples should fill sample bottles and be sealed with a Teflon cap, leaving no headspace, and be refrigerated until analysis proceeds (Brooks et al., 1980). Sediment samples should also fill sampling containers and they should be frozen. Alternatively, sediments can be transferred immediately to containers holding "poisoned" (e.g., sodium azide) hydrocarbon-free seawater, the container headspace flushed with helium or nitrogen and the container inverted at near-freezing temperature (Bernard et al., 1978). Biota samples should be frozen until subsampled for purgeable organics (Environmental Protection Agency, 1980).

Sample Preservation

There is a general lack of information on the longevity of petroleum hydrocarbons in stored, unextracted samples of all types. Thus, the procedures described are based, in most cases, on first principles, with regard to minimizing processes that will alter the compounds of interest.

All samples (sorbents, filters, sediments, tissues) should be frozen at -10° to -20°C after collection. Water samples, however, are impractical to freeze and can be solvent extracted aboard ship or preserved in the dark with a bacterial retardant (chloroform, methylene chloride, mercuric chloride, sodium azide). However, care should be exercised in the choice of preservation technique. Samples obtained for multiple use in chemical and biological studies should be preserved in a manner that does not mitigate against certain measurements; e.g., sodium azide would not be acceptable for samples to be used in a variety of biochemical or physiological studies. Volatilization of hydrocarbon components and microbial and photochemical oxidation of organic matter in samples are the primary concerns to be addressed in postsampling preservation. ASTM Method D 3325-78 presents a standard method for storing waterborne oil samples. The effects of long term (months to years) storage of samples under "preserved" conditions is largely unknown, although Medeiros and Farrington (1974) determined that, after 18 months of storage of oil-spiked cod liver lipid extract, analytical results for some major hydrocarbons were unchanged.

SPILLED OIL CHARACTERIZATIONS

As the behavior and environmental fate of spilled oil are dependent on the physical and chemical properties of the oil and the meteorological/ oceanographic conditions, there is a need for full characterization of an authentic sample of the source of oil and a series of oil samples from the water's surface and from oiled beaches. These oil samples will serve as reference materials for environmental analyses and also may be used in damage assessment studies and in judicial proceedings. In addition, rapid analytical information should be obtained during offshore spill scenarios to predict the physical, chemical, and toxicological properties of oils after being waterborne and as they may impact sensitive shorelines. Offshore and shoreline countermeasure strategies often hinge on the knowledge of the physical properties of spilled oil, actual and predicted.

Sample Collection and Preservation

The original 1975 NRC collection guidelines should be adhered to and supplemented by ASTM Methods D 3325-78 and D 3694-78, U.S. Coast Guard (1977) considerations of collection, sample documentation, and chain-of-custody procedures, and sample preservation.

Several authentic cargo samples should be collected in all cases along with waterborne oil samples. Replication is important, as floating oil patches exhibit significant heterogeneity. If possible, floating oil patches or slicks should be marked with buoys and sampled periodically until dissipation or landfall. Samples should be taken from small boats or helicopters, as it is often impractical for large ships to enter large oil patches. Cross-contamination should be avoided, especially while sampling in areas of heavy contamination wherein gear and clothing may become oiled. Gloves, protective clothing, and activated charcoal trap respirators should be used while working in heavy oil, and personnel should be monitored by a trained medical staff.

The samples should be taken in sufficient quantities to permit replicate physical and chemical analyses. One hundred milliliters of sample are needed for some physical tests (e.g., viscosity), so wherever possible, liter-sized jars should be filled with sample. Sample documentation should be made on prespecified, durable, water-proof tags (e.g., U.S. Coast Guard, 1977) to include information on collection location, date, time, name of collector, and sampling device. All collections should be logged in a master log and given a unique sample number. Consecutive numbering (National Oceanic and Atmospheric Administration, 1980) using collector codes has proven extremely efficient in sample collection operations, and avoids ambiguous situations which occur during all collections when several people or groups are sampling concurrently.

Preservation of oil samples involves the containment of low boiling components and the retardation of degradation through postsampling photochemical and microbial degradation.

Analytical Methods

Physical and chemical information should be obtained as soon as possible after the spill occurs.

Field Information

The existence, extent, and mapping of subsurface oil concentrations may be acquired during spill events through the use of in situ (towed) fluorometers (Environmental Devices Company, 1977; Calder et al., 1978) or continuous pumping through shipboard fluorometers (e.g., Boehm and Fiest, 1980b). Several important physical measurements, such as the determination of water content of oil (i.e., emulsification state) and the specific gravity of oil, can be made using simple devices (National Oceanic and Atmospheric Administration, 1977). This information is valuable to countermeasure strategies (i.e., use of dispersants, application of booms, estimations of cleanup efficiency).

Laboratory Information (Short Time Frame: Days to Weeks)

Samples shipped to the laboratory should be subjected to a series of routine physical property tests to determine the oil's characteristics and behavior. These include accurate specific gravity, viscosity, pour point, and fractional distillation temperatures. ASTM procedures exist for all of these measurements (R.C. Clark and Brown, 1977). In addition, useful parameters associated with the emulsification process are the asphaltene and wax contents of whole oil.

Ideally, chemical testing in the laboratory should include class separation to obtain information on the initial and changing relative proportions of saturated hydrocarbon, aromatic hydrocarbon, and polar and asphaltic fractions. Oils should initially be dissolved in methylene chloride, or similar solvent with water removed by phase separation and drying over sodium sulfate. The extract is then deasphalted by precipitation by pentane addition (ASTM Method D 893-80), and a portion of the pentane is charged to and eluted on silica gel, silica gel/ alumina, or other column (see Measurements and Detailed Analysis of Environmental Samples section). A class separation and characterization sequence based on initial normal phase high pressure liquid chromatography (HPLC) (equivalent to silica gel column chromatography) followed by detailed capillary GC analysis (Gas Chromatography section) and analytical HPLC (High Pressure Liquid Chromatography section) has been described by Crowley et al. (1980). Laboratory-derived data should include GC analysis, preferably capillary GC, of the hydrocarbon fractions so as to determine the boiling range and overall composition of the oil.

Laboratory Information (Long Time Frame: Weeks to Months)

Techniques of petroleum characterization include those that derive detailed compositional information as well as those that obtain information used to match waterborne oils with suspected cargoes through IR (infrared spectrometry), UV/F (ultraviolet fluorescence spectrometry), GC (gas chromatography), FID (flame ionization detector) element specific detectors, and trace metal (Ni/V) measurements (U.S. Coast Guard, 1977; ASTM Methods D 3415-79, D 3414-79, D 3650-78, D 3328-78, D 3327-79).

Gas chromatography with flame ionization and sulfur- or nitrogen-specific detectors yields considerable information on the molecular weight range of hydrocarbon components, and is one of the more powerful methods for broadly characterizing crude oils (Crowley et al., 1980; Rasmussen, 1976; Clark and Jurs, 1979) and refined products (e.g., Ury, 1981). Graphical plots of the relative saturated and aromatic compositions of oil samples (Patton et al., 1981; Atlas et al., 1981; Boehm and Fiest, 1980b) complement specific calculated parameter ratios in describing the oil's chemical properties.

IR measurements, in addition to having forensic use, can be used to characterize major compound groups and to evaluate weathering in a gross way by the appearance of carboxyl and hydroxyl functional groups (Rashid, 1974; Blumer et al., 1973; W.E. Reed, 1977).

Mass spectrometric (MS) class and group (or subclass) analyses providing quantitative information on some 25 molecular types have proven very useful in comparing oil types and in readily evaluating the chemical characteristics of fresh and weathered oils (Robinson and Cook, 1969; Petrakis et al., 1980; ASTM Method D 2786-71).

GC/MS techniques have been used to identify fresh and weathered oils based on detailed compositions (Hood and Erickson, 1980; Albaiges and Albrecht, 1979; Atlas et al., 1981; W.E. Reed, 1977; Calder et al., 1978; Overton et al., 1980b; DeLeon et al., 1980; Schmitter et al., 1981).

HPLC is another technique for characterizing oils on the basis of their aromatic hydrocarbon content (e.g., Crowley et al., 1980). A combination of IR and HPLC analyses, to quantify and characterize saturated and aromatic petroleum hydrocarbons, respectively, has been used in conjunction with GC for analysis (Riley and Bean, 1979).

Further long-time-frame characterizations of spilled oils include the techniques of carbon and sulfur isotope ratios (Koons et al., 1971; Hartman and Hammond, 1981; Sweeney et al., 1980), proton and ^{13}C nuclear magnetic resonance spectroscopy (Petrakis et al., 1980), and elemental (C, H, N, S) analysis (e.g., W.E. Reed, 1977; National Research Council, 1975). Additionally, many of the analytical techniques used by petroleum chemists may effect more detailed characterizations (Terrell, 1981). Examples of detailed multiple-technique characterizations of oils are given by W.E. Reed (1977) for weathered tars, W.E. Reed and Kaplan (1977) for marine petroleum seeps, and Overton et al. (1980b) for Ixtoc I oil.

MEASUREMENTS AND DETAILED ANALYSIS OF ENVIRONMENTAL SAMPLES

General

The analysis of a particular sample of water, sediment, tissue, air, etc., for petroleum hydrocarbons must be preceded by matching the particular informational need with the proper analytical technique. For example, information may be needed on the gross amount of oil in the dosing system of a toxicological study or on concentrations of an individual aromatic toxicant (e.g., naphthalene) and its metabolites (e.g., naphthol) in a marine fish.

Single analytical techniques (e.g., UV, GC) can be used for certain applications when the analytical end is to examine absolute levels or compound assemblages (nonpoint sources), but multiple techniques (e.g., UV + GC + IR) are required for forensic purposes in matching environmental compositions of petroleum to specific point sources. Figure 3-1 illustrates various analytical options for environmental samples. The proper choices of separation and analytical techniques are at the heart of environmental petroleum hydrocarbon chemistry.

In general, the less chemically specific techniques require less sample processing and manipulation. With increased processing, the level of analytical detail, and hence compositional and quantitative information, increases.

The field of oil pollution chemistry has expanded rapidly in the past 5-10 years without great attention to intercomparability of measurements between different laboratories using similar techniques and between different analytical techniques used to generate data. The generation of analytical data continues at a rapid pace at different

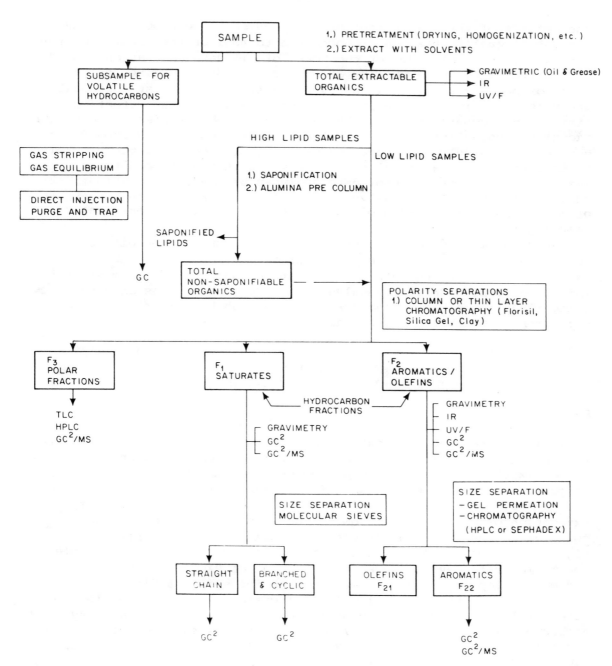

FIGURE 3-1 Analytical options for analysis of petroleum compounds in sediment, tissue, particulate matter, and water.

levels of sophistication. Recently the ACS Subcommittee on Environmental Analytical Chemistry published its "Guidelines on Data Acquisition and Data Quality Evaluation," which expressed three interwoven strategies of modern trace analysis: (1) the development of sensitive, specific, and validated methods; (2) the use of protocols that describe the details of the measurement process and sampling procedures; and (3) the use of quality assurance procedures to monitor the quality of the data as it is developed. At the heart of all data generation should be procedures of rigorous quality assurance including routine determinations of procedural blanks, instrument calibration and standardization, analytical precision on replicates, recovery of spikes, detection limits, and comparison of results with other laboratories (intercalibrations) (MacDougall and Crummett, 1980). These guidelines should become part of all petroleum hydrocarbon studies. In addition, the precision of environmental analytical measurements has three components: (1) instrumental variation (replicate analyses of the same solution), (2) analytical variability (analysis of replicates of the same homogenate, or subsamples), and (3) sampling variability (replicate analyses of sampling replicates).

Numerous methodologies have been used in conjunction with oil pollution studies, and the efficacy of the various methods used, for example, in extracting and fractionating organic matter from sediment and in performing detailed analysis of hydrocarbons, has only recently (since 1975) come under rigorous study through both intralaboratory experiments and thorough intercalibration exercises.

Extraction of Organic Matter
(High Molecular Weight, $C_{11}+$) Hydrocarbons

Sediments

Several different solvent extraction methods are commonly used for the extraction of petroleum hydrocarbons from sediments. No standard method exists, but most methods involve the combined use of polar and nonpolar solvents to effect an efficient extraction of organic matter. Geochemical and oil spill sediment samples differ in the ease of extraction of hydrocarbons from the sediment matrix, the latter containing loosely bound petroleum hydrocarbons. Thus while one of the rigorous extraction procedures is necessary to extract, for example, low to moderate levels (less than 10 μg/g) of PAH from a silt/clay sediment, simpler techniques may suffice for spill samples. As it is often important to discern levels of incremental addition of low to moderate levels of oil to sediments containing some prior history of anthropogenic pollution, the rigorous solvent extraction methods (e.g., Soxhlet, tumbler/shaker) are most appropriate for all environmental samples.

Sediment extraction techniques include organic solvent extractions (e.g., D.W. Brown et al., 1980), alkali digestions followed by solvent extractions (Environmental Protection Agency, 1980; Farrington and Tripp, 1975), headspace gas stripping (May et al., 1975), and steam distillation (Veith and Kiwus, 1977; Bellar et al., 1980). Solvent extractions employ (1) the use of the Soxhlet extractor with a combination of polar and nonpolar solvents (e.g., Hites et al., 1980; Farrington and Tripp, 1975; Lake et al., 1980; Environmental Protection Agency, 1980), (2) the reflux of sediment with organic solvents (e.g.,

Lake et al., 1980), or (3) ambient temperature extractions using the shaker table or ball mill tumbler (D.W. Brown et al., 1979, 1980; Boehm et al., 1981a). Van Vleet and Quinn (1978) utilized a methanol:toluene reflux of wet sediment to remove "unbound" lipoidal material followed by combined alkaline digestion solvent extraction to remove additional "bound" material. Farrington and Quinn (1973) and Boehm and Quinn (1978) employed simultaneous saponification/extraction using a methanolic KOH:toluene reflux to remove hydrocarbons from sediment. In headspace analysis, Wise et al. (1978) and May et al. (1975) combined dynamic headspace sampling of hydrocarbons volatilized at 70°C for 18 hours with a coupled-column reversed phase enrichment step to recover volatile compounds up to 4- and 5-ringed aromatic compounds. The volatilized hydrocarbons were trapped onto a Tenax GC trap, which was then desorbed directly onto a capillary GC column. Tan (1979) utilized ultrasonic solvent extraction using cyclohexane of freeze-dried sediment to reproducibly extract PAH material. A very rapid extraction technique using a process called "flow blending" utilizes solvent extraction of material in a flow-through cell (Radke et al., 1978). Using this technique, extraction efficiencies comparable to Soxhlet extraction were achieved for a wide range of sample types. While several of the recently published methods included the use of benzene (Boehm and Quinn, 1978; Lake et al., 1980), this solvent is now restricted from use in the laboratory due to health effects. Methylene chloride and toluene have replaced benzene as solvents of intermediate polarity, having good extraction efficiency alone or in azeotropic mixture with other solvents.

Comparisons of several of these extraction techniques have been reported by Farrington and Tripp (1975), Rohrbach and Reed (1976), Lake et al. (1980), D.W. Brown et al. (1980), Bellar et al. (1980), M.K. Wong and Williams (1980), and Templeton and Chasteen (1980). Present evidence suggests that Soxhlet and tumbler or shaker techniques are the most thorough extraction procedures and are most widely used. Use of different extraction techniques by different laboratories can yield comparable analytical results for individual n-alkanes and PAH at 0.01-to 1 μg/g concentrations in sediments (MacLeod et al., 1981a). The ambient temperature tumbler method described by D.W. Brown et al. (1980) allows for processing of large numbers of samples without the use of the expensive glassware, boiling solvents, hood space, and running water required by the Soxhlet extraction method.

Tissues

Petroleum hydrocarbons present in marine plants and animals include both those loosely bound to the tissue matrix and those occurring intracellularly. Several methods have been utilized extensively to extract these compounds. High natural lipid contents of many organisms present unique problems of extraction and analysis. Wet tissues can be extracted after homogenization by alkali digestion/saponification using aqueous or alcoholic KOH followed by addition of water or saturated NaCl and partitioning into ethyl ether, pentane, hexane or isooctane

(Warner, 1976; Dunn, 1976; Pancirov and Brown, 1981). Wet tissue may be freeze dried and the dried tissue extracted using methanol followed by a nonpolar solvent (e.g., hexane, benzene, methylene chloride) in a Soxhlet apparatus (Farrington et al., 1976a; Ehrhardt and Heinemann, 1975). Simultaneous high speed homogenization/extraction using anhydrous sodium sulfate and nonpolar solvent in a high speed homogenization apparatus can be utilized, followed by phase separation of the organic layer (Gay et al., 1980; Farrington and Medeiros, 1977). Alternatively, high speed homogenization/extraction using acetonitrile has been used in conjunction with pH adjustments followed by hexane extraction to obtain both hydrocarbon and other organic pollutants comprising the Environmental Protection Agency (1980) "priority pollutant" compounds, which include PAH compounds. Steam distillation of tissue homogenates to obtain an organic extract (Ackman and Noble, 1973; Veith and Kiwus, 1977; Boehm et al., 1983) is an attractive technique in that the extraction procedure is simplified and many polar lipid interferences are removed at the time of extraction. Lawler et al. (1978) described a solvent reflux technique using hexane followed by benzene for extracting mallard tissues. R.C. Clark and Brown (1977) presented a further discussion on tissue extraction methodologies.

Farrington and Medeiros (1977) and Gritz and Shaw (1977) have compared several of the more commonly used techniques--alkaline digestion/extraction, Soxhlet extraction, and high speed homogenization extraction--and found the alkaline digestion/extraction methods to be the most efficient. However, absolute recovery data on all of the tissue methods are generally lacking.

The organic extracts from any of the tissue procedures, even those involving direct saponification (i.e., alkaline digestion), contain interfering polar lipids and biogenic hydrocarbons along with petroleum hydrocarbons. Further saponification of tissue extracts facilitates subsequent separation and isolation techniques (Farrington et al., 1976a; National Research Council, 1975). Extraction of alkaline digestates with nonpolar solvent usually results in gel formations which probably are comprised of lipoproteins. These compounds may be removed by use of an alumina precolumn (Warner, 1976) or gel permeation chromatography (see Size Fractionation section).

Seawater

The precise choice of extraction techniques used to isolate petroleum hydrocarbons from seawater depends on the nature of the sampling device used (see Sampling and Sample Preservation section) and the analytical method to be employed. The extraction methods used depend on whether a water sample must be extracted or whether the organics in the water have been previously sorbed to an accumulator resin (e.g., XAD, polyurethane, reversed phase columns). Boehm (1980a) extracted a large volume (90 L) of filtered seawater in a continuous stainless steel liquid-liquid extractor for 6 hours and the corresponding glass fiber filter in sequential methylene chloride and hexane solvent reflux. R.A. Brown et al. (1975) and R.A. Brown and Huffman (1976) used batch (separatory funnel) carbon tetrachloride extractions of unfiltered 3- to 5-L samples to isolate petroleum hydrocarbons. Boehm's analytical goal was detailed GC and GC/MS analyses; the latter investigators were making infrared measurements of total hydrocarbons. Gordon et al. (1978) made batch extractions of 36 L of unfiltered seawater with doubly distilled pentane for subsequent UV/F and GC analyses. Clearly, the solvent employed must suit the analytical method. In all cases, scrupulous attention to solvent and reagent purities and blanks is required due to low levels expected (0.1-10 μg/L) (Boehm, 1980a; R.A. Brown et al., 1975).

ASTM Method D 3326-78 describes standard practices for the extraction of high levels of waterborne oil, including chloroform or methylene chloride extraction and centrifugation.

The use of accumulator columns (e.g., macrorecticular resins) to isolate hydrocarbons from seawater is becoming increasingly important, although definitive data on sorption and desorption efficiencies are generally not available for these promising techniques. Extraction of the absorbent is performed using a combination of polar and nonpolar boiling solvents either by direct column elution (Bean et al., 1980; Harvey and Giam, 1976) or in a Soxhlet extractor (Basu and Saxena, 1978). Reversed phase HPLC precolumns (e.g., C_{18} Bondapak) have been used to adsorb hydrocarbons from seawater, with the resulting column then being coupled to an analytical column and eluted into a liquid chromatograph for analysis (e.g., May et al., 1975; Ogan et al., 1979; Saner et al., 1979).

Concentration of Extract

Analysts must exercise caution when concentrating solvent extracts to small volume on dryness to avoid excessive loss of compounds in the extract. This is especially a problem for compounds in the C_{10}-C_{13} molecular weight range, and most analysts report data for compounds of molecular weight C_{14} or C_{15} and greater.

Sample Cleanup and Fractionation

In order to increase the precision and discrimination of all analytical techniques, samples may be treated by any one or a series of several types of sample cleanup and fractionation techniques which fall into three basic categories: polarity, size, and chemical separations. These techniques separate petroleum hydrocarbons from lipoidal material, fractionate the hydrocarbons into one or several parts to facilitate analysis, and can isolate nonhydrocarbon components (e.g., azaarenes, phenols) from the petroleum hydrocarbons. The application of one or more separation techniques depends both on the nature of the sample (i.e., amount of polar material) and on the focus of the analysis. Several illustrative sequences are shown in Figure 3-2.

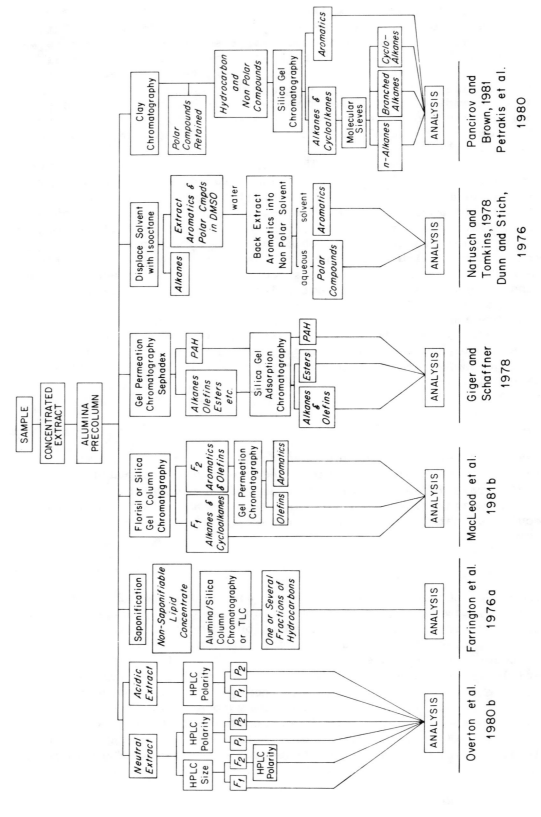

FIGURE 3-2   Representative petroleum hydrocarbon separation techniques.

Polarity Fractionation

Suggested references to some recent applications of polarity fractionation techniques are ASTM Methods D 2007-80 and D 2549-76, Warner (1976,
1978), Giger and Blumer (1974), MacLeod et al. (1981b), Environmental
Protection Agency (1980), J.N. Gearing et al. (1978), Daisey and Leyko
(1979), and Overton et al. (1980a,b).

Size Fractionation

Suggested references to more recent applications of size fractionation
techniques are Albaugh and Talarico (1972), Ramos and Prohaska (1981),
Giger and Schaffner (1978), Warner et al. (1980), Overton et al.
(1980b), Pym et al. (1975), Dastillung and Albrecht (1976), McTaggart
and Luke (1978), and Albaiges and Albrecht (1979).

Chemical Techniques

Sample extracts containing high concentrations of lipoidal material
(i.e., sediments of high organic content, and tissues) are often
saponified (Farrington et al., 1976a) prior to any fractionation or
analytical steps to convert solvent soluble waxes and glycerides to
their fatty acid salts and alcohols. The polar lipids can then be
partitioned into water, leaving a less complex extract containing
nonsaponifiable lipids, which include the petroleum hydrocarbons.
Saponification does, however, preclude the analysis of some compounds
such as the family of DDT pesticides, as the alkaline hydrolysis
converts DDT and DDD to DDE.
    A variety of separation techniques based on chemical interactions
of petroleum hydrocarbon and nonhydrocarbon molecules with resins,
complexing agents, and solvents has been used alone or, more commonly,
in conjunction with other procedures to achieve postextraction
fractionations to facilitate analyses. For example, removal of
elemental sulfur from sediment extracts by reaction with activated
copper prior to analysis (Blumer, 1957) is accomplished either on the
total sample extract or on the aromatic silica gel fraction. The
presence of elemental sulfur greatly interferes with mass spectral
determinations, and removal from sediment extracts prior to analysis is
recommended. Dimethyl sulfoxide (DMSO) has been used to partition PAH
compounds selectively from sample extracts (Dunn and Stich, 1976;
Natusch and Tomkins, 1978). Other chemical techniques have been used
to isolate nonhydrocarbon components of petroleum and petroleum-
containing samples. Basic nitrogen heterocyclics in an organic extract
may be partitioned into an aqueous acidic extract (Jewell, 1980).
Oxidation and reduction ($LiAlH_4$) reactions have been used to
facilitate the isolation of nitrogen and sulfur heterocyclics (Jewell,
1980; Willey et al., 1981). Pentane-soluble nonhydrocarbons such as
pyridines, phenols, carbazoles, and amides can be isolated from
nonpolar hydrocarbons by means of their chemical reaction in ion

exchange resins or transition metal salts (Petrakis et al., 1980). Aromatic hydrocarbons can be isolated from sample extracts via charge transfer complexation reactions (Giger and Blumer, 1974) in conjunction with polarity and/or size separations.

Analytical Methods for C_{11}+ Hydrocarbons (Nonvolatiles)

A variety of analytical methods is available to analyze the levels and composition of hydrocarbons in the solvent extract of an environmental sample. Since the mid- to late-1960s, investigators have realized that techniques to measure petroleum hydrocarbons (PHC) in the marine environment must differentiate incremental additions of newly spilled oil from background levels, both from a quantitative and a qualitative view. Such techniques must adequately measure the concentrations of PHC at levels 10 times or more lower than the traditional "oil and grease" (Environmental Protection Agency, 1979) measurement, and techniques must be available to resolve and quantify individual petroleum components at concentrations less than biogenic hydrocarbon concentrations. Techniques are now available that make the nonspecific oil and grease measurement and then fractionate the extract into one or more PHC fractions analyzable by different methods. The range of measurements go from (1) nondiagnostic gravimetric measurements, to (2) the gross concentration measurements of IR, UV absorption, UV/F spectrometry (with some diagnostic power, albeit concentrations are based on "oil equivalents" and thus are not "absolute concentration"), to (3) the middle to high resolution techniques of high performance (pressure) liquid chromatography, glass capillary gas chromatography (GC^2) and computer-assisted glass capillary gas chromatography/mass spectrometry (GC^2/MS). These techniques and several others encompass the methods available to arrive at petroleum hydrocarbon data. They are often used in a hierarchical manner where "screening" is followed by detailed analyses (e.g., Figure 3-3).

Gravimetric Methods

If sufficient material is available from the extraction processes, an aliquot of the extract may be weighed, using an electrobalance, by evaporating the solvent on the weighing pan and weighing the residue. As little as 5 µg of material can be weighed.

Gravimetry and microgravimetry have been applied to gross determinations of both total extractable material, i.e., oil and grease (Environmental Protection Agency, 1979; ASTM Method D 2778-70; Eganhouse and Kaplan, 1981), and of levels of petroleum hydrocarbons (Farrington et al., 1976a). The former determination does not differentiate the hydrocarbon from the nonhydrocarbon material in a given sample, thus lumping together hydrocarbons and the more abundant nonhydrocarbon lipoidal material. The oil and grease measurement may yield useful information vis-a-vis petroleum-related contamination in samples in which the hydrocarbons are far more abundant than background lipophilic

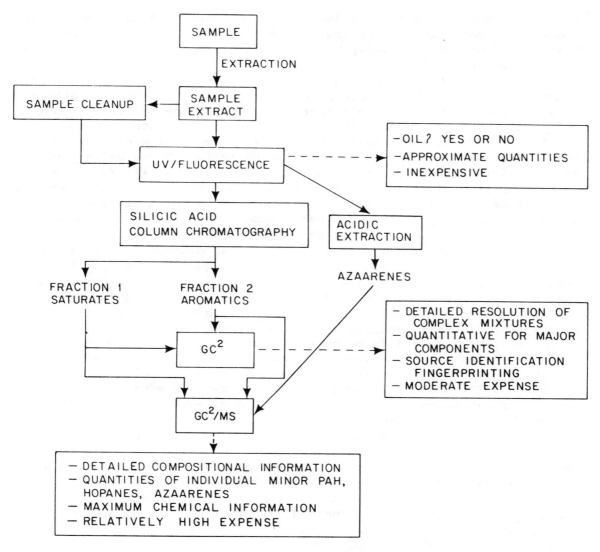

FIGURE 3-3 Hierarchical scheme for analyses of petroleum hydrocarbon in environmental samples.

material (e.g., oil spills), in which levels of hydrocarbons are greater than the method's detection limit (~10-50 µg/L in a 1-L sample, using an electrobalance; minimum absolute weight determination of 5 µg), and in which more detailed compositional information is also obtained.

Infrared Spectrometry

Infrared spectrometry is well suited as an analytical technique for the identification of waterborne oil and matching with suspected cargos of crudes and fuel oils (i.e., fingerprinting). Identifications are based

on comparisons of the intensities of spectral bands in an unknown with
those of suspect cargos over the entire spectrum, or using specific
bands (ASTM Method D 3414-79; U.S. Coast Guard, 1977; Bentz, 1980; J.W.
Anderson et al., 1980).

As a quantitative technique, IR has been used to monitor concentrations of oil in laboratory dosing of seawater systems through calibration and measurement of sample extract IR absorbance at 2,290-2,930
cm^{-1} (-CH_2-stretch) (e.g., Rice et al., 1976). Concentrations of
oil in environmental samples have been determined using similar IR
measurements on CCl_4 extracts of sediments following the Santa Barbara
blowout (Kolpack et al., 1971) and the Amoco Cadiz spill (Marchand and
Caprais, 1981). However, application of quantitative IR measurement to
sediment samples is only suited to situations in which large amounts of
oil (>1 µg/g) are present above background levels due to the method's
inability to (1) distinguish fresh oil from background (chronic) pollution in coastal sediments, and (2) distinguish petroleum from biogenic
material. Furthermore, if samples are not subjected to a fractionation
step separating hydrocarbons from those polar lipids prevalent in all
samples, then the method is totally inadequate to distinguish small
amounts of hydrocarbons in the presence of much larger quantities of
lipids, both of which contain -CH_2 bonds. Furthermore, measurement
of the -CH_2-stretch yields measurements of only saturated molecular
systems, giving no information on aromatics and not taking into account
postspill differential chemical weathering of the source oil. Riley
and Bean (1979) used IR in conjunction with HPLC (see High Pressure
Liquid Chromatography section) to monitor sedimentary PHC for saturated
and aromatic hydrocarbons.

R.A. Brown and Huffman (1976, 1979), R.A. Brown et al. (1975),
Gruenfeld (1975), and Gruenfeld and Frederick (1977) have used IR
measurements to determine concentrations of hydrocarbons in seawater
after performing a silica gel separation of hydrocarbons from "total
extractable organics." These measurements are better suited than the
oil and grease (i.e., unfractionated) IR data, but in nonspill situations they still suffer greatly from difficulties of interpretation and
accurate quantification of petroleum hydrocarbons.

IR analysis has been used for qualitative assessments of hydrocarbon oxidation through measurement of the C = O-stretch (e.g., Rashid,
1974; Wade et al., 1976; Blumer et al., 1973) and for the quick screening of the efficacy of nonpolar/polar compound separation techniques
(see Sample Cleanup and Fractionation section).

Ultraviolet-Visible Fluorescence (UV/F) Spectrometry

Also known as spectrofluorometry, this method has achieved widespread
use for the determination of background levels of petroleum hydrocarbons
in seawater (Levy, 1971, 1980; Keizer and Gordon, 1973; Law, 1981),
sediments (Boehm and Fiest, 1980c; Wakeham, 1977), and tissues (Fong,
1976; Zitko, 1975), and in oil spill sampling as well (e.g., Eaton and
Zitko, 1979; Law, 1978; Mackie et al., 1978; Fiest and Boehm, 1981).
UV/F has been used in conjunction with other analytical techniques to

match waterborne oil with suspected cargo in forensic studies (Bentz, 1980). UV/F spectrometry or spectrofluorometry involves both the ultraviolet (200-400 nm) and visible (350-600 nm) regions of the electromagnetic spectra. Fluorescence procedures are geared to the determination of aromatic hydrocarbon molecular structures, although other conjugated molecules (e.g., polyolefinic compounds) and hetero-cyclic aromatics will fluoresce.

UV/F spectrometry offers advantages of greater sensitivity and selectivity for aromatic molecules than the UV absorption techniques (Farrington et al., 1976a). The fluorescence spectra of crude oils and petroleum products have been under intense investigation in recent years (John and Soutar, 1976; ASTM Method D 3650-78; U.S. Coast Guard, 1977; Eastwood et al., 1978; Eastwood, 1981) as the diversity of the aromatic compositions of crude and refined oils has resulted in "finger-printable" spectral characteristics. Excitation, emission, and synchronous modes of fluorometry have all been utilized for these studies, but the latter two are of greatest interest. Emission spectrofluorometry involves a fixed-sample excitation wavelength and a scan of emission wavelengths.

The UV/F of complex mixtures is provided with greater resolution by use of synchronous scanning (Lloyd, 1971; Vo-Dinh, 1978; Gordon et al., 1976; Hargrave and Phillips, 1975; Keizer et al., 1977; Talmi et al., 1978), and both cryogenic low temperature (Shpol'skii) luminescence (Fortier and Eastwood, 1978; Colmsjo and Ostman, 1980) and "total luminescence" methods (Hornig, 1974) have great potential for further spectral detail. The various possible scan modes are illustrated in Figure 3-4 (Giering and Hornig, 1977).

Although a rather simple technique, room temperature UV/F is subject to severe errors and limitations, only some of which may be eliminated by attention to matters of calibration and matrix interferences. The keys to all of the quantification methods are (1) choice of the proper standard and (2) measurement in the dilution range where self-absorption and quenching effects are minimal. Ideally, the standard should be identical in molecular composition to the samples, but most often in nonspill scenarios the source(s) of contamination is unknown. In spill situations the most appropriate standard is, of course, the spilled oil itself. However, composition changes rapidly due to weathering upon addition of oil to the marine environment. In nonspill cases, measure-ments of "equivalent oil concentrations" (i.e., crude oil equivalents) result, which do not necessarily agree with other, more accurate quantification methods, e.g., GC (Keizer et al., 1977; Zsolnay, 1978b; Hoffman et al., 1979).

In monitoring subsurface concentrations of oil from an offshore blowout using synchronous UV/F, Fiest and Boehm (1981) encountered two entirely different spectral types (Figure 3-5), one similar to Ixtoc oil and the other to its water-soluble fraction. The authors used gravimetric measurements of oil-in-water samples to correct equivalent oil concentrations to absolute levels for the two spectral types. Thus, if the source of the hydrocarbons is known, the ease and speed of the technique make it quite useful for "hot spot" determinations and environmental monitoring (e.g., Eaton and Zitko, 1979; Law, 1978;

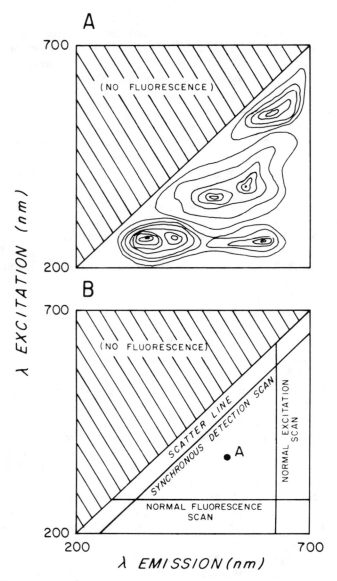

FIGURE 3-4   UV/F scan modes.

SOURCE: Adapted from Giering and Hornig (1977).

Mackie et al., 1978), with the caveat that accuracy and more detailed compositional information are sacrificed.

Baseline seawater measurements generally achieve estimated concentrations in crude oil equivalents (Levy and Moffatt, 1975; Law, 1981). These relative measurements are of some value in examining temporal and spatial trends if quantifying methodology is consistent (e.g., International Oceanic Commission/World Meteorological Organization, 1981).

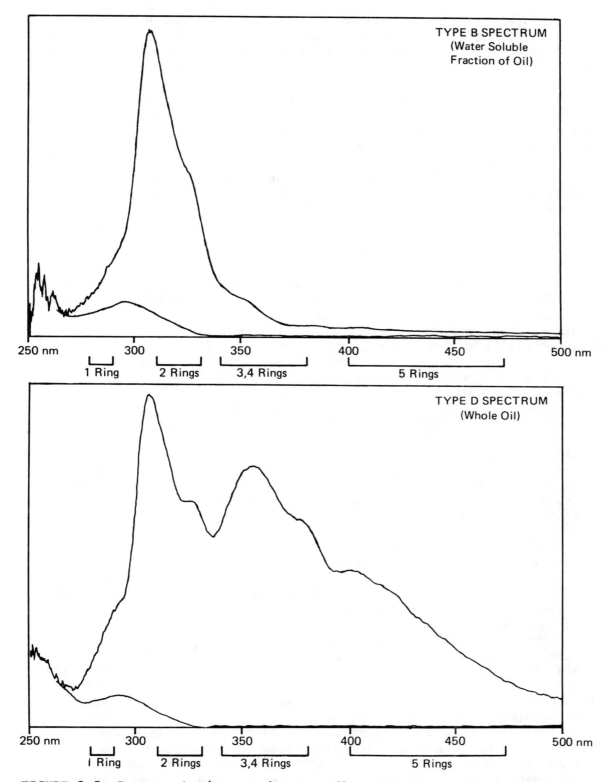

FIGURE 3-5 Representative synchronous fluorescence spectra of water samples collected near the Ixtoc I blowout.

The use of UV/F as a screening technique for establishing an incremental addition of oil to sediment and tissue samples has yet to be adequately tested. However, interference from compounds indigenous to the samples is anticipated to be more severe than for seawater extracts.

As a field spill assessment technique, continuous fluorometric analyses of water pumped through an underwater towed system, using fixed excitation and emission wavelengths, has proven quite useful in real-time measurements of relative quantities of oil in the water column during spill events (Calder et al., 1978; Calder and Boehm, 1981; Turner, 1979; Environmental Devices Company, 1977), especially in conjunction with other measurements such as acoustical reflection (Boehm and Fiest, 1980b).

Thin Layer Chromatography (TLC)

In addition to its use as a separation/fractionation tool (see Sample Cleanup and Fractionation section), TLC has been used as an analytical tool for both the rapid qualitative detection of the presence of oil in sediment samples (L.R. Brown et al., 1975) and the identification of spilled oil (U.S. Coast Guard, 1977). There is an extensive literature on the separation of aromatic hydrocarbons by TLC, followed by measurements directly on the plate or after spot elution (Dunn and Young, 1976; Hunter, 1975; DiSalvo et al., 1975).

Recently, TLC techniques have been used in conjunction with studies of aromatic hydrocarbon metabolism (Gruger et al., 1981; Varanasi and Gmur, 1981b; Malins et al., 1979) to determine the identity of polar metabolites in fish. Two-dimensional TLC, using two-solvent systems (hexane:diethyl ether, 95:5; and toluene:ethanol, 9:1), was used in conjunction with measurements of radioactivity of (labeled) compound metabolites to determine the identity and concentration of specific metabolites (Varanasi and Gmur, 1981b).

High Pressure Liquid Chromatography (HPLC)

HPLC techniques have advanced rapidly in the past 5 years. Although not even discussed in the 1975 NRC report, HPLC is now a major emerging method for the analysis of aromatic hydrocarbons, labile metabolites, and petroleum-derived nitrogen bases in marine samples in many cases (Colin et al., 1981; Wakeham et al., 1981; R.F. Lee et al., 1978, 1981a; Warner et al., 1980; Dunn and Armour, 1980; Malins et al., 1979; Chmielowiec and George, 1980; Ogan et al., 1978; Wise et al., 1977, 1980). As the technique involves nondestructive detection, HPLC may be used as a preparative mode and the column eluates saved (Wakeham et al., 1981), or in analytical mode using fixed and variable wavelength ultraviolet absorption, or fluorescence detectors singularly or in series (T.R. Smith and Strickler, 1980; Christensen and May, 1978; Das and Thomas, 1978). Refractive index detectors may also be used to measure nonfluorescent compounds at high concentrations (Riley and

Bean, 1979). Microparticulate columns can be used in the normal mode using silica columns (Wakeham et al., 1981), or in the reversed phase mode, wherein columns are packed with silica whose surface is coated with a chemically bonded phase (e.g., bonded octadecyl silane columns; Dunn and Armour, 1980). HPLC elution solvent systems range from single mixed solvent systems (e.g., water/methanol) to quaternary systems consisting of two pairs of mixed solvents which are aqueous and nonaqueous binary mixtures (T.R. Smith and Strickler, 1980), and which are run with progressively increasing amounts of the nonpolar solvent mixture (i.e., gradient elution).

HPLC offers major advantages in the analysis of thermally labile or nonvolatile polar compounds (i.e., metabolites; Krahn et al., 1980, 1981) and high-molecular-weight parent PAH (greater than 6 rings; Peaden et al., 1980). There are limitations for analyses of both these groups of compounds when using high resolution GC and GC/MS analytical techniques. HPLC is well suited in analyses where nondestructive techniques are required. However, its use suffers from several disadvantages. Fluorescence and UV absorption detection vary widely from compound to compound, and response factors change as wavelength settings are altered. Thus, a calibration curve of each compound must be determined. Consequently, the quantities of unknown compounds, or compounds for which standards are unavailable, can only be approximated. Additionally, partial fluorescence quenching may occur from interferences in the sample, and complex environmental samples may contain interferences due to nonhydrocarbon fluorescence or absorption.

HPLC shows promise for determining the degree of PAH alkylation in petroleum-contaminated samples. Presently, confused chromatograms often result, wherein alkylated PAH of a given ring size may coelute with the parent PAH of the next larger ring size, thus making HPLC more suited as an analytical tool for combustion-related rather than petroleum-derived aromatics (see Chapter 1). Preparative HPLC using aminosilane packing can be used to collect fractions based on ring size, which can then be analyzed by reversed phase columns to determine degrees of alkylation (Bartle et al., 1981; Wise et al., 1977, 1980; Chmielowiec and George, 1980; Hertz et al., 1980; Berthou et al., 1981). Thus, while GC^2 and GC^2/MS techniques are still preferable in terms of their resolving power, recent HPLC techniques have increased their applicability to environmental petroleum and PAH analyses. Advances in HPLC detectors and coupling HPLC with MS-computer systems offer great promise for future uses in petroleum pollution research and monitoring. Intercomparison of HPLC- and GC^2-derived data is an important priority for future intercalibration studies. Hertz et al. (1980) have provided an initial intercomparison study of GC, GC/MS, and HPLC analyses of PAH in shale oil.

Gas Chromatography (GC)

Great advances in GC column methodology have been responsible for a vastly increased ability to analyze nanogram levels of individual hydrocarbons routinely in complex environmental samples and to discern the overall hydrocarbon composition of marine samples (M.L. Lee and Wright, 1980; Cram and Yang, 1980; Grob and Grob, 1976, 1977; Bjorseth and Eklund, 1979). Stainless steel and glass columns packed with a liquid phase adsorbed onto a solid support phase have given way to stainless steel, borosilicate glass, and fused silica glass capillary columns in which the interior walls of the columns are coated directly with a liquid phase or coated with a support phase on which the liquid phase is bonded (Jennings, 1980). Increased resolution of complex hydrocarbon mixtures has resulted. In addition, the increased inertness of the glass capillary columns affords excellent GC resolution and response of large aromatic molecules (6 rings) and, in the case of nonhydrocarbon compounds, the resolution and response of nonderivatized polar molecules (e.g., triglycerides, sterols; Cram and Yang, 1980). Capillary columns are effective in distinguishing among different sources of hydrocarbons for the same sample. For example, background anthropogenic inputs, recent petroleum additions, and biogenic inputs comprise a typical composite source to marine samples from coastal regions. The existence of, and in some cases the relative contribution of, these assemblages to a particular sample's "hydrocarbon fraction" are discernible by high resolution glass capillary GC (or GC^2) (Overton et al., 1977; Farrington, 1980; Simoneit and Kaplan, 1980; Simoneit, 1978; Thompson and Eglinton, 1978a). Characterization of oils by high resolution GC^2 has been discussed by Rasmussen (1976), Crowley et al. (1980), and Cram and Yang (1980).

The concentrations of individual petroleum compounds are usually a small fraction of the total hydrocarbon concentration. GC^2 results can be used to obtain individual compound concentrations, obtain the sum of components (e.g., total n-alkanes), or obtain an estimate for a total concentration comparable to gravimetrically determined values by summing the total resolved plus unresolved hump. This hump, or unresolved complex mixture (UCM) (Farrington and Meyers, 1975; Farrington et al., 1976a; Simoneit, 1978), is a characteristic GC feature of some fresh oils and most weathered oils. As weathering proceeds and resolved components decrease in concentration, the UCM in both becomes more prominent. Additionally, most non-spill-related environmental samples containing anthropogenic inputs (Reed et al., 1977; Wakeham and Farrington, 1980; Farrington et al., 1980) and many geochemical samples (e.g., Thompson and Eglinton, 1978a,b) contain UCM material which may account for 80-90% of the total hydrocarbon weight. The UCM, which most likely consists of naphthenic, naphthenoaromatic, and other condensed ring structures, may be reduced in size but not eliminated by use of the more efficient capillary columns now available. There are techniques (Sampling and Sample Preservation section) that can separate, for example, PAH compounds from the UCM, and n-alkanes from UCM material. However, the shape of the UCM and its molecular weight, and the possible existence of bimodal UCM distributions, can provide information on the nature of the contaminant oil and postspill diagenetic, metabolic, and selective bioaccumulation processes. Indeed, Saxby (1978) and Butler (1975) have used UCM shapes and responses to obtain information on petroleum weathering processes.

Thus, GC^2 integration algorithms must be able to integrate peaks over changing baselines (i.e., the UCM).

In practice, a small amount of fractionated extract, as determined by the weight of a given extract, is injected into a GC inlet system or directly onto a glass capillary column (Grob and Grob, 1978a,b). The peaks recorded by the instrument are identified and quantified by one of several methods: (1) comparing retention times of the peaks with those of known standards, (2) determining the retention indices (Kovats and Keulemans, 1964; M.L. Lee et al., 1979) and comparing with those of known compounds, (3) coinjecting a sample with standards, or (4) using supplemental analytical equipment and techniques (see Gas Chromatography/Mass Spectrometry section). Use of retention times is prone to considerable variation from day to day on different instruments and even on the same instrument. Systems of retention indices have been developed to take these variations into account and to make data more reproducible within a given laboratory and between laboratories. There remains, however, the possibility that two components in the sample may have the same retention times and indices on a particular column (i.e., they coelute), even with high resolution techniques. Therefore, unless part of a predictable pattern (e.g., the n-alkanes, isoprenoids), the identity of a component should be verified by (1) injection of the sample on another GC column with a liquid phase of different polarity or (2) supplemental instrumental techniques (GC/MS).

Compound Quantification A variety of quantifying techniques has emerged in recent years. Sample components may be quantified by the use of an internal standard method. In this technique, one or more components with a molecular structure similar to the compounds being analyzed are spiked to the sample prior to extraction. Deuterated standards are becoming more widely used, as they eliminate the problem of finding a standard which, a priori, can be determined not to be in a sample (Cretney et al., 1980). The standard(s) and sample extract are thus carried through the entire analytical procedure. Peak areas of sample components are compared with those of the known quantity of internal standard. The external standard method involves spiking of standard(s) to the extract just prior to GC analysis. This quantification method must be corrected for the absolute recovery of the analytical procedure, which is usually accomplished by use of a spiking or recovery standard and which is, for all intents and purposes, equivalent to the internal standard. The instrumental calibration method involves GC analysis of an aliquot of the sample extract and comparison of instrumental response to that of a known amount of standard compound(s) run in a separate serial dilution to obtain a calibration curve of instrumental response versus amount injected. In this latter technique, both sample recovery and correction for the injection volume must be taken into account.

Selective Detectors A variety of GC detectors has been developed. They are presently in use, either singularly or in combination, to increase the sensitivity and discrimination power of the GC techniques (Hartmann, 1971; Hrivnac et al., 1976; Searl et al., 1979; Novotny et al., 1980a,b; Frame et al., 1979; ASTM Method D 3328-78; Overton et al., 1980b; D.A. Miller et al., 1981).

Data Acquisition, Handling, and Storage Given the present high level of detail of petroleum hydrocarbon GC determinations, many laboratories have instituted systems for acquiring, processing, calculating, storing, and retrieving data on marine environmental samples. Raw data from high resolution GC^2 analyses consist of sets of GC retention time and peak area data. Several independent systems are now in use in analytical laboratories for transmitting these pairs to an external computer and converting data to peak names (i.e., compound names and/or retention indices based on a standard) and concentrations, and for storing, retrieving, displaying, and manipulating data (e.g., Overton et al., 1978a,b; Reese, 1980a,b). However, due to the great abundance of data of this nature generated from high resolution GC^2 techniques, acquired data should be treated as a library of information on hydro-carbon and other compounds of potential use in (a) monitoring temporal changes of known pollutant compounds of concern, and (b) yielding information on compounds of which the significance has yet to be revealed. Further hardware and software advances will be needed to link the activities of compound identification, compound quantifi-cation, information storage, information editing, insertion into interlaboratory data files, and selective retrieval of data for display.

Gas Chromatographic Mass Spectrometry

Gas chromatographic mass spectrometry (GC/MS) is a technique wherein a mass spectrometer acts as the detector of a GC system, thus enabling the mass spectra of eluting components to be determined. Computer-assisted capillary GC/MS techniques (GC/MS/DS, where DS are data systems), are the most powerful tools available to confirm the identity of and to quantify trace levels of individual petroleum hydrocarbon components in environmental samples. The identity of an unknown compound can be determined by matching the compound's mass spectrum with that of the pure compound, either manually or via computer-assisted mass spectral library searches and probability-based matches. GC/MS computer systems have been used extensively in recent years for unambiguous identification and quantification of petroleum and combustion-derived aromatic hydrocarbon compounds in sediments (Teal et al., 1978; Hites et al., 1980; Youngblood and Blumer, 1975; Lake et al., 1980; Overton and Laseter, 1980), marine tissues (Grahl-Nielsen et al., 1978; Boehm et al., 1981b; Farrington et al., 1980, 1982a,b; Warner et al., 1980), and seawater (Boehm, 1980a). GC/MS has also produced valuable data on low levels of saturated hydrocarbon marker compounds in sediments (Simoneit and Kaplan, 1980; Atlas et al., 1981; Dastillung and Albrecht, 1976; Albaiges and Albrecht, 1979; Bieri et al., 1978) and in tissues (Anderlini et al., 1981). Novel molecular fingerprinting techniques using aromatic hydrocarbon and organosulfur compounds (Overton et al., 1981), acyclic isoprenoids, steranes, and triterprenoids (Albaiges, 1980) are rooted in GC/MS analyses. GC/MS is also the most powerful method for examining the identity of photochemical and biodegradative (Overton et al., 1980a) products and metabolites of parent hydrocarbon compounds. Indeed, GC/MS computer systems could provide a method for determining more components of the UCM GC^2 signal.

GC/MS is either performed in the chemical ionization (CI) mode (Warner, 1978) or in the more commonly utilized electron impact (EI) mode. In the latter case, the high energy (ionization voltage = 70 eV) of electron impact results in a high degree of fragmentation of nonaromatic hydrocarbons. CI, using methane as the ionizing gas, results in the preservation of the molecular ion (molecular weight) peak, thus facilitating identification of easily fragmented compounds such as the alkanes. In addition, specialized techniques such as plasma desorption probe CI (PD/CI/MS) have been developed for use in identifying thermally labile PAH metabolites (Krahn et al., 1980).

Basically, GC/MS systems are operated in one of several operational modes:

1. Acquisition of full range mass spectral data, which allows for total ion chromatograms (equivalent to GC/FID trace) and selected ion chromatograms or mass chromatograms to be reconstructed (i.e., computer searches and display of particular fragment or molecular ions characteristic of a compound). This mode is used to find information on unknown compounds or on a large number (>20) of compound spectra.

2. The selected ion monitoring (SIM) mode allows the mass spectrometer output to be scanned at preselected masses and the resultant mass chromatograms to be stored. SIM allows for greater sensitivity but permits only a limited number of masses to be scanned during a run.

3. In the probe distillation mode sample, mixtures are introduced directly into the ion source of the mass spectrometer. Increased mass dwell-time greatly increases the sensitivity of this mode so that compounds undetectable in the total ionization modes can be detected (Youngblood and Blumer, 1975; Hites et al., 1980). However, this mode is subject to interference by nontarget compounds present in the sample.

Quantitative GC/MS by mass fragmentography involves the computer integration of an ion current plot (molecular ion or fragment ion) for a compound derived from the mass chromatogram, comparison to an internal standard ion current, and correction based on relative mass spectral response factors. Although research has been conducted on applications of this technique for several years, extensive rigorous quality control and intercalibration are still needed.

More details of the various applications of GC/MS are found in Warner (1978), Warner et al. (1980), and Hites et al. (1980).

Other Methods

Less widely used analytical chemical methods have been applied to measurements of hydrocarbons in marine samples. Stable isotope ratios of carbon, sulfur, and hydrogen have been used to examine the hydro-carbon composition of oils and sediments (J.W. Miller, 1973; Sweeney et al., 1980) and to identify weathered oils (Hartman and Hammond, 1981). Nuclear magnetic resonance (NMR) spectroscopy has been used to examine the relative molecular composition of petroleum constituents and fractions (Petrakis et al., 1980; Petrakis and Edelheit, 1979).

Low-Molecular-Weight Hydrocarbons: Analytical Methods

Hydrocarbons in the C_1-C_{10} range are not amenable to routine solvent extraction techniques. Specialized techniques have been developed for isolating and analyzing C_1-C_{10} saturated and aromatic hydrocarbons in seawater (Lysyj et al., 1980; McAuliffe, 1971, 1980; Swinnerton and Linnenbom, 1967, 1976; Sackett and Brooks, 1975; Brooks et al., 1980; Sauer et al., 1978), sediments (May et al., 1975; Bernard et al., 1978), and marine biota (Chesler et al., 1978; L.C. Michael et al., 1980). The methods that all rely on GC and/or GC/MS as the final analytical tool fall into six categories: static headspace sampling (Friant and Suffet, 1979), dynamic headspace purge (May et al., 1975; Chesler et al., 1978; Michael et al., 1980), gas stripping (Sauer et al., 1978; Swinnerton and Linnenbom, 1967, 1976; Bellar and Lichtenberg, 1974; Lysyj et al., 1981), vacuum stripping or degassing (Brooks et al., 1973), multiple phase equilibrium (McAuliffe, 1971, 1980), and direct aqueous injection.

In the static headspace technique, a sample in a closed container is usually heated, and a subsample of the air is taken with a syringe and injected into a gas chromatograph. The method suffers from being suitable to analyze only high concentrations of compounds of high volatility, although the volatility limitation is improved by heating. Additionally, rigid control of temperature and headspace volume is required. Dynamic headspace techniques involve continuous movement of inert gas (e.g., helium) through the headspace of a heated flask con-taining the sample and the collection of the entrained volatiles on an accumulator column, e.g., Tenax GC (Michael et al., 1980; Chesler, 1978). This method is appropriate for use with biological samples but has also been used for sediments (Bernard et al., 1978; May et al., 1975). Nonvolatile hydrocarbons left behind may be analyzed by high-molecular-weight hydrocarbon methods. In gas stripping, the inert gas is bubbled through the sample, and the purged vapors are trapped on a sorbent (Tenax GC, Porapak Q, alumina or activated charcoal columns). The method is best suited for seawater analysis, as severe foaming limits its use for most biological samples (Michael et al., 1980). Swinnerton and Linnenbom (1967, 1976) and Sackett and Brooks (1975) have used gas stripping to isolate gaseous hydrocarbons from seawater, and Sauer et al. (1978) used the method to isolate volatile liquid (C_6-C_{14}) hydrocarbons from seawater. Lysyj et al. (1981) trapped sparged aromatic hydrocarbons from seawater, trapped them in activated charcoal, desorbed them with carbon disulfide, and directly injected the solution into the GC. The sensitivity was 0.1-0.2 µg/L. Given the proper design of adsorption and desorption systems, the gas stripping techniques are suitable for nL/L levels of low-molecular-weight hydrocarbons (Brooks et al., 1980).

Use of a vacuum stripping apparatus, "The Sniffer" by Brooks et al. (1973), involved continuous stripping of gaseous hydrocarbons from seawater by vacuum degassing. The apparatus is used by the petroleum industry for routine sampling to detect dissolved gases near suspected seep areas. The system consists of a tow body, a pump, and shipboard degassing apparatus, interfaced to provide an analysis at 90-s time intervals.

The multiple gas phase equilibrium approach of sample analysis has many applications to oil spill and other environmental studies (McAuliffe, 1980). While gas stripping techniques are partial equilibrium methods, the gas equilibration technique provides for transfer of volatile constituents from water to inert gas in proportion to the compound's vapor pressure and aqueous solubility (i.e., distribution coefficient). The inert gas is injected directly into a GC. The technique, as described by McAuliffe (1980), is well suited for spill-related studies and has been used extensively in experimental spill studies by McAuliffe et al. (1980) and in studies of New York harbor by McGowan (1975).

Direct aqueous injection onto a GC column is limited by high detection limits (mg/L) (ASTM Method D 2908-74), although the method may be used in conjunction with capillary GC to increase sensitivity.

All methods require specialized GC introduction systems for the efficient thermal desorption of trapped material (e.g., Michael et al., 1980) onto the head of a GC column, and for the selective removal of water from columns prior to desorption.

PETROLEUM HYDROCARBON INTERCALIBRATION/INTERCOMPARISON PROGRAMS

Interlaboratory intercalibration programs of various types have been undertaken with national and international scopes. The intent of these studies has been mainly to examine the variability of analytical results between laboratories, aid in the evaluation and comparisons of environmental data sets, evaluate the relative efficacy of different methodologies, and improve analytical methods so as to reduce interlaboratory discrepancies in data. In recent years, researchers have developed specialized techniques for analyzing hydrocarbons in marine samples. Comparison of analytical data developed in different studies has often been highly problematical, even among programs (data sets) utilizing similar basic analytical tools (e.g., capillary gas chromatography). The problem is, of course, more severe when one attempts to relate data sets which have utilized different analytical techniques (e.g., fluorescence spectroscopy and gas chromatography). Many laboratories are involved in the measurement of fossil fuel hydrocarbons in marine samples utilizing different techniques. There is currently little available information on the compatability of these data sets.

Three types of interlaboratory comparison exercises are possible: (1) sample splits involving relatively few laboratories, (2) calibration samples or sample extracts prepared for specific research or monitoring programs, and (3) standard certified reference materials. The most significant intercalibration exercises presently underway or previously undertaken address type (1) exercises, involving enough laboratories to enable statistical analysis of data. Type (3) materials with National Bureau of Standards (NBS) certification, containing known amounts of specified constituents, have been requested by scientists in environmental studies. To date, only one such sample has been prepared, due to uncertainties of sample homogeneity, storage stability and matrix effects, and definitive analytical methods. A new standard reference material (SRM 1580), "Organics in Oil Shale," is intended primarily for evaluating reliability of analytical methods for the determination of three PAH and two phenolic compounds in an oil matrix. Thus, most exercises involve type (2) programs. A summary of major petroleum hydrocarbon intercalibration studies undertaken in the 1976-1981 period is shown in Table 3-2. Interlaboratory precision has improved significantly over the past 5 years or so, as techniques for both analyzing samples and running intercalibration exercises have improved.

The roots of a well-conducted intercomparison program lie in the homogeneity of the sample and the comparability of data (i.e., the reporting of the same components by all participating laboratories on the same basis, corrected for recovery). During the last 5 years, the ability to conduct intercalibration exercises and to analyze samples rigorously and achieve comparable results have both improved markedly. Bearing in mind that there is no "right answer" in such exercises using environmental samples, a group of laboratories in the United States has obtained generally tightly grouped results based on GC^2 (and GC^2/MS) determined alkane and polynuclear aromatic hydrocarbon levels in sediments (MacLeod et al., 1981a). While statistical evaluations are still in progress, laboratories probably can achieve comparable (within a factor of 2 and often much better) analytical results. Coefficients of variation for individual aromatic hydrocarbon determinations in the Duwamish II study were, for example, +14% for fluorene, +17% for phenanthrene and fluoranthrene, and +39% for perylene, for the six data sets (MacLeod et al., 1981a) and were as good for n-alkane values.

The International Council for the Exploration of the Seas (ICES) intercalibration studies, while not as rigorously controlled as the Duwamish exercises (see Table 3-2), have yielded comparable fluorescence-based data on sediments with a coefficient of variation for "total petroleum" in the 10-30% range. This level of agreement was reached by using specified quantification methods, i.e., prescribed Integrated Global Ocean Station Systems (IGOSS) wavelengths. The ICES-sediment exercise yielded comparable UV-, IR-, and GC-based "total hydrocarbon" concentrations.

Intercalibrations on biological materials have posed more serious problems, with even UV-based data (ICES study) yielding poor results, probably due to both analytical problems and quantification techniques. The GC- and GC/MS-based EPA megamussel study currently under way (no final data available) specifies individual compounds and aromatic isomeric groupings for reporting.

The emerging view appears to be that, for the most part, comparability of petroleum hydrocarbon and PAH results is beginning to depend

TABLE 3-2 Summary of Interlaboratory Intercalibration Exercises, 1976-1981

Name of Study	Sponsoring Organization	Sample Type	Analytical Basis of Data	No. of Participating Laboratories	Reference	Results
Santa Barbara sediment	BLM	Sediment spiked with South Louisiana crude oil	1. Saturated (f_1) and aromatic (f_2) hydrocarbons by gravimetric and GC analyses 2. Individual $\underline{n}$-alkane concentrations 3. Individual 2-ring aromatics	12	Farrington (1978)	-Individual component concentrations vary by factors of 1-40 (generally 5-10) -Much less variability in gravimetric values for total fraction weights (1-4)
NBS sediment	NBS/EPA/ BLM	Alaskan intertidal sediments	1. Total f_1 and f_2 by GC 2. Most abundant aliphatic (f_1) and aromatic (f_2) components 3. Identity and amount of PAH 4 rings and larger	8	Hilpert et al. (1978)	-Large variability in all parameters reported ($1 = \pm 25\%$)
NBS mussel	NBS/BLM	Santa Barbara and Alaskan mussel	1. Total f_1 and f_2 by GC 2. Most abundant aliphatic (f_1) and aromatic (f_2) components 3. Identity and amount of PAH 4 rings and larger	8	Wise et al. (1980)	-Large variability in parameters ($1 = \pm 40\%$) especially individual aromatic and saturated compounds

TABLE 3-2 (continued)

Name of Study	Sponsoring Organization	Sample Type	Analytical Basis of Data	No. of Participating Laboratories	Reference	Results
Duwamish I	BLM/NOAA	Estuarine sediment	1. Individual aliphatic and aromatic components by GC (specified lists of components); statistical analysis	11	D.W. Brown et al. (1980); MacLeod et al. (1981a)	-Major improvements in consistency of parameters reported -Most values with a factor of 2
Duwamish II	BLM/NOAA	Estuarine sediment (finer grained)	1. Individual aliphatic and aromatic components by GC (specified lists of components); statistical analysis	9	McLeod et al. (1981a)	-Comparable results achieved in spite of use of different extraction procedures
ICES	ICES	Crude oil Dried sediment Mussel homogenate	1. Analytical and reporting requirements not specified for advance; GC, IR, and UV used 2. Fluorescence data specified using IOC/WHO methods	25	Law and Portman (1981)	-Good comparisons of sediment fluorescence data due to pre-prescribed quantification method -Good comparisons of methodology differences -Poor results for tissues
IKU	IKU/ Norway	Homogenized seawater sediments and biota from experimental spill	1. GC quantifications of alkanes and aromatic groupings	3	Haegh (undated)	-Data on sediments and organisms not comparable due to quantification method differences; data on seawater not reported on a consistent basis

EPA megamussel homogenate	Mussel	1. Individual alkane and aromatic components by GC	4+	Galloway et al. (1983)	-Shows good alkane and aromatic agreements (factor of 2)
IDOE 1,3,5	NSF	-1,3: cod liver oil spiked with fuel and crude oil -5: cod liver lipid extract spiked with distillate oil -tuna meal	1. Total petroleum by GC 2. Individual component, pristane	3	Farrington et al. (1976b)	-Good agreement on "total hydrocarbon" parameter ($\pm$12%) -Mixed results on quantification of individual components

NOTE: All acronyms appear in Appendix B.

more on the quantification process (i.e., how individual component GC peaks are quantified) than on the extraction and processing steps (i.e., several extraction procedures will suffice). This is true for the Duwamish I and II sediment studies, wherein differing extraction methodologies were used (D.W. Brown et al., 1980; MacLeod et al., 1981a), and may be emerging as the reason behind variability in the more difficult, interference-prone biotic measurements.

Clearly, further intercomparisons are required, addressing (1) comparability of results based on simpler, more universally available methods (i.e., UV fluorescence), (2) comparability of more rigorous techniques (i.e., GC and GC/MS), (3) intercomparability of the methods, and (4) the location within the analytical technique for discrepancy. Laboratories should be urged to participate in intercalibration programs in a nonthreatening atmosphere at the start of the environmental chemistry program, to enable the refinement of analytical techniques so as to achieve results within a determined statistical range. The NBS SRM oil shale, samples such as Duwamish I and II sediments, and the ICES sediment appear to be most appropriate for this purpose.

REMOTE DETECTION AND MEASUREMENT OF OIL SPILLS

Remote sensing devices used to monitor marine pollution are becoming more sensitive and reliable than they were just 5 years ago. The use of both airplanes and satellites as platforms for remote sensing devices has been explored. ICES and NOAA, as well as other organizations, have been involved in the development of satellite-carried equipment for sensing oceanographic parameters (Apel, 1978; Kniskern et al., 1975; Koffler, 1975; N.R. Anderson, 1980; Klemas, 1980). However, satellite monitoring is not without problems. Geosynchronous satellites do provide repeatable coverage, but the resolution is not great enough to be of practical use. The NASA ad hoc committee on remote sensing concluded that the physical parameter requirements for oil spill monitoring are at least an order of magnitude greater than the remote sensing data which are now available (Croswell and Fedors, 1979). In addition, Goldburg (1979) concluded that sensors in airplanes are more feasible and cost efficient than satellite remote sensing, thus, the focus on airborne sensors in this section.

The U.S. Coast Guard has developed remote sensing "packages" to aid in the detection of oil slicks. The two prototypes of the current package, AOSS I and AOSS II (Airborne Oil Surveillance Systems I and II), are described more fully in Bentz (1980), Maurer and Edgerton (1975), and G.P. White and Arecchi (1975). The third-generation aerial reconnaissance system, designated AIREYE (for aerial remote instrumentation), will be installed in Falcon 20-G jet planes and includes side-looking airborne radar (SLAR), an IR/UV scanner, a computerized data recording system, and an aerial reconnaissance camera (N.R. Anderson, 1980). By including sensors utilizing three portions of the electromagnetic spectrum, the number of false alarms due to kelp beds, wake scars, and the weather can be kept to a minimum (J.R. White et al., 1979).

Remote sensing devices can be divided into two categories: those based on passive (natural) reflectance and emission of some part of the electromagnetic spectrum, and those based on an active (man-induced) electromagnetic excitation of the ocean surface and the collection of reflected radiation. The passive group includes microwave, IR, and UV collectors. Those devices that depend on man-induced electromagnetic radiation include radar, UV fluorescence systems, and laser backscatter sensors. Table 3-3 (from N.R. Anderson [1980] and Maurer and Edgerton [1975]) reviews the types of remote sensing devices and the false alarms given by each.

Passive microwave systems measure radiation waves naturally emitted or reflected by the sea surface. Microwave brightness is a function of surface roughness and the dielectric constant of the surface. Thin oil films have a calming effect on the water surface, which results in a modification of the microwave structure and thus a lower brightness temperature. Thick films (>0.1 mm) emit more microwave energy than unpolluted water does; thus, the film thickness can be determined from the relative brightness temperature. Passive microwave systems can penetrate weather and are independent of lighting conditions. Disadvantages include coarse resolution and a limited swath (Maurer and Edgerton, 1975).

Infrared sensors detect apparent temperature differences between oil and water due to the physical properties of the two substances. Oil and water have different reflectance properties in the 2- to 4-μm spectral range (G.P. White and Arecchi, 1975). In the near IR range (0.6-1.1 μm), the radiance from an oil slick is 20-100% greater than the radiance from water, and at night, oil gives 50% greater radiance than water does (Catoe, 1972). Thermal IR (1.1-14 μm) sensing is limited to specific atmospheric windows where the atmosphere is transparent enough to allow the waves to pass through without significant absorption (Catoe, 1972). Thermal infrared sensing can also be used 24 hours a day, and IR waves can penetrate haze but not clouds. Odd local thermal structures (e.g., an upwelling) can cause false alarms (Maurer and Edgerton, 1975).

Passive ultraviolet collectors can detect oil patches because oil reflects more UV light than water does. The greater amount of UV radiation that water absorbs, the cooler it appears in relation to the oil slick it surrounds. Passive UV collectors require some ambient sunlight, but the light range can be extended if the collector is used in conjunction with a low light level television (LLLTV). False alarms from this system include kelp patches (Maurer and Edgerton, 1975), and atmospheric aerosols limit its use in hazy weather (Catoe, 1972).

One of the more widely used active sensing systems is radar. It is used with a great deal of success to detect offending ships and oil slicks on the sea surface. SLAR has a swath of up to 80 km (40 km on each side of the airplane). SLAR detects the capillary wave-damping effect caused by oil on the sea surface, so this technique becomes ineffective on flat, calm or extremely rough seas. Another disadvantage of SLAR is that it does not "see" a strip directly beneath the plane. An IR/UV line scanner is often used to overcome this problem (J.R. White et al., 1979).

TABLE 3-3 Oil Spill Detection by Remote Sensing: Sensors and Spectral Regions

Sensor Approach	Spectral Region	False Alarms[a]
Active reflectance	Microwave radar, 1.05-5 cm Laser backscatter UV fluorescence, 0.4 m	Natural organic slicks Wind slicks, ship wakes Pollutant organic slicks (detergents, sewage sludge) Kelp/debris Dense cloud cells Unrippled water under calm conditions
Passive reflectance	UV, 0.4 m	Natural organic slicks Suspended solids
	Visible 0.4-0.65 m	Natural organic slicks Pollutant organic slicks Suspended solids Shallow water Broken cloud deck
	Near IR, 0.65 m	Natural organic slicks Other pollutant slicks
Passive emission	Thermal IR, 3-14 m	Natural organic slicks Pollutant organic slicks Ship wakes Thermal discharges and effluents Upwelling
	Microwave, 0.2-1 cm	Foam patches Kelp/debris Dense cloud cells

[a]As all of the listed sensors detect oil on water, natural petroleum seeps would be a false target for each sensor.

SOURCE: N.R. Anderson (1980) and Mauer and Edgerton (1975).

A laser backscatter sensor (Dichromatic Lidar Polarimeter), which transmits at two coaxially aligned, vertically polarized wavelengths, has been developed (G.P. White and Arecchi, 1975). Depolarization occurs at the sea surface, and the two wavelengths are backscattered differentially. The backscatter is collected, and the magnitude of returned radiation and the depolarization ratios are used to determine the presence of oil. Hoge and Swift (1980) used a laser-induced water

Raman backscatter sensor to detect the presence of oil. They found that oil depressed the Raman backscatter, which returned to normal after the sensor was over water once again. Oil film thickness could also be determined using this method.

Probably the most promising remote sensing device currently being developed is the laser-induced UV fluorescence sensor. Laser-induced fluorescence systems not only differentiate oil from water but also can discriminate between oils as well (Kim and Hickman, 1973; Rayner et al., 1978; Fantasia et al., 1971; Fantasia and Ingrao, 1973; Horvath et al., 1971; O'Neil et al., 1975; Measures et al., 1975; Kung and Itzkan, 1976). A UV laser excites the sea surface, and the fluorescence return is collected. A photomultiplier tube converts the fluorescence to an electrical signal, and then a fluorescence spectrum can then be printed out.

Field trials by Fantasia et al. (1971), Horvath et al. (1971), and Rayner et al. (1978) have shown that, not only can oil fluorescence be detected over background fluorescence, but oil can be classified into three groups: diesel fuel, crude oil, and bunker fuel. O'Neil et al. (1980) reported that oil shows increased UV absorbance with decreasing excitation wavelength; thus, thinner oil layers can be detected. The shorter wavelengths also show greater structure in the fluorescence spectra, which gives greater discrimination power and allows classification of different oils.

Attempts have been made to detect oil in the water column using UV fluorescence sensors. These have been almost totally unsuccessful because there is so much nonpetroleum suspended organic matter in seawater and, because water absorbs so much UV light, there is very little fluorescence emitted (F.E. Hoge, personal communication, 1981).

MONITORING FOR PETROLEUM HYDROCARBONS

The success of any monitoring program depends on the proper selection of environmental parameters to be measured, the proper choice of analytical techniques to be used, the comparability of analytical results over time and between laboratories, and the statistical validity of the measurements (i.e., what level of sampling and analytical effort will detect change) (Risebrough et al., 1980). (See also the Introduction to this chapter.)

When the amounts of oil are large, simple analytical techniques (e.g., IR, gravimetry) or remote sensing may suffice. However, at low levels, analytical strategies become critical. A specific property of the oil such as UV/F may be determined and "equivalent oil concentrations" obtained. Alternatively, individual components in a single class of compounds (e.g., aromatic hydrocarbons) may be quantified. Measurements of specific properties, although more widely performable by more laboratories, rely on tenuous assumptions regarding calibration of the methods. Monitoring of individual compounds is more expensive and requires extensive quality control and intercalibration. However, much useful information for differentiation between hydrocarbon sources can be obtained, along with determination of the extent and severity of

pollution. If seawater is the targeted environmental compartment, then
UV/F may suffice due to low background levels. In cases where correlation analysis of hydrocarbon and other parameters is used as a monitoring tool, then these simpler techniques may differentiate impacted from
nonimpacted sediments (Boehm and Quinn, 1978). However, most monitoring
scenarios call for specific chemical component measurements, perhaps
guided by specific property techniques (see Figure 3-3).

Several far-reaching analytical monitoring programs have been
initiated in recent years which address two main concerns: (1)
detection of environmental change (i.e., environmental degradation or
improvement) due to petroleum hydrocarbon (and other pollutant) inputs
to the system, and (2) assessment of the temporal recovery of an oil
spill stressed system. A third concern only loosely being addressed
due to constraints of time and data handling is the identification of
"new pollutants." One example of the former type of program is the
U.S. EPA Mussel Watch program (National Academy of Sciences, 1980;
Farrington et al., 1983), which utilizes the sentinel organism approach.
Mussels on the mid-Atlantic, northeast, and west coasts, and oysters on
the southern and Gulf coasts are analyzed for specific petroleum
hydrocarbons and other pollutants, the rationale being that mussels
reflect the water quality over an integrated time scale. Emphasis in
the hydrocarbon program is on analysis of specific aromatic compounds
(currently up to 4 rings) and alkylated aromatics to determine absolute
levels of these compounds, their changing levels, and sources of
observed hydrocarbons (i.e., whether from pyrolytic or petroleum
sources). Intercalibrations have been underway in this program
(Galloway et al., 1983).

NOAA's Northeast (U.S.) Monitoring Program attempts to link chemical
to biological parameters over time. The focus is on the analysis of
sediments as a major sink for pollutants, and a selected set of organisms for individual PAH (and polychlorinated biphenyls (PCB) and metals)
compounds. This program attempts to utilize several preexisting data
bases (BLM-Benchmark; NOAA-MESA [New York Bight]), although in the past
no uniform techniques of measurement have been utilized nor intercalibrations stressed.

ICES monitoring programs, in existence since 1977, have focused on
metal and organochlorine residues in sediments and several fish and
invertebrate species. Petroleum hydrocarbon information is beginning
to be derived from this program, mainly based on specified UV/F
analysis, but presumably to be complemented by high resolution techniques as well. Residue levels are evaluated in terms of human health
concerns. The ICES "coordinated" monitoring programs include part of
NOAA's Northeast (U.S.) program as well. This program now proposes to
keep the following regions under annual surveillance: Irish Sea; German
Bight, Southern Bight of the North Sea; the Estuaries of the Forth,
Thames, Rhine, Scheldt, and Clyde; the Skagerrak, Kattegat, and Oslo
fjords; plus certain parts of the Gulf of Saint Lawrence and New York
Bight. The ICES program has three monitoring rationales: (1) the
provision of a continuing assurance of the quality of marine foodstuffs
with respect to human health, (2) the provision, over a wide geographical area, of an indication of the health of the marine environ-

ment in the entire ICES North Atlantic area, and (3) to provide an analysis of trends in pollutant concentrations. Intercalibration exercises for petroleum (see Petroleum Hydrocarbon Intercalibration/ Intercomparison Programs section) are underway, although many discrepancies in methodology need to be resolved.

Monitoring for the recovery of systems following oil spills has been conducted for many spills. Once a choice of sampling stations and measurements has been made, the same concerns face these programs as well as the "baseline-type" programs. Examples of spill monitoring programs are: Arrow spill (Keizer et al., 1978), West Falmouth spill (Teal et al., 1978), Tsesis spill (Linden et al., 1980; Boehm et al., 1981b), Amoco Cadiz spill (Atlas et al., 1981), and Iranian Crude-Norway spill (Grahl-Nielson et al., 1978). All relied on detailed chemical measurements of sediment and/or biota to monitor recovery based on the decrease and/or modification of petroleum residues.

CONCLUSIONS AND RECOMMENDATIONS

Conclusions

No single method of analysis provides a measure of total petroleum in water, sediment, or tissue because of the extreme complexity of the composition of petroleum. Unfortunately, apparent economic necessity has often forced analysts to the less expensive and less discriminating methods of analysis with attendant generation of a substantial amount of data which can only be interpreted with large uncertainty.

However, improved methodology for measuring fossil fuel compounds has been rapidly developed or applied since the 1975 NRC report. The range of selectivity and sensitivity makes it essential to choose the correct methods for a particular problem and to recognize the interpretation limits for the data.

Recommendations

Quality Control and Intercomparison of Data

The rapid increase in the number of analysts and the demand for larger sets of data require careful quality control and intercomparison of data, now even more than at the time of the 1975 NRC report.

We recommend that rigorous quality assurance protocols be integrated into the analysis of hydrocarbon and other fossil fuel compounds in environmental samples. The value of standard solutions, spiked samples, spiked extracts, and sample homogenates for quality control and intercomparison has been demonstrated in a few studies.

Identification of Sources of Input

Many studies of petroleum inputs or distribution in the marine environment have not applied analytical techniques to identify sources more exactly. The terms "petroleums" and "petroleum hydrocarbons" are often used incorrectly and too loosely when describing data resulting from less discriminating analyses. This is especially true in regard to inclusion of pyrogenic source hydrocarbons within the data for petroleum.

Application of Analytical Methods

We recommend the application of analytical methods with sufficient sensitivity and resolution to identify the various sources of input, e.g., high resolution glass capillary/gas chromatography/mass spectrometry/computer systems analysis or high performance liquid chromatography analysis coupled with mass spectrometry computer systems.

Nonhydrocarbon Compounds in Petroleum

Because many of the nonhydrocarbon compounds in petroleum are biologically active, we recommend a more concerted effort to measure these compounds in studies of inputs, fates, and effects.

Metabolites and Photochemical Reaction Products

The concern about the biological activity of several metabolites and photochemical reaction products as indicated in the fates and effects sections leads us to recommend research into methods for measuring these compounds in samples from laboratory and field studies. These methods would be used in studies of biogeochemical processes acting on fossil fuel compounds and in studies of biological effects. We do not advocate extensive analytical chemistry data-gathering exercises in monitoring program measurements of metabolites and reaction products until such time as research has clearly demonstrated the usefulness of such an approach. Rather, we recommend the investigation of biochemical or physiological parameters as potentially more useful for determining where biologically active compounds have been or are present.

Remote Sensing

Sensors of various types have been tested from aircraft and show promise for providing useful information in the measurement of the areal extent and thickness of slicks. We recommend further testing in conjunction with sea truth measurements to evaluate this concept further.

Remote sensing of dissolved fossil fuel compounds in seawater appears to be several years in the future and should naturally emerge from the more general basic research into remote sensing of chemical and biological components of the oceans.

Sampling Techniques

The bias introduced by various sampling protocols should be recognized explicitly. The mechanics of sampling needs close attention in order to maximize useful data. For example, many grab samples and gravity cores taken after oil spills did not contain the "floc" layer at the sediment-water interface, thereby introducing severe doubt as to whether or not petroleum compounds had reached the benthic ecosystem.

Large volume water samplers designed to avoid contamination during the sampling process have been developed and should be more extensively used to obtain useful samples.

In situ, or on deck, water pumping systems capable of obtaining samples of water for analyses of dissolved and particulate compounds should be subject to further deployments in a variety of oceanic and coastal regions to further evaluate their usefulness. Initial tests are quite favorable and suggest that these systems will prove useful to studies of fossil fuel compounds in the water columns.

PART B

Biological Methods

INTRODUCTION

An impressive amount of research has been done during the past decade on uptake and effects of petroleum by single species of organisms under controlled laboratory conditions. In fact, the methods for exposing organisms are now technically sophisticated in some cases. However, relatively few experiments have been conducted in the field to validate laboratory findings. Because of the inadequate comparison of results of laboratory experiments and postspill field investigations, the specific knowledge needed for predicting the impact of acute petroleum pollution in the marine environment is not yet available. Fortunately, the study of marine mesocosms, i.e., scaled living models of natural ecosystems, is a promising new means for developing the needed comparisons.

Physiological, Behavioral, Population, and Ecosystem Effects

Fundamental to the study of populations, communities, and their habitats is the identification of species. Because species names are a key to the biological literature, it is important to know exactly which animals, plants, and microorganisms are involved in any given study.

In general, physiological data are meaningful only when associated with reliably identified species. Although costly and time consuming, identification is a primary objective for both field and laboratory studies. Specimen depositories permit verification of identifications made in field surveys, before and after spills, and are vital for physiological and biochemical analyses during and subsequent to spills. In fact, depositories may provide the only means of establishing validity of the data gathered at the time of a given event.

Not only is it important to know the species involved in a given test system or event, the chemistry of the toxicant causing an impact has to be thoroughly understood. Changes in composition and concentration during exposure need to be monitored to establish cause-and-effect relationships.

After establishing the species and toxicant, the choice of methods for any biological study is determined by the goals of the investigation, availability of instruments, familiarity of the investigators with those instruments and methods, and the cost of the overall project.

The biological methods summarized in this report may be used in connection with the following objectives: to measure effects on physiology and behavior (a) on individual organisms, i.e., primarily acute (short term lethal or sublethal) effects, and (b) on the life cycles of organisms, including energy budgets, i.e., primarily chronic effects on growth, development, and reproduction; to assess population changes including (c) acute impacts and (d) subsequent changes in populations; and to obtain information at the ecosystem level (e) for experimental systems and (f) for natural systems.

Items (a), (b), (c), (d), and (f) are usually investigated in connection with accidents or chronic discharges. Item (a), but particularly (b) and (e), are principal approaches for searching out causal and possible predictive relationships or explanations for the effects of crude oils or their constituents. Although goal (d) is an important environmental consideration, a more fundamental concern is to maintain marine ecosystems (f) in a condition that allows them to be utilized as society deems appropriate.

Problems of Exposure, Type of Oil, Weathering, and Exposure Medium

Because of changes in composition that begin as soon as oil is spilled on the sea (see Chemical Methods section), experiments using unweathered oils do not indicate those responses expected when the same organisms are exposed to aged oils. Experiments designed to assess the impact of oil must take this disparity into account.

The relative immiscibility of oil and seawater makes the quantitative monitoring of petroleum in aqueous bioassay solutions difficult. As a result, several methods have been proposed and employed for the preparation of oil-water bioassay mixtures or for simulating the type of exposure to oil an organism might encounter in nature (also see Laboratory Exposure Methods section).

Behavioral, bioassay, and bioaccumulation studies using organisms exposed to weathered petroleum in the laboratory are meaningful if they improve or broaden our understanding of the biological responses of organisms in their natural habitats. The goal of such research should be to measure biological effects of a specific compound, or mixture of compounds of known concentration, on organisms under a prescribed set of environmental conditions. Often, however, assessment of laboratory results in terms of field situations is difficult because of the complexity of the environmental factors involved (G.V. Cox, 1974). Nevertheless, laboratory studies are important to explore potential damage caused by various concentrations and exposure times for pollutants and to assist in designing field studies.

METHODS FOR ASSESSING TOXICITY OF PETROLEUM TO MARINE ORGANISMS

A full appreciation of petroleum hydrocarbon concentrations that might actually occur in a given water column, sediment, and/or food found in different oil-contaminated marine environments is valuable in designing effects studies. (See Chapter 4.)

In the laboratory, test organisms are best exposed to petroleum hydrocarbon concentrations similar to those that might realistically be expected to occur in a contaminated marine environment. A wide range of exposure concentrations is used whenever possible, including environmentally realistic concentrations and concentrations up to about 10-20 times higher than the latter. Higher concentrations are helpful in eliciting obvious biological effects and are useful in estimating a safety factor (difference between lowest concentrations eliciting a response and expected environmental concentration), when environmentally realistic concentrations do not elicit a measurable response.

Acute Lethal Toxicity Bioassay

The usual first step in evaluating the toxicity of petroleum and specific petroleum hydrocarbons for marine organisms is the acute lethal bioassay (Sprague, 1978). This is a rapid screening method, designed to provide an estimate of the relative toxicity of crude or refined oils or specific hydrocarbons, to different species and life stages of marine organisms. It is a rough predictor of the maximum concentration of pollutant material that can be present in the marine environment for an extended period of time without causing damage to sensitive organisms and/or ecosystems (Sprague, 1971; Wilson, 1975; Perkins, 1979). Chronic bioassays, in which organisms are exposed for longer periods (most of their life cycle or even for several genera- tions), and studies of effects of chronic exposure to low concentrations of pollutant materials on various biochemical, physiological, and behavioral parameters in the test organisms, are more useful for deriving maximum acceptable concentrations of pollutant materials. If results of acute lethal bioassays show that a given pollutant is toxic, studies of chronic, life-cycle, and sublethal effects are a useful follow-up, as well as mesocosm studies, as appropriate, to establish maximum acceptable concentrations.

As methods for acute lethal toxicity bioassay protocols are improved, eventually they will be standardized. At the present time, several manuals and reviews are available in which such protocols are described in sufficient detail to measure the toxicity of petroleum for marine organisms (American Public Health Association, 1977; American Society for Testing and Materials, 1980; Environmental Protection Agency, 1975a,b; Environmental Protection Agency/Corps of Engineers, 1977).

Flow-through, as opposed to static, petroleum bioassays are preferred if resources and constraints peculiar to the organisms of choice allow. Several flow-through systems have been designed for use in petroleum bioassays (Hyland et al., 1977; Vanderhorst et al., 1977b). It is imperative in flow-through and static bioassays and in chronic effects studies that the petroleum hydrocarbons in the aqueous phase in contact with the test organisms be characterized and measured at regular intervals.

The LC_{50} (median lethal concentration) is currently the term most often used to report results of acute marine bioassays. The LC_{50} and its error may be estimated by simple graphical methods (American Public Health Association, 1977; Lichtfield and Wilcoxon, 1949), more precise probit (Finney, 1971), logit (Ashton, 1972; Hamilton et al., 1977), and nonparametric (Stephan, 1977) methods and by methods that make use of computer capabilities, e.g., BMD 03S Fortran program (Dixon, 1970).

H.J.W. Anderson et al. (1980) recommended use of the product of LC_{50} and exposure time as a toxicity index, to compare toxicity of different oils or sensitivity of different species. This method is still under study.

If log time is plotted versus log LC_{50}, a straight line can be produced which can then be extrapolated to predict mortality for exposure intervals likely to be encountered during an oil spill.

Chronic and Sublethal Effects Studies

To study chronic and sublethal effects, employment of full life-cycle assays are desirable but not always practical. In a life-cycle bioassay, test organisms are exposed over a complete life cycle, or a major portion of it, to sublethal concentrations of a given pollutant. Biological parameters usually measured include mortality, growth rate, time to maturation, fecundity, offspring survival, and physiological or genetic adaptation. The most sensitive stages in the life cycle of an organism are detected and effects of the pollutant on sensitive and ecologically important parameters such as growth and reproduction are determined (Hansen et al., 1978; Nimmo et al., 1977; Reish, 1980). Life-cycle bioassays using petroleum have been performed with polychaete worms (Carr and Reish, 1977; Rossi and Anderson, 1978) and crustaceans (Laughlin et al., 1978; W.Y. Lee, 1978).

Besides the mentioned biological parameters, others have been measured in an effort to define sublethal concentrations of petroleum causing deleterious responses to marine organisms (J.W. Anderson, 1977a,b; Malins, 1977; Neff and Anderson, 1981). The behavior of marine animals has been shown to be highly sensitive to petroleum-induced perturbation. Methods are cited therein for monitoring behavior, chemosensing, locomotory, and feeding responses, among others.

Change in respiration (oxygen consumption) has been used frequently as a criterion of sublethal response of marine organisms to oil pollution. Results have been highly variable because a great many endogenous and exogenous factors, other than pollution, influence respiratory rate. A fruitful approach is to combine respiration rate with other biological parameters (food consumption, growth, and excretion) to construct an energy budget for the animal (Bayne et al., 1976, 1979; Widdows, 1978). Several indices of stress can be derived from the energy budget, including scope for growth (energy available for growth and reproduction) and growth efficiency.

The ratio of oxygen consumed to nitrogen excreted (O:N ratio) can also be used as an index of pollutant stress, although there can be considerable variability arising from other environmental factors. Nevertheless, it provides an estimate of the proportion of metabolic energy derived from catabolism of proteins and amino acids. Bioenergetics methods, or variations of them, of which the O:N ratio is an example, have been used in several recent investigations of the effects of sublethal concentrations of petroleum on marine animals (Capuzzo and Lancaster, 1981; Edwards, 1978; Gilfillan and Vandermeulen, 1978; Johns and Pechenik, 1980; Stekoll et al., 1980).

Biochemical enzyme assays, primarily of blood serum, are a powerful diagnostic tool in human clinical medicine. Many of these enzyme assays have been applied to tissue samples of marine animals in an effort to detect changes in enzyme activity attributable to pollutant exposure, but such efforts have met with only limited success. This is a growing field and offers great promise.

The activity of the microsomal cytochrome P450 mixed function oxygenase (MFO) system of fish and, possibly, marine polychaete worms is increased (induced) by exposure of the animal to petroleum and selected aromatic hydrocarbons (Neff, 1979; Stegeman, 1981; Varanasi and Malins, 1977). Because it is induced by exposure to petroleum, the hepatic MFO in fish has been recommended as an index of petroleum contamination in the marine environment (J.F. Payne, 1976; Walton et al., 1978; J.F. Payne and Fancey, 1982). Several other pollutants, including heavy metals and chlorinated hydrocarbons, as well as natural environmental and biological factors, may influence MFO activity. It must be used with caution as a specific index of petroleum pollution, because of the effects of other pollutants.

Acute or chronic exposure to petroleum may cause a variety of tissue lesions, increased incidence of parasitism, or increased susceptibility to bacterial or viral disease. These can be detected and evaluated by examination using the light or electron microscope (Hodgins et al., 1977; Sinderman, 1979). Some success has been achieved using the light and electron microscope for histopathology and histochemistry to detect sublethal damage in laboratory and field populations of marine fish (Blanton and Robinson, 1973; Hawkes, 1977; DiMichele and Taylor, 1978; Payne et al., 1978; Hawkes et al., 1980; Eurell and Haensly, 1981; Haensly et al., 1982). These methods could, indeed, be useful for diagnosing characteristics of damage in marine invertebrates and fish caused by pollutants, provided they can be related directly to the pollutant.

Field Studies

There is a growing interest in adapting physiological, biochemical, and histopathological methods, such as those described above, for diagnosing the state of health of field populations of marine animals in the vicinity of oil spills or chronic oil inputs to the marine environment. Such methods, if validated and adapted for field use, could be useful for monitoring petroleum contamination of the marine environment. For example, two large interdisciplinary programs currently involve developing and evaluating field monitoring methods. These include the Pollutant Responses in Marine Animals (PRIMA) Program supported by the National Science Foundation and NOAA's Ocean Pulse Program, now the Northeast Monitoring Program. Suites of biochemical, physiological, and histopathological tests provide a diagnostic profile of the health of the test animal. Such characterization may prove more useful than any single test for assessing pollutant stress in populations of marine animals (McIntyre and Pearce, 1980). There are problems in such an approach, however, For example, plaice (_Pleuronectes_ _platessa_) from two estuaries heavily contaminated with oil from the _Amoco_ _Cadiz_ crude oil spill were examined for histopathology, and a wide variety of biochemical changes over a 2-year period were recorded. A progression of biochemical changes and pathological lesions was observed, which indicated an initial decline in the health of the fish, followed by improvement in these indices 27 months after the spill (Haensly et al., 1982). However, it is not clear that the effects were, in fact, due to the oil alone. Thus, one must be confident that a cause-and-effect relationship has been established.

Selection of Test Organisms

Several criteria are important in selecting test organisms for laboratory toxicity and accumulation/release studies. Because marine organisms vary widely in their sensitivity to oil and ability to metabolize and excrete petroleum hydrocarbons (Neff et al., 1976; Craddock, 1977; Varanasi and Malins, 1977; Neff and Anderson, 1981), several species, representing different major taxonomic groups provide a more useful system. Most frequently used are microalgae, polychaete worms, bivalve mollusks, crustaceans, and fish.

Ideally, test species should meet several criteria. However, the criteria for selection of a test species will depend on the question being asked. At the minimum, the test species ought to be available in large numbers, occur over an extended geographic range, represent important members of the ecosystem, and come from, or represent, marine habitats likely to be severely impacted by oil spills. Species used for hydrocarbon accumulation/release studies ought to include taxa possessing different types of hydrocarbon metabolizing ability or response.

Several lists of marine species have been described in the literature which fulfill some or all of these criteria, and several of these species have been recommended as standard bioassay/biological effects test organisms, including the microalga Skeletonema costatum, the copepod Acartia tonsa, the opossum shrimp Mysidopsis bahia and the cyprinodont fishes Fundulus heteroclitus and Cyprinodon variegatus (Becker et al., 1973; Environmental Protection Agency, 1975a,b; American Public Health Association, 1977; Environmental Protection Agency/Corps of Engineers, 1977; Reish, 1980). In many cases, however, it is preferable to use species indigenous to, or representative of, habitats of particular concern, such as coral reefs, fishing banks or continental shelves, estuaries, and arctic regimes.

In the design of field studies, the choice of suitable experimental species may be limited by what is available locally at a field site. Many of the criteria used to select laboratory subjects can be applied here also. In most cases, particular species are especially suitable for use in answering a particular environmental question; for example, bivalve mollusks are good subjects for studying petroleum contamination of biota because, in general, they accumulate hydrocarbons readily.

Preparation of Oil-Water Solutions

Test organisms can be exposed to petroleum in the laboratory in the form of water-soluble fractions, oil-in-water dispersions, surface slicks, oil-contaminated food, or oil-contaminated sediments. No single method of exposure to petroleum is applicable for all marine organisms. Experiments using microorganisms require different approaches from uptake studies with marine macroorganisms. The latter, in turn, need different methods applied than those used with birds or marine mammals.

The following discussion describes how petroleum and its components have been presented to a variety of marine organisms, recognizing, of course, that specific methods often are needed for different organisms or when different experimental approaches are applied.

Preparations of petroleum solutions should represent situations that can occur in the environment as a result of an accidental discharge of petroleum or from chronic inputs. Many methods used in preparing petroleum solutions for laboratory exposures can also be used for flow-through systems, particularly when larger organisms held in aquaria or tanks are to be exposed. Similarly, birds and marine mammals require different approaches for exposure studies; the former has been reviewed by Holmes and Cronshaw (1977) and the latter by Geraci and Smith (1977). See Chapter 5.

Water-Soluble Fractions: Static

When oil is mixed with seawater, the oil can form macroparticles (droplet dispersions), microparticles (collodial dispersions and oil-in-water emulsions), and single-phase, homogeneous mixtures (water-soluble fractions) of hydrocarbons. There are no definitive demarcations between these states of dissolution, although arbitrarily, decisions have been made, such as using filters having 0.45- and 1.2-μm pore size to differentiate between particulate oil (retained on the filter) from subparticulate and soluble oil (passing through the filter) (Gordon et al. 1973; Wells and Sprague, 1976). However, reaggregation may occur after filtering. Recent developments resulting in improved chemical analyses have permitted a more critical distinction between states of dissolution.

Published accounts of laboratory exposure studies conducted through the mid-1970s frequently described a test solution as a "water-soluble fraction" (WSF). Unfortunately, many of the reports contain no description of the exposure medium, whereas in others, an attempt was made to define the water-soluble fraction by reporting chemical analyses of only the major hydrocarbon compounds, providing limited data on oil particle sizes, and results only of visual examinations of the clarity of the fractions. Such information is inadequate because oil particles of 100 μm diameter or less are not readily discernible to the human eye (Nelson-Smith, 1973), and oil droplets smaller than 1-2 μm in diameter remain suspended in seawater for hours or days (Parker et al. 1971)—much longer than the settling period used routinely in preparing water-soluble fractions. In addition to the problems cited above, it is difficult to determine whether water-soluble fractions used in the tests reported in the early literature were truly single-phase solutions, dispersions of fine droplets of oil in water, or a combination of these, described as "accommodated" by Gordon et al. (1973) and R.C. Clark and MacLeod (1977). Unfortunately, an additional difficulty is that most dispersions of oil and seawater are unstable over time.

A water-soluble fraction is an artificial mixture and cannot be used to simulate precisely the conditions of hydrocarbon composition and concentration in a water column when oil is spilled in the marine environment. Equilibration conditions in nature may be quite different from those used to produce water-soluble fractions in the laboratory. The water-soluble fraction represents a compromise, a means of generating a highly reproducible and relatively stable oil-in-water mixture and is, therefore, very useful when comparing the relative toxicity of different crude and refined petroleums for marine organisms.

Laboratory studies have employed low-molecular-weight aromatic hydrocarbons because of their relatively high, short term toxicity for marine organisms; however, in the case of oil spills the partitioning of the simultaneously volatile and soluble low-molecular-weight hydrocarbons is a dynamic process, dependent upon a set of parameters unique to each spill (e.g., water and atmospheric mixing energies, temperature, salinity, presence of natural and human-contributed polar materials in the seawater, and type of petroleum. See Chapter 4 for details.

Table 3-4 provides a summary of data for four American Petroleum Institute (API) reference oils and Prudhoe Bay crude oil employed in many studies in recent years. The compositional analysis of the whole reference oils is compared to that of the water-soluble fraction, prepared by mixing one part of oil with nine parts of seawater for 20 hours. Analyses of water from the Prudhoe Bay crude oil exposures are quite different, since a flowing exposure system was used in extracting the latter. The n-paraffin compounds, which range in carbon chain length from 12 to 24, represent a large amount of the total measured components in the oils, even though their contribution to the water extracts is relatively small.

Measurements of monoaromatic compounds present in the oils and their extracts are not always quantitative, and in fact, low-boiling components frequently are not measured in the oil. However, the contribution of monoaromatics to the total concentrations of hydrocarbons in water-soluble fractions is significant if they occur in fresh oil; this is particularly true for crude oils that have not undergone any refining process. From an examination of the concentrations present in the water extract, the contribution of compounds higher in molecular weight than the alkylnaphthalenes is very small and may not be significant in producing acute toxicity. One could conclude that either the mono-aromatics or the diaromatics are the major contributors to the acute toxicity associated with these extracts.

Table 3-5 was prepared to summarize the data in Table 3-4 as percent of the classes of compounds, relative to the total amounts of hydrocarbon actually measured. The percent composition of individual hydrocarbons routinely measured in the oil by environmental chemists is relatively low (5-15%) in comparison with the numbers of compounds present (Malins, 1980).

Water-soluble fractions have been prepared by stirring varying ratios of petroleum compounds and experimental media for varying periods of time and allowing these to stand so as to arrive at a stabilized water-soluble fraction. Stirring times range from several hours to days, e.g., 12 hour stirring (Kauss and Hutchinson, 1975) and 72 hour stirring (Mahoney and Haskin, 1980). Following an equilibrium period of several minutes (e.g., Winters et al., 1977; Pulich et al., 1974) to several hours (e.g., Kauss and Hutchinson, 1974) for separation of the aromatic and aqueous phases, the aqueous phase may be filtered through materials which range from glass wool to 0.45-μm Millipore(R) filters. Subsequent dilution with filtered or unfiltered seawater or media provides a range of concentrations. Soto et al. (1975a,b) determined that the type of stirring affects the composition and, therefore, the biological effects of a petroleum extract, whereas Wells and Sprague (1976) determined that the type of stirring affected the concentration of the extractable organics measured by UV and fluorescence (i.e., aromatics) and, hence, influenced the toxicity of the preparations.

If mixing conditions are carefully standardized, highly reproducible results can be obtained in preparing water-soluble fractions (Linden et al., 1980). However, chemical and physical characteristics of the petroleum will affect the actual composition and concentrations of hydrocarbons in the water-soluble fraction preparations (J.W. Anderson et al., 1974; Neff and Anderson, 1981). Inasmuch as the oil-water partition coefficients of hydrocarbons favor retention in the oil phase, and evaporation and solubilization are competing processes, the aqueous phase of a water-soluble fraction never becomes saturated with hydro-

TABLE 3-4 Concentrations of C$_{12}$–C$_{24}$ n-Paraffins and Aromatic Hydrocarbons in Reference Oils and the 10% Water-Soluble Fractions (WSFs) Prepared From Them

Name of Compound	S. Louisiana Whole Oil (%)	S. Louisiana WSF (ppb)	Kuwait Whole Oil (%)	Kuwait WSF (ppb)	No. 2 Fuel Oil Whole Oil (%)	No. 2 Fuel Oil WSF (ppb)	Bunker C Whole Oil (%)	Bunker C WSF (ppb)	Prudhoe Bay Whole Oil (%)	Prudhoe Bay WSF[a] (ppb)
n-paraffins										
C$_{14}$	0.44	10	0.46	<0.5	0.82	5.0	0.11	0.8	0.36	b
C$_{15}$	0.48	10	0.41	<0.5	1.06	7.0	0.11	0.9	0.37	
C$_{16}$	0.54	12	0.43	0.6	1.20	8.0	0.15	1.2	0.36	
C$_{17}$	0.41	9	0.42	0.8	0.98	6.0	0.12	1.9	0.34	
C$_{18}$	0.30	7	0.28	0.5	0.60	4.0	0.10	1.0	0.30	
Total C$_{12}$–C$_{24}$ n-paraffins	3.98	89	4.00	2.9	7.38	47	1.26	12	4.40	2.93
Aromatics										
Benzene	c	6,750	c	3,360	c	550	c	40	c	36.1
Toluene		4,130		3,620		1,040		80	0.93	55.5
Alkylbenzenes[d]		760		730		970		110	3.19	53.8
Naphthalene	0.04	120	0.04	20	0.40	840	0.10	210	0.16	1.6
1-methylnaphthalene	0.08	60	0.05	20	0.82	340	0.28	190	0.21 ⎱	⎱ 2.4
2-methylnaphthalene	0.09	50	0.07	8	1.89	480	0.47	200	0.25 ⎰	⎰
Dimethylnaphthalenes[d]	0.36	60	0.20	20	3.11	240	1.23	200	0.80	1.4
Trimethylnaphthalenes[d]	0.27	8	0.19	3	1.84	30	0.88	100	0.48	0.6
Biphenyls[d]	<0.01	2	<0.01	1	0.16	28	<0.01	1	c	c
Fluorenes[d]	0.02	3	<0.01	3	0.36	20	0.24	11	c	c
Phenanthrenes[d]	0.06	4	0.04	3	0.53	20	1.11	23	0.39	0.3
Dibenzothiophenes[d]	0.02	1	0.01	1	0.07	4	<0.01	1	0.13[d]	c
Total aromatics	0.94	11,948	0.60	7,789	9.18	4,562	4.31	1,166	6.54	151.7
Total hydrocarbons measured	4.92	12,037	4.60	7,792	16.56	4,609	5.57	1,178	10.94	154.6
Total hydrocarbons present (IR analysis)		19,800		10,400		8,700		6,300		141

NOTE: Hydrocarbon concentrations in the oils are given as percent (g/100 mL) and in the WSFs as parts per billion (μg/L) in 20 o/oo seawater.

[a] 30 o/oo seawater in a flowing system.
[b] Individual n-paraffin concentrations were very low.
[c] Not measured.
[d] Total of several isomers.

SOURCE: J.W. Anderson et al. (1974), R.D. Anderson (1975), and Bean et al. (1980).

TABLE 3-5 Compositional Comparison of Reference Oils on a Basis of Those Hydrocarbons Measured

Composition	S. Louisiana[a]		Kuwait		No. 2 Fuel Oil		Bunker C		Prudhoe Bay	
	Whole Oil	WSF	Whole Oil	WSF	Whole Oil	WSF	Whole Oil	WSF	Whole Oil	WSF[b]
Percent n-paraffins	81	0.7	87	0.4	45	1.0	23	1.0	40	1.9
Percent monoaromatics	a	97	a	99	a	56	a	20	38	94
Percent diaromatics and triaromatics	19	3	13	1	54	43	77	79	22	4
Percent aromatics	--	100	--	100	--	99	--	99	60	98

NOTE: WSF is water-soluble fractions.

[a]Not measured.
[b]Flowing extract (30 o/oo seawater) of fresh oil.

SOURCE: J.W. Anderson et al. (1974) and Bean et al. (1980).

carbons. Thus, although naphthalene has a solubility of about 20 mg/L in seawater (Rossi and Neff, 1978) and No. 2 fuel oil (API reference oil) contains 4,000 mg/L naphthalene, a water-soluble fraction prepared from No. 2 fuel oil contains only 0.84 mg/L naphthalene (Neff and Anderson, 1981). Salinity of the test medium has little effect on the rate of loss, but temperature has a marked effect (Laughlin and Neff, 1979). For example, half times for naphthalene loss from a 15% water-soluble fraction vary from 3-5 days at 20°C to 2-4 days at 30°C. Loss of naphthalene from the aqueous phase is caused primarily by volatilization. It is clear that hydrocarbons in static tests are lost from the test solution (Vandermeulen and Ahern, 1976). To solve this problem, one can turn to flow-through systems.

Water-Soluble Fractions: Flow Through

Several attempts have been made to design a continuous flow system for use in petroleum exposure studies. The extreme compositional complexity of petroleum and its unusual behavior in water make the design of such a system difficult. A major problem has been, as in static systems, to maintain a constant hydrocarbon concentration and composition in the exposure tanks. Oil slicks, insoluble oil residues, or emulsions may accumulate in parts of the system over time, altering properties of the dispersed or soluble aqueous fraction. Weathering of oil and buildup of oil-degrading bacteria in the flow-through system may also pose problems.

The simplest flow-through systems utilize stock water-soluble fractions, prepared daily or more frequently, which are infused at a constant rate into and mixed with the inflowing seawater. By controlling flow rate of dilution water and water-soluble fraction, one can achieve precise exposure concentrations. Preparing the water-soluble fraction frequently makes possible the maintenance of relatively constant hydrocarbon concentration and composition. Examples of recent investigations using this approach include those of Moore et al. (1980) and Capuzzo and Lancaster (1981).

An elaborate flow-through solubilizer system was devised by Roubal et al. (1977a) to prepare water-soluble fractions of crude oil for months-long exposures of marine organisms. The system gave an uninterrupted flow of seawater extracts of Prudhoe Bay crude oil for test periods of up to 5 weeks, with a total hydrocarbon range from 5 to 6 mg/L. The diluted seawater extracts exhibited no Tyndall Effect and were clear. The water-soluble fraction was free of alkanes, which implied the virtual absence of suspended droplets of undissolved oil (Roubal et al., 1977a).

Another system, built by Krugel et al. (1978), produced a saturated solution of the water-soluble components of a jet fuel oil (JP8), free from droplets of undissolved oil, for flow-through bioassay tests. Water, as droplets, flowed through five consecutive vertical columns containing the oil. One of the advantages of the apparatus was its simplicity: the entire system was gravity fed with no pumps or regulators needed.

Giam et al. (1980) developed a dosing method for exposing marine organisms to low levels of poorly water-soluble compounds in a recirculating system using a column containing sand coated with a contaminant.

Another type of solubilizer, described by Nunes and Benville (1979a,b) and Benville et al. (1981), consisted of an oil reservoir (Figure 3-6a), oil pumps, modified solubilizer glass bottle (Figure 3-6b), and oil waste reservoir. The water-soluble components of crude oil were dissolved without serious loss of more volatile compounds and without the formation of emulsions or oil droplets, and the apparatus could be assembled as a recirculating system (Figure 3-6c).

Benville and Korn (1974) and Maynard and Weber (1981) described simple devices for metering low-molecular-weight hydrocarbons into seawater for use in either static or flow-through experiments. The devices produced water-soluble fractions with constant concentrations of low-molecular-weight aromatic hydrocarbons and reduced emulsion formation.

Oil-in-Water Dispersions

Oil-in-water dispersions were used in the early 1970s, particularly for short term static exposures and acute toxicity bioassays. These provide higher concentrations of oil in water because the solubility of oil can be exceeded. Interpretation of data from this early research was often difficult because of uncontrollable variations in exposure conditions or lack of reproducible results (Varanasi and Malins, 1977). Methods for preparing oil-in-water dispersions included mixing oil with seawater and shaking, stirring, blending, ultrasonically emulsifying, passing through baffle plates, and turbulent mixing in a water jet. The stability of oil-in-seawater dispersions is dependent upon a number of factors, including the size of the droplets and the presence of air or vapor space above the dispersions. For instance, very fine (<1 μm) oil droplets may stay in suspension for long periods, although some loss occurs from volatilization. Different hydrocarbon classes and molecular weight ranges are selectively partitioned between aqueous and oil phases (Boylan and Tripp, 1971). For example, the low-molecular-weight aromatics (e.g., benzenes and naphthalenes) are more highly accommodated in the aqueous phase than higher weight aromatics. Consequently, in static exposures, organisms are initially exposed to relatively high concentrations of aromatic hydrocarbons in the water, but because of their volatile nature, these aromatics are soon lost through evaporation.

Emulsified oil-in-water preparations can be prepared by shaking oil in seawater on a mechanical shaker (J.W. Anderson et al., 1974; Winters et al., 1977), by homogenizing without ultrasonification (C.K. Wong et al., 1981), or with ultrasonification (Wells and Sprague, 1976; Gruenfeld and Frederick, 1977). A continuous flow-through dosing apparatus for preparing water-accommodated fractions of oil was described by Hyland et al. (1977). No. 2 fuel oil was metered into a horseshoe-shaped oil separation chamber in which a constant input of

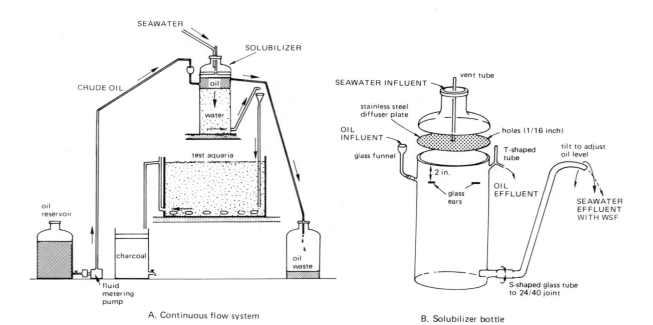

A. Continuous flow system

B. Solubilizer bottle

C. Recirculating system using the solubilizer

FIGURE 3-6 Solubilizer systems for exposing organisms to water-soluble fractions of crude oil: (a) continuous flow system (Nunes and Benville, 1979a), (b) solubilizer bottle (Whipple et al., 1978; Benville et al., 1981), (c) recirculating system (Nunes and Benville, 1979b).

oil and unfiltered seawater was turbulently mixed and then allowed to separate over a series of baffles. The water-accommodated fraction was removed at the end of the chamber at mid-depth and distributed to the experiment in exposure tanks. Clement et al. (1980) combined features of the designs of both Hyland et al. (1977) and Roubal et al. (1977a) to simulate chronic exposure to dispersions, using an apparatus capable of simultaneously delivering dispersions with nominal crude oil concentrations in water of 30, 300, and 3,000 µg/L.

Only a relatively small fraction of the total oil added to seawater becomes dispersed as fine droplets in the water column during preparation of an oil-in-water dispersion. For example, vigorous mixing of 100,000 µL/L of South Louisiana crude oil (API reference oil) with seawater produces a dispersion containing only 81.2 µL/L total hydrocarbons (J.W. Anderson et al., 1974). The hydrocarbon composition of the dispersion can be expected to resemble that of the parent oil, at least initially, as most of the hydrocarbons are still present in the form of dispersed oil droplets.

Oil-in-water dispersions are usually unstable under static bioassay conditions (vide infra). The concentration of total hydrocarbons in the water dispersions of South Louisiana crude oil decreases almost 90% within 24 hours during gentle aeration (J.W. Anderson et al., 1974). Similar results have been reported by other investigators. Concentrations of aliphatic hydrocarbons in dispersions decrease more rapidly than concentrations of aromatic hydrocarbons because of a much lower aqueous solubility of the former and as a result of their presence in oil droplets rising to the surface.

Petroleum dispersions in seawater that remain stable over periods of several months have been produced in an apparatus described by Vanderhorst et al. (1977b) (Figures 3-7a, b, and c). The oil and seawater are mixed as they are introduced through a funnel into the first compartment of the contactor. The mixture passes through a small hole in the side of the compartment into a second larger compartment, where the undispersed floating oil is separated by a baffle and discarded. The dispersion is directed to a second tank (Figure 3-7d), where baffles allow for further removal of undispersed and suspended oil. Finally, the solution is directed to a metering tank (Figure 3-7e) where the flow and dilution factors for the exposure tanks are controlled by the size of the tubing and by gravity. Composition and concentration of individual and total hydrocarbons in dispersions produced by this system have been measured (Bean and Blaylock, 1977; Bean et al., 1980; J.W. Anderson et al., 1980). Monoaromatic hydrocarbon concentrations varied somewhat over time, but maintenance of relatively constant hydrocarbon concentrations in the water column over several days was quite good. Chemically dispersed oil can be prepared in this apparatus by adding the appropriate chemical at the initial mixing funnel (J.W. Anderson et al., 1981).

Water-accommodated extracts prepared by hand or by mechanical shaking exhibit a similar range of variability in preparation as that observed for water-soluble extracts. However, shaking or turbulent stirring yields an extract which includes oil in particulate form (Gordon et al., 1973). Thus, the composition and concentration may

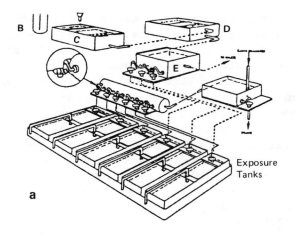

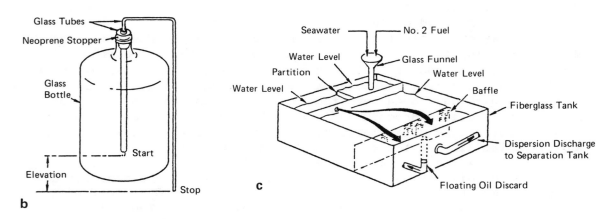

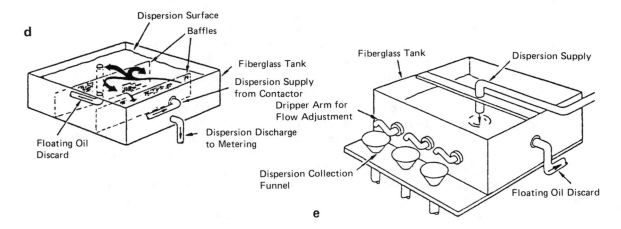

FIGURE 3-7 Fuel oil dispersion and bioassay apparatus: (a) arrangement of components, (b) fuel oil metering, (c) contactor, (d) separation tank, and (e) dispersion metering tank.

SOURCE: Vanderhorst et al. (1977b).

differ considerably as a result of inconsistent equilibrium times. As specific fractions of oil (aromatics) appeared to elicit pronounced cellular and physiological responses from phytoplankton populations, Winters et al. (1977), Pulich et al. (1974), Karydis (1979), and Batterton et al. (1978), among others, obtained and tested specific fractions of a variety of hydrocarbons by either distilling and collecting the required fractions or driving off specific fractions, such as volatile aromatics by heating or bubbling with air.

Subsequent handling of the oil-water preparation, such as preparing a dilution series, may also result in change in composition and concentration. There are other factors to be considered, aside from the physical, such as avoiding contamination of preparations, achieved by use of axenic (bacteria-free) cultures. Prouse et al. (1976) found a significant change in the C_{17}:pristane ratio within 30 minutes of adding algae to the media--suggesting bacteria degradation of the n-alkane, relative to the isoprenoid.

Such variation in oil-water preparation methods and subsequent handling, reported above, make interpretations of the presence or absence of effects difficult. As pointed out by Vandermuelen and Ahern (1976), Prouse et al. (1976), O'Brien and Dixon (1976), and Corner (1978), among others, the specific details of the methodology used in each experiment must be known--specifically, the method of extract preparation, whether sterilization or filtration was used, and if the actual concentration and composition of both the extracts and the mixtures used for experiments were determined analytically.

Filtering, heating, or autoclaving whole oils or their extracts in an attempt to obtain sterile preparations for use with axenic cultures further modify composition and concentration (Vandermuelen and Ahern, 1976), although Prouse et al. (1976) reported that autoclaving whole oils in bulk did not appreciably alter the aromatic content or relative concentrations of n-alkanes.

Surface Slicks: Laboratory

In the case of surface slick exposure studies and bioassays the oil is usually poured onto the seawater surface in the exposure chamber, and the oil components are then allowed to leach into the underlying seawater without mixing. The seawater underneath the slick may not necessarily be changed over the duration of the exposure (Clark and Finley, 1974a; Wells and Sprague, 1976). It may be exchanged in a flow-through system where the seawater contaminated by the slick is slowly exchanged or diluted with uncontaminated seawater (Bott et al., 1976; Taylor and Karinen, 1977; Shaw et al., 1977; Payne et al., 1978). Another type of flow-through system allows fresh seawater to fall through the surface slick and contaminated seawater to be drained from the bottom of the exposure tank (Eisler, 1975; Rinkevich and Loya, 1979).

Direct additions of crude oil as a surface slick were used by Shiels et al. (1973) with natural phytoplankton populations. Lacaze (1974) added crude oils with and without emulsifiers to outdoor mesocosms containing natural populations. Lacaze and Villedon de Naide (1976) made direct additions of fresh crude oil, but also aged or weathered the oil in the light and dark in open and closed flasks as a means of assessing effects on growth arising from photooxidation and/or losses of volatile fractions. Again, in water slick exposure experiments, quantification of the hydrocarbons entering the water phase is essential.

Oil in Food

Various investigators have incorporated petroleum, refined products, or fractions thereof into food for the organisms; exposure is via consumption of contaminated food (Corner et al., 1973; Hardy et al., 1974; Roubal et al., 1977b; Whittle et al., 1977; Varanasi et al., 1979). Another means is to introduce oil through the food web via an intermediate step, such as feeding oysters containing radiolabeled hydrocarbons to crabs (R.F. Lee et al., 1977), or dimethylnaphthalene-contaminated detritus to benthic deposit-feeding organisms (Roesijadi et al., 1978). In another example of food web transfer (Malins and Roubal, 1982), a tritiated 2,6-dimethylnaphthalene, accumulated by Fucus sp. algae from seawater without conversion to metabolites, was fed to sea urchins. Waldon et al. (1978) have used oiled food in MFO induction studies in fish.

Oiled Sediment Studies

Starting in the early 1970s, laboratory research was initiated to study the uptake of hydrocarbons from oiled sediments that serve both as a habitat and as a food-containing segment of the food web. Various devices have been used for exposing bottom-dwelling organisms to oiled sediments, ranging from simple beakers (Prouse and Gordon, 1976; Wells and Sprague, 1976; Gordon et al., 1978) and aquaria (Taylor and Karinen, 1977; Howgate et al., 1977), polyvinylchloride trays (J.W. Anderson et al., 1977), and aluminum pans suspended in aquaria (Shaw et al., 1977), to systems containing aquaria within aquaria or water tables, often with provisions for tidal simulation (McCain et al., 1978; J.W. Anderson et al., 1978; Roesijadi and Anderson, 1979).

Studies have been conducted in which externally oiled sediments were placed in the test tanks (Taylor and Karinen, 1977; J.W. Anderson et al., 1977, 1979; Rossi, 1977; Howgate et al., 1977; Busdosh et al., 1978; McCain et al., 1978; Gordon et al., 1978; Roesijadi and Anderson, 1979; Varanasi and Gmur, 1981a,b) or where clean sediments were contaminated by oil slicks deposited by simulated falling tides (Prouse and Gordon, 1976; Taylor and Karinen, 1977; Shaw et al., 1977).

McCain et al. (1978) described a typical laboratory system for exposing flatfish to oiled sediments for extended periods of time. Crude oil was mixed into wet sediment as a sediment-oil-seawater slurry in a commercial cement mixer. The sediment, initially containing 2,000 mg/L oil was placed on the bottom of an aquarium situated inside a second aquarium. Uncontaminated seawater was percolated through the oiled sediment. The concentration of oil dropped to 700 mg/L after an overnight flushing in running seawater before the addition of test fish. After 4 months under running seawater, the concentration in the sediment stabilized to ca. 350 mg/L (total extractable petroleum-derived hydrocarbons), at which time the fish were introduced. In another approach to preparing contaminated sediments, J.W. Anderson et al. (1979) first emulsified the oil in seawater using a blender at "high" speed before adding the suspension to the sediment in a cement mixer.

Solid intertidal substrata (e.g., rock, natural floating materials, artificial substrates) can be removed, complete with the attached organisms and placed in a laboratory environment in order to expose sessile macroorganisms to petroleum pollutants. The size of the exposure chamber limits the amount of substrate used. Further, care must be exercised in transporting the attached organisms to minimize physiological stress.

BACTERIA, YEASTS, AND FILAMENTOUS FUNGI

Methods for Estimating Microbial Numbers and Biomass

The microbial population size, i.e., numbers or biomass of microbes, must be measured so that changes can be normalized for total microbial numbers or biomass to quantify changes in microbial populations resulting from interaction with oil. This is particularly important for natural populations.

Microbial interactions with oil may increase or decrease the number of petroleum-transforming microorganisms (PTM). In general, lethal events are not measured when aerobic heterotrophic microorganisms interact with oil, especially after the oil has aged and the lighter solvent fractions have evaporated. The immediate response of a microbial population to addition of oil is an enrichment of PTM, both in relative and absolute terms. In fact, such an increase in number of PTM has been used to practical advantage for locating oils (Atlas, 1981).

Epifluorescent direct counts are useful for estimating total numbers of bacteria and fungi (Zimmerman and Meyer-Reil, 1974; Daley and Hobbie, 1975; Porter and Feig, 1980). Results of direct counts provide a reference to which PTM counts can be compared. Epifluorescent techniques may also be used to assess growth of bacteria on a specific carbon source, e.g., petroleum, in the presence of nalidixic acid (Kogure et al., 1979). The resulting elongated cells indicate the number of cells capable of growth on the substrate, in this case, petroleum. However, precise interactions of fluorescein dyes and nalidixic acid with petroleum are unknown.

Jones (1981) successfully employed measurement of adenosine triphosphate (ATP) to estimate the biomass of pure cultures associated with petroleum, and Griffiths et al. (1981) did similar work with mixed cultures. The effects of petroleum associated with losses of ATP from stress alone have not yet been quantified, creating a potential source of error in this measurement.

The most probable number (MPN) technique and the techniques listed below provide indirect estimates of PTM after samples of water or sediment are inoculated into media containing petroleum as the sole carbon and energy source. The MPN technique appears to have been first applied to the estimation of PTM by Gunkel and Trekel (1967). J.D. Walker and Colwell (1976a) modified this technique by using antibiotic-supplemented media to provide separate estimates of counts for bacteria and fungi growing on petroleum. They also compared results obtained using the MPN method with those of several other methods listed below. The ^{14}C-hydrocarbon method for estimating PTM has been used to estimate numbers as well as activity of PTM (Caparello and LaRock, 1975; Atlas, 1979).

A silica gel-oil (SGO) medium was developed independently by Seki (1973) and J.D. Walker and Colwell (1975b) as a method of estimating PTM growing on petroleum as the sole carbon and energy source. The advantages and disadvantages of agar plate procedures have been discussed, and the silica gel medium was developed to eliminate problems encountered with growth of non-PTM on oil-agar. Oil-agar was first described by Baruah et al., (1967) but was modified and employed by Atlas and Bartha (1972) and J.D. Walker and Colwell (1973, 1975a) to estimate PTM in estuarine samples. Use of oil-agar as a medium in estimating PTM is not encouraged. If a solid medium is desired, SGO is recommended. However, problems of sineresis and liquefaction may occur if the medium is not prepared fresh before use. Recently, a medium was developed that permits detection of microorganisms using a given hydrocarbon as the sole carbon and energy source, but the medium is limited to those hydrocarbons, e.g., phenanthrene, around which a hydrolytic zone is produced upon growth of microorganisms (Shiaris and Cooney, 1982).

As cited above, the number obtained and the biomass calculated from the numbers of cells enumerated must be interpreted, with the understanding that the entire population is not cultured, and the organisms in culture are not capable of using oil as a sole carbon and energy source.

Enumeration and biomass estimates of PTM are valuable as a means of comparing differences among localities and times, particularly if differences observed are of several orders of magnitude. It is especially important to note that, as indicators of microbial response, the number of PTM normalized to the total population of aerobic heterotrophic microbes is more meaningful than the number of PTM alone.

Methods for Estimating Metabolic Effects of Oil on Microorganisms

Not a great deal is known about the more subtle microbe-oil interactions. Some information exists on the effects of oil on the chemoreceptor system of bacteria and on the effects of oil on the biochemistry of sediment bacteria.

Methods employed to date for measuring metabolic changes in bacteria associated with oil are essentially modifications of standard methods employed in microbial ecology. Similarly, variation in population species composition can be determined using taxonomic methods and analyses.

Microbial processes commonly used in oil spill assessment include the changes in SO_4 reduction, sulphide oxidation, N_2 fixation, nitrification, denitrification, methane production, activity of enzymes (hydrolytic, chitinase, chitosanase, cellulase, and amylase), and ribonucleic acid (RNA) synthesis. Lipid content has also been examined.

Physiological activity of microorganisms associated with oil can be estimated using geomicrobiological methods or one of the following: (1) release of $^{14}CO_2$ from labeled glutamate; (2) heterotrophic potential, with changes in v_{max} measures; (3) change in specific activity of selected key enzyme(s); (4) relative rate of change in RNA or DNA synthesis; (5) changes in adenylate pools, and/or change in ATP cell; and (6) alteration of a measurable function such as N_2 fixation or N_2O production.

Methods for Obtaining Indirect and Direct Measurements of Oil Degradation

Measurement of oxygen uptake, carbon dioxide evolution, and ^{14}C-hydrocarbon degradation can be used to estimate degradation of oil indirectly. These methods do not require extraction of oil. Measurement of oxygen uptake is a simple method that provides a rapid estimate of microbial activity for samples containing large numbers of microorganisms and can be carried out by Winkler titration, or with a suitable respirometer, or oxygen electrode (Bridie and Bos, 1971; Gibbs, 1972).

Measurement of CO_2 evolution is also a simple method providing a rapid estimate of the activity of samples containing large numbers of microorganisms. CO_2 can be quantified by titration of $BaCO_3$ or by infrared gas analysis (Atlas and Bartha, 1972). Use of oxygen uptake and CO_2 evolution for estimating short term activity (i.e., minutes or hours) creates the problem of determining effects of oil, or oil degradation products, on endogenous respiration. Use of ^{14}C hydrocarbons eliminates this problem, because the fate of products resulting from the metabolism of the labeled substrate can be measured (e.g., evolution of $^{14}CO_2$) by incorporation of ^{14}C into microbial cells and metabolic products and persistence of unreacted ^{14}C hydrocarbon (J.D. Walker and Colwell, 1976b). Each method of determining activity is very much dependent upon the experimental conditions employed, including type of oil and the physical state of the oil being utilized and/or degraded.

A variety of chemical methods can be used to monitor microbial degradation of oil. These include high pressure liquid chromatography, infrared spectrophotometry, ultraviolet and fluorescence spectrophotometry, gas and/or column chromatography, and mass spectrometry. These are described elsewhere in this methods chapter.

PLANKTON

Spatial and temporal distribution of plankton in the environment is
constantly changing. Thus, demonstration of oil pollution effects is
difficult in the field. Sampling design must be rigorously defined
(Venrick, 1978a,b; Wiebe et al., 1973) if results are to be unequivocal.
In every case, the concentration of the oil and its components should
be determined. Furthermore, a defined control community is essential.
In many field situations, rigorous experimental design may be impossible
logistically, or prohibitive in cost. Nevertheless, for valid ecological predictions, laboratory experiments should be linked with field
investigations (Wilson et al., 1974). General references providing
useful methods for the study of phytoplankton include the publications
of Sournia (1978) and for zooplankton, Edmondson and Winberg (1971) and
Steedman (1975).

Plankton studies of an acute spill are desirable but will usually
require extensive resources for adequate sampling and analysis. Both
phytoplankton and zooplankton are notoriously patchy in distribution,
with variations often approaching an order of magnitude in normal,
unstressed situations. A few samples taken in oiled and control areas
are unlikely to yield statistically valid information concerning
relative standing crops. Continuous records of horizontal distribution
are highly desirable for preliminary surveys. However, there is some
question as to whether instruments ordinarily used for this purpose
(towed fluorometers for phytoplankton, Hardy-Longhurst samplers or
electronic particle counters for zooplankton) will operate effectively
in an oiled situation. This problem needs study. If continuous records
cannot be obtained, intensive spot sampling of species composition and
physiological properties may yield useful information on the effects of
an oil spill.

After the initial survey, resources should be focused at the population level, as well as on assessing changes in species composition of
the community. Derived indices, such as diversity, should not be used
alone as estimators of community health, in the absence of supporting
data. More attention paid to effects at the cellular level and on
modes of action of petroleum hydrocarbons on planktonic organisms will
improve interpretation and prediction of environmental changes (Wells,
1982).

Many previous laboratory studies utilizing physiological responses
of phytoplankton and zooplankton to assess the impact of oil have been
undertaken under varying environmental conditions. Thus attention must
be focused on maintaining known environmental conditions, such that the
organisms being studied have a known history. A discussion of methods
follows, but some of the methods cited may not be expedient in a spill,
and judgment must be used.

The ranges of size within each of the algal and animal plankton and
nekton are such that one type of sampling device will not catch representative species of any of the three groups. Thus, field samples from
a single sampling device cause an immediate bias. Moreover, the ranges
of size and density of bacteria, algae, and animals will overlap, not
only with each other but also with that of particulate matter (i.e.,

organic and inorganic detritus, or tripton). Hence, only the smallest free-living bacteria, the larger zooplankton, and the nekton fall into size classes distinct enough to be estimated on the basis of mechanical separation. In contrast, samples of phytoplankton will always be heavily contaminated by bacteria, zooplankton, and nonliving organic matter. Nevertheless, the quantity of algae can be crudely estimated by chlorophyll measures, and their activity can be measured by their photosynthesis. Microscope analyses permit the separation and enumeration of algae, zooplankton, and tripton; allow estimates of biomass; and provide the means for identifying the species under study.

Phytoplankton

Field Methods

Sampling procedures should be quantitative. Therefore, nets should not be used to collect all size classes of phytoplankton. Large species can be collected quantitatively with nets, provided clogging does not occur (Tangen, 1978). In working under oil slicks, nets and open water bottles generally employed for quantitative sampling can be subject to contamination when lowered through the slick. A method that was tried to avoid the surface slick problem was oblique towing (Grose and Mattson, 1977). Sampling gear that can be opened below the surface should be employed if samples are to be analyzed for accumulation of petroleum products in plankton.

Most routine laboratory analytical methods are amenable to field situations, given ideal working conditions (i.e., sufficient electrical power, space, and stability of the platform). However, serious consideration should be given to methods in which initial sample processing involves a minimum of equipment outlay and produces samples amenable to storage, with final analyses done in the laboratory. Historically, field work has included taking measurements of community composition from cell counts (Ignatiades and Mimicos, 1977; Wilhm and Dorris, 1966, 1968; R.F. Lee and Takahashi, 1977; Federle et al., 1979; G.A. Vargo et al., 1981), pigment concentration, and photosynthesis, principally by ^{14}C uptake.

Community composition and biomass determinations are critical for evaluating long term effects in phytoplankton communities, because changes in populations will result in changes throughout the food web (R.F. Lee and Takahashi, 1977; Federle et al., 1979; Elmgren et al., 1980a,b). Chlorophyll measurements are useful, but species counts can be used for estimating biomass. However, the lack of reliability of species counts should be recognized. Methods and precautions for sampling procedures, preservation and counting techniques, biomass conversions, and lists of taxonomic literature can be found in the Phytoplankton Manual prepared by Sournia (1978).

Measurement of photosynthesis, using either the oxygen evolution method or ^{14}C uptake, has been the most widely used method for assessing community response to petroleum hydrocarbons (Gordon and Prouse, 1973; Shiels et al., 1973; Bender et al., 1979; Hsiao et al.,

1978). Samples can be incubated in on-deck, seawater-cooled boxes retained in the sunlight and equipped with neutral density light screens, or attached to anchored and buoyed lines (Johansson, 1980). Maximum photosynthetic rates (P_{max}) normalized to chlorophyll will provide the most meaningful data. The index is independent of irradiance variability to some extent and permits rather unambiguous comparison between stations and samples. Alternatively, measurement of photosynthesis below light saturation, i.e., the initial slope, normalized to chlorophyll and light, is also recommended. Productivity rates, expressed as incorporation rate per volume and time (mg Cm^{-3} hr^{-1}), are less useful than normalized rates. Additional information concerning responses by different size categories of phytoplankton within the community can be gained by size fractionation (Malone and Chervin, 1979; O'Reilly and Thomas, 1980; Lannergren, 1978).

Field comparisons of productivity rates between control and impacted regions, whether based on oxygen or ^{14}C methods, must take into consideration potential bias arising from microbial activity and containment effects, a result of differential growth rates, bottle size, and duration of incubation (Venrick et al., 1977; Gieskes et al., 1979). More recently, Carpenter and Lively (1980) demonstrated that the type of glass bottle used for incubation, its UV transmission characteristics, and trace metal contamination from ^{14}C stock solutions and other equipment can seriously affect ^{14}C uptake rates, particularly in oligotrophic waters.

Cage cultures (dialysis encapsulation and Nucleopore® filter or nitex mesh cages), for either laboratory or field studies, offer another method potentially useful for evaluating responses of cultured or natural populations. Jensen et al. (1976) and Eide et al. (1979) employed cage cultures to monitor heavy metal pollution, while O'Connors et al. (1978) evaluated effects of PCB on unialgal and natural populations. Dialysis encapsulation (Jensen et al., 1972) and other types of cage cultures (Owens et al., 1977) may provide a unique method for studying in situ effects of petroleum hydrocarbons.

Laboratory Methods

Statistical comparison of growth rates for control and treated populations is more reliably done using cultures in the exponential growth phase. Regressions of exponential growth, compared by suitable statistics, e.g., covariance analysis, have been used by Hsiao (1978) and Prouse et al. (1976). Problems associated with extrapolating results obtained from laboratory cultures to the field (Braarud, 1961) have increased since it is now known that clonal cultures of the same species vary in physiological response to environmental and chemical perturbants (Eppley et al., 1969; Hargraves and Guillard, 1974; Fisher, 1977; Mahoney and Haskin, 1980; Murphy and Belastock, 1980). Therefore, results obtained using cultures of one genotype may not necessarily be representative of a species response.

Photosynthesis and Respiration   Photosynthesis has been the most common physiological index used to measure response of phytoplankton populations to petroleum hydrocarbons. Both the oxygen evolution and the ^{14}C uptake methods have been used extensively.

Oxygen evolution can be measured by the standard Winkler titration (Strickland and Parsons, 1972), as detailed by Shiels et al. (1973), or by using a Clark-type electrode (Pulich et al., 1974; Armstrong and Calder, 1978; Kusk, 1978; Batterton et al., 1978). Effects of heavy metals on oxygen evolution should be considered. Potentiometer end point determinations for Winkler titrations can increase reproducibility (Vargo and Force, 1981).

The ^{14}C method offers greater sensitivity, but concerns raised with respect to trace metal problems in the field (vide supra) also apply to laboratory use and also for O_2 evolution.

Clark-type electrodes and Gilson respirometry have been used to measure respiration rate in the presence of petroleum (Kusk, 1978; Karydis, 1979). Axenic cultures are required, and species with demonstrated heterotrophic metabolism are best selected for studies of oil effects on dark respiration. Electron-transport-estimated respiratory activity (Christensen and Packard, 1979) has not yet been used to assess response to oil.

Preexposure periods, with and without oil present (Shiels et al., 1973; Trudel, 1978; Hsiao et al., 1978), vary from short term, i.e., 0.5-15 min (Pulich et al., 1974; Batterton et al., 1978; Armstrong and Calder, 1978), to long term, i.e., 12-18 hours (Gordon and Prouse, 1973; T.R. Parsons et al., 1976). The duration of incubation must be considered along with exposure time because, with long incubation, the response elicited is an integration of effect over the entire incubation period. Short term incubations (i.e., minutes) with Clark-type electrodes (Kusk, 1978; Soto et al., 1975a,b) are recommended but are subject to problems of interpretation because of diurnal periodicity of photosynthesis and shock response. Long term incubations suffer from containment effects, and results are subject to changes arising from bacterial activity (Harris, 1978). Termination of ^{14}C estimated photosynthesis by addition of DCMU (3-(3,4-dichlorophyll)1,1-dimethyl urea, a photosynthetic inhibitor) (Lacaze and Villedon de Naide, 1976), neutralized formalin (Hsiao et al., 1978), or mercuric chloride (Trudel, 1978) needs further evaluation because cell lysis can occur when any of these methods is employed (Silver and Davoll, 1978).

Cellular Constituents and Cell Structure   Enzyme analyses and cell structural responses have received scant attention in hydrocarbon investigations. This is unfortunate because biochemical and electron microscope applications offer standardized methods which can be usefully employed to study effects of petroleum hydrocarbons, particularly to determine the site and mode of action of these compounds. Techniques have included scanning electron microscopy transmission, electron microscopy, and membrane studies (Van Overbeek and Blondeau, 1954; Goldacre, 1968; Baker, 1970; Boney, 1970). Electrolytes also leak from treated algal fronds (Reddin and Preudeville, 1981).

Standardized methods have been applied to assess petroleum effects on adenosine triphosphate (Vandermuelen and Ahern, 1976; Armstrong et al., 1981), alkaline phosphatase and phosphodiesterase activity (Armstrong et al., 1981).

Culture Systems Laboratory methods for determining effects of petroleum hydrocarbons on phytoplankton populations require suitable culture media. Recipes for solid, liquid, or biphasic media, using enriched seawater or defined media, and for the requirements of individual species or groups have been published by Nichols (1973) for fresh water and McLachlan (1973) for seawater systems.

Both "open" and "closed" containers have been used. Open systems, e.g., foam or cotton plug stoppers (Soto et al., 1975a,b), have the disadvantage of allowing the more volatile compounds to escape, thus reducing the hydrocarbon concentration during the course of an experiment (Kauss and Hutchinson, 1975; Dunstan et al., 1975; Vandermeulen and Ahern 1976). Closed systems, which retard losses of volatile compounds, may also potentially limit growth and final yield from CO_2 limitation, an effect not observed in seawater unless nutrient enrichment is very high (Dunstan et al., 1975; Pulich et al., 1974). Blankley (1973) discusses possible detrimental effects on the enclosed population arising from the composition and leachability of various types of materials used in the manufacture of stoppers and other types of closures.

Both unialgal and axenic cultures of marine and freshwater phytoplankton have been used in laboratory experiments to determine the effects of petroleum hydrocarbons. Mixed populations have not been studied. Thus, results obtained to date are indicative only of noncompeting populations. Use of unialgal cultures has been criticized, especially for determining effects of hydrocarbons on physiological responses, because bacteria are present and can interfere. In any case, axenic cultures are required if physiological responses (Kusk, 1978; Karydis, 1979; Soto et al., 1975a,b), enzyme analysis (Armstrong et al., 1981), or ATP (Vandermuelen and Ahern, 1976; Armstrong et al., 1981) are used as the indicator of stress.

Almost all studies employing marine species have been carried out using liquid media in batch culture. Many investigators, notably in earlier studies, ignored nutritional and irradiance prehistory of phytoplankton populations when determining effects of oil. Shifts in irradiance history, both intensity and photoperiod, as well as nutrient, temperature, and salinity regime of cultures, can elicit variable physiological responses. It is, therefore, critical that cultures be maintained under similar growth and environmental conditions both before and during an experiment. Sampling schedules should also be predicated on known diurnal rhythms or points on population growth curves. Both semicontinuous (i.e., turbidostat) and continuous (i.e., chemostat) culture systems can be employed to provide populations of known nutritional and growth history. However, the extensive time involved in chemostat work can make this approach less feasible. Batch cultures can be more helpful when working with several species, as opposed to the chemostat. Useful discussions of these methods have been published by Goldman and Davidson (1977) and Murphy and Belastock (1980).

A method for rapid screening for potential toxic, or inhibitory, effects of a particular petroleum compound has been employed by Pulich et al. (1974) and Winters et al. (1977). Termed the "algal lawn technique," cultures of the species to be tested are dispersed in a plate of molten agar, and a pad saturated with the substance to be tested is placed on the agar. Sensitivity is determined by a zone of inhibition. Usefulness of the method is limited to those species that can withstand the relatively high temperature of liquefied agar (approximately 40°C) and are able to grow on an agar medium. Low temperature gelling agar (approximately 26°C) is now available and can ameliorate this problem.

Accumulation and Depuration   Radiolabeled pure petroleum hydrocarbons offer the most direct method for establishing uptake, accumulation, and release rates (Kauss et al., 1973; Soto et al., 1975a,b). Improved and highly sensitive gas chromatography and combined gas chromatography/ mass spectometry have also been successfully employed to establish the presence of polycyclic aromatic hydrocarbons in epipelic diatoms.

Population Responses   Rate of growth or cell division, combined with generation time and final yield, can be considered integrators of factors influencing cellular metabolism. Population responses measured by visual counting methods also yield additional qualitative information on the "condition" of the population (i.e., color, movement, loss of flagellae). Reviews, such as those published by Guillard (1973) and Sournia (1978), should be consulted for proper choice of counting chambers, calibration, counting statistics, and limitations of each method. In addition, electronic particle counters provide rapid results, thus allowing increased replication of both counts and experiments; but problems of coincident counts, calibration for cell size, interference of nonliving matter, and counting of chain-forming species must be recognized and taken into account (Sheldon, 1978).

In vivo chlorophyll fluorescence is also easily measured and can provide a quantitative measurement of population increase, provided consideration is given to those factors which influence fluorescence yield, e.g., irradiance intensity, diel periodicity, dark exposure (see Kiefer, 1973; Loftus and Seliger, 1975). Enhancement of fluorescence by DCMU can provide an estimate of energy channeling within a cell and could be an indicator of potential photosynthesis, yielding a relatively quick and simple assay for detecting effect of petroleum products. Prezelin and Ley (1980), however, point out the inconsistencies of the method when it is used as an indicator of potential photosynthesis. Thus, further investigation is warranted.

Zooplankton

Field Methods

Field methods are the same as for the phytoplankton. Contamination of the zooplankton being sampled by the net itself is a significant problem. Also, care must be taken that the net does not pass through the surface slick and that it is rinsed with solvent between tows (R.C. Clark and Brown, 1977). Gelatinous zooplankton require special methods.

General ecological methods applied to zooplankton are also useful in assessing effects of petroleum. The least detailed parameters measured are biomass (Wiebe et al., 1973; Beers, 1976; Omori, 1978) and total abundance (Edmondson and Winberg, 1971). Live-dead counts may be done in conjunction with enumeration (Fleming and Coughlan, 1978; Seepersad and Crippen, 1978). Recognizing the heterogeneous distribution of zooplankton, species composition, age structure and sex ratios provides information on community structure, which may reveal changes more readily than biomass or total abundance. Species composition data permit calculation of diversity indices (Pielou, 1977), which may decline under pollutant stress (Copeland and Bechtel, 1971; Borowitzka, 1972) or may not decline (Elmgren et al., 1980a,b). Animal production (Edmondson and Winberg, 1971) of the community or individual species may also change under pollutant stress.

Biochemical composition of the zooplankton communities may change also (Samain et al., 1980). However, biochemical composition will vary with temperature and nutritional state, so that the effect of petroleum may be masked. Protein (Lowry et al., 1951; Packard and Dortch, 1975; Capuzzo and Lancaster, 1981), lipid (Marsh and Weinstein, 1966), carbohydrate (Dubois et al., 1956; Handa, 1966), fatty acid content (Morris and Culkin, 1976), and digestive enzyme activities (Samain et al., 1980) per unit weight may be altered due to stress. Enzymes may also be induced to deal with toxic substances, such as the induction of mixed function oxidases in response to oil (J.F. Payne, 1977; Walters et al., 1979), although not all species possess this capability. Interpretation of results is, therefore, difficult. In organisms with a mixed function oxidase, this system is important in metabolic functions. Hydrocarbons may be taken up by zooplankton (Harris et al., 1977b; Corner, 1978; Spooner and Corkett, 1979), and hydrocarbon content can serve as an indicator of oil pollution (Mackie et al., 1978).

Microscopic techniques for detecting oil contamination were employed by Conover (1971) and Polak et al. (1978), among others. In recent years, histopathological data (Yevich and Barsacz, 1977) have been used as indicators of long term sublethal stress. However, for the zooplankton community, there are few background descriptions of "normal" histological characteristics. Furthermore, there is no general consensus on which histopathological changes are reliable indicators of stress, although chromosomal studies of fish eggs have revealed that changes in chromosomal structure and mitotic division may be useful (Longwell, 1978).

Laboratory Methods

In laboratory studies, functional parameters, such as survival under both acute and chronic application of oil; recovery; physiological processes (respiration, excretion, reproduction, and growth); behavior (feeding and locomotion); and uptake, retention, and metabolism of oil have been used to measure the reaction of an organism to oil. Survival has been measured in acute toxicity bioassays (24-96 hours) in both static or flow-through systems (see Methods of Assessing Toxicity of Petroleum to Marine Organisms section). Such bioassays are helpful in ranking oils in order of toxicity but are of limited value for eco-logical prediction (Wilson, 1975). Chronic long term exposures are more useful because they allow detection of delayed mortality (Berdugo et al., 1977) and significant sublethal effects. Furthermore, physio-logical processes of an organism are more sensitive indicators of stress. Many physiological tests can be carried out on board ship, as well as in the laboratory. Respiration can be measured polarographi-cally (Edwards, 1978; W.Y. Lee et al., 1978; Gyllenberg and Lundqvist, 1976), chemically by the modified Winkler method (Carritt and Carpenter, 1966; Vargo and Force, 1981), or manometrically (Capuzzo and Lancaster, 1981). Manometric techniques are limited, however, in that not all zooplankton can withstand confinement within the small volume and shaking required. In all cases, suitable precautions must be taken to ensure that zooplankton are not stressed in several ways simultaneously, e.g., by crowding and lack of food (Ikeda, 1976, 1977).

Ammonium excretion is usually measured colorimetrically by the method of Strickland and Parsons (1972) or Solorazano (1969), although other nitrogenous compounds, such as urea and primary amines, may also be excreted and, therefore, measured (McCarthy, 1971; McCarthy et al., 1977). For the small size zooplankton fraction, the ^{15}N isotope dilution method can be useful (Caperon et al., 1979). Respiration and excretion rate measurements can be combined as an O:N ratio, providing an indication of the biochemical substrate utilized as energy reserves (Capuzzo and Lancaster, 1981; S. Vargo, 1981; Mayzaud, 1973) and provide some indication of disruption of normal energetic processes.

Reproduction and larval development may also be affected by oil, and several parameters have been used to measure such effects, i.e., number of eggs (Ustach, 1977; Berdugo et al., 1977; Ott et al., 1978), embryological development and hatching (Donahue et al., 1977b; Tatem, 1977), and larval development and survival (Donahue et al., 1977a; Laughlin et al., 1978; Wells and Sprague, 1976; Nicol et al., 1977; Bryne and Calder, 1977).

Growth, as a sum of physiological processes, can be a good indicator of effects of oil. Measurement of growth can be a simple biomass determination (Edwards, 1978) or more complex assessments (Edmondson and Winberg, 1971), with carbon or energy budgets also proven useful (Edwards, 1978).

Extensive investigations have been conducted on zooplankton feeding (Conover, 1978). Simple, short term feeding experiments can be carried out on board ship using ^{14}C-labeled phytoplankton and the larger size fraction of zooplankton. However, with longer incubation, the ^{14}C may be excreted, resulting in less reliable conclusions. Landry and Hassett (1982) have developed a method for evaluating grazing rates in whole water samples. In situ measurements have been made by Haney (1971). More laboratory-oriented methods include the use of cultures of single species of phytoplankton and zooplankton and microscopic counting techniques to evaluate those phytoplankton or zooplankton

(Wells and Sprague, 1976) that have been eaten. Feeding on several species of phytoplankton, or a natural assemblage, can be measured using electronic particle counters (Berman and Heinle, 1980). Fecal pellet production can also be used to estimate changes in feeding rates (Spooner and Corkett, 1974, 1979), but the test animals must produce fairly large, cohesive fecal pellets for this method to be practical.

Behavioral responses, such as feeding and locomotion, comprise integrated physiological and biochemical processes. Behavior appears to be altered quickly when the animal is under stress and, therefore, shows promise as an indicator (Wilson, 1975; Olla et al., 1980). However, not all behavioral responses can be easily quantified, and care must be exercised in selection of a behavioral response.

Locomotion includes at least two components: rate and pattern of movement. Narcotization by petroleum can result in cessation of movement (Gyllenberg and Lundqvist, 1976; Wells and Sprague, 1976; W.Y. Lee and Nicol, 1977; R.F. Lee et al., 1978). Changes in rate can be observed visually, but changes in the pattern of movement are more difficult to quantify. Video taping, coupled with computer analysis, shows promise (Lang et al., 1981), although methods currently employed are not suitable for species that accelerate rapidly. These methods are also time consuming and expensive. Locomotory responses to external stimuli, such as light, gravity, and pressure, may also be used to assess effects of oil (Bigford, 1977). Changes in locomotory responses provide a qualitative, rather than quantitative, indicator of effects of oil, and are difficult to interpret in terms of permanence of effect(s).

Uptake, retention, and metabolism of petroleum in planktonic organisms have been studied using ^{14}C-labeled hydrocarbons (Corner et al., 1976; Harris et al., 1977a,b; R.F. Lee et al., 1981a). Such measurements reveal rates of uptake, retention time, and presence or absence of ability to metabolize hydrocarbons.

ACCUMULATION AND MODIFICATION OF PETROLEUM BY MACROORGANISMS

Field Exposure Methods

Organisms can be exposed to petroleum in the field using any one or combination of water-soluble fractions, oil-in-water dispersions, surface slicks, oil-contaminated food, or oiled habitat. Table 3-6 lists examples of field exposure studies based on the physical form of the contaminant. Alternatively, field experiments can also be clas-sified according to method of exposure: (a) introduction of uncontami-nated organisms into contaminated areas (uptake studies) and vice versa (depuration studies), (b) sediment tray experiments where oil is mixed into a sediment in the laboratory and then returned to the field for a period of observation, (c) oiled enclosures, and (d) field surveys that compare environmentally exposed, wild samples with uncontaminated reference samples (monitoring schemes).

Introduction Experiments

One method of exposure consists of transplanting small marine organisms from uncontaminated areas to contaminated areas. DiSalvo and Guard (1975) designed a mussel-exposure apparatus made of wide-diameter plastic PVC pipe. The apparatus containing the test organisms was placed on pilings below the low tide level or hung under floats. Other methods include suspending mussels and other shellfish in mesh bags or cages (B.A. Cox et al., 1975; Whittle et al., 1978; Burns and Smith, 1978; Wolfe et al., 1981) or placing test animals in cages and suspending the cages off the bottom in the intertidal zone (Bender et al., 1977; Bieri et al., 1977, 1979). Benthic organisms have been placed directly on contaminated substrates, either in open-bottom trays (Shaw et al., 1976; J.W. Anderson et al., 1978; Roesijadi et al., 1978; Roesijadi and Anderson 1979; Augerfeld et al., 1980) or without enclosure (Lake and Hershner, 1977; Friocourt et al., 1981).

Sediment Tray Experiments

Trays of contaminated sediment, either free of organisms for recruitment studies or containing selected bivalve mollusks or worms, arranged carefully by hand, were placed in intertidal beaches. The bottom of each tray was fitted with a screen to allow for natural tidal flow of seawater.

In studies to determine the effect of oil-contaminated sediments on natural recruitment, sediment was collected and sieved in the field to standardized particle size. The sediment was exposed to three cycles of freezing and thawing in the laboratory to eliminate macroorganisms, after which it was poured into trays and immersed in a large aquarium. The flowing seawater aquarium was equipped with a mechanism to provide simulated tidal draining of the seawater. When the water level was 1 cm above the sediment, a 4% volume of oil was poured over the seawater, forming a uniform slick. The seawater was drained and the oil was allowed to remain on the sediment for 2 hours. The flow of seawater was reinstated and the excess, floating oil was skimmed off. Two additional tidal cycles were completed. The sediment-filled trays were then placed in holes dug in the beach (Anderson et al., 1978).

Alternatively: the oil was mixed with the sediment prior to filling the trays. Coating of the sediment was achieved by adding a blender-prepared, oil-water emulsion to the freshly sieved sediment and mixing it in a cement mixer (J.W. Anderson et al., 1977). Oil analyses should be done before, during, and after deployment in the field.

Enclosure Experiments

Bakke and Johnsen (1979) used sediment enclosures made of aluminum with tops of polyethylene sheets. Divers embedded the enclosures to a depth of 0.1 m in the sandy bottom. The polyethylene sheets were mounted while the frames were in place on the bottom. The benthic organisms in

TABLE 3-6 Experimental Field Methods for Exposing Marine Macroorganisms to Petroleum Contaminants

Habitat	Petroleum Material	Duration	Type of Exposure	Organism	Reference
Water-Soluble Fractions					
Coastal and estuarine subtidal, Calif.	Suspected fossil fuel contamination	1-4 mo.	Wire mesh baskets 1 m, transplants	Blue mussel, coastal mussel	DiSalvo et al. (1975)
Coastal and estuarine subtidal, Calif.	Suspected fossil fuel contamination	1 week	Plastic biosampler	Blue mussel	DiSalvo and Guard (1975)
Open sea, North Sea	Ekofisk oil well blowout	<5 days	Cages off bottom, transplants	Blue mussel	Whittle et al. (1978)
Estuarine intertidal, South Australia	Suspected fossil fuel contamination	7 weeks	Polyethylene mesh bags, transplants	Blue mussel	Burns and Smith (1978)
Estuarine subtidal Brittany, France	Amoco Cadiz crude oil spill	4 weeks	Shrimp cages, 1 m transplants	Blue mussel, cockle	Wolfe et al. (1981)
Estuarine intertidal Brittany, France	Amoco Cadiz crude oil spill	8-16 mo.	On bottom, transplants	Oysters (2)	Friocourt et al. (1981)
Oil-in-Water Dispersions					
Estuarine embayment, Trinity Bay, Tex.	Oil separator platform	3 mo.	On bottom	Clam	Fucik et al. (1977)
Sandy bottom community, Western Norway	Ekofisk crude oil	9 mo.	Plastic and aluminum enclosures; divers	Benthic organisms	Bakke and Johnsen (1979)
Surface Slick					
Shrimp pond, Tex.	No. 2 fuel oil[a]	3+ mo.	Midwater cages	Clam, oyster, shrimp	B.A. Cox et al. (1975)

Location	Oil type	Duration	Method	Organisms	Reference
Estuarine marsh pond, Miss.	Empire Mix crude oil	15 mo.	On bottom, un-restrained	Estuarine marsh flora and fauna	Lytle (1975)
Subarctic intertidal, Port Valdez, Alaska	Prudhoe Bay crude oil	2 mo.	On bottom	Clam	Shaw et al. (1976)
Estuarine marsh, Chesapeake Bay, Va.	No. 2 fuel oil	2,6,15 weeks	On bottom	Mussel, oyster	Lake and Hershner (1977), Hershner and Lake (1980)
Estuarine marsh, Chesapeake Bay, Va.	South Louisiana crude oil, fresh and weathered	10 mo.	Off bottom trays, on bottom, cages	Oyster, clam, fish, plants	Bieri et al. (1977, 1979), Bender et al. (1977)
Estuarine marsh, Chesapeake Bay, Va.	No. 2 fuel oil	3 weeks	Off bottom trays	Oyster, clam	Bieri and Stamoudis (1977)
Estuarine intertidal, Sequim, Wash.	Prudhoe Bay crude oil	10-15 mo.	In bottom trays	Benthic invertebrates	J.W. Anderson et al. (1978)
Contaminated Sediment (By Mixing)					
Arctic shallow subtidal, Beaufort Sea, Alaska	Prudhoe Bay crude oil	8 mo.	Bottom trays	Benthic invertebrates	Atlas et al. (1978), Busdosh et al. (1978)
Estuarine intertidal, Sequim, Wash.	Prudhoe Bay crude oil	2 mo.	In bottom trays	Clams (2), worm	Roesijadi et al. (1978), Roesijadi and Anderson (1979), Augerfeld et al. (1980)
Estuarine intertidal, Sequim, Wash.	Prudhoe Bay crude oil	15 mo.	In bottom trays	Mollusks, poly-chaetes, and crustaceans	Vanderhorst et al. (1981)
Artificial Substrates					
Estuarine intertidal Sequim, Wash.	Prudhoe Bay crude oil	1 mo.	Concrete bricks	Intertidal epifauna	Vanderhorst et al. (1981)

NOTE: Table is in roughly chronological order by method, since the preparation of the 1975 NRC report.

[a] American Petroleum Institute reference oil (R.C. Clark and Brown, 1977).

the enclosure were exposed to 30-35 L of an oil-in-seawater dispersion of Ekofisk crude oil. Experiments took place over a 9-month period.

Oil can be poured onto the surface of an enclosure, such as a pond or a cordoned-off segment of a bay. Bottom-dwelling and free-swimming organisms are exposed to various forms of oil, depending on the natural or human-induced mixing conditions, tidal fluctuations, etc. (B.A. Cox et al., 1975; Lake and Hershner, 1977; Bieri et al., 1977, 1979). The enclosure can be as large as an embayment, as in the case of a controlled oil spill (Blackall et al., 1981; Blackall and Sergy, 1981).

Artificial Substrate Experiments

In studies of epifaunal recovery, concrete construction bricks have been used as experimental substrates to represent oiled rock habitats (Vanderhorst et al., 1981). Bricks of uniform size, shape, and porosity are readily available in large quantities and are easily placed on a beach without requiring a physical support system to keep them in place. In the study cited above, the bricks were preconditioned in flowing laboratory seawater for 2 weeks and then treated with a surface-borne slick of Prudhoe Bay crude oil to simulate repeated exposure of an intertidal rock for 5 days of tides. After another day in flowing, clean seawater, the bricks were placed in the intertidal zone for 1 month. At the end of the experiment, the animal species present on the top of each brick were identified and counted. Analyses of the oil (carbontetrachloride-soluble) adhering to the bricks were conducted using infrared spectrophotometry and glass capillary gas chromatography.

Field Surveys

A less well-controlled method of study involves comparison of contaminated wild populations with uncontaminated reference populations. Some large scale monitoring projects, e.g., Mussel Watch (National Academy of Sciences, 1980) and Oyster Continental Shelf Environmental Assessment Program, were intended, in part, to establish reference characteristics of petroleum contamination in various wild populations of marine macroorganisms. The information obtained has been useful in evaluating petroleum uptake, distribution, and discharge for selected organisms. The common problems associated with such methods are lack of scientific control over the wild populations and a lack of detailed knowledge of the history of contaminant-organism interactions.

One difficulty in comparing hydrocarbon uptake data from chronically polluted areas with data from reference areas is filtering out the natural "noise" or variability in biological systems. Kwan and Clark (1981) described a system of pattern recognition analysis for ranking mussels collected at sites of differing contaminant input using paraffin hydrocarbon parameters. They were able to rank mussels collected from different locations in Puget Sound according to their degree of chronic and acute pollution exposure. In the case of acute petroleum pollution (i.e., major oil spills), an extensive discussion on the philosophy and application of sampling for benthic and pelagic organisms was given by G.V. Cox (1980).

Laboratory Exposure Methods

General methods for preparing petroleum solutions for use in studying uptake and behavior modification of petroleum by macroorganisms have been described. Our brief review is presented in the Preparation of Oil-Water Solution section. In addition to laboratory exposure of macroorganisms to water-soluble fractions, dispersions, surface slicks, oil-contaminated food, and oiled habitats, intraperitoneal injection of pure hydrocarbons has been employed in a few studies, as a direct method of exposing individual animals to a known concentration of hydrocarbon, and is useful for studying pathways of metabolic conversion (Varanasi et al., 1979). Frequently, radiolabeled hydrocarbons containing tritium (^{3}H) or carbon-14 (^{14}C) are used, a method offering a sensitive and specific mode of detection, as well as a means of studying metabolite formation. Interestingly Roubal et al. (1977b) reported that it was not possible to correlate feeding experiments directly with rates of depletion of aromatic hydrocarbons or their metabolites from tissues in intraperitoneal injection studies in salmon.

Table 3-7 lists the principal "state-of-the-art" techniques used for laboratory studies of uptake and effect of hydrocarbons by pelagic and benthic marine macroorganism (excluding marine mammals and birds). Several of the methods described in Table 3-7 permit the application of dispersant-petroleum mixture and testing in a fashion similar to petroleum exposure. Initially, static laboratory tests were used to establish acute toxicity of dispersants in the absence of petroleum. Several methods developed for acute toxicity testing of dispersant-petroleum mixtures are also applicable for uptake.

The "sea" test (Norton et al., 1978) developed in the United Kingdom employs a propeller-mixing device inside a plastic cylinder inside a transparent plastic exposure tank containing the test animals. The oil can be added to the mixing vortex within the central cylinder by syringe or to the inner cylinder, prior to agitation, followed by syringe-delivered dispersant, a short waiting period, and finally, initiation of mixing. This procedure has been used to expose free-swimming organisms (e.g., the brown shrimp, Crangon crangon).

The "beach" test provides a means of assessing toxicity and uptake of dispersant on semimobile, intertidal organisms such as limpets (Norton et al., 1978). The organisms are placed on transparent plastic test plates in flowing seawater until they have attached. After several days of conditioning in a simulated tidal seawater system, the plates containing the organisms are hand sprayed either with dispersants or oil. For toxicity tests, the number of limpets detaching immediately after exposure and at 24 and 48 hours after starting the test are counted.

TABLE 3-7 Laboratory Methods for Exposing Marine Macroorganisms to Hydrocarbon Contaminants

Method	Petroleum Material	Exposure[a] Type	Duration	Organism	Comments	Reference
Water-soluble Fractions						
Surface slick, simulated tidal pumping in aquarium	No. 2 fuel oil No. 5 fuel oil	S	Sh	Blue mussel	Recirculating seawater system	Clark and Finley (1974b)
Stirred for 24 h, depurated for 58 d	^{14}C-benzo[a]-pyrene	S	Sh	Estuarine clam	Isotope added directly to exposure tank	Neff and Anderson (1975a)
Saturated benzene	Benzene	F	Sh	Striped bass	Method of Benville and Korn (1974)	Meyerhoff (1975)
Hydrocarbon added to artificial seawater	^{14}C-benzo[a]-pyrene	S	Sh	Clam	30 ppb	Neff et al. (1976)
Hydrocarbon added to seawater in ethanol solution, no mixing, 1-3 d exposure	^{3}H-benzo[a]-pyrene ^{3}H-methyl-chloanthene ^{14}C-fluorene	S	Sh	Blue crab	2-L aquarium containing 1 25-40 g crab; approx. 10% of isotope lost in 48 h	R.F. Lee et al. (1976)
Saturated benzene solution by shaking in separatory funnel	^{14}C-benzene	S	Sh	Northern anchovy striped bass	48-h exposure	Korn et al. (1977)
Same as previous	^{14}C-benzene ^{14}C-toluene	S	Sh	Pacific herring	100 ppb concentr. initially	Korn et al. (1977)
Oil-in-Water Dispersion						
Stirred for 20 h, stand for 1-6 h;	No. 2 fuel oil,[b] Bunker C fuel oil[b]	S	Sh	Mysids, grass and brown shrimp	See discussion in text	J.W. Anderson et al. (1974)
20-100,000 ppm (v/v) oil in artificial seawater, 200 cycles/min, 5 min, stand 30-60 min	South Louisiana crude oil,[b] Kuwait crude oil,[b] No. 2 fuel oil [b]	S	Sh	Mysids, grass & brown shrimp, sheephead minnow, silversides, killifish	Mixed in exposure container	J.W. Anderson et al. (1974)

Method	Oil			Species	Details	Reference
Two-cycle outboard run in 260 L seawater for 100 min	10% outboard motor effluent	F	L	Blue mussel, oyster	18-hp, 1,200 1,800 rpm, 6 times/10 d	R.C. Clark et al. (1974)
Ultrasonic emulsification, 25-150 ppm (v/v)	No. 2 fuel oil, South Louisiana crude oil	S	Sh	Soft shell clam	Oil-soluble dye (Oil Red O) used	Stainken (1975)
Blender for 2 min, 20-500 ppm oil in seawater	Waste motor oil	S	L	Oyster, scallop, Atlantic silversides	Recirculating seawater system	Gardner et al. (1975)
Method of J.W. Anderson et al. (1979)	South Louisiana crude oil,[b] Kuwait crude oil,[b] No. 2 fuel oil,[b] Bunker C fuel oil[b]	S	Sh, L	Oyster	1 L jars and aquaria, 4- to 7-d exposures	R.D. Anderson (1975)
Water agitation: mixed by streams of oil and water, stand 1 h	Prudhoe Bay crude oil	S	L	Pink salmon	0.7-5.7 ppm, 10-d exposure	Rice et al. (1975)
Method of J.W. Anderson et al. (1974)	No. 2 fuel oil[b]	S	Sh	Shrimp, clam, oyster, fish	IR and UV analysis in water	Neff et al. (1976)
Mixed in centrifugal metering pump	No. 2 fuel oil[b]	F	Sh	Clam, oyster	Alkane enrichment in water	Neff et al. (1976)
Method of Vanderhorst et al. (1977b)	No. 2 fuel oil[b]	F	Sh	Coon stripe shrimp	See text for apparatus, 0.4-1.7 ppm by analysis	Vanderhorst et al. (1976)
Oil mixed by aeration in seawater in tank	No. 2 fuel oil[b]	S	Sh	Polychaete worm	Clean sediment in tank	Prouse and Gordon (1976)
Method of J.W. Anderson et al. (1974)	No. 2 fuel oil[b]	S	Sh	Clam, oyster	Various exposure times	Neff et al. (1976)
Shaken mechanically 1 h, room temperature	No. 2 fuel oil[b] n-hexadecane Phenanthrene	S	Sh	Hard clam	8-h exposure	Boehm and Quinn (1973)
Shaken 300 times in 6-L separatory funnel, stand 1 h; 90-380 ppb	Kuwait crude oil	F	L	Soft shell clam	10 d in sediment with running seawater	W.C. Wong (1976)

TABLE 3-7 (continued)

Method	Petroleum Material	Exposure[a] Type	Duration	Organism	Comments	Reference
Water-suspension of kaolin metered with peristaltic pump	Diesel fuel oil	F	L	Blue mussel	75 mL oil/100 g kaolin	Fossato and Canzonier (1976)
Ultrasonic emulsification by method of Gruenfeld and Fredrick (1977)	No. 2 fuel oil	S	L	Soft shell clam	28-d exposure, 10, 50, 100 ppm oil initially	Stainken (1977)
Method of J.W. Anderson et al. (1974)	South Louisiana crude oil,[b] Kuwait crude oil,[b] No. 2 fuel oil,[b] Bunker C fuel oil[b]	S	Sh	Oyster	4 d at 1% oil dispersion	J.W. Anderson (1977a)
Ultrasonic emulsification	^{3}H-n-Tricosane	S	Sh	Blue crab, mullet	1- to 2-d exposure	Geizler et al. (1977)
Method of Vanderhorst et al. (1977b), field colonized bricks	No. 2 fuel oil[b]	F	L	83 species of intertidal plants and animals	0.12-0.62 ppm by analysis	Vanderhorst et al. (1977a)
Surface oil in tank mixed by pumping	Arabian Light crude oil	S	L	Eel	Water replaced periodically	Ogata et al. (1977)
Stirred for 20 h, stand for 3 h, 10°-12°C, 1% oil	Prudhoe Bay crude oil	F	L	Deposit-feeding clam	Recirculating water and sediment in tray	Taylor and Karinen (1977)
Method of J.W. Anderson et al. (1974)	Prudhoe Bay crude oil	S	Sh	Sipunculid worm	Seawater in fingerbowls	J.W. Anderson et al. (1977)
Method of J.W. Anderson et al. (1974)	South Louisiana crude oil,[b] Kuwait crude oil,[b] No. 2 fuel oil,[b] Bunker C[b]	F	L	Polychaete worm	14- and 28-d exposure, 2 concentrations	J.W. Anderson et al.
Method of J.W. Anderson et al. (1974)	No. 2 fuel oil	S	Sh	Killifish	Few hours of exposure	Dixit and Anderson (1977)

Method	Oil	Flow	Stage	Species	Conditions	Reference
Dissolved fuel oil by shaking	No. 2 fuel oil	F	L	Marine fish	Daily addition to maintain 170 ppb	Kurelec et al. (1977)
Ultrasonic emulsification, method of Gruenfeld and Frederick (1977)	No. 2 fuel oil	S	L	Soft shell clam	28-d exposure, 10, 50, 100 ppb dispersion	Stainken (1978)
Method of Hyland et al. (1977)	No. 2 fuel oil	F	L	Winter flounder	10 and 100 ppb dispersion	Kuhnhold et al. (1978)
Shaken vigorously with Gallenkamp bottle shaker for 20 min	Toluene	S	Sh	Marine isopod	5.5:10,000 toluene to seawater	Bakke and Skjoldal (1979)
Method of Vanderhorst et al. (1977b)	Prudhoe Bay crude oil	F	L	Benthic amphipod	4-, 5-, 27-d exposures	J.W. Anderson et al. (1979)
Method of J.W. Anderson et al. (1974)	Prudhoe Bay crude oil	S	Sh	Benthic amphipod	4-d exposure	J.W. Anderson et al. (1979)
Surface oil in tank mixed by pumping	Arabian Light and Zubea crude oils (4:1)	S	L	Eel, short necked clam	Eel: 1,000 ppm, 15 d; clam: 50 ppm, 20 d	Ogata et al. (1979)
Stirred for 20 h, stand for 3 h, 1% oil, 4°-8°C	Cook Inlet crude oil, No. 2 fuel oil	S	Sh	30 species subarctic marine fish and invertebrates	Method of Taylor and Karinen (1977)	Rice et al. (1979)
Surface oil in tank mixed by pumping	Arabian Light and Zubea crude oils (4:1)	S	L	Eel	250 ppm, 5 d	Ogata and Miyake (1979)
Method of Hyland et al. (1977)	No. 2 fuel oil	F	Sh	Mud snail	See text for description	Hyland and Miller (1979)
Emulsified oil method, 4 ppm	Caucasus crude oil	S	L	Mussel	Seawater changed daily	Mironov and Shckekaturina (1979)
Surface oil in tank mixed by pumping, dispersion changed daily	Arabian Light and Subea crude oils (4:1)	S	L	Eel, short necked clam	Eel: 250 ppm, 16 d; clam: 59 ppm, 8 d	Ogata et al. (1980)
Method of Vanderhorst et al. (1977b)	Prudhoe Bay crude oil	F	Sh, L	Grass and coon stripe shrimp, mysids	300 ppb hydrocarbons by analysis	J.W. Anderson et al. (1980)

TABLE 3-7 (continued)

Method	Petroleum Material	Exposure[a] Type	Duration	Organism	Comments	Reference
Combination of methods of Hyland et al. (1977) and Roubal et al. (1977a)	Prudhoe Bay crude oil	F	L	Deposit-feeding clam	180-d exposure, 30, 300, 3,000 ppb levels	Clement et al. (1980)
Mixing in baffle-filled glass tube, separation in tanks like Vanderhorst et al. (1977b)	Solvent refined coal liquids	F	L	Freshwater systems	Total organic carbon and phenols analyzed	Dauble et al. (1981)
Method of J.W. Anderson et al. (1974)	Benzene, toluene, benzo[a]pyrene	S	Sh	Coastal mussel	UV spectrometry	Sabourin and Tullis (1981)
Mixed in tubing and in incubation vessels by magnetic stirrers	Naphthalene	F	Sh	Oyster	90 ppb for 3 d, start at 0 concentr.	Riley et al. (1981)
Method of Vanderhorst et al. (1977b), added at initial mixing funnel	Prudhoe Bay crude oil and dispersant	F	L	Coon stripe shrimp	Oil only: 141 ppb; oil and dispersant: 1,400 ppm	J.W. Anderson et al. (1981)
Method of J.W. Anderson et al. (1974)	Kuwait crude oil[b]	F	Sh	Ghost crab	4-d exposure	Jackson et al. (1981)
Surface Slick						
Tidal aquarium, recirculating seawater	No. 2 fuel oil, No. 5 fuel oil	S	Sh	Blue mussel	Exposed twice daily for 2-3 d	R.C. Clark and Finley (1974b)
Inflowing seawater sprayed into slick	Iranian and Sinai crude oils and dispersant	F	L	15 species of Red Sea marine organisms	2 m tall x 1 m dia. tanks	Eisler (1975)
Simulated tidal cycle with oil added on water surface and allowed to drain into sediment	No. 2 fuel oil, Nigerian crude oil, waste motor oil	F	L	Mixed benthic algae	Low-salinity communities in laboratory microcosm	Bott et al. (1976)

Simulated tideflat aquaria; oil added 5 successive d on falling tide	Prudhoe Bay crude oil	F	L	Deposit-feeding clam	60-d exposure, 1.2, 2.4, 5.0 cm^{-2} d^{-1} oil	Taylor and Karinen (1977)
Simulated tidal cycle with oil added on water surface and allowed to drain into sediment, 1 oiling, 0.32 cm thick	Prudhoe Bay crude oil	F	L	Sipunculid worm	13 d before adding worms, 14-d exposure	J.W. Anderson et al. (1977)
Oiled seawater percolated through sediment in trays once daily for 3 d	Prudhoe Bay crude oil	S	L	Deposit-feeding clam	500 ppm oil in seawater, 30-d recovery	Shaw et al. (1977)
Oil retained in ring placed in exposure tank, air mixing	No. 2 fuel oil	S	L	Marine fish	Described as saturated with oil in 90 min	Kurelec et al. (1977)
Oil (150 mL) layered on surface (1 m^2), weathered oil removed weekly	Venezuelan crude oil	F	L	Fish	6-mo. exposure	J.F. Payne et al. (1978)
Inflowing seawater sprayed into slick, 3 ppm equivalent	Iranian crude oil	F	L	Red Sea coral	Oil added weekly for 1 d, coral removed for 2-6 mo.	Rinkevich and Loya (1979)

Oil in Food

Soft core extruded from tube into foregut	Naphthalene	F	L	Crab	In lard matrix	Corner et al. (1973)
Cod liver oil capsules in squid diet	Kuwait crude oil (high boiling ends)	F	L	Cod	1 mg/d oil for 175 d	Hardy et al. (1974)
Isotope dissolved in ethanol and injected into frozen shrimp	^{3}H-benzo[a]-pyrene ^{14}C-fluorene ^{14}C-heptadecane ^{14}C-dotriacontane ^{14}C-hexadecane ^{14}C-2-methylnaphthalene ^{14}C-naphthalene	S	Sh	Blue crab	Analyzed 2 d after feeding	R.F. Lee et al. (1976)

TABLE 3-7 (continued)

Method	Petroleum Material	Exposure[a] Type	Duration	Organism	Comments	Reference
Live oysters maintained in seawater with isotope for 1 d	^{14}C-benzo[a]pyrene ^{14}C-hexadecane	S	Sh	Blue crab	Oysters fed to crabs	R.F. Lee et al. (1976)
Isotope added to Oregon moist pellet; 5 μCi of each fed	^{14}C-benzene ^{14}C-naphthalene ^{14}C-anthracene	F	L	Coho salmon	Sampled after 1, 3, 7, 14 d	Roubal et al. (1977b)
Detritus (powdered alfalfa, 5 mg/d) impregnated with 10-15 ppm isotope	^{14}C-2-methylnaphthalene	S	L	Polychaete worm	16-d exposure	Rossi (1977)
Isotope evaporated onto cubes of squid, then coated with gelatin	^{14}C-n-hexadecane ^{14}C-benzo[a]pyrene	S	Sh	Herring	43-45 h before sacrificing	Whittle et al. (1977)
Isotope placed on food Isotope dissolved in corn oil and placed in esophagus of fish	^{3}H-n-nonadecane ^{3}H-n-tricosane	S S	L L	Blue crab Mullet	15-d exposure	Geizler et al. (1977)
Dissolved in cod liver oil and injected into gut via mouth	^{14}C-naphthalene	S	Sh	Killifish	0.5 and 1.0 μCi, sacrificed after 2, 4, 8 h	Dixit and Anderson (1977)
Force-fed gelatin capsule with isotope dissolved in salmon oil	^{14}C-naphthalene	F	Sh	Coho salmon	5.5 μCi each, sacrificed after 8 and 16 h	Collier et al. (1978)
Dry sediment and oil and seawater (1:2:10), 1 h on oscillating shaker, rinsed once	Prudhoe Bay crude oil	F	L	Deposit-feeding clam	0.10-0.50 g/cm² dose, 30-d exposure	Taylor and Karinen (1977)
Mixed in cement mixer using blender-prepared emulsion	Prudhoe Bay crude oil	F	L	Sipunculid worm	500-750 ppm oil by analysis	J.W. Anderson et al. (1977)

Method	Oil			Organism	Exposure detail	Reference
Water-soluble fraction (J.W. Anderson et al., 1974) mixed with sediment	No. 2 fuel oil	F	L	Polychaete worm	28-d exposure	Rossi (1977)
Dry beach sand and oil in pentane mixed in cement mixer, 0.5 cm thick	North sea crude oil, Kuwait crude oil	F	L	Plaice, Norwegian lobster, brown shrimp	Exposed to 10 d	Howgate et al. (1977)
Sediment-oil-seawater slurry in cement mixer	Prudhoe Bay crude oil	F	L	English sole	4-mo. exposure, 5 cm thick, 0.2% oil	McCain et al. (1978)
Oil and sediment mixed by hand, 2-L beakers filled to 2 cm from top	South Louisiana crude oil,[b] Kuwait crude oil,[b] No. 2 fuel oil,[b] Bunker C fuel oil,[b] weathered Bunker C (Arrow)	F	Sh, L	Polychaete worm	35-2,000 ppm oil analyzed	Gordon et al. (1978)
Mechanical mixing, stand 96 h, rinsed and placed in trays to 4 cm	Prudhoe Bay crude oil, fresh and weathered	S	L	Benthic amphipods	1:5 oil to clean sandy sediment	Busdosh et al. (1978)
Method of J.W. Anderson et al. (1977), simulated tidal flux	Prudhoe Bay crude oil	F	L	Deposit-feeding clam	1 mg oil/g sediment, 55-d exposure	Roesijadi and Anderson (1979)
Detritus shaken with aqueous solution of oil, isotope, and ethyl ether, forms 1- to 2-mm layer on bottom of tank	^{14}C-phenanthrene ^{14}C-chrysene ^{14}C-dimethylbenz[a]-anthracene ^{14}C-benzo[a]pyrene and Prudhoe Bay crude oil	F	L	Deposit-feeding clam	2,000 ppm total petroleum hydrocarbons by analysis, 10 µCi and 0.033 mL oil added	Roesijadi et al. (1978)
Force-fed gelatin capsule with isotope dissolved in salmon oil	^{3}H-naphthalene	F	L	Starry flounder, rock sole	56 µCi each, sacrificed after 1, 2, 7 d	Varanasi et al. (1979)
Force-fed gelatin capsule with isotope evaporated on salmon diet	^{14}C-phenanthrene	F	L	Coal fish	1.0 µCi each, sacrificed 1/2 h to 28 d	Solbakken et al. (1979)
Intragastric administration by syringe	^{14}C-phenanthrene	F	L	Norwegian lobster	Sacrificed 1-28 d	Palmork and Solbakken (1980)
Intragastric administration by syringe	^{14}C-phenanthrene	F	L	Spiny dogfish, rainbow trout	Sacrificed 1-28 d	Solbakken and Palmork (1980)

TABLE 3-7 (continued)

Method	Petroleum Material	Exposure[a] Type	Duration	Organism	Comments	Reference
Hydrocarbon-contaminated food (Artemia salina) in artificial seawater in aquaria	2,6-dimethylnaphthalene	S	L	Grass shrimp	32-d exposure, 0.24 μg/g hydrocarbon in food	Dillon (1981)
Hydrocarbon-contaminated food (Fucus sp. algae) in aquarium	^{3}H-2,6-dimethyl-napthalene	F	L	Sea urchin	14-d exposure, freshly contaminated algae added daily	Malins and Roubal (1982)

Habitat: Sediment

Method	Petroleum Material	Exposure[a] Type	Duration	Organism	Comments	Reference
Mechanically mixed placed in 2-L beakers	No. 2 fuel oil,[b] fresh and weathered	F	L	Polychaete worm	<250 mg oil/g sediment	Prouse and Gordon (1976)
Hydrocarbon added to artificial seawater	^{14}C-naphthalene	S	Sh	Polychaete worm	1-d exposure	Rossi (1977)
Sonicated hydrocarbons in 8-L seawater	^{14}C-tricosane	S	L	Blue crab, mullet	80-160 ppb, 14-d exposure	Geizler et al. (1977)
Solubilizer of Roubal et al. (1977a)	Prudhoe Bay crude oil	F	L	Starry flounder, coho salmon	See text for apparatus	Roubal et al. (1978)
Electrocardiograph, recirculating fish chamber	Oil spill dispersants (8)	S	Sh	Fish	Sublethal levels	Kiceniuk et al. (1978)
Solubilizer of Benville et al. (1981)	Cook Inlet crude oil	F	L	Starry flounder	See discussion in text	Whipple et al. (1978)
Solubilizer of Benville et al. (1981)	Cook Inlet crude oil	F	Sh, L	Manila clam	See text for apparatus	Nunes and Benville (1979a)
Solubilizer of Roubal et al. (1977a)	Prudhoe Bay crude oil	F	L	Spot shrimp	110 ppb total hydrocarbons	Sanborn and Malins (1980)
Hydrocarbons added directly to seawater	^{3}H + ^{14}C-naphthalene	F	Sh	Spot shrimp	^{3}H:^{14}C was 48:1	Sanborn and Malins (1980)

Method	Material	Exposure type[a]	Exposure duration[a]	Species		Reference
Sand dosing column	Hexachlorobenzene 0.04–0.45 µg/L	F	L	Killifish	Recirculating system	Giam et al. (1980)
Hydrocarbons added to seawater	^{14}C-phenanthrene	S	Sh	Horse mussel	10 mussels in 4 L seawater	Palmork and Solbakken (1981)
Flow-through system of glass jars in mobile lab	Ballast-water treatment effluent	F	Sh, L	Pink salmon, kelp shrimp	Effluent analyzed twice daily by gas chromatography	Rice et al. (1981)
Method of J.W. Anderson et al. (1977), 3 cm thick in aquaria	Prudhoe Bay crude oil	F	L	Benthic amphipod	290 ppm total hydrocarbons by analysis	J.W. Anderson et al. (1979)
Mechanically mixed, 2-L in 17-L jar	Prudhoe Bay crude +^3H-benzo[a]pyrene; Prudhoe Bay crude +^3H-benzo[a]pyrene +^{14}C-naphthalene	F	Sh	Starry flounder	1% oil + 20 nmol/g, 1% oil + 1.8 nmol/g; nmol/g + 5.9	Varanasi and Gmur (1981a)
Mechanically mixed, 2-L in 17-L jar, Method of McCain et al. (1978)	Prudhoe Bay crude +^3H-benzo[a]pyrene +^{14}C-naphthalene	F	L	English sole	1% oil + 3,640 µCi + 140 µCi, respectively	Varanasi and Gmur (1981b)
Injections						
Intraperitoneal, in ethanol	^{14}C-benzene, ^{14}C-napthalene, ^{14}C-anthracene	F	L	Coho salmon	2.5 µCi	Roubal et al. (1977b)
Intraperitoneal, in salmon oil[b]	^3H-naphthalene	F	L	Starry flounder, rock sole	56 µCi	Varanasi et al. (1979)

NOTE: Table is in roughly chronological order by method, since the preparation of the 1975 NRC report.

[a]Exposure type: S = static, F = flow-through, Exposure duration, Sh = short (96 hr or less), L = long.

[b]American Petroleum Institute reference oils (R.C. Clark and Brown, 1977).

COMMUNITIES

The success of surveys or monitoring depends upon careful attention to sampling design, methods, and analysis. Many investigators have described methods for sampling (Holme and McIntyre, 1971; A.D. Michael et al., 1979) and for measuring ecological variables useful in investigation of pollution (Gray et al., 1980), as well as study design and analysis (R.H. Green, 1979).

Sampling Methods

Subtidal Habitats

The trend in the development of sampling methods is toward ensuring complete capture of organisms within a unit area of seabed or habitat surface (Holme and McIntyre, 1971). Because grab samplers may not sample to a uniform depth under the area covered and may vary in depth of penetration, the preferred sampling device for soft substrates is a corer. In intertidal or very shallow water habitats, hand-operated corers are simple and efficient. In deeper waters, however, box-coring devices for sampling a larger surface area to the desired depth in the sediment are required, especially in soft sediments. Large ships and heavy handling equipment are required to operate such devices. Spade box corers are the most commonly employed large volume coring devices and, properly equipped, can be used to collect comparatively undisturbed sediment columns to a depth of 20-50 cm or more. Frequently adequate penetration and core capture are difficult to achieve in sandy sediments.

Larger, and consequently less common, benthic animals must be sampled by a trawl or dredge dragged over the bottom. Accurate and precise assessment of population density is, therefore, difficult. Relatively sedentary large animals may, indeed, be an important and susceptible component of the biota. Short tows of dredges and photographic or televideo instruments or direct counts by divers or from submersibles are, currently, the most reliable methods for assessing such populations.

Collection and adequate preservation of sediment samples for determination of sediment granulometry and petroleum hydrocarbons are important. More samples can be collected than may be feasible to analyze, thus allowing for archiving samples, should they be needed for subsequent analysis. Samples are best collected as a vertical integral of the surface sediment to a prescribed uniform depth. However, some investigators prefer that cores be sectioned as finely as possible and the sections be analyzed separately. If this is not possible, the uppermost layers are analyzed.

For macrobenthos and meiobenthos, variability in sample processing can be a significant source of error. The very small mesh openings of sieves needed for separating animals from sediments reduce the loss of the target populations, but present some practical limitations. In the West Falmouth spill study, Sanders et al. (1980) used a mesh of 0.3 mm and collected essentially all the adult macrobenthos. Historically, 1.0-mm sieves are used, but these may lose a sizable portion of the macrobenthos. Furthermore, they are susceptible to variation in efficiency, due to length of sieving, amount of debris, etc. Careful use of a 0.5-mm mesh sieve results in capture of most of the adult macrofaunal taxa and is generally an acceptable compromise for studies in coastal and continental shelf habitats. In deeper water, where the macrofauna are small, the mesh should be 0.3 mm or finer.

Meiobenthos, i.e., metazoans which, as adults, pass through a 0.5-mm mesh sieve, require specialized sample processing procedures (Hulings and Gray, 1971). They are commonly removed from sediment collected in a small diameter corer by elutriation or density separation. Animals elutriated are retained on a sieve of opening size between 40 and 100 μm (commonly 63 μm). The sensitivity of meiobenthos to petroleum is not yet well documented, largely because these constituents are difficult to identify. In general, only the abundance of a higher taxon (e.g., total nematodes) is assessed (Wormald, 1976; Grassle et al., 1981; Elmgren et al., 1980a,b). In cases of less than catastrophic pollution, it is probably more realistic to expect to observe effects only on individual species, rather than on the total density of a higher taxon. However, because of difficulties in species identification of the dominant meiobenthic taxon, Nematoda, few studies have actually been done to assess the effects of petroleum on dominant component species of the meiobenthos (Gier, 1979; Boucher, 1980). Species and population analysis of subdominant taxa, in particular harpacticoid copepods, is, in general, relatively more easily accomplished.

Intertidal Habitats

Because of environmental changes which occur in the intertidal zone, including both rocky and sediment-covered substrates, stratified random sampling or sampling along line or belt transects through the intertidal zone is advisable (A.D. Michael et al., 1979). For hard substrates, a line intercept method may be used to estimate populations. This approach to measurement of occupied surface area allows nondestructive sampling and provides the advantage of monitoring the same individuals to assess long term effects on recovery.

Although attention should be focused on the primary space occupied on structure-creating macroalgae and animals of rocky shores, associated epibiota should not be ignored. For example, although no effect of the Tsesis oil spill on the dominant alga Fucus vesiculosus could be measured, most of the associated fauna, particularly the small Crustacea, were in fact adversely affected (Notini, 1980).

Microalgae living in intertidal sediments or on hard substrates are often important primary producers and may be subjects of concern if the habitat is affected by petroleum. Although their biomass can be approximated simply by measuring chlorophyll concentrations of surface sediments, chlorophyll concentration demonstrates, in nature, a high degree of temporal and spatial variability.

Vegetated Habitats

Biomass and productivity of marsh grass are best assessed by clipped quadrats placed in a random manner within habitat strata, e.g., high Spartina, low Spartina, salt meadow, etc. (J. Michael et al., 1978). Some benthic fauna can be sampled using corers and seines. Hand seines and baited traps should also be employed to sample marsh benthos. Epifauna, such as marsh snails, can be sampled along with the grass. However, the more difficult estimation of the burrower population, such as fiddler crabs, necessitates excavation of the substrate. Other physiological and functional measurements of salt marsh communities which may prove to be relevant in oil pollution studies, can also be made (Pomeroy and Wiegert, 1981).

Sea grass biomass and production can be assessed, much as is done with marsh grasses, except that defoliated leaves are rapidly lost from the bed, complicating production estimates. Phillips and McRoy (1979) described a variety of methods used to assess sea grass productivity and other aspects of the biota of the sea grass community. Because of their ecological importance in trophic interactions and their suspected sensitivity to oil, the epibiota (including the diverse crustaceans) are best sampled by placing a fine mesh net over the grass and clipping the grass or collecting by suction methods. The concentration of the epibiota can be expressed both in unit area of seabed and biomass of grass.

For mangroves, biomass and primary productivity of trees in the mangrove swamps are difficult to measure. Attention, instead, should be placed on measurement of the degree of defoliation and, subsequently, on the recovery of defoliated trees. For faunal studies, the same conditions occur as in salt marshes, except that the epibiotic community of trunks and prop roots may provide useful information.

Coral Reefs

Many methods used in oceanography can be adapted for the study of coral reefs (see also Chapter 5); however, special consideration should be addressed to shallow areas, i.e., reef lagoons, reef flats, and algal ridges. Wave surge, shallow water depth, and tidal variations will dictate that specialized sampling methods be used. Stoddart and Johannes (1978) provide a detailed source of information for quantitative sampling of various biological, chemical, and geological phenomena of coral reefs.

Calcareous algae and scleractinian corals are major primary producers and are particularly important to study because they are more susceptible to toxic effects of oil pollution. In a recent review, Loya and Rinkevich (1980) discussed field and laboratory studies on the effects of petroleum in coral reef communities.

Analysis and Interpretation of Data

Typical data resulting from benthic community sampling form a matrix consisting of a limited array of environmental variables and an imposing list of species counts for each of many collections. The approach often taken in interpretation of such data is to describe the distribution of a few abundant, or apparently diagnostic, species and relate these to environmental factors, thus effectively discarding a large portion of the data very often obtained at great expense. Alternatively, or in addition, an attempt may be made to simplify the complex array of data by computing a diversity index or some other statistic designed to represent the structure of the community, which discounts the majority of the information content of the sample, as well, including the fundamentally important information of what kinds of organisms were present. Thus, the results are difficult to interpret and the loss of information from such data reduction can have a significant effect on the conclusions of the study.

Derived Community Indices

Statistical methods for studying joint abundances of species, using probability distributions fitted to abundance data and diversity indices, can be valuable for certain applications. Of course, the underlying assumptions must be understood and, where possible, the conclusions tested by other means. Clearly, the statistical properties of some indices make them more useful than others, depending on the application. From a pragmatic point of view, a single number for certain types of inferences, or monitoring, might be desirable, as decision makers might prefer the parsimony of an index number. It is a well-established statistical fact, however, that one number or index may be insufficient for summarizing data; even so simple a distribution as the normal requires two numbers (a mean and a variance) for accurate description.

The bewildering complexity of data resulting from surveys of macrobenthos often prompts investigators to simplify results and report a derived index. Diversity indices have been particularly popular because of the presumed relationship between species diversity and environmental quality (Wilhm and Dorris, 1968), although Green (1979) and Smith et al. (1979a,b) state that this may not necessarily be a valid generalization.

Diversity indices should not be used alone to assess impact but should be coupled with population or multivariate analyses that reflect qualitative community composition. Any index involves an inevitable loss of information, compared to the data from which it was calculated. In the case of species diversity measurement, the information lost includes the identity of the species in the community and the species history documented in the literature, both of which are characteristics of obvious importance in evaluating the consequences of any alteration in community structure. If diversity patterns are to be assessed, they should be expressed primarily as species richness and evenness components, both of which are intuitively more meaningful.

Graphical Representation of Community and Population Data

The distribution of abundance in a biotic assemblage can often be more effectively represented graphically than by index. For population data, density plots and life stage histograms (sex, age) are particularly useful.

A commonly used graphical approach in community analyses is to plot, as a logarithmic ordinate, the density or biomass of all component species arranged sequentially from the most abundant to the least (Spies et al., 1980). A different representation of the distribution of abundance involves plotting the frequency of species in geometrically increasing size classes of abundance (Gray, 1980; Gray and Mirza, 1979).

Sanders et al. (1980) prepared similar graphical plots in the analysis of community variability with time by plotting not the abundance of each species, but a measure of species variability (e.g., coefficient of variation), representing species ranked by this variability. Such plots clearly represent the greater temporal population variability characteristic of disturbed communities (Figure 3-8).

Multivariate Analyses

Two samples can possess identical diversity indices and yield similar dominance curves, for example, and yet not share any taxa in common. Multivariate analyses, particularly numerical classification and ordination, allow the simplification of complex data sets, wherein information regarding the taxonomic composition of samples being compared is retained (Clifford and Stephenson, 1975; Boesch, 1977; Pielou, 1977; Orloci, 1978; Whittaker, 1978; R.H. Green, 1979; Orloci et al. 1979; van der Maarel, 1980). Too often practitioners will use a particular classification approach because it is available or was previously used. This frequently results in weak clustering and generally uninterpretable results (e.g., Bender et al., 1979). Criteria for the design of apropriate classification algorithms are given by Boesch (1977). Insight gained from application of numerical classification can be greatly enhanced by postclustering analyses, i.e., two-way or nodal analysis (Boesch, 1977; Straughan, 1980).

Among ordination techniques, the most satisfactory methods for ecological analyses are reciprocal averaging (correspondence analysis) and nonmetric, multidimensional scaling (Hill and Gaugh, 1980). Ordination is generally more instructive than classification when the data are not too heterogeneous and when it is useful to view patterns as gradients, such as along a pollution gradient or in a time series. Multiple discriminant analysis (Green and Vascotto, 1978) is a useful technique which allows the investigation of environmental factors that best discriminate among previously defined groups (e.g., exposed to oil or not exposed to oil).

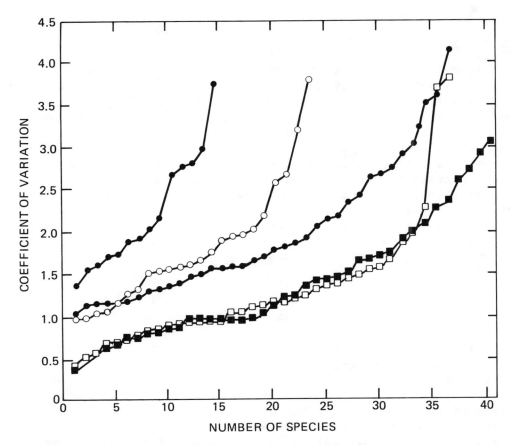

FIGURE 3-8 Plots of coefficients of variation for species populations in communities of macrobenthos near West Falmouth, Massachusetts. Stations are indicated where sediments were contaminated with No. 2 fuel oil.

SOURCE: Sanders et al. (1980).

Species and Populations

Certain benthic species exist in polluted environments or tolerate other disturbances. It is often appealing to interpret results of an impact assessment by examining the distribution of species known to be pollution tolerant, but great care must be taken in selection of a species, notably with regard to its identification and distribution. For example, Bender et al. (1979), in an evaluation of the effects of petroleum production in coastal Louisiana, observed that "if there had been a buildup of hydrocarbons or other pollutants, then one would expect a reduced benthic fauna and the appearance of pollution-tolerant species, such as _Capitella capitata_, in large numbers in affected localities." On the other hand, Sanders and Jones (1981) noted that in this case the benthos was of comparatively low density and was characterized by pronounced dominance by two highly opportunistic species, the polychaet <u>Spiochaetopterus</u> <u>oculatus</u> and the bivalve <u>Mulinia</u> <u>lateralis</u>. Gray et al. (1980) pointed out that species such as <u>Capitella</u> respond to many forms of disturbance and concluded that the use of so-called indicator species is not reliable for ecological monitoring. Unfortunately, <u>Capitella</u> has been shown by Grassle and Grassle (1976) to comprise a group of sibling species, formerly thought to be one species. It is very difficult to separate members of <u>Capitella</u> <u>capitata</u> into their constituent species by morphology. Thus, in its composite, <u>C</u>. <u>capitata</u>, may not be a reliable indicator species even though its sibling species, individually, may prove to be finely tuned to different stresses. The opportunistic species cited by Sanders and Jones (1981) may be considered characteristic of predator-dominated estuarine habitats (Virnstein, 1977) such as those represented in coastal Louisiana. In general, it must be stated that reliance on one measure of change, disturbance, or stress is unwise and suites of measures are best employed because they offer greater reliability over the long term.

Experimental Approaches Beyond the Laboratory

Experiments conducted in the field or in mesocosms offer promise in advancing knowledge of the effects of petroleum on marine benthos. For example, Grassle et al. (1981) followed the response of macrobenthos and meiobenthos in large volume mesocosms during and following chronic addition of No. 2 fuel oil. The behavior of the communities transplanted in the control tanks followed the natural communities in the adjoining bay reasonably well and, by comparison of oiled treatments and controls, effects could be related to the buildup of petroleum hydrocarbons in the sediment.

In another valuable experimental approach using enclosed spills, Bender et al. (1980) described effects of weathered and unweathered crude oil on the benthos of a mesohaline salt marsh using controlled release of oil in intertidal communities. A different approach that has been pursued is the colonization by benthos of azoic sediments placed in trays in petroleum-contaminated and control environments. Thus, rates of colonization and the composition of the recolonizing biota can be compared with azoic controls (Grassle and Grassle, 1974). Also, the settlement of larvae onto contaminated field plots is under intensive study (Vanderhorst et al., 1981) and settlement has been observed in the field on several occasions (Woodin et al., 1972; Straughan, 1971, 1972).

The use of such experimental approaches in the field for assessment of effects of petroleum on benthos greatly extends our ability to detect and understand these effects and, however, allows the development and testing of valid hypotheses regarding effects (e.g., Spies and Davis, 1979; Spies et al., 1978, 1980).

FISH, SEABIRDS, AND MAMMALS

Fish

Some relatively unique methods have been used to measure effects of petroleum on fish. Because fish provide diverse and biologically complex subjects for study, complete coverage of every method applied to fish is not feasible.

Acute toxicity tests provide good preliminary knowledge concerning upper tolerance limits of fish to petroleum hydrocarbons. Fish generally respond rapidly to oil exposure, and therefore, bioassays need to extend only a few days in duration, in contrast to bioassays of many invertebrates requiring longer exposure periods to respond. Flow-through assays are preferable to static tests because of the high metabolic activity of fish and the resulting greater need for oxygen replenishment and metabolic waste removal. The oxygen level is critical for many species of fish, e.g., salmon require >5.0 ppm oxygen for maintenance. The LC_{50}s obtained from acute assays can subsequently be used to determine hydrocarbon concentrations for determining sublethal effects.

Methods for measuring hydrocarbon uptake, distribution in tissues, metabolism, and excretion are similar, in many cases, to methods used with invertebrates. However, methods unique for fish include those devised to determine the pathway of oil hydrocarbons metabolized by fish, and the metabolic potential of fish.

Three exposure media are commonly employed to introduce hydrocarbons of petroleum to fish: food, water, and sediment. To determine accumulation via the gut, contaminated food may be prepared for ingestion by mixing the food with oil or oil components, which may be radiolabeled if the experiment dictates. The food containing the hydrocarbon may be fed directly, encapsulated in gelatin capsules and force fed (Solbakken and Palmork, 1980; R.E. Thomas and Rice, 1981), or fed by gastric lavage (Nava and Engelhardt, 1980). A useful method is to expose living, natural prey organisms to the hydrocarbons and allow the fish to feed on the prey, but the hydrocarbon concentration in the food organisms is difficult to control. Exposure of fish to hydrocarbons via water can be done by placing the fish in exposure solutions amended with radiolabeled or nonlabeled hydrocarbons or oil WSF for a given time interval, after which the fish are transferred to hydrocarbon-free water for depuration. The exposure medium may also be sediment mixed with hydrocarbon and placed in the tank with the fish.

Internal hydrocarbon concentrations can be monitored by collecting subsamples of groups of exposed fish at predetermined intervals. The fish are sacrificed and samples of selected tissues and fluids, viz., liver, gall bladder, muscles, and blood, are analyzed. Although extraction procedures vary, they permit separation of metabolites from parent compound(s). For example, Nava and Engelhardt (1980) treated homogenates with methanol-benzene (1:1) for 24 hours on a rotator at 4°C, followed by centrifugation. The organic and aqueous phases were tested separately for radioactivity to measure crude-oil-derived fractions and total metabolites, respectively. Another useful method involves digestion of half the sample in papain, to separate polar metabolites from nonpolar parent compounds, followed by extraction of carbon-14, employing 90% formic acid overlaid with 5-10 mL hexane (Roubal et al., 1977a,b). Total radioactivity can then be determined for the other half of the sample. Detection techniques include liquid scintillation counting, gas-liquid chromatography, and thin layer chromatography.

Metabolism in fish occurs mainly through the inducible, mixed function oxidase (MFO) system to form oxidized metabolites. These more polar derivatives are discharged by diffusion across membranes or conjugated with serum components and excreted (Burns, 1976). Induction of MFO enzymes, such as aryl hydrocarbon hydroxylase (AHH) can serve as sensitive physiological indicators of oil contamination (J.F. Payne and Penrose, 1975).

MFO activity is measured in tissue homogenates by quantifying the conversion of a substrate to a metabolite, with different substrates requiring different methods for analysis. Individual methods are not covered in detail, but the following are some of the useful methods: cytochrome P450 (Omura and Sato, 1964), cytochrome b5 (Ernster et al., 1962), NADPH cytochrome c reductase (Ernster et al., 1962; Masters et al., 1967), NADPH dichloropenolindophenol (Ernster et al., 1962), reductase (Masters et al., 1967), NADH cytochrome c reductase (Masters et al., 1967), aryl hydrocarbon hydroxylases (Dehnen et al., 1973; Depierre et al., 1975), aryl-4 monooxygenase (Lowry et al., 1951), and nitro reductase (Lowry et al., 1951).

Determination of the pathway and excretion routes of petroleum hydrocarbons in fish has been accomplished (R.E. Thomas and Rice, 1981) by employing a split box arrangement, whereby a fish is fitted with a rubber dam attached posterior to the gills so that water could be sampled from sealed anterior (gills) and posterior (feces, urine) chambers.

Many different kinds of sublethal measurements can be made in fish exposed to oil, similar to those used with invertebrates. In general, fish are more easily excited than most invertebrates, and measurements of respiration, heart rate, blood cortisol levels, etc., must be done with this in mind. A group of fish, once sampled, may not be ready for another sampling for 24 hours or more.

In general, sublethal measurements on fish have been done to measure stress and the energy utilized after exposure to hydrocarbons. For example, heart rates have been measured by implanting electrodes and recording EKG (Wang and Nicol, 1977). Opercular rhythm and coughing have been monitored by measuring pressure changes in the buccal cavity (Barnett and Toews, 1978) or by external electrodes (Spoor et al., 1971; P. Thomas and Rice, 1979). Energy reserves can be measured by determining rates of lipogenesis (Stegeman and Sabo, 1976) or total lipid content. Growth rates are measured by recording weight and length (Moles et al., 1981). Respiration can also be monitored by analysis of oxygen consumption rates, taking care to maintain proper PO_2 levels in the water (Brocksen and Bailey, 1973). The effect of hydrocarbon on the stress response hormone, cortisol, also has been measured (P. Thomas et al., 1980) as well as the effects of petroleum on ability of fish to handle stress, using stamina tunnels that assess the ability of fish to swim under measured water velocities (Beamish, 1978).

Growth of fish is relatively easy to monitor, but fish require long exposure times before significant differences can be detected, compared with controls. In contrast, energy budgets (scope for growth) provide a sensitive method for early detection of sublethal responses in fish, but are labor intensive. Energy input and output are determined by measuring caloric intake, respiration, and excretion. Scope for growth (P) is the difference between caloric intake and energy expended for respiration (Widdows, 1978a,b):

$$P = C \times [(F-E)/(1-E)F] - R$$

C (cal/day) = ([mg] food consumed/day) x calories/mg
F   = (ash-free dry weight/dry weight) of ingested food
E   = (ash-free dry weight/dry weight) of excrement
R (cal/day) = (mL O_2 consumed/day) x 4.86

Effects on reproduction are measured by observing gamete condition (Whipple et al., 1981) and development of sexual dimorphism (Hedtke and Puglishi, 1980), and by quantifying egg fertility, egg survival, timing of hatch, and hatching success.

Observations of behavioral responses yield information that may help link laboratory research with response and impact in the natural environment. For example, observations can be made on schooling ability in large circulating tanks fitted with a rotating gantry of fish (Partridge, 1982). Food selection tests, allowing fish to choose between contaminated and uncontaminated food, will indicate preference for or against (or neither) contaminated food (Blackman, 1974). Avoidance tests also can be done to determine whether fish prefer, avoid, or neither prefer nor avoid water containing hydrocarbons (Rice, 1973; Weber et al., 1981). The behavioral response of larvae is usually monitored by simple visual observation of given attributes.

Seabirds

Experimental Methods

Much of the experimental methodology for assessing oil effects on seabirds is derived from standard freshwater or terrestrial bird studies, although some specialized techniques have been developed (Peakall et al., 1979). For example, techniques used for studying blood chemistry, growth, nutrition, respiration, and so forth are routinely applied to seabirds with equal success. However, there are difficulties in maintaining wild seabirds in captivity, especially adult birds, that have led researchers to use more tolerant species (e.g., mallard ducks). Depending on the questions studied, the use of such substitute species can provide answers relevant to oiling problems in seabirds.

Marine Birds   The most meaningful information, especially on growth, nutrition, osmoregulation, and blood chemistry, has come from work done with seabirds. Here the approach has been to work either with captive seabirds or in the wild with nesting colonies. Not all seabirds are maintained that easily in captivity. Adult seagulls especially are very difficult to keep in good health. Nestlings, on the other hand, including nestling seagulls, are much easier to handle, provided a good supply of food (fish) is available. Seabirds, however, can be unexpectedly choosy in their food. Young puffins have been raised successfully in captivity by housing them in individual burrows, although their maintenance is very time consuming (Leighton et al., 1983).

Work with wild colonies has largely involved either eggs and nestlings, or the young of burrow-nesting seabirds. Manipulating either eggs or young on the nest, in the wild, does not appear to cause any problems. While there may be some desertion of eggs earlier in the breeding season, the experience is that the adults are less likely to desert the nest after hatching. Work with fledglings depends on the species. Seagull chicks can be experimentally handled in the nestling stage (e.g., R.G. Butler and Lukasiewicz, 1979), but once they become mobile and begin to explore their surroundings outside the nest, they become difficult to handle. In this respect, burrow-nesting species such as puffins are much more tractable (e.g., Peakall et al., 1980a). Of the burrow-nesting species, the most useful have been the auks (puffins and black guillemots) and Leach's petrel (Peakall, 1980b; Ainley et al., 1981; C.H. Walker and Knight, 1981). Thus, in one study (Peakall et al., 1980b), nestling herring gulls (Larus argentatus) and black guillemots (Cepphus grylle) and adult Leach's petrels (Oceanodroma leucorhoa) were collected and studied on islands off the coast of Maine. Both the petrels and the guillemots were sampled directly in their burrows or taken from their burrows and brought into the laboratory. On the other hand, the gull nestlings had to be banded at the beginning of the experiments and spotted and captured at each sampling time.

One major drawback with wild birds, and a criticism of some experimental studies, is the absence of nutritional history. While some seabirds show marked preferences for certain foods, many feed on a range of prey. Any variability due to such nutritional differences can be exacerbated in colonies near human settlements and industrial centers where the birds may, in addition, be exposed to a range of man-introduced chemicals (e.g., Knight and Walker, 1982a,b). These problems can be overcome, in part, by using suitable nestlings or fledglings raised in captivity, where nutrition can be controlled.

Freshwater and Terrestrial Birds   Frequently circumstances are such that wild seabirds are not readily available, and one has to turn to other species. Bird eggs, in terms of their susceptibility to oiling, are not highly species specific, and chicken-eggs, for example, could be used to study effects of oiling on hatching processes (e.g., Albers and Gay, 1982; Macko and King, 1980; Wootton et al., 1979).

However, when seabirds are not available, relevant results are likely to be obtained using either similar species or seawater-adaptable birds. In this respect, valuable results have been obtained using mallards (e.g., Holmes et al., 1978; Gorsline et al., 1981). Although a mallard duck may not be as good an osmoregulator as a seabird, its capability to induce the nasal gland and the osmoregulatory system makes this species a useful and readily available test animal for oiling studies of seabirds.

<u>Geographic Variations</u> With increasing oil and gas exploratory activity in the polar regions, attention will in the future be directed to studying effects of oil on seabirds in cold climates. Fortunately, the common seabird species have a wide range in their distribution and also are found in more temperate areas. Thus, oil effect studies done on temperate species are probably applicable to northern forms, and data can be extrapolated with some degree of confidence and relevance. It is important to consider the added factor of temperature or cold stress in oiling studies of species of polar regions (e.g., Erasmus et al., 1981).

Census Studies

Probably the greatest hindrance to assessing the effects of oil on seabirds at the population level is the scarcity of information on population sizes. The most accurate census can be taken at seabird colonies during the breeding season. In both the eastern United States and Britain, these counts date back to ca. 1900, yielding an 80-year census history. However, for the rest of North America and elsewhere, census records date back only to ca. 1970 (e.g., R.G.B. Brown et al., 1975). Therefore, in most cases it is difficult to assess any apparent shift in colony sizes within the context of long term population trends. Even when there is a long history of counts, lack of standardized of methods often makes it virtually impossible to detect all but the largest changes.

Fortunately, refinement in census techniques (e.g., Nettleship, 1976) in North America and Europe within the last decade will allow for statistically replicable census taking of colonies based on a system of standardized plots. Adequate census taking at colonies of burrow-nesting species such as storm petrels and many of the auks, especially when the species are also nocturnal, will remain extremely difficult.

Such population estimates cannot take account of the large numbers of adolescent birds that may spend several years away from their colonies before returning to breed. Counting of birds at sea, from both ships and aircraft, has been developed and refined within the last years, and the approaches allow better statistical treatment of the relative abundance of seabirds in different zones of the open ocean, e.g., relative abundances over cold water continental shelves versus warm, offshore tropical seas. It is difficult to convert these figures into absolute estimates of the numbers of birds at sea and relate back to absolute numbers in the colonies.

While ship and, especially, aircraft surveys allow more extensive census taking, they also have certain limitations. Shipboard surveys can cover only relatively small areas and are difficult to repeat with regularity. Ships necessarily avoid hazardous zones of shallow water, close inshore, where seabirds and sea ducks are often abundant. Aerial surveys, on the other hand, can cover all marine zones along sampling tracks with repeated regularity. Several surveys have produced absolute estimates of bird numbers at sea. Unfortunately, absolute identification of birds from the air is not always possible. For example, many of the species of alcids, seabirds especially vulnerable to oil pollution, cannot be identified while they are in the air, a serious limitation to aerial surveys off both the West and East coasts of North America (e.g., R.G.B. Brown, 1980, p. 13). In addition, the problem of ground-truthing aerial surveys, i.e., both census number and bird identification, has not yet been solved satisfactorily for most species. Finally, aerial surveys are restricted by range and safety requirements of the aircraft, as well as by expense.

Oil-induced mortality of seabirds is difficult to establish with certainty, (R.G.B. Brown, 1982). For several reasons, all mortality figures are gross underestimates because they are based mainly on the number of birds that drift ashore. Most oiled bird counts are based on shoreline surveys, including a body count per kilometer and extrapolation to the length of oiled beach. However, many slicks never reach land and birds encountering such offshore slicks are never enumerated. Similarly, many oiled birds sink and drown before reaching coastlines. One estimate (Hope-Jones et al., 1970) suggests that only 20% of oiled birds reach land; the percentage is thought to be smaller for west Atlantic waters, where prevailing winds tend to carry corpses of the birds out into the Atlantic (R.G.B. Brown, 1982).

Marine Mammals

Aside from oil itself, other activities and by-products of oil exploration and exploitation (e.g., noise, debris, shipping movements) can affect marine mammals. Because of their movements on and through the sea surface, marine mammals are highly vulnerable to contact with oil. Notwithstanding these concerns, probably less is known of how oil affects marine mammals than any other group of marine organisms. Only recently has work begun to address this issue experimentally, i.e., with suitable controls, instead of relying on field observations and news accounts, which may be conflicting and, in some cases, imprecise. There are several reasons for this gap in understanding, mainly that the behavior, physiology, etc., of marine mammals are not well understood even without the added dimension of effects of oil. Also, many of these animals are too large, or otherwise unsuited, for captive or field studies. There are marked differences among the 9 groups of marine mammals, which comprise a total of more than 130 different species. Thus, it is difficult to select representative marine mammals or to measure effects of oil, noise, and other factors on each species. Because research on marine mammals attracts public interest, experimental studies are vulnerable to public criticism. Nonetheless, much can be learned of how oil affects marine mammals if intelligent choice of study animals and appropriate clinical methods is made.

Toxicological Studies

Conventional toxicology studies, e.g., establishing acute lethal toxicities, on marine mammals are not popular and receive wide public attention and criticism. Also, such tests provide little information on how oil interacts with marine mammals because such tests are rarely designed to provide the information that is needed (e.g., Sprague, 1971) and only a small number of animals usually are available. Abundant information gathered for other species is available and can be used to derive conclusions concerning the toxicity of oil for marine mammals (Geraci and St. Aubin, 1982).

Standard clinical methods, coupled with chemical analytical techniques, can be used with good success to examine uptake and metabolism of petroleum hydrocarbons in marine mammals, with relatively little harm to the animals. For example, controlled oil uptake studies have now been done on captive marine mammals, using small doses of radio-labeled hydrocarbons subsequently detected in biopsied tissue, blood, urine, and feces (Engelhardt et al., 1977). Similarly, the physio-logical impact of oiling on these animals can be monitored and studies done using standard clinical methods (Geraci and Smith, 1976; Engelhardt, 1982).

Surface Oiling

Contact of oil with the skin or hide of marine mammals can cause problems for the animals. Traditionally, direct immersion procedures have been employed. There are several disadvantages with this approach. Rarely are enough animals available to yield statistically significant results. Only one petroleum product can be tested at a given time and exposure time on the skin cannot be easily controlled. Again, such methods are subject to intense public criticism.

An alternative approach that is promising is analogous to skin allergy tests for humans. The method is relatively harmless to the test animal, permits controlled testing of a variety of compounds, and yields far more information (Geraci and St. Aubin, 1982). It has been used repeatedly on captive and live-stranded cetaceans and provides information on biochemical and ultrastructural changes in skin cells. Such an approach is particularly appropriate for skin-sensitivity studies of large mammals such as the mysticetes.

Oil Avoidance

Information on the ability of cetaceans to detect and avoid surface oil can be obtained from direct observations in the field or using trained captive mammals. Both have provided some data, but much more information is needed, since some of the observations reported to date are contradictory, pointing out the difficulties in extrapolating from experimental "in captivity" studies to responses in the field. Certain cetaceans, such as the bottlenose dolphin Tursiops truncatus, can be readily trained to detect and respond to surface oil slicks as limited as 1 mm thick, apparently by echolocation (T.G. Smith et al., 1983, Geraci et al., 1983). This approach, apparently harmless to the test animals, provides useful information, specifically on oil avoidance and in general on echolocation. Possibly this approach could be extended to other captive species to determine whether findings for bottlenose dolphins can be extrapolated to other odontocetes.

Greater difficulty is met in (1) obtaining similar observations from wild marine mammals in the field and (2) extrapolating from the experimental data to oiling incidents in the wild. Observations have been made on wild gray whales swimming through natural oil seep areas off the coast of California. The advantage to this approach is that observation of several parameters related to behavior under natural oiling conditions can be made, including aerial surveys. However, under such conditions it is difficult to obtain a large number of observations on any single test animal. Also, one has little control over the amount or chemical composition of oil to which the animal is exposed.

Interference With Feeding

Little information is available on interference with or blocking of feeding in oiled marine mammals. Potential fouling of the feeding mechanism in baleen whales has gained attention. However, direct oiling of the baleen apparatus in live animals, whether in the wild or in captivity, is subject to public scrutiny. The problem can be attacked by seeking evidence of oil contamination of the baleen apparatus in stranded or commercially hunted specimens and by carrying out studies of isolated baleen in specially designed water flumes (Geraci and St. Aubin, 1982). The latter offers significantly greater control over experimental conditions and, with appropriate modifications, should be useful for studies of subtle effects of oiling on the baleen structure.

Spills of Opportunity

In the final analysis, valuable information can be gathered from systematic studies carried out during accidental oil spills at sea, both by direct observation of mammal behavior in oiled waters and by gathering autoptic data on stranded oiled animals (e.g., J. Parsons et al., 1980).

Only limited information is available concerning the behavior of marine mammals found in the vicinity of an oil slick, whether at sea or stranded on shorelines. Stranded animals provide useful information on effects of the physiology and, therefore, samples should be collected and analyzed to confirm results of studies done antemortem. In the past, potentially useful information was lost or missed because systematic reporting and collection of samples of stranded marine mammals during spills was not done. To correct this, and perhaps reduce the need to work with live animals in captivity, federal and regional officials should collect and freeze, where possible, whole carcasses, or at least be familiar with methods for collection of skin, blubber, blood, visceral organ, and stomach content samples.

CYTOGENIC AND MUTAGENIC METHODS

A wide range of cytogenic and mutagenic assays are now available, with both new developments as well as improvements on existing techniques being reported in the scientific literature (Stich and San, 1981). Most have been developed for bacterial or mammalian cell lines. Very few are available for use with marine tissues or cell lines.

The methods are essentially designed to measure or score chromosomal abnormalities, i.e., cytogenic or cytologic aberrations, and mutational changes, i.e., inheritable changes in nucleic materials in daughter cells.

Chromosomal Aberrations

Cytogenetic methods have not been used extensively for marine systems. Thus, it is not surprising that chromosomes have not been examined to any extent, with respect to petroleum pollution.

Chromosomal aberrations can be structural (breaks or gaps) or can involve changes, either in the number of chromosomes or chromosome sets, resulting from faulty separation or division.

Consequences of such aberrations are generally deleterious, not beneficial or neutral. Abnormal chromosomes are especially prone to further misdivision and breakage, resulting in additional deficiencies and duplications. Chromosome aberrations in somatic cells are involved in some stage of tumorigenesis. When the aberration occurs in the germ line, it is transmittable to the next generation, provided it is compatible with reasonably normal embryo development. Unbalanced genetic complements (duplication for one segment or deficiency for another) usually lead to death early in embryonic development. Certain reciprocal translocations can result in sterility in both animals and plants (Russel and Matter, 1980; Sparrow and Woodwell, 1963).

The methods described with respect to marine species are treated in detail in the literature (Hollaender, 1971, 1973, 1976; Hollaender and deSerres, 1976; Kilbey et al., 1977).

Direct Examination of the Chromosome Karyotype

Direct examination of the chromosome karyotype is practical only when the organism has a suitable complement of chromosomes, that is, of a given size range and number. The direct chromosome method can be applied only to certain tissues and cell types that lend themselves well to the spreading of individual chromosomes and possess a good mitotic index. Organisms with such karyotypes can provide detailed, qualitative information on chromosome mutation. Even when species with less than ideal karyotypes have been studied intensely, careful study of the chromosomes has been productive, as for example in the mouse and human systems.

Marine Invertebrates Little cytogenetic work has been done on marine invertebrates. Nonvertebrates present special problems as to method and source of tissue to be examined. Marine invertebrates and temperate and cold water fishes have significantly lower rates of mitotic turnover, in general, than mammals and certain terrestrial plant tissues, and their gametogenesis is seasonal. Since chromosomes are directly studied cytogenetically only in selected stages of mitosis or meiosis, difficulties arise in obtaining sufficient numbers of cells for chromosome analysis on a year-round basis. Externally fertilized, freely developing zygotes, however, offer a good source of material.

Among the invertebrates, the Ostreidae, species of which are commercially valuable, possess 10 pairs of chromosomes. The chromosomes of several oyster species have been well described, and a full karyotype analysis has been published for the American oyster Crassostrea virginica (Longwell et al., 1967). The chromosomes and freely spawned, externally fertilized eggs of shellfish make them ideal subjects for studies of oocyte chromosomes in meiosis, chromosomes of the male in fertilization, and early developing embryos (Longwell and Stiles, 1968). Hatchery cultivation of the oyster in aquaculture makes this shellfish attractive for such work, especially since oysters play a significant role in nongenetic assays of water quality.

Recently, the marine polychaete worm Neanthes arenaceodentata was reported to have an ideal karyotype for direct chromosome study. Also, Neanthes has been used to examine sister-chromatid exchange (Pesch and Pesch, 1980).

Plants A significant amount of cytologic information is available from the examination of terrestrial plants, notably meiosis in the development of the male gamete (microsporogenesis), along with root tip meristem cells.

Fish In general, fish possess large numbers of small chromosomes. Although fish comprise the largest of the vertebrate groups, the chromosomes of only a few species have been counted or described, and even less are marine. Complete chromosome karyotypes have been characterized for only about 2-3% of the 20,000-23,000 living species of fish (Chen, 1969; Gold, 1979). In contrast, 30% of the living species of eutherian mammals has been studied, some extensively. Recently there is renewed interest in the chromosomes of fish with respect to studies of their evolution, particularly in the Salmonidae and also with respect to genetic engineering applications for the breeding of aquaculture strains (Gold, 1979; Ojima, 1980).

Basic cytogenetics and gene mutations of certain groups of freshwater fish have been studied (Schröder, 1973; Ojima, 1980), and a few freshwater fish have been shown to possess ideal karyotypes, desirable for direct studies of chromosome mutation, viz., the mud minnows Umbra limi and Umbra pygmaea (Kligerman et al., 1975; Kligerman and Bloom, 1977; Hooftman and Vink, 1981).

Prime tissue sources of fish cells for chromosome analysis are epithelial cells of gills, fin, scales, and cornea and hematopoietic tissue of the anterior kidney, testis, intestine, and the early embryo (Denton, 1973; Wolf and Quimby, 1969). Lymphocytes of circulating blood in fish do not, in general, respond well to mitotic stimulation in vitro, as do those of human lymphocytes. However, there have been some reports of success using this method. It has been confirmed by autoradiographic observation that the intestinal epithelium of fish is a typical cell renewal system (Hyodo-Taguchi, 1968; T.S. Johnson et al., 1970). The yolk-sac membrane of pelagic fish eggs collected from sea surface waters in plankton provides good material, in the early egg development stages, for direct chromosome study (Longwell and Hughes, 1980). Larval tissues of fish may prove more useful than those of adults because of a higher mitotic turnover.

Marine Mammals Direct examination for chromosomal aberrations in mammals is usually done using tissue culture cells, many lines of which are of embryo origin, and hematopoietic cells of bone marrow, or lymphocytes of circulating blood stimulated to undergo a simple mitosis in vitro. However, restrictions and difficulties in sampling marine mammals create special problems for cytologic studies, somewhat analogous to the human, for which special methods to circumvent these difficulties had to be developed. Future studies could well be done on tissue-cultured cells employing biopsies, aborted fetuses, and maternal or fetal membranes and monitoring by culturing samples of circulating blood as is done in human studies.

Birds Methods for chromosome analysis are available for avian embryos and can be applied to marine birds. Chromosome analysis of chicken embryos has been proposed as a method for monitoring of chemical-induced mutagenicity (Bloom, 1978). The major difficulty faced in studying marine birds is collection of samples in sufficient number to obtain statistically significant results. Work with affected bird populations after major oil spills would provide some opportunities for this cytogenetic work if its potential significance were recognized.

Micronucleus Test and Related Mitotic Assays

This test has found extensive use in mammalian and plant studies. With proper modifications, it appears to be a promising approach for marine tissues (Schmid, 1976, 1977). In mammals the test is usually applied to the polychromatic erythrocyte of the bone marrow. Several laboratories have shown the micronucleus test comparable in sensitivity to direct chromosome examination, as well as offering substantial savings in time and cost (Kliesch and Adler, 1980; Jenssen and Ramel, 1976, 1980).

The test was developed specifically for chemical mutagenicity (Countryman and Heddle, 1976) and has since been used effectively to measure cytogenetic aberrations in plants (Ma, 1979), rat spermatids (Lahdetie and Parvinen, 1981), mammalian liver cells, prenatal mammals (Stoyel and Clark, 1980; Cole et al., 1981), and mammalian meiosis (Lahdetie and Parvinen, 1981). This assay, or modifications thereof, may be adapted to studies of adult, larval, and fish embryo in bioassays of petroleum and dispersants, in monitoring and in field studies of the effect of oil spills, to certain invertebrate tissues of sufficiently high mitotic turnover, or to marine mammals.

For several years, Soviet researchers have measured the effects of incorporated radionuclides on fish embryo cells by scoring micronuclei of cells in mitosis, telophase bridges of translocated chromosomes, and laggard or fragmented chromosomes, as well as micronuclei in nondividing or interphase cells (AEC, 1968, 1972). This approach differs from the micronucleus test applied to mammalian bone marrow only in that it is used on dividing as well as nondividing cells, and scores also mitotic events leading to micronucleus formation as well as micronuclei. Using such a suite of cytological criteria, Longwell and Hughes (1980) measured mitotic-chromosome abnormalities in developing eggs of the Atlantic mackerel collected by plankton in the New York Bight, employing both the embryo itself and the yolk-sac membrane of the egg.

Hagström and Lönning (1967) used mitotic disturbances, among other factors, in a cytological study of the effect of lithium on the development of the sea urchin embryo. Sea urchin eggs have been proposed as general test objects in pollution bioassays, oil, and related studies (Kobayashi, 1974; Lönning, 1977).

Dominant Lethal Gene Mutations

This test is a measure of heritable chromosomal abnormalities in the male germ line which kills the embryo early in development. The basis for the test is the incompatibility of gross chromosome changes with successful development. Recent cytogenetic studies in mice and hamsters have quantitatively related induced dominant lethal mutation frequencies to the incidence of broken chromosomes at early cleavage metaphases of the embryo (see Röhrborn and Hansmann, 1977). It is likely that the absence of one or more chromosomes of the several chromosome pairs is a prime cause of genetic death of many early embryos (Russell and Matter, 1980). For all its shortcomings, especially when the female is treated, the dominant lethal test has been widely applied in the past as the only practical mammalian in vivo test for heritable mutation in the germ line.

The production of dominant lethal mutations has been studied in Tilapia (Hemsworth and Wardhaugh, 1978; Wardhaugh, 1981), the guppy (Woodhead, 1977), trout (Newcombe and McGregor, 1967), and medaka (Egami and Hyodo-Taguchi, 1973). Embryo deformity was usually manifested initially during gastrulation. The large majority of morphologically defective embryos were nonviable. Further information on the applicability of the test has come from studies on fish sperm treated with known mutagens, e.g., carp, rainbow trout, peled (Tsoi, 1969; Tsoi et al., 1974). Fertilizability of the sperm was not affected; however, effects were expressed at subsequent developmental stages.

The dominant lethal gene mutation assay, notably cytogenetic modifications of it, appear to be applicable to marine species that are culturable through at least gastrulation, particularly for the oyster where it can be combined readily with cytogenetic examination of early cleavage chromosomes.

Sister-Chromatid Exchange

Staining methods yielding differential staining of the two chromatids of the chromosome allow detection and scoring of sister-chromatid exchange events in a modification of the chromosome spread procedure used to detect chromosome aberrations. This type of test procedure has some of the same limitations as the direct chromosome examination procedure, but fewer metaphase need be examined.

It is attractive, however, as comparative studies of chromosome aberrations and sister-chromatid exchange often show the latter to occur in the absence of chromosome breakage. In cultured Chinese hamster cells, sister-chromatid exchanges were observed at considerably lower concentrations of mutagens than those needed to induce aberrations (Natarajan et al., 1976). Sister-chromatid exchange has proved very sensitive for the detection of many mutagenic chemicals, both in vivo and in vitro, including those, such as the polycyclic aromatic hydrocarbons, that require metabolic activation to be mutagenic.

In vitro (tissue culture) applications of this method are much more easily conducted than in vivo, even though the assay has been successfully employed in vivo using some plant species (Kihlman and Kronberg, 1975), the chick embryo (Bloom and Hsu, 1975), and two species of mammals (for example, Stetka and Wolff, 1976).

Barker and Rackham (1979) successfully applied sister-chromatid exchange to cultured cells of the freshwater fish Ameca splendens. When the response to mutagens was compared to that of cultured mammalian cells, fish cells seemed less sensitive. This method has also notably been applied to the freshwater mud minnow (Kligerman and Bloom, 1977). Valentine and Bishop (1980) found intraperitoneal injection of two known direct mutagens to induce dose-dependent increases in sister-chromatid exchange in intestinal metaphase chromosomes of the same fish.

Sister-chromatid exchange was also tested on 8- to 10-day-old embryos of the marine polychaete worm Neanthes. Effects of a known mutagen, Mitomycin C, exhibited the expected dose response at the level of sensitivity exhibited by mammalian systems (Pesch and Pesch, 1980).

Sperm Mutation Test

Certain mutants in mice and particular mouse strains are regularly characterized by abnormal shapes of the mature sperm heads and tails. It has recently been determined that irregularities in shape of sperm heads are produced by known mutagens, but not by nonmutagens. As a

result, sperm abnormalities have been proposed as an assay for mutagenesis. Such a test would have the advantage of providing direct
evidence for gene toxic activity within the germ cells in vivo (Sobels,
1977).

The sperm-shape abnormality assay, which appears to detect gene,
not chromosome, mutation as in the dominant lethal gene mutation test,
has been used to detect mutagens, carcinogens, and teratogens (Heddle
and Bruce, 1977; Topham, 1980a,b; Wyrobek and Bruce, 1978). Recently,
Bruce and Heddle (1979) reported an excellent overlap in the mutagen
detection capacity between the Salmonella bacterial assay and the
abnormal sperm assay. Polycyclic hydrocarbons are positive. Effects
on the progeny of mice with induced mutations affecting sperm head
shape have been assessed by measuring the incidence of abnormal sperm
in offspring (Topham, 1980a).

The sperm of several fish has been described (Ginsberg, 1972) and
these compare favorably to mammalian sperm, with respect to usefulness
in the sperm mutation test. Oyster sperm are somewhat smaller but can
be scored for abnormalities rapidly using the light microscope.

Hollstein et al., (1979) point out that this test could be of value
in monitoring genetic effects of environmental and industrial chemicals,
along with cytogenetic examinations of oocytes and spermatocytes. The
sperm mutation test should be applicable to field, laboratory, and
monitoring studies of marine fish and mollusks, provided mutation
therein also affects abnormalities in sperm morphology.

In Vitro Cell Transformation

Assays employing cell transformation (Hollstein et al., 1979) are
carcinogenicity tests rather than tests of chromosome mutagenicity.
Also, they require considerable technical skill, experienced judgment,
and time and demonstrate a strong correlation between carcinogenicity
and mutagenicity. Petroleum-induced abnormal tissue growth in crabs
(Albeaux-Fernet and Laur, 1970), bryozoans (Powell et al., 1970), and
fish (Kezic et al., 1980) has been reported.

Typically, the test concerns changes in morphological nature and
division rate of cultured cells after incubation with the suspect
chemical. Colonies of transformed cells are sometimes demonstrated to
be malignant when injected into host animals, unlike their nontrans-
formed, precursor cell types. For example, polycyclic hydrocarbons
were reported as having both cytotoxic and transforming effects on
Syrian hamster fetal lung cultures (Richter-Reichhelm et al., 1979).
Similar studies might be conducted on cultured fish cells. In a
modification of the in vitro transformation assay, mammalian embryos
were exposed to chemicals in vivo before culturing (DiPaolo et al.,
1973). An analogous test can be devised for intact, freely spawned
early embryos of marine species, the larval development of many of
which is adversely affected by petroleum hydrocarbons (National
Research Council, 1975).

Mutagenesis Assays

Methods for measuring mutagenic potential in the strictest sense, i.e., those that score changes in phenotype, are based largely on the use of specialized bacterial strains, certain yeasts, and mammalian cell lines. Among the bacteria most commonly used are Salmonella typhimurium (Ames et al., 1975) and Escherichia coli. The latter, E. coli WP2, was used successfuly to detect mutagens in mussels from the polluted Lagoon of Venice (Parry and Al-Mossawi, 1979). Best known, however, is the "Ames test," based on a number of specialized S. typhimurium mutant test strains that revert readily, via point mutations, to wild type upon exposure to mutagens.

The Ames Assay

The Ames assay is rapid, easily applied to a wide range of potential mutagens, and extremely reproducible. It consists very simply of mixing one of the several tester strains that are available with soft agar and an amount of the test substance, prior to plating the mixture onto a histidine-deficient (minimal medium) plate. The soft agar contains sufficient histidine to allow approximately two divisions of the S. typhimurium tester strain. After the histidine has been utilized, the culture ceases growth, except for spontaneous mutants which continue multiplication (ca. 50/109 cells). However, if the culture is exposed to a mutagen (e.g., UV or benzopyrene), the number of mutants increases, increasing the colony count, ideally in a direct dose-response proportion.

In many cases when the response to a suspect mutagen is negative, a positive response can be elicited by the inclusion of a microsomal extract (S9) of rat liver in the bacteria/soft agar mixture. The explanation is that some compounds are not mutagenic themselves, e.g., benzo(a) pyrene, but mutagenicity resides in their epoxide derivative. Usually such "activation" is done using the S9 fraction of induced liver homogenates obtained from animals exposed to a known mutagen. A positive response can also, in some instances, be enhanced by pre-incubating the Ames tester strain with S9 soft agar mixture before overlaying onto the minimal medium plate. For applications and limitation of the Ames assay, the literature should be consulted, viz., Nagao et al. (1977), Muller et al., (1980), and Klekowski and Barnes (1979). It should be pointed out that, apparently, some complex environmental mixtures contain antimutagenic components, as in the case of acetone extracts of lyophilized Mytilus edulis tissue from the Dutch coast, which were reportedly able to completely abolish the mutagenicity of benzo(a)pyrene in the Ames test (Mattern et al., 1981, cited by Odense, 1982).

Modifications on the Ames Assay

Recently Bjorseth et al. (1982) incorporated the Ames test into thin layer chromatography (TLC), by layering the mixture of soft agar, bacterial tester strain, and S9 directly onto a chromatographed TLC plate. Thus, mutagens or promutagens requiring activation and occurring in the different fractions (TLC spots) were identified directly.

A second modification involves use of a liquid incubation medium employing blue in plates with 50 or 96 wells, allows statistical evaluation of the mutagenic response by scoring the number of acidified wells (Hubbard et al., 1981). A third version, the host-mediated assay, involves injection of both the suspect mutagen and the Ames bacterial tester strain into a suitable host (mice, rats) (Frezza et al., 1982). The advantage of this modification is that the mutagen is "activated" naturally by the host S9 system instead of by commercially prepared S9 in a test tube mixture, a method found useful in detecting mutagens in mussels collected from a lagoon in Venice.

Other Systems

A limitation of the Ames test is that, while it appears generally to be good for detecting mutagenicity using bacteria, a possible result does not always correlate with mutagenicity in animals. Thus, eukaryotic cells, i.e., yeasts or mammalian cell lines, are more applicable for detecting mutations based on chromosome damage (e.g., Von Borstel, 1981; McCormick and Maher, 1981). Little work has been done with marine algae, but one approach potentially applicable to marine phytoplankton is that of Vandermeulen and Lee (1977), who employed Chlamydomonas reinhardtii to examine the potential mutagenicity of several oils.

Applicability to Petroleum Studies

One hindrance to the experimental testing of mutagenic effects of petroleum (and other substances) on marine plants and animals is the lack of genetically uniform (inbred) test organisms. The availability of highly inbred animals would permit testing a smaller number of animals, yet yield higher precision. Some strains of freshwater fish species are being bred for such purposes, but the marine research needs have not yet been met.

No single test of mutagenicity or cytogenic aberrations employing a single tissue or life stage can be expected to provide full information on chromosome mutation risk. Uptake, metabolism of the chemical, and transport will all affect the response measured using a particular tissue, cell type, or phase of the life cycle. Also, the sensitivity of chromosomes to mutation will vary according to stage of mitosis or meiosis. Unfortunately many marine species have less than ideal chromosome complements for direct, detailed study of individual chromosomes, i.e., karyotyping. However, the newer, indirect measures have the advantage over simpler tests on prokaryotes in that they measure directly the effects of a substance on chromosomes in the eukaryotic cell and organism of concern and are not merely predictors of mutagenic potential.

ECOSYSTEMS

An ecosystem study is that which assesses interacting habitats, involves determining primary and secondary productivity, standing crops of organisms in major compartments, key species within those compartments, and the cycling of carbon and nutrients through those compartments as a function of sunlight, temperature, current regimes, and biological and physical-chemical interactions over at least one annual cycle and, for some questions, over a time scale of decades. Assessment of the impact of oil on an ecosystem should include measurements relating changes in the biological components and functions of those components with changes in chemical composition of the oil and its metabolites and reaction products. All types of oil over several orders of magnitude of concentration will have impacts (Hyland and Schneider, 1976).

Field studies have the advantage of examining the natural ecosystem, even though there may be disadvantages, such as a lack of necessary background data and/or suitable controls, uncertainty as to what should be measured and over what period of time, as well as considerable expense. Mesocosm studies offer the advantages of controlled experiments with a greater representation of the ecosystem and are, in general, less expensive than field studies. They offer particular value for the study of chronic spills, rather than acute events.

From accumulated experience with oil spills (Gundlach and Hayes, 1978; Glemarec and Hussenot, 1981; Sanders et al., 1980), vulnerability indices have been developed. These indices have been applied to the Amoco Cadiz oil spill in Brittany, to the lower Cook Inlet, Alaska, and to the Peck Ship spill in Puerto Rico (Gundlach and Hayes, 1978; Michael et al., 1978; Davis et al., 1980). The approach comprises geomorphological aspects, species assemblages, and persistence of oil.

The residence time of oil clearly is an important factor in ecosystem analyses. The residence time of oil in the water column, in general, is in the order of hours to days, unless new discharges occur, whereas oil residence time is much more prolonged for the benthos. Recovery time may be in the order of hours to days for phytoplankton and zooplankton, depending on the extent and duration of the event. Fish eggs and larvae may be destroyed in the immediate vicinity of an oil spill during the short period of time that oil concentrations may be lethal. However, it should be pointed out that there is no known instance of an oil spill covering an entire spawning area for a particular fish species, although it is possible that, for localized coastal spawning populations, such as salmon and herring, this could be a net result.

The admonition regarding the changing composition of oil as it weathers and is degraded must be repeated (see Chapter 4). As the oil is altered, its toxicity will also change, and this is important in assessing ecosystem effects.

Even in areas of chronic hydrocarbon input, effects of hydrocarbons on the ecosystem will be difficult to differentiate from inputs of nutrients, metals, and other organic compounds. Habitats most impacted by an oil spill, assessed by the persistence of oil in that habitat and according to vulnerability indices (Gundlach and Hayes, 1978), are the ones that should be studied over the long term. Most often, these will comprise benthic studies of depositional areas on the continental shelf, in estuaries, or in salt marshes.

Mesocosms are useful principally for studying ecosystem inter-relationships of plankton and small benthic invertebrates of the sort that are normally collected in core samples. However, mesocosms are not able to provide complete simulation of the entire ecosystems, i.e., including fishes and large benthic epifauna. Yet they are especially effective for analyzing problems of chronic pollution, where plankton interrelationships and the effect of perturbations of the plankton on benthic fauna are important and need to be monitored. Mesocosms are less useful for studying oil spill situations because only in few cases have serious effects on plankton been demonstrated.

An important future use for mesocosms may well prove to be in the simulation of ecosystem effects in bays and estuaries subject to inputs of multiple pollutants such as petroleum, other organic contaminants, and trace metals. The potential synergistic effects these various substances may produce are poorly understood at the present time and merit attention. A variety of experimental procedures will be required, since the relationships are too complex to be dealt with definitively by field investigations alone.

Studies of the plankton and a portion of the benthos can be carried out in properly scaled mesocosms. Mesocosm experiments recently have been reviewed in a volume edited by Grice and Reeve (1982). At the Marine Ecosystems Research Laboratory (MERL), mesocosms have been scaled to mimic a temperate pelagic estuarine water column with a heterotrophic soft-bottom benthic community in order to conduct studies on the fate and effects of pollutants in estuaries (Pilson et al., 1979). In experiments with No. 2 fuel oil, unexpected changes in biological compartments in the water column and benthos were detected quantitatively (Elmgren et al., 1980a,b). It is possible these changes would have been only qualitatively perceived in field studies of oil spills (Elmgren and Frithsen, 1981). In addition, the biogeochemical behavior of individual hydrocarbon compounds, including their role in degradation and the products thereby produced, can be traced and budgeted through the ecosystem when mesocosms are employed (Hinga et al., 1980; Gearing and Gearing, 1982a,b; R.F. Lee et al., 1981b). Mesocosm experiments, however, are not a panacea. In fact, over time scales of a metazoan generation or longer, proper dynamic scaling may be impossible.

RECOMMENDATIONS

One of the difficulties in assessing the significance of many of the laboratory petroleum exposure studies reported during the past decade, and in comparing results obtained by different laboratories, has been the lack of standardization of methods. Each group of investigators has proceeded to develop methods to suit its own needs, abilities, and goals. Consequently there is no universally acceptable set of experi-mental parameters, such as test species, seawater medium, temperature, duration of exposure, contaminant concentrations, mode of contaminant introduction, means of determining rates of uptake or loss, and measurements of physiological change. Experimental parameters that can be used under given conditions have been described (Becker et al., 1973; G.V. Cox, 1974, 1980) and are useful to consider in connection with the design of the experiment. The following represent some specific recommendations arising out of our considerations.

1. Bacteria and fungi provide the major mechanisms for oil degradation. Experiments with higher plants and animals must take into account the fact that organisms under study interact with a constantly changing substrate, as a result of concomitant microbial activity. Substrates altered by microbial processes may be more toxic than the original oil.

2. The demonstration of oil pollution effects on planktonic communities is hampered by the constantly changing spatial and temporal distribution of plankton in the environment. The enormous resources required for adequate sampling and analysis of postspill planktonic communities usually render such large scale studies prohibitive. Resources should be focused on community species composition and biomass determinations.

3. Both enzyme analyses and studies of structural distortions of cellular components in response to hydrocarbon contamination are promising areas of research that, as yet, have received little attention. Furthermore, histopathological methods for detecting long term sublethal stress suffer from a lack of background descriptions of normal histological characteristics. A general consensus should be developed as to which histopathological changes best serve as reliable indicators of stress.

4. The use of both semicontinuous and continuous culture systems, necessary to provide populations of known nutritional and growth history but heretofore untried for studies of hydrocarbon effects, should be given greater attention. Fluorescence enhancement in phytoplankton by DCMU is a promising area of research but requires further investigation.

5. Benthic plants and animals are often the biota most in contact with deposited petroleum and consequently should be studied in relation to petroleum effects. As benthos occur in such a variety of habitats, each varying in susceptibility to oil contamination, studies on benthos should focus on areas where contamination may persist, such as depositional areas in coastal wetlands, estuaries and continental shelves. After an initial survey of community composition, resources should be focused on populations of specific taxa. Summary indices such as species abundance, diversity, taxa ratios, and total biomass (density) should not be used as sole measures of community health. The prevailing use of isolated species diversity indices should also be discouraged.

6. In soft-bottom assemblages subject to an acute oil spill, benthic crustaceans should be studied because of their known sensitivity to perturbations, including petroleum contamination, and their trophic importance. Chronic, low level effects in such habitats, especially on

growth and reproduction, should be studied with long lived, low mobility species when available. In hard substrates, primary space occupiers, key species, or structure-building (framework) species should be the major taxa studied.

7. Whenever possible, acute toxicity bioassays should utilize flow-through exposure systems of proven design. Weathered and photooxidized oils should be investigated. Toxicity of petroleum metabolites produced by marine bacteria and animals should be evaluated. Major emphasis in laboratory effects studies should be placed on life-cycle and chronic sublethal effects. Studies of effects of petroleum on reproduction, bioenergetics, and histopathology should receive greater attention.

8. Suites of physiological and biochemical diagnostic tests should be developed, validated, and adopted for field use for monitoring impacts of pollution and recovery in the marine environment.

Long term monitoring, while desirable before a spill, is not a necessary prerequisite for impact assessment of acute spills. Control (reference) sites are requisite for comparisons, even if baseline data exist.

Whole-ecosystem studies of the fates and effects of oil or its constituents are desired and can be effectively conducted in meso-cosms. Such studies may reveal unexpected effects on phytoplankton, zooplankton, and benthic fauna not observable under field conditions. Determinations of the rates of disappearance and biodegradation of oil compounds in the water column and in the sediments can also be made in mesocosm experiments. It is important also, however, to be aware of problems in dynamic physical scaling.

A pressing need exists for the development of suitable field experiments to confirm many of the results obtained in the laboratory. The goal is the capability to predict the impacts of chronic and acute petroleum contamination on the various levels of biological organization (cellular, organ, organism, population, community, ecosystem, and biome) in the marine environment.

REFERENCES

AEC. 1968. Problems of biological oceanography. AEC-TR-6940. National Technical Information Service, U.S. Department of Commerce, Springfield, VA 22151.

AEC. 1972. Marine radioecology. AEC-TR-7299. National Technical Information Service U.S. Department of Commerce, Springfield, VA 22151.

Ackman, R.G., and D. Noble. 1973. Steam distillation: a simple technique for recovery of petroleum hydrocarbons from tainted fish. J. Fish. Res. Board Can. 30:711-714.

Adlard, E.R., L.F. Creaser, and P.H.D. Matthews. 1972. Identification of hydrocarbon pollutants on sea and beaches by gas chromatography. Anal. Chem. 44:64-73.

Ainley, D.G., R. Grau, T.E. Roundybush, S.H. Morrell, and J.M. Utts. 1981. Petroleum ingestion reduces reproduction in Cassin's auklets. Mar. Pollut. Bull. 12:314-317.

Aizenshtat, A. 1973. Perylene and its geochemical significance. Geochim. Cosmochim. Acta 37:559-567.

Albaiges, J. 1980. Fingerprinting petroleum pollutants in the Mediterranean Sea, pp. 69-81. In J. Albarges, ed. Analytical Techniques in Environmental Chemistry. Pergamon, New York.

Albaiges, J., and P. Albrecht. 1979. Fingerprinting marine pollutant hydrocarbons by computerized gas chromatography-mass spectrometry. Intern. J. Environ. Anal. Chem. 6:171-190.

Albaiges, J., and J. Borbon. 1981. Gas chromatographic-mass spectrometric identification of geochemically significant isoalkane hydrocarbons. J. Chromatogr. 204:491-498.

Albaugh, E.W., and P.C. Talarico. 1972. Identification and characterization of petroleum and petroleum products by gel permeation chromatography with multiple detectors. J. Chromatogr. 74:233-253.

Albeaux-Fernet, M., and C.M. Laur. 1970. Influence de la pollution par le mazout sur les testicules de crabes (etude histologique). Natl. Acad. Sci. Paris 270:170.

Albers, P.H. and M.L. Gay. 1982. Unweathered and weathered aviation kerosene: Chemical characterization and effect on hatching success of duck eggs. Bull. Environ. Contam. Toxicol. 28:430-434.

American Petroleum Institute. 1958. Determination of volatile and non-volatile oily material. Infrared spectrometric method number 733-58. Washington, D.C.

American Petroleum Institute. 1977. Oil spill studies: strategies and techniques. API Publication No. 4286. Washington, D.C. August 1977.

American Public Health Association. 1977. Standard methods for the examination of water and wastewater. Washington, D.C. 874 pp.

American Society for Testing and Materials. 1980. Standard practice for conducting static acute toxicity tests with larvae of four species of bivalve molluscs. Designation: E724-80. Philadelphia, Pa. 27 pp.

American Society for Testing and Materials. ASTM Method D 893-80. Test for insolubles used in lubricating oils. 1980 Annual Book of ASTM Standards, Part 23.

American Society for Testing and Materials. ASTM Method D 2007-80. Groups in rubber extender and processing oils by the clay-gel adsorption chromatographic method. 1980 Annual Book of ASTM Methods, Part 24.

American Society for Testing and Materials. ASTM Method D 2549-76. Test for separation of representative aromatic and nonaromatic fractions of high boiling oils by elution chromatography. 1980 Annual Book of ASTM Methods, Part 24.

American Society for Testing and Materials. ASTM Method D 2778-70. Solvent extraction of organic matter from water. 1980 Annual Book of ASTM Standards, Part 31.

American Society for Testing and Materials. ASTM Method D 2786-71. Hydro-carbon types analysis of gas-oil saturate fractions by high ionizing voltage mass spectrometry. 1980 Annual Book of ASTM Standards, Part 24.

American Society for Testing and Materials. ASTM Method D 2908-74. Measuring volatile organic matter in water by aqueous-injection gas chromatography. 1980 Annual Book of ASTM Methods, Part 31.

American Society for Testing and Materials. ASTM Method D 3239-76. Aromatic types analysis of gas-oil aromatic fractions by high ionizing voltage mass spectrometry. 1980 Annual Book of ASTM Methods, Book 25.

American Society for Testing and Materials. ASTM Method D 3325-78. Preservation of waterborne oil samples. 1980 Annual Book of ASTM Standards, Book 31.

American Society for Testing and Materials. ASTM Method D 3326-78. Preparation of samples for identification of waterborne oils. 1980 Annual Book of ASTM Standards, Part 31.

American Society for Testing and Materials. ASTM Method D 3327-79. Analysis of selected elements in waterborne oils. 1980 Book of ASTM Methods, Book 31.

American Society for Testing and Materials. ASTM Method D 3328-78. Comparison of waterborne petroleum oils by gas chromatography. 1980 Annual Book of ASTM Standards, Part 31.

American Society for Testing and Materials. ASTM Method D 3414-79. Infrared analysis of waterborne oils. 1980 Annual Book of ASTM Standards, Part 31.

American Society for Testing and Materials. ASTM Method D 3415-79. Identification of waterborne oils. 1980 Annual Book of ASTM Standards, Part 31.

American Society for Testing and Materials. ASTM Method D 3650-78. Comparison of waterborne petroleum oils by fluorescence analysis. 1980 Annual Book of ASTM Standards, Part 31.

American Society for Testing and Materials. ASTM Method D 3694-78. Practices for preparation of sample containers and for preservation of organic constituents. 1980 Annual Book of ASTM Standards, Part 31.

Ames, B.N., J. McCann, E. Yamasaki. 1975. Methods for detecting carcinogens and mutagens with the Salmonella/mammalian microsome mutagenicity test. Mut. Res. 31:347-364.

Anderlini, V.C., L. Al-Hormi, B.W. deLappe, R.W. Risebrough, W. Walker II, B.R.T. Simoneit, and A.S. Newton. 1981. Distribution of hydrocarbons in the oyster Pinctada margaratifera, along the coast of Kuwait. Mar. Pollut. Bull. 12:57-62.

Anderson, J.W. 1977a. Effects of petroleum hydrocarbons on the growth of marine organisms. Rapp. P.-v. Reun. Cons. Int. Explor. Mer 171:157-165.

Anderson, J.W. 1977b. Responses to sublethal levels of petroleum hydrocarbons: are they sensitive indicators and do they correlate with tissue contamination?, pp. 95-114. In D.A. Wolfe, ed. Fate and Effects of Petroleum Hydrocarbons in Marine Organisms and Ecosystems. Pergamon, New York.

Anderson, J.W., J.M. Neff, B.A. Cox, H.E. Tatem, and G.M. Hightower. 1974. Characteristics of dispersions and water-soluble extracts of crude and refined oils and their toxicity to estuarine crustaceans and fish. Mar. Biol. 27:75-88.

Anderson, J.W., L.J. Moore, J.W. Blaylock, D.L. Woodruff, and S.L. Kiesser. 1977. Bioavailability of sediment-sorbed naphthalenes to the sipunculid worm, Phascolosoma agassizii, pp. 276-285. In D.A. Wolfe, ed. Fate and Effects of Petroleum Hydrocarbons in Marine Organisms and Ecosystems. Pergamon, New York.

Anderson, J.W., R.G. Riley, and R.M. Bean. 1978. Recruitment of benthic animals as a function of petroleum hydrocarbon concentrations in the sediment. J. Fish. Res. Board Can. 35:776-790.

Anderson, J.W., S.L. Kiesser, and J.W. Blaylock. 1979. Comparative uptake of naphthalenes from water and oiled sediment by benthic amphipods, pp. 579-584. In Proceedings, 1979 Oil Spill Conference (Prevention, Behavior, Control, Cleanup). American Petroleum Institute, Washington, D.C.

Anderson, J.W., S.L. Kiesser, and J.W. Blaylock. 1980. The cumulative effect of petroleum hydrocarbons on marine crustaceans during constant exposure. Rapp. P.-v. Reun. Cons. Int. Explor. Mer 179:62-70.

Anderson, J.W., S.L. Kiesser, R.M. Bean, R.G. Riley, and B.L. Thomas. 1981. Toxicity of chemically dispersed oil to shrimp exposed to constant and decreasing concentrations in a flowing system, pp. 69-75. In Proceedings, 1981 Oil Spill Conference (Prevention, Behavior, Control, Cleanup). American Petroleum Institute, Washington, D.C.

Anderson, N.R. 1980. The Potential Application of Remote Sensing in Chemical Oceanographic Research. National Science Foundation, Washington, D.C.

Anderson, R.D. 1975. Petroleum hydrocarbons and oyster resources of Galveston Bay, Texas, pp. 541-548. In Proceedings, 1975 Conference on the Prevention and Control of Oil Pollution. American Petroleum Institute, Washington, D.C.

Apel, J.R. 1978. Past, present and future capabilities of satellites relative to the needs of ocean sciences, p. 7. Bruun Memorial Lectures, 1977. IOC Technical Series 19. UNESCO, Paris.

Armstrong, J.E., and J.A. Calder. 1978. Inhibition of light-induced pH increase and O_2 evolution of marine microalgae by water soluble components of crude and refined oils. Appl. Environ. Microbiol. 35:858-862.

Armstrong, J.E., W.G. Fehler, and J.A. Calder. 1981. Effects of petroleum hydrocarbons on the growth and energetics of marine microalgae, pp. 449-466. In F.J. Verberg et al., eds. Biological Monitoring of Marine Pollutants. Academic Press, New York.

Ashton, W.D. 1972. The logit transformation with special reference to its use in bioassay. Hafner Publishing, New York. 88 pp.

Atema, J. 1976. Sublethal effects of petroleum fractions on the behavior of the lobster Homarus americanus, and the mud snail Nassarius obsoletus, pp. 302-312. In M. Wiley, ed. Estuarine Processes. Vol. 1. Uses, Stresses and Adaptation to the Estuary. Academic Press, New York.

Atlas, R.M. 1979. Measurement of hydrocarbon biodegradation potentials and enumeration of hydrocarbon-utilizing microorganisms using carbon-14 hydrocarbon-spiked crude oil. pp. 196-204. In J.W.

Costerton and R.R. Colwell, eds. Native Aquatic Bacteria: Enumeration, Activity and Ecology. ASTM STP 695. American Society for Testing and Materials, Philadelphia, Pa.

Atlas, R.M. 1981. Microbial degradation of petroleum hydrocarbons: an environmental perspective. Microb. Rev. pp. 180-209.

Atlas, R.M., and R. Bartha. 1972. Degradation and mineralization of petroleum by two bacteria isolated from coastal waters. Biotech. Bioeng. 14:297-308.

Atlas, R.M., A. Horowitz, and M. Busdosh. 1978. Prudhoe crude oil in arctic marine ice, water, and sediment ecosystems: degradation and interactions with microbial and benthic communities. J. Fish. Res. Board Can. 35:585-590.

Atlas, R.M., P.D. Boehm, and J.A. Calder, 1981. Chemical and biological weathering of oil from the Amoco Cadiz oil spillage, within the littoral zone. Estuarine Coastal Mar. Sci. 12:589-608.

Augerfeld, J.M., J.W. Anderson, D.L. Woodruff, and J.L. Webster. 1980. Effects of Prudhoe Bay crude oil-contaminated sediments on Protothaca staminea (Mollusca:Pelecypoda): hydrocarbon content, condition index, free amino acid level. Mar. Environ. Res. 4:135-143.

Avignan, J., and M. Blumer. 1968. On the origin of pristane in marine organisms. J. Lipid Res. 9:350-352.

Backlund, L. 1979. Airborne oil spill surveillance systems in Sweden, pp. 305-312. In Proceedings, 1979 Oil Spill Conference (Prevention, Behavior, Control, Cleanup). American Petroleum Institute, Washington, D.C.

Baker, J.M. 1970. The effects of oils on plants. Environ. Pollut. 1:27-44.

Bakke, T., and T.M. Johnsen. 1979. Response of a subtidal sediment community to low levels of oil hydrocarbons in a Norwegian fjord, pp. 633-639. In Proceedings, 1979 Oil Spill Conference (Prevention, Behavior, Control, Cleanup). American Petroleum Institute, Washington, D.C.

Bakke, T., and H.R. Skjoldal. 1979. Effects of toluene on the survival, respiration, and adenylate system of a marine isopod. Mar. Pollut. Bull. 10:111-115.

Barker, C.J., and B.D. Rackham. 1979. The induction of sister-chromatid exchanges in cultured fish cells (Ameca splendens) by carcinogenic mutagens. Mut. Res. 68:381.

Barnett, J., and D. Toews. 1978. The effect of crude oil and the dispersant, Oilsperse 43, on respiration and coughing rates in Atlantic salmon (Salmo salar). Can. J. Zool. 56:307-310.

Bartle, K.D., M.L. Lee, and S. A. Wise. 1981. Modern analytical methods for environmental polycyclic aromatic compounds. Chem. Soc. Rev. 10:113-158.

Baruah, J.N., Y. Alroy, and R.I. Mateles. 1967. Incorporation of liquid hydrocarbons into agar media. Appl. Microbiol. 15:961.

Basu, D.K., and J. Saxena. 1978. Monitoring of polynuclear aromatic hydrocarbons in water. II. Extraction and recovery of six representative compounds in polyurethane foams. Environ. Sci. Technol. 12:791-794.

Batterton, J.C., K. Winters, and C. Van Baalen. 1977. Toxicity of crude oils and fuel oils presented directly to microalgae. J. Phycol. 13 (Suppl):6.

Batterton, J.C., K. Winters, and C. Van Baalen. 1978. Sensitivity of three microalgae to crude oils and fuel oils. Mar. Environ. Res. 1:31-41.

Bayne, B.L., D.R. Livingstone, M.N. Moore, and J. Widdows. 1976. A cytochemical and a biochemical index of stress in Mytilus edulis L. Mar. Pollut. Bull. 7:221-224.

Bayne, B.L., M.N. Moore, J. Widdows, D.R. Livingstone, and P. Salkeld. 1979. Measurement of the responses of individuals to environmental stress and pollution: studies with bivalve molluscs, pp. 165-180. In H.A. Cole, ed. The Assessment of Sublethal Effects of Pollutants in the Sea. The Royal Society, London.

Beamish, F.W.H. 1978. Swimming capacity, pp. 110-172. In W.S. Hoar and D.J. Randall, eds. Fish Physiology. Vol. 7. Academic Press, New York.

Bean, R.M., and J.W. Blaylock. 1977. Characterization of volatile hydrocarbons in flowing seawater suspensions of No. 2 fuel oil, pp. 397-403. In D.A. Wolfe, ed. Fate and Effects of Petroleum Hydrocarbons in Marine Organisms and Ecosystems. Pergamon, New York.

Bean, R.M., J.W. Blaylock, and R.G. Riley. 1980. Application of trace analytical techniques to a study of hydrocarbon composition upon dispersion of petroleum in a flowing seawater system, pp. 235-236. In L. Petrakis and F.T. Weiss, eds. Petroleum in the Marine Environment. Advances in Chemistry Series 185. American Chemical Society, Washington, D.C.

Beardmore, J.A., C.J. Barker, B. Battaglia, R.J. Berry, A.C. Longwell, J.F. Payne, and A. Rosenfield. 1980. The use of genetical approaches to monitoring biological effects of pollution. Rapp. P.-v. Reun. Cons. Int. Explor. Mer 179:299.

Becker, C.D., J.A. Lichatowich, M.J. Schneider, and J.A. Strand. 1973. Regional survey of marina biota for bioassay standardization of oil and oil dispersant chemicals. Research Report Publication 4167. American Petroleum Institute, Washington, D.C. 106 pp.

Beers, J.R. 1976. Determination of zooplankton biomass, pp. 35-74. In H.F. Steedman, ed. Zooplankton Fixation and Preservation. UNESCO, Paris.

Bellar, T.A., and J.L. Lichtenberg. 1974. Recovery of organic compounds from environmentally contaminated bottom materials, pp. 57-70. In R.A. Baker, ed. Contaminants and Sediments. Vol. 2. Ann Arbor Scientific Publishers, Ann Arbor, Mich.

Bellar, T.A., J.J. Lichtenberg, and S.C. Lonneman. 1980. Determining volatile organics at the _g/l levels by gas chromatography. J. Am. Water Works Assoc. 66:739-744.

Bender, M.E., E.A. Shearls, R.P. Ayres, C.H. Hershner, and R.J. Huggett. 1977. Ecological effects of experimental oil spills on eastern coastal plain estuarine ecosystems, pp. 505-509. In Proceedings, 1977 Oil Spill Conference (Prevention, Behavior, Control, Cleanup). American Petroleum Institute, Washington, D.C.

Bender, M.E., D.J. Reish, and C.H. Ward. 1979. Re-examination of the offshore ecology investigation. Rice Univ. Stud. 65(4&5):35-116.

Bender, M.E., E.A. Shearls, L. Murray, and R.J. Huggett. 1980. Ecological effects of experimental oil spills in eastern coastal plain estuaries. Environ. Intern. 3:121-133.

Bentz, A.P. 1980. Oil spill identification and remote sensing. In L. Petrakis and F.T. Weiss, eds. Petroleum in the Marine Environment. Advances in Chemistry Series 185. American Chemical Society, Washington, D.C.

Benville, P.E., Jr., and S. Korn. 1974. A simple apparatus for metering volatile liquids into water. J. Fish. Res. Board Can. 31:367-368.

Benville, P.E., Jr., T.G. Yocum, P. Nunes, and J.M. O'Neill. 1981. Simple, continuous-flow systems for dissolving the water-soluble components of crude oil into seawater for acute or chronic exposure of marine organisms. Water Res. 15:1197-1204.

Berdugo, V., R.P. Harris, and S.C. O'Hara. 1977. The effect of petroleum hydrocarbons on reproduction of an estuarine planktonic copepod in laboratory cultures. Mar. Pollut. Bull. 8(6):138-143.

Berman, M., and D.R. Heinle. 1980. Modification of the feeding behavior of marine copepods by sublethal concentrations of water-accommodated fuel oil. Mar. Biol. 56(1)59-64.

Bernard, B.B., J.M. Brooks, and W.M. Sackett. 1978. Light hydrocarbons in recent Texas Continental Shelf and slope sediments. J. Geophys. Res. 83(8):4053-4061.

Berthou, F., Y. Gourmelun, Y. Dreano, and M.P. Friocourt. 1981. Application of gas chromatography on gas capillary columns to the analysis of hydrocarbon pollutants from the Amoco Cadiz oil spill. J. Chromatogr. 203:279-292.

Bieri, R.H., and V.C. Stamoudis. 1977. The fate of petroleum hydrocarbons from a No. 2 fuel oil spill in a seminatural estuarine environment, pp. 332-344. In D.A. Wolfe, ed. Fate and Effects of Petroleum Hydrocarbons in Marine Organisms and Ecosystems. Pergamon, New York.

Bieri, R.H., V.C. Stamoudis, and M.K. Cueman. 1977. Chemical investigations of two experimental oil spills in an estuarine ecosystem, pp. 511-515. In Proceedings, 1977 Oil Spill Conference (Prevention, Behavior, Control, Cleanup). American Petroleum Institute, Washington, D.C.

Bieri, R.H., M.K. Cueman, C.L. Smith, and C.W. Su. 1978. Polynuclear aromatic and polycyclic aliphatic hydrocarbons in sediments from the Atlantic Outer Continental Shelf. Intern. J. Environ. Anal. Chem. 5:293-310.

Bieri, R.H., V.C. Stamoudis, and M.K. Cueman. 1979. Chemical investigations of two experimental oil spills in an estuarine ecosystem, Part II, pp. 693-697. In Proceedings, 1979 Oil Spill Conference (Prevention, Behavior, Control, Cleanup). American Petroleum Institute, Washington, D.C.

Bigford, T.E. 1977. Effects of oil on behavioral responses to light, pressure and gravity in larvae of the rock crab Cancer irroratus. Mar. Biol. 43:137-148.

Bjorseth, A., and G. Eklund. 1979. Analysis of PAH by glass capillary gas chromatography using simultaneous flame ionization and electron cpature detection. J. High Resolution Capillary Column Chromatogr. 2:22-26.

Bjorseth, A., G. Eidsa, J. Gether, L. Landmark, and M. Moller. 1982. Detection of mutagens in complex samples by the Salmonella assay applied directly to thin-layer chromatography plates. Science 215:87-89.

Blackall, P.J., and G.A. Sergy. 1981. The BIOS Project--Frontier oil spill countermeasures research, pp. 167-172. In Proceedings, 1981 Oil Spill Conference (Prevention, Behavior, Control, Cleanup). American Petroleum Institute, Washington, D.C.

Blackall, P.J., G.A. Sergy, and D.E. Thornton. 1981. The BIOS Project-1980 review. Spill Technol. Newsletter 6:21-36.

Blackman, R.A.A. 1974. Effects of sunken oil on the feeding of plaice on brown shrimps and ther benthos. Intern. Counc. Explor. Sea 7 pp.

Blackman, R.A.A., and R.J. Law. 1980. The Eleni V oil spill: fate and effects of the oil over the first twelve months. Part I. Oil in waters and sediments. Mar. Pollut. Bull. 11:199-204.

Blankley, W.F. 1973. Toxic and inhibitory materials associated with culturing, pp. 207-229. In J.R. Stein, ed. Handbook of Phycological Methods-Culture Methods and Growth Measurements. Cambridge University Press, New York.

Blanton, W.G., and M.C. Robinson. 1973. Some acute effects of low boiling petroleum fractions on the cellular structures of fish gills under field conditions, pp. 265-273. In D.G. Ahearn and S.P. Meyers, eds. The Microbial Degradation of Oil Pollutants. Publication LSU-SG-73-01. Louisiana State University, Center for Wetland Resources, Baton Rouge, La.

Bloom, S.E. 1978. Chick embryos for detecting environmental mutagens. In A. Hollaender, and F.J. de Serres, eds. Chemical Mutagens--Principles and Methods for Their Detection. Vol. 5. Plenum, New York.

Bloom, S.E., and T.C. Hsu. 1975. Differential fluorescence of sister chromatids in chicken embryos exposed to 5-bromodeoxyuridine. Chromosoma 51:261.

Blumer, M. 1957. Removal of elemental sulfur from hydrocarbon fractions. Anal. Chem. 29:1039-1041.

Blumer, M. 1967. Hydrocarbons in digestive tract and liver of a basking shark. Science 156:390-391.

Blumer, M. 1969. Oil pollution of the ocean, pp. 5-13. In D.P. Hoult ed. Plenum, New York.

Blumer, M. 1976. Polycyclic aromatic compounds in nature. Sci. Am. 234:34-45.

Blumer, M., and J. Sass. 1972. The West Falmouth oil spill. Reference 7219. Woods Hole Oceanographic Institute, Woods Hole, Mass.

Blumer, M., and D.W. Thomas. 1965a. Phytodienes in zooplankton. Science 147:1148-1149.

Blumer, M., and D.W. Thomas. 1965b. Zamene, isomeric C_{19} monoolefins from marine zooplankton, fishes, and mammals. Science 148:370-371.

Blumer, M., J.C. Robertson, J.E. Gordon, and J. Sass. 1960. Phytol-derived C_{19} di- and tri-olefinic hydrocarbons in marine zooplankton and fishes. Biochemistry 8:4067-4074.

Blumer, M., M.M. Mullin, and D.W. Thomas. 1963. Pristane in zooplankton. Science 140:974.

Blumer, M., J. Gordon, J.C. Robertson, and J. Sass. 1969. Phytol-derived C_{19} and di- and tri-olefinic hydrocarbons in marine zooplankton and fishes. Biochemistry 8:4067-4074.

Blumer, M., M.M. Mullin, and R.R.L. Guillard. 1970. A polyunsaturated hydrocarbon (3,6,9,12,15,18-heneicosahexane) in the marine food web. Mar. Biol. Berlin 6:226-235.

Blumer, M., R.R.L. Guillard, and T. Chase. 1971. Hydrocarbons of marine phytoplankton. Mar. Biol. 8:183-189.

Blumer, M., M. Ehrhardt, and J.H. Jones. 1973. The environmental fate of stranded crude oil. Deep Sea Res. 20:239-259.

Blyth, C.R. 1972. On Simpson's paradox and the sure-thing principle. J. Am. Statist. Assoc. 67:364-366.

Boehm, P.D. 1978. Hydrocarbon chemistry. In New England Benchmark Study. Vol. II. Draft Final Report. Contract AA550-CT6-51. Bureau of Land Management, New York.

Boehm, P.D. 1980a. The decoupling of dissolved, particulate and surface microlayer hydrocarbons in northwestern Atlantic continental shelf waters. Mar. Chem. 9:255-281.

Boehm, P.D. 1980b. Gulf and Atlantic survey--Cape Hatteras to Gulf of Maine survey for selected organic pollutants in finfish and benthic animals. Final Report. NOAA Contract NA-80-FA-C-00046. NOAA/NMFS Northeast Fisheries Center, Sandy Hook, N.J.

Boehm, P.D. 1981. Investigations on pollutant organic chemical fluxes in the Hudson Raritan estuarine and New York Bight coastal systems. Final Report. NOAA Grant NA-80-AA-D-00062. NOAA, Office of Marine Pollution Assessment, Rockville, Md.

Boehm, P.D., and T.F. Dorsey. 1981. A method for the analysis of nitrogen and sulfur heterocyclic compounds in marine organisms (in preparation).

Boehm, P.D., and D.L. Fiest. 1978. Analyses of water samples from the Tsesis oil spill and laboratory experiments on the use of Niskin bacteriological sterile bag sampler. Final Report. NOAA Contract 03-A01-9-4178. NOAA, Office of Marine Pollution Assessment, Rockville, Md.

Boehm, P.D., and D.L. Fiest. 1980a. Aspects of the transport of petroleum hydrocarbons to the benthos during the Ixtoc I blowout in the Bay of Campeche, pp. 207-236. In Proceedings of the Conference on the Preliminary Scientific Results from the Researcher/Pierce Cruise to the Ixtoc I Blowout. NOAA, Office of Marine Pollution Assessment, Rockville, Md.

Boehm, P.D., and D.L. Fiest. 1980b. Surface water column transport and weathering of petroleum hydrocarbons during the Ixtoc I blowout in the Bay of Campeche and their relation to surface oil and microlayer compositions, pp. 169-185. In Proceedings of the Conference on the Preliminary Scientific Results from the Researcher/Pierce Cruise to the Ixtoc I Blowout. NOAA, Office of Marine Pollution Assessment, Rockville, Md.

Boehm, P.D., and D.L. Fiest. 1980c. Determine hydrocarbon composition and concentration in major components of the marine ecosystem. Vol. VI. In W.B. Jackson and G.M. Faw, eds. Biological/Chemical Survey of Texoma and Capline Sector Salt Dome Brine Disposal Sites off Louisiana, 1978-1979. NOAA Technical Memorandum NMFS-SEFC-30. National Technical Information Service, Springfield, Va. 136 pp.

Boehm, P.D., and J.G. Quinn. 1973. Solubilization of hydrocarbons by the dissolved organic matter in sea water. Geochim Cosmochim. Acta 37:2459-2477.

Boehm, P.D., and J.G. Quinn. 1978. Benthic hydrocarbons of Rhode Island Sound. Estuarine Coastal Mar. Sci. 6:471-494.

Boehm, P.D., M. Elia, and T. Dorsey. 1983. Methodology development for analyzing sulfur and nitrogen heterocyclics in biological matrices. Final Report. Contract NA80-GA-C-00065 submitted by Energy Resources Company to National Marine Fisheries Service, National Seafood Quality and Inspection Laboratory, University of Southern Mississippi, Hallesburg.

Boehm, P.D., D.L. Fiest, and A. Elskus. 1981a. Comparative weathering patterns of hydrocarbons from the Amoco Cadiz oil spill observed at a variety of coastal environments, pp. 159-173. In Proceedings, Amoco Cadiz: Fate and Effects of the Oil Spill, November 19-22, 1979. Centre Nationale pour l'Exploitation des Oceans, COB, Brest, France.

Boehm, P.D., J.E. Barak, D.L. Fiest, and A.A. Elskus. 1981b. A chemical investigation of the transport and fate of petroleum hydrocarbons in littoral and benthic environments: the Tsesis oil spill. Mar. Environ. Res. (in press).

Boehm, P.D., D.L. Fiest, D. Mackay, and S. Paterson. 1981c. Physical-chemical weathering of petroleum hydrocarbons from the Ixtoc I blowout--chemical measurements and a weathering model. In Proceedings, 1981 Oil Spill Conference (Prevention, Behavior, Control, Cleanup). Publication 4334. American Petroleum Institute, Washington, D.C.

Boesch, D.F. 1977. Application of numerical classification in ecological investigation of water pollution. Ecological Research Series, EPA-600/3-77-033. U.S. Environmental Protection Agency, Washington, D.C. 114 pp.

Boesch, D.F., and R. Rosenberg. 1981. Response to stress in marine benthic communities, pp. 179-200. In G.M. Barrett and R. Rosenberg, eds. Stress Effects on Natural Ecosystems. John Wiley & Sons, New York.

Boney, A.D. 1970. Toxicity studies with an oil spill emulsifier and the green alga Prasinocladus marinus. J. Mar. Biol. Assoc. U.K. 50:461-473.

Borneff, J., F. Selenka, H. Kunte, and A. Maximos. 1968a. Die synthese von 3,4-benzpyren and anderen plyzyklischen, aromatischem kohnlenwasserstoffen in pflanzen. Arch. Hyg. Bakt. 152:279-282.

Borneff, J., F. Selenka, H. Kunte, and A. Maximos. 1968b. Experimental studies on the formaton of polycyclic aromatic hydrocarbons in plants. Environ. Res. 2:22-29.

Borowitzka, M.A. 1972. Intertidal algae species diversity and the effect of pollution. Aust. J. Mar. Freshwater Res. 23:73-84.

Bott, T.L., K. Rogenmuser, and P. Thorne. 1976. Effect of No. 2 fuel oil, Nigerian crude oil and used crankcase oil on the metabolism of benthic algal communities, pp. 373-393. In Sources, Effects and Sinks of Hydrocarbons in the Aquatic Environment. American Institute of Biological Sciences, Arlington, Va.

Boucher, G. 1980. Impact of Amoco Cadiz oil spill on intertidal and sublittoral meiofauna. Mar. Pollut. Bull. 11:95-101.

Bowden, W.B. 1977. Comparison of two direct-count techniques for enumerating aquatic bacteria. Appl. Environ. Microbiol. 33:1229-1232.

Box, G.E.P., and G.C. Tiao. 1975. Intervention analysis with application to economic and environmental problems. J. Am. Statist. Assoc. 71:71-79.

Boylan, D.B., and B.W. Tripp. 1971. Determination of hydrocarbons in seawater extracts of crude oil and crude oil fractions. Nature 230:44-47.

Braarud, T. 1961. Cultivation of marine organisms as a means of understanding environmental influences on populations, pp. 271-298. In M. Sear, ed. Oceanography. Publication 67. AAAS, Washington, D.C.

Bridie, A.L., and J. Bos. 1971. Biological degradation of mineral oil in seawater. J. Inst. Petrol. London 57:270-277.

Brocksen, R.W., and H.T. Bailey. 1973. Respiratory response of juvenile chinook salmon and striped bass exposed to benzene, a water-soluble component of crude oil, pp. 783-791. In Proceedings, Joint Conference on Prevention and Control of Oil Spills. American Petroleum Institute, Washington, D.C.

Brockway, D.L., J. Hill IV, J.R. Mandsley, and R.R. Lassiter. 1979. Development, replicability and modeling of naturally derived microcosms. Intern. J. Environ. St. 13:149-158.

Brooks, J.M., A.D. Fredericks, W.M. Sachett, and J.W. Swinnerton. 1973. Baseline concentrations of light hydrocarbons in Gulf of Mexico. Environ. Sci. Technol. 7:639-642.

Brooks, J.M., D.A. Wiesenberger, R.A. Burke, and M. Kennicutt. 1980. Gaseous and volatile hydrocarbons in the Gulf of Mexico following the Ixtoc I blowout, pp. 53-85. In Proceedings of Conference on the Preliminary Results from the Researcher/Pierce Cruise to the Ixtoc I Blowout. NOAA, Office of Marine Pollution Assessment, Rockville, Md.

Brown, D.W., L.S. Ramos, A.J. Friedman, and W.D. MacLeod. 1979. Analysis of trace levels of petroleum hydrocarbons in marine sediments using a solvent/slurry extraction procedure, pp. 161-167. In Trace Organic Analysis: A New Frontier in Analytical Chemistry. Special Publication 519. National Bureau of Standards, Washington, D.C.

Brown, D.W., L.S. Ramos, M.Y. Uyeda, A.J. Friedman, and W.D. MacLeod, Jr. 1980. Ambient temperature contamination of hydrocarbons from marine sediment--comparison with boiling solvent extractions, pp. 313-326. In L. Petrakis and F.T. Weiss, eds. Petroleum in the Marine Environment. Advances in Chemistry Series 185. American Chemical Society, Washington, D.C.

Brown, L.R., G.S. Pabst, and M. Light. 1975. A rapid field method for detecting oil in sediments. Mar. Pollut. Bull. 9:81-82.

Brown, R.A., and H.L. Huffman, Jr. 1976. Hydrocarbons in open ocean waters. Science 191:847-849.

Brown, R.A., and H.L. Huffman, Jr. 1979. Extractable organics and hydrocarbons in the Mediterranean Sea. Mar. Pollut. Bull. 10:291-298.

Brown, R.A., and F.T. Weiss. 1978. Fate and effects of polynuclear aromatic hydrocarbons in the aquatic environment. Publication 4297. American Petroleum Institute, Environmental Affairs Department, Washington, D.C.

Brown, R.A., J.J. Elliott, J.M. Kelliher, and T.D. Searl. 1975. Sampling and analysis of non-volatile hydrocarbons in ocean water, pp. 172-187. In T.R.P. Gibbs, Jr., ed. Analytical Methods in Oceanography. Advances in Chemistry Series 147. American Chemical Society, Washington, D.C.

Brown, R.G.B. 1980. Seabirds as marine mammals, pp. 1-39. In J. Burger, B.L. Olla, and H.E. Winn, eds. Behavior of Marine Animals. Vol. 4. Plenum, New York.

Brown, R.G.B. 1982. Birds, oil and the Canadian environment, pp. 105-112. In J.B. Sprague, J.H. Vandermeulen and P.G. Wells, eds. Oil and Dispersants in Canadian Seas--Research Appraisal and Recommendations. Economic and Technical Review Report EPS 3-EC-82-2. Environment Canada.

Brown, R.G.B., D.N. Nettleship, P. Germain, C.E. Tull, and T. Davis. 1975. Atlas of Eastern Canadian Seabirds. Canada Wildlife Service, Ottawa, Ontario. 220 pp.

Bruce, W.R., and J.A. Heddle. 1979. The eutogenic activity of 61 agents as determined by the micronucleus, Salmonella, and sperm abnormality assays. Can. J. Genetics Psych. 21:319.

Bryant, R.A., D.J.A. Williams, and A.E. James. 1980. A sampler for cohesive sediment in the benthic boundary layer. Limnol. Oceanogr. 25:572-576.

Bryne, C.J., and J.A. Calder. 1977. Effect of the water-soluble fractions of crude, refined, and waste oils on the embryonic and larvae stages of the quahog clam Mercenaria sp. Mar. Biol. 40:225-231.

Burns, K.A. 1976. Microsomal mixed function oxidases in an estuarine fish, Fundulus heteroclitus, and their induction as a result of environmental contamination. Comp. Biochem. Physiol. 53B:443-446.

Burns, K.A., and J.L. Smith. 1978. Biological monitoring of ambient water quality: the case for the use of mussels as indicators of certain organic pollutants. Unpublished manuscript. Marine Chemistry Unit, Ministry for Conservation, 7B Parliament Place, Melbourne, Victoria 3002, Australia. (Cited in NAS 1980.)

Burns, K.A., and J.M. Teal. 1979. The West Falmouth oil spill: hydrocarbons in the salt marsh ecosystem. Estuarine Coastal Mar. Sci. 8:349-360.

Burwood, R., and G.C. Speers. 1974. Photooxidation as a factor in the environmental dispersal of crude oil. Estuarine Coastal Mar. Sci. 2:117-135.

Busdosh, M., K.W. Dobra, A. Horowitz, S.E. Neff, and R.M. Atlas. 1978. Potential long-term effects of Prudhoe Bay crude oil in arctic sediments on indigenous benthic invertebrate communities, pp. 856-874. In Proceedings of the Conference on Assessment of Ecological Impacts of Oil Spills. American Institute of Biological Sciences, Arlington, Va.

Butler, J.N. 1975. Evaporative weathering of petroleum residues: the age of pelagic tar. Mar. Chem. 3:9-21.

Butler, R.G., and P. Lukasiewicz. 1979. A field study of the effect of crude oil on herring gull chick growth. Auk 96:809-812.

Calder, J.A. and P.D. Boehm. 1981. The chemistry of Amoco Cadiz oil in the l'Aber Wrach, pp. 149-158. In Amoco Cadiz: Fate and Effects of the Oil Spill. CNEXO, Paris.

Calder, J.A., J. Lake, and J. Laseter. 1978. Chemical composition of selected environmental and petroleum samples from the Amoco Cadiz oil spill, pp. 21-84. In W.H. Hess, ed. The Amoco Cadiz Oil Spill. A Preliminary Scientific Report. NOAA, EPA, Washington, D.C.

Caparello D.M., and P.A. LaRock. 1975. A radioisotope assay for the quantification of hydrocarbon biodegradation potential in environmental samples. Microb. Ecol. 2:28-42.

Caperon, J., D. Schell, J. Husta, and E. Laws. 1979. Ammonium excretion rates in Kanoehe Bay, Hawaii, measured by a 15N isotope dilution technique. Mar. Biol. 54:33-40.

Capuzzo, J.M., and B.A. Lancaster. 1981. Physiological effects of south Louisiana crude oil on larvae of the American Lobster (Homarus americanus), pp. 405-424. In F.J. Vernberg, A. Calabrese, F.P. Thurberg and W.B. Vernberg, eds. Biological Monitoring of Marine Pollutants. Academic Press, New York.

Carpenter, E.J., and J.S. Lively. 1980. Review of estimates of algal growth using C tracer techniques, pp. 161-178. In P.G. Falkowski, ed. Primary Production in the Sea. Brookhaven Symposia on Biology 31. Plenum, New York.

Carr, R.S., and D.J. Reish. 1977. Effect of petroleum hydrocarbons on the survival and life history of polychaetous annelids, pp. 168-173. In D.A. Wolfe, ed. Fate and Effects of Petroleum Hydrocarbons in Marine Organisms and Ecosystems. Pergamon, New York.

Carritt, D.C., and J.H. Carpenter. 1966. Comparison and evaluation of currently employed modifications of the Winkler method for determining dissolved oxygen in sea water: a NASCO report. J. Mar. Res. 24(3):286-318.

Catoe, C.E. 1972. The applicability of remote sensing techniques for oil slick detection. Paper presented at the Fourth Annual Offshore Technology Conference, Houston, Texas, May 1-3, 1972.

Chen, T.R. 1969. Karyological heterogamety of deep-sea fishes Postilla. No. 130. Peabody Museum, Yale University, March 18, 1969.

Chesler, S.N., B.H. Bump, H.S. Hertz, W.E. May, S. Dyszel, and D.P. Enagonio. 1976. Trace hydrocarbon analysis: The National Bureau of Standards Prince William Sound/N.E. Gulf of Alaska baseline study. Technical Note 889. National Bureau of Standards, Washington, D.C.

Chesler, S.N., B.H. Bump, H.S. Hertz, W.E. May, and S.A. Wise. 1978. Determination of trace level hydrocarbons in marine biota. Anal. Chem. 50:805-810.

Chmielowiec, J., and A.E. George. 1980. Polar bonded-phase sorbents for HPLC separations of polycyclic aromatic hydrocarbons. Anal. Chem. 52:1154-1157.

Christensen, J.C., and T.T. Packard. 1979. Respiratory electron transport activity in marine phytoplankton and bacteria: comparison of methods. Limnol. Oceanogr. 24(3):576-583.

Christensen, R.G., and W.E. May. 1978. Detectors for liquid chromatographic analysis for polynuclear aromatic hydrocarbons. J. Liq. Chromatogr. 1(3):385-399.

Clark, H.A., and P.C. Jurs. 1979. Classification of crude oil gas chromatograms by pattern recognition techniques. Anal. Chem. 51(6):616-622.

Clark, R.C., Jr. 1966. Occurrence of normal paraffin hydrocarbons in nature. Unpublished manuscript. Technical Report 66-34. Woods Hole Oceanographic Institution, Woods Hole, Mass.

Clark, R.C., Jr., and M. Blumer. 1967. Distribution of n-paraffins in marine organisms and sediment. Limnol. Oceanogr. 12:79-87.

Clark, R.C., Jr., and D.W. Brown. 1977. Petroleum: properties and analyses in biotic and abiotic systems, pp. 1-89. In D.C. Malins, ed. Effects of Petroleum on Arctic and Subarctic Marine Environments and Organisms. Vol. 1. Nature and Fate of Petroleum. Academic Press, New York.

Clark, R.C., Jr., and J.S. Finley. 1974a. Tidal aquarium for laboratory studies of environmental effects on marine organisms. Progressive Fish-Culturist 36:134-137.

Clark, R.C., Jr., and J.S. Finley. 1974b. Uptake and loss of petroleum hydrocarbons by the mussel, Mytilus edulis, in laboratory experiments. Fish Bull. 73:508-515.

Clark, R.C., Jr., and J.S. Finley. 1982. Occurrence and impact of petroleum on arctic environments, pp. 295-341. In L. Rey, ed. The Arctic Ocean: The Hydrographic Environment and the Fate of Pollutants. Macmillan, London.

Clark, R.C., Jr., and W.D. MacLeod, Jr. 1977. Inputs, transport mechanisms, and observed concentrations of petroleum in the marine environment, pp. 91-223. In D.C. Malins, ed. Effects of Petroleum on Arctic and Subarctic Marine Environments and Organisms. Vol. 1. Nature and Fate of Petroleum. Academic Press, New York.

Clark, R.C., Jr., J.S. Finley, and G.G. Gibson. 1974. Acute effects of outboard motor effluent on two marine shellfish. Environ. Sci. Technol. 8:1009-1014.

Clement, L.E., M.S. Stekoll, and D.G. Shaw. 1980. Accumulation, fractionation and release of oil by the intertidal clam Macoma balthica. Mar. Biol. 57:41-50.

Cleveland, W.L., and B. Kleiner. 1975. A graphical technique for enhancing scatter plots with moving statistics. Technometrics 17:447-454.

Clifford, H.T., and W. Stephenson. 1975. An Introduction to Numerical Classification. Academic Press, New York. 229 pp.

Cochran, W.G. 1951. The comparison of percentages in matched samples. Biometrika 37:256-266.

Cochran, W.G. 1968. The effectiveness of subclassification in removing bias in observational studies. Biometrics 24:295-313.

Cole, R.J., N. Taylor, J. Cole, and C.F. Arlett. 1981. Short-term tests for transplacentally active carcinogens. I. Micronucleus formation in fetal and maternal erythroblasts. Mut. Res. 80:141.

Colin, H., J.M. Schmitter, and G. Guiochon. 1981. Liquid chromatography of azaarenes. Anal. Chem. 53:625-631.

Collier, T.K., L.C. Thomas, and D.C. Malins. 1978. Influence of environmental temperature on disposition of dietary naphthalene in coho salmon (Oncorhynchus kisutch): isolation and identification of individual metabolites. Comp. Biochem. Physiol. 61C:23-28.

Colmsjo, A.L., and C.E. Ostman. 1980. Selectivity properties in Shpol'skii fluorescence of polynuclear aromatic hydrocarbons. Anal. Chem. 52:2093-2095.

Conover, R.J. 1971. Some relations between zooplankton and Bunker C oil in Chedabucto Bay following the wreck of the tanker Arrow. J. Fish. Res. Board Can. 28:1327-1330.

Conover, R.J. 1978. Transformation of organic matter, pp. 221-499. In O. Kinne, ed. Marine Ecology. Vol. IV, Dynamics. Wiley-Interscience, New York.

Cooney, J.J., and M.P. Shiaris. 1982. Utilization and co-oxidation of aromatic hydrocarbons by estuarine microorganisms. Dev. Ind. Microbiol. 23:177-185.

Copeland, B.J., and T.J. Bechtel. 1971. Species diversity and water quality in Galveston Bay, Texas. Water Air Soil Pollut. 1:89-105.

Corner, E.D.S. 1978. Pollution studies with marine plankton. Part I. Petroleum hydrocarbons and related compounds. Advances Mar. Biol. 15:289-380.

Corner, E.D.S., C.C. Kilvington, and S.C.M. O'Hara. 1973. Qualitative studies on the metabolism of naphthalene in Maia squinado (Herbst). J. Mar. Biol. Assoc. U.K. 53:819-832.

Corner, E.D.S., R.P. Harris, C.C. Kilvington, and S.C.M. O'Hara. 1976. Petroleum compounds in the marine food web: short-term experiments on the fate of naphthalene in Calanus. J. Mar. Biol. Assoc. U.K. 56:121-133.

Countryman, P.I., and J.A. Heddle. 1976. The production of micronuclei from chromosome aberrations in irradiated cultures of human lymphocytes. Mut. Res. 41:321.

Cox, B.A., J.W. Anderson, and J.C. Parker. 1975. An experimental oil spill: the distribution of aromatic hydrocarbons in the water, sediment and animal tissues within a shrimp pond, pp. 607-612. In Proceedings, 1975 Conference on the Prevention and Control of Oil Pollution. American Petroleum Institute, Washington, D.C.

Cox, G.V. 1974. Marine Bioassays Workshop Proceedings. Marine Technology Society, Washington, D.C. 308 pp.

Cox, G.V. 1977. Information management, Chap. 2. In A. Barnett, ed. Oil Spill Studies: Strategies and Techniques. Publication 4286. American Petroleum Institute, Washington, D.C.

Cox, G.V. 1980. Oil Spill Studies: Strategies and Techniques. American Petroleum Institute, Washington, D.C. 150 pp.

Craddock, D.R. 1977. Acute toxic effects of petroleum on arctic and subarctic marine organisms, pp. 1-93. In D.C. Malins, ed. Effects of Petroleum on Arctic and Subarctic Marine Environments and Organisms. Vol. 2, Biological Effects. Academic Press, New York.

Cram, S.P., and F.J. Yang. 1980. High resolution gas chromatography: an overview, pp. 105-121. In L. Petrakis and F.T. Weiss, eds. Petroleum in the Marine Environment. Advances in Chemistry Series 185. American Chemical Society, Washington, D.C.

Cretney, W.J., P.A. Chrhistensen, B.W. McIntyre, and B.R. Fowler. 1980. Quantification of polycyclic aromatic hydrocarbons in marine environmental smaples, pp. 315-330. In B.K. Afghan and D. Mackay, eds. Hydrocarbons and Halogenated Hydrocarbons in the Aquatic Environment. Plenum, New York.

Crisp, D.G., A.O. Christi, and A.F.A. Ghobashy. 1967. Narcotic and toxic action of organic compounds on barnacle larvae. Comp. Biochem. Physiol. 22:629-649.

Croswell, W.F., and J.C. Fedors. 1979. An ongoing assessment of the use of space technology in monitoring oil spills, pp. 313-316. In Proceedings, 1979 Oil Spill Conference (Prevention, Behavior, Control, Cleanup). American Petroleum Institute, Washington, D.C.

Crowley, R.J., S. Siggia, and P.C. Uden. 1980. Class separation and characterization of shale oil by liquid chromatography and capillary column gas chromatography. Anal. Chem. 52:1224-1228.

Daisey, J.M., and M.A. Leyko. 1979. Thin-layer gas chromatographic method for the determination of polycyclic aromatic and aliphatic hydrocarbons in airborne particulate matter. Anal. Chem. 51(1):24-26.

Daley, R.J., and J.E. Hobbie. 1975. Direct counts of aquatic bacteria by a modified epifluorescence technique. Limnol. Oceanogr. 20:875-882.

Daniels, E.W., A.C. Longwell, J.M. McNiff, and R.W. Wolfgang. 1971. Ultrastructure of spermatozoa from the American oyster Crassostrea virginica. Trans. Am. Micros. Soc. 90:275.

Das, B.S., and G.H. Thomas. 1978. Fluroescence detection in high performance liquid chromatographic determination of polycyclic aromatic hydrocarbons. Anal. Chem. 50(7):967-973.

Dastillung, M., and P. Albrecht. 1976. Molecular test for oil pollution in surface sediments. Mar. Pollut. Bull. 7:13-15.

Dauble, D.D., E.W. Lusty, W.E. Fallon, and R.H. Gray. 1981. Mixing and separation device for continuous flow bioassays with coal liquids. Bull. Environ. Contam. Toxicol. 26:717-723.

Davis, J.B. 1968. Paraffinic hydrocarbons in the sulfate-reducing bacterium Desulfovibrio desulfuricans. Chem. Geol. 3:155-160.

Davis, W.P., G.I. Scott, C.D. Getter, M.O. Hayes, and E.R. Grundlach. 1980. Methodology for environmental assessments of oil and hazardous substance spills. Helgolander Meeresunters 33:246-256.

Degens, E.T. 1969. Biogeochemistry of stable carbon isotopes, pp. 304-329. In G. Eglinton and M.T.J. Murphy, eds. Organic Geochemistry. Springer-Verlag, New York.

Dehnen, W., R Tomingas, and D.J. Roos. 1973. A modified method for the assay of benzo(a)pyrene hydroxylase EC-1.14.1.1 Anal. Biochem. 53(2):373-383.

deLappe, B.W., R.W. Risebrough, A.M. Spinger, T.T. Schmidt, J.C. Shropshire, E.F. Letterman, and J.R. Payne. 1980. The sampling and measurement of hydrocarbons in natural waters, pp. 29-68. In D.K. Afghan and D. Mackay, eds. Halogenated Hydrocarbons in the Aquatic Environment. Environment Research Series. Plenum, New York.

DeLeon, I.R., L.V. McCarthy, C.K. Raschke, E.B. Overton, and J.L. Laseter. 1980. Characterization of azaarenes in Ixtoc I oil by gas chromatography and gas chromatography-mass spectrometry. Extended abstract 2 presented before the Division of Environmental Chemistry, American Chemical Society, Washington, D.C., August 24-29, 1980. Vol. 20.

Denton, T.E. 1973. Fish Chromosome Methodology. C.C. Thomas, Springfield, Ill.

DePierre, J.W., M.S. Moron, K.A.M. Johannese, and D.L. Ernster. 1975. A reliable sensitive and convenient radioactive assay for benzopyrene monooxygenase. Anal. Biochem. 63(2):470-484.

Dicks, B. 1976. The importance of behavioral patterns in toxicity testing and ecological prediction, pp. 303-320. In J.M. Baker, ed. Marine Ecology and Oil Pollution. John Wiley & Sons, New York.

Dillon, T.M. 1981. Effects of dimethylnaphthalene and fluctuating temperatures on estuarine shrimp, pp. 79-85. In Proceedings, 1981 Oil Spill Conference (Prevention, Behavior, Control, Cleanup). American Petroleum Institute, Washington, D.C.

DiMichele, J., and M.H. Taylor. 1978. Histopathological and physiological responses of Fundulus heteroclitus L. to naphthalene exposure. J. Fish. Res. Board Can. 35:1060-1066.

Di Paolo, J.A., R.L. Nelson, P.J. Donovan, and C.H. Evans. 1973. Host-mediated in vivo--in vitro assay for chemical carcinogens. Arch. Path. 95:380.

DiSalvo, L.H., and H.E. Guard. 1975. Hydrocarbons associated with suspended particulate matter in San Francisco Bay waters, pp. 169-173. In Proceedings, 1975 Conference on Prevention and Control of Oil Pollution. American Petroleum Institute, Washington, D.C.

DiSalvo, L.H., H.E. Guard, and L. Hunter. 1975. Tissue hydrocarbon burden of mussels as potential monitor of environmental hydrocarbon insult. Environ. Sci. Technol. 9:247-251.

Dixit, D., and J.W. Anderson. 1977. Distribution of naphthalenes within exposed Fundulus similus and correlations with stress behavior, pp. 633-636. In Proceedings, 1977 Oil Spill Conference (Prevention, Behavior, Control, Cleanup). American Petroleum Institute, Washington, D.C.

Dixon, W.J., ed. 1970. Biomedical Computer Programs. University of California Press, Berkeley. 600 pp.

Donahue, W.H., M.F. Welch, W.Y. Lee, and J.A.C. Nicol. 1977a. Toxicity of water-soluble fractions of petroleum oils on larvae of crabs, pp. 77-94. In C.S. Giam, ed. Pollutant Effects on Marine Organisms. Lexington Books, Lexington, Mass.

Donahue, W.H., R.T. Wang, M. Welch, and J.A.C. Nicol. 1977b. Effects of water-soluble components of petroleum oils and aromatic hydrocarbons on barnacle larvae. Environ. Pollut. 13:187-202.

D'Ozouville, L., M.O. Hayes, E.R. Gundlach, W.J. Sexton, and J. Michel. 1979. Occurrence of oil in offshore bottom sediments at the Amoco Cadiz oil spill site, pp. 187-192. In Proceedings, 1979 Oil Spill Conference (Prevention, Behavior, Control, Cleanup). American Petroleum Institute, Washington, D.C.

Dubois, M., K.A. Gilles, J.D. Hamilton, P.A. Rebers, and F. Smith. 1956. Colorimetric method for determination of sugars and related substances. Anal. Chem. 28:350-356.

Dunn, B.P. 1976. Techniques for determination of benzo(a)pyrene in marine organisms and sediments. Environ. Sci. Technol. 10:1018-1021.

Dunn, B.P. 1979. Benzo(a)pyrene in the marine environment: analytical techniques and results. In Proceedings of the International Symposium on the Analysis of Hydrocarbons and Halogenated Hydrocarbons in the Aquatic Environment. Plenum, New York. (in press).

Dunn, B.P., and Armour, R.J. 1980. Sample extraction and purification for determination of polycyclic aromatic hydrocarbons by reversed phase chromatography. Anal. Chem. 52:2027-2031.

Dunn, B.P., and H.F. Stich. 1976. Monitoring procedures for chemical carcinogens in coastal waters. J. Fish. Res. Board Can. 33:2040-2046.

Dunn, B.P., and D.R. Young. 1976. Baseline levels of benzo(a)pyrene in southern California. Mar. Pollut. Bull. 7:231-234.

Dunstan, W.M., L.P. Atkinson, and J. Natale. 1975. Stimulation and inhibition of plankton growth by low molecular weight hydrocarbons. Mar. Biol. 31:305-310.

Eastwood, D. 1981. Use of luminescence spectroscopy in oil identification. In. E.L. Wehyr, ed. Modern Fluorescence Spectroscopy. Vol. 4. Plenum, New York.

Eastwood, D., S.H. Fortier, and M.S. Hendrick. 1978. Oil identification--recent developments in fluorescence and low temperature luminescence. American Laboratory (March).

Eaton, P., and V. Zitko. 1979. Polycyclic Aromatic Hydrocarbons in Marine Sediment and Shellfish Near Creosoted Wharf Structures in Eastern Canada. International Council for the Exploration of the Seas, Charlottenlund, Denmark.

Edmondson, W.T., and G.G. Winberg. 1971. Secondary productivity in fresh waters. IBP Handbook 17. Blackwell Scientific Publications, Oxford. 358 pp.

Edwards, R.R.C. 1978. Effects of water-soluble oil fractions on metabolism, growth, and carbon budget of the shrimp Crangon crangon. Mar. Biol. 46:259-265.

Egami, N., and Y. Hyodo-Taguchi. 1973. Dominant lethal mutations in the fish, Oryzias latipes, irradiated at various stages of gametogenesis. In J.H. Schroder, ed. Genetics and Mutagenesis of Fish. Springer-Verlag, New York.

Eganhouse, R.P., and I.R. Kaplan. 1981. Extractable organic matter in municipal wastewaters. I. Petroleum hydrocarbons--temporal variations and mass emission rates to the ocean. Environ. Sci. Technol. (in press).

Ehrhardt, M. 1976. A veratile system for the accumulation of dissolved, non-polar organic compounds from seawater. Meteor. Forsch-Ergebnisse A(18):9-12.

Ehrhardt, M. 1978. An automatic sampling buoy for the accumulation of dissolved and particulate organic material from seawater. Deep Sea Res. 25:119-126.

Ehrhardt, M., and J. Heinemann. 1975. Hydrocarbons in blue mussels from the Kiel Bight. Environ. Pollut. 9:263-281.

Eide, I., A. Jensen, and S. Melsom. 1979. Application of in situ cage cultures of phytoplankton for monitoring heavy metal pollution in two Norwegian fjords. J. Exp. Mar. Biol. Ecol. 37:71-286.

Eisenbeiss, F., H. Hein, R. Joester, and G. Naundorf. 1978. The separation by L.C. and determination of polycyclic aromatic hydrocarbons in water using an antegrated enrichment step. Chromatogr. Newsletter 6(1):8-12.

Eisler, R. 1975. Toxic, sublethal, and latent effects of petroleum on Red Sea macrofauna, pp. 535-540. In Proceedings, 1975 Conference on the Prevention and Control of Oil Pollution. American Petroleum Institute, Washington, D.C.

Elder, S.R., W.O. Thompson, and R.H. Myers. 1981. Properties of composite sampling procedures. Technometrics 22:179-186.

Elinton, G., and R.J. Hamilton. 1963. The distribution of alkanes, pp. 187-217. In T. Swain, ed. Chemical Plant Taxonomy. Academic Press, New York.

Elmgren, R., and J.B. Frithsen. 1981. The use of experimental ecosystems for evaluating the environmental impact of pollutants: a comparison of an oil spill in the Baltic and a chronic oil addition experiment in microcosms. In G.D. Grice and M.R. Reeve, eds. Marine Mesocosms: Biological and Chemical Research in Experiment Ecosystems. Springer-Verlag, New York (in press).

Elmgren, R., J.F. Grassle, J.P. Grassle, D.R. Heinle, G. Langlois, S.L. Vargo, and G.A. Vargo. 1980a. Trophic interactions in experimental marine ecosystems perturbed by oil, pp. 779-800. In J.P. Gisey, ed. Microcosms in Ecological Research. DOE Symposium Series. CONF 781101. National Technical Information Service, Springfield, Va.

Elmgren, R., S. Hansson, U. Larsson, and B. Sundelin. 1980b. Impact of oil on deep soft bottoms, pp. 97-126. In J.J. Kineman, R. Elmgren, and S. Hansson, eds. The Tsesis Oil Spill. National Oceanic and Atmospheric Administration, Boulder, Colo.

Engelhardt, F.R. 1982. Hydrocarbon metabolism and cortisol balance in oil-exposed ringed seal, Phoca hispida. Comp. Biochem. Physiol. 72C:133-136.

Engelhardt, F.R., J.R. Geraci, and T.G. Smith. 1977. Uptake and clearance of petroleum hydrocarbons in the ringed seal, Phoca hispida. J. Fish. Res. Board Can. 34(8):1143-1147.

Environmental Devices Company. 1977. Oil fluorescence data and current measurements performed during February 1977 for the Buzzards Bay oil spill. Final Report. NOAA, Boulder, Colo.

Environmental Protection Agency. 1975a. Methods for acute toxicity tests with fish, macroinvertebrates, and amphibians. EPA-660/3-75-009. EPA, Corvallis, Oreg.

Environmental Protection Agency. 1975b. Bioassay procedures for the ocean disposal permit program. EPA-600/9-78-010. EPA, Gulf Breeze, Fla.

Environmental Protection Agency. 1979. Method 418.1 petroleum hydrocarbons, total recoverable (spectrometric, infrared). In Methods for Chemical Analysis of Water and Wastes. EPA, Environmental Monitoring and Support Laboratory, Cincinnati, Ohio.

Environmental Protection Agency. 1980. Interim methods for the sampling and analysis of priority pollutants in sediments and fish tissue. EPA, Environmental Monitoring and Support Laboratory, Cincinnati, Ohio.

Environmental Protection Agency/Corps of Engineers. 1977. Ecological evaluation of proposed discharge of dredged material into ocean waters. Environmental Effects Laboratory, U.S. Army Engineer Waterways Experiment Station, Vicksburg, Miss.

Eppley, R.W., J.N. Rogers, and J.J. McCarthy. 1969. Half-saturation constants for uptake of nitrate and ammonium by marine phytoplankton. Limnol. Oceanogr. 14:912-920.

Erasmus, T., R.M. Randall, and B.M. Randall. 1981. Oil pollution, insulation and body temperatures in the jackass penguin. Comp. Biochem. Physiol. 69(A):169-171.

Ernster, L., P. Siekevitz, and G.E. Palade. 1962. Enzyme-structure relationships in the endoplasmic reticulum of rat liver: a morphological and biochemical study. J. Cell Biol. 15(3):541-562.

Eurell, J.A., and W.E. Haensly. 1981. The effects of exposure to water-soluble fractions of crude oil on selected histochemical parameters of the liver of Atlantic croaker, Micropogon undulatus L. J. Fish Dis. 4:187-194.

Fantasia, J.F., and H.C. Ingrao. 1973. The development of an experimental airborne laser oil spill remote sensory system, pp. 101-106. In Proceedings, 1973 Oil Spill Conference, American Petroleum Institute, Washington, D.C.

Fantasia, J.F., T.M. Hard, and H.G. Ingrao. 1971. An investigation of oil fluorescence as a technique for remote sensing of oil spills. Report TSC-USCG-71-7. DOT Transportation Systems Center, U.S. Coast Guard, Washington, D.C.

Farrington, J.W. 1978. Santa Barbara sediment--South Louisiana crude oil intercalibration. Final Report. Contract AA550-CT6-43. Bureau of Land Management, Washington, D.C.

Farrington, J.W. 1980. An overview of the biogeochemistry of fossil fuel hydrocarbons in the marine environment, pp. 1-22. In L. Petrakis and F.T. Weiss, eds. Petroleum in the Marine Environment. Advances in Chemistry Series 185. American Chemical Society, Washington, D.C.

Farrington, J.W., and G.C. Medeiros. 1977. Evaluation of some methods of analysis for petroleum hydrocarbons in marine organisms, pp. 115-122. In Proceedings, 1975 Oil Spill Conference. American Petroleum Institute, Washington, D.C.

Farrington, J.W., and P.A. Meyers. 1975. Hydrocarbons in the marine environment, pp. 109-136. In G. Eglinton, ed. Environmental Chemistry. Vol. I. The Chemical Society, London.

Farrington, J.W., and J.G. Quinn. 1973. Petroleum hydrocarbons in Narragansett Bay. I. Survey of hydrocarbons in sediments and clams (Mercenaria mercenaria). Estuarine Coastal Mar. Sci. 1:71-79.

Farrington, J.W., and B.W. Tripp. 1975. A comparison of methods of analysis for hydrocarbons in surface sediments, pp. 267-289. In T.M. Church, ed. Marine Chemistry in the Coastal Environment. Symposium Series 18. American Chemical Society, Washington, D.C.

Farrington, J.W., and B.W. Tripp. 1977. Hydrocarbons in western North Atlantic surface sediments. Geochim. Cosmochim. Acta 41:1627-1641.

Farrington, J.W., J.M. Teal, and P.L. Parker. 1976a. Petroleum hydrocarbons. In E.D. Goldberg, ed. Strategies for Marine Pollution Monitoring. John Wiley & Sons, New York.

Farrington, J.W., J. Teal, G.C. Medeiros, K.A. Burns, E.A. Robinson, Jr., J.G. Quinn, and T.L. Wade. 1976b. Intercalibration of gas chromatographic analyses for hydrocarbons in tissues and extracts of marine organisms. Anal. Chem. 48:1711-1716.

Farrington, J.W., J. Albaiges, K.A. Burns, B.P. Dunn, P. Eaton, J.L. Laseter, P.L. Parker, and W. Wise. 1980. Fossil fuels, pp. 7-77. In The International Mussel Watch. National Academy of Sciences, Washington, D.C.

Farrington, J.W., B.W. Tripp, J.M. Teal, G. Mille, K. Tjessem, A.C. Davis, J.B. Livramento, N.A. Hayward, and N.M. Frew. 1982a. Biogeochemistry of aromatic hydrocarbons in the benthos of microcosms. Toxicol. Environ. Chem. 5:331-346.

Farrington, J.W., R.W. Risebrough, P.L. Parker, A.C. Davis, B. deLappe, J.K. Winters, D. Boatwright, and N.M. Frew. 1982b. Hydrocarbons, polychlorinated biphenyls and DDE in mussels and oysters from the U.S. coast 1976-1978--The Mussel Watch. Technical Report WHOI 82-42. Woods Hole Oceanographic Institution, Woods Hole, Mass.

Farrington, J.W., E.D. Goldberg, R.W. Risebrough, J.H. Martin, and V.T. Bowen. 1983. U.S. "Mussel Watch" 1976-1978: An overview of the trace metal, DDE, PCB, hydrocarbon, and artificial radionuclide data. Environ. Sci. Tech. 17:490-496.

Federle, T.W., J.R. Vestal, G.R. Hater, and M.C. Miller. 1979. Effects of Prudhoe Bay crude oil on primary production and zooplankton in arctic tundra thaw ponds. Mar. Environ. Res. 2:3-18.

Fiest, D.L., and P.D. Boehm. 1981. Subsurface distributions of petroleum from an offshore well blowout, the Ixtoc I blowout, Bay of Campeche. Environ. Sci. Technol. (in press).

Finney, D.J. 1971. Probit Analysis. 3rd ed. Cambridge University Press, Cambridge, England. 333 pp.

Fisher, N.S. 1977. On the differential sensitivity of estuarine and open ocean diatoms to exotic chemical stress. Am. Naturalist 111(981):871-895.

Fleming, J.H., and J. Coughlan. 1978. Preservation of vitally stained zooplankton for live/dead sorting. Estuaries 1(2):135-137.

Fong, W.C. 1976. Uptake and retention of Kuwait crude oil and its effects on oxygen uptake by the soft shell clam Mya arenaria. J. Fish. Res. Board Can. 33:2774-2780.

Fortier, S.H., and D. Eastwood. 1978. Identification of fuel oils by low-temperature luminescence spectrometry. Anal. Chem. 50(2):334.

Fossato, V.U., and W.J. Canzonier. 1976. Hydrocarbon uptake and loss by the mussel Mytilus edulis. Mar. Biol. 36:243-250.

Frame, G.M., G.A. Flanigan, and D.C. Carmody. 1979. Application of gas chromatography, using nitrogen-selective detection, to oil spill identification. J. Chromatogr. 168:365-376.

Frank, U. 1975. Identification of petroleum oils by fluorescence spectroscopy, pp. 87-91. In Proceedings of Joint Conference on Prevention and Control of Oil Spills. San Francisco, Calif.

Freegarde, M., C.G. Hatchard, and C.A. Parker. 1971. Oil spilt at sea: its identification, determination and ultimate fate. Lab. Practice 20(1):35-40.

Frezza, D., B. Pegoraro, and S. Presciuttini. 1982. A marine host-mediated assay for the detection of mutagenic compounds in polluted sea waters. Mut. Res. 104:215-223.

Friant, S.L., and I.H. Suffet. 1979. Interactive effects of temperature, salt concentration, and pH on head space analysis for isolating volatile trace organics in aqueous environmental samples. Anal. Chem. 51:2167-2172.

Friocourt, M.P., Y. Gourmelun, F. Berthou, R. Cosson, and M. Marchand. 1981. Effets de la pollution de l'Amoco Cadiz sur l'ostréiculture en Bretagne Nord: Suivi chimique de la pollution, de l'epuration et de l'adaptation, pp. 617-631. In Amoco Cadiz: Fate and Effects of the Oil Spill. CNEXO, Paris.

Fucik, K.W., H.W. Armstrong, and J.M. Neff. 1977. Uptake of naphthalenes of the clam, Rangia cuneata, in the vicinity of an oil separator platform in Trinity Bay, Texas, pp. 637-640. In Proceedings, 1977 Oil Spill Conference (Prevention, Behavior, Control, Cleanup). American Petroleum Institute, Washington, D.C.

Galloway, W.B., J.L. Lake, D.K. Phelps, P.F. Rogerson, V.T. Bowen, J.W. Farrington, E.D. Goldberg, J.L. Laseter, G.C. Lawler, J.H. Martin, and R.W. Risebrough. 1983. The Mussel Watch: intercomparison of trace level constituent determinations. Environ. Toxicol. Chem. 2:395-410.

Galtshoff, P.S., H.F. Prytherch, R.O. Smith, and V. Koehring. 1935. Effect of crude oil pollution on oysters in Louisiana waters. Bur. Fish. Bull. 18:143-209.

Gardner, G.R., P.P. Yevich, and P.F. Rogerson. 1975. Morphological anomalies in adult oyster, scallop, and Atlantic silversides exposed to waste motor oil, pp. 473-477. In Proceedings, 1975 Conference on the Prevention and Control of Oil Pollution. American Petroleum Institute, Washington, D.C.

Gardner, W.D. 1980. Field assessment of sediment traps. J. Mar. Res. 38(1):41-52.

Gay, M.L., A.A. Belishe, and J.F. Patton. 1980. Quantification of petroleum-type hydrocarbons in avian tissue. J. Chromatogr. 187:153-160.

Gearing, J.N., P.J. Gearing, T.F. Lytle, and J.S. Lytle. 1978. Comparison of thin-layer and column chromatography for separation of sedimentary hydrocarbons. Anal. Chem. 50(13):1833-1836.

Gearing, P.J., and J.N. Gearing. 1982a. Behavior of No. 2 fuel oil in the water column of controlled ecosystems. Mar. Environ. Res. 6:115-132.

Gearing, P.J., and J.N. Gearing. 1982b. Transport of No. 2 fuel oil between water column, surface microlayer, and atmosphere in controlled ecosystems. Mar. Environ. Res. 6:133-143.

Gearing, P.J., J.N. Gearing, R.J. Pruell, T.L. Wade, and J.G. Quinn. 1980. Partitioning of No. 2 fuel oil in controlled estuarine ecosystems: sediments and suspended particulate matter. Environ. Sci. Technol. 14(9):1129-1136.

Geizler, P.C., B.J. Grantham, and G.J. Blomquist. 1977. Fate of labeled n-alkanes in the blue crab and stripped mullet. Bull. Environ. Contam. Toxicol. 17:463-467.

Gelpi, E., D.W. Nooner, and J. Oro. 1970. The ubiquity of hydrocarbons in nature: aliphatic hydrocarbons in dust samples. Geochim. Cosmochim. Acta 34:421-425.

Geraci, J.R., and T.G. Smith. 1976. Direct and indirect effects of oil on ringed seals (Phoca hispada) of the Beaufort Sea. J. Fish. Res. Board Can. 33:1976-1984.

Geraci, J.R., and T.G. Smith. 1977. Consequences of oil fouling on marine mammals, pp. 399-410. In D.C. Malins, ed. Effects of Petroleum on Arctic and Subarctic Marine Environments and Organisms. Vol. II, Biological Effects. Academic Press, New York.

Geraci, J.R., and D.J. St. Aubin. 1982. Study of the effects of oil on cetaceans. Final Report. Contract AA 551-CT9-29. U.S. Department of the Interior, Bureau of Land Management, Washington, D.C. 274 pp.

Geraci, J.R., D.J. St. Aubin, and R.J. Reisman. 1983. Bottlenose dolphins, Tursiops truncatus, can detect oil. Can. J. Fish. Aquat. Sci. (submitted).

Giam, C.S., H.E. Murray, L.E. Ray, and S. Kira. 1980. Bioaccumulation of hexachlorobenzene in killifish (Fundulus similis). Bull. Environ. Contam. Toxicol. 25:891-897.

Gibbs, C.F. 1972. A new approach to the measurement of oxidation of crude oil in sea water systems. Chemosphere 3:119-124.

Gier, O. 1979. The impact of oil pollution on intertidal meiofauna, field studies after the La Coruna spill, May 1976. Can. Biol. Mar. 20:231-251.

Giering, L.P., and A.W. Hornig. 1977. Total Luminescence Spectroscopy, A Powerful Technique for Mixture Analysis. American Laboratory.

Gieskes, W.W.C., G.W. Kraay, and M.A. Baars. 1979. Current ^{14}C methods for measuring primary production: gross underestimates in oceanic waters. Neth. J. Sea Res. 13(1):58-78.

Giger, W., and M. Blumer. 1974. Polycyclic aromatic hydrocarbons in the environment: isolation and characterization by chromatography, visible, ultraviolet and mass spectrometry. Anal. Chem. 46:1663-1671.

Giger, W., and C. Schaffner. 1978. Determination of polycyclic aromatic hydrocarbons in the environment by glass capilllary gas chromatography. Anal. Chem. 50:243-249.

Gilfillan, E.S., and J.H. Vandermeulen. 1978. Alternation in growth and physiology of soft-shell clams, Mya arenaria, chronically oiled with Bunker C from Chedabucto Bay, Nova Scotia, 1970-76. J. Fish. Res. Board Can. 35:630-636.

Ginsberg, A.S. 1972. Fertilization in Fishes and the Problem of Polyspermy. National Technical Information Service, Springfield, VA 22151.

Glass, V., V.L. Willson, and J.M. Gottman. 1975. Design and analysis of time-series experiments. Colorado Associated University Press, Boulder.

Glemarec, M., and E. Hussenot. 1981. Definition d'une succession ecologique en milieu meuble anormalement enrichi en matières organiques a la suite de la catastrophe de l'Amoco Cadiz. In Amoco Cadiz: Fate and Effects of the Oil Spill. CNEXO, Paris.

Golary, M.J.E. 1958. Gas chromatography. In V.J. Coates, H.J. Noebels, and I.S. Fagerson, eds. Academic Press, New York.

Gold, J.R. 1979. Cytogenetics. In W.S. Hoar, D.J. Randall, and J.R. Brett, eds. Fish Physiology. Vol. 8. Academic Press, New York.

Goldacre, R.J. 1968. Effects of detergents and oils on the cell membrane, pp. 131-137. In Field Study Council, ed. The Biological Effects of Oil Pollution on Littoral Communities. Vol. II. Classey, London.

Goldburg, E.D., ed. 1979. Remote sensing and problems of the hydrosphere. Conference Publication 2109. NASA, Washington, D.C.

Goldman, J.C., and J.A. Davidson. 1977. Physical model of marine phytoplankton chlorination at coastal power plants. Environ. Sci. Technol. 11:908-913.

Gordon, D.C., Jr., and P.D. Keizer. 1974. Hydrocarbon concentrations in sea water along the Halifax-Bermuda Section: lessons learned regarding sampling and some results, pp. 113-115. In Marine Pollution Monitoring (Petroleum). NBS Special Publication No. 409. National Bureau of Standards, Gaithersburg, Md.

Gordon, D.C., Jr., and N.J. Prouse. 1973. The effects of three oils on marine phytoplankton photosynthesis. Mar. Biol. 22:329-333.

Gordon, D.C., Jr., P.D. Keizer, and N.J. Prouse. 1973. Laboratory studies of the accommodation of some crude and residual fuel oils in sea water. J. Fish. Res. Board Can. 30:1611-1618.

Gordon, D.C., Jr., P.D. Keizer, W.R. Hardstaff, and D.G. Aldous. 1976. Fate of crude oil spilled in seawater contained in outdoor tanks. Environ. Sci. Technol. 10:580-585.

Gordon, D.C., Jr., J. Dale, and P.D. Keizer. 1978a. Importance of sediment working by the deposit-feeding polychaete Arenicola marina on the weathering rate of sediment-bound oil. J. Fish. Res. Board Can. 35:591-603.

Gordon, D.C., Jr., P.D. Keizer, and J. Dale. 1978b. Temporal variations and probable origins of hydrocarbons in the water column of Bedford Basin, Nova Scotia. Estuarine Coastal Mar. Sci. 7:243-256.

Gorsline, J., W.N. Holmes, and J. Cronshaw. 1981. The effects of ingested petroleum on the gasoline metabolizing properties of liver tissue in seawater-adapted mallard ducks, Anas platyrhynchos. Environ. Res. 24:377-390.

Goutx, M., and A. Saliot. 1980. Relationship between dissolved and particulate fatty acids and hydrocarbons, chlorophyll A and zooplankton biomass in Villefranche Bay, Mediterranean Sea. Mar. Chem. 8:299-318.

Grahl-Nielsen, O., J.T. Staveland, and S. Wilhelmsen. 1978. Aromatic hydrocarbons in benthic organisms from coastal areas polluted by Iranian crude oil. J. Fish. Res. Board Can. 35:615-623.

Grassle, J.F., and J.P. Grassle. 1974. Opportunities, life histories and genetic systems in marine benthic polychaetes. J. Mar. Res. 32:253-284.

Grassle, J.F., R. Elmgren, and J.P. Grassle. 1981. Response of benthic communities in MERL experimental ecosystems to low level, chronic additions of No. 2 fuel oil. Mar. Environ. Res. 4:279-297.

Gray, J.S. 1980. The measurement of effects of pollutants on benthic communities. Rapp. P.-v. Reun. Cons. Int. Explor. Mer 179:188-193.

Gray, J.S., and F.B. Mirza. 1979. A method for the detection of pollution-induced disturbances on marine benthic communities. Mar. Pollut. Bull. 10:142-146.

Gray, J.S., D. Boesch, C. Heip, A.M. Jones, J. Lassig, R. Vanderhorst, and D. Wolfe. 1980. The role of ecology in marine pollution monitoring. Rapp. P.-v. Reun. Cons. Int. Explor. Mer 179:237-252.

Green, D.R. 1978. Sampling sea water for trace hydrocarbon determination: the state of the art in 1977. Publication NRCC 16565. Marine Analytical Chemistry Standards Program, National Research Council of Canada, Halifax, Nova Scotia.

Green, R.H. 1979. Sampling Design and Statistical Methods for Environmental Biologists. Wiley-Interscience, New York. 257 pp.

Green, R.H., and G.L. Vascotto. 1978. Analysis at environmental factors controlling spatial patterns of species composition. Water Res. 12:583-590.

Grice, G.D., and M.R. Reeve. 1982. Marine Mesocosms. Biological and Chemical Research in Experimental Ecosystems. Springer-Verlag, New York. 430 pp.

Grice, G.D., G.R. Harvey, V.T. Brown, and R.H. Backus. 1972. The collection and preservation of open ocean marine organisms for pollution analysis. Bull. Environ. Contam. Toxicol. 7:125-132.

Griffiths, R.P., B.A. Caldwell, W.A. Broich, and R.Y. Morita. 1981. Long-term effects of crude oil on uptake and respiration of glucose and glutamate in arctic and subarctic marine sediments. Appl. Environ. Microbiol. 42:792-801.

Grimmer, G., and D. Duval. 1970. Investigations of biosynthetic formation of polycyclic hydrocarbons in higher plants. Naturforsch. 25b:1171-1175.

Gritz, R.L., and D.G. Shaw. 1977. A comparison of methods for hydrocarbon analysis of marine biota. Bull. Environ. Contam. Toxicol. 17(4):408-415.

Grob, K., and G. Grob. 1976. Glass capillary gas chromatography in water analysis. How to initiate use of the method. In L.H. Keith, ed. Identification and Analysis of Organic Pollutants in Water. Ann Arbor Scientific Publishers, Ann Arbor, Mich.

Grob, K., and G. Grob. 1977. Practical capillary gas chromatography--a systematic approach. J. High Resolut. Chromatogr. Chromatogr. Commun. (March 1979):113-117.

Grob, K., and G. Grob. 1979. Practical capillary gas chromatography, a systematic approach. J. High Resolut. Chromatogr. Chromatogr. Commun. 2:109-117.

Grob, K., and K. Grob, Jr. 1978a. On-column injection on the glass capillary columns. J. Chromatogr. 151:311-320.

Grob, K., and K. Grob, Jr. 1978b. Splitless injection and the solvent effect. J. High Resolut. Chromatogr. Chromatogr. Commun. (July 1978):57-64.

Grose, T.L., and J.M. Mattson (Eds.) 1977. NOAA Special Report. 113 pp. U.S. Department of Commerce, Washington, D.C.

Gruenfeld, M. 1973. Extraction of dispersed oils from water for quantitative analysis by infrared spectrophotometry. Environ. Sci. Technol. 7:636-639.

Gruenfeld, M. 1975. Quantitative analysis of petroleum oil pollutants by infrared spectrophotometry. In Water Quality Parameters. ASTM STP 573. American Society for Testing and Materials, Philadelphia, Pa.

Gruenfeld, M., and R. Frederick. 1977. The ultrasonic dispersion, source identification, and quantitative analysis of petroleum oils in water. Rapp. P.-v. Reun. Cons. Int. Explor. Mer 171:33-38.

Gruger, E.H., Jr., J.V. Schenell, P.S. Fraser, D.W. Brown, and D.C. Malins. 1981. Metabolism of 2,6-dimethyl-napthalene in starry flounder (Platichthys stellatus) exposed to napthalene and p-cresol. Aquat. Toxicol. 1:37-48.

Guillard, R.R.L. 1973. Division rates, pp. 283-311. In J.R. Stein, ed. Handbook of Phycological Methods. Culture Methods and Growth Measurements. Cambridge University Press, New York.

Gundlach, E., and M. Hayes. 1978. Investigations of beach processes, pp. 85-96. In W.N. Hess, ed. The Amoco Cadiz Oil Spill. NOAA/EPA Special Report. Superintendent of Documents, U.S. Government Printing Office. Washington, D.C. 20402.

Gunkel, W., and H.H. Trekel. 1967. On the method of quantitative determination of oil-decomposing bacteria in oil-polluted sediments and soils, oil-water mixtures, oils and tarry substances. Helgolander Wiss. Meeresunters 16:336-348.

Gyllenberg, G., and G. Lundqvist. 1976. Some effects of emulsifiers and oil on two copepod species. Acta Zool. Fenn. 148:1-24.

Haegh, T., P.P. Rosmanith, R.G. Lichtenthaler, J. Boler, F. Oreld, O. Grahl-Nielsen, and K. Westrheim. Inter-laboratory comparison of determinations of petroleum hydrocarbons in sea water, marine sediments and organisms. Continental Shelf Institute, Trondheim, Norway.

Haensly, W.E., J.M. Neff, A.C. Morris, M.F. Bodgood, and P.D. Boehm. 1981. Histopathology of the plaice Pleuronectes platessa from Aber Wrac'h and Aber Benoit, Brittany, France: long term effects of the Amoco Cadiz crude oil spill. L. J. Fish. Dis. 5:365-391.

Hagstrom, B.E., and S. Lonning. 1967. Cytological and morphological studies of the action of lithium on the development of the sea urchin embryo. Roux Archiv Entwicklungsmechanik. 158:1.

Hamilton, M.A., R.C. Russo, and R.V. Thurston. 1977. Trimmed SpearmanKarber method for estimating median lethal concentrations in toxicity bioassays. Environ. Sci. Technol. 11:714-719.

Han, J., and M. Calvin. 1969. Hydrocarbon distribution of algae and bacteria and microbiological activity in sediments, pp. 436-443. In Proceedings, National Academy of Sciences. Vol. 64. Washington, D.C.

Han, J., E.D. McCarthy, M. Calvin, and M.M. Benn. 1968. Hydrocarbon constituents of the bluegreen algae Nostoc uscorum, Anacystis nidulans, Phormidium luridum, and Chlorogloea fritschii. J. Chem. Soc. (C):2785-2791.

Handa, N. 1966. Examination on the applicability of phenol sulfuric acid method to the determination of dissolved carbohydrates in sea water. J. Oceanogr. Soc. Japan 2:79-86.

Haney, T. 1971. An in situ method for the measurement of zooplankton grazing rates. Limnol. Oceanogr. 16(6):970-977.

Hansen, D.J., P.R. Parrish, S.C. Schimmel, and L.R. Goodman. 1978. Life-cycle toxicity test using sheepshead minnows (Cyprinodon variegatus), pp. 109-117. In Bioassay Procedures for the Ocean Disposal Permit Program. EPA-600/9-78-010. EPA, Gulf Breeze, Fla.

Hardy, R., P.R. Mackie, K.J. Whittle, and A.D. McIntyre. 1974. Discrimination in the assimilation of n-alkanes in fish. Nature 252:577-578.

Hargrave, B.T., and G.A. Phillips. 1975. Estimates of oil in aquatic sediments by fluorescence spectroscopy. Environ. Pollut. 8:193-215.

Hargraves, P.E., and R.R.L. Guillard. 1974. Structural and physiological observations on some small marine diatoms. Phycologia 13:163-172.

Harris, R.P. 1978. Photosynthesis, productivity and growth: the physiological ecology of phytoplankton. Arch. Hydrobiol. Beih. Ergebn. Limnol. 10:1-171.

Harris, R.P., V. Berdugo, S.C.M. O'Hara, and E.D.S. Corner. 1977a. Accumulation of ^{14}C-1-naphthalene by an oceanic and an estuarine copepod during long-term exposure to low level concentrations. Mar. Biol. 42:187-195.

Harris, R.P., V. Berdugo, E.D.S. Corner, C.C. Kilvington, and S.C.M. O'Hara. 1977b. Factors affecting the retention of a petroleum hydrocarbon by marine planktonic copepods, pp. 286-304. In D.A. Wolfe, ed. Fate and Effects of Petroleum Hydrocarbons in Marine Ecosystems and Organisms. Pergamon, New York.

Hartman, B., and D. Hammond. 1981. The use of carbon and sulfur isotopes as correlation parameters for the source identification of beach tar in the southern California borderland. Geochim. Cosmochim. Acta 45:309-319.

Hartmann, C.H. 1971. Gas chromatography detectors. Anal. Chem. 43:113A.

Harvey, G.R., and C.S. Giam. 1976. Polychlorinated biphenyls and DDT compounds, pp. 35-46. In E.D. Goldberg, ed. Strategies for Marine Pollution Monitoring. John Wiley & Sons, New York.

Hase, A., and R.A. Hites. 1976. On the origin of polycyclic aromatic hydrocarbons in recent sediments: biosynthesis by anaerobic bacteria. Geochim. Cosmochim. Acta 40:549-555.

Hauser, T.R., and J.N. Pattison. 1972. Analysis of aliphatic fraction of air particulate matter. Environ. Sci. Technol. 6:549-555.

Hawkes, J.W. 1977. The effects of petroleum hydrocarbon exposure on the structure of fish tissues, pp. 115-128. In D.A. Wolfe, ed. Fate and Effects of Petroleum Hydrocarbons in Marine Organisms and Ecosystems. Pergamon, New York.

Hawkes, J.W., E.H. Gruger, Jr., and O.P. Olson. 1980. Effects of petroleum hydrocarbons and chlorinated biphenyls on the morphology of the intestine of chinook salmon (Oncorhynchus tshawytscha). Environ. Res. 23:149-161.

Heddle, J.A., and W.R. Bruce. 1977. Comparison of the micronucleus and sperm assays for mutagenicity with the carcinogenic activities of 61 different agents. In H.H. Hiatt, J.D. Watson, and J.A. Winsten, eds. Cold Spring Harbor Symposium, Origins of Human Cancer. Book C. Cold Spring Harbor, Ill.

Hedtke, S.F., and F.A. Puglishi. 1980. Effects of waste oil on the survival and reproduction of the American flagfish, Jordanella floridae. Can. J. Fish. Aquat. Sci. 37:757-764.

Hemsworth, B.N., and A.A. Wardhaugh. 1978. The induction of dominant lethal mutations in Tilapia mossambica by alkane sulphonic esters. Mut. Res. 58:263-268.

Hershner, C., and J. Lake. 1980. Effects of chronic oil pollution on a saltmarsh grass community. Mar. Biol. 56:163-173.

Hertz, H.S., J.M. Brown, S.N. Chesler, F.R. Guenther, L.R. Hilpert, W.E. May, R.M. Paris, and S.A. Wise. 1980. Determination of individual organic compounds in shale oil. Anal. Chem. 52:1650-1657.

Hill, M.O., and H.G. Gaugh. 1980. Detrended correspondence analysis: an improved ordination technique. Vegetatio 42:47-58.

Hilpert, L.R., W.E. May, S.A. Wise, S.N. Chesler, and H.S. Hertz. 1978. Interlaboratory comparison of determinations of trace level petroleum hydrocarbons in marine sediments. Anal. Chem. 50:458-463.

Hinga, K.R., M.E.Q. Pilson, R.F. Lee, J.W. Farrington, K. Tjessem, and A.C. Davis. 1980. Biogeochemistry of benzanthracene in an enclosed marine ecosystem. Environ. Sci. Technol. 14:1136-1143.

Hites, R.A., and W.G. Bieman. 1975. Identification of specific organic compounds in a highly anoxic sediment by gas chromatographic-mass spectrometry and high resolution mass spectrometry, pp. 188-201. In R.P. Gibbs, ed. Analytical Methods in Oceanography. Advances in Chemistry Series 147. American Chemical Society, Washington, D.C.

Hites, R.A., R.E. Laflamme, and J.G. Windsor, Jr. 1980. Polycyclic aromatic hydrocarbons in marine/aquatic sediments, pp. 289-311. In L. Petrakis and F.T. Wiess, eds. Petroleum in the Marine Environment. Advances in Chemistry Series 185. American Chemical Society, Washington, D.C.

Hodgins, H.O., B.B. McCain, and J.W. Hawkes. 1977. Marine fish and invertebrate diseases, host disease resistance, and pathological effects of petroleum, pp. 95-173. In D.C. Malins, ed. Effects of Petroleum on Arctic and Subarctic Marine Environments and Organisms. Vol. II, Biological Effects. Academic Press, New York.

Hoffman, E.J., and J.G. Quinn. 1979. Gas chromatographic analyses of Argo Merchant oil and sediment hydrocarbons at the wreck site. Mar. Pollut. Bull. 10:20-24.

Hoffman, E.J., J.G. Quinn, J.R. Jadamec, and S.H. Fortier. 1979. Comparison of UV fluorescence and gas chromatographic analyses of hydrocarbons in sediments from the vicinity of the Argo Merchant wreck site. Bull. Environ. Contam. Toxicol. 23:536-543.

Hoge, F.E., and R.N. Swift. 1980. Oil film thickness measurement using airborne laser induced water Raman backscatter. Appl. Opt. 19:3269-3281.

Hollaender, A., ed. 1971. Chemical Mutagens, Principles and Methods for Their Detection. Vols. 1 and 2. Plenum Press, New York.

Hollaender, A. 1973. Chemical Mutagens, Principles and Methods for Their Detection. Vol. 3. Plenum Press, New York.

Hollaender, A. 1976. Chemical Mutagens, Principles and Methods for Their Detection. Vol. 4. Plenum Press, New York.

Hollaender, A., and F.J. de Serres, eds. 1976. Chemical Mutagens, Principles and Methods for Their Detection. Vol. 5. Plenum Press, New York.

Hollstein, M., J. McCann, F.A. Angelosanto, and W.W. Nicholls. 1979. Short-term tests for carcinogens and mutagens. Mut. Res. 65:133.

Holme, N., and A.D. McIntrye, eds. 1971. Methods for the Study of Marine Benthos. Blackwell Scientific Publishers, London.

Holmes, W.N., and J. Cronshaw. 1977. Biological effects of petroleum on marine birds, pp. 359-398. In D.C. Malins, ed. Effects of Petroleum on Arctic and Subarctic Marine Environments and Organisms. Vol. II, Biological Effects. Academic Press, New York.

Holmes, W.N., J. Cronshaw, and J. Gorsline. 1978. Some effects of ingested petroleum on seawater adapted ducks, Anus platyrhynchos. Environ. Res. 17:177-190.

Hood, L.V.S., and C.M. Erikson. 1980. Analysis of Alaskan crude oils by glass capillary gas chromatography/mass spectrometry. J. High Resolut. Chromatogr. Chromatogr. Commun. 3:516-520.

Hooftman, R.N., and G.J. Vink. 1981. Cytogenetic effects on the eastern mud-minnow, Umbra pygmaea, exposed to ethyl methanesulphonate, benzo(a)pyrene and river water. Ecotoxicol. Environ. Safety (in press).

Hope-Jones, P., G. Howells, E.I.S. Rees, and J. Wilson. 1970. Effect of Hamilton Trader oil on birds in the Irish Sea in May 1970. Brit. Bds. 63:97-110.

Hornig, A.W. 1974. Identification, estimation, and monitoring of petroleum in marine waters by luminescence methods, pp. 135-144. In Marine Pollution Monitoring (Petroleum). NBS Special Publication 409. National Bureau of Standards, Gaithersburg, Md.

Horvath, R., W.L. Morgan, and S.R. Stewart. 1971. Optical remote sensing of oil slicks: signature analysis and systems evaluation. Final Report. U.S. Coast Guard Project 724104.2/1. Willow Run Laboratories, University of Michigan, Ann Arbor.

Howgate, P., P.R. Mackie, K.J. Whittle, J. Farmer, A.D. McIntyre, and A. Eleftheriou. 1977. Petroleum tainting in fish. Rapp. P.-v. Reun. Cons. Int. Explor. Mer 171:143-146.

Hrivnac, M., W. Frischknecht, and L. Cechoa. 1976. Gas chromatographic multidector coupled to a glass capillary column. Anal. Chem. 48:937-940.

Hsiao, S.I.C. 1978. Effects of crude oils on the growth of arctic marine phytoplankton. Environ. Pollut. 17:93-107.

Hsiao, S.I.C., D.W. Kittle, and M.G. Fox. 1978. Effects of crude oil and the oil dispersant Corexit on primary production of arctic marine phytoplankton and seaweed. Environ. Pollut. 15:209-221.

Hubbard, S.A., M.H.L. Green, and J.W. Bridges. 1981. Detection of carcinogens using the fluctuations test with S9 or with hepatocyte activation, pp. 296-305. In H.F. Stich, and R.H.C. San, eds. Short Term Tests for Chemical Carcinogens. Springer-Verlag, New York.

Huitema, B.E. 1979. An analysis of covariance Markov model for interrupted time series experiments. Commun. Statist. Theor. Meth. A8:789-797.

Hulings, N., and J.S. Gray, eds. 1971. A manual for the study of meiofauna. Smithson. Contr. Zool.

Hunter, L. 1975. Quantitation of environmental hydrocarbons by thin-layer chromatography, gravimetry/densitometry comparison. Environ. Sci. Technol. 9:241-246.

Hyland, J.L., and D.C. Miller. 1979. Effects of No. 2 fuel oil on chemically evolved feeding behavior of the mud snail, Ilyanassa obsoleta, pp. 603-607. In Proceedings, 1979 Oil Spill Conference (Prevention, Behavior, Control, Cleanup). American Petroleum Institute, Washington, D.C.

Hyland, J.L., and E.D. Schneider. 1976. Petroleum hydrocarbons and their effects on marine organisms, populations, communities, ecosystems, pp. 464-506. In Sources, Effects and Sinks of Hydrocarbons in the Aquatic Environment. American Institute of Biological Science, Arlington, Va.

Hyland, J.L., P.F. Rogerson, and G.R. Gardner. 1977. A continuous flow bioassay system for the exposure of marine organisms to oil, pp. 547-550. In Proceedings, 1977 Oil Spill Conference (Prevention, Behavior, Control, Cleanup). American Petroleum Institute, Washington, D.C.

Hyodo-Taguchi, Y. 1968. Rate of development of intestinal damage in the goldfish after X-irradiation and mucosal cell kinetics at different temperatures. In M.F. Sullivan, ed. Gastrointestinal Radiation Injury. Excerpta Medica Foundation, Amsterdam.

Ignatiades, L., and N. Mimicos. 1977. Ecological responses of phytoplankton to chronic oil pollution. Environ. Pollut. 13:109-118.

Ikeda, T. 1976. The effect of laboratory conditions on the extrapolation of experimental measurements to the ecology of marine zooplankton. I. Effect of feeding condition on the respiration rate. Bull. Plankton Soc. Japan 23(2):51-60.

Ikeda, T. 1977. The effect of laboratory conditions on the extrapolation of experimental measurements to the ecology of marine zooplankton. II. Effect of oxygen saturation on the respiration rate. Bull. Plankton Soc. Japan. 24(1):19-28.

International Council for the Exploration of the Sea. 1977. The Ekofisk Bravo blowout. Compiled Norwegian Contributions. Fisheries Improvement Committee, Ref.: Shellfish and Benthos Committee. ICES C.M. 1977/E:55. Charlottenlund, Denmark. (Note: ICES ruling requires permission of the authors to cite these papers. Most have been cited by other authors already, and none have been published in full elsewhere.)

IOC/WMO. 1976. Guide to operational procedures for the IGOSS project on marine pollution (petroleum) monitoring. UNESCO, Paris.

Jackson, L., T. Bidleman, and W. Vernberg. 1981. Influence of reproductive activity on toxicity of petroleum hydrocarbons to ghost crabs. Mar. Pollut. Bull. 12:63-65.

Jannasch, H.W., O.C. Zafiriou, and J.W. Farrington. 1980. A sequencing sediment trap for time-series studies of fragile particles. Limnol. Oceanogr. 25(5):939-943.

Jennings, W. 1980. Gas Chromatography with Glass Capillary Columns. 2nd ed. Academic Press, New York.

Jensen, A., B. Rystad, and S. Melsom. 1976. Heavy metal tolerance of marine phytoplankton, Part 2. Copper tolerance of 3 species in dialysis and batch cultures. J. Exp. Mar. Biol. Ecol. 22(3):249-256.

Jensen, A., B. Rystad and L. Skoglund. 1972. The use of dialysis culture in phytoplankton studies. J. Exp. Mar. Biol. Ecol. 8:241-248.

Jenssen, D., and C. Ramel. 1976. Dose response at low doses of X-irradiation and MMS on the induction of micronuclei in mouse erythroblasts. Mut. Res. 41:311.

Jenssen, D., and C. Ramel. 1980. The micronucleus test as part of a short-term mutagenicity test program for the prediction of carcinogenicity evaluated by 143 agents tested. Mut. Res. 75:191.

Jewell, D.M. 1980. The role of nonhydrocarbons in the analysis of virgin and biodegraded petroleum, pp. 219-235. In L. Petrakis and F.T. Weiss, eds. Petroleum in the Marine Environment. Advances in Chemistry Series 185. American Chemical Society, Washington, D.C.

Johansson, S. 1980. Impact of oil on the pelagic ecosystem, pp. 61-80. In J.J. Kineman, R. Elmgren, and S. Hansson, eds. The Tsesis Oil Spill. National Oceanic and Atmospheric Administration, Boulder, Colo.

Johansson, S., U. Larsson, and P. Boehm. 1980. The Tsesis oil spill. I. Impact of the pelagic ecosystem. Mar. Pollut. Bull. 11:284-293.

John, P., and I. Soutar. 1976. Identification of crude oils by synchronous excitation spectrofluorometry. Anal. Chem. 48:520-524.

Johns, D.M., and J.A. Pechenik. 1980. Influence of the water-accommodated fraction of No. 2 fuel oil on energetics of Cancer irroratus larvae. Mar. Biol. 55:247-254.

Johnson, J.C., C.D. McAuliffe, and R.A. Brown. 1978. Physical and chemical behavior of small crude oil slicks in the ocean. In McCarthy, Lindblom, and Walter, eds. Chemical Dispersants for the Control of Oil Spills. ASTM STP 659. American Society for Testing and Materials, Philadelphia, Pa.

Johnson, T.S., R.E. Nakatani, and F.P. Conte. 1970. Influence of temperature and X-irradiation on the cellular dynamics of the intestinal epithelium in coho salmon, Oncorhynchus kisutch. Radiat. Res. 42:129.

Jones, H. 1981. The use of adenosine triphosphate as a measure of bacterial biomass in cultures grown on crude oil. Masters thesis. University College of North Wales, Department of Biology, Marine Science Laboratories, Menai Bridge, Anglesey, Gwynedd, U.K. 51 pp.

Karydis, M. 1979. Short-term effects of hydrocarbons on the photosynthesis and respiration of phytoplankton species. Bot. Marina 22(5):281-285.

Kauss, P.B. and J.C. Hutchinson. 1974. Studies on the susceptibility of An kistrodesmus species to crude oil components. In Proceedings of the 19th Congress of the International Association of Theoretical and Applied Limnology. Winnipeg, Canada.

Kauss, P.B., and J.C. Hutchinson. 1975. The effects of water soluble petroleum components on the growth of Chlorella vulgaris Beijerinck. Environ. Pollut. 9:157-174.

Kauss, P.B., J.C. Hutchinson, O. Soto, J. Hellebust, and M. Griffiths. 1973. The toxicity of crude oil and its components to freshwater algae, pp. 703-714. In Proceedings, Joint Conference on Prevention and Control of Oil Spills. American Petroleum Institute, Washington, D.C.

Keizer, P.D., and D.C. Gordon, Jr. 1973. Detection of trace amounts of oil in sea water by fluorescence spectroscopy. J. Fish. Res. Board Can. 30:1039-1046.

Keizer, P.D., D.C. Gordon, Jr., and J. Dale. 1977. Hydrocarbons in eastern Canadian marine waters determined by fluorescence spectroscopy and gas-liquid chromatography. J. Fish. Res. Board Can. 34:347-353.

Keizer, P.D., T.P. Ahern, J. Dale, and J.H. Vandermeulen. 1978. Residues of Bunker C oil in Chedabucto Bay, Nova Scotia, 6 years after the Arrow spill. J. Fish. Res. Board Can. 35:528-535.

Kezic, N., M. Rijavec, and B. Kurelec. 1980. Frequency of neoplasia in fish from the river Sava. Mut. Res. 74:195.

Kiceniuk, J.W., W.R. Penrose, and W.R. Squires. 1978. Oil spill dispersants cause bradycardia in a marine fish. Mar. Pollut. Bull. 9:42-45.

Kiefer, D.A. 1973. Chlorophyll fluorescence in marine centric diatoms: responses of chloroplasts to light and nutrient stress. Mar. Biol. 23:39-46.

Kieth, L.H., W. Crummet, J. Deegan, Jr., R.A. Libby, J.K. Taylor, and G. Wentler. 1983. Principles of environmental analysis. Anal. Chem. 55:2210-2218.

Kihlman, B.A., and D. Kronberg. 1975. Sister chromatid exchanges in Vicia faba. I. Demonstration by a modified fluorescent plus Giemsa (FPG) technique. Chromosoma 51:1.

Kilbey, B.J., M. Legator, W.W. Nichols, and C. Ramel, eds. 1977. Handbook of Mutagenicity Test Procedures. Elsevier, New York.

Kim, H., and G. Hickman. 1973. An airborne laser fluorosensor for the detection of oil on water. In Proceedings of Second Conference on Environmental Quality Sensors F-5. July 10-11, 1973. Las Vegas, Nevada.

Klekowski, E.J., and W. Barnes. 1979. The detection of mutagenic pollutants in aquatic and marine environments: studies of the Millers River and Barton Harbor. Water Resources Research Center, University of Massachusetts, Amherst.

Klemas, V. 1980. Remote sensing of coastal fronts and their effects on oil dispersion. Int. J. Remote Sensing 1:11-28.

Kliesch, V., and I.D. Adler. 1980. Sensitivity comparison of chromosome analysis and micronucleus test in mouse bone marrow. Mut. Res. 74:160.

Kligerman, A.D., and S.E. Bloom. 1977. Rapid chromosome preparations from solid tissues of fishes. J. Fish. Res. Board Can. 34:266.

Kligerman, A.D., S.E. Bloom, and W.M. Howell. 1975. Umbra limi, a model for the study of chromosome aberrations in fishes. Mut. Res. 31:225.

Knight, G.C., and C.H. Walker. 1982a. A study of hepatic microsomal epoxide hydroxylase in seabirds. Comp. Biochem. Physiol. 73(B):463-467.

Knight, G.C. and C.H. Walker. 1982b. A study of hepatic microsomal monooxygenase of seabirds and its relationship to organochlorine pollutants. Comp. Biochem. Physiol. 73(C):211-221.

Kniskern, F.E., et al. 1975. NOAA operational products in oceanography derived from satellite data. Paper presented to the International Council for the Exploration of the Sea, Charlottenlund, Denmark.

Kobayashi, N. 1974. Marine pollution bioassay by sea urchin eggs. An attempt to enhance accuracy. Publ. Seto Mar. Biol. Lab. 21:377.

Koffler, R. 1975. Use of NOAA environmental satellites to remotely sense ocean phenomena, pp. 835-839. In IEEE Ocean '75.

Kogure, K., U. Simidu, and N. Taga. 1979. A tentative direct microscopic method for counting living marine bacteria. Can. J. Microbiol. 25:415-420.

Kolattukudy, P.E., and T.J. Walton. 1972. The biochemistry of plant citicular lipids. In R.T. Holman, ed. Progress in the Chemistry of Fats and Other Lipids. Pergamon, New York.

Kolpack, R.L., J.S. Mattson, J.B. Mark, Jr., and T.C. Tu. 1971. Hydrocarbon content of Santa Barbara Channel sediments. In R.L. Kolpack, ed. Biological and Oceanographical Survey of the Santa Barbara Channel Oil Spill 1969-1970. Vol. II, Physical, Chemical, and Geological Studies. Allan Hancock Foundation, University of Southern California, Los Angeles.

Koons, C.B. 1973. Chemical composition: a control on the physical and chemical processes acting on petroleum in the marine environment. In Background Papers for a Workshop on Inputs, Fates, and Effects of Petroleum in the Marine Environment. Vol. II. National Academy of Sciences, Washington, D.C.

Koons, C.B., G.W. Jaieson, and L.S. Giereszko. 1965. Normal alkane distribution in marine organisms: possible significance to petroleum origin. Bull. Am. Assoc. Petrol. Geol. 49:301-304.

Koons, C.B., P.H. Monaghan, and G.C. Bayliss. 1971. Pitfalls in oil spill characterization: needs for multiple parameter approach and direct comparison with specific parent oils. Paper presented at Southwest Regional ACT Meeting, San Antonio, Tex., December 1971.

Korn, S., N. Hirsch, and J.W. Struhsaker. 1976. Uptake, distribution, and depuration of ^{14}C benzene in northern anchovy, Engraulis mordax, and striped bass, Horone saxatilis. Fish. Bull. 74:545-551.

Korn, S., N. Hirsch, and J.W. Struhsaker. 1977. The uptake, distribution, and depuration of ^{14}C benzene and ^{14}C toluene in Pacific herring, Clupea harengus pallasi. Fish. Bull. 75:633-636.

Kovats, E., and A.I.M. Keulemans. 1964. The Kovats retention index system. Anal. Chem. 36:31A-41A.

Krahn, M.M., D.W. Brown, T.K. Collier, A.J. Friedman, R.G. Jenkins, and D.C. Malins. 1980. Rapid analysis of naphthalene and its metabolites in biological systems: determination by high-performance liquid chromatography/fluorescence detection and by plasma desorption/chemical ionization mass spectrometry. J. Biochem. Biophys. Methods 2:233-246.

Krahn, M.M., J.V. Schnell, M.Y. Uyeda, and W.D. MacLeod, Jr. 1981. Determination of mixtures of benzo(a)pyrene, 2,6-dimethylnaphthalene and their metabolites by high-performance liquid chromatography with fluorescence detection. Anal. Chem. 112.

Krogh, F., and J. Haldorsen. 1978. OILSIM--a computer program simulating the fate of oil spills, pp. 1-9. Norwegian Maritime Research No. 4/1978.

Krugel, S., D. Jenkins, and S.A. Klein. 1978. Apparatus for the continuous dissolution of poorly water-soluble compounds for bioassays. Water Res. 12:269-272.

Kuhnhold, W.W., D. Everich, J.J. Stegeman, J. Lake, and R.E. Wolke. 1978. Effects of low levels of hydrocarbons on embryonic larval and adult winter flounder (Pseudopleuronectes americanus), pp. 677-711. In Proceedings of the Conference on Assessment of Ecological Impacts of Oil Spills. NTIS No. AD-A072 859. National Technical Information Service, Springfield, Va.

Kung, R.T.V., and I. Itzkan. 1976. Absolute oil fluorescence conversion efficiency. Appl. Opt. 15:409-415.

Kurlelec, B., S. Britvic, M. Rijavec, W.E.G. Müller, and R.K. Zahn. 1977. Benzo(a)pyrene monooxygenase induction in marine fish--molecular response to oil pollution. Mar. Biol. 44:211-216.

Kusk, K.O. 1978. Effects of crude oil and aromatic hydrocarbons on the photosynthesis of the diatom Nitzchia palea. Phys. Plant. 43:1-6.

Kwan, P.W., and R.C. Clark, Jr. 1981. Assessment of oil contamination in the marine environment by pattern recognition analysis of paraffinic hydrocarbon content of mussels. Anal. Chim. Acta 133:151-168.

Lacaze, J.C. 1974. Exotoxicology of crude oil, and the use of experimental marine ecosystems. Mar. Pollut. Bull. 5:153-156.

Lacaze, J.C., and O. Villedon de Naide. 1976. Influence of illumination on phytotoxicity of crude oil. Mar. Pollut. Bull. 7:73-76.

Laflamme, R.E., and R.A. Hites. 1978. The global distribution of polycyclic aromatic hydrocarbons in recent sediments. Geochim. Cosmochim. Acta 42:289-304.

Lahdetie, J., and M. Parvinen. 1981. Meiotic micronuclei induced by X-rays in early spermatids of the rat. Mut. Res. 81:103.

Lake, J.L., and O. Hershner. 1977. Petroleum sulfur-containing compounds and aromatic hydrocarbons in the marine mollusks Modiolus demissus and Oras virginica, pp. 627-632. In Proceedings, 1977 Oil Spill Conference (Prevention, Behavior, Control, Cleanup). American Petroleum Institute, Washington, D.C.

Lake, J.L., C.W. Dimock, and C.B. Norwood. 1980. A comparison of methods for the analysis of hydrocarbons in marine sediments, pp. 343-360. In L. Petrakis and R.T. Weiss, eds. Advances in Chemistry Series 185. American Chemical Society, Washington, D.C.

Landry, M.R., and R.P. Hassett. 1982. Estimating the grazing impact of marine micro-zooplankton. Mar. Biol. 67:283-288.

Lang, W.H., D.C. Miller, P.J. Ritacco, and M. Marcy. 1981. The effects of copper and cadmium on the behavior and development of barnacle larvae, pp. 165-203. In F.J. Vernberg, A. Calabrese, F.P. Thurberg, and W.B. Vernberg, eds. Biological Monitoring of Marine Pollutants Academic Press, New York.

Lannergren, C. 1978. Net and nanoplankton: effects of an oil spill in the North Sea. Bot. Marina 21(6):353-356.

Larson, R.A., L.L. Hunt, and D.W. Blankenship. 1977. Formation of toxic products from a No. 2 fuel oil by photooxidation. Environ. Sci. Technol. 11(5):492-496.

Laughlin, R.B., Jr., and J.M. Neff. 1979. The interactive effects of polynuclear aromatic hydrocarbons, salinity, and temperature on the survival and development rate of the larvae of the mud crab, Rhithropanopeus harrisii. Mar. Biol. 53:281-291.

Laughlin, R.B., Jr., L.G.L. Young, and J.M. Neff, 1978. A long-term study of the effects of water-soluble fractions of No. 2 fuel oil on the survival, development rate, and growth of the mud crab Rhithropanopeus harrisii. Mar. Biol. 47:87-95.

Law, R.J. 1978. Petroleum hydrocarbon analyses conducted following the wreck of the supertanker Amoco Cadiz. Mar. Pollut. Bull. 9:293-296.

Law, R.J. 1981. Hydrocarbon concentrations in water and sediments from U.K. marine waters, determined by fluorescence spectroscopy. Mar. Pollut. Bull. 12:153-157.

Law, R.J., and J.E. Portman. 1981. Report on the first ICES intercomparison exercise on petroleum hydrocarbons. Draft.

Lawler, G.C., W.-A. Loong, and J.C. Laseter. 1978. Accumulation of saturated hydrocarbons in tissues of petroleum-exposed mallard ducks (Anas platyrhynchos). Environ. Sci. Technol. 12:47-50.

Lee, M.L., and B.W. Wright. 1980. Capillary column gas chromatography of polycyclic aromatic compounds: a review. J. Chromatogr. Sci. 18:345-358.

Lee, M.L., G.P. Prado, J.B. Howard, and R.A. Hites. 1977. Source identification of urban airborne polycyclic aromatic hydrocarbons by gas chromatographics mass spectrometry and high resolution mass spectrometry. Biomed. Mass. Spec. 4:182-186.

Lee, M.L., D.K. Vassilonos, C.M. White, and M.N. Novotny. 1979. Retention indices for programmed temperature capillary column gas chromatography of polycyclic aromatic hydrocarbons. Anal. Chem. 51:768-775.

Lee, R.F. 1977. Fate of petroleum components in estuarine waters of the Southeastern United States, pp. 611-616. In Proceedings, 1977 Oil Spill Conference (Prevention, Behavior, Control, Cleanup). American Petroleum Institute, Washington, D.C.

Lee, R.F., and M. Takahashi. 1977. The fate and effect of petroleum in controlled ecosystem enclosures. Rapp. P.-v. Reun. Cons. Int. Explor. Mer 171:150-156.

Lee, R.F., J.C. Nevenzel, G.A. Paffenhofer, A.A. Benson, S. Parron, and R.E. Kavanagh. 1970. A unique hexadecane hydrocarbon from a diatom (Skeletonema costatum). Biochim. Biophys. Acta 202:386-388.

Lee, R.F., C. Ryan, and M.L. Neuhauser. 1976. Fate of petroleum hydrocarbons taken up from food and water by the blue crab, Callinectes sapidus. Mar. Biol. 37:363-370.

Lee, R.F., M. Takahashi, J.H. Beers, W.H. Thomas, D.R.L. Sieibert, P. Koeller, and D.R. Green. 1977. Controlled ecosystems: their use in the study of the effects of petroleum hydrocarbons on plankton, pp. 323-342. In F.-J. Vernberg, A. Calabrese, F.P. Thurberg, and W.B. Vernberg, eds. Physiological Responses of Marine Biota to Pollutants. Academic Press, New York.

Lee, R.F., W.S. Gardner, J.W. Anderson, J.W. Blaylock, and J. Barwell-Clarke. 1978. Fate of polycyclic aromatic hydrocarbons in controlled ecosystem enclosures. Environ. Sci. Technol. 12:832-838.

Lee, R.F., D. Lehsau, M. Madden, and W. Marsh. 1981a. Polycyclic aromatic hydrocarbon in oysters (Crassostrea virginica) from Georgia coastal waters, analyzed by high pressure liquid chromatography, pp. 341-345. In Proceedings, 1981 Oil Spill Conference (Prevention, Behavior, Control, Cleanup). American Petroleum Institute, Washington, D.C.

Lee, R.F., B. Dornseif, F. Gonsoulin, K. Tenore, and R. Hanson. 1981b. Fate and effects of a heavy fuel oil spill on a Georgia salt marsh. Mar. Environ. Res. 5:125-143.

Lee, W.Y. 1978. Chronic sublethal effects of the water soluble fractions of No. 2 fuel oils on the marine isopod Sphaeroma quadridentatum. Mar. Environ. Res. 1:5-17.

Lee, W.Y., and J.A.C. Nicol. 1977. The effects of the water soluble fractions of No. 2 fuel oil on the survival and behavior of coastal and oceanic zooplankton. Environ. Pollut. 12:279-292.

Lee, W.Y., K. Winters, and J.A.C. Nicol. 1978. The biological effects of the water-soluble fractions of a No. 2 fuel oil on the planktonic shrimp, Lucifer faxoni. Environ. Pollut. 15:167-183.

Levin, D.E., M. Hollstein, M.F. Christman, E.A. Schwiers, and B.N. Ames. 1982. A new Salmonella tester strain (TA102) with A.T. basepairs at the site of mutation detects oxidative mutagens. P.N.A.S. 79:7445-749.

Levy, E.M. 1971. The presence of petroleum residues off the east coast of Nova Scotia, in the Gulf of St. Lawrence and the St. Lawrence River. Water Res. 5:723-733.

Levy, E.M. 1979a. Intercomparison of Niskin and Blumer samplers for the study of dissolved and dispersed petroleum residues in sea-water. J. Fish. Res. Board Can. 36:1513-1516.

Levy, E.M. 1979b. Concentration of petroleum residues in the waters and sediments of Baffin Bay and the eastern Canadian Arctic. Report Series BI-R-79-3. Bedford Institute of Oceanography, Dartmouth, Nova Scotia. 34 pp.

Levy, E.M. 1980. Background levels of petroleum residues in Baffin Bay and eastern Canadian Arctic, pp. 277-294. In Petromar 80: Petroleum and the Marine Environment. Association Europeenne Oceanique, Monaco.

Levy, E.M., and J.D. Moffatt. 1975. Floating petroleum residues in the North Atlantic (1971-1974). Data Series Bl-D-75-9. Bedford Institute of Oceanography, Dartmouth, Nova Scotia.

Lichtfield, J.T., and F. Wilcoxon. 1949. A simplified method for evaluating dose effects experiments. J. Pharmacol. Exp. Therap. 96:99-113.

Linden, O., R. Elmgren, and P. Boehm. 1980a. The Tsesis oil spill: its impact on the coastal ecosystem of the Baltic Sea. Ambio. 8:244-253.

Linden, O., R. Laughlin, Jr., J.R. Sharp, and J.M. Neff. 1980b. The combined effect of salinity, temperature and oil on the growth pattern of embryos of the killifish, Fundulus heteroclitus Walbaum. Mar. Environ. Res. 3(2):129-144.

Lindley, D.V., and M.R. Norick. 1981. The role of exchangeability in inference. Annals Statist. 41:45-58.

Lloyd, J.B.F. 1971. The nature and evidential value of the luminescence of automobile engine oils and related materials. J. Forensic Sci. Soc. 11:83-94,153-10,235-253.

Loftus, M.E., and H.H. Seliger. 1975. Some limitations of the in vivo fluorescence technique. Chesapeake Sci. 16(2):79-92.

Longwell, A.C. 1977. A genetic look at fish eggs and oil. Oceanus 20:45.

Longwell, A.C. 1978. Field and laboratory measurements of stress responses at the chromosome and cell levels in planktonic fish eggs and the oil problem, pp. 116-125. In M.P. Wilson, J.G. Quinn, and K. Sherman, eds. In the Wake of the Argo Merchant. Center for Ocean Management Studies, University of Rhode Island, Kingston.

Longwell, A.C., and J.B. Hughes. 1980. Cytologic, cytogenetic and developmental state of Atlantic mackerel eggs from sea surface water of the New York Bight, and prospects for biological effects monitoring with ichthyoplankton. Rapp. P.-v. Reun. Cons. Int. Explor. Mer 179:274.

Longwell, A.C., and S.S. Stiles. 1968. Fertilization and completion of meiosis in spawned eggs of the American oyster, Crassostrea virginica. Caryologia 21:65.

Longwell, A.C., S.S. Stiles, and D.G. Smith. 1967. Chromosome complement of the American oyster, Crassostrea virginica, as seen in meiotic and cleaving eggs. Can. J. Genet. Cytol. 9:845.

Lönning, S. 1977. The sea urchin as a test object in oil pollution studies. Rapp. P.-v. Reun. Cons. Int. Explor. Mer 171:186.

Lowry, O.H., N.A. Rosenbrough, A.L. Farr, and R.J. Randall. 1951. Protein measurement with Folin phenol reagent. J. Biol. Chem. 193:265-275.

Loya, Y., and B. Rinkevich. 1980. Effects of oil pollution on coral reef communities. Mar. Ecol. Prog. Ser. 3:167-180.

Lysyj, I., and E.C. Russell. 1974. Dissolution of petroleum-derived products in water. Water Res. 8:863-868.

Lysyj, I., R. Rushworth, R. Melvold, and E.C. Russell. 1980. A scheme for analysis of oily waters, pp. 247-266. In L. Petrakis and F.T. Weiss, eds. Petroleum in the Marine Environment. Advances in Chemistry Series 185, American Chemical Society, Washington, D.C.

Lysyj, I., G. Perkins, and J.S. Farlow. 1981. Trace analysis for aromatic hydrocarbons in natural waters. Environ. Intern. (in press).

Lytle, J.S. 1975. Fate and effects of crude oil on an estuarine pond, pp. 595-600. In Proceedings, 1975 Conference on Prevention and Control of Oil Pollution. American Petroleum Institute, Washington, D.C.

Ma, Te-Hsiu. 1979. Micronuclei induced by X-rays and chemical mutagens in meiotic pollen mother cells of _Tradescantia_, a promising mutagen test system. Mut. Res. 64:307.

MacDougall, D., and W.B. Crummett. 1980. Guidelines for data acquisition and data quality evaluation in environmental chemistry. Anal. Chem. 52:2242-2249.

Mackay, D., and P.J. Leinonen. 1977. Mathematical model of the behavior of oil spills on water with natural and chemical dispersion. Economic and Technical Review Report EPA-3-EC-77-19. Fisheries and Environment Canada.

Mackie, P.R., and R. Hardy, E.I. Butler, P.M. Holligan, and M.F. Spooner. 1978. Early samples of oil in water and some analyses of zooplankton. Mar. Pollut. Bull. 9(11):296-297.

MacLeod, W.D., P.G. Prohaska, D.D. Gennero, and D.W. Brown. 1981a. Inter-laboratory comparisons of selected trace hydrocarbons found in marine sediments. NOAA National Analytical Facility, Northwest Alaska Fisheries Center, Environmental Conservation Division, 2725 Montlake Boulevard East, Seattle, WA 98112.

MacLeod, W.D., Jr., L.S. Ramos, A.J. Friedman, D.G. Burrows, P.G. Prohaska, D.L. Fisher, and D.W. Brown. 1981b. Analysis of residual chlorinated hydrocarbons, aromatic hydrocarbons, and related compounds in selected sources, sinks, and biota of New York Bight. NOAA/MESA Program, New York Bight Project, Stonybrook, New York.

Macko, S.A., and S.N. King. 1980. Weathered oil: effect on hatchability of heron and gull eggs. Bull. Environ. Anal. Chem. 25:316-320.

Mahoney, B.M., and H.H. Haskin. 1980. The effects of petroleum hydrocarbons on the growth of phytoplankton recognized as food forms for the eastern oyster, _Crassostrea virginica_ Omelin. Environ. Pollut., Ser. A 22(2):123-132.

Malins, D.C., ed. 1977. Effects of Petroleum on Arctic and Subarctic Marine Environments and Organisms. Vol. II, Biological Effects. Academic Press, New York.

Malins, D.C. 1980. Pollution of the marine environment. Environ. Sci. Technol. 14:32-37.

Malins, D.C., and W.T. Roubal. 1982. Aryl sulfate formation in sea urchins (Strongylocentrotus droebachiensis) ingesting marine algae (Fucus distichus) containing 2:6-dimethylnaphthalene. Environ. Res. 27:290-297.

Malins, D.C., T.K. Collier, L.C. Thomas, and W.T. Ronbal. 1979. Metabolic fate of aromatic hydrocarbons in aquatic organisms: analysis of metabolites by thin-layer chromatography and high-pressure liquid chromatography. Intern. J. Anal. Chem. 6:55-66.

Malins, D.C., M.M. Krahn, D.W. Brown, W.D. MacLeod, Jr., and T.K. Collier. 1980. Analysis for petroleum products in marine environments. Helgolander Meeresunters. 33:257-271.

Malone, T.C., and M.B. Chervin. 1979. The production and fate of photoplankton size fractions in the plume of the Hudson River, New York Bight. Limnol. Oceanogr. 24(4):683-696.

Marchand, M., and M.P. Caprais. 1981. Suivi de la pollution de l'Amoco Cadiz dans l'eau de mer et les sediments marins, pp. 23-54. In Amoco Cadiz: Fate and Effects of the Oil Spill. CNEXO, Paris.

Marsh, J.B., and D.B. Weinstein. 1966. A simple charring method for determination of lipids. J. Lipid. Res. 1:574-576.

Masters, B.S.S., C.H. Williams, Jr., and H. Kamin. 1967. In R.W. Estabrook and M.E. Pullman, eds. Methods in Enzymology. Vol. X. ACA Press, New York. p. 565.

Mattern, I.E., W. Van Der Zwagn, and G.W. Vink. 1981. The antimutagenic effects of an extract of the mussel Mytilus edulis (meeting abstract). Mut. Res. 85:241-242.

Maurer, A., and A.T. Edgerton. 1975. Flight evaluation of U.S. Coast Guard airborne oil surveillance system, pp. 129-142. In Proceedings 1975 Oil Spill Conference. American Petroleum Institute, Washington, D.C.

May, W.E., S.N. Chesler, S.P. Cram, B.H. Bump, H.S. Hertz, D.F. Enagonio, and S.M. Dyszel. 1975. Chromatographic analysis of hydrocarbons in marine sediments and seawater. J. Chromatogr. Sci. 13:535-540.

May, W.E., S.N. Chesler, B.H. Bump, and H.S. Hertz. 1978. An analysis of petroleum hydrocarbons in the marine environment: results of an interlaboratory comparison exercise. J. Environ. Sci. Health A13(5&6):403-410.

Maynard, D.J., and D.D. Weber. 1981. Avoidance reactions of juvenile coho salmon (Oncorhynchus kisutch) to monocyclic aromatics. Can. J. Fish. Aquat. Sci. 38:772-778.

Mayzaud, P. 1973. Respiration and nitrogen excretion of zooplankton. II. Studies of the metabolic characteristics of starved animals. Mar. Biol. 21:19-28.

Mazaki, M., T. Ishii, and M. Uyeta. 1982. Mutagenicity of hydrolysates of citrus fruit juices. Mut. Res. 101:283-291.

McAuliffe, C.D. 1966. Solubility in water of paraffin, cycloparaffin, olefin, acetylene, cycloolefin and aromatic hydrocarbons. J. Phys. Chem. Wash. 70:1267-1275.

McAuliffe, C.D. 1971. GC determination of solutes by multiple phase equilibration. Chem. Technol. 1:46-51.

McAuliffe, C.D. 1976. Surveillance of the marine environment for hydrocarbons. Mar. Sci. Commun. 2:13-42.

McAuliffe, C.D. 1980. The multiple gas-phase equilibration method and its application to environmental studies, pp. 193-218. In L. Petrakis and F.T. Weiss, eds. Petroleum in the Marine Environment. Advances in Chemistry Series 185. American Chemical Society, Washington, D.C.

McAuliffe, C.D., J.C. Johnson, S.H. Greene, G.P. Canevari, and T.D. Searl. 1980. Dispersion and weathering of chemically treated crude oils on the ocean. Environ. Sci. Technol. 14:1509-1518.

McCain, B.B., H.O. Hodgins, W.D. Gronlund, J.W. Hawkes, D.W. Brown, M.S. Myers, and J.H. Vandermeulen. 1978. Bioavailability of crude oil from experimentally oiled sediments to English sole (Parophrys vetulus), and pathological consequences. J. Fish. Res. Board Can. 35:657-664.

McCarthy, J.J. 1971. The role of urea in marine phytoplankton ecology. B. Ammonium and urea excretion of zooplankton, pp. 50-70. Ph.D. dissertation. Scripps Institution of Oceanography, La Jolla, Calif.

McCarthy, J.J., W.R. Taylor, and J.L. Taft. 1977. Nitrogenous nutrition of the plankton in the Chesapeake Bay. I. Nutrient availability and phytoplankton preferences. Limnol. Oceanogr. 22:996-1011.

McCormick, J.J., and V.M. Maher. 1981. Mutagenesis studies in diploid human cells with different DNA repair capacities, p. 264. In H.F. Stich, and R.H.C. San, eds. Short Term Tests for Chemical Carcinogenesis. Springer-Verlag, New York.

McFadden, W. 1973. Techniques of combined gas chromatography/mass spectrometry. John Wiley & Sons, New York. 463 pp.

McGowan, W.E. 1975. Monitoring dissolved hydrocarbons as a function of tidal cycle (New York Harbor). NTIS No. ADA015882. National Technical Information Service, Springfield, Va.

McIntyre, A.D., and J.B. Pearce, eds. 1980. Biological Effects of Marine Pollution and the Problems of Monitoring. Proceedings from ICES Workshop held in Beaufort, N.C., 26 February-2 March, 1979. Rapp. P.V. Rev. Vol. 179.

McKinlay, S.M. 1975. The design and analysis of the observational study--a review. J. Am. Statist. Assoc. 71:513-523.

McLachlan, J. 1973. Growth media marine, pp. 25-51. In J.R. Stein, ed. Handbook of Phycological Methods. Cambridge University Press.

McTaggart, N.G., and L.A. Luke. 1978. Molecular sieves for the analysis of petroleum. Fresenius Z. Anal. Chem. 290:1-9.

Measures, R.M., J. Garleck, W.R. Houston, and D.G. Stephenson. 1975. Laser induced spectral signatures of relevance to environmental sensing. Can. J. Remote Sensing 1:95.

Medeiros, G.C., and J.W. Farrington. 1974. IDOE-5 intercalibration sample: results of analysis after sixteen months storage. NBS Special Publication. 409. Marine Pollution Monitoring (Petroleum), Proceedings of a Symposium and Workshop, May 13-17, 1974. National Bureau of Standards, Gaithersburg, Md.

Meyerhoff, R.D. 1975. Acute toxicity of benzene, a component of crude oil, to juvenile striped bass (Morone saxatilis). J. Fish. Res. Board Can. 32:1864-1866.

Michael, A.D., D. Boesch, C. Hershner, R.J. Livingston, K. Roas, and D. Straughan. 1979. Oil spill studies. Benthos. J. Environ. Pathol. Toxicol. 93:91-118.

Michael, J., M.O. Hayes, and P.J. Brown. 1978. Application of an oil spill index vulnerability index to the shoreline of lower Cook Inlet, Alaska. Environ. Geol. 2(2):107-117.

Michael, L.C., M.D. Erickson, S.P. Parks, and E.D. Pellizzari. 1980. Volatile environmental pollutants in biological matrices with a headspace purge technique. Anal. Chem. 52:1836-1841.

Milan, C.S., and T. Whelan III. 1978. Accumulation of petroleum hydrocarbons in a salt marsh ecosystem exposed to steady state of oil input, pp. 875-893. In Proceedings of the Conference on Assessment of Ecological Impacts of Oil Spills. American Institute of Biological Sciences, Arlington, Va.

Miller, D.A., K. Skogerboe, and E.P. Grimsrud. 1981. Enhancement of electron capture detector response to polycyclic aromatic and related hydrocarbons by addition of oxygen to carrier gas. Anal. Chem. 53:464-467.

Miller, D.S., D.B. Peakall, and W.B. Kinter. 1978. Ingestion of crude oil: sublethal effect in herring gull chicks. Science 199:315-317.

Miller, J.W. 1973. A multiparameter oil pollution source identification system, pp. 195-204. In Proceedings, 1973 Oil Spill Conference. American Petroleum Institute, Washington, D.C.

Mironov, O.G., and T.L. Shckekaturina. 1979. Oil change in excretory products of mussels (Mytilus galloprovincialis). Mar. Pollut. Bull. 10:232-234.

Moles, A., S. Bates, S.D. Rice, and S. Korn. 1981. Reduced growth of coho salmon fry exposed to two petroleum components, toluene and naphthalene, in fresh water. Trans. Am. Fish. Soc. 110:430-436.

Moore, M.N., D.R. Livingstone, P. Donkin, B.L. Bayne, J. Widdows, and D.M. Lower. 1980. Mixed function oxygenases and xenobiotic detoxication/toxication systems in bivalve molluscs. Helgolander Meeresunters. 35:278-291.

Morris, R.J. 1973. Uptake and discharge of petroleum hydrocarbons by barnacles. Mar. Pollut. Bull. 4:107-109.

Morris, R.J., and F. Culkin. 1976. Marine lipids: analytical techniques and fatty acid ester analyses. Oceanogr. Mar. Biol. Cinn. Rev. 14:391-433.

Moss, M.E., and G.D. Tasker. 1979. Progress in the design of hydrologic-data networks. Rev. Geophys. Space Phys. 17:1298-1316.

Mulholland, R.J., and C.M. Gowdy. 1977. Theory and application of the measurement of structure and the determination of function for laboratory ecosystems. J. Theoret. Biol. 69:321-341.

Muller, D., J. Nelles, E. Deparade, and P. Auni. 1980. The activity of S9-liver fractions from seven species in the Salmonella/ mammalian-microsome mutagenicity test. Mut. Res. 70:279-300.

Murphy, L.S., and R.D. Belastock. 1980. The effect of environmental origin on the response of marine diatoms to chemical stress. Limnol. Oceanogr. 25:160-165.

Nagao, M., Y. Yahagi, T. Sieno, T. Sugimura, and N. Ito. 1977. Mutagenicities of quinoline and its derivatives. Mut. Res. 42:335-342.

Natarajan, A.T., A.D. Tates, P.P.W. van Buul, M. Meijers, and N. De Vogel. 1976. Cytogenetic effects of mutagens/carcinogens after activation in a microsomal system in vitro. I. Induction of chromosome aberrations and sister chromatid exchanges by diethylnitrosamine (DEN) and dimethylnitrosamine (DMN) in CHO cells in the presence of rat-liver microsomes. Mut. Res. 37:83.

National Academy of Sciences. 1980. The International Mussel Watch. Washington, D.C. 248 pp.

National Oceanic and Atmospheric Administration. 1977. NOAA/USCG spilled oil research team operations manual. Unpublished document. NOAA, Office of Marine Pollution Assessment, Rockville, Md.

National Oceanic and Atmospheric Administration. 1980. Proceedings of symposium on the Researcher/Pierce cruise to the Ixtoc 1 oil spill wellhead. NOAA, Office of Marine Pollution Assessment, Rockville, Md.

National Research Council. 1975. Petroleum in the Marine Environment. National Academy of Sciences, Washington, D.C.

Natusch, D.F.S., and B.A. Tomkins. 1978. Isolation of polycyclic organic compounds by solvent extraction with dimethylsulfoxide. Anal. Chem. 50:1429-1434.

Nava, M.E., and F.R. Engelhardt. 1980. Compartmentalization of ingested labelled petroleum in tissues and bile of the American eel (Anguilla rostrata). Bull. Environ. Contam. Toxicol. 24:879-885.

Neff, J.M. 1979. Polycyclic Aromatic Hydrocarbons in the Aquatic Environment: Sources, Fates and Biological Effects. Applied Science Publishers, London. 262 pp.

Neff, J.M., and J.W. Anderson. 1975a. Accumulation, release, and distribution of benzo(a)pyrene-^{14}C in the clam Rangia cuneata, pp. 469-471. In Proceedings, 1975 Conference on the Prevention and Control of Oil Pollution. American Petroleum Institute, Washington, D.C.

Neff, J.M., and J.W. Anderson. 1975b. An ultraviolet spectrophotometric method for the determination of naphthalene and alkylnaphthalenes in the tissues of oil-contaminated marine animals. Bull. Environ. Contam. Toxicol. 14:122-128.

Neff, J.M., and J.W. Anderson. 1981. Response of Marine Animals to Petroleum and Specific Petroleum Hydrocarbons. Applied Science Publishers, London. 177 pp.

Neff, J.M., B.A. Cox, D. Dixit, and J.W. Anderson. 1976. Accumulation and release of petroleum-derived aromatic hydrocarbons by four species of marine animals. Mar. Biol. 38:279-289.

Nelson-Smith, A. 1973. Oil Pollution and Marine Ecology. Plenum, New York. 260 pp.

Nettleship, D.N. 1976. Census techniques for seabirds of arctic and eastern Canada. Occasional. Pap. 25. Canadian Wildlife Service, 31 pp.

Newcombe, H.B. and J.F. McGregor. 1967. Major congenital malformations from irradiations of sperm and heads. Mut. Res. 4:663.

Nichols, H.W. 1973. Growth media freshwater, pp. 7-24. In, J.R. Stein, ed. Handbook of Phycological Methods. Cambridge University Press, New York.

Nicol, J.A.C., W.H. Donahue, R.T. Wang, and K. Winters. 1977. Chemical composition and effects of water extracts of petroleum on eggs of the sand dollar Melitta quinquesperforata. Mar. Biol. 40:309-316.

Nimmo, D.R., L.H. Bahner, R.A. Rigby, J.M. Sheppard, and A.J. Wilson, Jr. 1977. Mysidopsis bahia: an estuarine species suitable for life-cycle toxicity tests to determine the effects of a pollutant, pp. 109-116. In F.L. Mayer and J.L. Mahelink, eds. Aquatic Toxicology and Hazard Evaluation. ASTM STP 634. American Society for Testing and Materials, Philadelphia, Pa.

Norton, M.G., F.L. Franklin, and R.A.A. Blackman. 1978. Toxicity testing in the the United Kingdom for the evaluation of oil slick dispersants, pp. 18-34. In L.T. McCarthy, Jr., G.P. Lindblom, and H.F. Walter, eds. Chemical Dispersants for the Control of Oil Spills. ASTM STP 659. American Society for Testing and Materials, Philadelphia, Pa.

Notini, M. 1980. Impact of oil on the littoral ecosystem, pp. 129-165. In J.J. Kineman, R. Elmgren, and S. Hansson, eds. The Tsesis Oil Spill. National Oceanic and Atmospheric Administration, Boulder, Colo.

Novotny, M., F.J. Schwende, M.J. Nartigan, and J.E. Purcell. 1980a. Capillary gas chromatography with ultraviolet spectrometric detection. Anal. Chem. 42(4):636-740.

Novotny, M., R. Kump, F. Merli, and L.J. Todd. 1980b. Capillary gas chromatography/mass spectrometric determination of nitrogen aromatic compounds in complex mixtures. Anal. Chem. 52:401-406.

Nunes, P., and P.E. Benville, Jr. 1979a. Effects of the water-soluble fraction of Cook Inlet crude oil on the marine alga, Dunalliela teritiolecta. Bull. Environ. Contam. Toxicol. 21:727-732.

Nunes, P., and P.E. Benville, Jr. 1979b. Uptake and depuration of petroleum hydrocarbons in the Manila clam, Tapes semidecussata Reeve. Bull. Environ. Contam. Toxicol. 21:719-726.

O'Brien, P.Y., and P.S. Dixon. 1976. The effects of oil and oil components on algae: a review. Br. Phycol. J. 11(2):115-142.

O'Conners, H.B., Jr., C.F. Worster, C.D. Powers, D.C. Biggs, and R.G. Rowland. 1978. Polychlorinated biphenyls may alter marine trophic pathways by reducing phytoplankton size and production. Science 201:737-739.

Odense, R.B. 1982. Measurement of the mutagenic potential in extracts of Mytilus edulis collected from polluted harbours. Marine Environment Studies thesis. Institute for Resource Environment Studies, Dalhousie University, Halifax, Nova Scotia. 108 pp.

Odum, E.P. 1971. Fundamentals of Ecology. W.B. Saunders, Co. 574 pp.

Ogan, K., E. Katz, and W. Slavin. 1978. Concentration and determination of trace amounts of several PAH in aqueous samples. J. Chromatogr. Sci. 16:517-522.

Ogan, K., E. Katz, and W. Slavin. 1979. Determination of polycyclic aromatic hydrocarbons in aqueous samples by reversed-phase liquid chromatography. Anal. Chem. 51:1315-1320.

Ogata, M., and Y. Miyake. 1979. Identification of organic sulfur compounds transferred to fish from petroleum suspension by mass chromatography. Water Res. 13:1179-1185.

Ogata, M., Y. Miyake, S. Kira, K. Matsunaga, and M. Imanaka. 1977. Transfer to fish of petroleum paraffins and organic sulfur compounds. Water Res. 11:333-338.

Ogata, M., Y. Miyake, and Y. Yamasaki. 1979. Identification of substances transferred to fish or shellfish from petroleum suspension. Water Res. 13:613-618.

Ogata, M., Y. Miyake, K. Fujisawa, S. Kira, and Y. Yoshida. 1980. Accumulation and dissipation of organosulfur compounds in short-necked clam and eel. Bull. Environ. Contam. Toxicol. 75:130-135.

Ojima, Y. 1980. Fish cytogenetics. In K. Fujino, ed. Genetics and Breeding of Aquatic Creatures. Proceedings, Special Symposium, Present Status of Genetics and Future Prospects of Breed Improvement. Japanese Society of Scientific Fisheries, Japan National Academy of Science (in press).

Olla, B.L., W.H. Pearson, and A.L. Studholme. 1980a. Applicability of behavioral measures in environmental stress assessment. Rapp. P.-v. Reun. Cons. Int. Explor. Mer 179:162-173.

Omori, H. 1978. Some factors affecting dry weight, organic weight and concentrations of carbon and nitrogen in freshly prepared and in preserved zooplankton. Int. Rev. Ges. Hydrobiol. 63:261-269.

Omura, T., and R. Sato. 1964. The carbon monoxide-bending pigment of liver microsomes. II. Solubilization, purification and properties. J. Biol. Chem. 239(7):2379-2385.

O'Neil, R.A., A.R. Davies, H.G. Gross, and S. Kruss. 1975. Special Technical Publication 573. American Society for Testing and Materials, Philadelphia, Pa. p. 424.

O'Neil, R.A., L. Buja-Bijunas, and D.M. Rayner. 1980. Field performance of a laser fluorosensor for the detection of oil spills. Appl. Opt. 19:863-870.

O'Reilly, J.E., and J.P. Thomas. 1980. A manual for the measurement of Cotal daily primary productivity on Mar Ap and Ocean Pulse cruises using simulated in situ sunlight incubation. Ocean Pulse Technical Manual 1. Report SHL 79-06. NOAA, Northeast Fisheries Center, Sandy Hook Laboratory, Highlands, N.J.

Orloci, L. 1978. Multivariate Analysis in Vegetarian Research. 2nd ed. Junk, The Hague. 451 pp.

Orloci, L., C.R. Rao, and W.M. Stiteler, eds. 1979. Multivariate Methods in Ecological Work. International Cooperative Publishing, Fairland, Md. 550 pp.

250

Oro, J., T.G. Tornabene, D.W. Nooner, and E. Gelpi. 1967. Aliphatic hydrocarbons and fatty acids of some marine and fresh water microorganisms. J. Bacteriol. 93:1811-1818.

Ott, F.S., R.P. Harris, and S.C.M. O'Hara. 1978. Acute and sublethal toxicity of naphthalene and three methylated derivatives to the estuarine copepod, Eurytemora affinis. Mar. Environ. Res. 1:49-58.

Ourisson, G., P. Albrecht, and M. Rohmer. 1979. The hopanoids: paleo-chemistry and biochemistry of a group of natural products. Pure Appl. Chem. 51:709-729.

Overton, E.B., and J.L. Laseter. 1980. Distribution of aromatic hdyrocarbons in sediments from selected Atlantic, Gulf of Mexico and Pacific Outer Continental and Shelf areas, pp. 327-341. In L. Petrakis and F.T. Weiss, eds. Petroleum in the Marine Environment. Advances in Chemistry Series 185. American Chemical Society, Washington, D.C.

Overton, E.B., J. Bracken, and J.L. Laseter. 1977. Application of glass capillary columns to monitor petroleum-type hydrocarbons in marine sediments. J. Chromatogr. Sci. 15:169.

Overton, E.B., C.F. Steele, and J.L. Laseter. 1978a. Computer reconstruction of high resolution gas chromatograms. J. High Resolut. Chromatogr. (August 1978):109-110.

Overton, E.B., C.F. Steele, and J.L. Laseter. 1978b. Improved data processing software for high resolution chromatography data. Paper presented at 29th Annual Pittsburgh Conference, March 1978.

Overton, E.B., J.R. Patel, and J.L. Laseter. 1979. Chemical characterization of mousse and selected environmental samples from the Amoco Cadiz oil spill, pp. 169-214. In Proceedings, 1979 Oil Spill Conference (Prevention, Behavior, Control, Cleanup). American Petroleum Institute, Washington, D.C.

Overton, E.B., J.L. Laseter, W. Mascarella, C. Raschke, I. Nuiry, and J.W. Farrington. 1980a. Photochemical oxidation of Ixtoc I oil, pp. 41-386. In Proceedings of the Conference on the Preliminary Scientific Results from the Researcher/Pierce Cruise to the Ixtoc I Blowout. NOAA, Office of Marine Pollution Assessment, Rockville, Md.

Overton, E.B., L.W. McCarthy, S.W. Mascarella, M.A. Maberry, S.R. Antoine, J.L. Laseter, and J.W. Farrington. 1980b. Detailed chemical analysis of Ixtoc I crude oil and selected environmental samples from the Researcher and Pierce cruises, pp. 439-480. In Proceedings of the Conference on the Preliminary Scientific Results from the Researcher/Pierce Cruise to the Ixtoc I Blowout. NOAA, Office of Marine Pollution Assessment, Rockville, Md.

Overton, E.B., J. McFall, S.W. Mascarella, C.F. Steele, S.A. Antoine, I.R. Politzer, and J.L. Laseter. 1981. Petroleum residue sources identification after a fire and oil spill, pp. 541-546. In Proceedings, 1981 Oil Spill Conference (Prevention, Behavior, Control, Cleanup). American Petroleum Institute, Washington, D.C.

Owens, O.V., P. Dresler, C.C. Crawford, M.A. Taylor, and H.H. Seliger. 1977. Phytoplankton cages for the measurement in situ of the growth rates of mixed natural populations. Chesapeake Sci. 18(4):325-333.

Packard, T.T., and Q. Dortch. 1975. Particulate protein-nitrogen in North Atlantic surface waters. Mar. Biol. 33:347-354.

Paine, R.T. 1966. Food web complexity and species diversity. Amer. Nat. 100:65-75.

Palmork, K.H., and J.E. Solbakken. 1980. Accumulation and elimination of radioactivity in the Norway lobster (Nephrops norvezicus) following intragastric administration of [9-^{14}C] phenanthrene. Bull. Environ. Contam. Toxicol. 25:668-671.

Palmork, K.H., and J.E. Solbakken. 1981. Distribution and elimination of [9-^{14}C] phenanthrene in the horse mussel (Modiola modiolus). Bull. Environ. Contam. Toxicol. 26:196-201.

Pamatmat, M. 1971. Oxygen consumption by the seabed. 4. Shipboard and laboratory experiments. Limnol. Oceanogr. 16:536-550.

Pancirov, R.J., and R.A. Brown. 1977. Polynuclear aromatic hydrocarbons in marine tissues. Environ. Sci. Tech. 11:989-992.

Pancirov, R.J., and R.A. Brown. 1981. The measurement of polynuclear aromatic hydrocarbons in heating oils, crude oils, marine tissues, foodstuffs, and sediments. Final Report (unpublished). American Petroleum Institute, Washington, D.C.

Pancirov, R.J., T.D. Seal, and R.A. Brown. 1980. Methods of analysis for polynuclear aromatic hydrocarbons in environmental samples, pp. 123-142. In L. Petrakis and F.T. Weiss, eds. Petroleum in the Marine Environment. Advances in Chemistry Series 185. American Chemical Society, Washington, D.C.

Parker, C.A., M. Freegarde, and C.G. Hatchard. 1971. The effect of some chemical and biological factors on the degradation of crude oil at sea, pp. 237-244. In P. Hepple, ed. Water Pollution by Oil. Institute of Petroleum, London.

Parry, J.M., and M.A.J. Al-Mossawi. 1979. The detection of mutagenic chemicals in the tissue of the mussel Mytilus edulis. Environ. Pollut. 19:175-186.

Parsons, J., J. Spry, and T. Austin. 1980. Preliminary observations on the effects of Bunker C fuel oil on seals on the Scotian Shelf, pp. 193-202. In J.H. Vandermeulen, ed. Scientific Studies During the Kurdistan Tanker Incident: Proceedings of a Workshop. Report Series BI-R-80-3. Bedford Institute of Oceanography, Dartmouth, Nova Scotia.

Parsons, T.R., W.K.W. Li, and R. Waters. 1976. Some preliminary observations on the enhancement of phytoplankton growth by low levels of mineral hydrocarbons. Hydrobiology. 51:85-89.

Partridge, B.L. 1982. The structure and function of fish schools. Sci. Am. 246(6):114-123.

Patton, J.S., M.W. Rigler, P.D. Boehm, and D.L. Fiest. 1981. The Ixtoc I oil spill: flaking of surface mousse in the Gulf of Mexico. Nature 290:235-238.

Payne, J.F. 1976. Field evaluation of benzo(a)pyrene hydroxylase induction as a monitor for marine petroleum pollution. Science 191:945-946.

Payne, J.F. 1977. Mixed function oxidase in marine organisms in relation to petroleum hydrocarbon metabolism and detection. Mar. Pollut. Bull. 8(5):112-116.

Payne, J.F., and L.L. Fancey. 1982. Effect of long-term exposure to petroleum on mixed function oxygenases in fish: further support for use of the enzyme system in biological monitoring. Chemosphere 11(2):207-213.

Payne, J.F., and W.R. Penrose. 1975. Induction of aryl hydrocarbon benzo(a)pyrene hydroxylase in fish by petroleum. Bull. Environ. Contam. Toxicol. 14:112-116.

Payne, J.F., J.W. Kiceniuk, and W.R. Squires. 1978. Pathological changes in a marine fish after a 6-month exposure to petroleum. J. Fish. Res. Board Can. 35:665-667.

Payne, J.R., G. Smith, P.J. Mankiewicz, R.F. Shokes, N.W. Flynn, V. Moreno, and J. Altamirano. 1980a. Horizontal and vertical transport of dissolved hydrocarbons from the Ixtoc I blowout, pp. 239-266. In Proceedings of Symposium on the Preliminary Results from the Researcher/Pierce Cruise to the Ixtoc I Blowout. NOAA, Office of Marine Pollution Assessment, Rockville, Md.

Payne, J.R., N.W. Flynn, P.J. Mankiewicz, and G.S. Smith. 1980b. Surface evaporation/dissolution partitioning of lower-molecular-weight aromatic hydrocarbons in a down-plume transect from the Ixtoc I wellhead. In Proceedings of the Conference on the Preliminary Scientific Results from the Researcher/Pierce Cruise to the Ixtoc I Blowout. NOAA, Office of Marine Pollution Assessment, Rockville, Md.

Peaden, P.A., M.L. Lee, Y. Hirata, and M. Novotny. 1980. High performance liquid chromatographic separation of high-molecular-weight polycyclic aromatic compounds in carbon black. Anal. Chem. 52:2268-2271.

Peakall, D.B., D.S. Miller, and W.B. Kinter. 1979. Physiological techniques for assessing the impact of oil on seabirds, pp. 52-60. In E.E. Kenaga, ed. Avian and Mammalian Wildlife Toxicology. ASTM STP 693. American Society for Testing and Materials, Philadelphia, Pa.

Peakall, D.C., D. Hallett, D.S. Miller, R.G. Butler, and W.B. Kinter. 1980a. Effects of ingested crude oil on black guillemots: a combined field and laboratory study. Ambio. 9:28-30.

Peakall, D.B., J. Tremblay, W.B. Kinter, and D.S. Miller. 1980b. Endocrine function in seabirds caused by ingested oil. Environ Res. 23:1124-1129.

Pearce, J.B. 1981. Review of archiving and storage techniques used in maintaining biological tissues and other environment samples for future contaminant analyses. Report for the Advisory Committee on Marine Pollution. Unpublished manuscript. International Council for the Exploration of the Seas, Charlottenlund, Denmark.

Pearson, W.H., and B.L. Olla. 1979. Detection of naphthelene by the blue crab, Collinectus sapidus. Estuaries 2:64-65.

Percy, J.A., and T.C. Mullin. 1977. Effects of crude oil on the locomotory activity of arctic marine invertebrates. Mar. Pollut. Bull. 8:35-40.

Perkins, E.J. 1979. The need for sublethal studies, pp. 27-41. In H.A. Cole, ed. The Assessment of Sublethal Effects of Pollutants in the Sea. The Royal Society, London.

Pesch, G.G., and C.E. Pesch. 1980. _Neanthes arenaceodentata_ (Polychaeta: Annelida), a proposed cytogentic model for marine gentic toxicology. Can. J. Fish. Aquat. Sci. 37:1225.

Petrakis, L., and E. Edelheit. 1979. The utilization of nuclear magnetic resonance spectroscopy for petroleum, coal, oil shale, petrochemicals, and polymers. Phenomenology, Paradigms of Applications, and Instrumentation. Appl. Spect. Rev. 15(2):195-260.

Petrakis, L., and F.T. Weiss, eds. 1980. Petroleum in the Marine Environment. Advances in Chemistry Series 185. American Chemical Society, Washington, D.C.

Petrakis, L., D.M. Jewell, and W.F. Benusa. 1980. Analytical chemistry of petroleum: an overview of practices in petroleum industry laboratories with emphasis on biodegradation, pp. 23-54. In L. Petrakis and F.T. Weiss, eds. Petroleum in the Marine Environment. Advances in Chemistry Series 185. American Chemical Society, Washington, D.C.

Phillips, R.C., and C.P. McRoy. 1979. Handbook of Seagrass Biology. Garland, New York.

Philp, R.P., J.R. Maxwell, and G. Eglinton. 1976. Environmental organic geochemistry of aquatic sediments. Sci. Prog. Oxford 63:521-545.

Pielou, E.C. 1977. An Introduction to Mathematical Ecology. 2nd ed. John Wiley & Sons, New York. 385 pp.

Pierce, R.C., and M. Katz. 1975. Determination of atmospheric isomeric polycyclic arenes by thin-layer chromatography and fluorescence spectrophotometry. Anal. Chem. 47:1743.

Pilson, M.E.Q., C.A. Oviatt, G.A. Vargo, and S.L. Vargo. 1979. Replicability of MERL microcosms: initial observations, pp. 359-381. In F.S. Jacoff, ed. Advances in Marine Environmental Research. Proceedings of a Symposium, June 1977. EPA-600/9-79-035. Environmental Protection Agency, Narragansett, R.I.

Piotrowicz, S.R., C.A. Hogan, R.A. Shore, and A.A.P. Pszenny. 1981. The variability in the distribution of weak acid leachable Cd, Cr, Cu, Fe, Ni, Pb, and Zn, in the sediments of the Georges Bank/Gulf of Maine region. Environ. Sci. Technol. (in press).

Polak, R., A. Filion, S. Fortier, J. Lanier, and K. Cooper. 1978. Observations on _Argo Merchant_ oil in zooplankton of Nantucket Shoals, pp. 109-115. In M.P. Wilson, J.G. Quinn, and K. Sherman, eds. In the Wake of the _Argo Merchant_. Center for Ocean Management Studies, University of Rhode Island, Kingston.

Pomeroy, L.R., and R.G. Wiegert. 1981. The Ecology of a Salt Marsh. Springer-Verlag, New York. 271 pp.

Porter, K.G., and Y.S. Feig. 1980. The use of DAPI for identifying and counting aquatic miroflora. Limnol. Oceanogr. 25:943-948.

Posthuma, J. 1977. The composition of petroleum. Rapp. P.-v. Reun. Cons. Int. Explor. Mer 171:7-16.

Powell, N.A., C.S. Sayce, and D.F. Tufts. 1970. Hyperplasia in an estuarine bryozoan attributable to coal tar derivatives. J. Fish. Res. Board Can. 27:2095.

Prezelin, B.B., and A.C. Ley. 1980. Photosynthesis and chlorophyll fluorescence rhythms of marine phytoplankton. Mar. Biol. 55:295-307.

Prouse, N.J., and D.C. Gordon, Jr. 1976. Interactions between the deposit feeding polychaete Arenicola marina and oiled sediment, pp. 407-422. In Sources, Effects and Sinks of Hydrocarbons in the Aquatic Environment. American Institute of Biological Sciences, Arlington, Va.

Prouse, N.J., D.C. Gordon, Jr., and P.D. Keizer. 1976. Effects of low concentrations of oil accommodated in sea water on the growth of unialgal marine phytoplankton cultures. J. Fish. Res. Board Can. 33:810-818.

Pulich, W.M., Jr., K. Winters, and C. Van Baalen. 1974. The effects of a No. 2 fuel oil and two crude oils on the growth and photosynthesis of microalgae. Mar. Biol. 28:87-94.

Pym, J.G., J.E. Ray, G.W. Smith, and E.V. Whitehead. 1975. Petroleum triterpane fingerprinting of crude oils. Anal. Chem. 47:1617-1622.

Radke, M., H.G. Sittardt, and D.H. Welte. 1978. Removal of soluble organic matter from rock samples with a flow-through extraction cell. Anal. Chem. 50(4):663-665.

Ramos and Prohaska. 1981. Sephadex LH-20 chromatography of extracts of marine sediment and biological samples for the isolation of polynuclear aromatic hydrocarbons. J. Chromatogr. (in press).

Randolph, D.J., and J.R. Payne. 1980. Fate and weathering of petroleum spills in the marine environment: a literature review and synopsis. Ann Arbor Science Publishers, Ann Arbor, Mich. 175 pp.

Rashid, M.A. 1974. Degradation of Bunker C oil under different coastal environments of Chedabucto Bay, Nova Scotia. Estuarine Coastal Mar. Sci. 2:137-144.

Rasmussen, D.V. 1976. Characterization of oil spills by capillary column gas chromatography. Anal. Chem. 48:1562.

Rayner, D.M., M. L. Lee, and A.G. Szabo. 1978. Effect of sea-state on the performance of laser fluorosensors. Appl. Opt. 17:2730.

Reddin, A., and G.N. Preudeville. 1981. Effect of oils on cell membrane permeability in Fucus serratus and Laminoria digitata. Mar. Pollut. Bull. 12(10):339-342.

Reed, M., and M.C. Spaulding. 1979. A fishery-oil spill interaction model, pp. 63-74. In Proceedings, 1979 Oil Spill Conference (Prevention, Behavior, Control, Cleanup). Publication 4308. American Petroleum Institute, Washington, D.C. 728 pp.

Reed, W.E. 1977. Molecular compositions of weathered petroleum and comparison with its possible source. Geochim. Cosmochim. Acta 41:237-247.

Reed, W.E., and I.R. Kaplan. 1977. The chemistry of marine petroleum seeps. J. Geochem. Explor. 7:255-293.

Reed, W.E., I.R. Kaplan, M. Sandstrom, and P. Mankiewicz. 1977. Petroleum and anthropogenic influence on the composition of sediments from the Southern California Bight, pp. 183-188. In Proceedings, 1977 Oil Spill Conference (Prevention, Behavior, Control, Cleanup). American Petroleum Institute, Washington, D.C.

Reese, C.E. 1980a. Chromatographic data acquisition and processing. 1. Data acquisition. J. Chromatogr. Sci. 18:201-206.

Reese, C.E. 1980b. Chromatographic data acquisition and processing. 2. Data manipulation. J. Chromatogr. Sci. 18:249-257.

Reish, D.J. 1980. The use of polychaetous annelids as test organisms for marine bioassay experiments, pp. 140-154. In Aquatic Invertebrate Bioassays. American Society for Testing and Materials. Philadelphia, Pa.

Rice, S.D. 1973. Toxicity and avoidance tests with Prudhoe Bay oil and pink salmon fry, pp. 667-670. In Proceedings, Joint Conference on Prevention and Control of Oil Spills. American Petroleum Institute, Washington, D.C.

Rice, S.D., D.A. Moles, and J.W. Short. 1975. The effect of Prudhoe Bay crude oil on survival and growth of eggs, alevins, and fry of pink salmon, Oncorhynchus gorbuscha, pp. 503-507. In Proceedings, 1975 Conference on the Prevention and Control of Oil Pollution. American Petroleum Institute, Washington, D.C.

Rice, S.D., J.W. Short, and J.F. Karianen. 1976. Toxicity of Cook Inlet crude oil and No. 2 fuel oil to several Alaskan marine fishes and invertebrates. In Proceedings of the Symposium on the Sources, Effects, and Sinks of Hydrocarbons in the Aquatic Environment. The American Institute of Biological Sciences, Arlington, Va.

Rice, S.D., R.E. Thomas, and J.W. Short. 1977. Effect of petroleum hydrocarbons on breathing and coughing rates and hydrocarbon uptake-depuration in pink salmon fry, pp. 259-277. In F. Vernberg, J.A. Calabrese, F.P. Thurberg, and W.B. Vernberg, eds. Proceedings, Physiological Responses of Marine Biota to Pollutants. Academic Press, New York. 462 pp.

Rice, S.D., A. Moles, T.L. Taylor, and J.F. Karinen. 1979. Sensitivity of 39 Alaskan marine species to Cook Inlet crude oil and No. 2 fuel oil, pp. 549-554. In Proceedings, 1979 Oil Spill Conference (Prevention, Behavior, Control, Cleanup). American Petroleum Institute, Washington, D.C.

Rice, S.D., S. Korn, C.C. Brodersen, S.A. Lindsay, and S.A. Andrews. 1981. Toxicity of ballast-water treatment effluent to marine organisms at Port Valdez, Alaska, pp. 55-61. In Proceedings, 1981 Oil Spill Conference (Prevention, Behavior, Control, Cleanup). American Petroleum Institute, Washington, D.C.

Richter-Reichhelm, H.-B., M. Emura, S. Matthei, and U. Mohr. 1979. Cytotoxic and transforming effects of polycyclic hydrocarbons (PAH) on Syrian hamster fetal lung cell cultures. Mut. Res. 42:46.

Riley, R.G., and R.M. Bean. 1979. Application of liquid and gas chromatographic techniques to a study of the persistence of petroleum in marine sediments. NBS Special Publication 519. Trace Organic Analysis: A New Frontier in Analytical Chemistry. Proceedings of the 9th Materials Research Symposium, April 10-13, 1978. National Bureau of Standards, Gaithersburg, Md.

Riley, R.T., M.C. Mix, R.L. Schaffer, and D.L. Bunting. 1981. Uptake and accumulation of naphthalene by the oyster Ostrea edulis, in a flow-through system. Mar. Biol. 61:267-276.

Rinkevich, B., and Y. Loya. 1979. Laboratory experiments on the effects of crude oil on the Red Sea coral Stylophora pistillata. Mar. Pollut. Bull. 10:328-330.

Risebrough, R., V. Anderlini, M. Berhard, D.V. Ellis, G. Polykarpov, W. Robertson IV, E. Schneider, and G. Topping. 1980. Monitoring strategies for the protection of the coastal zone, pp. 236-238. In The International Mussel Watch. National Academy of Sciences, Washington, D.C.

Robinson, C.J., and G.L. Cook. 1969. Low-resolution mass spectrometric determination of arometric fractions from petroleum. Anal. Chem. 41:1548-1553.

Roesijadi, G., and J.W. Anderson. 1979. Condition index and free amino acid content of Macoma inquinata exposed to oil-contaminated marine sediments, pp. 69-83. In W.B. Vernberg, A. Calabrese, F.P. Thurberg, and F.J. Vernberg, eds. Marine Pollution Functional Responses. Academic Press, New York.

Roesijadi, G., J.W. Anderson, and J.W. Blaylock. 1978. Uptake of hydrocarbons from marine sediments contaminated with Prudhoe Bay crude oil: influence of feeding type of test species and availability of polycyclic aromatic hydrocarbons. J. Fish. Res. Board Can. 35:608-614.

Rohrbach, B.G., and W.E. Reed. 1976. Evaluation of extraction techniques for hydrocarbons in marine sediments. Publication 1537. Institute for Geophysics and Planetary Physics, University of California, Los Angeles.

Rohrborn, G., and I. Hansmann. 1977. Chromosomal aberrations in the early embryogenesis of mice. Mut. Res. 42:46.

Rossi, S.S. 1977. Bioavailability of petroleum hydrocarbons from water, sediments, and detritus to the marine annelid, Neanthes arenaceodentata, pp. 621-625. In Proceedings, 1977 Oil Spill Conference (Prevention, Behavior, Control, Cleanup). American Petroleum Institute, Washington, D.C.

Rossi, S.S., and J.W. Anderson. 1978. Petroleum hydrocarbon resistance in the marine worm Neanthes arenaceodentata (Polychaeta:Annelida) induced by chronic exposure to No. 2 fuel oil. Bull. Environ. Contam. Toxicol. 20:513-521.

Rossi, S.S., and J.M. Neff. 1978. Toxicity of polynuclear aromatic hydrocarbons to the marine polychaete Neanthes arenaceodentata. Mar. Pollut. Bull. 9:220-223.

Rossi, S.S., G.W. Rommel, and A.A. Benson. 1979. Comparison of hydrocarbons in benthic fish from Coal Oil Point and Tanner Bank, California, pp. 573-577. In Proceedings, 1979 Oil Spill Conference (Prevention, Behavior, Control, Cleanup). American Petroleum Institute, Washington, D.C.

Roubal, W.T., D.H. Bovee, T.K. Collier, and S.I. Stranahan. 1977a. Flowthrough system for chronic exposure of aquatic organisms to seawater-soluble hydrocarbons from crude oil: construction and applications, pp. 551-555. In Proceedings, 1977 Oil Spill Conference (Prevention, Behavior, Control, Cleanup). American Petroleum Institute, Washington, D.C.

Roubal, W.T., T.K. Collier, and D.C. Malins. 1977b. Accumulation and metabolism of carbon-14 labeled benzene, naphthalene, and anthracene by young coho salmon (Oncorhynchus kisutch). Arch. Environ. Contam. Toxicol. 5:513-529.

Roubal, W.T., S.I. Stranahan, and D.C. Malins. 1978. The accumulation of low molecular weight aromatic hydrocarbons of crude oil by coho salmon (Oncorhynchus kisutch) and starry flounder (Platichthys stellatus). Arch. Environ. Contam. Toxicol. 7:237-244.

Russel, L.B., and B.E. Matter. 1980. Whole mammal mutagenicity tests: evaluation of five methods. Mut. Res. 75:279.

Sabourin, T.D., and R.E. Tullis. 1981. Effect of three aromatic hydrocarbons on respiration and heart rates of the mussel, Mytilus californianus. Bull. Environ. Contam. Toxicol. 26:729-736.

Sackett, M.W., and J.M. Brooks. 1975. Origin and distribution of low-molecular-weight hydrocarbons in the Gulf of Mexico coastal waters, pp. 211-230. In T.M. Church, ed. Marine Chemistry in the Coastal Environment. ACS Symposium Series 18. American Chemical Society, Washington, D.C.

Samain, J.F., J. Moal, A. Coum, J.R. LeCoz, and J.Y. Daniel. 1980. Effects of the Amoco Cadiz oil spill on zooplankton: a new possibility of ecophysiological survey. Helgolander Meeresunters. 33:225-235.

Sanborn, H.R., and D.C. Malins. 1980. The disposition of aromatic hydrocarbons in adult spot shrimp (Pandalus platyceros) and the formation of metabolites of naphthalene in adult and larval spot shrimp. Xenobiotica 10:193-200.

Sanders, H.L., and C.C. Jones. 1981. Oil, science and public policy, pp. 73-94. In T.C. Jackson and D. Reische, eds. Friends of the Earth, San Francisco, Calif.

Sanders, H.L., J.F. Grassle, G.R. Hampson, L.S. Morse, S. Garner-Price, and C.C. Jones. 1980. Anatomy of an oil spill: long-term effects from the grounding of the barge Florida off West Falmouth, Massachusetts. J. Mar. Res. 38:265-380.

Saner, W.A., J.R. Jadamec, R.W. Sager, and T.J. Killeen. 1979. Trace enrichment with hand-packed CO-PELL ODS guard columns and Sep-Pak C_{18} cartridges. Anal. Chem. 51:2180-2188.

Sauer, T.C., Jr., W.M. Sackett, and L.M. Jeffrey. 1978. Volatile liquid hydrocarbons in the surface coastal waters of the Gulf of Mexico. Mar. Chem. 7:1-16.

Saxby, J.D. 1978. Comparison of crude oils and their alteration products. Geochim. Cosmochim. Acta 42:215-217.

Scalan, R.S., and J.E. Smith. 1970. An improved measure of the odd-even predominance in the normal alkanes of sediment extracts and petroleum. Geochim. Cosmochim. Acta 34:611-620.

Schenck, P.A., and E. Eisma. 1964. Quantitative determination of n-alkanes in crude oils and rock extracts by gas chromatography, pp. 403-415. In U. Columbo and G.D. Hobson, eds. Advances in Organic Geochemistry. Macmillan, New York.

Schmid, W. 1976. The micronucleus test for cytogenetic analysis. In A. Hollaender, ed. Chemical Mutagens, Principles and Methods for their Detection. Vol. 4. Plenum, New York.

Schmid, W. 1977. Micronucleus test. In B.J. Kilbey, M.S. Legator, W.W. Nichols, and C. Ramel, eds. Handbook of Mutagenicity Test Procedures. Elsevier, New York.

Schmitter, J.M., Z. Vajta, and P.J. Arpine. 1981. Investigation of nitrogen bases from petroleum, pp. 67-76. In J. Maxwell and A. Douglas, eds. Advances in Organic Geochemistry. Macmillan, New York.

Schomburg, G., H. Behlau, R. Dielmann, F. Weeke, and H. Husman. 1977. Sampling techniques in capillary gas chromatography. J. Chromatogr. 142:87-102.

Schroder, J.H. 1973. Genetics and Mutagenesis of Fish. Springer-Verlag, New York.

Searl, T.D., W.K. Robbinsm, and R.A. Brown. 1979. Determination of polynuclear aromatic hydrocarbons in water and wastewater by a gas chromatographic ultraviolet spectrophotometric method. Special Technical Publication 686. American Society for Testing and Materials, Philadelphia, Pa.

Seepersad, B., and R.W. Crippen. 1978. Use of Aniline blue for distinguishing between live and dead freshwater zooplankton. J. Fish. Res. Board Canada 34(10):1362-1366.

Seki, H. 1973. Silica gel medium for enumeration of petroleumlytic microorganisms in the marine environment. Appl. Microbiol. 26:318-320.

Shaw, D.G., A.J. Paul, L.M. Cheek, and H.M. Feder. 1976. Macoma balthica: an indicator of oil pollution. Mar. Pollut. Bull. 7:29-31.

Shaw, D.G., A.J. Paul, and E.R. Smith. 1977. Responses of the clam Macoma balthica to Prudhoe Bay crude oil, pp. 493-494. In Proceedings, 1977 Oil Spill Conference (Prevention, Behavior, Control, Cleanup). American Petroleum Institute, Washington, D.C.

Sheldon, R.W. 1978. Sensing-zone counters in the laboratory, pp. 202-214. In A. Sournia, ed. Phytoplankton Manual. Monographs on Oceanographic Methodology 6. UNESCO, Paris.

Shiels, W.E., J.J. Goering, and D.W. Hood. 1973. Crude oil phytotoxicity studies. In D.W. Hood, W.E. Shiels, and E.J. Kelley, eds. Environmental Studies of Port Valdez. Occasional Publication 3. Institute of Marine Sciences, University of Alaska, Fairbanks.

Silver, M.W., and P.J. Davoll. 1978. Loss of ^{14}C activity after chemical fixation of phytoplankton: error source for autoradiography and other productivity measurements. Limnol. Oceanogr. 23(2):326-368.

Simoneit, B.R.T. 1978. Organic chemistry of marine sediments, pp. 233-311. In R. Chester and J.P. Riley, eds. Chemical Oceanography. Vol. 7. Academic Press, New York.

Simoneit, B.R.T., and I.R. Kaplan. 1980. Triterpenoids as molecular indicators of paleoseepage in recent sediments of the Southern California Bight. Mar. Environ. Res. 3:113-128.

Sinderman, C.J. 1979. Pollution-associated diseases and abnormalities of fish and shellfish: a review. Fish. Bull. 76:717-749.

Smeach, S.J., and R.W. Jernigan. 1977. Further aspects of a Markovian sampling policy for water quality monitoring. Biometrics 33:41-46.

Smith, T.G., J.R. Geraci, and D.J. St. Aubin. 1983. The reaction of bottlenose dolphins, Tursiops truncatus, to controlled oil spills. Can. J. Fish. Aquat. Sci. (submitted).

Smith, T.R., and V.A. Strickler. 1980. A quaternary solvent system for the reverse-phase liquid chromatographic separation of polycyclic aromatics. J. High Resolut. Chromatogr. Chromatogr. Commun. 3:634-640.

Smith, W. 1979. An oil spill sampling strategy. In R.M. Cormack, G.P. Patil, and D.S. Robson, eds. Sampling Biological Populations. International Cooperative Publishing, Fairland, Md.

Smith, W., V.R. Gibson, L.S. Brown-Leger, and J.F. Grassle. 1979a. Diversity as an indicator of pollution: cautionary results from microcosm experiments, pp. 269-277. In J.F. Grassle, G.P. Patil, W. Smith, and O. Taillie, eds. Ecological Diversity in Theory and Practice. International Cooperative Publishing, Fairland, Md.

Smith, W., D. Kravitz, and J.F. Grassle. 1979b. Confidence intervals for similarity measures using the two sample jackknife, pp. 253-262. In L. Orloci, C.R. Rao, and W.M. Stiteler, eds. Multivariate Methods in Ecological Work. International Cooperative Publishing, Fairland, Md.

Smith, W., V.R. Gibson, and J.F. Grassle. 1981. Replication in controlled marine systems: presenting the evidence. In G.D. Grice and M.R. Reeve, eds. Marine Mesocosms. Springer-Verlag, New York (in press).

Sobels, F.H. 1977. Some problems associated with the testing for environmental mutagens and a perspective for studies in "comparative mutagenesis." Mut. Res. 46:245.

Solbakken, J.E., and K.H. Palmork. 1980. Distribution of radioactivity in the Chondrichthyes Squalus acanthias and the Steichthyes Salmo gairdneri following intragastric administration of (9-^{14}C) phenanthrene. Bull. Environ. Contam. Toxicol. 25:902-908.

Solbakken, J.E., K.H. Palmork, T. Neppelberg, and R.R. Scheline. 1979. Distribution of radioactivity in coa-Cish (Pollachius virens) following intragastric administration of (9-1) phenanthrene. Bull. Environ. Contam. Toxicol. 23:100-103.

Solorazano. L. 1969. Determination of ammonia in natural waters by the phenol-hypochlorite method. Limnol. Oceanogr. 14:799-801.

Soto, C., J.A. Helleburst, T.C. Hutchinson, and T. Sawa. 1975a. Effect of naphthalene and aqueous crude oil extracts on the green flagellate Chlamydomonas angulosa. I. Growth. Can. J. Bot. 53:109-117.

Soto, C., J.A. Helleburst, and T.C. Hutchinson. 1975b. Effect of naphthalene and aqueous crude oil extracts on the green flagellate Chlamydomonas angulosa. I. Growth. Can J. Bot. 53:109-117.

Sournia, A., ed. 1978. Phytoplankton Manual. Monographs on Oceanographic Methodology 6. UNESCO, Paris. 337 pp.

Sparrow, A.H., and G.M. Woodwell. 1963. Prediction of the sensitivity of plants to chronic gamma irradiation. In V. Schultz and A.W. Klement Jr., eds. Radioecology. Reinhold Publishers, New York.

Spies, R.B., and P.H. Davis. 1979. The infaunal benthos of a natural oil seep in the Santa Barbara Channel. Mar. Biol. 50:227-237.

Spies, R.B., P.H. Davis, and D.H. Stuermer. 1978. The infaunal benthos of petroleum-contaminated sediments: study of a community at a natural oil seep, pp. 735-755. In Proceedings of the Conference on Assessment of Ecological Impacts of Oil Spills. American Institute of Biological Sciences, Arlington, Va.

Spies, R.B., P.H. Davis, and D.H. Stuermer. 1980. Ecology of a submarine petroleum seep off the California coast, pp. 229-263. In

R.A. Geyer, ed. Marine Environmental Pollution. Vol. I, Hydrocarbons. Elsevier, New York.

Spooner, M.F., and C.J. Corkett. 1974. A method for testing the toxicity of suspended oil droplets on planktonic copepods used at Plymouth, pp. 69-74. In L.R. Benyon and E.B. Cowells, eds. Ecological Aspects of Toxicity Testing of Oils and Dispersants. Applied Science Publishers, Barking, Essex, England. 149 pp.

Spooner, M.F., and C.J. Corkett. 1979. Effects of Kuwait oils on feeding rates of copepods. Mar. Pollut. Bull. 10:197-202.

Spoor, W.A., T.W. Neiheisel, and R.A. Drummond. 1971. An electrode chamber for recording respiratory and other movements in free-swimming animals. Trans. Am. Fish. Soc. 1:22-28.

Sprague, J.B. 1971. Measurement of pollutant toxicity to fish. III. Sublethal effects and "safe" concentrations. Water Res. 5:245-266.

Stainken, D.M. 1975. Preliminary observations on the mode of accumulation of #2 fuel oil by the soft shell clam, Mya arenaria, pp. 463-468. In Proceedings, 1975 Conference on Prevention and Control of Oil Pollution. American Petroleum Institute, Washington, D.C.

Stainken, D.M. 1977. The accumulation and depuration of No. 2 fuel oil by the soft shell clam, Mya arenaria L., pp. 313-322. In D.A. Wolfe, ed. Fate and Effects of Petroleum Hydrocarbons in Marine Organisms and Ecosystems, Pergamon, New York.

Stainken, D.M. 1978. Effects of uptake and discharge of petroleum hydrocarbons on the respiration of the soft-shell clam, Mya arenaria. J. Fish. Res. Board Can. 35:637-642.

Steedman, H.F., ed. 1975. Zooplankton fixation and preservation. Monographs on Oceanographic Methodology 4. UNESCO, Paris. 350 pp.

Stegeman, J.J. 1981. Polynuclear aromatic hydrocarbons and their metabolism in the marine environment, pp. 1-60. In H.V. Gelboin and P.O.P. Ts'O, eds. Polycyclic Hydrocarbons and Cancer. Vol. 3. Academic Press, New York.

Stegeman, J.J., and D.J. Sabo. 1976. Aspects of the effects of petroleum hydrocarbons on intermediary metabolism and xenobiotic metabolism in marine fish, pp. 423-431. In Proceedings, Sources Effects and Sinks of Hydrocarbons in the Aquatic Environment, ERDA, EPA, BLM, and API. American Institute of Biological Sciences, Arlington, Va.

Stekoll, M.S., L.E. Clement, and D.G. Shaw. 1980. Sublethal effects of chronic oil exposure on the intertidal clam Macoma balthica. Mar. Biol. 57:51-60.

Stephan, C.E. 1977. Methods for calculating an LC_{50}, pp. 65-84. In F.L. Mayer and J.L. Hamelink, eds. Aquatic Toxicology and Hazard Evaluation. ASTM STP 634. American Society for Testing and Materials. Philadelphia, Pa.

Stetka, D.G., and S. Wolff. 1976. Sister chromatid exchange as an assay for genetic damage induced by mutagen-carcinogens. II. In vitro test for compounds requiring metabolic activation. Mut. Res. 41:343.

Stich, H.F., and R.H.C. San, eds. 1981 Short-Term Tests for Chemical Carcinogenesis. Springer-Verlag, New York.

Stoddart, D., and R. Johannes, eds. 1978. Coral reef: research methods. Monograph on Oceanographic Methodology 5. UNESCO, Paris. 581 pp.

Stoyel, C.J., and A.M. Clark. 1980. The transplacental micronucleus test. Mut. Res. 73:393.

Straughan, D. 1971. Breeding and larval settlement of certain intertidal invertebrates in the Santa Barbara Channel following pollution by oil, pp. 223-244. In Biological and Oceanographical Survey of the Santa Barbara Channel Oil Spill 1969-1970. Vol. I. Allan Hancock Foundation, University of Southern California, Los Angeles.

Straughan, D. 1972. Biological effects of oil pollution in the Santa Barbara Channel. In M. Ruivo, ed. Marine Pollution and Sea Life. Fishing News (Books), Surrey. 355 pp.

Straughan, D. 1980. Analysis of mussel (_Mytilus californianus_) communities in areas chronically exposed to natural oil seepage. Publication 4319. American Petroleum Institute, Washington, D.C. 115 pp.

Strickland, J.D.H., and T.R. Parsons. 1972. A practical handbook of seawater analysis. 2nd ed. Bulletin 167. Fisheries Research Board of Canada. 310 pp.

Sweeney, R.E., R.I. Haddad, and I.R. Kaplan. 1980. Tracing the dispersal of the _Ixtoc_ I oil using C, H, S and N stable isotope ratios, pp. 89-118. In Proceedings, Symposium on the Preliminary Results from the _Researcher/Pierce_ Cruise to the _Ixtoc_ I Blowout. NOAA, Office of Marine Pollution Assessment, Rockville, Md.

Swinnerton, J.W., and V.J. Linnenbom. 1967. Determination of C_1 to C_4 hydrocarbons in sea water by gas chromatography. J. Gas Chromatogr. 5:570-573.

Swinnerton, J.W., and V.J. Linnenbom. 1976. Gaseous hydrocarbons in sea water: determination. Science 156:1119-1120.

Talmi, Y., D.C. Baker, J.R. Jadamec, and W.A. Saner. 1978. Fluorescence spectrometry with optoelectronic image detectors. Anal. Chem. 50:930A-952A.

Tan, Y.L. 1979. Rapid simple sample preparation technique for analyzing polynuclear aromatic hydrocarbons in sediment by gas chromatography-mass spectrometry. J. Chromatogr. 176:319-327.

Tangen, K. 1978. Nets, pp. 50-58. In A. Sournia, ed. Phytoplankton Manual. Monograph on Oceanographic Methodology 6. UNESCO, Paris.

Tatem, H.E. 1977. Accumulation of naphthalenes by grass shrimp: effects on respiration, hatching, and larval growth, pp. 201-207. In D.A. Wolfe, ed. Fate and Effects of Petroleum Hydrocarbons in Marine Organisms and Ecosystems. Pergamon, New York.

Taylor, T.L., and J.F. Karinen. 1977. Response of the clam, _Macoma balthica_ (Linnaeus), exposed to Prudhoe Bay crude oil as unmixed oil, water-soluble fraction, and oil-contaminated sediment in the laboratory, pp. 229-237. In D.A. Wolfe, ed. Fate and Effects of Petroleum Hydrocarbons in Marine Organisms and Ecosystems. Pergamon, New York.

Teal, J.M., K. Burns, and J. Farrington. 1978. Analyses of aromatic hydrocarbons in intertidal sediments resulting from two spills of No. 2 fuel oil in Buzzards Bay, Mass. J. Fish. Res. Board Can. 35:510-520.

Templeton, G.D., III, and N.D. Chasteen. 1980. Evaluation of extraction schemes for organic matter in anoxic estuarine sediments. Mar. Chem. 10:31-46.

Terrell, R.E. 1981. Petroleum. Anal. Chem. 53:88R-142R.

Thomas, P., and S.D. Rice. 1979. The effect of exposure temperatures on oxygen consumption and opercular breathing rates of pink salmon fry exposed to toluene, naphthalene, and water-soluble fractions of Cook Inlet crude oil and No. 2 fuel oil, pp. 39-52. In W.B. Vernberg, A. Calabrese, F.P. Thurberg, and F.J. Vernberg, eds. Marine Pollution: Functional Responses. Academic Press, New York.

Thomas, R.E., and S.E. Rice. 1981. Excretion of aromatic hydrocarbons and their metabolites by freshwater and seawater Dolly Varden char, pp. 425-448. In F.J. Vernberg, A. Calabrese, F.P. Thurberg, and W.B. Vernberg, eds. Biological Monitoring of Marine Pollutants Academic Press, New York.

Thomas, P., B.R. Woodin, and J.M. Neff. 1980. Biochemical responses of the striped mullet Mugil dephalus to oil exposure. I. Acute responses-interrenal activations and secondary stress responses. Mar. Biol. 59:141-149.

Thompson, S., and G. Eglinton. 1978a. Composition and sources of pollutant hydrocarbons in the Severn Estuary. Mar. Pollut. Bull. 9:133-136.

Thompson, S., and G. Eglinton. 1978b. The fractionation of a recent sediment for organic geochemical analysis. Geochim. Cosmochim. Acta 42:199-207.

Tietjen, G.L., and R.J. Beckman. 1974. On duplicate measurements in the chemical laboratory. Technometrics 16:53-56.

Topham, J.C. 1980a. The detection of carcinogen-induced sperm head abnormalities in mice. Mut. Res. 69:149.

Topham, J.C. 1980b. Chemically-induced transmissible abnormalities in sperm-head shape. Mut. Res. 70:109.

Torgrimson, G.M. 1981. A comprehensive model for oil spill simulation, pp. 423-428. In Proceedings, 1981 Oil Spill Conference. American Petroleum Institute, Washington, D.C.

Tripp, B.W., J.W. Farrington, and J.M. Teal. 1981. Unburned coal as a source of hydrocarbons in surface sediments. Mar. Pollut. Bull. 12:122-126.

Troy, B., Jr., and J. Hollinger. 1977. The measurement of oil spill volume by a passive microwave imager. NRL Memorandum Report 3515. Naval Research Lab., Washington, D.C.

Trudel, B.K. 1978. The effect of crude oil and crude oil/Corexit 9527 suspensions on carbon fixation by a natural marine phytoplankton community. Spill Tech. Newsletter 3(2):56-64.

Tsoi, R.M. 1969. Effect of nitrosomethyl urea and dimethyl sulfate on sperm of rainbow trout (Salmo irideus Gibb.) and peled (Coregonus peled Gmel.). Dokl. Biol. Sci. 189:849.

Tsoi, R.M., A.I. Men'shova, and Y.V.F. Golodov. 1974. Soviet Genetics. Translated from Genetika 10:68.

Turner, G. 1979. Fluorometry and marine environmental monitoring. Sea Technology (July 1979):30-33.

Ury, G.B. 1981. Automated gas chromatographic analysis of gasolines for hydrocarbon types. Anal. Chem. 53:481-485.

U.S. Coast Guard. 1977. Oil spill identification system. U.S. Department of Transportation Report CG-D-52-77. Accession No. ADA044750. National Technical Information Service, Springfield, Va.

Ustach, J.F. 1977. Effects of sub-lethal oil levels on the reproduction of a copepod, Nitocra affinis. Sea Grant Publication UNC-SG 76-10.

Valentine, L.C., and W.E. Bishop. 1980. Use of the central mudminnow, Umbia limi, in the development and evaluation of the sister-chromatid exchange test for detecting mutagens in vivo. Paper presented at Fifth Symposium on Aquatic Toxicology. American Society of Testing and Materials, Philadelphia, Pa.

Vanderhorst, J.R., C.I. Gibson, and L.J. Moore. 1976. Toxicity of No. 2 fuel oil to coon stripe shrimp. Mar. Pollut. Bull. 7:106-108.

Vanderhorst, J.R., R.M. Bean, L.J. Moore, P. Wilkinson, C.I. Gibson, and J.W. Blaylock. 1977a. Effects of a continuous low-level No. 2 fuel dispersion on laboratory held intertidal colonies, pp. 557-561. In Proceedings, 1977 Oil Spill Conference (Prevention, Behavior, Control, Cleanup). American Petroleum Institute, Washington, D.C.

Vanderhorst, J.R., C.I. Gibson, L.J. Moore, and P. Wilkinson. 1977b. Continuous-flow apparatus for use in petroleum bioassay. Bull. Environ. Contam. Toxicol. 17:577-584.

Vanderhorst, J.R., J.W. Blaylock, P. Wilkinson, M. Wilkinson, and G. Fellingham. 1981. Effects of experimental oiling on recovery of Strait of Juan de Fuca intertidal habitats. NOAA-MESA/EPA Report EPA-600/7-81-088. Office of Environmental Engineering and Technology, Environmental Protection Agency, Washington, D.C. 129 pp.

Vandermeulen, J.H., and T.P. Ahern. 1976. Effect of petroleum hydrocarbons on algal physiology: review and progress report, pp. 107-125. In A.P. Lochwood, ed. Effects of Pollutants on Aquatic Organisms. Cambridge University Press, New York.

Vandermeulen, J.H., and R.W. Lee. 1977. Absence of mutagenicity due to crude and refined oils in the alga Chlamydomonas reinhardtii. ICES CM 1977/E:69. International Council for the Exploration of the Seas, Charlottenlund, Denmark.

Van Overbeek, J., and R. Blondeau. 1954. Mode of action of phytotoxic oils. Weeds 3:55-65.

Van Vleet, E.S., an J.G. Quinn. 1978. Contribution of chronic petroleum inputs to Narragansett Bay and Rhode Island Sound sediments. J. Fish. Res. Board Can. 35:536-543.

Varanasi, U., and D.J. Gmur. 1980. In vivo metabolism of naphthalene and benzo[a]pyrene by flatfish. In A.J. Dennis and M. Cook, eds. Proceedings, Fifth International Symposium on Polynuclear Aromatic Hydrocarbons. Battelle Press, Columbus, Ohio. (in press).

Varanasi, U., and D.J. Gmur. 1981a. In vivo metabolism of naphthalene and benzo[a]pyrene by flatfish, pp. 367-376. In M. Cooke and A.J. Dennis, eds. Chemical Analysis and Biological Fate: Polynuclear Aromatic Hydrocarbons. Battelle Press, Columbus, Ohio.

Varanasi, U., and D.J. Gmur. 1981b. Hydrocarbons and metabolites in English sole (Parophrys vetulus) exposed simultaneously to (^{3}H) benzo(a)pyrene and (^{14}C) napthalene in oil-contaminated sediment. Aquat. Toxicol. 1:49-67.

Varanasi, U., and D.O. Malins. 1977. Metabolism of petroleum hydrocarbons: accumulation and biotransformation in marine organisms, pp. 175-270. In D.C. Malins, ed. Effects of Petroleum on Arctic and Subarctic Marine Environments and Organisms. Vol. II, Biological Effects. Academic Press, New York.

Varanasi, U., D.J. Gmur, and P.A. Treseler. 1979. Influence of time and mode of exposure on biotransportation of naphthalene by juvenile starry flounder (Platichthys stellatus) and rock sole (Lepidopsetta bilineata). Arch. Environ. Contam. Toxicol. 8:673-692.

Vargo, G.A., M. Hutchins, and G. Almquist. 1981. The effect of low, chronic levels of No. 2 fuel oil on natural phytoplankton assemblages in microcosms. I. Species composition and seasonal succession. Mar. Environ. Res. 6:245-264.

Vargo, S. 1981. The effects of chronic low concentrations of No. 2 fuel oil on the physiology of a temperate estuarine zooplankton community in the MERL microcosms, pp. 295-322. In F.J. Vernberg, A. Calabrese, F.P. Thurberg, and W.B. Vernberg, eds. Biological Monitoring of Marine Pollutants. Academic Press, New York.

Vargo, S., and K. Force. 1981. A simple photometer for precise determination of dissolved oxygen concentration by the Winkler method with recommendations for improving respiration rate measurements in aquatic organisms. Estuaries 4(1):70-74.

Veith, G.D., and L.M. Kiwus. 1977. An exhaustive steam distillation and solvent extraction unit for pesticides and industrial chemicals. Bull. Environ. Contam. Toxicol. 17:631-636.

Venkatesan, M.I., P. Mankiewicz, W.K. Ho, R.E. Sweeney, and I.R. Kaplan. 1980. Determination of petroleum contamination in marine sediments by organic geochemical and stable sulfur isotope analyses. In G.W. Ernst, ed. Ruby Colloquium on Marine Processes (in press).

Venrick, E.L. 1971. Recurrent groups of diatoms in the North Pacific. Ecology 52(4):614-625.

Venrick, E.L. 1978a. Systematic sampling in a planktonic ecosystem. Fishery Bull. 76(3):617-627.

Venrick, E.L. 1978b. Sampling strategies, pp. 7-16. In A. Sournia, ed. Phytoplankton Manual. Monographs on Oceanographic Methodology 6. UNESCO, Paris.

Venrick, E.L., J.R. Seers, and J.F. Heinbokel. 1977. Possible consequences of containing microplankton for physiological rate measurements. J. Exp. Mar. Biol. Ecol. 26:58-76.

Virnstein, R.W. 1977. The importance of predation by crabs and fishes on benthic infauna in Chesapeake Bay. Ecology 58:1199-1217.

Vo-Dinh, T. 1978. Multicomponent analysis by synchronous luminescence spectrometry. Anal. Chem. 50:396-401.

Von Borstel, R.C. 1981. The yeast Saccharomyces cervesiae: an assay organism for environmental mutagens, p. 161. In H.F. Stich and R.H.C. San, eds. Short-Term Tests for Chemical Carcinogenesis. Springer-Verlag, New York.

Wade, T.L., and J.G. Quinn. 1980. Incorporation, distribution and fate of saturated petroleum hydrocarbons in sediments from a controlled marine ecosystem. Mar. Environ. Res. 3:15-33.

Wade, T.L., J.G. Quinn, W.T. Lee, and C.W. Brown. 1976. Source and distribution of hydrocarbons in surface waters of the Sargasso Sea, pp. 271-286. In Proceedings, Symposium on Sources, Effects, and Sinks of Hydrocarbons in the Aquatic Environment. American Institute of Biological Sciences, Arlington, Va.

Wakeham, S.G. 1977. Synchronous fluorescence spectroscopy and its application to indigenous and petroleum-derived hydrocarbons in lacustrine sediments. Environ. Sci. Technol. 11:272-276.

Wakeham, S.G., and R. Carpenter. 1976. Aliphatic hydrocarbons in sediments of Lake Washington. Limnol. Oceanogr. 21(5):711-723.

Wakeham, S.G., and J.W. Farrington. 1980. Hydrocarbons in contemporary aquatic sediments, pp. 3-32. In R.A. Baker, ed. Contaminants and Sediments. Vol. 1. Ann Arbor Science Publishers, Ann Arbor, Mich.

Wakeham, S.G., J.W. Farrington, R.B. Gagosian, C. Lee, H. DeBaar, G.E. Nigrelli, B.W. Tripp, S.O. Smith, and N.M. Frew. 1980. Organic matter fluxes from sediment traps in the Equatorial Atlantic Ocean. Nature 286(5775):798-800.

Wakeham, S.G., C. Schaffner, and W. Giger. 1981. Diagenic polycyclic aromatic hydrocarbons in recent sediments: structural information obtained by high performance liquid chromatography, pp. 353-363. In J. Maxwell and A. Douglas, eds. Advances in Organic Geochemistry. Macmillan, New York.

Waldon, D.E., W.R. Penrose, and S.M. Greene. 1978. The petroleum-induced mixed function oxidase of cunner (Dautogloabrus adspirus), some characteristics relevant to hydroxization monitoring. J. Fish. Res. Board Can. 35:1847-1582.

Walker, C.H., and G.C. Knight. 1981. The hepatic microsomal enzyme of seabirds and their interaction with lipo-soluble pollutants. Aquat. Toxicol. 1:343-354.

Walker, J.D., and R.R. Colwell. 1973. Microbial ecology of petroleum utilization in Chesapeake Bay, pp. 605-691. In Proceedings, Joint Conference on Prevention and Control of Oil Spills. American Petroleum Institute, Washington, D.C.

Walker, J.D., and R.R. Colwell. 1975a. Factors affecting enumeration and isolation of actinomycetes from Chesapeake Bay and Southeastern Atlantic Ocean sediments. Mar. Biol. 30:193-201.

Walker, J.D., and R.R. Colwell. 1975b. Degradation of hydrocarbons and mixed hydrocarbon substrate by microorganisms from Chesapeake Bay. Proceedings of the 1973 International Conference on water Pollution. Progress Water Technol. 7:781-783.

Walker, J.D., and R.R. Colwell. 1976a. Enumeration of petroleum-degrading microorganisms. Appl. Environ. Microbiol. 31:188-207.

Walker, J.D., and R.R. Colwell. 1976b. Measuring potential activity of hydrocarbon-degrading bacteria. Appl. Environ. Microbiol. 31:189-197.

Walker, J.D., R.R. Colwell, and L. Petrakis. 1975. Microbial petroleum degradation: application of computerized mass spectrometry. Can. J. Microbiol. 21:1760-1767.

Walters, J.M, R.B. Cain, I.J. Higgins, and E.D.S. Corner. 1979. Cell-free benzo[a]pyrene hydroxylase activity in marine zooplankton. J. Mar. Biol. Assoc. U.K. 59:553-563.

Walton, D.G., W.R. Penrose, and J.M. Green. 1978. The petroleum-inducible mixed-function oxidase of cunner (Tautogolabrus adspersus Walbaum 1972): some characteristics relevant to hydrocarbon monitoring. J. Fish. Res. Board Canada 35:1547-1552.

Wang, R.T., and J.A.C. Nicol. 1977. Effects of fuel oil on sea catfish: feeding activity and cardiac responses. Bull. Environ. Contam. Toxicol. 18:170-176.

Ward, D.M., R.M. Atlas, P.D. Boehm, and J.A. Calder. 1980. Microbial biodegradation and chemical evolution of oil from the Amoco spill. Ambio 9:277-283.

Wardhaugh, A.A. 1981. Dominant lethal mutations in Tilapia mossambica (Peters) elicited by Myleran. Mut. Res. 88:191.

Warner, J.S. 1976. Determination of aliphatic and aromatic hydrocarbons in marine organisms. Anal. Chem. 48:578-583.

Warner, J.S. 1978. Chemical characterization of marine samples. Publication 4307. American Petroleum Institute, Washington, D.C.

Warner, J.S., R.M. Riggin, and T.M. Engel. 1980. Recent advances in the determination of aromatic hydrocarbons in zooplankton and macrofauna, pp. 87-104. In L. Petrakis and F.T. Weiss, eds. Petroleum in the Marine Environment. Advances in Chemistry Series 185. American Chemical Society, Washington, D.C.

Weber, D., D.J. Maynard, W.D. Gronlund, and V. Konchin. 1981. Avoidance reactions of migrating adult salmon to petroleum hydrocarbons. Can. J. Fish. Aquat. Sci. (in press).

Wells, P.G. 1982. Background Papers/Petroleum in the Marine Environment Update NAS/NRC Ocean Sciences Board Workshop, November 9-13, 1981.

Wells, P.G., and J.B. Sprague. 1976. Effects of crude oil on American lobster (Homarus americanus) larvae in the laboratory. J. Fish. Res. Board Can. 33:1604-1614.

Whipple, J.A., T.G. Yocom, D.R. Smart, and M.H. Cohen. 1978. Effects of chronic concentrations of petroleum hydrocarbons on gonadal maturation in starry flounder (Platichthys stellatus [Pallas]), pp. 756-806. American Institute of Biological Sciences, Arlington, Va.

Whipple, J.A., M.B. Eldridge, and P. Benville, Jr. 1981. An ecological perspective of the effects of monocyclic aromatic hydrocarbons on fishes, pp. 483-551. In F.J. Vernberg, A. Calabrese, F.P. Thurberg, and W.B. Vernberg, eds. Biological Monitoring of Marine Pollutants. Academic Press, New York.

White, G.P., and A.V. Arecchi. 1975. Local area pollution surveillance systems: a summary of the Coast Guard's research and development activities, pp. 123-128. In Proceedings, 1975 Oil Spill Conference. American Petroleum Institute, Washington, D.C.

White, J.R., R.E. Schmidt, and W.E. Plage. 1979. The AIREYE remote sensing system for oil spill surveillance, pp. 301-304. In Proceedings, 1979 Oil Spill Conference (Prevention, Behavior, Control, Cleanup). American Petroleum Institute, Washington, D.C.

Whittaker, R.H., ed. 1978. Ordination of Plant Communities. Junk, The Hague.

Whittle, K.J., J. Murray, P.R. Mackie, R. Hardy, and J. Farmer. 1977a. Fate of hydrocarbons in fish. Rapp. P.-v. Reun. Cons. Int. Explor. Mer 171:139-142.

Whittle, K.J., P.R. Mackie, R. Hardy, A.D. McIntyre, and R.A.A. Blackman. 1977b. The alkanes of marine organisms from the United Kingdom and surrounding waters. Rapp. P.-v. Reun. Cons. Int. Explor. Mer 171:72-78.

Whittle, K.J., P.R. Mackie, J. Farmer, and R. Hardy. 1978. The effects of the Ekofisk blowout on hydrocarbon residues in fish and shellfish, pp. 540-559. In Proceedings of the Conference on Assessment of Ecological Impacts of Oil Spills. American Institute of Biological Sciences, Arlington, Va.

Widdows, J. 1978a. Combined effects of body size, food concentration and season on the physiology of Mytilus edulis. J. Mar. Biol. Assoc. U.K. 58:109-124.

Widdows, J. 1978b. Physiological indices of stress in Mytilus edulis. J. Mar. Biol. Assoc. U.K. 58:125-142.

Wiebe, P.H., G.D. Grice, and E. Hoagland. 1973. Acid-iron waste as a factor affecting the distribution and abundance of zooplankton in the New York Bight. II. Spatial variations in the field and implications for monitoring studies. Estuarine Coastal Mar. Sci. 1:51-64.

Wilhm, J.L., and T.C. Dorris. 1966. Species diversity of benthic macroinvertebrates in a stream receiving domestic and oil refinery effluents. Am. Midland Naturalist 76:427-449.

Wilhm, J.L., and T.C. Dorris. 1968. Biological parameters for water quality criteria. Bioscience 18:477-481.

Willey, C., M. Iwao, R.N. Castle, and M.L. Lee. 1981. Determination of sulfur heterocyclics in coal liquids and shale oils. Anal. Chem. 53:400-407.

Wilson, K.W. 1975. The laboratory estimation of the biological effects of organic pollutants. Proc. R. Soc. London, Ser. B 189:459-477.

Wilson, K.W., E.B. Cowell, and L.R. Beynon. 1974. The toxicity testing of oils and dispersants: a European view, pp. 129-141. In L.R. Beynon and E.B. Cowell, eds. Ecological Aspects of Toxicity Testing of Oils and Dispersants. John Wiley & Sons, New York.

Windsor, J.G., Jr., and R.A. Hites. 1979. Polycyclic aromatic hydrocarbons in Gulf of Maine sediments and Nova Scotia soils. Geochim. Cosmochim. Acta 43:27-33.

Winters, K., C. Van Baalen, and J.A.C. Nicol. 1977. Water soluble extractives from petroleum oils: chemical characterization and effects on microalgae and marine animals. Rapp. P.-v. Reun. Cons. Int. Explor. Mer 171:166-174.

Wise, S.A., S.N. Chesler, H.S. Hertz, L.R. Hilpert, and W.E. May. 1977. Chemically-bonded aminosilane stationary phase for the high performance liquid chromatography separation of polynuclear aromatic compounds. Anal. Chem. 49:2306-2310.

Wise, S.A., S.N. Chesler, H.S. Hertz, L.R. Hilpert, and W.E. May. 1978. Methods for polynuclear hydrocarbon analysis in the marine environment. In P.W. Jones and R.I. Freudenthal, eds. Carcinogenesis. Vol. 3, Polynuclear Aromatic Hydrocarbons. Raven Press, New York.

Wise, S.A., S.N. Chesler, F.R. Guenther, H.S. Hertz, L.R. Hilpert, W.E. May, and R.M. Parris. 1980. Interlaboratory comparison of determination of trace level hydrocarbons in mussels. Anal. Chem. 52:1828-1833.

Wolf, K., and M.C. Quimby. 1969. Fish cell and tissue culture. In W.S. Hoar and D.J. Randall, eds. Fish Physiology. Vol. 3. Academic Press, New York.

Wolfe, N.A., R.C. Clark, Jr., C.A. Foster, J.W. Hawkes, and W.D. MacLeod, Jr. 1981. Hydrocarbon accumulation and histopathology in bivalve molluscs transplanted to the Baie de Morlaix and the Rade de Brest, pp. 599-616. In Amoco Cadiz: Fates and Effects of the Oil Spill. CNEXO, Paris.

Wong, C.K., F.R. Englehardt, and J.R. Strickler. 1981. Survival and fecundity of Daphnia pulex on exposure to particulate oil. Bull. Environ. Contam. Toxicol. 36:606-612.

Wong, M.K., and P.J. leB. Williams. 1980. A study of three extraction methods for hydrocarbons in marine sediment. Mar. Chem. 9:183-190.

Wong, W.C. 1976. Uptake and retention of Kuwait crude oil and its effect on oxygen uptake by the soft-shell clam, Mya arenaria. J. Fish. Res. Board Can. 33:2774-2780.

Woodhead, D.S. 1977. The effect of chronic irradiation on the breeding performance of the guppy, Poecilia reticulata (Osteichthyes: Teleostei). Int. J. Radiat. Biol. 32:1.

Woodin, S.A., C.F. Nyblade, and F.S. Chia. 1972. Effect of diesel oil spill on invertebrates. Mar. Pollut. Bull. 3:139-143.

Wootton, T.A., G.R. Grau, and T.E. Roundybush. 1979. Reproductive responses of quail to Bunker C oil fractions. Arch. Environ. Contam. Toxicol. 8:457-463.

Wormald, A.P. 1976. Effects of a spill of marine diesel oil on the meiofauna of a sandy beach at Picnic Bay, Hong Kong. Environ. Pollut. 11:117-130.

Wyrobek, A.J., and W.R. Bruce. 1978. The induction of sperm-shape abnormalities in mice and humans. In A. Hollaender, A. and F.J. de S-rres, eds. Chemical Mutagens, Principles and Methods for Their Detection. Vol. 5. Plenum, New York.

Yang, W.C., and H. Wang. 1977. Modeling of oil evaporation in aqueous environment. Water Res. 11:879-887.

Yevich, P.P., and C.A. Barsacz. 1977. Neoplasms in soft-shell clam, Mya arenaria, collected from oil-impacted sites. Annals N.Y. Acad. Sci. 298:409-426.

Yost, R.W., L.S. Ettre, and R.D. Conlon. 1980. Practical liquid chromatography: an introduction. Perkin-Elmer, Norwalk, Conn. 255 pp.

Young, R.H., and A.J. Sethi. 1975. Compositional changes of a fuel oil from an oil spill due to natural exposure. Water Air Soil Pollut. 5:195-205.

Youngblood, W.W., and M. Blumer. 1973. Alkanes and alkenes in marine benthic algae. Mar. Biol. 21:163-172.

Youngblood, W.W., and M. Blumer. 1975. Polycyclic aromatic hydrocarbons in the environment: homologous series in soils and recent marine sediments. Geochim. Cosmochim. Acta 39:1303-1314.

Youngblood, W.W., M. Blumer, R.L. Guillard, and F. Fiore. 1971. Saturated and unsaturated hydrocarbons in marine benthic algae. Mar. Biol. 8:190-201.

Zimmerman, R., and L.A. Meyer-Reil. 1974. A new method for fluorescence staining of bacterial populations on membrane filters. Kiel. Meeresforsch. 30:24-27.

Zitco, V. 1975. Aromatic hydrocarbons in aquatic fauna. Bull. Environ. Contam. Toxicol. 14:621-631.

Zitko, V., and W.V. Carson. 1970. The characterization of petroleum oils and their determination in the aquatic environment. Fish. Res. Board Can. Tech. Rep. 217:29.

Zsolnay, A. 1978a. Caution in the use of Niskin bottles for hydrocarbon samples. Mar. Pollut. Bull. 9:23-24.

Zsolnay, A. 1978b. Lack of correlation between gas-liquid chromatographic and UV absorption indicators of petroleum pollution in organisms. Water Air and Soil Pollut. 9:45-51.

Zurcher, F., and M. Thuer. 1978. Rapid weathering processes of fuel oil in natural waters: analyses and interpretations. Environ. Sci. Technol. 12:838-843.

4
Fates

INTRODUCTION

Petroleum introduced to the marine environment goes through a variety of physical, chemical, and biological transformations during its transport by the advective and spreading processes discussed below. This section identifies the major factors controlling each of these processes, reviews the relevant experimental and field evidence for quantitative evaluation of the effect of these various processes on petroleum, and estimates the amount of petroleum hydrocarbons in the marine environment at the present time. Although much of the subsequent discussion deals with the fate of oil spills, this source of oil in the marine environment only accounts for about 15% of the annual input, with chronic discharges being of much greater significance (see Chapter 2, Table 2-22). The latter are subject to essentially the same kinds of fates but are sometimes more difficult to study owing to the dispersed nature of the inputs and lower concentrations of petroleum compared to oil spills.

Advection and spreading begin immediately after introduction of petroleum to the ocean and cause a rapid increase in the exposure area of the oil to subsequent "weathering" processes. These include evaporation, dissolution, vertical dispersion, emulsification, and sedimentation. Involved in all of these processes are chemical factors determined by the specific composition of the petroleum in question. Additionally, photochemical oxidation of some of the components of petroleum can be induced by sunlight. Dark or autooxidation may also occur. The products of these processes include hydrocarbon fractions and reaction products introduced to the atmosphere, slicks and tar lumps on the surface of the ocean, dissolved and particulate hydrocarbon materials in the water column, and similar components in the sediments.

While physical and chemical processes are occurring, biological processes also act on the different fractions of the original petroleum in various ways. The biological processes considered include degradation of petroleum by microorganisms to carbon dioxide or organic components in intermediate oxidation stages, uptake by larger organisms and subsequent metabolism, storage, or discharge.

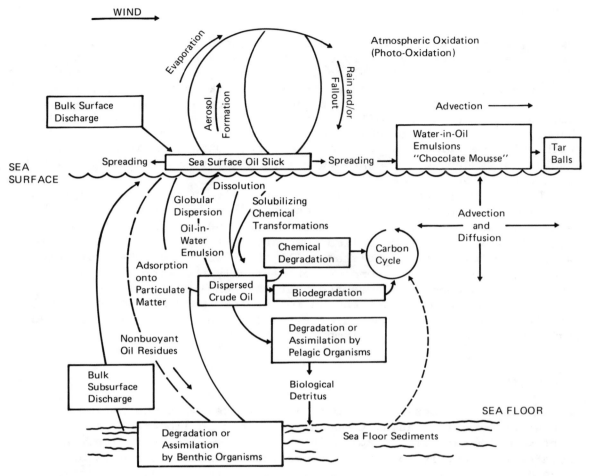

FIGURE 4-1 Schematic of physical, chemical, and biological processes.

SOURCE: Adapted from Burwood and Speers (1974).

Figure 4-1 presents many of these processes in a simple schematicized form.

PHYSICAL AND CHEMICAL FATES

Physical and Chemical Characteristics of Petroleum

The chemical composition of petroleum was discussed in detail in Chapter 3. There are, however, several critical physical properties (given below) that are important when considering the fate of petroleum in the marine environment.

Density

Density of spilled oil increases as evaporation removes the lighter
constituents, but only rarely does the density reach that of seawater.
This effect is partially balanced by a density decrease with increasing
temperature. The effective density of a slick tends to increase due to
weathering, but more significant increases are attributable to (1) the
uptake of water by many oils to form emulsions ("mousse"), which have
higher densities (approaching that of seawater), and (2) association
with suspended minerals or organic matter. Oxidation also may cause a
density increase, but the products may be quite water soluble and will
thus migrate out of the oil.

Density plays an obviously important role in the fate of spilled
oil, for the density difference between oil and water determines the
extent to which the slick is submerged and the residence time of oil
droplets which may be propelled downward in the water column by breaking
waves. Following the Kurdistan spill there were anecdotal but unsub-
stantiated accounts of submerged or neutrally buoyant oil (Vandermeulen,
1981). It is generally accepted that the density of most weathered
oils does not become great enough for neutral buoyancy to occur and
result in significant amounts of particles and pancakes in suspended
equilibrium in the water column.

Viscosity and Pour Point

Spill viscosity (resistance to flow) increases with weathering and
decreases with increasing temperature. This is important, as it
controls the rate of spreading in the gravity-viscous regime. A
related property is pour point temperature for oils which is often
invoked as an "equivalent to melting point" for organic chemicals.
Phenomena associated with the rapid increase in viscosity as the pour
point is approached are not well understood. Probably more important
is the effect of emulsified water on the bulk viscosity of emulsions.

Oils usually have non-Newtonian rheologies (flow) characteristics;
thus a viscosity measurement is meaningful only in the context of a
particular rheological model if the shear conditions are defined. It
is appropriate to measure and record low shear rate viscosities using,
for example, an Ostwald viscometer.

Vapor Pressure

Vapor pressure controls evaporation rate and air concentrations of
hydrocarbons and, therefore, the fire hazard in the vicinity of
spills. Vapor pressures can be estimated using Raoult's law (vapor
pressure of a solution equals the product of the vapor pressure of the
solvent and the mole fraction of the solvent) if the composition of the
mixture is known—which is usually not the case. The use of
"pseudo-components" or analytical expressions for vapor pressure is
discussed in the Evaporation section below.

TABLE 4-1 Henry's Law Constants for Selected Hydrocarbons[a]

Compound (at 25°C)	Molecular Weight (g/mol)	Vapor or Partial Pressure (atm)	Aqueous Solubility (g/m³)	Henry's Law Constant (atm m³/mol)
Methane	16	1.0	24.1	0.67
n-Butane	58	1.0	61.4	0.95
n-Hexane	86	0.2	9.5	1.81
n-Octane	114	0.019	0.66	3.28
n-Decane	148	0.0017	0.052	4.83
Cyclohexane	84	0.13	55	0.19
1-Hexane	84	0.24	50	0.41
Benzene	78	0.13	1780	0.0055
Toluene	92	0.038	515	0.0067
o-Xylene	106	0.0087	175	0.0052
Naphthalene	128	0.00011	34	0.00043
Biphenyl	154	0.000013	7.5	0.00027

[a]For gaseous solutes the solubility is at 1.0 atm pressure.

SOURCE: After McKay, 1981.

Aqueous Solubility

Henry's law, CP=HC, where p is pressure in atmospheres, C is concentration in solution, and H is Henry's law constant, can be invoked, although some error may be introduced because there is evidence that mixtures are more soluble than is expected from a direct mole fraction dependence (Leionen et al., 1971), a phenomenon that is at least partially due to the presence of dissolved natural humic-like matter in seawater (Boehm and Quinn, 1973). Henry's law constants for selected hydrocarbons in distilled water are given in Table 4-1.

The solubilities of hydrocarbons in seawater are generally some 60-80% of the distilled water values owing to the "salting out" effect. This can be characterized by the Setchenow equation (Aquan-Yuen et al., 1979).

Processes

Advection and Spreading

Transport of oil spilled onto the sea surface is due to two processes: advection and spreading. Advection is due to the influence of overlying winds and/or underlying currents. This process may be subdivided further depending on the causes of motion. For example, there may be advection by Stokes drift, Ekman currents, Langmuir circulation, geostrophic currents, or even turbulent flow. Descriptions and mathematical treatments of these various advection processes can be found in texts on general and physical oceanography.

The other transport process is spreading, a phenomenon resulting from a dynamic equilibrium between the forces of gravity, inertia, friction, viscosity, and surface tension.

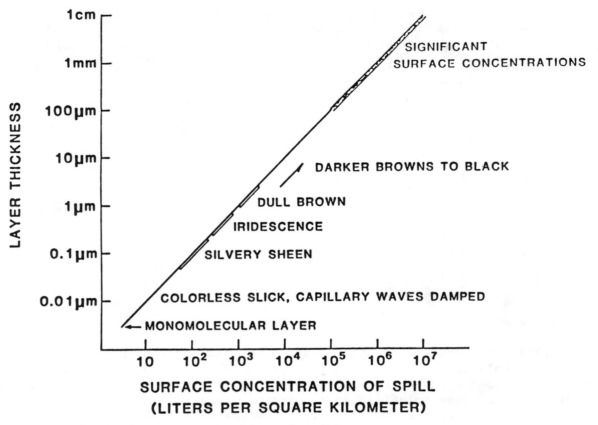

FIGURE 4-2 Surface concentration of spill.

SOURCE: Barger et al. (1974).

Oil on the sea surface manifests itself as slicks of variable
thickness. An approximate classification of these slicks is "thin"
slicks, less than 10 µm thick, to "thick" slicks, often millimeters
or even centimeters thick. Generally, the area of thin slick exceeds
that of the thick, but most of the oil volume usually resides in the
thick slick. Figure 4-2 is a labeled plot of thickness versus surface
concentration.

Observations of water may show first the evidence of oil by damping
of capillary waves: the surface becomes less "rough" and more
"glassy," but no oil is visible. As the slick thickens to 1 µm,
light interference effects become apparent, often giving irridescent
colors. Further increase in thickness to approximately 10 µm gives
darker films.

The behavior of thin films is dominated by surface tension (or
interfacial energy) effects; spreading is promoted when the sum of the
oil-water and oil-air interfacial tensions is less than that of the
water-air interfacial tension. Behavior in this regime is complicated
by the presence of natural organic surface layers on the ocean surface,
especially in quiescent and biologically productive areas. Although

these interfacial tensions can be measured, reliable deduction of behavior is not possible because (1) as the oil spreads it evaporates and dissolves, and the interfacial tension changes; (2) oxidation (probably photolytic in origin) alters the composition of the oil, especially by forming oxygenated compounds with low interfacial tensions; (3) as hydrocarbons dissolve in water they alter the water-air interfacial tension; and (4) spreading induces a change in composition of the oil by selective dissolution and evaporation of certain components.

Under relatively quiescent conditions, slicks of thickness greater than 10 μm tend to be surrounded by thin slicks; thus, they do not experience a surface tension force to induce spreading. Accordingly, the thick slicks tend to spread more slowly, at a rate controlled by a balance of hydrostatic, viscous, and inertial forces. This fluid flow process can be described mathematically if certain simplifying assumptions are made. However, the results probably will not have general utility because (1) solutions are very complex; (2) the rheology (flow) of the oil is often complex, i.e., the viscosity is not constant; (3) wave action stretches and compresses the oil slick; (4) water-in-oil emulsions may form; (5) usually the slick is wind driven relative to the water; (6) the presence of natural surface convergences or divergences will cause the oil slick to separate or accumulate; (7) oil composition (as distinct from viscosity or surface properties) appears to influence spreading (Fazal and Milgram, 1979); and (8) the entire spreading process is likely to be profoundly influenced by sea state, especially under rough conditions in which oil may be carried by spray.

In recent years there have been many attempts to model the fate of oil spills. For example, more than 35 different models are described in a comprehensive report by Huang and Monastero (1982). Because advective processes are the principal controls for the fate of a spill, they are the most frequently modeled. The general consensus of modeling experts is that there is no one universal model that will generally yield predictions that are realistic or undistorted. The modelling of the many dispersed smaller slicks is a major unsolved problem.

Wind causes surface water drift at a velocity of a few percent of the wind speed. Oil behaves similarly, the consensus being that the drift velocity is 3-4% of the wind speed. G.L. Smith (1977) has treated this in some detail, arriving at a drift factor of 3.64 ± 0.51%. An important observation is that the drift factor of the thick slick exceeds that of the thin; thus, the thicker region tends to accumulate at the leading edge of the slick, with the thinner region trailing.

Calculation of drift is essential in oil spill trajectory models, but is complicated by (1) the possible influence of Coriolis forces on the slick (tending to cause diversion to the "right" in the northern hemisphere and "left" in the southern hemisphere), (2) by residual and tidal ocean currents which provide an additional vector, (3) by Stokes surface drift associated with gravity waves (Lange and Hufnerfuss, 1978), and (4) the reduction by floating oil of wind stress transmitted to the sea.

Comparison of computed and actual trajectories of slicks such as those from the <u>Argo</u> <u>Merchant</u>, <u>Ixtoc</u> blowout, <u>Kurdistan</u>, or <u>Amoco</u> <u>Cadiz</u> suggests that the major sources of uncertainties are (1) lack of reliable data on wind speed and direction (due to distance from weather stations) and (2) lack of detailed ocean surface current data.

Evaporation

Evaporation, which may be responsible for the loss of from one- to two-thirds of an oil spill mass in a period of a few hours or a day (Jordan and Payne, 1980), causes considerable changes in chemical composition and physical properties of the oil. Calculation of evaporation rates is difficult because the rate depends on a number of factors, all of which may change with time. Observations of evaporation rate and attempts to predict that rate have been reported by Kreider (1971), McAuliffe (1977), Mackay et al. (1980b), Butler (1976), and Harrison et al. (1975) and are generally reviewed by Jordan and Payne (1980).

The rate of evaporative loss from a given volume of oil depends on (1) the area exposed, which tends to increase continuously as the slick spreads; (2) the oil phase component vapor pressures, which are a function of oil temperature and composition, and which fall as the lighter components are depleted from the slick; (3) the oil-air mass transfer coefficient, which depends primarily on the wind speed but also on the hydrocarbon vapor diffusivity; and (4) the possible presence of diffusive barriers such as a water-in-oil emulsion or a "skin" on the oil surface.

Thus the "half-lives" for the various hydrocarbon components in the slick cannot be determined, although approximate values can be suggested for defined conditions. The rate of evaporation from a thick, cold slick under calm conditions may be orders of magnitude slower than from a thin, warm slick under stormy conditions.

There are two general approaches to calculating evaporation rates. First is a pseudocomponent approach in which the oil is postulated to consist of a number of components or pseudocomponents of defined volatility and with proportions selected to give a mixture with volatility characteristics similar to that of the oil. As evaporation proceeds, the change in oil composition is computed and the falling vapor pressure is calculated from Raoult's law at the desired temperature. This approach has been used by Mackay and Leionen (1975), Yang and Wang (1977), and Mackay and Paterson (1981).

The second approach is to postulate an analytical expression for the amount evaporated as a function of time and composition as attempted by Butler (1976) and Mackay et al. (1980b). In the latter case, a method was proposed by which oil distillation data could be used to predict vapor pressures and, hence, evaporation rates.

Evaporation rates and composition changes can be measured by simple pan evaporation experiments, either outdoors or in wind tunnels, with an attempt to extrapolate the results to oceanic conditions. There remains a need to improve the prediction of oil evaporation rates and to characterize oil volatility characteristics more accurately by means of information obtained from pan evaporation experiments, distillation temperature data, and evaporation by a controlled air flow bubbled through the oil. Such information probably can be used to estimate the oil fractions evaporated under various defined conditions and to calculate the fractional retention of specific hydrocarbons at various times. Such a capability would be invaluable as a means of calculating changes in density or viscosity, assessing changing toxicity, and improving identification of slick samples for legal purposes. Although parts of this overall capability are in place, a comprehensive treatment is still lacking.

Hydrocarbons may evaporate from true solution in surface water quite rapidly--often with half-lives of an hour or less. This is illustrated by the analytical data reported for samples collected under dispersed oil slicks in which there was evidence of substantial removal of volatiles from the water column (McAuliffe et al., 1980). In the case of high-molecular-weight hydrocarbons of low solubility, most may be in colloidal or accommodated form and are not immediately available for evaporation. This topic has been reviewed recently by Mackay et al. (1981a), using calculations based on previous work by Mackay and Leionen (1975) and Liss and Slater (1974).

Dissolution

Dissolved hydrocarbon concentrations in water are particularly important because of their potentiality for exerting a toxic effect on biological systems.

They are less important from the viewpoint of the mass lost by the spill, for dissolution of even a few percent of the spill is unlikely. Dissolution is believed to be directly from the slick to the water column and from dispersed oil drops to the water column. In analyzing spill behavior a prediction of dissolution rate is unnecessary because the mass dissolved is negligible compared with that removed by droplet entrainment and can be subsumed in the dispersion rate expression.

The extent of dissolution is obviously influenced by the oil's aqueous solubility which, for a crude oil, is typically 30 mg/L. Most of the dissolved hydrocarbons are the more soluble low molecular weight aromatics such as benzene, toluene, and the xylenes. As the oil evaporates, these hydrocarbons are removed; thus the oil solubility drops and the dissolution rate falls to a negligible value. Some illustrative solubility data for fresh and weathered crude oils are given by Mackay and Shiu (1975).

Calculations of the rate of dissolution are imprecise, and only Cohen et al. (1980) and Butler (1976) have attempted to make estimates. The most soluble hydrocarbons, which are also the most volatile, are likely to be preferentially removed by evaporation, which is typically orders of magnitude faster. Even when hydrocarbons do dissolve, many are likely to be removed by subsequent evaporation from the water, provided they have sufficient volatility.

It should be reemphasized that in the subsequent chapter on Effects the simplest aromatic compounds are shown to be among the most toxic compounds of crude and refined oil, and as they are also the most soluble, their impact on the marine environment is greater than simple mass balance considerations would imply.

Dispersion/Vertical Transport

The lifetime of an oil slick on an ocean surface is often controlled by the dispersion or vertical transport of small particles of oil or oil-in-water emulsions into the water column (Mackay et al., 1980a). This lifetime usually determines whether a given slick is likely to impact on a particular shoreline that may be, for example, several days drift time from the site of the spill. Dispersion also results in exposure of subsurface marine organisms to particulate and dissolved oil. These organisms, in turn, may mediate the sedimentation of some of the oil through incorporation into fecal pellets.

The nature of the fluid mechanics of the event resulting in natural vertical dispersion is not well understood and is undoubtedly complex. Breaking or surface turbulence waves probably cause the oil to be driven into the water column, thus forming a swarm of oil droplets. The larger particles probably rise and coalesce with the slick, while the smaller oil droplets are conveyed with water eddies vertically downward to become permanently incorporated into the water column. These smaller droplets, which do not rise to the surface, and their associated water medium may be classified as an oil-in-water emulsion. This emulsion formation is only a part of the overall dispersion process.

Expressions for natural dispersion rates have been assembled by Mackay et al. (1980a), Spaulding et al. (1978), Garver and Williams (1978), and Aravamudan et al. (1981). The simplest approach for including dispersion in an oil spill model is that used in the SLICKTRAC model by Blaikley et al. (1978), who tabulated estimated vertical dispersion rates expressed as a percentage of the oil per day as a function of sea state and duration of the spill. This tabulation is undoubtedly an oversimplification of a complex phenomenon. A similar approach has been used by Audunson et al. (1980).

Experimental wind-wave tank measurements and a mathematical treatment of this process have been made by Mackay et al. (1981a). Equations were proposed for transport rates as a function of the oil slick thickness, the oil-water interfacial tension, the sea state, and in particular, the fraction of the sea covered by breaking waves. Although there are some data on this fraction, it is only for seas in the absence of oil. As is well known, oil reduces the incidence of breaking waves. Also, the dispersion process is believed to occur even when there are no breaking waves, a possible mechanism being "folding" of the oil when short waves of relatively high amplitude and short wavelength pass through the oil layer.

When the surface layer of water is well mixed, vertical eddy diffusion presumably causes further transport downward, and hypothetically,

Langmuir circulation cells may be even more important. Sutcliffe et al. (1963) reported sinking rates of water in the convergences of 2.7-5.7 cm/s with moderate wind speeds. This should be sufficient to overcome the buoyancy of some oil droplets that otherwise would not sink, but direct observations are lacking.

Further research on the problem of vertical dispersion is justified. An adequate set of equations cannot be developed until the basic mechanisms are better understood.

Emulsification/Mousse Formation (Water-in-Oil)

Laboratory studies to evaluate water-in-oil emulsion formation for different crude oils and petroleum products have demonstrated a dependence on the unique chemical compositions of each of the materials tested (Payne, 1984, and references therein). Heavier crudes with high viscosities are, in general, found to form the more stable emulsions (Bocard and Gatellier, 1981), and the presence of asphaltenes and higher-molecular-weight waxes have been found to be positively correlated with mousse stability (Berridge et al., 1968a,b; Davis and Gibbs, 1975; MacGregor and McLean, 1977; Mackay et al., 1979, 1980a; Twardus, 1980; Bocard and Gatellier, 1981; Bridie et al., 1980). Slightly differing results have been obtained in different investigations, but generally these materials act together in the emulsification process, although the asphaltenes do appear to play a more significant role (Bridie et al., 1980; Berridge et al., 1968a,b). The crystallizing properties of the component waxes (near the pour points of the oils tested) are believed to be important in affecting the internal oil-mousse structure and viscosity, and the asphaltenes are believed to act as surfactants, preventing water-water coalescence in the more stable mixtures (Berridge et al., 1968c; Canevari, 1969; Mackay et al., 1973; Bridie et al., 1980; Cairns et al., 1974). Other indigenous surface-active agents such as metalloporphyrins and nitrogen, sulfur, and oxygen compounds are believed to be equally important.

The products of photochemical and microbial oxidation have also been identified as having an important role as stabilizing agents (Bocard and Gatellier, 1981; Klein and Pilpel, 1974; Burwood and Spears, 1974; Zajic et al., 1974; Friede, 1973; Guire et al., 1973). In several instances, mousse could only be formed with photochemically or microbially weathered oils which were also subject to evaporation/dissolution processes. Brega, Nigerian, Zarzatine, and light Arabian crude oils have all been shown to exhibit this behavior in laboratory studies (Berridge et al., 1968b; Bocard and Gatellier, 1981). The formation of a stable mousse at the Ixtoc I wellhead was also observed to be delayed until after these processes had been operative for 24-48 hours on the oil released during that blowout (Payne, 1981).

No stable mousses could be formed in laboratory studies at any temperature with light petroleum distillates such as gasoline, kerosenes and several diesel fuels (Berridge et al., 1968a,b; Twardus, 1980) and could only be obtained with several light lube oils when they are fortified with wax and asphaltene mixtures obtained from known mousse-forming oils such as Kuwait crude (Bridie et al., 1980). This asphaltene mixture could also contain other surface-active agents of higher molecular weights.

Temperature is also a factor in mousse formation, and in several instances at low temperatures approaching the pour point of the heavier oils, stable emulsions have been generated regardless of wax or asphaltene content. Conversely, if stable water-in-oil emulsions are repeatedly exposed to freeze-thaw cycles, some destabilization and separation of water and oil have been noted (Dickens et al., 1981; and Twardus, 1980). Similar results have been obtained when laboratory generated and real spill water-in-oil emulsions were subjected to prolonged heating or removal from the water column.

The absolute amount of water content and the size of water droplets incorporated into various mixtures of mousse also significantly affect their stability and viscosity (Berridge et al., 1968a,b; Mackay et al., 1980a; Twardus, 1980; Bocard and Gatellier, 1981). Positive correlations of percent water versus mousse stability and viscosity have been noted for several of the crude oils studied (Mackay et al., 1979, 1980a). In general, with many oils, maximum stability is achieved with a water content in the range of 20-80%; however, at an oil-specific critical point, significant destabilization of the emulsions occurred (Berridge et al., 1968a,b; Twardus, 1980). Presumably, this reflects enhanced water-water contact and coalescence resulting in ultimate phase separation.

In most of the laboratory studies, the presence and/or absence of bacteria and suspended particulate material did not appear to affect emulsion behavior (Berridge et al., 1968a,b; Davis and Gibbs, 1975). Bacterial growth was generally limited to the surface of the mousse products tested, and is believed to have been inhibited by limited oxygen and nutrient diffusion into the mousse. Toxic materials inherent to the oils themselves may also be responsible for these observations, although water content (and in particular the size of the water droplets encapsulated within the mixtures) has also been correlated with the presence of bacteria in the less stable mousses (Berridge et al., 1968a,b). In several laboratory studies significant bacterial utilization of the mousse only occurred after treatment with dispersants, resulting in the break-up of the material, with concomitant increased surface-to-volume ratios (Bocard and Gatellier, 1981).

Physical Properties of Water-in-Oil Emulsions   The physical properties of stable emulsions are different from those of the starting crudes, and increases in specific gravity and viscosity have been observed to affect spreading, dispersion, and solution rates (Berridge et al., 1968b; Davis and Gibbs, 1975; MacGregor and McLean, 1977; Mackay et al., 1979, 1980a; Twardus, 1980). Some evidence has also suggested that evaporation of hydrocarbons of lower molecular weight (C_9-C_{12}) is affected by the emulsion (Twardus, 1980; Nagata and Kondo, 1977). In general, these effects are most significant in the emulsions containing greater than 50% water. Water-in-oil emulsions with less water usually have pour points, spreading properties, and viscosities which proportionately resemble those of the starting oils (Twardus, 1980; Mackay et al., 1980a).

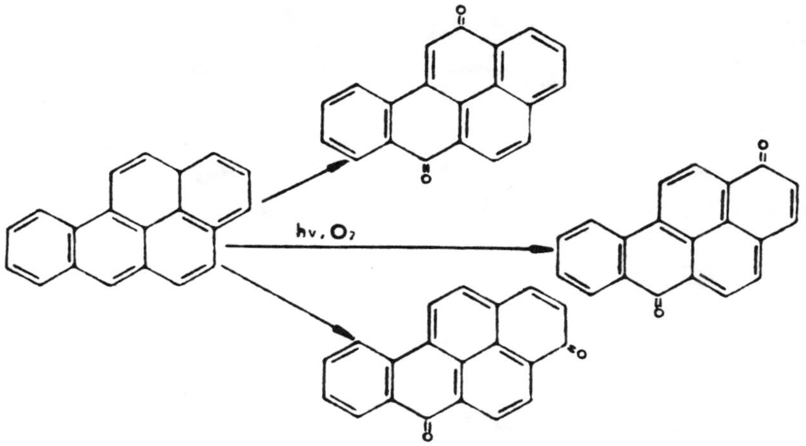

FIGURE 4-3 Photooxidation of benzo(a)pyrene to the 6,12-, 1,6-, and 3,6-diones.

SOURCE: National Research Council (1972, pp. 67-69).

The flash points and burn points of the water-in-oil emulsions studied have been found to vary significantly with water content, and for medium crudes, in situ combustion was significantly inhibited when the water content reached 79% (Twardus, 1980). For heavier crudes, significant combustion inhibition occurred when the water content reached 30% (Twardus, 1980). Water-dependent increases in viscosity also affect cleanup procedures, for skimming, mopping, and pumping of such mixtures becomes more difficult. The sorption capacity of various commercially available sorbant materials has also been observed to decrease as the water content in the mousse mixtures increases. This behavior is believed to be inherent to the hydrophobic properties of the sorbant materials examined (Twardus, 1980).

Photooxidation/Autooxidation

Photooxidation About 25% of the average oil spill evaporates and, in the gaseous state, is almost certainly all oxidized photochemically by OH radical and other species in hours or days to CO, CO_2, oxygenated organics, "secondary" aerosols, etc. (Altshuler and Bufalini, 1971; Heicklen, 1976). A proposed reaction for the photooxidation of benzo-[a]pyrene to the 6,12-, 1,6-, and 3,6-drones is illustrated in Figure 4-3. These processes prevent oil's reentry into the sea as petroleum.
The dissolved fraction of petroleum, such as the aromatics and polar compounds, is also subject to photochemical oxidation (Zafiriou, 1977; Zika, 1980). Mill et al. (1980) showed that cumene, an alkylbenzene, could be photooxidized by mechanisms involving the absorption of light by humic substances in natural waters, while PAH absorb sunlight directly and may be oxidized both by direct photolysis and as

TABLE 4-2 Petroleum Photooxidation Summary

REPORTED PRODUCTS:[a]

Carbon dioxide	Aldehydes and ketones
Organic acids, esters, lactones	Hydroperoxides
Diacids	Alcohols
Phenols and polyphenols	Sulfoxides

REPORTED EFFECTS:[b]

Darkening of oil
Increased ease of emulsification
Changes in oil-on-water spreading
Increased formation of soluble organics/carbons
Increased toxicity of petroleum or WSF to various organisms
Decreased primary production

NOTE: References to data in table appear in text.
[a]Not in order of importance.
[b]See section above on emulsification/mousse formation and Chapter 5 section on the toxicity of chemical products.

the indirect pathway (Zepp and Cline, 1977; Zepp, 1978; Zepp and Baughman, 1978).

The photooxidation of nondissolved oil, such as slicks, tar balls, films, "sheen," droplets and microdroplets, and colloidally dispersed oil, has been studied intensively since the 1973 NRC workshop. Table 4-2 presents a summary of photooxidation products and effects that have been reported in the literature. Other reviews that have considered oil photooxidation are given by Clark and MacLeod (1977), McAuliffe (1977), Wheeler (1978), Sprague et al. (1981), and Jordon and Payne (1980).

Laboratory and Simulated Environmental Studies   A series of laboratory and field studies, beginning with Berridge et al. (1968a), has elucidated the photolytic conversion under various conditions of petroleum hydrocarbons to a broad range of secondary and tertiary reaction products such as 3- to 4- ring PAH, thiacyclane oxides, methyl esters, salicyclic and phthalic acids, alcohols, ketones, hydroperoxides, lactones, hydroxyaromatic acids, and polyphenols (Frankenfeld, 1973; Burwood and Speers, 1974; Klein and Pilpel, 1974; Hansen, 1975, 1977; Larson et al., 1977, 1979; Bocard and Gatellier, 1981; Brunnock et al., 1968).

Environmental Case Histories   The U.S. group studying the Amoco Cadiz spill was able to link photooxidation products found in environmental samples very closely with those formed in laboratory studies

(Calder et al., 1978; Patel et al., 1979; Overton et al., 1979). A slick sample less than 10 days old showed a series of C_1-C_3 benzothiophene sulfoxide homologs, corresponding to the benzothiophenes that were present in the original oil. Using a tungsten lamp with uranium glass filter, Overton et al. (1979) photolysized a similar crude suspended in aqueous saline solution with the visible analog of sunlight (without much UV). This experiment showed again that for some oils, visible and long wave UV light can be important in oil photooxidation, in agreement with Freegarde et al. (1971) and contrasting with Hansen's (1975) conclusions that wavelengths below 350 nm are needed. Product identification in the Amoco Cadiz studies confirmed in molecular detail the prediction of Berridge et al. (1968a), that the reactivity of oxidized sulfur compounds in a case in which the compounds that are oxidized--benzothiophenes--are much less reactive than typical thioorganics.

There are several reports and papers involving oil photooxidation in the Ixtoc I event. These include the only case in which experienced scientists, observing the occurrence of the event at close range for long periods of time reported seeing apparently light-dependent changes in oil behavior. Atwood and Ferguson (1982) stated that in zone 4 (roughly from 10 to 20 nautical miles from the blowout) the light-brown emulsion visibly darkened and formed into black streaks. "This was assumed to result from oxidation of the oil, and the rate seemed to be dependent on sunlight intensity. . . The extent of the zone and rate of mousse formation apparently were dependent on sunlight intensity and wind stress." Reports by Mackay et al. (1981a,b) and Boehm and Fiest (1980) give varying interpretations of the importance of photooxidation of the oil spilled from Ixtoc I. Generally, they agree that photooxidation is minor with regard to material balance considerations but that it produced significant changes in the oil at 50 km and more from the wellhead.

Overton et al. (1980) exposed very fresh Ixtoc I oil to natural sunlight, air, and synthetic seawater. The exposures were under a quartz plate, with the oil being added as a 1:10 heptane solution/ dispersion and the heptane being allowed to evaporate. The temperature was controlled, and 4-day Louisiana sunlight exposures were used. Each experiment had a "dark" control. Photographs documented the conversion of the oil to a crusted material which broke up into tarry flakes, confirming experimentally the involvement of photolysis suggested by the anecdotal observations of Atwood and Ferguson (1982) and the report by Patton et al. (1981) that at 750-1,000 km from the well-head on sheen-covered water, the primary material was gram-sized pancakes and milligram-sized flakes--both covered with a 5- to 10-μm thick "skin." The flakes were enriched in polar compounds and were denser than pancakes.

The polar aromatic fraction of Overton et al. (1980), contained an unresolved mixture as determined by high pressure liquid chromatography (HPLC). This mixture was absent in unexposed material. The acid fraction, on methylation, yielded numerous peaks, including large amounts of $\underline{n}$-C_9 to $\underline{n}$-C_{11} fatty acid methyl esters (FAMES). Branched FAMES and C_1- and C_2-alkylated naphthols were also found, as well as substituted 1- to 3-ring aromatic and heteroaromatic acids. The authors stressed that a separate microbial oxidation simulation experiment showed many of the same product types, but with different ratios than those found in the photochemical experiment.

Autooxidation The thermodynamic instability of reduced carbon compounds in the presence of oxygen is not always manifested in a significant reaction rate. When it is, the noncombustion mechanisms involved most commonly are multistep, free radical reactions that are notoriously complex in mechanism and detailed behavior. They are subject to induction periods, and rates are often very sensitive to both inhibitory and acceleratory effects of trace components and to the physicochemical environment.

Relevant background material is widely available; basic reviews are given by Fallab (1967) and Nonhebel and Walton (1974), among others. Many constituents of oil are active in these reactions (e.g., branched hydrocarbons, benzylic C-H bonds, aromatic rings). Petroleum also contains potential inhibitors (amines, metastable radicals, sulfur compounds) and accelerators (transition metal complexes, selected organics). The issue is therefore entirely one of rate, and basic theory cannot be of much help in assessing environmental rates. However, it can predict the probable products and help to rationalize those actually found.

There are only a few relevant observations on dark autooxidations of petroleum in the environment (Kawahara, 1969; Burwood and Speers, 1974; Brunnock et al., 1968; Lysyj and Russell, 1974; Larson et al., 1977; McLean and Betancourt, 1973). The chemical mechanisms involved in oil autooxidation and photooxidation are probably similar. The formation of virtually all the reported breakdown products (largely oxygenated compounds) can be rationalized by known types of photoreactions and autooxidation (Howard and Ingold, 1966; Fallab, 1967; Nonhebel and Walton, 1974; Hendry et al., 1976). In contrast, we are lacking both good chemical evidence for, and good mechanisms to explain, apparent "polymerization" to asphaltenes, to skins or to "semisolid flakes." These mechanisms probably involve both free radical pathways and singlet molecular oxygen (Larson and Hunt, 1978). Whatever the detailed chemical mechanisms involved, in some cases they appear to give products similar to those formed in biological processes (Overton et al., 1980; Dowty et al., 1974; Patel et al., 1978).

Sedimentation

Laboratory Studies The various forms of oil in seawater can be sorbed onto settling particles and delivered to the bottom sediments. Meyers and Quinn (1973) suggested that as hydrocarbons become more soluble with increasing temperature and/or interaction with dissolved organic matter, correspondingly smaller amounts are available to associate with suspended mineral particles in seawater. These authors studied the sorption of n-alkanes and polycyclic aromatic hydrocarbons onto bentonite clay in saline solutions. The heats of sorption indicated physical adsorption of the Van der Waals type. The extent of association decreased with increasing solubility of the hydrocarbons from n-alkanes to aromatics. Several other clay minerals were also investigated and showed differing amounts of hydrocarbon association. In more recent studies, Meyers and Oas (1978) investigated the adsorption of hydrocarbons by smectite clay in saline solutions. They found that association increased linearly with increasing n-alkane concentration and with increasing carbon chain length. The degree of association of aromatic hydrocarbons was generally low, and isoalkanes were more effectively adsorbed than n-alkanes containing the same number of carbon atoms. Button (1976) reported that when the solubility of n-dodecane in saline solution was not exceeded, this n-alkane was readily metabolized by microorganisms but not sorbed by clays. He concluded that findings of the type reported by Meyers and Quinn (1973) may represent the affinity of clays for small, oil-phase particles.

Removal of indigenous organic matter from sediment particles by hydrogen peroxide treatment increased the uptake of n-alkanes and No. 2 fuel oil hydrocarbons from saline solutions as compared to that by untreated particles (Meyers and Quinn, 1973). Meyers and Quinn suggested that sedimentary organic matter interfered with oil uptake by masking sorption sites in the sediment and/or by binding small particles together, thus reducing the effective surface area. Another possibility is that some of the sedimentary organic matter was released from the particles to the solution, thereby increasing the solubility of the oil as reported by Boehm and Quinn (1973, 1974). In contrast to the results of this study are the findings of Karickoff et al. (1979), who examined the sorption of the aromatic hydrocarbons (especially pyrene) in saline solutions by pond and river sediments. The partition coefficients were directly related to organic carbon content for a given sediment particle size. The sand fraction was a considerably less effective adsorbent than the fine fractions, and differences in the sorption between the silt and clay fractions were largely due to organic carbon content. The authors concluded that a reasonable estimation of the sorption behavior can be made from a knowledge of the particle size distribution, associated organic matter in the sediment, and the octanol-water distribution coefficient of the hydrocarbons.

Bassin and Ichiye (1977) studied the flocculation behavior of suspended sediments and crude oil emulsions in both fresh and brackish waters. They demonstrated that oils and clays formed colloids or colloidal electrolytes in the presence of dissolved salts. Oil sedimentation seemed to be caused by adsorption of oil films onto clay particles which were subsequently flocculated by the shrinking of the double-layer charge on the collodial clay particles. Thus, a significant quantity of oil may be adsorbed to the clays and sedimented with them. They further suggested that the sedimentation of oils by clays in coastal areas is due more to electrolytic flocculation than to the affinity between the oil and the clays.

The weathering processes affecting No. 2 fuel oil in saline solutions were investigated by Zürcher and Thüer (1978). They concluded that partial dissolution, adsorption, dispersion, and agglomeration play important roles as initial processes in weathering of oil in natural aquatic systems and result in the fractionation of the original oil mixture. Alkylated benzenes and naphthalenes were enriched in water phase; some high-molecular-weight aliphatic hydrocarbons were adsorbed by clays, and oil droplets were associated with suspended minerals. The latter process depended on the formation of dispersed oil droplets through turbulence and interfacial tension of the oil, and this played a major role in oil sedimentation.

Model Ecosystems   Gearing et al. (1980) studied the partitioning of No. 2 fuel oil among seawater, suspended particulate matter, and sediments at the Marine Ecosystems Research Laboratory (MERL) located at the University of Rhode Island. The fuel oil was added as an oil-water dispersion in semiweekly doses to three systems over a 4-month period, and samples were analyzed for various hydrocarbon fractions. Transport from the water column was examined in relation to their physical and chemical properties. The amount of hydrocarbons in the water column associated with suspended particulate matter was inversely proportional to their water solubility. The result was a fractionation of saturated and aromatic hydrocarbons. The eventual settling of the suspended material carried about 50% of the relatively insoluble saturated components and less than 20% of the more soluble aromatic hydrocarbons to the sediments. Once in the sediments, these components were slowly mixed down through the zone of bioturbation (3-4 cm), and 10-20% of the hydrocarbons originally delivered to the sediments persisted for at least 1 year after the end of the oil additions.

Hinga et al. (1980) used the MERL facility to study the biogeo-chemistry of ^{14}C-labeled benzo[a]anthracene. They reported that this polycylic aromatic hydrocarbon, having a low water solubility and low rate of metabolism in the water column, became associated with sediments either through direct adsorption on the bottom after turbulent mixing at the sediment-water interface, or by adsorption onto or incorporation into suspended particulate matter, followed by subsequent deposition. Once incorporated into surface sediments, the hydrocarbon and its metabolites were mixed deeper into the sediments by benthic animal activity. R.F. Lee et al. (1978) studied the fate of polycyclic aromatic hydrocarbons added to ecosystem enclosures (CEPEX) suspended in Saanich Inlet, British Columbia, Canada. Their results indicated that aromatic hydrocarbons have a short residence time (on the order of a few days) in marine waters and, because of their low solubility in water, the higher-molecular-weight components are associated with par-ticles in the water column. After sedimentation, biological degradation became an important factor in their removal. In areas with low concen-trations of suspended material, rates of hydrocarbon sedimentation would be low.

Field Investigations   In the case of actual oil spills in the marine environment, there are several mechanisms (including those studied under laboratory and enclosed or controlled mesocosm conditions) by which petroleum can reach the sediments. For the purpose of this discussion, the most important of these are (1) sorption of oil by suspended particles (these particles include terrigenous minerals,

plankton and detrital particles, and resuspended bottom sediments),
(2) ingestion of oil by zooplankton and incorporation into fecal
pellets, (3) weathering of oil by physical/chemical processes, and (4)
direct mixing of oil and sediments. The following field investigations
illustrate each of the above mechanisms.

Sorption of Oil The impact of the Tsesis oil spill in 1977 on the
pelagic ecosystem of the Baltic Sea was the subject of a detailed study
by Johansson et al. (1980). Sediment traps were used to demonstrate
the importance of sedimentation as a mechanism for the removal of this
No. 5 fuel oil from the water column. Up to 0.7% of the sedimented
material collected during the second week after the spill was petroleum
hydrocarbons, and the estimated total sedimentation in the impacted
areas was 30-60 tons. This accounted for 10-15% of the approximately
300 tons of oil spilled. The most probable mechanism for the rapid
sedimentation of oil was through adsorption to detritus or clay particles, for particulate levels were high due to wind-induced resuspension of bottom sediments.

Another study by McAuliffe et al. (1975) estimates the fates of
35,000-65,000 bbl of crude oil discharged by Chevron Production
platform C, Man Pass Block 41 in the Gulf of Mexico. By measuring the
concentrations of hydrocarbons in the sediments and subtracting background values, they calculated that less than 1% of the spilled oil was
found in the sediments within a 5 mile radius of the platform and
suggested that sorption of oil droplets on settling particles was the
probable mechanism for sedimenting the hydrocarbons.

In addition to sorption reaction in the water column, hydrocarbons
can associate with particles before entering estuarine and coastal
waters. For example, Van Vleet and Quinn (1977) reported that about
95% of the hydrocarbons discharged from a municipal wastewater secondary
treatment plant were associated with suspended solids. This plant is
located on the Providence River at the head of Narragansett Bay and is
a major source of the suspended petroleum hydrocarbons found in the
water column of the bay (Schultz and Quinn, 1977). The material is
transported throughout the bay via tidal currents and eventually settles
to the bottom, resulting in decreasing sedimentary hydrocarbon concentrations from the river to the mouth of the bay (Hurtt and Quinn,
1979). DiSalvo and Guard (1975) reported that at least 13.5 tons of
pollutant hydrocarbons were present in association with suspended
particulate matter in San Francisco Bay at any given time during their
sampling period. The source of the particulate hydrocarbons was
thought to be suspended material from sewage effluents. Analysis of
water samples suggested that hydrocarbons were associated with large
particles or with flocculated smaller particles which were able to
settle from suspension in less than 16 hours and which accumulated in
the shoal areas of the bay.

Incorporation of Oil Into Zooplankton Fecal Pellets Forrester
(1971) found small (5 μm to 2 mm size) droplets of Bunker C oil in
the water column following the 1970 wreck of the tanker Arrow in
Chedabucto Bay, Nova Scotia. These droplets were formed by surf action

on oiled beaches and by wave action in oil-covered water, and were stirred into the water column to a depth of 80 m in some places. The size range of the oil particles included that of natural foods consumed by zooplankton. Conover (1971) found that zooplankton ingested small particles of the oil that were dispersed throughout the water column. He reported that up to 10% of the oil was associated with the zooplankton and their feces contained up to 7% Bunker C. As much as 20% of the oil was delivered to the bottom sediments as fecal pellets in addition to the particulate oil which was exported from the area by hydrodynamic processes.

Weathering of Oil Patton et al. (1981) studied the Ixtoc I oil spill in the Gulf of Mexico and reported that weathering processes first formed a cracked, scaly surface on oil pancakes, which then flaked in turbulent seas. This process exposed fresh oil within the pancakes, which produced a surface sheen until a new skin formed. After weathering, the skin flaked off on agitation, and the process was repeated. The flaking of these particles seemed to be a significant intermediate in the dispersion of this oil spill, and the increased density of the chemically weathered flakes relative to the pancakes suggested that this process may ultimately result in the sedimentation of oil.

Direct Mixing of Oil and Sediments The barge Florida ran aground off West Falmouth, Massachusetts, in September 1969 and spilled about 600 tons of No. 2 fuel oil into Buzzards Bay. Strong winds churned the oil into an oil-water emulsion and drove it into West Falmouth Harbor and Wild Harbor. Oil was incorporated into the sediments under a water depth of at least 10 m due to the intense mixing of oil and water by gale force winds. The effects of this spill have been studied in detail by several workers (e.g., Blumer et al., 1970b; Blumer and Sass, 1972). Most recently, Sanders et al. (1980) reported that 5 years after the spill, partly degraded fuel oil was still present in the sediments of Wild Harbor River and Estuary.

A spill of Bunker C oil in San Francisco Bay in 1971 was studied by Conomos (1975). The spilled oil increased in density by evaporative loss of its lighter fractions and/or solution of its low-molecular-weight components, and was mixed through the water column by strong tidal currents and wind. Some of the oil globules carried to the bottom by the turbulence mixed with sandy and gravelly sediments and remained near the bottom. The oil was eventually moved, beaching in eastern San Pablo Bay after being transported landward by the near-bottom water currents.

In regard to oil stranded on coastlines, the behavior of spilled oil in different environments is primarily dependent on the porosity of sediments and the energy of the waves acting on the coastline. Rocky shores tend to "self-clean" within a matter of months, whereas soft-sediment lagoons or mangrove swamps act as long term (years to decades) petroleum sinks. On cobble and sandy beaches, oil can sink deeply into the sediments and remain longer than on bare rocks. Pools of oil are likely to collect in hollows among rocks, protected by a skin of weathered oil and may remain essentially unchanged for a long time. Tidal pumping is the active factor causing penetration into the sediments. Sediment grain size and compaction control the rate of penetration. In muddy sediments, penetration is minimal, and only the upper few centimeters are affected.

The breakup of the supertanker Amoco Cadiz (spilling 223,000 tons of crude oil) in March 1978, off the Brittany coast, coincided with the annual rebuilding phase of beaches in which tons of sand are transported onto the shallow winter beach slope (Hess, 1978). This process resulted in the stranding of oil and mousse on the beaches, followed by transport and burial within these beaches. Long et al. (1981) report that beaches having thick sand layers became long term reservoirs in which the oil moved slowly and continuously downward until it reached a level near the water table where it was somewhat stabilized, having a residence time of more than 3 years. For beaches with a thin sand layer overlying an impermeable basement, the oil moved laterally along the bedding plane and could be seen "washing out" at outcrops.

Summary and Recommendations

Advection and spreading are the most important processes affecting the fate of spilled oil. Predictions based on complex mathematical modeling of these processes are unreliable because of the wide spectrum of oil types and the changing environmental conditions occurring during a spill. The best estimate possible at the present time is that the drift velocity is 3-4% of wind speed.

Evaporation of oil is the next most important process, accounting for up to one- to two-thirds of the mass lost. Evaporation of various hydrocarbons from aqueous solution is also important. However, the predictability of evaporative behavior is difficult due to the complexity of the oil and uncertainties in solubility and other thermodynamic data for individual compounds. Dissolution is considerably less important than evaporation in determining the fate of spilled oil because of the low aqueous solubility of most components.

In order to be able to develop the complex models which will yield the predictive capability required for future spillages of petroleum at sea, better and more comprehensive thermodynamic data are needed on individual hydrocarbons as well as on nitrogen-, sulfur-, and oxygen-containing compounds.

The movement of oil into the water column is important because it determines the lifetime of a slick. The primary mechanism for the process is believed to be propulsion by surface turbulence of oil into the water column as a "shower" of oil droplets. Modeling awaits an understanding of the exact mechanism for this process. Thus, research leading to a better understanding of the mechanisms for vertical dispersion of oil is recommended.

Our present knowledge of atmospheric photochemistry suggests that almost all of the oil that evaporates is photochemically oxidized in the atmosphere. In surface water, photochemical oxidation may be important, taking place within a time scale of minutes to days. In addition, reaction products may be more or less toxic than their precursors. The significance of photooxidation and autooxidation and their products is unknown and needs to be determined.

Emulsification and mousse formation arise from the physical mixing due to wave turbulence when oil is released into the sea and involves surface-active compounds (possibly asphaltenes, porphyrins, and other nitrogen, sulfur, and oxygen compounds). Products of photochemical and microbial oxidation also can serve as surfactants, but mousse formation may slow bacterial action. In addition, emulsification can be initiated and dispersants added to spills to curb or reduce impacts. Research in this area should be focused on the relationship between chemical composition and formation and the stability of oil-water emulsions, including the role of photochemical and biochemical reaction products.

Sedimentation of spilled oil takes place primarily through sorption on particulates or by incorporation into fecal matter. Weathering processes increase the density of floating oil and, when this occurs, incorporation into particles will eventually cause an increase in density above that of seawater so that the oil then sinks below the surface into the water column and, in some cases, eventually to the sediments. A better understanding of interactions of petroleum with particulates in the water column and sediments is needed.

BIOLOGICAL FATES

Introduction

The biodegradation of petroleum is seen by most workers as one of the principal mechanisms for removal of petroleum from the marine environment. This applies particularly to the nonvolatile components of crude oil or refined products. The various compounds differ widely in terms of their biodegradability. Thus alkanes and alkenes and the simpler monoaromatics are biodegraded quite readily, but the tars and resins are virtually impervious to biological attack.

The pathways used for biodegradation of petroleum tend to fall into two distinct approaches: that used by bacteria, and that of the eukaryotic invertebrate and vertebrate systems. Insufficient data are available from green plants to make broad generalizations on their ability to biodegrade petroleum hydrocarbons. Microorganisms (bacteria, yeasts, fungi) are important in the degradation of petroleum in surface films, slicks, the water column, and sediments. Phytoplankton may degrade hydrocarbons in the water column, but little is known of this. Zooplankton are known to aid in the sedimentation of oil droplets and oil associated with particulate matter through their ingestion of microparticulate oil from the water column, followed by excretion of what is apparently unmodified oil in the feces (Andrews and Floodgate, 1974; Conover, 1971; among others). Benthic invertebrates such as polychaetes, which normally play an important part in the oxidation and recycling of sediment organic matter, also have a significant role in the degradation of sediment-bound oil (e.g., Gardner et al., 1979; Gordon et al., 1978b; Lee et al., 1979).

Fish, marine mammals, and birds can become contaminated through uptake of oil from the water column and through ingestion of oiled food, or in the case of marine mammals and seabirds, through cleaning and preening of oiled fur or plumage. Therefore, these animals can also contribute to the overall biodegradation of petroleum in the marine environment.

The understanding of petroleum biodegradation is far from complete because of the complexity of both petroleum and the various metabolic processes. Of the various classes that make up petroleum, most of the attention has been on the hydrocarbons. Some information is available also on the degradation of sulfur-containing compounds, e.g., dibenzothiophene. The accumulation and metabolism of various sulfur-, nitrogen-, and oxygen-containing compounds present in petroleum, by marine algae and higher invertebrates and vertebrates, are largely unstudied.

Microbial Biodegradation

When hydrocarbons become available to a microbial community in a complex mixture such as petroleum, biodegradation of most petroleum compounds occurs simultaneously, but at widely differing rates. Generally the biodegradation of the n-alkanes is most rapid, followed closely by the simple aromatic components. The isoalkanes, cyclo-alkanes, and condensed aromatics are degraded more slowly. Various hydrocarbon components may also influence each other's degradation indirectly through the phenomena of cometabolism or "diauxie." In the first process, a normally refractory hydrocarbon may be degraded in the presence of a second readily degraded hydrocarbon. In the case of diauxie, the presence of a more easily utilized hydrocarbon represses enzyme induction necessary for metabolism of the second hydrocarbon. The latter is degraded only after the first is exhausted.

The following summary of hydrocarbon biodegradation by microbes is based primarily on recent reviews of the current literature (Bartha and Atlas, 1977; Atlas and Bartha, 1981).

Aliphatic Hydrocarbons

The biodegradation of normal and branched alkanes was reviewed by McKenna (1971) and Rathledge (1978). Pirnik (1977) reviewed some specific problems related to the biodegradation of methyl-branched alkanes.

Alkanes of the C_{10}-C_{22} range are the most readily and frequently utilized hydrocarbon substrates. The gaseous alkanes (C_1-C_4) are degraded by certain groups of microorganisms; because of low solubility, the C_5-C_9 alkanes are attacked by relatively few hydrocarbon degraders. The n-alkanes above C_{22} are not readily biodegraded because they are only slightly soluble at temperatures within the range normally found in the ocean. Nevertheless, biodegradation of n-alkanes up to C_{44} in length has been demonstrated (Haines and Alexander, 1974), albeit slowly, particularly at low temperatures.

Isoalkanes are less readily utilized in comparison to n-alkanes. Methyl branching in the 2- or 3-position is a hindrance to betaoxidation, and relatively few alkane degraders possess mechanisms to bypass such blockage. Further branching, resulting in quaternary carbon atoms, may render an isoalkane completely resistant to microbial biodegradation.

Olefins tend to be more toxic and, at least under aerobic conditions, are less readily utilizable than the corresponding alkanes. Theoretically, olefins should be less stable under anaerobic conditions than alkanes, as they can be hydroxylated without a need for oxygenase enzyme systems.

The most common type of primary metabolic attack by microorganisms on n-alkanes is mediated by mixed function oxidases (monooxygenases) that, acting on the terminal carbon, convert the hydrocarbon molecule to a primary alcohol. A cytochrome P450 and a rubredoxin system have been characterized as mediating such oxidations, both resulting in the same primary alcohol product.

Although in the great majority of cases the initial attack is directed at the terminal carbon atom of the hydrocarbon molecule (Figure 4-4a), some microorganisms attack hydrocarbons subterminally, converting them to secondary alcohols (Markovetz, 1971). Oxidation continues to the keto and ester stage. The ester, most commonly a formate or acetate ester, is hydrolyzed, yielding formic or acetic acid and a primary alcohol.

The primary alcohols, whether derived from terminal or subterminal oxidations, are further oxidized to aldehydes and fatty acids. The fatty acids are subsequently shortened by C_2 units by betaoxidation. In some cases, however, especially when betaoxidation is hindered by branching, the fatty acid is attacked at the other terminal carbon by the process called omegaoxidation. Alternatively, the blockage posed by the methyl branch can be eliminated by the mechanism elucidated by Seubert and Fass (1964). This pathway essentially elongates the methyl branch by a carboxylation step, and the resulting C_2 unit is released as acetic acid.

Alkanes may be attacked, either as the alkanes at a saturated terminal carbon, or may be oxidized directly at the double bond with formation of an epoxy compound. This is hydrated to a diol, which in turn, is oxidized and cleaved to yield a fatty acid and a primary alcohol.

Alicyclic Hydrocarbons

Low-molecular-weight cycloalkanes, such as cyclohexane and decalin, exhibit considerable solvent-type membrane toxicity and serve as growth substrates for microbes only in exceptional cases. At low concentrations, in mixed enrichments, and in the marine environment, cycloalkanes are degraded at moderate rates. Initial cometabolic attack followed by commensal utilization of the products is the main mechanism of biodegradation (Perry, 1977, 1979; Trudgill, 1978; Beam and Perry, 1974). The metabolic sequence, as illustrated for cyclohexane, is shown in Figure 4-4b. The oxidase responsible for converting the cycloalkane to the cyclic alcohol and the monooxygenase that lactonizes the ring apparently are seldom present in the same microorganism, necessitating the synergistic degradation sequence.

Aromatic and Condensed Polyaromatic Hydrocarbons

Monoaromatic hydrocarbons have considerable membrane toxicity because of their solvent properties, but in low concentrations they are rapidly utilized by a considerable number of microorganisms. Condensed poly-aromatics having 2-4 rings are somewhat less toxic and are biodegradable at rates that decrease with the level of condensation. Condensed polyaromatics with 5 and more rings fail to serve as growth substrates and are eliminated from the environment very slowly. The initial metabolic transformation steps, if any, are cometabolic. The microbial metabolism of aromatic hydrocarbons has been subject to several updated reviews (Gibson, 1968, 1971, 1977; Hopper, 1978). Using benzene to illustrate the sequence of events (Figure 4-4c), initial bacterial oxidation occurs by dioxygenase attack. The postulated dioxetane product is first reduced to cis-1,2-dihydroxy-dihydrobenzene and is oxidized, in turn, to catechol, regenerating NADH (hydrogen form of nicotinamide-adenine dinucleotide) in the process. The catechol ring is opened by either orthocleavage or metacleavage, yielding in the first case, cis, cis-muconic acid, beta-ketoadipic acid, and the succinate plus acetate fragments. In the second case the cleavage yields 2-hydroxy-cis, cis-muconic semialdehyde, and subsequently the pyruvate plus 2-keto-4-pentenoic acid fragments are produced.

As reviewed by Cripps and Watkinson (1978), condensed polyaromatic hydrocarbons having 2 or more fused aromatic rings command special interest because some compounds in this group are potential carcino-gens, or may be transformed to carcinogens by microbial metabolism. Two- and three-ring condensed aromatic hydrocarbons such as naphtha-lene, anthracene, and phenanthrene are degraded by successive opening of the aromatic rings, essentially by the mechanism described for benzene. More highly condensed polycyclic aromatic hydrocarbons such as benzo(a)pyrene and benzo(a)anthracene (Gibson, 1975, 1976) are cooxidized to dihydrodiols and thus are activated to carcinogens. They apparently are not extensively degraded by pure cultures and are mineralized to CO_2 in the environment only at extremely slow rates (R.F. Lee and Ryan, 1976; R.F. Lee, 1977a; Herbes and Schwall, 1978). Recently, Wu and Wong (1981) reported microbial methyl hydroxylation of 7,12-dimethylbenzo(a)anthracene, resulting also in carcinogenic activation.

Alkylaromatic hydrocarbons with short alkyl varieties such as toluene may be degraded by the mechanisms described for benzene (Figure 4-4c). Alternatively, the initial attack may occur at the methyl group with a conversion, in several steps, to benzoic acid. Oxidative decar-boxylation leads to catechol that is subject to ring cleavage. Phenyl-

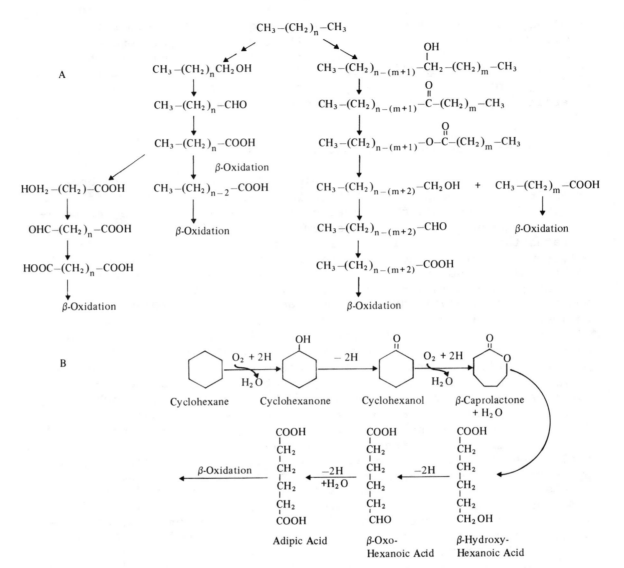

FIGURE 4-4 Degradative pathways of petroleum hydrocarbons. (a) n-alkanes: (left) diterminal or omegaoxidation, (center) monoterminal betaoxidation, and (right) subterminal oxidation (Atlas and Bartha, 1973a). (b) An example of metabolism of alicyclic hydrocarbons (Atlas and Bartha, 1981). (c) Microbial metabolism of the aromatic ring (simplified) by meta or ortho cleavage, as shown for cyclohexane (Atlas and Bartha, 1981).

alkanes with long alkyl chains are regularly metabolized, starting at the terminal carbon of the alkyl moiety (omegaoxidation). Successive betaoxidation steps shorten the alkyl chain to benzoic acid (in the case of odd carbon numbers) or to phenylacetic acid (even carbon numbers). Benzoate is easily degraded as outlined above, but phenylacetic acid is more persistent and, in pure culture experiments, often accumulates as an end-product.

C

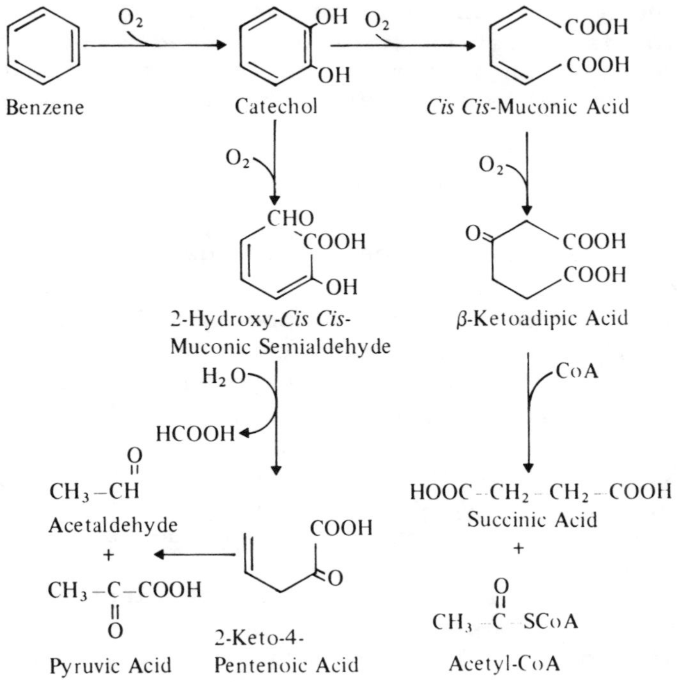

Benzene → O_2 → Catechol (OH, OH) → O_2 → Cis Cis-Muconic Acid (COOH, COOH)

Catechol → O_2 → 2-Hydroxy-Cis Cis-Muconic Semialdehyde (CHO, COOH, OH)

Cis Cis-Muconic Acid → O_2 → β-Ketoadipic Acid (O, COOH, COOH)

2-Hydroxy-Cis Cis-Muconic Semialdehyde → H_2O → HCOOH

$$CH_3-CH \quad \overset{O}{\underset{}{\|}}$$
Acetaldehyde
+
$$CH_3-C-COOH \quad (\|\,O)$$
Pyruvic Acid

← 2-Keto-4-Pentenoic Acid (COOH, O)

β-Ketoadipic Acid → CoA → $HOOC-CH_2-CH_2-COOH$ Succinic Acid
+
$$CH_3-C-SCoA \quad (O)$$
Acetyl-CoA

FIGURE 4-4 (continued)

Asphaltenes and Resins

Asphaltenes are a heterogeneous and poorly characterized assortment of compounds with high molecular weights and low volatility and solubility. Analytical techniques are in general inadequate to define the individual chemical structures of asphaltenes and are even less able to follow their fate in the environment (viz., Chapter 3). However, "tar," high in asphaltenes, is widely distributed throughout the marine environment (Butler et al., 1973; e.g., physical fates section). As well, both laboratory and practical experience show that these compounds are highly resistant to biodegradation (e.g., Traxler et al., 1965).

Resins comprise the polar and often heterocyclic NSO compounds (compounds containing N, S, O as constituents). When not highly condensed they may be available to limited microbial metabolism. This includes the lower-molecular-weight resin fraction such as phenols, cresols, thiols, thiophenes, pyridines, and pyrroles. The latter have considerable toxicity toward microorganisms, but at least some of them are likely to be biodegraded at low concentrations. Very little work has been published in this area, and available information is restricted to the condensed dibenzothiophene (Yamada et al., 1968; Nakatani et al., 1968; Kodama et al., 1970, 1973; Laborde and Gibson, 1977).

Somewhat paradoxical is the microbially mediated production of long chain alkanes (waxes) during biodegradation of petroleum (Walker and Colwell, 1976b). These are produced only as a consequence of biodegradation and not by nonbiological weathering. The mechanism of their formation is as yet unexplored. A head-to-head condensation of reactive biodegradation intermediates (e.g., free radicals) is considered to be a possible explanation for their appearance.

Phytoplankton and Marine Algae

The evidence suggests that unicellular algae are able to take up and metabolize both aliphatic and aromatic hydrocarbons. However, the extent to which this occurs is only poorly understood. Much less work has been done with macroalgae, except to show that such genera as Fucus, Enteromorpha, and Phylospadix, when exposed to an oil spill, will take up petroleum hydrocarbons (Clark et al., 1973, 1975; Burns and Teal, 1979; Vandermeulen and Gordon, 1976).

The uptake and metabolism of aliphatic hydrocarbons to fatty acids by the diatom, Chaetoceros simplex calcitrans, was investigated by Boutry et al. (1977). A green alga, Scenesdesmus, was reported to metabolize the alkane, heptadecane, in the light but not in the dark (Masters and Zajic, 1971). Since that time, increasing awareness of the importance of photooxidation leaves their result open to question.

The ability to metabolize the simple aromatic hydrocarbon, naphthalene, appears to be fairly widespread. Thus Cerniglia et al. (1980a) found that nine species of blue-green algae, five green algae, one brown alga, and two diatoms were able to oxidize naphthalene under photoautotrophic conditions, with at least six metabolites produced. The alga, Prototheca zopfii, which lacks chlorophyll, was reported to degrade both the aromatic and aliphatic portions of crude oil (Walker et al., 1975). Blue-green algae have also been reported to metabolize biphenyl and are thought to be capable of metabolizing other aromatic hydrocarbons (Cerniglia et al., 1980b). Work in the last 3 years (Cerniglia et al., 1980b) has led to further understanding of metabolic pathways in phytoplankton (Figure 4-5).

Cerniglia et al. (1979) reported that a culture of cyanobacteria exposed to ^{14}C-naphthalene metabolized 1.4% of the substrate in hours. The major product was 1-naphthol, with preliminary evidence presented for the formation of both cis- and trans-diols. This would suggest that blue-green algae appear to have attributes of both bacteria (which metabolize naphthalene to cis-diols) and higher organisms (which produce trans-diols) (Hopper, 1978).

A single study has examined, in phytoplankton, the fate of a polycyclic aromatic hydrocarbon. After introduction of ^{14}C-benzo(a)-anthracene into a marine mesocosm, some of the radioactivity was subsequently found associated with a phytoplankton fraction (Hinga et al., 1980). Most of the fraction was still in the form of benzo(a)-anthracene, but there were significant amounts of polar metabolites reported also. However, the possibility cannot be dismissed that associated bacteria were involved. Axenic cultures are needed for this kind of work.

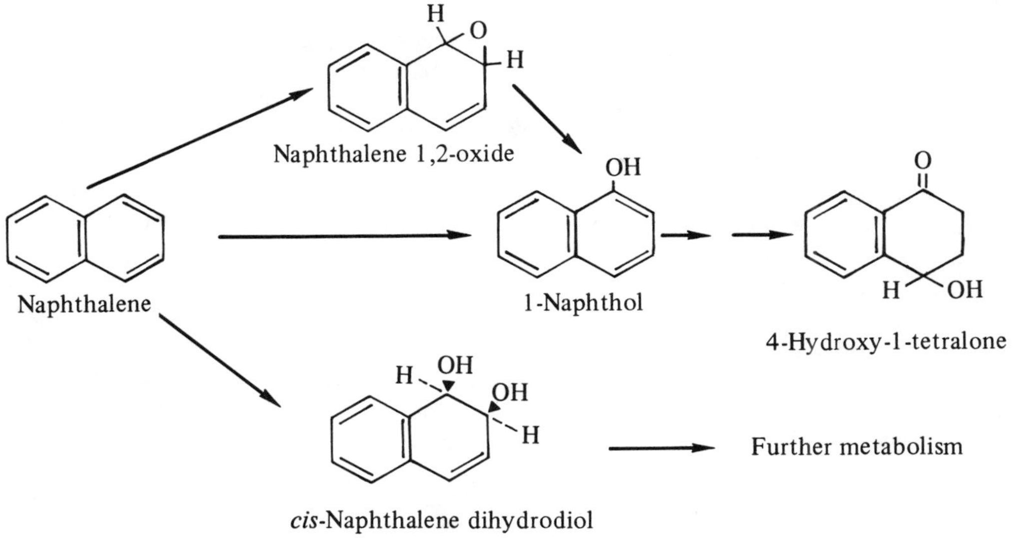

FIGURE 4-5 Proposed pathways for the metabolism of naphthalene by *Oscillatoria* sp., strain JCM.

SOURCE: Adapted from Cerniglia et al. (1980b).

Invertebrates and Vertebrates

General Patterns of Hydrocarbon Uptake and Tissue Contamination

Unlike microorganisms, animals tend not to utilize petroleum hydrocarbons as a carbon source, but generally oxidize and conjugate the products, rendering the end-products more water soluble, thereby facilitating their elimination via the usual modes of excretion of dissolved substances.

All animal groups tested have been capable of taking up petroleum hydrocarbons from either the water column directly or via their food. This uptake can occur directly through the general body integument, across respiratory surfaces (gills, lungs, or other gas-exchange surfaces), and via the gut. Although the precise mechanism of availability of hydrocarbons is a topic about which we still know little, uptake may be simple, nonmediated transport across epithelial layers (Kotyk, 1973).

Bioavailability depends to a considerable degree on whether the hydrocarbons are dissolved in the water column, sorbed by or bound onto particulate sediments or organic material, or bound up in food (see Sedimentation section). The organic content of sediments or particles, for example, can determine the sorption characteristics for hydrocarbons (Means et al., 1979) and therefore the amount of hydrocarbon in solution in natural waters.

In bivalves the sorption of specific hydrocarbons and their apparent bioavailability have been found to vary with hydrophobicity (Dunn, 1980). In fish and some invertebrates, factors related to solubility of hydrocarbons may well be responsible for the greater accumulation or retention of alkylated aromatics as compared to the unsubstituted forms (Roubal et al., 1977; Melancon and Lech, 1979; Neff, 1979).

The degree of correlation between accumulation of some lipophilic foreign compounds by fish and octanol-water partition coefficients (Veith et al., 1979) supports the idea that partitioning into and uptake via the gills (Hunn and Allen, 1974) is a major pathway in these animals. The nature of the compound may also dictate the absorption of hydrocarbons via the gut. Indeed, in mammals the absorption of aliphatic hydrocarbons by the gut was found to be dependent on carbon number (Albro and Fishbein, 1970). In two species of marine fish, the absorption of hexadecane from contaminated food differed markedly from that of benzo(a)pyrene, and the patterns of absorption for the two compounds were quite different in various species (Whittle et al., 1977). The bases for such differences are not yet apparent. The role of bacteria in the guts in metabolizing ingested hydrocarbons in marine vertebrates is also unknown.

Macroinvertebrates All invertebrates studied to date readily take up petroleum components, and a majority also metabolize them fairly readily, although little is known of the various metabolic pathways involved when compared to the body of knowledge available on microbial metabolism of hydrocarbons.

Experimental studies have demonstrated that the process of elimination of hydrocarbons is initiated within minutes or hours of their uptake, although less is known of metabolite formation and their eventual fate.

The route of uptake varies with the organism and its feeding habits. Thus in copepods (Calanus helgolandicus), Harris et al. (1977) and Corner et al. (1976b) demonstrated that dietary uptake of naphthalene was more important than uptake from the water. However, in blue crabs (Callinectes sapidus), hydrocarbon in the food was not accumulated rapidly and was quickly voided in the feces (R.F. Lee and Neuhauser, 1976). While there is no doubt that hydrocarbon contamination may be available from reservoirs within oiled sediments, in many instances the main route appears to be via the water column. Thus Rossi (1977) reported that most of the aromatic hydrocarbons accumulated by the polychaete Neanthes arenaceodentata were derived from water and not from sediments. Soft-shell clams (Mya arenaria) in oiled sediments appear to behave similarly (Vandermeulen et al., 1981, 1982).

Judging from the few studies that have addressed the question of tissue distribution of hydrocarbons, petroleum becomes readily distributed throughout the exposed animals, but storage of hydrocarbon apparently occurs in lipid-rich tissues, and concentrations of hydrocarbons are generally found to be higher in lipid-rich animals.

Most analytical work has been done with readily accessible intertidal invertebrates such as crabs and bivalves. As a consequence the data base for pelagic and offshore benthic invertebrates is slim, leaving a gap in our understanding of oil distribution in marine organisms generally. In many instances, extrapolation appears to be valid, but comparative corroboration is needed.

Bivalves, many of which filter large volumes of water while feeding, can take up and concentrate petroleum hydrocarbons from the water, whether in solution, absorbed to suspended particles, or as finely dispersed oil globules (Anderson, 1975; Boehm and Quinn, 1977; Clement et al., 1980; Disalvo et al., 1975; Dobroski and Epifano, 1980; Farrington and Quinn, 1973; Farrington et al., 1982; Fossato and Canzonier, 1976; Fucik and Neff, 1977; Hansen et al., 1978; R.F. Lee et al., 1972a, 1978; Neff et al., 1976; Nunes and Benville, 1979; Palmork and Solbakken, 1981; Stainken, 1977; W.C. Wong, 1976). Reviews of the literature on the uptake and discharge of petroleum hydrocarbons by bivalves have been presented (R.F. Lee, 1977a; Neff, 1979; National Research Council, 1980).

Numerous studies have shown that bivalves can accumulate hydro-carbons to a level several orders of magnitude above the concentration in the water (Table 4-3). Table 4-4 is a summary of the accumulation of petroleum hydrocarbons by marine bivalves taken from areas con-taminated by spills or chronic pollution.

The maximum concentration of petroleum hydrocarbons in bivalves exposed to oil under laboratory or field conditions was between 300 and 400 μg/g (Tables 4-3 and 4-4). Clams, oysters, and mussels differed in their rates of hydrocarbon uptake, possibly due to differences in filtering rates and amounts of lipids (Clark and Finley, 1974; Neff et al., 1976). Stegeman and Teal (1973) noted that oysters with high lipid content took up more fuel oil (314 μg/g wet weight) from the water than others with less lipid content (161 μg/g). Burns and Smith (1977) reported that mussel and oyster tissues appeared to be saturated at approximately 30 mg of hydrocarbons per gram of body lipid. In oiled areas, burrowing bivalves such as Mya arenaria or Modiolus demissus have much higher hydrocarbon concentrations than attached epibenthic bivalves such as Mytilus edulis or Crassostrea virginica (R.F. Lee et al., 1981b; Vandermeulen and Gordon, 1976). Detritus-feeding bivalves accumulate more hydrocarbons than suspension feeders (Augenfeld et al., 1981; Roesijadi et al., 1978).

Several factors can affect tissue accumulation of hydrocarbons and their subsequent elimination. For example, hydrocarbons accumulated in lipid-rich gametes will be discharged during gamete release. Thus, the seasonal reproductive cycle is an important factor in hydrocarbon accumulation. Maxima for benzo(a)pyrene and perylene concentrations in Mytilus edulis from Laguna Veneta, Italy, occurred in January with minima in May (Fossato et al., 1979). Spawning took place from March to April. Temperature and salinity are also parameters affecting uptake. Uptake of polynuclear aromatic hydrocarbons by clams was greater at reduced temperatures, while changes in salinity had little or no effect (Fucik and Neff, 1977). The discharge rate was not affected by temperature or salinity.

Depuration, i.e., elimination, of hydrocarbons by bivalves is not yet completely understood. Depuration does occur, but it depends in part on the manner of contamination and, in most instances, appears to occur only incompletely. Petroleum hydrocarbons accumulated by bivalves maintained under laboratory conditions generally had a half-life of only a few days (Table 4-3); however, mussels collected from heavily

Table 4-3 Uptake and Loss of Petroleum Hydrocarbons in Bivalves Under Laboratory or Microcosm Conditions

Species	Petroleum Source	Hydrocarbons Analyzed	Hydrocarbon Concn. in Water[a] (µg/L)	Exposure Time (days)	Maximum Hydrocarbon Concn. in Animals (µg/g)	Hydrocarbon Half-Life (days)	Depuration Experiments		Reference
							Depuration Period (days)	Hydrocarbon Concn. in Animals (µg/g)	
Oysters Crassostrea virginica	No. 2 fuel oil	aliphatics, aromatics	106	50	334	5	30	30	Stegeman and Teal (1973)
	crude oil in microcosms	naphthalenes	170	8	120	2	23	0.0	R.F. Lee et al., (1978)
		fluoranthene	7	8	5	5	23	0.4	R.F. Lee et al., (1978)
		benzo(a)-anthracene	5	8	3	9	23	0.3	R.F. Lee et al., (1978)
		benzo(a)-pyrene	2	8	0.4	18	23	0.1	R.F. Lee et al., (1978)
	No. 2 fuel oil	diaromatics	n.d.	4	412	5	13	2.2	Anderson (1975)
Mussels Mytilus edulis	No. 2 fuel oil	n-paraffins	n.d.	2	110	4	35	8	Clark and Finley (1975)
	diesel fuel	aliphatics	200-400	41	400	3	35	50	Fossato and Canzonier (1976)
Clams Rangia cuneata	No. 2 fuel oil	naphthalene	300	1	10	2	9	0.2	Anderson (1975)
		dimethyl-naphthalene	200	1	1.2	3	9	0.3	Anderson (1975)
	benzo(a)pyrene	benzo(a)pyrene	30	1	7.2	2-5	20	0.1	Neff and Anderson (1975)

[a]n.d., not determined.

SOURCE: Adapted from R.F. Lee (1977); see also Point Source Distributions section.

TABLE 4-4   Amounts of Petroleum Hydrocarbon in Bivalves from Contaminated Areas

Species	Contamination Source	Hydrocarbons Analyzed	Hydrocarbon Concentration in Animals (μg/g)	Hydrocarbon Half-Life[a] (days)	Reference
Oysters					
Crassostrea virginica	spill, No. 2 fuel oil	aliphatics, aromatics	70	--	Blumer et al. (1970b)
	chronic	aliphatics, aromatics	236	--	Ehrhardt (1972)
Mussels					
Mytilus edulis	spill, Bunker C	aromatics	77-103	--	Zitko (1971)
	spill, No. 2 fuel oil	n-paraffins	1.4	--	Clark and Finley (1973)
	chronic	n-paraffins	1.0	--	Clark and Finley (1973)
	spill	aliphatics, aromatics	400	2	DiSalvo et al. (1975)
	chronic	aromatics	6.75	48-60	DiSalvo et al. (1975)
	chronic	aromatics	8	--	Ehrhardt and Heineman (1975)
	chronic	aliphatics	250	4	Fossato (1975)
	chronic	benzo(a)pyrene	0.05	16	Dunn and Stich (1976)
	spill, No. 2 fuel oil	pristane	8	2	Farrington et al. (1982)
		alkylnaphthalenes	8	1	Farrington et al. (1982)
		phenanthrene	0.5	2	Farrington et al. (1982)
Clams					
Mercenaria mercenaria	sewage effluent	C_{16-32}	4-69	--	Farrington and Quinn (1973)
Mya arenaria	spill, No. 2 fuel oil	aliphatics, aromatics	26	--	Blumer et al. (1970a)

[a] Determined by exposure of animals to clean water and measuring loss of hydrocarbons from tissue.

SOURCE: Adapted from R.F. Lee (1977a).

contaminated areas sometimes required a much longer period for depuration (Table 4-4). There also appear to be inherent species differences; thus, some studies have shown a difference between Mercenaria mercenaria and Mytilus edulis from similarly polluted areas. A weakened physiological state may well affect the rate of depuration. Possibly there are "conservative tissues" or "stable compartments" from which depuration is slow (Stegeman and Teal, 1973).

The fate of hydrocarbons in other invertebrates is much the same as in bivalves, dependent on the same environmental and physiological factors for uptake, residence time, tissue distribution, and depuration/ elimination. A major difference that may exist relates to the metabolic fate of hydrocarbons, in particular, those in which metabolism is mediated by cytochrome P450-dependent enzyme systems (mixed function oxygenases). Most invertebrates examined to date appear to have this capability (e.g., crabs, lobster, polychaetes, and bivalves).

The metabolism of petroleum hydrocarbons in invertebrates has not received the attention needed, but a range of metabolites has been described including derivatives of simple aromatic hydrocarbons in crabs (e.g., Corner et al., 1976b). Gordon et al. (1978) provided evidence for uptake and removal of hydrocarbons from chronically oiled sediments by the polychaete Arenicola marina. They pointed out, however, that the precise role of the polychaete, as distinct from that of possible microbial interaction, had not been defined. Polychaetes are now known to have mixed function oxidases (R.F. Lee, 1981), but possible effects of their excretion and bioturbation on bacterial activity have received little attention.

Fish The nature of the compounds and possible association with blood serum components (Plack et al., 1979) may influence deposition in various tissues. Association of hydrocarbons with cell membranes in coho salmon (Oncorhynchus kisutch) was indicated in at least one study (Roubal, 1974), but details of transfer processes through membranes or cells have not been demonstrated for hydrocarbons in aquatic vertebrates.

Although in general the levels of hydrocarbons in exposed fish are greatest in the liver, in experiments with a variety of species the levels in neural tissues of fish have equaled or even exceeded those in the liver (Neff et al., 1976; Roubal et al., 1977; Collier et al., 1980). This would seem to be consistent with the high lipid content and vascularization of these tissues. The observed distribution may be related to the molecular size of the hydrocarbons, for Roubal et al. (1977) observed that aromatics with a high molecular weight were retained more readily in brain tissues of the coho salmon (O. kisutch) than compounds with a smaller molecular weight. Similar observations have also been obtained in other tissues.

Efficiency of uptake of hydrocarbons from the food may be low in fish in some circumstances (Whittle et al., 1977). While the mass of material accumulated via drinking of water in marine fish under most circumstances would be minimal, the uptake via gills can be very important (Lee et al., 1972b). Benthic fish, particularly the Pleuronectidae, readily take up hydrocarbons from the sediment (Varanasi et al.,

1981), possibly via desorption from sedimentary particles, but the route is uncertain.

The contamination of animals in early development deserves close attention, for they may be more susceptible to toxic effects at these stages in their life cycle than as adults. Contamination of fish or invertebrate eggs is either direct, or via exposure of females during oocyte maturation (e.g., Kuhnhold et al., 1979; Hose et al., 1981; Rossi and Anderson, 1977; Longwell, 1958).

The elimination of hydrocarbons from contaminated animals is affected by numerous factors and is highly variable. Disposition may be accomplished by several routes. One often suggested for fish is direct partitioning through the gills into water. The significance of this pathway has not been established for hydrocarbons in general, but Thomas and Rice (1981) offered evidence that substantial proportions of naphthalene and toluene could be discharged directly via the gills. Whether this is as important as excretion via the kidneys remains to be seen.

Birds and Marine Mammals   The significance of various pathways in birds and mammals will differ from that in fish. Entry of hydrocarbons via the respiratory epithelium in lungs of birds and mammals is, barring aspiration of water, restricted mainly to those volatile compounds transported in the atmosphere, whether particulate or in a true gas phase. Thus hydrocarbons in aerosols in the vicinity of an oil spill may be an important source to birds. Sorption to particles apparently is an important factor in the delivery of hydrocarbons to the lungs of terrestrial mammals. Absorption of hydrocarbons by seals immersed in contaminated water has been clearly demonstrated (Engelhardt et al., 1977) but whether by lung or skin was not confirmed. However, the same studies indicated that dietary as well as nondietary routes are quite important in seals and, presumably, in other marine mammals. Absorption via the gut (in part by preening) can occur in seabirds (Grau et al., 1977) as in other waterfowl (Lawler et al., 1978a,b). A blood-brain barrier may exclude saturated aliphatic but not aromatic hydrocarbons in birds (Lawler et al., 1978).

The exposure of avian eggs to hydrocarbons may be direct or maternal. Exposure can occur through direct transfer of oil on the plumage to egg shells in the nest, and small amounts (50-100 μL) have been shown to be toxic to embryos. Contamination of terrestrial bird (quail) eggs with Bunker C or No. 2 fuel oil via maternal routes resulted in reduced egg production and reduced egg viability (Grau et al., 1977). Transplacental contamination of developing marine mammals has not been demonstrated.

Factors Influencing Petroleum Biodegradation Rates

Some of the factors that determine biodegradation rates are inherent to the polluting oil; others are environmental and subject to variation. Organismic factors (abundance of hydrocarbon degraders and their substrate range) and certain environmental parameters (pH, salinity) that are nearly uniform and in a favorable range throughout the marine environment (Tait and De Santo, 1972) are not considered here.

Composition and Weathering of Petroleum

The biodegradable portion of various crude oils ranges from 11 to 90% (Colwell and Walker, 1977). A low percentage of biodegradation may result from a high amount of volatile components, for these ordinarily evaporate before significant biodegradation can take place. Because of this, the biodegradation percentage is often related to the "topped" (preevaporated) crude rather than the intact one. Low percentages of biodegradation can result also from high proportions of condensed polyaromatic, condensed cycloparaffinic, and asphaltic petroleum components, because these compounds are biodegraded at extremely slow rates if at all.

Toxicity of certain petroleum components can delay or prevent the biodegradation of susceptible ones. Atlas and Bartha (1972b) and Atlas (1975) noted such action by volatile components of certain petroleums in the environment. Toxic and lipophilic substances such as pesticides (Seba and Corcoran, 1969; Hartung and Klinger, 1970), polychlorinated biphenyls (Sayler and Colwell, 1976), and mercury (Walker and Colwell, 1976a; Sayler and Colwell, 1976) can be concentrated in the oil slick 10^2-10^5 times above their ambient concentration in the water and may inhibit biodegradation of the petroleum. Photooxidation (Burwood and Speers, 1974) may remove methyl branches that block biodegradation, but in high concentrations photooxidation products may become toxic to microorganisms (Van der Linden, 1978). Formation of mousse (Berridge et al., 1968c) reduces the surface area and availability of mineral nutrients and O_2, thus hindering biodegradation (Atlas et al., 1980; Colwell et al., 1978).

Temperature

The nature of the marine environment restricts petroleum biodegradation to the mesophilic and psychrophilic organisms. Hydrocarbon biodegradation has been reported at temperatures below 0°C (ZoBell and Agosti, 1972). Because of arctic and subarctic oil exploration, this has led to a substantial interest in psychrophilic and psychrotrophic hydrocarbon degraders (Malins, 1977). Generally, the rate and extent of hydrocarbon biodegradation was severely restricted at low water temperatures (Gunkel, 1968; ZoBell, 1969; Mulkins-Phillips and Stewart, 1974). The temperature dependence of hydrocarbon biodegradation rates can be expressed in terms of Q_{10} values (Q_{10} = increases in rate per 10° change in temperature). Gibbs et al. (1975) and Gibbs and Davis (1976) determined an average Q_{10} value of 2.7 over the 6°-26°C range.

Hydrostatic Pressure

Some crude oils exceed the specific weight of water and others may do so at an advanced stage of weathering. Thus, hydrocarbons can enter the deep-sea environment and, consequently, the effect of hydrostatic pressure on oil biodegradation is of interest. Using an enrichment culture obtained from 4,940-m depth, Schwarz et al. (1974a,b, 1975) found that, at 20° and 25°C, 500-atm pressure delayed biodegradation only moderately. The same pressure at 4°C reduced metabolism by more than 1 order of magnitude as compared to a 1-atm control incubated at the same temperature. The authors concluded that the biodegradation of any petroleum residue that reaches the deep-sea environment will be exceedingly slow (Schwarz et al., 1975).

Oxygen

The initial attack on hydrocarbons is commonly performed by oxygenases. There have been several reports on anaerobic conversion of hydrocarbons (Senez and Azoulay, 1961; Choteau et al., 1962; Iizuka et al., 1969; Traxler and Bernard, 1969; Parekh et al., 1977). These were all in vitro studies with isolated cultures. Thus, a pathway appears to exist for the anaerobic utilization of alkanes, with sulfate or nitrate serving as electron acceptors. Nevertheless, anaerobic hydrocarbon biodegradation in the environment is either undetectable, or orders of magnitude lower than aerobic biodegradation (Bailey et al., 1973; Ward and Brock, 1978; Delaune et al., 1980; Ward et al., 1980).

Oxygen limitation of petroleum biodegradation is unlikely in the case of surface slicks that are in direct contact with atmospheric oxygen. When oil is dispersed in the water column, oxygen limitation may occur. Complete oxidation of 1 L of oil would exhaust the dissolved oxygen in 320,000-400,000 L of seawater (ZoBell, 1969). Whether or not oxygen actually becomes limiting depends on the oil concentration, the rate of biodegradation, and oxygen replenishment by turbulence (wave and current action). Oxygen limitation of petroleum biodegradation in the water column may occur sometimes, but mineral nutrients are more likely to become limiting before the oxygen is depleted.

Marine sediments commonly are anaerobic below a thin surficial layer. The thickness of this layer depends on grain size, organic content, and the degree of physical or biological disturbance of the sediment. Petroleum that becomes incorporated in anaerobic marine sediments is essentially immune to biodegradation until some disturbance releases the oil or oxygenates the sediment. The main reason for increased petroleum persistence in fine-grained sediments as compared to coarse-grained ones appears to be a lower rate of oxygen diffusion.

Mineral Nutrients

Petroleum-degrading microorganisms need to obtain mineral nutrients from seawater. Considering the composition of seawater (Tait and De Santo, 1972) in relation to mineral nutrient requirements, actual levels of phosphorus, nitrogen, and iron are likely to approach limiting concentrations. Other essential elements appear to be present in sufficient or excess concentration. In vitro experiments employing relatively high oil-to-seawater ratios have convincingly demonstrated phosphorus and nitrogen limitation of petroleum biodegradation (Atlas and Bartha, 1972b). Iron limitation was confirmed in clear offshore seawater but not in sediment-rich coastal seawater (Dibble and Bartha, 1976). The need for phosphorus and nitrogen supplements for optical oil biodegradation activity in seawater was noted also by Bridie and Bos (1971), Reisfeld et al. (1972), Gibbs (1975), and LePetit and N'Guyen (1976). Based on the data of Atlas and Bartha (1972b) and Reisfeld et al. (1972), 1.5-2.5% N and 0.2% P (w/w) addition (calculated on the basis of petroleum that was actually degraded) allowed maximal petroleum biodegradation in these in vitro experiments. Bridie and Bos (1971) found somewhat higher and Gibbs (1975) somewhat lower requirements, explainable by differing experimental conditions and petroleum-to-seawater ratios. A summary discussion of the relation of mineral nutrient requirements for biodegradation of oil pollutants was provided by Floodgate (1979) with the conclusion that the scarcity of mineral nutrients in seawater is often limiting for petroleum biodegradation, not only in vitro but also under closely simulated environmental conditions.

Rates of Petroleum Biodegradation in the Marine Environment

Microbial Degradation

Considering the multitude of factors that influence petroleum biodegradation and the variety of methods employed in its measurement, generalized statements about rates are difficult. Nevertheless, rates are of central interest in terms of the self-purification capacity of the marine environment and need to be defined at least with the crude accuracy of an order of magnitude. Petroleum biodegradation rates have been estimated in four types of experimental systems in ways that might be applied to spill conditions in temperate waters: (1) fermentation studies, (2) seawater enrichments under optimized conditions, (3) in situ marine enrichments, and (4) in situ petroleum degradation potentials. These are shown in Table 4-5. System 1 provides an estimate of optimal rate and is shown only for comparison. The rates of system 2 may be approached in situ using nutrient-stimulated biodegradation as described by Atlas (1977). Conditions equal to system 3 occur upon prolonged exposure of a marine environment to petroleum pollutants, whereas system 4 measures the potential of a marine environment for petroleum biodegradation before the existing microbial population has an opportunity to shift in response to the

TABLE 4-5 Estimates Under Temperate Conditions of Hydrocarbon
Biodegradation Rates

System	Degradation Rates $(g/m^3/day)$	References
1. High density monocultures in fermenters, optimal conditions	10,000-100,000	6,8
2. Seawater community under nutrient-enriched conditions	5-2,500	3,5,7,12
3. In situ marine oil additions (long incubation times)	0.5-60	1,2,3,4
4. In situ marine unoiled condition (short incubation times)	0.001-0.030	9,10,11

The cited references served as a basis for calculations; the above
listed figures do not necessarily appear in the papers in this particular form. Key to the references: (1) Atlas and Bronner (1981),
(2) Atlas and Bartha (1973b), (3) Atlas et al. (1980), (4) Atlas et al.
(1981), (5) Caparello and La Rock (1975), (6) Dibble and Bartha (1976),
(7) Coty and Leavitt (1971), (8) Kanazawa (1975), (9) Lee (1977a), (10)
Robertson et al. (1973), (11) Seki (1976), and (12) Walker et al.,
(1976).

petroleum exposure. This potential is greatly influenced not only by
the prevailing environmental conditions but also by the previous history
of exposure of the site to petroleum pollutants.

The relevance of the four systems in Table 4-5 to the biodegradation of oceanic petroleum spills is as follows: The initial values are
expected to be in the range observed in system 4. On prolonged contact
with the petroleum, population shifts can be expected to occur, and
biodegradation rates can approach the values of system 3.

Cytochrome P450-Dependent Metabolism

One principal route of metabolism of hydrocarbons in metazoans is that
initiated by cytochrome P450-dependent polysubstrate monooxygenases
(MO) or mixed function oxidases (MFO). The metabolism of hydrocarbons
by this route is essentially a means whereby animals can convert these
lipophilic compounds to derivatives that are more water soluble and,
hence, more readily excreted, thus profoundly affecting the nature,
disposition, and effects of hydrocarbon residues in the animal. The
scientific literature is replete with numerous studies and several

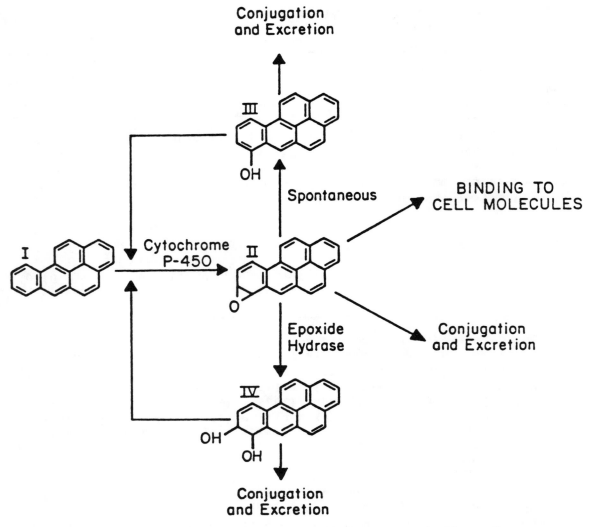

Conjugation
and Excretion

BINDING TO
CELL MOLECULES

Cytochrome
P-450

Spontaneous

Epoxide
Hydrase

Conjugation
and Excretion

Conjugation
and Excretion

FIGURE 4-6 Diagram of cytochrome P450-dependent metabolism of benzo(a)pyrene.

reviews of P450 systems and their function in xenobiotic metabolism (for example, Johnson, 1979; Gelboin, 1980; Mannering, 1981). Briefly, the P450 complex initially forms epoxide derivatives of aromatic hydrocarbons, which can then be further metabolized by other enzymes (for example, epoxide hydrolase) and various conjugating enzymes. This subsequent conjugation usually results in detoxification of the epoxide intermediates, which by themselves are reactive and may be mutagenic. A scheme describing the biochemical production, cycling, and fate of products of primary and secondary metabolism of an important model aromatic hydrocarbon, benzo(a)pyrene, is depicted in Figure 4-6.

Patterns of P450-Mediated Metabolism

The characteristics, functions, and induction of P450 systems in aquatic, principally marine, species have been recently reviewed in detail (Bend and James, 1978; Stegeman, 1981; Lech et al., 1981; R.F. Lee, 1981). In fish the liver is the primary site of hydrocarbon metabolism, while the hepatopancreas apparently serves this function in invertebrates. MFO activity has been found in fish species whenever sought. There have been fewer studies to determine the presence or absence of MFO systems in marine invertebrates. To date, 18 marine invertebrate species, belonging to 4 phyla (Annelida, Arthropoda, Echinodermata, and Mollusca) are known to contain MFO activity in their hepatopancreas, digestive gland, and other tissues (R.F. Lee, 1981). Marine invertebrates belonging to such phyla as Porifera, Platyhel-minthes, Rhynchocoela, Rotifera, Gastrotricha, Nematoda, Echiurida, Brachiopoda, Phoronida, and Chaetogratha have not been examined for MFO activity. Numerous studies have estimated the activity of P450-dependent enzyme systems responsible for hydrocarbon metabolism, using benzo(a)pyrene as a model substrate (Table 4-6). However, a study comparing activity of the hepatic microsomes of coho salmon (Oncorhynchus kisutch) in vitro indicated that rates of metabolism of different aromatic hydrocarbons are not always equal (Schnell et al., 1980). Also there are multiple forms of cytochrome P450, and the metabolism of different types of compounds may depend on the relative abundance of specific forms of the enzyme.

Seabirds and some other waterfowl possess P450 systems (Bend et al., 1977a; Knight et al., 1981; Shackelford and Kahn, 1981; Peakall et al., 1981), but there have been few estimates of the capacity for hydrocarbon metabolism. Similarly, such activity has been little assessed in marine mammals (Engelhardt, 1981). The only known examples are included in Table 4-6.

The rates in some extrahepatic tissues are appreciable (Pohl et al., 1974; Stegeman et al., 1979; Singer and Lee, 1977). The mass transformed per gram of fish in most tissue is usually only a small fraction of that in liver, yet the significance of metabolism in a given organ to the health of that organ may be great. Aromatic hydro-carbon metabolism has also been demonstrated in embryonic stages of marine fish, even during early development before the appearance of the liver (Binder, 1981). In at least one species (Fundulus heteroclitus) such activity in embryos is very low until hatching, when within 24 hours there is a 10-fold increase in "constitutive" activity.

Metabolism In Vivo

As the major site of metabolism of hydrocarbons is the liver, the gall bladder is accordingly a major route of excretion of metabolites in fish (Lee et al., 1972b; Melancon and Lech, 1978; Collier et al., 1978; Solbakken et al., 1980; Varanasi and Gmur, 1981). Thus the analysis of bile might serve to monitor exposure of fish to various foreign chemicals (Statham et al., 1976). Most of the metabolites appearing in

TABLE 4-6 Microsomal Benzo(a)pyrene Metabolism in Some Marine Animals

Animal	Tissue	Benzo(a)pyrene Hydroxylase (nmol/min/mg protein)	Cytochrome P450 (nmol/mg microsomal protein)	Reference
Marine Invertebrates				
Panulirus argus (spiny lobster)	hepatopancreas	--	0.91±39 (45)	James et al. (1979)
Homarus americanus (lobster)	hepatopancreas	--	0.043 to 0.13	Èlmamlouk et al. (1974)
Callinectes sapidus (blue crab)	stomach	0.037±.013 (4)	0.14	Singer et al. (1979)
	hepatopancreas	--	0.064±0.009 (5)	S.C. Singer (unpublished data, 1982)
	hepatopancreas	--	0.18±0.08 (11)	James et al. (1979)
Nereis virens (sand worm)	intestine	0.021±0.004 (3)	0.089±0.024 (4)	Lee and Singer (1980) S.C. Singer (unpublished data)
Balanus eburneus (barnacle)	digestive gland	0.043±0.001 (3)	0.11±0.01 (3)	Stegeman and Kaplan (1981)
	intestine	0.016±0.004 (3)	0.09±03 (4)	Stegeman and Kaplan (1981)

TABLE 4-6 (continued)

Species	Tissue	BP Hydroxylase Activity nmol/min/mg	nmol/min/nmol P450	Reference
Elasmobranchs				
Dogfish (Squalus acanthias)	liver	0.014±0.001[a]	0.065	Stegeman (1981)
Little skate (Raja erinacea)	liver	0.009±0.009[a]	0.028	James and Bend (1980)
Stingray (Dasyatis sabina)	liver	0.039±0.020[a]	0.078	James and Bend (1980)
Teleosts				
Scup (Stenotomus chrysops)	liver	0.685±0.040[b]	1.117	Stegeman et al. (1979)
Fundulus heteroclitus	liver	0.280±0.048[a]	0.721	Stegeman (1979)
Winter flounder (Pseudopleuronectes americanus)	liver	0.213±0.015[a]	0.237	Stegeman et al. (1982)
Sheephead (Archosargus probatocephalus)	liver	0.155±0.090[b]	0.574	James and Bend (1980)
Southern flounder (Paralichthys dentatus)	liver	0.012±0.004[b]	0.019	James and Bend (1980)
Birds				
Gull (Larvus argentus)	liver	4.99±0.68[c]	--	Peakall et al. (1981)
Mammals				
Ringed seal (Phoba hispida)	liver	0.006[a]	--	Engelhardt (1981)

[a] nmoles 3-OH-BP equivalents produced per min per mg microsomal protein.
[b] nmoles (^{3}H)-BP metabolites produced per min per mg microsomal protein.
[c] Fluorescence units (due to OH-BP) produced per mg microsomal protein. These values may be artifically low, as the data were obtained using frozen tissue.

TABLE 4-7 Comparison of Metabolites of Naphthalene in Liver, Muscle, and Bile of Starry Flounder (Platichthys stellatus)

Metabolites	Liver	Muscle	Bile
Total conjugates	62.3 ± 8.8[a]	46.1 ± 0.3	91.5 ± 0.3
Total nonconjugates	37.7 ± 8.8	53.9 ± 0.3	8.5 ± 0.3
Glucuronides	16.6 ± 1.6	23.7 ± 1.4	81.7 ± 0.3
Mercapturic acids	10.5 ± 6.6	8.0 ± 0.9	8.8 ± 0.1
Sulfate/glucosides	34.8 ± 14.8	14.4 ± 0.5	0.9 ± 0.1
Dihydrodiol (1,2-isomer)	12.4 ± 11.7	23.9 ± 0.9	3.0 ± 0.1
Naphthols (1-a 2-)	11.5 ± 3.9	21.6 ± 0.6	2.9 ± 0.2
Uncharacterized	14.0 ± 7.1	8.4 ± 0.4	2.6 ± 0.1

NOTE: Data at 12°C and 7 days.
[a]Data are expressed as percent of total metabolites.

SOURCE: Varanasi et al. (1981).

bile are conjugated derivatives of oxygenated forms of hydrocarbon. Excretion of metabolites of aromatic hydrocarbons by other routes, including gills, urine, and skin, has also been indicated for fish and invertebrates. The patterns of hepatic metabolism of hydrocarbons in vivo are largely similar to those in vitro, at least as judged from identity of polar derivatives of hydrocarbons isolated from bile. Biliary metabolites of naphthalene, methylnaphthalene, phenanthrene, and benzo(a)pyrene freed from conjugates have included dihydrodiol, phenolic, and quinone derivatives (Melancon and Lech, 1978; Solbakken et al., 1980; Varanasi and Gmur, 1981). Examples of principal metabolites of some compounds reportedly found in fish bile are listed in Table 4-7. In crabs and shrimp exposed to naphthalene, the metabolites produced included conjugates with glucose and sulfate, diols, and phenolic derivatives (Corner et al., 1973; Lee et al., 1976; Sanborn and Malins, 1980). Rates of metabolism in vivo cannot yet be inferred from in vitro rates. In many cases the rates of biotransformation in vivo can be expected to match the intake, and little accumulation of parent compound will be evident.

When the parent compound is retained at low levels, metabolites may be retained in various tissues for longer periods than the parent (e.g., Varanasi et al., 1979; Melancon and Lech, 1979). The identity of naphthalene metabolites extracted from tissues (muscle and/or liver) of starry flounder (Platichthys stellatus) included conjugates as well as phenols and dihydrodiols (Varanasi et al., 1981), but the profile of tissue metabolites differed from that in bile (Table 4-8). Metabolites of benzo(a)pyrene (B(a)P) extracted from liver of English sole

TABLE 4-8 Induction by Hepatic Benzo(a)pyrene Hydroxylase in Marine Fish by Aromatic Hydrocarbons

Species	Treatment	BP Hydroxylase	
		Control	Treated
Little skate[a]	dibenzanthracene (intraperitoneal)	0.009 ± 009[c]	0.396 ± 0.194
Sheepshead[a]	3-methylcholanthrene (i.p.)	0.070 ± 0.025[c]	0.550 ± 0.190
Croaker[b]	benzo(a)pyrene (i.p.)	0.026 ± 0.002[d]	0.420 ± 0.021

[a] James and Bend [1981].
[b] Stegeman [1981].
[c] nmoles 3-OH-BP equivalent/min/mg microsomal protein.
[d] nmoles [^{3}H]-BP metabolites/min/mg microsomal protein.

(Parophrys vetulus) included a suite of metabolites like those seen in vitro (Varanasi and Gmur, 1981). Moreover, a portion of B(a)P in liver in that study was found to be in cellular constituents, ostensibly the result of formation of active intermediates in vivo. The origin of metabolites in many extrahepatic tissues is not fully known. Given that both oxidative capacity (Stegeman et al., 1979) and conjugating enzymes (James et al., 1979) occur in most tissues of fish, the various types of extrahepatic metabolites possibly are produced in situ. The export of metabolites from the liver to other tissues is also possible, as suggested by studies in mammals. The effects of these metabolites and their further transformation in various tissues are not known. The total flux of metabolites by the various excretory routes is unknown for any marine species and needs to be addressed.

Factors Influencing Metabolism

The rates of hydrocarbon metabolism and elimination can be influenced by a variety of environmental and physiological factors. Paramount among these is the induction of cytochrome P450 by hydrocarbons and other environmental pollutants. Polynuclear aromatic hydrocarbons, crude oil, and refined petroleum products, as well as polychlorinated biphenyls and polybrominated biphenyls, are among those compounds that have the capacity to induce increased levels of P450 and increased rates of hydrocarbon metabolism in liver and some extrahepatic tissues of fish and some invertebrates. The phenomenon has been described in numerous marine and freshwater fish species (e.g., Payne and Penrose, 1975; Bend et al., 1977), including embryonic and larval forms (Binder and Stegeman, 1980) as well as polychaetes (R.F. Lee, 1981). Induction is usually described as an increase in the rate of B(a)P metabolism in vitro (e.g., Table 4-8), but metabolism of other hydrocarbons is also elevated, as indicated by the higher rates of methylnaphthalene metabolism in vivo in induced trout (Statham et al., 1978). The increased rates of B(a)P metabolism in fish appear to be attributable to the synthesis of novel forms of cytochrome P450 (Elcombe et al., 1979a,b; James and Bend, 1980; Stegeman et al., 1981). The presence of new cytochrome P450s in crabs has also been suggested after exposure to pollutants (R.F. Lee et al., 1982).

There is growing evidence for induction in fish in the environment. This includes induction of hydrocarbon metabolism in fish by spilled oil in the environment (Payne, 1976; Kurlelec et al., 1977; Stegeman, 1978). This is not a petroleum-specific phenomenon, however. There is increasing evidence for widespread induction of P450 in fish by chemicals of unknown origin in the environment (e.g., Bend, 1980; Dunn, 1980; Stegeman et al., 1981). The causes of such induction have not been established, although in one case (Dunn, 1980) there was a correlation with PAH in the sediment. In the polychaete, Capitella capitata, exposed to crude oil, the third generation had much higher MFO activity than the first or second generation (R.F. Lee, 1981). Grassle and Grassle (1976, 1977) showed that C. capitata is actually a complex of at least six sibling species based on electrophoretic patterns. Exposure to oil thus may result in selection for species or strains that are resistant to oil because of high MFO activity.

Induction of hydrocarbon metabolism in the liver of seabirds has been noted in one study of oiled birds from the Amoco Cadiz. where elevated levels of MFO were observed. Similar induction in mammals has not been demonstrated but is virtually certain to occur, based on comparison with responses to foreign chemicals seen in their terrestrial counterparts. Elevated B(a)P hydroxylase activity was reported in the kidneys of seals fed crude oil (Engelhardt, 1981). Gulls fed Prudoe Bay crude oil did not show induction 8 days after treatment, but any induction response had probably disappeared by that time. Another type of induction, by DDT (which induces different types of P450 from those of aromatic hydrocarbons), has been demonstrated in puffins (Bend et al., 1977).

There are several studies that have disclosed an effect of temperature on disposition of naphthalene in vivo. In both coho salmon (Oncorhynchus kisutch) and starry flounder (Platichthys stellatus) there was a pronounced increase in the retention of naphthalene and its metabolites in the tissues with a decrease in temperature (Collier et al., 1978; Varanasi et al., 1981). Moreover, lower temperature also effected a shift in the pattern of naphthalene metabolism in starry flounder, resulting in a substantially greater proportion of glucuronides and 1,2-dihydrodiol in liver at the lower temperature a week after exposure (Varanasi et al., 1981). Metabolites in muscle did not follow the same pattern, however. The exact basis for and consequences of such responses are not known and clearly warrant investigation.

In addition to effects on disposition and elimination of hydrocarbons, low temperature also can cause an attenuation of the induction of P450 in fish (Stegeman, 1979). Further evidence also indicates seasonal and sex-linked responses to inducers (Stegeman and Chevion, 1980, Forlin, 1980). Moreover, the many different types of pollutant compounds present in the environment can influence the metabolism, disposition, and effects of each other. In a study by Gruger et al. (1981) the pattern of metabolism of 2,6-dimethylnaphthalene in vivo in starry flounder was substantially altered by the presence of either naphthalene or cresol in the animal. Such results have implications for the toxicity of mixed chemicals in the environment and also for extrapolations based on studies of biotransformation of single compounds.

Formation of Mutagenic Metabolites

The metabolism of aromatic hydrocarbons can result in their activation to toxic, and in some cases mutagenic and carcinogenic derivatives. The activity of the metabolites will depend, in part, on which part of the molecule has been metabolized. The patterns of metabolism of some aromatics by fish liver in vitro have been studied, but the principal effort has been on benzo(a)pyrene (e.g., Ahokas et al., 1979; Bend et al., 1979; Stegeman, 1981; Varanasi and Gmur, 1981). The principal phenols, dihydrodiols, and quinone metabolites of benzo(a)pyrene formed by aquatic species have been reviewed (Stegeman, 1981). In general, metabolism of B(a)P by teleost fish liver preparations results in formation of high percentages of benzo-ring (7,8- and 9,10-0 dihydrodiols) but little K region (4,5-dihydrodiol). This is also true of extrahepatic and embryonic tissues. The major metabolite with crabs was shown to be 3-hydroxy benzo(a)pyrene, with minor amounts of diols and other phenols also being produced (Singer et al., 1979). Metabolism of benzo(a)pyrene in vitro by fish produces derivatives that are mutagenic (Ahokas et al., 1979; Stegeman, 1977) and that bind to DNA (Ahokas et al., 1979; Varanasi et al., 1981). The patterns of in vitro metabolism of some other aromatics by fish have also been established, e.g., the metabolism of 2-methylnaphthalene by trout liver (Breger et al., 1981). Similar studies on patterns of hydrocarbon metabolism by marine birds and mammals have yet to be done.

In short, induction can increase rates of metabolism and disposition, but it also increases rates of formation of mutagenic derivatives. Thus, hydrocarbon-induced P450s are known to be efficient in forming mutagenic derivatives of carcinogenic hydrocarbons (Wood et al., 1976). Moreover, a strong induction in fish proceeds with little or no change in the activity of conjugating enzymes that detoxify reactive products (e.g., Statham et al., 1978). This could shift the steady-state level of toxic derivatives in vivo. The relationships of metabolism, disposition, and toxic action of many xenobiotics have recently been reviewed (Lech and Bend, 1980), but for hydrocarbons much remains to be determined.

Conclusions and Recommendations

Microbial degradation is a major mechanism for elimination of petroleum pollutants from the aerobic marine environment. The environmental constraints that influence the rate of biodegradation have been defined. Substantial progress has been made toward determining the rate of biodegradation in various marine environments, but further refinement and standardization of methodology are required before reliable rate projections can be made.

Evidence to date indicates that uptake of hydrocarbons from food and/or water is a universal phenomenon in animals, with partitioning from water or from particles or sediments after desorption the key process. The levels of specific compounds in different species or different tissues do not accurately reflect exposure, and this is due to differences in partitioning and in metabolic factors. Metabolic transformation of hydrocarbons occurs in most groups of animals but at widely differing rates. Equilibrium concentrations in tissues are dependent on their ability to metabolize various compounds as well as physical-chemical processes, and in many cases metabolic rates may balance uptake rates with little apparent bioaccumulation. Intermediary metabolites are found in many tissues and may be retained longer than parent compounds.

Although considerable progress has been made on the biological fate of petroleum, numerous gaps remain:

1. The fate of hydrocarbons in green plants has received insufficient attention.

2. In order to understand the fate of oil more completely, more work is needed on the metabolism of heterocyclic compounds and non-hydrocarbon compounds in petroleum. Little is known about the slow but possibly significant biodegradation of the hydrocarbons of high molecular weight and the metabolites that might be formed.

3. Some basic cellular processes remain obscure: for example, movement of hydrocarbons across cell membranes and their potential interaction with cellular nucleic acids. Further elucidation of metabolic pathways in animals is needed.

4. Study of the distribution and fate of metabolites is recommended, including the question of whether polar compounds that are produced have a significant effect on mousse formation, as described in the physical fates section.

AMOUNTS OF HYDROCARBONS IN THE MARINE ENVIRONMENT

Introduction

A considerable amount of data has been gathered on the concentration of hydrocarbons in the oceanic water column, sediments, and biota since the 1975 NRC report. In many studies, however, differentiation between biological, petroleum, pyrogenic, and other sources of these hydrocarbons was not unequivocal because of the lack of definitive diagnostic parameters and analytical problems associated with handling the trace quantities of hydrocarbons found in most natural samples. The quality of the hydrocarbon data given in the following sections should be viewed in the context of these analytical difficulties, which have been discussed in the chapter on chemical methods.

Dissolved Petroleum Hydrocarbons

Low-Molecular-Weight Hydrocarbons (C_1-C_4)

Ambient Concentrations   Low-molecular-weight hydrocarbon (LMWH) concentrations in most of the ocean are largely influenced by natural processes (e.g., air-sea exchange, seepage or diffusion across the sea-sediment interface, in situ biological production) which produce near-surface concentrations of methane, ethane, propane, and the butanes in the open ocean of 40-150, 0.2, 0.2, and <0.2 nL/L, respectively. Methane in the upper water column of the ocean has been shown by several investigators (Lamontagne et al., 1971, 1973, 1974; MacDonald, 1976; Swinnerton and Lamontagne, 1974; Brooks and Sackett, 1973, 1977; Brooks et al., 1973, 1981a; Scranton and Brewer, 1977; Scranton and Farrington, 1977) to be supersaturated with respect to the partial pressure of methane in the atmosphere. Most of this supersaturation apparently results from in situ biological production (Lamontagne et al., 1973; Seiler and Schmidt, 1974; Scranton and Brewer, 1977; Scranton and Farrington, 1977; Brooks et al., 1981a) and advection from coastal areas (Sackett and Brooks, 1975; Brooks, 1979). Methane supersaturation appears to be a permanent feature of the mixed layer in the world ocean, except in regions of strong upwelling such as the Yucatan Shelf (Brooks et al., 1973) and in some ice-covered areas of the Antarctic (Lamontagne et al., 1974). Most of the published profiles of methane in the open ocean have a subsurface maximum of at least twice surface levels. This phenomenon may be due to a combination of factors including outgasing of surface water and higher production rates of biogenic methane at intermediate depths. Methane levels in the deep ocean are below equilibrium with the atmosphere in areas of deep-water formation, apparently due to methane degradation during advection (Scranton and Brewer, 1978). Less is known about the biological origin and distribution of C_2-C_4 LMWH in the ocean.

Sackett and Brooks (1975), Cline and Holmes (1977a,b), and Reitsema et al. (1978) have shown that concentrations of LMWH are significantly higher in near-bottom waters on the continental shelves and in the vicinity of structural highs and gas seeps. Brooks (1979) reported massive deep methane maxima in the northwest Caribbean Sea which were attributed to submarine seepage off the Jamaica Ridge system. From another sea floor seep region, Norton Sound, Alaska, Cline and Holmes (1977a,b) observed C_2-C_4 hydrocarbon concentrations elevated by a factor of 10 or more.

Point Source Distributions   Because LMWH are abundant constituents of crude oils and natural gas, their presence in the water column is a sensitive indicator of petroleum inputs. Through thousands of analyses from hydrocarbon "sniffing" and discrete sampling programs, ports and estuaries--with their associated commercial, petrochemical, and transportation activities and offshore petroleum operations--have been identified as the major man-derived sources of LMWH in the Gulf of Mexico. The water column for at least one example of each of these types of inputs has shown LMWH concentrations to be several orders of magnitude higher than the water column in the open Gulf. The underwater venting of waste gases and brine discharges, both associated with offshore platforms, was the major source of nonmethane LMWH in coastal surface waters, and was apparently responsible for an increase of 2 orders of magnitude in most Louisiana shelf waters over levels in the open Gulf. Average concentrations of 3,100, 31, and 22 nL/L for methane, ethane, and propane, respectively, were observed (Brooks et al., 1973, 1977, 1979, 1981b; Brooks, 1976; Brooks and Sackett, 1973, 1977; Sackett and Brooks, 1974, 1975; Sackett, 1977; Wiesenburg et al. 1981b). Weisenburg et al. (1981b) found that the average 2,970 µL/L LMWH in brine from the Buccaneer Gas and Oil Field in the northwest Gulf of Mexico caused discernible increases in LMWH levels in surface waters only within 300 m of the platform. Brooks et al. (1978, 1981b) also reported very high LMWH concentrations in the vicinity of the Ixtoc I well blowout in the Gulf of Mexico.

Volatile Liquid Hydrocarbons (C_5-C_{12})

Ambient Concentrations Although ca. 30-40% of crude oil consists of volatile liquid hydrocarbons (VLH), there is little information available on their distribution in the marine environment. This is primarily a result, until the past few years, of the difficult methodologies involved in VLH measurements. Sauer (1978, 1980), Sauer and Sackett (1980), and Sauer et al. (1978) found that open ocean, nonpetroleum-polluted surface waters contain VLH concentrations of approximately 60 ng/L, while heavily polluted Louisiana shelf and coastal waters reached over 500 ng/L. Aromatics accounted for about 60-85% of the total VLH in surface waters. Cycloalkane concentrations were <1.0 ng/L in open ocean water and from 60 to 110 ng/L in polluted waters (about 20% of the total VLH). Total alkanes increased from 15 ng/L in open ocean water to as much as 40 ng/L in polluted shelf waters. These latter elevated levels were shown to be a direct result of the large amounts of brine discharges and underwater venting of waste gases associated with offshore production on the Louisiana shelf. An approximately linear relationship existed between anthropogenic gaseous and volatile hydrocarbons.

Although Schwarzenbach et al. (1979) used a somewhat different technique (i.e., solvent extraction off charcoal versus heat desorption from Tenax-GC), their studies of two coastal stations--one in Vineyard Sound and the second a tidal creek in Massachusetts--yielded similar results. They identified approximately 50 compounds with individual abundances of ≈1-100 ng/L, although concentrations above 20 ng/L were rare. The total VLH recovered from these very near-shore stations ranged from 0.2 to 1.0 µgC/L. Gschwend et al. (1982) and Mantoura et al. (1982) have followed the preliminary report of Schwarzenbach et al. (1979) with further data from sites near Woods Hole, Massachusetts. At one site, alkylbenzenes and aldehydes were the major VLHs observed. The alkylbenzenes were dominated by anthropogenic inputs and air-sea exchange, with selective biodegradation effecting minor changes in the summer. Offshore mixing and adsorption on particulates appeared minor. The alkylnaphthalenes showed a contrasting pattern to the alkylbenzenes, which was explained by a dominant winter time source such as space heating oil. Alkanes were frequently petroleum derived, but penta-decane, heptadecane, and pristane showed evidence of strong biogenic sources as well. Gschwend et al. (1980) have reported concentrations of VLHs in the Peru upwelling region of the Pacific.

Point Source Distributions Weisenburg et al. (1981a) reported mean VLH concentrations of 1.9 µg/L composed of more than 80% light aromatics (benzene and naphthalenes) around the Buccaneer Gas and Oil Field in the Gulf of Mexico. Sauer (1981b) also reported VLH concentrations in the same field as well as in vents in the Gulf of Mexico. Brooks et al. (1978) found VLH concentrations as high as 19 µg/L around a gas well blowout on the Texas shelf. VLH concentrations of as much as 400 µg/L were reported in the immediate vicinity of the Ixtoc I blowout on the Campeche shelf as a result of dispersed oil in the water column (Brooks et al., 1981b). At 6 and 12 miles downplume, VLH concentrations had decreased to 63 and 4 µg/L, respectively. Payne et al. (1980) also reported VLH in the water column around the Ixtoc I blowout: benzene and toluene concentrations were 49 and 97 µg/L, respectively, 6 miles from the blowout, with most values further downplume in the 1- to 4-µg/L range. Lysyj et al. (1981) reported aromatic VLH concentrations as high as 120 µg/L in Port Valdez, Alaska, as a result of discharges from a ballast treatment plant.

High-Molecular-Weight Hydrocarbons

In general, studies on the quantities and sources of high-molecular-weight hydrocarbons (HMWH) in the water column have indicated that the hydrocarbon composition is a function of both biosynthetic and anthro-pogenic sources (Barbier et al., 1973; Brown et al., 1973; Gordon et al., 1974, 1978a, Iliffe and Calder, 1974; Brown and Huffman, 1976; Keizer et al., 1977; Calder, 1977; Boehm et al., 1979; Boehm, 1980). Reported concentrations of petroleum-derived HMWH in seawater are difficult to interpret. This is due to the radically different methods of quantification (i.e., gravimetric, GC, GC/MS, IR, fluorescence, etc.) used in various studies and to the problem of contamination during sampling and processing. Another problem encountered is the wide range of compounds grouped together in the hydrocarbon fraction (i.e., alkanes, olefins, isoprenoids, cycloalkanes, and aromatics). Many methods of isolation and quantification will bias the results for or against one or more of these groups of compounds (see Chapter 3, Chemical Methods section). In addition, water processing varies from study to study and can range from no filtration and batch extraction to filtration plus adsorption and reverse phase LC analysis. Concentrations of total HMWH mentioned in the literature typically range from 0.2 to 100 µg/L, although the lack of uniform analytical techniques and reporting formats makes comparisons difficult. In most of these studies, certain quantification of the relative proportions of natural and anthropogenic hydrocarbons is impossible.

HMWH have been reported for numerous areas of the Atlantic. A preponderance of measurements has given median concentrations of 1-10 µg/L for large areas of the North Atlantic (Levy, 1971; Levy and Walton, 1973; Brown et al., 1973, 1975; Monaghan et al., 1973; Gordon and Keizer, 1974; Barbier et al., 1973; Hardy et al., 1975; Brown and Huffman, 1976; Zsolnay, 1977a; Boehm et al., 1979; Boehm, 1980). Barbier et al. (1973) reported values up to 140 µg/L with an average of 40 µg/L at two stations near Dakar, Africa; Wade and Quinn (1975) observed a range of 13-239 (average 73) µg/L in the Sargasso Sea. Grahl-Neilson (1978) examined the petrogenic hydrocarbons, naphthalene, phenanthrene, and dibenzothiophene in the North Sea, using fluorescence, and concluded petroleum hydrocarbons were below detection limits ($\sim$20 ng/L). Mackie et al. (1974) reported n-alkanes from 0.2-4.9 µg/L off Scotland. Mackie et al. (1976) and Hardy et al. (1977) have reported average n-alkane concentrations of 0.166 and 4.5 µg/L, respectively, in the North Sea. Law (1981) using fluorescence in the United Kingdom coastal waters found petroleum hydrocarbons in the 1.1- to 74-µg/L range (mean values were 1.3, 2.5, and 2.6 µg/L in the northern North Sea, southern North Sea, and Irish Sea, respectively). Keizer et al. (1977) and Gordon et al. (1974) have reported n-alkanes by GC and petroleum HMWH by fluorescence from <20 to 145 ng/L and 1 µg/L, respectively, in the Sargasso Sea.

Boehm (1980) and Boehm et al. (1979) found anthropogenic hydrocarbons were ubiquitous in the Georges Bank area, except in zooplankton. Petroleum HMWH were apparently from the Argo Merchant oil spill, chronic inputs from ballast washings, and normal ship traffic. Dissolved concentrations were generally in the 0.1- to 2-µg/L range throughout the year, but were above 10 µg/L in the 4 months after the Argo Merchant spill in late 1976. Individual polynuclear aromatics were persistent during the year at levels of 1-10 ng/L, but were elevated in concentration (10-50 ng/L) in early 1977. Variations were patchy and were not related to differences in water mass characteristics (Boehm et al., 1979). Unresolved complex mixtures (UCM) generally accounted for 60-80% of the total determined HMWH. Similar distributions have been observed by Iliffe and Calder (1974), Barbier et al. (1973), and Keizer et al. (1977).

Unusually high concentrations of HMWH (3-12 mg/L) were reported by Harvey et al. (1979) for six samples collected from a depth of about 200 m at five stations on a transect from the eastern Caribbean into the southwest North Atlantic. On the basis of limited chemical characterization of this material the authors suggested that it was a "biochemically weathered oil that has not undergone evaporative weathering" and that it probably originated from a natural submarine seep on the Venezuelan shelf. A "conservative estimate of its dimension" was used by them to calculate that the layer contained more than 1 megaton of oil, an amount approximating that released in the Ixtoc I spill, and considerably more than the 0.2 mta best estimate given in Table 2-1 of Chapter 2.

In several Baltic Sea studies, 57.2 µgC/L of nonolefinic dissolved hydrocarbons and 157 ng/L of aromatic hydrocarbons were reported by Zsolnay (1972, 1973). A significant correlation between aromatic and saturated hydrocarbons was found, suggesting a common anthropogenic source. A correlation with chlorophyll was noted for waters off northwest Africa and Nova Scotia (Zsolnay, 1974, 1977b). Average hydrocarbon concentrations of 4.9 µg/L were observed off Nova Scotia and 4.6 µg/L off northwest Africa (Zsolnay, 1977b). Zsolnay (1979) reported HMWH averaging 6.9-25.8 µg/L in the Mediterranean Sea, with the Alboran Sea and the area off Libya having the greatest concentrations. Monaghan et al. (1973) and Brown et al. (1975) also have presented IR data for HMWH from the Mediterranean Sea in the range of <1-195 µg/L.

Petroleum hydrocarbons were not typically observed in the water column in the South Texas and Florida outer continental shelf areas of the Gulf of Mexico (Parker et al., 1972; Jeffrey, 1979); however, Boehm and Feist (1979) and Shokes et al. (1979a,b) have found petrogenic hydrocarbons in both whole and filtered seawater from near-shore Louisiana. Total HMWH in seawater at two sites averaged 37 and 5.8 µg/L. Literature values for total hydrocarbons in the open Gulf range from 0.1 µg/L to approximately 76 µg/L (Parker et al., 1972; Koons and Monaghan, 1973; Calder, 1977; Jeffrey, 1979; Iliffe and Calder, 1974; Brown et al., 1973). Parker et al. (1972) found n-paraffins of ∿1 µg/L in the Gulf of Mexico and Caribbean. There is generally a large decrease in HMWH between the surface and a depth of about 10 m (Brown et al., 1973; Parker et al., 1972), below which values are generally <1 µg/L. This may suggest that hydrocarbons are present as particulate matter rather than in true solution. Koons and Monaghan (1973) reported hydrocarbon concentrations from 1 to 8 µg/L in the Gulf of Mexico.

Indications of chronic petroleum pollution in the open Gulf of Mexico include aromatic concentrations ranging from 1 to 3 µg/L (Brown et al., 1973) and HMWH concentrations of up to 75 µg/L which were attributed to tanker traffic in the Florida Straits (Iliffe and Calder, 1974). In a further study from the northeast Gulf of Mexico, Calder (1977) estimated dissolved and particulate hydrocarbon concentrations at 0.4 and 0.3 µg/L, respectively. Aromatic hydrocarbons were estimated by Hiltrabrand (1978) at proposed deep-water port sites in the Gulf of Mexico at 23-24 µg/L in surface water and only 0.6-1.3 µg/L at 16 m. The aromatic hydrocarbons showed a sharp decrease with depth in these profiles. Kennicutt (1980) found approximately 9-99 ng/L n-alkanes; 1 ng/L aromatics; and 6 ng/L pristane, phytane, olefins, and cycloalkanes for both the dissolved and particulate fractions in the northeast Gulf of Mexico. The major source of these hydrocarbons was inferred to be biological.

Less is known about hydrocarbon distributions in the Pacific. Cretney and Wong (1974) reported some fluorescence in the northeast Pacific off Vancouver Island (0.011-0.027 µg/L as chrysene), and

Koons and Brandon (1975) reported values in a hydrocarbon seep region off California (0.4-16 µg/L). In an extensive survey (ca. 350 water samples collected along 17,000 miles of tanker routes in the Pacific), Brown and Searl (1976) reported HMWH had a median concentration of 1.6 µg/L for the surface and 0.9 µg/L at 10 m. HMWH appeared to be a mixture of both biogenic and petrogenic compounds; approximately 17-23% of the HMWH were aromatics. In the Indian Ocean, San Gupta et al. (1980) found 0.6 and 26.5 µg/L of petroleum hydrocarbons by UV absorbance, with highest values along tanker routes.

Petroleum Hydrocarbons in the Surface Microlayer

Several studies have reported on the isolation and composition of hydrocarbons in surface films (Garrett, 1967; Duce et al., 1972; Keizer and Gordon, 1973; Ledet and Laseter, 1974; Morris, 1974; Hardy et al., 1975; Wade and Quinn, 1975; Marty and Saliot, 1976; Boehm et al., 1979; Boehm, 1980). Garrett (1967) first identified hydrocarbons in surface films, and Duce et al. (1972) reported that hydrocarbons (ca. 8.5 µg/L) were concentrated in the surface film by a factor of 1.4 over subsurface concentrations. Morris (1974) measured alkanes in surface film samples from the Mediterranean and concluded that they contained petroleum hydrocarbons. Ledet and Laseter (1974) found that alkanes at the air-sea interface off Louisiana and Florida were of a mixed biogenic-anthropogenic origin, while Wade and Quinn (1975) observed HMWH from 14 to 599 µg/L with an average of 155 µg/L in the micro-layer of the Sargasso Sea and suggested that the major source of hydrocarbons in the surface waters of the Sargasso Sea was small particles of weathered pelagic tars. Concentrations averaged only 73 µg/L at 20-30 cm below the surface. Marty and Saliot (1976) found an enrichment factor averaging 50 in the eastern Atlantic and Mediter-ranean. Dissolved n-alkanes varied from 15-114 µg/L in the micro-layer compared to 0.1-5.7 µg/L for underlying water. The particulate n-alkanes varied from 0.3 to 7.2 µg/L in the underlying water and from 3.3 to 1,214 µg/L in the microlayer. Dissolved n-alkanes were generally higher than in particulates, except in polluted waters. Boehm (1980) and Boehm et al. (1979) reported a surface film enrichment of 1.4-90 times that of subsurface water in the Georges Bank area. The surface microlayer contained HMWH (5- to 76-µg/L range) that were mainly petroleum. There have been extensive problems with sampling just the surface microlayer and some values may be low due to entrain-ment of subsurface waters.

Particulate Petroleum Hydrocarbons (Tar Balls)

Most particulate petroleum residues in the sea are found floating at or near the surface and are called "tar balls." They are the residues remaining after various physical and chemical processes have acted on floating oil for varying periods of time (see Physical and Chemical Fates section). Butler et al. (1973), Butler (1975), and Morris (1971) have estimated that 10-30% of oil discharged to the ocean remains in the form of tar balls and has an estimated residence time on the order of a year.

Butler (1975) concluded that small tar lumps are formed by fragmentation of much older petroleum residues. Tar balls or lumps range in size from less than a millimeter to many centimeters in diameter, although most are quite small (1-10 mm in diameter). Their texture varies from soft to very hard. Some have living organisms on their surfaces [Horn et al. (1970) found barnacles as old as 4 months], as well as incorporated mineral particles, organisms, shells, and/or organic debris. Tar balls found near shore or in the littoral zone are often broken into smaller pieces and have incorporated mineral material.

The abundance of floating tar has been studied more intensively in the North Atlantic, Mediterranean, Gulf of Mexico, and Caribbean than for the remainder of the oceanic areas. Most of the data collected since 1973 are for the same areas already covered in the 1975 NRC report. The estimate for total tar in the ocean is reduced from 318 to 277 x 10^3 tons. The average amounts of pelagic tar concentrations and estimated total tar observed in various portions of the world's oceans are summarized in Table 4-9. There are little or no reported data for two-thirds of the world's oceans (e.g., South Atlantic, Antarctic, most of the Arctic, large areas of the Pacific and Indian, and portions of the North Atlantic).

The highest reported concentrations continue to be for the Sargasso and Mediterranean seas, with mean values of 10 mg/m^2. These two seas and the Gulf of Mexico provide the best data for estimating temporal variations in tar concentrations. For example, Morris and Butler (1975) reported a reduction in mean floating tar concentrations in the Ionian Sea, but in other portions of the Mediterranean (e.g., Alboran and Tyrrheian seas) there appeared to be an increase from 1969 to 1975. Knap et al. (1980), by analyzing beach tars in Bermuda (in the Sargasso Sea), concluded that there had been no decrease in the inputs of petroleum residues to the Atlantic between 1971 and 1979. In the Gulf of Mexico, Jeffrey (1980) determined that the average floating tar concentration was 1.35 mg/m^2 (range 0-10 mg/m^2), based on 220 neuston tows taken on nine cruises between 1972 and 1976. No apparent change in tar ball concentrations was observed during this period. In another temporal study in the eastern Gulf of Mexico, Van Vleet et al. (1982a) reported that tar concentrations increased off the southwestern Florida shore from winter to summer due to the increasing influence of the loop current, which has relatively high concentrations of tar--due, apparently, to tanker traffic through the Caribbean and Yucatan Strait. This study supports the concensus by most investigators that there is a strong correlation between high levels of tar concentrations and tanker routes.

A few investigators have analyzed the composition of pelagic tars (Butler et al., 1973; Jeffrey et al., 1976; Ehrhardt and Derenbach, 1977; Van Vleet et al., 1982b). For the most part, pelagic tars have been classified as having the general sources of tanker sludge residues, weathered crude oil, fuel (Bunker) oil residues, and highly weathered residues of indeterminate origin. The classifications are based on gas

TABLE 4-9 Tar Densities on the World Oceans

Location and Reference	Area (10^{12} m^2)	Tar (mg/m^2) Max.	Mean	Total Tar (10^3 t)
NW Atlantic Marginal Sea (Morris, 1971; Morris and Butler, 1973; McGowan et al., 1974)	2	2.4	1	2
East Coast Continental Shelf (Attaway et al. 1973; Sherman et al., 1973, 1974; van Dolah et al., 1980; Cordes et al., 1980)	1	10	1	1
Caribbean (Jeffrey, 1973; Sherman et al., 1973; Jeffrey et al., 1974; Geyer and Giammona, 1980; Sleeter et al., 1976)	2	13.4	1	2
Gulf of Mexico (Jeffrey, 1973; Sherman et al., 1973; Jeffrey et al., 1974; Light, 1977; Geyer and Giammona, 1980; Koons and Monaghan, 1973; Pequegnat et al., 1979; Van Vleet et al., 1982a,b)	2	10	1.4	2.8
Gulf Stream (Morris, 1971; Morris and Butler, 1973; Sherman et al., 1974; Levy, 1977)	8	10	2.2	18
Sargasso Sea (Polikarpov et al., 1971; Attaway et al., 1973; Morris and Butler, 1973; Sherman et al., 1973, 1974; McGowan et al., 1974; Sleeter et al., 1974)	7	91	10	70
Canary and North Equatorial Current (Heyerdahl, 1971a,b; Ehrhardt and Derenbach, 1977; Sleeter et al., 1976)	3	2,270	8	24
Rest of Northeast Atlantic (McGowan et al., 1974)	8	10.7	0.5	4
North Sea (Oppenheimer et al., 1977; Smith, 1976)	3	12.1	0.3	0.9
Baltic	1		?	?
South Equatorial Current	3		?	?
South Atlantic (Eagle et al., 1979)	50	232	0.5	25
Mediterranean (Horn et al., 1970; Morris, 1971, 1974; Morris and Culkin, 1974; Morris and Butler, 1975; Morris et al., 1976; Zsolnay et al., 1978)	2.5	540	10	25
Indian Ocean	75	?	?	? (large)
Southwest Pacific (R.A. Lee, 1973; C.S. Wong et al., 1976)	45		0.0003	0.01

TABLE 4-9 (continued)

Location and Reference	Area (10^{12} m^2)	Tar (mg/m^2) Max.	Mean	Total Tar (10^3 t)
Kuroshio System (C.S. Wong et al., 1974b, 1976)	10	14	2.1	21
Rest of Northwest Pacific (C.S. Wong et al., 1976)	30		0.4	12
Northeast Pacific (C.S. Wong et al., 1974b, 1976)	40	3	0.03	1.2
Arctic (C.S. Wong et al., 1974a; Smith, 1976)	13	3	0.1	1.3
Antarctic	10		?	?
Area accounted for	227 (63%)			210
TOTAL AREA	361			277[a]

[a]Assuming density 0.5 mg/m^2 for areas unaccounted.

SOURCE: Modified from National Research Council (1975).

chromatographic analyses, sulfur contents, stable isotopes, vanadium and nickel concentrations, and gel permeation. Butler et al. (1973) and Morris and Butler (1973), using gas chromatography, concluded that most of the floating tars collected in the North Atlantic were of a tanker sludge origin. In samples from the northeast Atlantic, Ehrhardt and Derenbach (1977) found that 61% were from crude oil sludges, 35% from crude oil and Bunker oil residues, and 4% of unknown origin. Using molecular characterization by gas chromatography and percentage sulfur of floating tars in the Gulf of Mexico and Caribbean, Jeffrey et al. (1976) found that approximately 30% of the tar balls analyzed were tanker sludge residues, based on a bimodal UCM and a high percentage of high-molecular-weight alkanes. Only 2% of the tar balls were identified as fuel oil residues, whereas 65% were crude oils of many origins. Jeffrey (1980) postulated that a significant fraction of the tar collections in the western Gulf originates from young seep oil in the southwestern Gulf. Koons and Monaghan (1973) also suggested that some of the tar balls they analyzed originated from natural seepage.

Special mention should be made here of the report Global Oil Pollution by Levy et al. (1981), which gives results of the IGOSS marine pollution (petroleum) monitoring pilot project (MAPMOPP). The report summarizes data on "over 85,000 visual observations of oil slicks and other floating pollutants, 4000 collections of floating tar balls, 3100 collections of beach tar, and almost 3000 measurements of dissolved/dispersed hydrocarbons" obtained during the period 1975-1978. Some of these data, which are available in other published reports, e.g., C.S. Wong et al. (1976) are included in Table 4-9. Generally, the data and conclusions of the IGOSS pilot project are consistent with those presented here.

Petroleum Hydrocarbons in Marine Sediments

Petroleum and hydrocarbons of biosynthetic origin have been investigated by many workers in a wide variety of marine sediments. Table 4-10 summarizes some of the results that have appeared in the reviewed literature between 1977 and mid-1981. A similar table covering earlier work has been prepared by Clark and MacLeod (1977a).

The studies referenced in Table 4-10 had a variety of objectives and used numerous analytical approaches. Virtually all of these studies provided detailed information about concentrations of individual constituent molecules of petroleum, in addition to the data on total hydrocarbon concentration. Totals for unpolluted marine sediments seldom exceed 50 μg/g. Much more information about the extent of petroleum contamination can be obtained from examination of the kinds and amounts of individual hydrocarbon molecules present in a sediment. This ability to recognize and quantify sedimentary petroleum in the presence of hydrocarbons from other sources is the result of a continuing evolution in analytical techniques and increased knowledge of compounds from specific sources. Many of the areas classified as having little, if any, petroleum pollution reveal, upon detailed examination, small but distinct concentrations of petroleum hydrocarbons. Recent applications of methods for detailed interpretation of hydrocarbons in marine sediments include Atlas et al. (1981), Barrick et al. (1980), Venkatesan et al. (1980), and numerous others.

A suite of criteria, i.e., total hydrocarbons, UCM, and n-alkanes; pristane and phytane contents; triterpanes; aromatics; and chlorinated hydrocarbons, was used to distinguish the pollutant histories of these sediments. Individually, none of these criteria is definitive. Each, as discussed by Venkatesan et al. (1980), and the references they cite, has limitations and requires careful application; yet together, they present a cohesive story (see also Chapter 3, Chemical Methods section for more details).

Table 4-10 shows that observed hydrocarbon concentrations vary by more than 6 orders of magnitude on a global basis, with ranges of 3 orders of magnitude common within a given locality. Much of this variation is the result of regional and temporal differences in petroleum input rates combined with variability in sediment type and depositional history. Major inputs take the forms of catastrophic oil spills, of repetitive small spills (typically in harbors), and of petroleum associated with sewage and urban runoff discharged into coastal waters. Because these inputs are all initially added to the water column and are later transferred to sediments, variations in local sedimentation patterns also strongly influence the distribution of hydrocarbons in sediments. The interplay of inputs and sedimentation patterns can be seen in numerous well-studied coastal systems.

The coastal system of Narragansett Bay/Rhode Island Sound is an area of complex sedimentation patterns (McMaster, 1960) and substantial petroleum inputs, mostly from the Providence area at the head of the bay. The sedimentary hydrocarbons of this system have been extensively studied (Farrington and Quinn, 1973; Zafiriou, 1973; Van Vleet and Quinn, 1978; Hurtt and Quinn, 1979). Sewage treatment plants discharge

TABLE 4-10 Hydrocarbons in Marine Sediments

Location	Sediment Depth (cm)	Water Depth (m)	Condition of Area[a]	Number of Samples	Concentration[j] (μg/g dry wt.)	Hydrocarbon Type	Reference
Chedabucto Bay, 7-8 years after spill	5-15[b]	intertidal	P	27	7-1,280 (41)	aliphatic and aromatic	Keizer et al. (1978a)
Scotian Shelf	surface	--	N	20	0.03-21.1 (0,9)	aliphatic	Keizer et al. (1978h)
Gulf of Maine	0-88[b]	--	N	6	2.7-19.0[c]	Total HC	Van Vleet and Quinn (1978)
	0-88[b]	214	N	3	24-57	aliphatic and aromatic	Farrington et al. (1977)
Searsport, Maine, 5 years after spill	0-15	intertidal	P	14	6-250 (105)	aliphatic and aromatic	Mayo et al. (1978)
Buzzards Bay, MA	0-72[h]	--	N	11	10-149	aliphatic and aromatic	Farrington et al. (1977)
	0-49[b]	intertidal	P	20	20-20,600 (1880)	aromatics	Teal et al. (1978)
Wild Harbor, 2-6 years after spill	0-120[b]	intertidal	P	20	0.4-4,307	aliphatic and aromatic	Burns and Teal (1979)
Georges Bank	surface	--	N	124	0.2-20	total HC	Boehm et al. (1979)
Argo Merchant spill site 2 and 7 mo. after spill	0-14[b]	--	P	>34	0.1-327	total HC	Hoffman and Quinn (1979)
Narragansett Bay	0-50[h]	<50	N-P	28	4.0-1,650 (366)	total HC	Hurtt and Quinn (1979)
	0-40[b]	<50	N-P	20	20-5,410[c] (570)	total HC	Van Vleet and Quinn (1978)
Rhode Island Sound	surface	30	N-P	21	1.0-301 (29.7)	total HC	Boehm and Quinn (1978)
	0-38[b]	--	N-P	4	1.9-24.9[c]	total HC	Van Vleet and Quinn (1978)
Hudson Canyon and Channel	surface	54-3,785	N-P	11	14-560 (60)	aliphatic and aromatic	Farrington and Tripp (1977)
	0-94[b]	986	N	4	7.6-59	aliphatic and aromatic	Farrington et al. (1977)
Continental Slope and Abyssal Plain, Northwest Atlantic	surface	190-5,465	N	9	1.2-16 (7.5)	aliphatic and aromatic	Farrington and Tripp (1977)

TABLE 4-10 (continued)

Location	Sediment Depth (cm)	Water Depth (m)	Condition of Area[a]	Number of Samples	Concentration[l] (μg/g dry wt.)	Hydrocarbon Type	Reference
New York Bight	surface / surface	-- / 23-39	P / P	35 / 4	6-6,530 (97) / 35-2,900 (1,224)	Total C_{15+} aliphatic and aromatic	Koons and Thomas (1979) / Farrington and Tripp (1977)
Bermuda Platform	0-13[b]	--	P	26	0.04-0.55[d]	aliphatic	Sleeter et al. (1979)
Bahia Sucia, Puerto Rico 4 years after spill	5-15	intertidal	P	18	11-60,240 (632)	aliphatic and aromatic	Page et al. (1979)
5-6 years after spill	surface	intertidal	P	37[e]	249-186,500 (8090)	aliphatic and aromatic	Gilfillan et al. (1981)
Northeast Gulf of Mexico	0-10	~27-55	N-P	60	1.5-11.7[f]	total HC	Gearing et al. (1976)
Exploratory oil rig, Gulf of Mexico	0-10	~25	N	50	18.3-41.8[f]	aliphatic and aromatic	Lytle and Lytle (1979)
Southern California Bight San Pedro Basin	0-1 / 0-13.0[b]	-- / --	N-P / P	11 / 8	13.0-764 (78.9) / 20-3,900[c]	aliphatic / aliphatic and aromatic	Reed et al. (1977) / Venkatesan et al. (1980)
San Nicolas Basin	0-13.5[b]	--	N	7	22-97[c]	aliphatic and aromatic	Venkatesan et al. (1980)
Puget Sound	8	--	P	71	3-360	aliphatic	Barrick et al. (1980) Barrick and Hedges (1981)
Strait of Juan de Fuca	0-3	intertidal	N-P	16	25.5-1,500 (106)	aliphatic and aromatic	MacLeod et al. (1977)
Port Valdez, Alaska	0-2	110-260	N	8	0.64-26.2 (1.66)	aliphatic and aromatic	Shaw and Baker (1978)
Eastern Bering Sea	surface	--	N	59	0.8-240.9 (8.2)	aliphatic and aromatic	Venkatesan et al. (1980)
Beaufort Sea, Arctic Ocean	surface	<11	N	20	0.1-19.8 (4.6)	aliphatic and aromatic	Shaw and Wiggs (1979)
Severn Estuary (England)	1-5	intertidal	P	1	114	aliphatic and aromatic	Thompson and Eglinton (1978)
Fawley marsh, Southampton Water, England	surface	intertidal	P	14	100-5,640 (2,150)	aliphatics	Dicks and Iball (1981)

Location	Depth		Pollution[a]		Range (mean)	Type	Reference
Brittany, France 5–15 mo. after Amoco Cadiz spill	surface	≤70	P	370	0–>500	total HC[h]	Beslier et al. (1980)
3 mo. after Amoco Cadiz spill	surface	--	P	11	4.1–73.3 (38.9)	total HC[i]	Law (1978)
1–16 mo. after Amoco Cadiz spill	0–3	intertidal	P	13	132–19,150 (275)	aliphatic and aromatic	Atlas et al. (1981)
King Edward Cove, South	0–19[b]	<20	P	15	0.55–15.65 (7.6)	n-alkanes	Mackie et al. (1978)
	0–36[b]	≈20	P	19	1.8–6.6	n-alkanes	Platt and Mackie (1980)
Westernport Bay, Australia	0–1	--	N-P	26	1.8–5,271 (10.2)	total C$_{14}$–C$_{34}$	Burns and Smith (1977)

[a] N=nonpolluted, P=polluted, N-P=ranging from nonpolluted to polluted; all classifications are generalizations.
[b] Core(s) analyzed in segments.
[c] Decreasing with depth.
[d] Concentrations of petroleum only, does not include biogenic hydrocarbons.
[e] Not the same stations as sampled 4 years after spill.
[f] Means of subgroups.
[g] ^{210}Pb dated recent cores; data presented by age, not depth.
[h] Estimated by IR and UV fluorescence.
[i] Estimated by UV fluorescence, wet weight basis.
[j] Range; median of surface values, where applicable.

upward of 10^3 metric tons of hydrocarbons annually into Narragansett Bay and Rhode Island Sound. Most of this is of fossil origin, and a substantial fraction is surely petroleum derived, although coal-burning (Tripp et al., 1981) and other combustion products (Lake et al., 1979) also contribute. These hydrocarbons are weakly bound to detrital particles and, consequently, sediment throughout the bay. Approximately half of the suspended hydrocarbons are deposited in the vicinity of the discharge and half are dispersed toward the mouth of the bay. In part, this reflects distance from the primary source. It also reflects changes in the sedimentation pattern: sediments tend to become coarser and lower in organic carbon toward the mouth of the bay.

Segmentally analyzed cores show that fossil hydrocarbons in Narragansett Bay are essentially a twentieth century phenomenon. Although these sediments are extensively reworked after deposition by burrowing organisms, most show decreasing hydrocarbon concentrations with depth and differences in the molecular composition of the hydrocarbons, which indicate that those in the deeper sediments are biogenic. A few cores show subsurface maxima in petroleum hydrocarbons. These suggest discrete pollution events, possibly oil spills or dumping of polluted dredge spoils.

Because it is farther from the city of Providence and has coarser sediments, Rhode Island Sound is lower in sedimentary hydrocarbons than Narragansett Bay. During the period from 1968 through 1970, highly petroleum-polluted dredge spoil from the Providence River, containing 10^4 metric tons of hydrocarbons, was dumped in the sound (Boehm and Quinn, 1978). In the vicinity of the dump site, hydrocarbons having the molecular characteristics of degraded petroleum were observed at elevated concentrations. Evidence of dredge spoil hydrocarbons was limited, however, to within 2 km of the dump site, probably due in part to the fact that the finest, most polluted spoil was subsequently covered with coarser, less contaminated material, which tended to prevent dispersal.

The trends outlined above for the Narragansett Bay/Rhode Island Sound system appear repeatedly elsewhere (Wakeham and Farrington, 1980). Detailed comparison, however, shows many differences. Differences in present pollution rates and pollution histories, as well as sedimentation patterns and other environmental variables, make each area distinct. At present, the accurate, quantitative prediction of petroleum accumulation in sediments which will result from a particular pollution event at a particular location is simply not possible.

The rate of loss of hydrocarbons from chronically polluted sediments is not easily measured. In areas affected by acute oil spills, hydrocarbon half-lives in sediments range from months in exposed locations to decades in sheltered areas (Hoffman and Quinn, 1979; Gilfillan et al., 1981; and others). While similar processes of physical flushing and biochemical degradation must be at work in chronically polluted sediments, direct experimental evidence that defines the rates of these processes is not available.

Petroleum Hydrocarbons in Marine Organisms

Pelagic Biota Very little information is available on petroleum
hydrocarbons in pelagic marine organisms. Organisms themselves produce
various isoprenoids, particularly pristane, olefins, and straight-chain
saturated paraffins, and these have been studied in some detail (Blumer
et al., 1963, 1969, 1970a, 1971; Burns and Teal, 1973). By contrast,
cycloalkanes and aromatic hydrocarbons are not normally found in
organisms, and therefore the implication of oil pollution as a source
of hydrocarbons in marine biota is usually based on their presence
along with an UCM. In the open ocean, concentrations of hydrocarbons
in organisms are low and the origin of the hydrocarbons is not always
easily determined. In areas where large inputs of hydrocarbons have
occurred, HMWH in organisms can be directly related to petroleum
pollution.

For plankton, Boehm (1980) and Bieri (1979) found that zooplankton
appeared free of petroleum hydrocarbons in the Georges Bank and Florida
shelf areas, respectively, although at least in the Georges Bank area
other components of the system contained significant petroleum. Morris
et al. (1976) and Burns and Teal (1973) reported petroleum hydrocarbon
concentrations in two phytoplankters from the Sargasso Sea at 10-20
µg/L. In the Gulf of Mexico, petrogenic HMWH have been observed in
zooplankton samples, but have been attributed to the incorporation of
tar balls in the samples (Calder, 1976; Parker et al., 1972) or contamination by offshore production activities (Middleditch et al.,
1979a,b).

For higher trophic levels, metabolic processes determine the
internal concentrations of petroleum hydrocarbons. Processes of
ingestion, metabolism, excretion, and lipid storage, which are both
active and selective, lead to petroleum concentrations which are
strongly influenced, not only by the organism's external environment,
but also by its behavior, physiology, and biochemistry (see Biological
Fates section). Factors such as these increase the variability of
hydrocarbon concentrations in organisms and make elucidation of the
causes of that variability more difficult. Horn et al. (1970) found
tar balls in the stomachs of saury collected in the Mediterranean.
Teal (1976) found evidence of tar ball ingestion in a galatheid from
the Nares Abyssal Plain in the North Atlantic.

Morris et al. (1976) found petroleum contamination in most levels
of the Sargasso Sea community. Petroleum-derived HMWH ranged from 16
to 3,230 µg/g dry weight in crabs, fish, and shrimp from the Sargasso
area. Middleditch et al. (1977, 1979b) found n-alkanes ranged from 0
to 16 µg/g in muscle and from 0 to 13 µg/g in livers of fish around
a production platform with some evidence of petroleum-derived HMWH.
Bieri (1979) found no evidence of petroleum hydrocarbons in fish from
the Florida shelf, except in one sample which correlated with evidence
of petroleum in underlying sediments.

Benthic Biota Hydrocarbons are naturally present in most, if not all,
marine organisms as a result of endogenous biosynthesis and dietary
exposure. In addition, many marine organisms have been found to contain

TABLE 4-11 Petroleum Hydrocarbons in Bottom-Dwelling Marine Organisms

Organism	Location	Area Type[a]	Concentration Range (μg g^{-1})[b]	Median	Hydrocarbon Type	Reference
Green macroalga Entermorpha clathrata	Buzzards Bay, Mass.	D	429 (wet)		total petroleum	Burns and Teal (1979)
Red macroalga Polysiphonia fibrillosa	Buzzards Bay, Mass.	D	6.3 (wet)		total petroleum	Burns and Teal (1979)
Brown macroalga Fucus distichus	Cook Inlet, Alaska	C	40.1-153 (dry)		saturated	Shaw and Wiggs (1979)
Marsh succulent Salicornia sp.	Buzzards Bay, Mass.	D	13.2 (wet)		total petroleum	Burns and Teal (1979)
Marsh grass Spartina patens	Buzzards Bay, Mass.	D	15.2 (wet)		total petroleum	Burns and Teal (1979)
Eelgrass Zostera marina	Chedabucto Bay, Nova Scotia	D	17.13±8.18 (wet)		Bunker C	Vandermeulen and Gordon (1976)
Sponge Halichondria panicea	Norwegian Coast	D	<1-14 (wet)[c]		aromatic	Grahl-Nielson et al. (1979)
Anemone Tealia felina	Norwegian Coast	D	<1-14 (wet)[c]		aromatic	Grahl-Nielson et al. (1979)
Mussel Mytilus californianus	Washington Coast, USA	D	7.9-17 (dry)	5	n-alkanes	Clark et al. (1978)
M. edulis	Cook Inlet, Alaska	C	2.9 (dry)		saturated	Shaw and Wiggs (1980)
M. edulis	Norwegian Coast	D	1-28 (wet)[c]		aromatic	Grahl-Nielson et al. (1979)
M. edulis	Westernport Bay, Australia	C	0-3220 (dry)	126	total petroleum	Burns and Smith (1977)
Modiolus demissus	Buzzards Bay, Mass.	D	218 (wet)		total petroleum	Burns and Teal (1979)
Clam Mercenaria mercenaria	Narragansett Bay, Rhode Island	C	41.9 (wet)		total hydrocarbons	Boehm and Quinn (1977)
Mya arenaria	Chedabucto Bay, Nova Scotia	D	191-350 (wet)	262	total petroleum	Gilfillan and Vandermeulen (1978)
M. arenaria	Chedabucto Bay, Nova Scotia	D	14.21±6.02[d] (wet)		Bunker C	Vandermeulen and Gordon (1976)
M. arenaria	Chedabucto Bay, Nova Scotia	D	12.40±4.03[e] (wet)		Bunker C	Vandermeulen and Gordon (1976)
Oyster Pinctada margaratifera	Kuwait Coast	B	39.2-348 (dry)	88	saturated and unsaturated	Anderlini et al. (1981)

Limpet						
Patella vulgata	Norwegian Coast	D		<1-7 (wet)[c]	aromatic	Grahl-Nielson et al. (1979)
Collisella pelta	Cook Inlet, Alaska	C		1692 (dry)	saturated and unsaturated	Shaw and Wiggs (1980)
Periwinkle						
Litorina littorea	Norwegian Coast	D		1-22 (wet)[c]	aromatic	Grahl-Nielson et al. (1979)
Goose Barnacle						
Nitella polymerus	Washington Coast, USA	D		1.1-30 (dry)	n-alkanes	Clark et al. (1978)
Crab						
Cancer pagurus	Norwegian Coast	D		<0.1-0.8 (wet)[c]	aromatic	Grahl-Nielson et al. (1979)
Uca pugnax	Buzzards Bay, Mass.	D	259	183-287 (wet)	total petroleum	Burns (1976), Burns and Teal (1979)
Seastar						
Asterias rubens	Norwegian Coast	D		<1-8 (wet)[c]	aromatic	Grahl-Nielson et al. (1979)
Sea cucumber						
Holothurian	Nares Abyssal Plain	A		22.2 (wet)	saturated	Teal (1976)

[a] Area types: A-oceanic, B-chronic coastal pollution, C-chronic harbor pollution, D-spill.
[b] Wet, lipid, or dry weight basis stated.
[c] Decreasing with time.
[d] Upper beach.
[e] Lower beach.

hydrocarbons from petroleum and other pollution sources. Table 4-11 summarizes some of the information about petroleum in benthic organisms that has been reported in the reviewed literature between 1977 and 1981. A similar table covering earlier work has been prepared by Clark and Macleod (1977a). These tables show that petroleum has been detected in numerous taxa, in a variety of locations representing different marine environmental types, and resulting from several kinds of pollution sources. Other work (not tabulated) shows that marine organisms in many parts of the world ocean are free of detectable petroleum.

Petroleum has been found in association with several genera of marine benthic algae and vascular plants: Enteromorpha, Polysiphonia, Fucus, Salicornia, Spartina, and Zostera. In the Case of Fucus collected in a small boat harbor (Shaw and Wiggs, 1979), petroleum was only an external coating. For Enteromorpha, Polysiphonia, Salicornia, and Spartina from a heavily polluted marsh (Burns and Teal, 1979), however, petroleum was incorporated into the plant tissues.

Several workers have investigated the occurrence of petroleum in bivalve molluscs (Table 4-11, mussels, clams, oysters). An extensive investigation of petroleum hydrocarbons, as well as other organic and inorganic pollutants in marine bivalves, especially mussels (Mytilus sps.) and oysters (Crassostrea and Ostrea sps.), has been carried out in several countries. In the United States, scientists have examined the hydrocarbons in tissues of mussels (Mytilus edulis and M. californianus) and oysters (Crassostrea virginica and Ostrea equestris) along the coastline of the 48 contiguous states (Goldberg et al., 1978; Farrington et al., 1980, 1982). To quote Farrington et al. (1980, p. 17), "A general consensus from these studies is that elevated concentrations as much as two orders of magnitude above background in remote areas are found near known or suspected sources of inputs of fossil fuel compounds. . . . Detailed analyses often make possible identification of probable major sources for the fossil fuel compounds." Both petroleum and pyrogenic inputs are suggested by the data (Farrington et al., 1982).

Hydrocarbons in the oyster Pinctada margaratifera from Arabian Gulf waters of coastal Kuwait were determined by Anderlini et al. (1981). Animals from all six stations sampled showed evidence of petroleum (UCM, phytane) as well as biogenic hydrocarbons. The highest petroleum concentration (approximately 300 µg/g dry weight) was observed at the station closet to Kuwait's major oil loading and refinery facilities. This concentration is comparable to those observed by the U.S. Mussel Watch for bivalves collected near urban areas.

Grahl-Neilson (1978) monitored the concentrations of individual aromatic hydrocarbons in seven species of benthic invertebrates for 1 year following a spill of 2,000 tons of crude oil near the coast of Norway. All species showed substantial reductions in total hydrocarbon concentrations with time. Within this general trend, however, there were marked differences apparently related to species, pollution history of sampling sites, and molecular identity of the various aromatic hydrocarbons.

Summary and Recommendations

Several U.S. agencies, including the Bureau of Land Management, the National Oceanic and Atmospheric Administration, and the Environmental Protection Agency, as well as those of other countries, have supported studies on the amount of petroleum in the water column, sediments, and organisms in the ocean. Results have been accumulated during the past 10 years so that several conclusions can be drawn. For the water column, PHC concentrations vary by several orders of magnitude and are related to the proximity to petroleum sources, e.g., offshore and shore-based coastal production and refining activities as well as transportation practices and accidents. The amount of particulate-bound petroleum, primarily floating oil residues (viz., tar balls), also varies by orders of magnitude, with the highest concentration associated with tanker shipping lanes. Significant decreases have not been observed in amounts of petroleum recorded in the most intensely studied areas, such as the Mediterranean Sea, north central Atlantic Ocean, and the Gulf of Mexico.

In marine sediments, elevated PHC concentrations are related to the proximity of sewage and industrial outfalls, offshore dumping sites, and accidental discharges. The extent and history of pollution sedimentation, and related environmental factors are characteristic for each geographical area. Thus, a primary research recommendation is studies of the rate of loss of PHCs from chronically polluted sediments need to be done so that accurate accounting of these compounds can be obtained.

Relatively little information is available on PHC concentrations in pelagic organisms, mainly because of analytical problems, that is, in differentiating between PHCs and hydrocarbons produced by organisms in nature and micro tar balls caught in plankton net tows made in heavily traversed oceanic areas. PHCs are usually detected in samples of benthic organisms collected from polluted areas, but not from areas free of spills or accidents.

REFERENCES

Ahokas, J.T., H. Saarni, D.W. Nebert, and O. Pelkonen. 1979. The in vitro metabolism and covalent binding of benzo(a)pyrene to DNA catalyzed by trout liver microsomes. Chem. Biol. Interact. 25:103-111.

Albro, P.W., and L. Fishbein. 1970. Absorption of aliphatic hydrocarbons in rats. Biochim. Biophys. Acta 219:437-446.

Altshuler, A.P. and J.J. Bufalini. 1971. Photochemical aspects of air pollution. Rev. Environ. Sci. Technol. 5:39-64.

Anderlini, V.C., L. Al-Hormi, B.W. DeLappe, R.W. Risebrough, W. Walker II, B.R.T. Simoneit, and A.S. Newton. 1981. Distribution of hydrocarbons in the oyster Pinctada margaratifera, along the coast of Kuwait. Mar. Pollut. Bull. 12:57-62.

Anderson, J.W. 1975. Laboratory studies on the effects of oil on marine organisms: an overview. Publication 4349. American Petroleum Institute, Washington, D.C.

Andrews, A.R., and G.D. Floodgate. 1974. Some observations on the interactions of marine protozoa and crude oil residues. Mar. Biol. 25:7-12.

Aquan-Yuen, M., D. Mackay, and W.Y. Shiu. 1979. Solubility of hexane, phenanthrene, chlorobenze and p-dichlorobenzene in aqueous solutions. J. Chem. Eng. Data 24:30.

Aravamudan, K.S., P.K. Ray, and G. Marsh. 1981. Simplified models to predict the breakup of oil on rough seas pp. 153-159. In Proceedings, 1981 Oil Spill Conference, American Petroleum Institute, Washington, D.C.

Atlas, R.M. 1975. Effects of temperature and crude oil composition on petroleum biodegradation. Appl. Microbiol. 30:396-403.

Atlas, R.M. 1977. Stimulated petroleum biodegradation. Crit. Rev. 5:371-386.

Atlas, R.M., and R. Bartha. 1972a. Biodegradation of petroleum in seawater at low temperatures. Can. J. Microbiol. 18:1851-1855.

Atlas, R.M., and R. Bartha. 1972b. Degradation and mineralization of petroleum in seawater: limitation by nitrogen and phosphorus. Biotechnol. Bioeng. 14:309-317.

Atlas, R.M., and R. Bartha. 1973a. Fate and effects of oil pollution in the marine environment. Residue Rev. 49:49-85.

Atlas, R.M., and R. Bartha. 1973b. Stimulated biodegradation of oil slicks using oleophilic fertilizers. Environ. Sci. Technol. 7:538-541.

Atlas, R.M., and R. Bartha. 1981. Microbial Ecology--Fundamentals and Applications. Addison-Wesley, Reading, Mass.

Atlas, R.M., and A. Bronner. 1981. Microbial hydrocarbon degradation within intertidal zones impacted by the Amoco Cadiz oil spillage, pp. 251-256. In Proceedings of the International Symposium on the Amoco Cadiz: Fate and Effects of the Oil Spill. Centre Oceanologique de Bretagne, Brest, France. UNESCO.

Atlas, R.M., G. Roubal, A. Bronner, and J. Haines. 1980. Microbial degradation of hydrocarbons in mousse from Ixtoc I, pp. 1-24. In Proceedings of Conference on Researcher/Pierce Ixtoc I Cruises. National Oceanographic and Atmospheric Administration, AOMI, Miami, Fla.

Atlas, R.M., P.D. Boehm, and J.A. Calder. 1981. Chemical and biological weathering of oil from the Amoco Cadiz spillage within the littoral zone. Estuararine Coastal Mar. Sci. 12:589-608.

Attaway, D., J.R. Jadamec, and W. McGowan. 1973. Rust in floating petroleum found in the marine environment. U.S. Coast Guard, Groton, Conn.

Atwood, D.K., and R. Ferguson 1982. Example study of the weathering of spilled petroleum in a tropical marine environment--Ixtoc I. Bull. Mar. Sci. 32:1-13.

Audunson, T., V. Dalen, J.P. Mathison, J. Haldorsen, and F. Krogh. 1980. SLICKFORCAST--A simulation program for oil spill emergency tracking and long-term contingency planning. Proc. Petromar 1980, Monaco, May 26-30. Continental Shelf Institute.

Augenfeld, J.M., R.G. Riley, B.L. Thomas, and J.W. Anderson. 1981. The fate of polyaromatic hydrocarbons in an intertidal sediment exposure system. I. Bioavailability to Macoma inquinata (Mollusca: Pelecypoda) and Abarenicola pacifica (Annelida:Polychaeta). Mar. Environ. Res. (in press).

Bailey, N.J.L., A.M. Jobson, and M.A. Rogers. 1973. Bacterial degradation of crude oil: comparison of field and experimental data. Chem. Geol. 11:203-221.

Barbier, M., D. Joly, A. Saliot, and D. Tourres. 1973. Hydrocarbons from sea water. Deep Sea Res. 20:305-314.

Barger, W.R., D.R. Sherard, and W.D. Garrett. 1974. Determination of the Thickness of Petroleum Films on Water. Memorandum Report 2883. Naval Research Laboratory, Washington, D.C.

Barrick, R.C., J.I. Hedges, and M.L. Peterson. 1980. Hydrocarbon geochemistry of the Puget Sound region. I. Sedimentary acyclic hydrocarbons. Geochim. Cosmochim. Acta 44:1349-1362.

Bartha, R., and R.M. Atlas. 1977. The microbiology of aquatic oilspills. Adv. Appl. Microbiol. 22:225-226.

Bassin, N.J., and T. Ichiye. 1977. Floculation behavior of suspended sediments and oil emulsions. J. Sed. Petrol. 47:671-677.

Beam, H.W., and J.J. Perry. 1974. Microbial degradation of cycloparaffinic hydrocarbons via cometabolism and commensalism. J. Gen. Microbiol. 82:163-169.

Bend, J.R. 1980. Induction of drug-metabolizing enzymes by polycyclic aromatic hydrocarbons: mechanism and some implications in environmental health research, p. 93. In Environmental Chemicals, Enzymic Mechanisms and Human Disease. Ciba Foundation Symposium No. 76. Excerpta medica, Amsterdam.

Bend, J.R., and M.O. James. 1978. Xenobiotic metabolism in marine and freshwater species, p. 128. In D.C. Malins and J.R. Sargent eds. Biochemical and Biophysical Perspectives in Marine Biology. Vol. 4. Academic Press, New York.

Bend, J.R., D.S. Miller, W.B. Kinter, and D.B. Peakall. 1977a. DDE-induced microsomal mixed-function oxidases in the puffin (Fratercula arctica). Biochem. Pharmacol. 26:1000-1001.

Bend, J.R., M.O. James, and P.M. Dansette. 1977b. In vitro metabolism of xenobiotics in some marine animals. Ann. N.Y. Acad. Sci. 298:505-521.

Bend, J.R., L.M. Ball, T.H. Elmamlouk, M.O. James, and R.M. Philpot. 1979. Microsomal mixed-function oxidation in untreated and polycyclic aromatic hydrocarbon-treated fish. In M.A.Q. Khan, J.J. Lech, and J.J. Menn, eds. Pesticide and Xenobiotic Metabolism in Aquatic Organisms, p. 297. American Chemical Society, Washington, D.C.

Berridge, S.A., R.A. Dean, R.G. Fallows, and A. Fish. 1968a. The properties of persistent oils at sea. J. Instit. Petr. 54:300-309.

Berridge, S.A., R.A. Dean, R.G. Fallows, and A. Fish. 1968b. The properties of persistent oils at sea, pp. 2-11. In Peter Hepple ed. Proceedings of a Symposium, Scientific Aspects of Pollution of the Sea by Oil. Institute of Petroleum, London.

Berridge, S.A., M.T. Thew, and A.G. Loriston-Clarke. 1968c. The formation and stability of emulsions of water in crude petroleum and similar stocks, pp. 35-59. In P. Hepple, ed. Scientific Aspects of Pollution of the Sea by Oil. Institute of Petroleum, London.

Beslier, A., J.L. Birrien, L. Cabioch, C. Larsonneur, and L. Le Borgne. 1980. La pollution des Baies de Morlaix et de Lannion por les hydrocarbures de l' Amoco Cadiz: Repartition sur les fonds et evolution. Heloglander Meeresunters 33:209-224.

Bieri, R. 1979. Hydrocarbons in demersal fish, macro-epifauna, and zooplankton, Vol. II, Chapter 9. MAFLA final report prepared by Dames and Moore. Contract AA550-CT7-34. Bureau of Land Management, Washington, D.C.

Binder, R.L. 1981. Xenobiotic monooxygenase activity and the response to inducers of cytochrome P-450 during embryonic and larval development in fish. Ph.D. thesis, WHOI/MIT Joint Program in Oceanography.

Binder, R.L., and J.J. Stegeman. 1980. Induction of aryl hydrocarbon hydroxylase activity in embryos of an estuarine fish. Biochem. Pharmacol. 29:949-951.

Blikely, D.R., G.F.L. Dietzel, A.W. Glass, P.J. van Kleef. 1977. Slicktrak--A computer simulation of offshore oil spills, cleanup, effect, and associated costs, pp. 45-53. In Proceedings, 1977 Oil Spill Conference. American Petroleum Institute, Washington, D.C.

Blumer, M., M. M. Mullin, and D. W. Thomas. 1963. Pristane in zooplankton. Science 140:974.

Blumer, M., J.C. Robertson, J.E. Gordon, and J. Sass. 1969. Phytol-derived C_{19} di- and tri-olefinic hydrocarbons in marine zooplankton and fishes. Biochemistry 8:4067-4074.

Blumer, M., and J. Sass. 1972. Oil pollution: Persistence and degradation of spilled fuel oil. Science 176:1120-1122.

Blumer, M., G. Souza, and J. Sass. 1970a. Hydrocarbon pollution of edible shellfish by an oil spill. Mar. Biol. 5:195-202.

Blumer, M., J. Sass, G. Souza, H. Sanders, F. Grassle, and G. Hampson. 1970b. The West Falmouth oil spill. Reference 70-44. Woods Hole Oceanographic Institution, Woods Hole, Mass.

Blumer, M., M.M. Mullin, and D.W. Thomas. 1970c. Pristane in the marine environment. Heloglander Wissenschaftliche Merresuntersuchungen 10:187-201.

Blumer, M., R.R.L. Guillard, and Y. Chase. 1971. Hydrocarbons of marine phytoplankton. Mar. Biol. 8:183-189.

Bocard, C. and C. Gatellier. 1981. Breaking of fresh and weathered emulsions by chemicals, pp. 601-607. In Proceedings, 1981 Oil Spill Conference. Publication 4334. American Petroleum Institute, Washington, D.C.

Boehm, P.D. 1980. Evidence for the decoupling of dissolved, particulate, and surface microlayer hydrocarbons in northwestern Atlantic continental shelf waters. Mar. Chem. 9:255-281.

Boehm, P.D., and D.L. Fiest. 1979. Determine hydrocarbon composition and concentrations of major components of the marine ecosystem, Vol. VI. In W.B. Jackson and G.W. Faw, eds. Biological/Chemical

Survey of Texamo and Caplin Sector Salt Dome Brine Disposal Sites off Louisiana, 1978-1979. Technical Memorandum NMFS-SEFC-32. NOAA, Rockville, Md.

Boehm, P.D. and D.L. Feist. 1980. Surface water column transport and weathering of petroleum hydrocarbon during the Ixtoc I blowout in the Bay of Campeche and their relation to surface oil and microalga compositions, pp. 169-185. In Proceedings of the Conference on the Preliminary Scientific Results from the Researcher/Pierce Cruise to the Ixtoc I Blowout. NOAA, Office of Marine Pollution Assessment, Rockville, Md.

Boehm, P.D. and J.G. Quinn. 1973. Solubilization of hydrocarbons by the dissolved organic matter in sea water. Geochim. Cosmochim. Acta 37:2459-2477.

Boehm, P.D. and J.G. Quinn. 1974. The solubility behavior of No. 2 fuel oil in sea water. Mar. Pollut. Bull. 5(7):101-105.

Boehm, P.D., and J.G. Quinn. 1976. The effect of dissolved organic matter in seawater on the uptake of mixed individual hydrocarbons and No. 2 fuel oil by a marine filter-feeding bivalve (Mercenaria mercenaria). Estuarine and Coastal Mar. Sci. 4:93-105.

Boehm, P.D., and J.G. Quinn. 1977. The persistence of chronically accumulated hydrocarbons in the hard shell clam Mercenaria mercenaria. Mar. Biol. 44:227-233.

Boehm, P.D. and J.G. Quinn. 1978. Benthic hydrocarbons of Rhode Island Sound. Estuarine Coastal Mar. Sci. 6:471-494.

Boehm, P.D., W.G. Steinhauer, D.L. Fiest, N. Mosesman, J.E. Barak, and G.H. Perry. 1979. A chemical assessment of the present levels and sources of hydrocarbon pollutants in the Georges Bank region, pp. 333-341. In Proceedings, 1979 Oil Spill Conference. American Petroleum Institute, Washington, D.C.

Boutry, L.-C., M. Bordes, A. Feurier, M. Barbier, and A. Saliot. 1977. La diatomee marine Chaetoceros simplex calcitrans Paulsen et son environment. IV. Relations avec le milieu de culture: etude des hydrocarbures. J. Exp. Mar. Biol. Ecol. 28:41-51.

Breger, R.K., R.B. Franklin, and J.J. Lech. 1981. Metabolism of 2-methylnaphthalene to isomeric dihydrodiols by hepatic microsomes of rat and rainbow trout. Drug Metab. Disp. 9(2):88-93.

Bridie, A.L., and J. Bos. 1971. Biological degradation of mineral oil in seawater. J. Inst. Pet. London 57:270-277.

Bridie, A.L., Th.H. Wanders, W. Zegveld, and H.B. Van der Heijde. 1980. Formation, prevention, and breaking of sea water in crude oil emulsions "Chocolate Mousses." Mar. Pollut. Bull. 2:434-348.

Brooks, J.M. 1976. Flux of light hydrocarbons into the Gulf of Mexico via runoff, pp. 135-200. In H. Windom and R.A. Duce, eds. Marine Pollutant Transfer. D.C. Heath and Co., Lexington, Mass.

Brooks, J.M. 1979. Deep methane maxima in the northwest Caribbean Sea: possible seepage along the Jamaica Ridge. Science 206:1060-1071.

Brooks, J.M., and W.M. Sackett. 1973. Sources, sinks and concentrations of light hydrocarbons in the Gulf of Mexico. J. Geophys. Res. 78:5248-5258.

Brooks, J.M. and W.M. Sackett. 1977. Significance of low-molecular-weight hydrocarbons in marine waters, pp. 445-448. In R. Campos and J. Goni, eds. Proceedings of the 7th International Meeting on Organic Geochemistry. Espanola de Micropaleontologia, Madrid.

Brooks, J.M., A.D. Fredericks, W.M. Sackett, and J.W. Swinnerton. 1973. Baseline concentrations of light hydrocarbons in the Gulf of Mexico. Environ. Sci. Technol. 7:639-642.

Brooks, J.M., B.B. Bernard, and W.M. Sackett. 1977. Input of low molecular-weight-hydrocarbons from petroleum operations into the Gulf of Mexico, pp. 373-384. In D.A. Wolfe, ed. Fate and Effects of Petroleum Hydrocarbons in Marine Ecosystems and Organisms. Pergamon, New York.

Brooks, J.M., B.B. Bernard, T.C. Sauer, Jr., and H. Abdel-Reheim. 1978. Environmental aspects of a well-flowout in the Gulf of Mexico. Environ. Sci. Technol. 12:695-703.

Brooks, J.M., B.B. Bernard, W.M. Sackett, and J.R. Schwarz. 1979. Natural gas seepage on the south Texas shelf, pp. 371-378. Paper No. OTC 3411 Offshore Technology Conference Houston, Tex.

Brooks, J.M., D.F. Reid, and B.B. Bernard. 1981a. Methane in the upper water column of the northwest Gulf of Mexico. J. Geophys. Res. 86:11029-11040.

Brooks, J.M., D.A. Wiesenburg, R.A. Burke, Jr., and M.C. Kennicutt. 1981b. Gaseous and volatile hydrocarbons input from a subsurface oil spill in the Gulf of Mexico. Environ. Sci. Technol. 15:91-959.

Brown, R.A., and H.L. Huffman. 1976. Hydrocarbons in open ocean waters. Science 191:847-849.

Brown, R.A., and T.D. Searl. 1976. Nonvolatile hydrocarbons along tanker routes of the Pacific Ocean, pp. 259-274. Paper No. 2448, Vol. 1. Offshore Technology Conference, Houston, Tex.

Brown, R.A., T.D. Searl, J.J. Elliott, B.G. Phillips, D.E. Brandon, and P.H. Monaghan. 1973. Distribution of heavy hydrocarbons in some Atlantic Ocean waters, pp. 505-519. In Proceedings, Joint Conference on Prevention and Control of Oil Spills. American Petroleum Institute, Washington, D.C.

Brown, R.A., J.J. Elliot, and T.D. Searl. 1974. Measurement and characterization of nonvolatile hydrocarbons in ocean water, pp. 131-133. In Marine Pollution Monitoring (Petroleum). NBS Special Publication 409. National Bureau of Standards, Gaithersburg, Md.

Brown, R.A., J.J. Elliot, J.M. Kellcher, and T.D. Searl. 1975. Sampling and analysis of nonvolatile hydrocarbons in ocean water, pp. 172-187. In T.R.P. Gibb, Jr., ed. Analytical Methods in Oceanography. Advances in Chemistry Series 147. American Chemical Society, Washington, D.C.

Brunnock, J.V., D.F. Duckworth, and G.G. Stephens. 1968. Analysis of beach pollutants. J. Inst. Petrol. 54:30-325.

Burns, K.A. 1976. Hydrocarbon metabolism in the intertidal fiddler crab Uca pugnax. Mar. Biol. 36:5-11.

Burns, K.A., and J.L. Smith. 1977. Distribution of petroleum hydrocarbons in Westernport Bay (Australia): results of chronic low level inputs, pp. 442-453. In D. Wolfe, ed. Fate and Effects of Petroleum Hydrocarbons in Marine Ecosystems and Organisms. Pergamon, New York.

Burns, K.A. and J.M. Teal. 1973. Hydrocarbons in the pelagic Sargasso community. Deep Sea Res. 20:207-211.

Burns, K.A., and J.M. Teal. 1979. The West Falmouth oil spill: hydrocarbons in the salt marsh ecosystem. Estuarine Coastal Mar. Sci. 8:349-360.

Burwood, R., and G.C. Speers. 1974. Photooxidation as a factor in the environmental dispersal of crude oil. Estuarine Coastal Mar. Sci. 2:117-135.

Butler, J.N. 1975. Evaporative weathering of petroleum residues: the age of pelagic tar. Mar. Chem. 3:9-21.

Butler, J.N. 1976. Transfer of petroleum residues from sea to air: evaporative weathering. In H.L. Windon and R.A. Duce, eds. Marine Pollutant Transfer Lexington Books, Lexington, Mass.

Butler, J.N., B.F. Morris and J. Sass. 1973. Pelagic tar from Bermuda and the Sargasso Sea, Special Publication 10. Bermuda Biological Station for Research, St. George's West.

Button, D.K. 1976. The influence of clay and bacteria on the concentration of dissolved hydrocarbons in saline solution. Geochim. Cosmochim. Acta 40:435-440.

Cairns, R.J.R., D.M. Grist, and E.L. Neustadter. 1974. The effect of crude oil-water interfacial properties on water-crude oil emulsion stability, pp. 135-151. In A.L. Smith, ed. Theory and Practice of Emulsion Technology. Academic Press, New York.

Calder, J.A. 1976. Hydrocarbons from zooplankton of the eastern Gulf of Mexico. In Proceedings of a Symposium on Sources, Effects, and Sinks of Hydrocarbons in Aquatic Environments. American Institute of Biological Sciences, Arlington, Va.

Calder, J.A. 1977. Seasonal variations of hydrocarbons in the water column of the MAFLA lease area, pp. 432-441. In D.A. Wolfe, ed. Fate and Effects of Petroleum Hydrocarbons in Marine Ecosystems and Organisms. Pergamon, New York.

Calder, J.A., J. Lake, and J. Laseter. 1978. Chemical composition of selected environmental and petroleum samples from the Amoco Cadiz oil spill. In W.H. Hess, ed. The Amoco Cadiz Oil Spill. A Preliminary Scientific Report. NOAA, EPA, Washington, D.C.

Canevari, G.P. 1969. General dispersant theory, pp. 171-177. In Proceedings, Joint Conference on Prevention and Control of Oil Spills. NTIS Report PB 194-395. National Technical Information Service, Springfield, Va.

Caparello, D.M., and P.A. La Rock. 1975. A radioisotope assay for the quantification of hydrocarbon biodegradation potential in environmental samples. Microb. Ecol. 2:28-42.

Cerniglia, C.E., D.T. Gibson, and C. Van Baalen. 1979. Algal oxidation of aromatic hydrocarbons; formation of 1-naphthol from naphthalene by Agmencllum quadruplicatum, strain PR-6. Biochem. Biophys. Res. Commun. 88:50-58.

Cerniglia, C.E., D.T. Gibson, and C. Van Baalen. 1980a. Oxidation of naphthalene by cyanobacteria and microalgae. J. Gen. Microbiol. 116:495-500.

Cerniglia, C.E., C. Van Baalen, and D.T. Gibson. 1980b. Oxidation of biphenyl by the cyanobacterium, Oscillatoria sp., strain JCM. Arch. Microbiol. 125:203-207.

Chouteau, J., E. Azoulay, and J.E. Senez. 1962. Anaerobic formation of n-heptene-1 from n-heptane by resting cells of Pseudomonas aeruginosa. Nature 194:576-578.

Clark, R.C., and J.S. Finley. 1973. Parraffin hydrocarbon pattern in petroleum polluted mussels. Mar. Pollut. Bull. 4:172-176.

Clark, R.C., and J.S. Finley. 1974. Acute effects of outboard motor effluent on two marine shellfish. Environ. Sci. Technol. 8:1009-1014.

Clark, R.C., and J.S. Finley. 1975. Uptake and loss of petroleum hydrocarbons by mussels, Mytilus edulis, in laboratory experiments. Fishery Bull. 73:508-515.

Clark, R.C., and W.D. MacLeod. 1977a. Inputs, transport mechanisms, and observed concentrations of petroleum in the marine environment, pp. 91-223. In D.C. Malins, ed. Effects of Petroleum on Arctic and Subarctic Environments and Organisms Vol. 1, Nature and Fate of Petroleum Academic Press, New York.

Clark, R.C., Jr., and W.D. MacLeod, Jr. 1977b. Photochemical modification, pp. 129-132. In D.C. Malins, ed. Effects of Petroleum on Arctic and Subarctic Marine Environments and Organisms. Vol. 1, The Nature and Fate of Petroleum. Academic Press, New York.

Clark, R.C., Jr., J.S. Finley, B.G. Patten, D.F. Stefoni, and E.E. DeNike. 1973. Interagency investigations of a persistent oil spill on the Washington coast, pp. 793-808. In Proceedings, Joint Conference on Prevention and Control of Oil Spills. American Petroleum Institute, Washington, D.C.

Clark, R.C., J.S. Finley, B.G. Patten, and E.E. DeNike. 1975. Long-term chemical and biological effects of a persistent oil spill following the grounding of the General M.C. Meigs. In 1975 Conference on Prevention and Control of Oil Pollution American Petroleum Institute, Washington, D.C.

Clark, R.C., Jr., B.G. Patten, and E.E. DeNike. 1978. Observations of a cold-water intertidal community after 5 years of a low level, persistent oil spill from the General M.C. Meiggs. J. Fish. Res. Board Can. 35:754-765

Clement, L.E., M.S. Stekol, and D.G. Shaw. 1980. Accumulation, fractionation and release of oil by the intertidal clam Macoma balthica. Mar. Biol. 57:41-50.

Cline, J.D. and M.L. Holmes. 1977a. Submarine seepage of natural gas in Norton Sound, Alaska. Science 198:1149-1153.

Cline, J.D. and M.L. Holmes. 1977b. Anomalous gaseous hydrocarbons in Norton Sound: biogenic or thermogenic? Offshore Technology Conference, Houston, Texas. Paper No. OTC 3052, 1:81-86.

Cohen, Y., D. Mackay, and W.Y. Shiu. 1980. Mass transfer rates between oil slicks and water. Can J. Chem. Eng. 58:569.

Collier, T.K., L.C. Thomas, and D.C. Malins. 1978. Influence of environmental temperature on disposition of dietary naphthalene in coho salmon (Oncorhynchus kisutch): isolation and identification of individual metabolites. Comp. Biochem. Physiol. 61C:23-28.

Collier, T.K., M.K. Krahn, and D.C. Malins. 1980. The disposition of naphthalene and its metabolities in the brain of rainbow trout (Salmo gairdneri). Environ. Res. 23:35-41.

Colwell, R.R., and J.D. Walker. 1977. Ecological aspects of microbial degradation of petroleum in the marine environment. Crit. Rev. Microbiol. 5:423-445.

Colwell, R.R., A.L. Mills, J.D. Walker, P. Garcia-Tello, and V. Campos-P. 1978. Microbial ecology studies of the Metula spill in the Straits of Magellan. J. Fish. Res. Board Can. 35:573-580.

Conomos, T.J. 1975. Movement of spilled oil as predicted by estuarine nontital drift. Limmol. Oceanogr. 20:159-173.

Conover, R. J. 1971. Some relations between zooplankton and Bunker C oil on Chedabucto Bay following the wreck of the tanker Arrow. J. Fish. Res. Board Can. 28:1327-1330.

Cordes, C., L. Atkinson, R. Lee, and J. Blanton. 1980. Pelagic tar off Georgia and Florida in relation to physical processes. Mar. Pollut. Bull. 11:315-317.

Corner, E.D.S., C.C. Kilvington, and S.C.M. O'Hara. 1973. Qualitative studies on the metabolism of naphthalenes in Maia squinado (Herbst). J. Mar. Biol. Assoc. U.K. 53:819-832.

Corner, E.D.S., R.P. Harris, K.J. Whittle, and P.R. Mackie. 1976a. Hydrocarbons in marine zooplankton and fish, pp. 71-105. In A.P.M. Lockwood, ed. Effects of Pollutants on Aquatic Organisms. Cambridge University Press, London.

Corner, E.D.S., R.P. Harris, C.C. Kilvington, and S.C.M. O'Hara. 1976b. Petroleum compounds in the marine food web: short-term experiments on the fate of naphthalene in Calanus. J. Mar. Biol. Assoc. U.K. 56:121-133.

Coty, V.F., and R.I. Leavitt. 1971. Microbial protein from hydrocarbons. Devel. Ind. Microbiol. 12:61-71.

Cretney, W.J. and C.S. Wong. 1974. Fluorescence monitoring study at ocean weather station "P," pp. 175-180. In Marine Pollution Monitoring (Petroleum). NBS Special Publication 409. National Bureau of Standards, Gaithersburg, Md.

Cripps, R.E., and R.J. Watkinson. 1978. Polycyclic aromatic hydrocarbons: metabolism and environmental aspects, pp. 113-134. In J.R. Watkinson, ed. Developments in Biodegradation of Hydrocarbons. Applied Science Publishers, London.

Crisp, P.T., S. Brenner, M.I. Venaktesan, E. Ruth, and I.R. Kaplan. 1979. Organic chemical characterization of sediment-trap particulates from San Nicholas, Santa Barbara, Santa Monica, and San Pedro Basins, California. Geochim. Cosmochim. Acta 43:1791-1801.

Davis, S.J. and C.F. Gibbs. 1975. The effect of weathering on a crude oil exposed at sea. Water Res. 9:275-289.

Delaune, R.D., G.A. Hambrick III, and W.H. Patrick, Jr. 1980. Degradation of hydrocarbons in oxidized and reduced sediments. Mar. Pollut. Bull. 11:103-106.

Dibble, J.T., and R. Bartha. 1976. The effect of iron on the biodegradation of petroleum in seawater. Appl. Environ. Microbiol. 31:544-550.

Dickens, D.F., I.A. Buist, and W.M. Pistruzak. 1981. Dome's petroleum study of oil and gas under sea ice, pp. 183-189. In Proceedings, 1981 Oil Spill Conference. Publication 4334. American Petroleum Institute, Washington, D.C.

Dicks, B., and K. Iball. 1981. Ten years of saltmarsh monitoring--the case history of a Southampton Water saltmarsh and a changing refinery effluent discharge, pp. 361-374. In Proceedings, 1981 Oil Spill Conference. Publication 4334. American Petroleum Institute, Washington, D.C.

DiSalvo, L.H. and J.E. Guard. 1975. Hydrocarbons associated with suspended particulate matter in San Francisco Bay waters, pp. 169-173. In Proceedings, 1975 Conference on Prevention and Control of Oil Pollution. American Petroleum Institute, Washington, D.C.

DiSalvo, L.H., H.E. Guard, and L. Hunter. 1975. Tissue hydrocarbon burden of mussels as potential monitor of environmental hydrocarbon insult. Environ. Sci. Technol. 9:247-251.

Dobroski, C.J., and C.E. Epifano. 1980. Accumulation of benzo(a)pyrene in a larval bivalve via trophic transfer. Can. J. Fish. Aquat. Sci. 37:2318-2322.

Dowty, B.J., N.E. Brightwell, J.L. Laseter, and G.W. Griffin. 1974. Dye-sensitized photo-oxidation of phenanthrene. Biochem. Biophys. Res. Comms. 57:452-455.

Duce, R.A., J.G. Quinn, C.E. Olney, S.R. Piotrowicz, B.J. Ray, and T.L. Wade. 1972. Enrichment of heavy metals and organic compounds in the surface micro-layer of Narragansett Bay, Rhode Island. Science 176:161-163.

Dunn, B.P. 1980. Polycyclic aromatic hydrocarbons in marine sediments, bivalves, and seaweeds: analysis by high-pressure liquid chromatography, pp. 367-377. In A. Bjorseth and A.J. Dennis, eds. Proceedings of the Fourth International Symposium on Polynuclear Aromatic Hydrocarbons. Batelle Press, Columbus, Ohio.

Dunn, B.P., and H.F. Stich. 1976. Release of the carcinogen benzo(a)pyrene from environmentally contaminated mussels. Bull. Environ. Contam. Toxicol. 15:398-401.

Eagle, G.A., A. Green, and J. Williams. 1979. Tar ball concentrations in the ocean around the Cape of Good Hope before and after a major oil spill. Mar. Pollut. Bull. 10:321-325.

Ehrhardt, M. 1972. Petroleum hydrocarbons in oysters from Galveston Bay. Environ. Pollut. 3:257-271.

Ehrhardt, M., and J. Derenbach. 1977. Composition and weight per area of pelagic tar collected between Portugal and south of the Canary Islands. Meteor. Forsch-Ergebnisse A(19):1-9.

Ehrhardt, M., and J. Heinemann. 1975. Hydrocarbons in blue mussels from the Kiel Bight. Environ. Pollut. 9:263-282.

Elcombe, C.R., R.B. Franklin, and J.J. Lech. 1979a. Induction of hepatic microsomal enzymes in rainbow trout, p. 319. In M.A.Q. Khan, J.J. Lech, and J.J. Menn, eds. Pesticide and Xenobiotic Metabolism in Aquatic Organisms. American Chemical Society, Washington, D.C.

Elcombe, C.R., R.B. Franklin, and J.J. Lech. 1979b. Induction of microsomal hemoprotein(s) P-450 in the rat and rainbow trout by polyhalogenated biphenyls. Ann. N.Y. Acad. Sci. 320:193.

Elmamlouk, T.H., T. Gessner, and A.C. Brownie. 1974. Occurrence of cytochrome P-450 in hepatopancreas of Homarus Americanus. Comp. Biochem. Physiol. 48B:419.

Engelhardt, F.R. 1981. Hydrocarbon metabolism and cortisol balance in oil-exposed ring seals, _Phoca hispida_. Comp. Biochem. Physiol. (in press).

Engelhardt, F.R., J.R. Geraci, and T.G. Smith. 1977. Uptake and clearance of petroleum hydrocarbons in the ringed seal, _Phoca hispida_. J. Fish. Res. Board Can. 34(8):1143-1147.

Fallab, S. 1967. Reactions with molecular oxygen. Agnew. Chem. Int. Edition 6:496-507.

Farrington, J.W., and J.G. Quinn. 1973. Petroleum hydrocarbons in Narragansett Bay. I. Survey of hydrocarbons in sediments and clams (_Mercenaria mercenaria_). Estuarine Coastal Mar. Sci. 1:71-79.

Farrington, J.W., and B.W. Tripp. 1977. Hydrocarbons in western North Atlantic surface sediments. Geochim. Cosmochim. Acta 41:1627-1641.

Farrington, J.W., N.M. Frew, P.M. Gschwend, and B.W. Tripp. 1977. Hydrocarbons in cores of northwestern Atlantic coastal and continental margin sediments. Estuarine Coastal Mar. Sci. 5:793-808.

Farrington, J.W., J. Albaiges, K.A. Burns, B.P. Dunn, P. Eaton, J.L. Laseter, P.L. Parker, and S. Wise. 1980. Fossil fuels, pp. 7-77. In The International Mussel Watch. National Academy of Sciences, Washington, D.C.

Farrington, J.W., A.C. Davis, N.W. Frew, and K.S. Rabin. 1981. No. 2 fuel oil compounds in _Mytilus edulis_: retention and release after an oil spill. Mar. Biol. 66:15-26.

Fazal, R.A., and H.J. Milgram. 1979. The effect of surface phenomena on the spreading of oil on water. Report MITSG 79-31. MIT Sea Grant College Program.

Floodgate, G.D. 1979. Nutrient limitation, pp. 107-118. In A.W. Bourquin and P.H. Pritchard, ed. Proceedings of Workshop, Microbial Degradation of Pollutants in Marine Environments. EPA-66019-79-012. Environmental Research Laboratory, Gulf Breeze, Fla.

Forlin, L. 1980. Effects of Clophen A50, 3-methylcholanthrene, pregnenolone 16a-carbonitrile and phenobarbital on the hepatic microsomal cytochrome P-450 dependent monooxygenase system in rainbow trout, _Salmo gairdneri_, of different age and sex. Toxicol. Appl. Pharmacol. 54:420.

Forrester, W.D. 1971. Distribution of suspended oil particulates following the grounding of the tanker _Arrow_. J. Mar. Res. 29:151-170.

Fossato, V.U. 1975. Elimination of hydrocarbons by mussels. Mar. Pollut. Bull. 6:7-10.

Fossato, V.U., and W.J. Canzonier. 1976. Hydrocarbon uptake and loss by the mussel _Mytilus edulis_. Mar. Biol. 36:243-250.

Fossato, V.U., C. Nasci, and F. Dolci. 1979. 3,4-benzopyrene and perylene in mussels, _Mytilus_ sp., from the Laguna Veneta, Northeast Italy. Mar. Environ. Res. 2:47-53.

Frankenfeld, J.S. 1973. Factors governing the fate of oil at sea: variations in the amounts and types of dissolved or dispersed materials during the weathering process. In Proceedings, Joint Conference on Prevention and Control of Oil Spills. American Petroleum Institute, Washington, D.C.

Freegarde, M., C.G. Hatchard, and C.A. Parker. 1971. Oil spilt at sea: its identification, determination and ultimate fate. Lab. Practice 20(1):35-40.

Friede, J.D. 1973. The isolation and chemical and biological properties of microbial and emulsifying agents for hydrocarbons. Progress Report Ad 770-630. National Technical Information Service, U.S. Department of Commerce, Springfield, Va., 5 pp.

Fucik, K.W., and J.M. Neff. 1977. Effect of temperature and salinity of naphthalene uptake in the temperate clam, Rangia cuneata, and the boreal clam, Prototheca staminea, pp. 305-312. In D. Wolfe, ed. Fate and Effects of Petroleum Hydrocarbons in Marine Organisms and Ecosystems. Pergamon, New York.

Gardner, W.S., R.F. Lee, K.R. Tenore, and L.W. Smith. 1979. Degradation of selected polycyclic aromatic hydrocarbons in coastal sediments: importance of microbes and polychaete worms. Water Air Soil Pollut. 11:339-347.

Garrett, W.D. 1967. Stabilization of air bubbles at the air-sea interface by surface active material. Deep Sea Res. 14:666-672.

Garver, D.R., and G.N. Williams. 1978. Advances in oil spill trajectory modelling. Proc. Oceans '78. Marine Technology Society and IEEE, Washington, D.C., September 6-8, 1978.

Gearing, P.J, J.N. Gearing, T.F. Lytle, and J.S. Lytle. 1976. Hydrocarbons in 60 northeast Gulf of Mexico shelf sediments: a preliminary survey. Geochim. Cosmochim. Acta 40:1005-1017.

Gearing, P.J., J.N. Gearing, R.J. Pruell, T.L. Wade, and J.G. Quinn. 1980. Partitioning of No. 2 fuel oil in controlled estuarine ecosystems: sediments and suspended particulate matter. Environ. Sci. Technol. 14(9):1129-1136.

Gelboin, H.V. 1980. Benzo(a)pyrene metabolism, activation, and carcinogenesis: role and regulation of mixed-function oxidases and related enzymes. Physiol. Rev. 60:1107-1155.

Geyer, R.A., and C.P. Giammona. 1980. Naturally occurring hydrocarbons in the Gulf of Mexico and Caribbean Sea, pp. 37-106. In R.A. Geyer, ed. Marine Environmental Pollution. Vol. I, Hydrocarbons. Elsevier, New York.

Gibbs, C.F. 1975. Quantitative studies in marine biodegradation of oil. I. Nutrient limitation at ^{14}C. Proc. R. Soc. London, Ser. B 188:61-82.

Gibbs, C.F., and S.J. Davis. 1976. The rate of microbial degradation of oil in a beach gravel column. Microb. Ecol. 3:55-64.

Gibbs, C.F., K.B. Pugh, and A.R. Andrews. 1975. Quantitative studies in marine biodegradation of oil. II. Effects of temperature. Proc. R. Soc. London, Ser. B 188:83-94.

Gibson, D.T. 1968. Microbial degradation of aromatic compounds. Science 161:1093-1097.

Gibson, D.T. 1971. The microbial oxidation of aromatic hydrocarbons. Crit. Rev. Microbiol. 1:199-223.

Gibson, D.T. 1975. Oxidation of the carcinogens benzo(a)pyrene and benzo(a)anthracene to dihydrodiols by a bacterium. Science 189:295-297.

Gibson, D.T. 1976. Microbial degradation of polycyclic aromatic hydrocarbons, pp. 57-66. In J.M. Sharpley and A.M. Kaplan, eds. Proceedings of the Third International Biodegradation Symposium. Applied Science Publishers, London.

Gibson, D.T. 1977. Biodegradation of aromatic petroleum hydrocarbons, pp. 36-46. In D. Wolfe, ed. Fate and Effects of Petroleum Hydrocarbons in Marine Ecosystems and Organisms. Pergamon, New York.

Gilfillan, E.S., and J.H. Vandermuelen. 1978. Alternation in growth and physiology of soft-shell clams, Mya arenaria, chronically oiled with Bunker C from Chedabucto Bay, Nova Scotia, 1970-76. J. Fish. Res. Board Can. 35:630-636.

Gilfillan, E.S., D.S. Page, R.P. Gerber, S. Hansen, J. Cooley, and J. Hotham. 1981. Fate of the Zoe Colocotroni oil spill and its effects on infaunal communities associated with mangroves, pp. 353-360. In Proceedings, 1981 Oil Spill Conference. American Petroleum Institute, Washington, D.C.

Goldberg, E.D., V.T. Bowen, J.W. Farrington, G. Harvey, J.H. Martin, P.L. Parker, R.W. Risebrough, W. Robertson, E. Schneider, and E. Gamble. 1978. The Mussel Watch. Environ. Conserv. 5:101-125.

Gordon, D.C., Jr., and P.D. Keizer. 1974. Estimation of petroleum hydrocarbons in seawater by fluoescence spectroscopy: improved sampling and analytical methods. Technical Report 481. Fisheries and Marine Sciences, Environment Canada, Ottowa, Ontario.

Gordon, D.C., P.D. Keizer, and J. Dale. 1974. Estimates using fluorescence spectoscopy of the present state of petroleum hydrocarbons contamination in the water column of the northwest Atlantic Ocean. Mar. Chem. 2:251-261.

Gordon, D.C., P.D. Keizer, and J. Dale. 1978a. Temporal variations and probable origins of hydrocarbons in the water column of Bedford Basin, Nova Scotia. Estuarine Coastal Mar. Sci. 7:243-256.

Gordon, D.C., J. Dade, and P.D. Keizer. 1978b. Importance of sediment working by the deposit-feeding polychaete Arenicola marina on the weathering rate of sediment-bound oil. J. Fish. Res. Board Can. 35:591-603.

Grahl-Nielson, O. 1978. The Ekofish Bravo blowout: petroleum hydrocarbons in the sea, pp. 476-487. In Proceedings of the Conference on Assessment of Ecological Impacts of Oil Spills. American Institute of Biological Sciences. Arlington, Va.

Grahl-Nielsen, O., K. Westrheim, and S. Wilhelmsen. 1979. Petroleum hydrocarbons in the North Sea, pp. 629-632. In Proceedings, 1979 Oil Spill Conference. American Petroleum Institute, Washington, D.C.

Grassle, J.F., and J.P. Grassle. 1977. Temporal adaptations in sibling species of Capitella, pp. 177-189. In B.C. Coull, ed. Ecology of Marine Benthos. University of South Carolina Press, Columbia.

Grassle, J.P., and J.F. Grassle. 1976. Sibling species in the marine pollution indicator Capitella (polychaete). Science 192:567-659.

Grau, C.R., T. Roudybush, J. Dobbs, and J. Wathen. 1977. Altered yolk structure and reduced hatchability of eggs from birds fed single doses of petroleum oils. Science 195:779-781.

Gruger, E.H., Jr., J.V. Schnell, P.S. Fraser, D.W. Brown, and D.C. Malins. 1981. Metabolism of 2,6-dimethylnaphthalene in starry flounder (Platichthys stellatus) exposed to naphthalene and p-cresol. Aquat. Toxicol. 1:37-48.

Gschwend, P.M., O.C. Zafiriou, and R.B. Gagosian. 1980. Volatile organic compounds in seawater from the Peru upwelling region. Limnol. Oceanogr. 25:1044-1053.

Gschwend, P.M., O.C. Zafiriou, R.F. Mantoora, R.P. Swarzenbach, R.B. Gagosian. 1982. Volatile organic compounds at the coastal site; I. Seasonal variations. Environ. Sci. Technol. 16:31-37.

Guire, P.E., J.D. Friede, and R.K. Gholson. 1973. Production and characterization of emulsifying factors from hydrocarbonoclastic yeast and bacteria, pp. 229-231. In D.G. Ahern and S.P. Meyers, eds. The Microbial Degradation of Oil Pollutants. Publication LSU-Sg-73-01. Center for Wetland Resources, Louisiana State University, Baton Rouge.

Gunkel, W. 1968. Bacteriological investigations of oil-polluted sediments from the Cornish coast following the Torrey Canyon disaster. Helgolander Wiss. Meeresunters. 17:151-158.

Haines, J.K., and M. Alexander. 1974. Microbial degradation of high-molecular weight alkanes. Appl. Microbiol. 28:1084-1085.

Hansen, H.P. 1975. Photochemical degradation of petroleum hydrocarbon surface films on seawater. Mar. Chem. 3:183-195.

Hansen, H.P. 1977. Photodegradation of hydrocarbon surface films. Rapp. P.-v. Reun. Cons. Int. Explor. Mer 171:101-106.

Hansen, N., V.B. Jensen, H. Applequist, and E. Morch. 1978. The uptake and release of petroleum hydrocarbons by the marine mussel Mytilus edulis. Prog. Water Tech. 10:351-359.

Hardy, R., P.R. Mackie, K.J. Whittle, A.D. McIntyre, and R.A.A. Blackman. 1975. Occurrence of hydrocarbons in surface films, subsurface water and sediment around the U.K., p. 71. In A.D. McIntyre and K. Whittle, eds. Petroleum Hydrocarbons in the Marine Environment. Proceedings from the ICES Workshop held in Aberdeen, Sept. 9-12, 1975. International Council for the Exploration of the Sea. 171:71.

Hardy, R., P.R. Mackie, K.J. Whittle, A.D. McIntyre, and R.A.A. Blackman. 1977. Occurrence of hydrocarbons in surface films, subsurface water, and sediment around the United Kingdom. Proceedings from the ICES Workshop held in Aberdeen, Sept. 9-12, 1975. International Council for the Exploration of the Sea. 171:61-65.

Harris, R.P., V. Berdugo, E.D.S. Corner, C.C. Kilvington, and S.C.M. O'Hara. 1977. Factors affecting the retention of a petroleum hydrocarbon by marine planktonic copepods, pp. 286-304. In D.A. Wolfe ed. Fate and Effects of Petroleum Hydrocarbons in Marine Ecosystems and Organisms. Pergamon, New York.

Harrison, W., M.A. Winnick, P.T.Y. Kwang, and D. Mackay. 1975. Crude oil spills: disappearance of aromatic and aliphatic components from small sea-surface slicks. Environ. Sci. Technol. 9:231-234.

Hartung, R., and G.W. Klinger. 1970. Concentration of DDT by sedimented polluting oils. Environ. Sci. Technol. 4:407-410.

Harvey, G.R., A.G. Reguejo, P.A. McGillivary, and J.M. Tokar. 1979. Observations of a subsurface oil-rich layer in the open ocean. Science 205:999-1001.

Heicklen, J. 1976. Atmospheric Chemistry. Academic Press, New York, 406 pp.

Hendry, D.G., C.W. Gould, D. Scheutzle, M.G. Syz, and F.R. Mayo. 1976. Auto-oxidations of cyclohexane and its auto-oxidation products. J. Organ. Chem. 41(1):1-459.

Herbes, S.E., and L.R. Schwall. 1978. Microbial transformation of polycyclic aromatic hydrocarbons in pristine and petroleum contaminated sediments. Appl. Environ. Microbiol. 35:306-316.

Hess, W.N., ed. 1978. The Amoco Cadiz oil spill: a preliminary scientific report. NOAA/EPA Special Report. NTIS No. PB-285-805. National Technical Information Service, Springfield, Va.

Heyerdahl, T. 1971a. Atlantic Ocean pollution and biota observed by the "Ra" expedition. Biol. Conserv. 3:164-167.

Heyerdahl, T. 1971b. The Ra Expeditions, chaps. 8-11. Doubleday, New York.

Hiltrabrand, R.R. 1978. Estimation of aromatic hydrocarbons in seawater at proposed deepwater port (DWP) sites in the Gulf of Mexico. Mar. Pollut. Bull. 9:19-21.

Hinga, K.R., M.E.Q. Pilson, R.F. Lee, J.W. Farrington, K. Tjessem, and A.C. Davis. 1980. Biogeochemistry of benzanthracene in an enclosed marine ecosystem. Environ. Sci. Technol. 14:1136-1143.

Hoffman, E.J. and J.G. Quinn. 1979. Gas chromatographic analyses of Argo Merchant oil and sediment hydrocarbons at the wreck site. Mar. Pollut. Bull. 10:20-24.

Hopper, D.J. 1978. Microbial degradation of aromatic hydrocarbons, pp. 85-112. In J.R. Watkinson ed. Developments in Biodegradation of Hydrocarbons Vol. 1. Applied Science Publishers, London.

Horn, M.H., J.M. Teal, and R.H. Backus. 1970. Petroleum lumps on the surface of the sea. Science 168:245-246.

Hose, J.E., J.B. Hannah, M.L. Landolt, B.S. Miller, S.P. Felton, and W.T. Iwaoka. 1981. Uptake of benzo(a)pyrene by gonadal tissue of flatfish (family Pleuronectidae) and its effect on subsequent egg development. J. Toxicol. Environ. Health 7:991.

Howard, J.A., and K.U. Ingold. 1966. Absolute rate constants for hydrocarbon auto-oxidation. III. Methylstyrene, -methlstyrene and indene. Can. J. Chem. 44:1113-1118.

Huang, J.C., and F.C. Monastero. 1982. Review of the state-of-the-art of oil spill simulation models. Final report. American Petroleum Institute, Washington, D.C.

Hunn, J.B., and J.L. Allen. 1974. Movement of drugs across the gills of fishes. Ann. Rev. Pharmacol. 14:47-55.

Hurtt, A.C., and J.G. Quinn. 1979. Distribution of hydrocarbons in Narragansett Bay sediment cores. Environ. Sci. Technol. 13:829-836.

Iizuka, H., M. Ilida, and S. Fujita. 1969. Formation of -n-decene-1 from n-decane by resting cells of C. rugosa. Z. Allg. Mikrobiol. 9:223-226.

Iliffe, T.M., and J.A. Calder. 1974. Dissolved hydrocarbons in the eastern Gulf of Mexico Loop Current and the Caribbean Sea. Deep Sea Res. 21:481-488.

James, M.O., and J.R. Bend. 1980. Polycyclic aromatic hydrocarbon induction of cytochrome P-450 dependent mixed-function oxidases in marine fish. Toxicol. Appl. Pharmacol. 54:117.

James, M.O., E.R. Bowen, P.M. Dansette, and J.R. Bend. 1979. Epoxide hydrase and glutathione S-transferase activities with selected alkene and arene oxides in several marine species. Chem. Biol. Interactions 25:321-344.

Jeffrey, L.M. 1973. Preliminary report on floating tar balls in the Gulf of Mexico and Caribbean Sea. Sea Grant Project 53399. Texas A&M University, College Station.

Jeffrey, L.M. 1979. Water column particulates and dissolved hydrocarbons, Vol. II, Chap. 25. In MAFLA final report prepared for Dames and Moore. Contract AA550-CT7-34. Bureau of Land Management, Washington, D.C.

Jeffrey, L.M. 1980. Petroleum residues in the marine environment, pp. 163-179. In R.A. Geyer ed. Marine Environmental Pollution. Vol. I, Hydrocarbons. Elsevier, New York.

Jeffrey, L.M., W.E. Pequegnat, E.A. Kennedy, A. Vos, and B.N. Jane. 1974. Pelagic tar in the Gulf of Mexico and Caribbean Sea, pp. 233-235. In Marine Pollution Monitoring (Petroleum). Special Publication 409. National Bureau of Standards, Gaithersburg, Md.

Jeffrey, L.M., A. Vos, and N. Powell. 1976. Progress report on the chemistry of environmental tars of the Gulf of Mexico and Atlantic. Report for Naturally Occurring Hydrocarbons in the Gulf of Mexico. Project 55173. Texas A&M University, College Station.

Johansson, S., U. Larsson, and P. Boehm. 1980. The _Tsesis_ oil spill. Mar. Pollut. Bull. 11:284-293.

Johnson, E.F. 1979. Multiple forms of cytochrome P-450: criteria and significance, pp. 1-26. In E. Hodgson, J.R. Bend, and R.M. Philpot, eds. Reviews in Biochemical Toxicology.

Jordan, R.E., and J.R. Payne. 1980. Fate and weathering of petroleum spilled in the marine environment: a literature review and synopsis. Ann Arbor Science Publishers, Ann Arbor, Mich.

Kamahara, F.K. 1969. Identification and differentiation of heavy residual oil and asphalt pollutants in surface waters by comparative ratios of infrared absorbances. Environ. Sci. Technol. 3:150-153.

Kanazawa, M. 1975. Production of yeast from n-paraffins, pp. 438-453. In S.R. Tannenbaum and D.I.C. Wang, ed. Single Cell Protein, Vol. 2. MIT Press, Cambridge, Mass.

Karickhoff, S.W., D.S. Brown, and T.A. Scott. 1979. Sorption of hydrophobic pollutants on natural sediments. Water Res. 13:241-248.

Keizer, P.D., and D.C. Gordon. 1973. Detection of trace amounts of oil and seawater by fluorescence spectroscopy. J. Fish. Res. Board Can. 30:1039-1046.

Keizer, P.D., D.C. Gordon, Jr., and J. Dale. 1977. Hydrocarbons in eastern Canadian marine waters determined by fluorescence spectroscopy and gas-liquid chromatography. J. Fish. Res. Board Can. 34:347-353.

Keizer, P.D., T.P. Ahern, J. Dale, and J.H. Vandermuelen. 1978a. Residues of Bunker C oil in Chedabucto Bay, Nova Scotia, 6 years after the Arrow spill. J. Fish. Res. Board Can. 35:528-535.

Keizer, P.D., J. Dale, and D.C. Gordon, Jr. 1978b. Hydrocarbons in surficial sediments from the Scotian Shelf. Geochim. Cosmochim. Acta 42:165-172.

Kennicutt, M.D., II. 1980. Particulate and dissolved lipids in seawater. Ph.D. thesis. Texas A&M University, College Station.

Klein, A.E., and N. Pilpel. 1974. The effects of artificial sunlight upon floating oils. Water Res. 8:79-83.

Knap, A.H., T.M. Iliffe, and J.N. Butler. 1980. Has the amount of tar on the open ocean changed in the past decade? Mar. Pollut. Bull. 11:161-164.

Knight, G.C., C.H. Walker, D.C. Cabot, and M.P. Harris. 1981. The activity of two hepatic microsomal enzymes in sea birds. Comp. Biochem. Physiol. 68C:127-132.

Kodama, K., S. Nakatani, K. Umehara, K. Shimizu, Y. Minoda, and K. Yamada. 1970. Microbial conversion of petro-sulfur compounds. III. Isolation and identification of products from dibenzothiophene. Agric. Biol. Chem. 34:1320-1324.

Kodama, K.K., K. Umehara, S. Shimizu, K. Nakatani, Y. Minoda, and K. Yamada. 1973. Identification of microbial products from dibenzothiophene and its proposed oxidation pathway. Agric. Biol. Chem. 37:45-50.

Koons, C.B., and D.E. Brandon. 1975. Hydrocarbons in water and sediment samples from Coal Oil Point area, offshore California, pp. 513-552. Paper No. 2387. Offshore Technology Conference, Houston, Tex.

Koons, C.B., and P.H. Monaghan. 1973. Petroleum derived hydrocarbons in Gulf of Mexico waters. Trans. Gulf Coast Assoc. Geol. Soc. 16:170-181.

Koons, C.B., and J.P. Thomas. 1979. C_{15} + hydrocarbons in the sediments of the New York Bight, pp. 625-628. In Proceedings, 1979 Oil Spill Conference. American Petroleum Institute, Washington, D.C.

Kotyk, A. 1973. Mechanisms of non-electrolyte transport. Biochem. Biophys. Acta 300:183-210.

Kreider, R.E. 1971. Identification of oil leaks and spills. In Proceedings, Joint Conference on Prevention and Control of Oil Spills. American Petroleum Institute, Washington, D.C.

Kuhnhold, W.W, D. Everich, J.J. Stegeman, J. Lake, and R.E. Wolke. 1979. Effects of low levels of hydrocarbons on embryonic, larval and adult winter flounder (Pseudopleuronectes americanus), pp. 677-711. In Proceedings of the Conference on Assessment of Ecological Impacts of Oil Spills. NTIS No. AD-A072 859. National Technical Information Service, Springfield, Va.

Kurelec, B., S. Britvic, M. Rijavec, W.E.G. Muller, and R.K. Zahn. 1977. Benzo(a)pyrene monooxygenase induction in marine fish--molecular response to oil pollution. Mar. Biol. 44:211-216.

Laborde, A.L., and D.T. Gibson. 1977. Metabolism of dibenzothiophene by a Beijerinckia species. Appl. Environ. Microbiol. 34:783-790.

Lake, J.L., C. Norwood, C. Dimock, and R. Bower. 1979. Origins of polycyclic aromatic hydrocarbons in estuarine sediments. Geochim. Cosmochim. Acta 43:1847-1854.

Lamontagne, R.A., J.W. Swinnerton, and V.J. Linnenbom. 1971. Nonequilibrium of carbon monoxide and methane at the air-sea interface. J. Geophys. Res. 76:5117-5121.

Lamontagne, R.A., J.W. Swinnerton, V.J. Linnenbom, and W.D. Smith. 1973. Methane concentrations in various marine environments. J. Geophys. Res. 78:5317-5324.

Lamontagne, R.A., J.W. Swinnerton, and V.J. Linnenbom. 1974. C_1-C_4 hydrocarbons in the North and South Pacific. Tellus 16:71-77.

Lange, P., and H. Hufnerfuss. 1978. Drift response of monomolecular slicks to wave and wind action. J. Phys. Oceanogr. 8.

Larson, R.A. and L.L. Hunt. 1978. Photo-oxidation of a refined petroleum oil: inhibition by β-carotene and role of singlet oxygen. Photochem. Photobiol. 28:553-555.

Larson, R.A., L.A. Hunt, and D.W. Blakenship. 1977. Formation of toxic products from a No. 2 fuel oil by photooxidation. Environ. Sci. Technol. 11(5):492-496.

Larson, R.A., T.L. Bott, L.L. Hunt, and K. Rogenmuser. 1979. Photo-oxidation products of a fuel oil and their anti-microbial activity. Environ. Sci. Technol. 13:965-969.

Law, R.J. 1978. Petroleum hydrocarbon analysis conducted following the wreck of the supertanker Amoco Cadiz. Mar. Pollut. Bull. 9:293-296.

Law, R.J. 1981. Hydrocarbon concentrations in water and sediments from UK marine waters, determined by fluorescence spectroscopy. Mar. Pollut. Bull. 12:153-157.

Lawler, G.C., W.-A. Loong, and J.L. Laseter. 1978a. Accumulation of saturated hydrocarbons in tissues of petroleum-exposed mallard ducks (Anas platyrhynchos). Environ. Sci. Technol. 12:51-50.

Lawler, G.C., W-A. Loong, and J.L. Laseter. 1978b. Accumulation of aromatic hydrocarbons in tissues of petroleum-exposed mallard ducks (Anas platyrhynchos). Environ. Sci. Technol. 12:51-54.

Lech, J.J., and J.R. Bend. 1980. The relationship between biotransformation and the toxicity and fate of xenobiotic chemicals in fish. Environ. Health Perspect. 35:115.

Lech, J.J., M.J. Vodcnik, and C.R. Elcombe. 1981. Induction of monooxygenase activity in fish. In L. Weber, ed. Aquatic Toxicology. Raven Press, New York.

Ledet, E.J., and J.L. Laseter. 1974. Alkanes at the air-sea interface from offshore Louisiana and Florida. Science 186:261-263.

Lee, R.A. 1973. Uptake of petroleum hydrocarbons by marine copepods. Unpublished manuscript.

Lee, R.F. 1977a. Accumulation and turnover of petroleum hydrocarbons in marine organisms, pp. 60-70. In D. Wolfe, ed. Fate and Effects of Petroleum Hydrocarbons in Marine Organisms and Ecosystems. Pergamon, New York.

Lee, R.F. 1977b. Fate of petroleum components in estuarine waters of the Southeastern United States, pp. 611-616. In Proceedings, 1977 Oil Spill Conference. American Petroleum Institute, Washington, D.C.

Lee, R.F. 1981. Mixed function oxygenases (MF) in marine invertebrates. Mar. Biol. Letts. 2:87-105.

Lee, R.F., and M.L. Neuhauser. 1976. Fate of petroleum hydrocarbons taken up from food and water by the blue crab, _Callinectes sapidus_. Mar. Biol. 37:363-370.

Lee, R.F., and C. Ryan. 1976. Biodegradation of petroleum hydrocarbns by marine microbes, pp. 119-126. In J.M. Sharpley and A.M. Kaplan ed. Proceedings of the Third International Biodegradation Symposium. Applied Science Publishers, London.

Lee, R.F., and S.C. Singer. 1980. Detoxifying enzymes system in marine polychaetes. Increases in activity after exposure to aromatic hydrocarbons. Rapp. R.V. Reun. Cons. Int. Explor. Mer 179:29.

Lee, R.F., G. Sauerheber, and A.A. Benson. 1972a. Petroleum hydrocarbons: uptake and discharge by the marine mussel _Mytilus edulis_. Science 177:344-346.

Lee, R.F., R. Sauerheber, and G.H. Dobbs. 1972b. Uptake, metabolism, and discharge of polycyclic aromatic hydrocarbons by marine fish. Mar. Biol. 17:201.

Lee, R.F., C. Ryan, and M.L. Neuhauser. 1976. Fate of petroleum hydrocarbons taken up from food and water by the blue crab _Callinectes sapidus_. Mar. Biol. 37:363-370.

Lee, R.F., W.S. Gardner, J.W. Anderson, J.W. Blaylock, and J. Barwell-Clarke. 1978. Fate of polycyclic aromatic hydrocarbons in controlled ecosystem enclosures. Environ. Sci. Technol. 12(7):832-838.

Lee, R.F., S.C. Singer, K.R. Tenore, W.S. Gardner, and R.M. Philpot. 1979. Detoxification system in polychaete worms: importance in the degradation of sediment hydrocarbons, pp. 23-37. In W.B. Vernberg, F.P. Thurberg, A. Calabrese, and F.J. Vernberg, eds. Pollution and Physiology of Marine Organisms. Academic Press, New York.

Lee, R.F., J. Stolzenbach, S. Singer, and K.R. Tenore. 1981a. Effects of crude oil on growth and mixed-function oxygenase activity in polychaetes, _Nereis_ sp., pp. 323-334. In F.J. Vernberg, A. Calabrese, F.P. Thurberg, and W.B. Vernberg, eds. Biological Monitoring of Marine Organisms. Academic Press, New York.

Lee, R.F., B. Dornseif, F. Gonsoulin, K. Tenore, and R. Hanson. 1981b. Fate and effects of a heavy fuel oil spill on a Georgias alt marsh. Mar. Environ. Res. 5: 125-43.

Lee, R.F., S.C. Singer, and D.S. Page. 1982. Responses of cytochrome P-450 systems in marine crabs and polychaetes to organic pollutants. Aquat. Toxicol. In press.

Leinonen, P.J., D. Mackay, and C.R. Phillips. 1971. A correlation for the solubility of hydrocarbons in water. Can. J. Chem. Eng. 49:747-752.

LePetit, J., and M.-H. N'Guyen. 1976. Besoins en phosphore des bacteries metabolisant les hydrocarbures en mer. Can. J. Microbiol. 22:1364-1373.

Levy, E.M. 1971. The presence of petroleum residues off the coast of Nova Scotia, in the Gulf of St. Lawrence and the St. Lawrence River. Water Res. 5:723-733.

Levy, E.M. 1972. Evidence for the recovery of waters off the coast of Nova Scotia from the effects of a major oil spill. Water Air Soil Pollut. 1:144-148.

Levy, E.M. 1977. The geographical distribution of tar in the North Atlantic. Rapp. P.-v. Reun. Cons. Int. Explor. Mer 171:55-60.

Levy, E.M., and A. Walton. 1973. Dispersed and particulate petroleum residues in the Gulf of St. Lawrence. J. Fish. Res. Board Can. 30:261-267.

Levy, E.M., M. Ehrhardt, K. Kohnke, E. Sobtchenko, T. Suzuoki, and A. Tokuhiro. 1981. Global oil pollution. Results of MAPMOPO, the IGOSS pilot project on marine pollution (petroleum) monitoring. Intergovermental Oceanographic Commission, UNESCO, Paris.

Light, M. 1977. Report on Gulf of Mexico 1976 fall cruise (DWP Oceano, FY 77-1) CTC Acushnet. U.S. Coast Guard, Groton, Conn.

Liss, P.S., and P.G. Slater. 1974. Flux of gases across the air-sea interface. Nature 247:181-184.

Long, B.F.N., J.H. Vandermeulen, and T.P. Ahern. 1981. The evolution of stranded oil within sandy beaches, pp. 519-524. In Proceedings, 1981 Oil Spill Conference. American Petroleum Institute, Wasington, D.C.

Lysyj, I., and E.C. Russell. 1974. Dissolution of petroleum-derived products in water. Water Res. 8:863-868.

Lysyj, I., G. Perkins, J.S. Farlow, and R.W. Morris. 1981. Distribution of aromatic hydrocarbons in Port Valdez, Alaska, pp. 47-54. In Proceedings, 1981 Oil Spill Conference. American Petroleum Institute, Washington, D.C.

Lytle, T.F., and J.S. Lytle. 1979. Sediment hydrocarbons near an oil rig. Estuarine Coastal Mar. Sci. 9:319-330.

MacDonald, R.W. 1976. The distribution of low-molecular-weight hydrocarbons in the southern Beaufort Sea. Environ. Sci. Technol. 10:1241-1246.

MacGregor, C., and A.Y. McLean. 1977. Fate of crude oil spilled in a simulated arctic environment, pp. 461-463. In Proceedings, 1977 Oil Spill Conference. Publication 4284. American Petroleum Institute, Washington, D.C.

Mackay, D., and P.J. Leionen. 1975. The rate of evaporation of low solubility contaminants from water bodies. Environ. Sci. Technol. 9:1178-1180.

Mackay, D., and S. Paterson. 1981. A mathematical model of oil spill behavior. Report EE7. Department of Environment, EPS, Ottawa, Ontario.

Mackay, D., and W.Y. Shiu. 1975. The aqueous solubility of weathered crude oils. Bull. Environ. Contamin. and Toxicol. 15:101.

Mackay, D., I. Buist, R. Mascarenhas, and S. Paterson. 1979. Experimental studies of dispersion and emulsion formation from oil slicks, pp. 1.17-1.40. In Workshop on the Physical Behavior of Oil in the Marine Environment, Princeton University. Prepared for the National Weather Service, Silver Spring, Md.

Mackay, D., I. Buist, R. Mascarenhas, and S. Paterson. 1980a. Oil spill processes and models. Report submitted to Environmental Emergency

Branch, Environmental Impact Control Directorate, Environment Protection Service, Environment Canada. Ottawa, Ontario K1A 1C8.

Mackay, D., S. Paterson, and S. Nadeau. 1980b. Calculation of the evaporation rate of volatile liquids, p. 361. In Proceedings of the National Conference on Control of Hazardous Material Spills, Louisville, Ky.

Mackay, D., I. Buist, R. Mascarenhas and S. Paterson. 1981a. Oil spill processes and models. Report EE8. Department of Environment, Environment Protection Service, Ottawa, Ontario K1A 1C8.

Mackay, D., S. Paterson, P.D. Boehm, and D.L. Feist. 1981b. Physical-chemical weathering of petroleum hydrocarbons from the Ixtoc 1 blowout--chemical measurements and a weathering model. In Proceedings, 1981 Oil Spill Conference. American Petroleum Institute, Washington, D.C.

Mackay, G.C.M., A.Y. McLean, O.J. Betancourt, and B.C. Johnson. 1973. The formation of water-in-oil emulsions subsequent to an oil spill. V. J. Inst. Petroleum 59(568):164-172.

Mackie, P.R., K.J. Whittle, and R. Hardy. 1974. Hydrocarbons in the marine environment. I. n-alkanes in the Firth of Clyde. Estuarine Coastal Mar. Sci. 2:359-374.

Mackie, P.R., R. Hardy, K.J. Whittle, D.V.P. Conway, and A.D. McIntyre. 1976. Relationship between hydrocarbons, chlorophyll, and particulate carbon in the area between Firth and Forth and the Fortes oil field. ICES CM Fish. Improvement Comm. 1976/E:42. International Council for the Exploration of the Seas, Charlottenlund, Denmark.

Mackie, P.R., H.M. Platt, and R. Hardy. 1978. Hydrocarbons in the marine environment. II. Distribution of n-alkanes in the fauna and environment of the sub-antarctic island of South Georgia. Estuarine Coastal Mar. Sci. 6:301-313.

MacLeod, W.D., D.W. Brown, R.G. Jenkins, and L.S. Ramos. 1977. Intertidal sediment hydrocarbon levels at two sites on the Strait of Juan de Fuca, pp. 385-396. In D.A. Wolfe, ed. Fate and Effects of Petroleum Hydrocarbons in Marine Ecosystems and Organisms. Pergamon, New York.

Malins, D.C., ed. 1977. Effects of Petroleum on Arctic and Subarctic Marine Environments and Organisms. Vol. 1, Nature and Fate of Petroleum. Academic Press, New York.

Mannering, G.J. 1981. Hepatic cytochrome P-450 linked drug-metabolizing systems, pp. 53-166. In P. Jenner and B. Testa, eds. Part B. Concepts in Drug Metabolism. Marcel Dekker, New York.

Mantoura, R.F., P.M. Gschwend, O.C. Zafiriou, and K.R. Clark. 1982. Volatile organic compound at a coastal site: ill, short-term variations. Environ. Sci. Technol. 16:38-45.

Markovetz, A.J. 1971. Subterminal oxidation of aliphatic hydrocarbons by microorganisms. Crit. Rev. Microbiol. 1:225-237.

Marty, J.C., and A. Saliot. 1976. Hydrocarbons (normal alkanes) in the surface microlayer of seawater. Deep Sea Res. 23:863-873.

Masters, M.J., and J.E. Zajic. 1971. Myxotrophic growth of algae on hydrocarbon substrates. Dev. Ind. Microbiol. 12:77-86.

Mayo, D.W., D.S. Page, J. Cooley, E. Sorenson, F. Bradley, E.S. Gilfillan, and S.A. Hanson. 1978. Weathering characteristics of petroleum hydrocarbons deposited in fine clay sediments, Searsport, Maine. J. Fish. Res. Board Can. 35:552-562.

McAuliffe, C.D. 1977. Dispersal and alteration of oil discharged on a water surface. Chap. 3. In P.D.A. Wolfe, ed. Fate and Effects of Petroleum Hydrocarbons in Marine Organisms and Ecosystems. Pergamon, New York.

McAuliffe, C.D., J.C. Johnson, S.H. Greene, G.P. Canevari, and T.D. Searl. 1980. Dispersion and weathering of chemically treated crude oils in the oceans. Environ. Sci. Technol. 14:1509-1518.

McGowan, W.E., W.A. Saner, and G.L. Hufford. 1974. Tar ball sampling in the western N. Atlantic, pp. 83-84. In Marine Pollution Monitoring (Petroleum). NBS Special Publication 409. National Bureau of Standards, Gaithersburg, Md.

McKenna, E.J. 1971. Microbial degradation of normal and branched alkanes, pp. 73-97. In Degradation of Synthetic Organic Molecules in the Biosphere. National Academy of Sciences, Washington, D.C.

McLean, A.Y., and O.J. Betancourt. 1973. Physical and Chemical changes in spilled oil weathering under natural conditions pp. 1249-1256 Paper No. OTC-1784. Offshore Technology Conference, Houston, Tex.

McMaster, R.L. 1960. Sediments of Narragansett Bay system and Rhode Island Sound, Rhode Island. J. Sediment Petrol. 30:249-274.

Means, J.C., J.J. Hassett, S.G. Wood, and W.L. Banwart. 1979. Sorption properties of energy-related pollutants and sediments, p. 327. In P.W. Jones and P. Leber eds. Polynuclear Aromatic Hydrocarbons. Ann Arbor Science Publishers, Ann Arbor, Mich.

Melancon, M.J., and J.J. Lech. 1978. Distribution and elimination of naphthalene and 2-methylnaphthalene in rainbow trout during short- and long-term exposures. Arch. Environ. Contam. Toxicol. 7:207-220.

Melancon, M.J., Jr., and J.J. Lech. 1979. Uptake, biotransformation, disposition, and elimination of 2-methylnaphthalene and naphthalene in several fish species, pp. 5-22. In L.L. Marking and R.A. Kimerele eds. Aquatic Toxicology. ASTM STP 667. American Society for Testing and Materials, Philadelphia, Pa.

Meyers, P.A., and T.G. Oas. 1978. Comparison of associations of different hydrocarbons with clay particles in simulated seawater. Environ. Sci. Technol. 12:934-937.

Meyers, P.A., and J.G. Quinn. 1973. Association of hydrocarbons and mineral particles in saline solution. Nature 244:23-24.

Middleditch, B.S., B. Basile, and E.S. Chang. 1977. Environmental effects of offshore oil production: alkanes in the region of the Buccaneer oil field. J. Chromatogr. 142:777-785.

Middleditch, B.S., E.S. Chang, and B. Basile. 1979a. Alkanes in plankton from the Buccaneer oil field. Bull Environ. Contam. Toxicol. 21:421-427.

Middleditch, B.S., E.S. Chang, B. Basile, and S.R. Missler. 1979b. Alkanes in fish from the Buccaneer oil field. Bull. Environ. Contam. Toxicol. 22:249-257.

Mill, T., D.G. Hendry, and H. Richardson. 1980. Free radical oxidants in natural waters. Science 207:886-887.

Monaghan, P.H., J.H. Seelinger, and R.A. Brown. 1973. The persistent hydrocarbon content of the sea along certain tanker routes. Preliminary report presented at API Tanker Conference, Hilton Head Island, S.C., May 7-9, 1973.

Morris, B.F. 1971. Petroleum: tar quantities floating in the northwestern Atlantic taken with a new quantitative neuston net. Science 173:430-432.

Morris, B.F. 1974. Lipid composition of surface films and zooplankton from the eastern Mediterranean. Mar. Pollut. Bull. 5:105-109.

Morris, B.F., and J.N. Butler. 1973. Petroleum residues in the Sargasso Sea and on Bermuda beaches, pp. 521-530. In Proceedings, Joint Conference on Prevention and Control of Oil Spills. American Petroleum Institute, Washington, D.C.

Morris, B.F., and J.N. Butler. 1975. Pelagic tar in the Mediterranean Sea 1974-1975. Environ. Conserv. 2:275-281.

Morris, B.F., and F. Culkin. 1974. Lipid chemistry of eastern Mediterranean surface layers. Nature 250:640-642.

Morris, B.F., J. Cadwallader, J. Geiselman, and J.N. Butler. 1976. Transfer of petroleum and biogenic hydrocarbons in the Sargasso community, pp. 235-259. In H.L. Windom and R.A. Duce, eds. Marine Pollutant Transfer. Lexington Books, Lexington, Mass.

Mulkins-Phillips, G.J., and J.E. Stewart. 1974. Effect of environmental parameters on bacterial degradation of Bunker C oil, crude oils and hydrocarbons. Appl. Microbiol. 28:915-922.

Nagata, S., and G. Kondo. 1977. Photo-oxidation of crude oils. In Proceedings, 1977 Oil Spill Conference. American Petroleum Institute, Washington, D.C.

Nakatani, S., T. Akasaki, K. Kodama, Y. Minoda, and K. Yamada. 1968. Microbial conversion of petro-sulfur compounds. II. Culture conditions of dibenzothiophene-utilizing bacteria. Agric. Biol. Chem. 32:1205-1211.

National Research Council. 1972. Particulate Polycyclic Organic Matter. Washington, D.C.

National Research Council. 1975. Pollution in the Marine Environment. National Academy of Sciences, Washington, D.C.

Neff, J.M. 1979. Polycyclic Aromatic Hydrocarbons in the Aquatic Environment: Sources, Fates and Biological Effects. Applied Science Publishers, London.

Neff, J.M., and J.W. Anderson. 1975. Accumulation, release and distribution of benzo(a)pyrene ^{14}C in the clam Rangia cuneata, pp. 469-471. In Proceedings, 1975 Conference on Prevention and Control of Oil Pollution. American Petroleum Institute, Washington, D.C.

Neff, J.M., B.A. Cox, D. Dixit, and J.W. Anderson. 1976. Accumulation and release of petroleum-derived aromatic hydrocarbons by four species of marine animals. Mar. Biol. 38:279-289.

Nonhebel, D.C., and J.C. Walton. 1974. Free Radical Chemistry. Cambridge University Press, New York. 572 pp.

Nunes, P., and P.E. Benville. 1979. Uptake and depuration of petroleum hydrocarbons in the Manila clam, Tapes semidecussata Reeve. Bull. Environ. Contam. Toxicol. 21:719-726.

Oppenheimer, C.H., W. Gunkel, and G. Gassman. 1977. Microorganisms and hydrocarbons in the North Sea during July-August 1975, pp. 593-610. In Proceedings, 1977 Oil Spill Conference. American Petroleum Institute, Washington, D.C.

Overton, E.B., J.R. Patel, and J.L. Laseter. 1979. Chemical characterization of mousse and selected environmental samples from the Amoco Cadiz oil spill. In Proceedings, 1979 Oil Spill Conference. American Petroleum Institute, Washington, D.C.

Overton, E.B., J.L. Laseter, W. Mascarella, C. Rashke, I. Noiry, and J.W. Farrington. 1980. Photochemical Oxidation of Ixtoc I Oil. Researcher/Pierce Ixtoc I Symposium.

Page, D.S., D.W. Mayo, J.F. Cooley, E. Sorenson, E.S. Gilfillan, and S.A. Hanson. 1979. Hydrocarbon distribution and weathering characteristics at a tropical oil spill site, pp. 709-712. In Proceedings, 1979 Oil Spill Conference. American Petroleum Institute, Washington, D.C.

Palmork, K.H., and J.E. Solbakken. 1981. Distribution and elimination of ($9-{}^{14}$C) phenanthrene in the horse mussel (Modiola modiolus). Bull. Environ. Contam. Toxicol. 26:196-201.

Parekh, V.C., R.W. Traxler, and J.M. Sobek. 1977. n-Alkane oxidation enzymes of a Pseudomonad. Appl. Environ. Microbiol. 33:881-884.

Parker, P.L., J.K. Winters, and J. Mortan. 1972. A Base-Line Study of Petroleum in the Gulf of Mexico, pp. 555-581. In Base-Line Studies of Pollutants in the Marine Environment (Heavy Metals, Halogenated Hydrocarbons and Petroleum). National Science Foundation, IDOE, Washington, D.C.

Patel, J.R., J.A. McFall, G.W. Griffin, and J.L. Laseter. 1978. Toxic photo-oxygenated products generated under environmental conditions from phenanthrene. Paper presented at the EPA. Symposium on Carcinogenic Polynuclear Aromatic Hydrocarbons in the Marine Environment. Pensacola Beach, Fla., August 14-18, 1978.

Patel, J.R., E.B. Overton, and J.L. Laseter. 1979. Environmental photo-oxidation of dibenzothiophenes following the Amoco Cadiz oil spill. Chemosphere 8:557-561.

Patton, J.S., M.W. Rigler, P.D. Boehm, and D.L. Fiest. 1981. Ixtoc 1 oil spill: flaking of surface mousse in the Gulf of Mexico. Nature 290:235-238.

Payne, J. 1984. Petroleum Spills in the Marine Environment. Butterworth Publishers, Woburn, Mass. (in press).

Payne, J.F. 1976. Field evaluation of benzo(a)pyrene hydroxylase induction as a monitor for marine pollution. Science 191:945-946.

Payne, J.F., and W.R. Penrose. 1975. Induction of aryl hydrocarbon hydroxylase in fish by petroleum. Bull. Environ. Contam. Toxicol. 14:112-116.

Payne, J.R., N.W. Flynn, P.J. Mankiewicz, and G.S. Smith. 1980. Surface evaporation/dissolution partitioning of lower-molecular-weight aromatic hydrocarbons in a down-plume transect from the

Ixtoc I wellhead, pp. 239-266. In Proceedings of the Conference on the Preliminary Scientific Results from the Researcher/Pierce Cruise to the Ixtoc I Blowout. NOAA, Office of Marine Pollution Assessment, Rockville, Md.

Peakall, D.B., and D.J. Hallett et al. 1981. Toxicity of Prudhoe Bay Crude Oil and its Aromatic Fractions to Nestling Herring Gulls. (in press).

Pequegnat. 1979. Pelagic tar concentrations in the Gulf of Mexico over the south Texas continental shelf. Contrib. Mar. Sci. 22:31-39.

Perry, J.J. 1977. Microbial metabolism of cyclic hydrocarbons and related compounds. Crit. Rev. Microbiol. 5:387-412.

Perry, J.J. 1979. Microbial cooxidations involving hydrocarbons. Microbiol. Rev. 43:59-72.

Pirnik, M.P. 1977. Microbial oxidation of methyl branched alkanes. Crit. Rev. Microbiol. 5:413-422.

Plack, P.A., E.R. Skinner, A. Rogie, and A.I. Mitchell. 1979. Distribution of DDT between lipoproteins of trout serum. Comp. Biochem. Physiol. 62C:119-126.

Platt, H.M., and P.R. Mackie. 1980. Distribution and fate of aliphatic and aromatic hydrocarbons in antarctic fauna and environment. Helogolander Meeresunters. 33:236-245.

Pohl, R.J., J.R. Bend, A.M. Guarino, and J.R. Fouts. 1974. Hepatic microsomal mixed-function oxidase activity of several marine species from coastal Maine. Drug Metab. Dis. 2:545-555.

Polikarpov, G.G., V.N. Yegorov, V.N. Ivanov, A.V. Tokareva, and I.A. Felepov. 1971. Oil areas as an ecological niche. Priroda 11 [transl. by N. Precoda], Pollut. Abstr. 3:72.

Rathledge, C. 1978. Degradation of aliphatic hydrocarbons, pp. 1-46. In J.R. Watkinson ed. Developments in Biodegradation of Hydrocarbons Vol. 1. Applied Science Publishers, London.

Reed, W.E., I.R. Kaplan, M. Sandstrom, and P. Mankiewicz. 1977. Petroleum and anthropogenic influence on the composition of sediments from the Southern California Bight, pp. 183-188. In Proceedings, 1977 Oil Spill Conference. American Petroleum Institute, Washington, D.C.

Reisfeld, A., E. Rosenberg, and D. Gutnick. 1972. Microbial degradation of oil: factors affecting oil dispersion in seawater by mixed and pure cultures. Appl. Microbiol. 24:363-368.

Reitsema, R.J., F.A. Lindberg, and A.J. Kaltenback. 1978. Light hydrocarbons in Gulf of Mexico waters: sources and relation to structural highs. J. Geochem. Explor. 10:139-151.

Robertson, B., S. Arhelger, P.J. Kinney, and D.K. Button. 1973. Hydrocarbon biodegradation in Alaskan waters, pp. 171-184. In D.G. Ahearn and S.P. Meyers, eds. The Microbial Degradation of Oil Pollutants. Publication LSU-SG-73-01. Center for Wetland Resources, Louisiana State University, Baton Rouge.

Roesijadi, G., J.W. Anderson, and J.W. Blaylock. 1978. Uptake of hydrocarbons from marine sediments contaminated with Prudhoe Bay crude oil: influence of feeding type of test species and availability of polycyclic aromatic hydrocarbons. J. Fish. Res. Board Can. 35:608:614.

Rossi, S.S. 1977. Bioavailability of petroleum hydrocarbons from water, sediments and detritus to the marine annelid Neanthes arenaceodentata, pp. 621-625. In Proceedings, 1977 Oil Spill Conference. American Petroleum Institute, Washington, D.C.

Rossi, S.S., and J.W. Anderson. 1977. Accumulation and release of fuel-oil derived diaromatic hydrocarbons by the polychaete Neanthes arenaceodentata. Mar. Biol. 39:51-55.

Roubal, W.T. 1974. Spin-labeling of living tissue--a method for investigating pollutant-host interaction, pp. 367-379. In F. Vernberg and W. Verberg eds. Pollution and Physiology of Marine Organisms. Academic Press, New York.

Roubal, W.T., T.K. Collier, and D.C. Malins. 1977. Accumulation and metabolism of cabon-14 labeled benzene, naphthalene, and anthracene by young coho salmon (Oncorhynchus kisutch). Arch. Environ. Contam. Toxicol. 5:513-529.

Sackett, W.M. 1977. Use of hydrocarbon sniffing in offshore exploration. J. Geochem. Explor. 7:243-254.

Sackett, W.M., and J.M. Brooks. 1974. Use of low-molecular-weight hydrocarbon concentrations as indicators of marine pollution, pp. 171-173. In Marine Pollution Monitoring (Petroleum). Special Publication 409. National Bureau of Standards, Gaithersburg, Md.

Sackett, W.M., and J.M. Brooks. 1975. Origin and distribution of low-molecular-weight hydrocarbons in Gulf of Mexico coastal waters, pp. 211-230. In T.M. Church, ed. Marine Chemistry in the Coastal Environment. ACS Symposium Series 18. American Chemical Society, Washington, D.C.

Sanborn, H.R., and D.C. Malins. 1980. The disposition of aromatic hydrocarbons in adult spot shrimp (Pandalus platyceros) and the formation of metabolites of naphthalene in adult and larval spot shrimp. Xenobiotica 10:193-200.

Sanders, H.L., J.F. Grassle, G.R. Hampson, L.S. Morse, S. Garner-Price, and C.C. Jones. 1980. Anatomy of an oil spill: long-term effects from the grounding of the barge Florida off West Falmouth, Massachusetts. J. Mar. Res. 38:265-380.

San Gupta, R., S.Z. Qasim, S.P. Fondekar, and R.S. Topgi. 1980. Dissolved petroleum hydrocarbons in some regions of the northern Indian Ocean. Mar. Pollut. Bull. 11:164-174.

Sauer, T.C., Jr. 1978. Volatile liquid hydrocarbons in the marine environment. Ph.D. thesis. Texas A&M University, College Station.

Sauer, T.C., Jr. 1980. Volatile liquid hydrocarbons in water of the Gulf of Mexico and Caribbean Sea. Limnol. Oceanogr. 25:338-351.

Sauer, T.C., Jr. 1981a. Volatile organic compounds in open ocean and coastal surface waters. Organ. Geochem. 3:91-101.

Sauer, T.C. 1981b. Volatile liquid hydrocarbons characterization of underwater hydrocarbon vents and formation waters from offshore production operations. Environ. Sci. Technol. 15:917-923.

Sauer, T.C., Jr., and W.M. Sackett. 1980. Gaseous and volatile hydrocarbons in marine environments with emphasis on the Gulf of Mexico, pp. 133-161. In R.A. Geyer, ed. Marine Environment Pollution Vol. I., Hydrocarbons. Elsevier, New York.

Sauer, T.C., Jr., W.M. Sackett, and L.M. Jeffrey. 1978. Volatile liquid hydrocarbons in the surface coastal waters of the Gulf of Mexico. Mar. Chem. 7:1-16.

Sayler, G.S., and R.R. Colwell. 1976. Partitioning of mercury and chlorinated biphenyl by oil, water and sediment. Environ. Sci. Technol. 10:1142-1145.

Schnell, J.V., E.H. Gruger, Jr., and D.C. Malins. 1980. Monooxygenase activities of coho salmon (Oncorhynchus kisutch) liver microsomes using three polycyclic aromatic hydrocarbon substrates. Xenobiotica 10:229-234.

Schultz, D.M., and J.G. Quinn. 1977. Suspended material in Narragansett Bay: fatty acid and hydrocarbon composition. Organ. Geochem. 1:27-36.

Schwarz, J.R., J.D. Walker, and R.R. Colwell. 1974a. Hydrocarbon degradation at ambient and in situ pressure. Appl. Microbiol. 28:982-986.

Schwarz, J.R., J.D. Walker, and R.R. Colwell. 1974b. Growth of deep-sea bacteria on hydrocarbons at ambient and in situ pressure. Dev. Ind. Microbiol. 15:239-249.

Schwarz, J.R., J.D. Walker, and R.R. Colwell. 1975. Deep-sea bacteria: growth and utilization on n-hexadecane at in situ temperature and pressure. Can. J. Microbiol. 21:682-687.

Schwarzenbach, R.P., R.H. Bromund, P.M. Gschwen, and O.C. Zafiriou. 1979. Volatile organic compounds in coastal seawater: preliminary results. J. Organ. Geochem. 1:45-61.

Scranton, M.I., and P.G. Brewer. 1977. Occurrence of methane in the near-surface waters of the western subtropical North Atlantic. Deep Sea Res. 24:127-138.

Scranton, M.I., and P.G. Brewer. 1978. Consumption of dissolved methane in the deep ocean. Limnol. Oceanogr. 23:1207-1213.

Scranton, M.I., and J.W. Farrington. 1977. Methane production in the waters of Walvis Bay. J. Geophys. Res. 82:4947-4953.

Seba, D.B., and E.F. Corcoran. 1969. Surface slicks as concentrators of pesticides in the marine environment. Pesticide Monit. J. 3:190-193.

Seiler, W., and V. Schmidt. 1974. Dissolved non-conservative gases in seawater, pp. 219-243. In E.D. Goldberg, ed. The Sea. Vol. V. Wiley Interscience, New York.

Seki, H. 1976. Method for estimating the decomposition of hexadecane in the marine environment. Appl. Environ. Microbiol. 31:439-441.

Senez, J.C., and E. Azoulay. 1961. Dehydrogenation of paraffinic hydrocarbons by resting cells and cell free extracts of Pseudomonas aeruginosa. Biochim. Biophys. Acta 47:307-316.

Seubert, W., and E. Fass. 1964. Untersuchugen uber den bakteriellen Abban von Isoprenoiden. V. Der Mechanismns des Isoprenoidabbanes. Biochem. Z. 341:35-44.

Shackelford, M.E., and M.A.Q. Khan. 1981. Hepatic mixed-function oxidase of the mallard duck (Anas platyrhychos). Comp. Biochem. Physiol. 70C.

Shaw, D.G., and B.A. Baker. 1978. Hydrocarbons in the marine environment of Port Valdez, Alaska. Environ. Sci. Technol. 12:1200-1205.

Shaw, D.G., and J.N. Wiggs. 1979. Hydrocarbons in Alaskan intertidal algae. Phytochem. 18:2025-2027.

Shaw, D.G., and J.N. Wiggs. 1980. Hydrocarbons in the intertidal environment of Kachemak Bay, Alaska. Mar. Pollut. Bull. 11:297-300.

Sherman, K., J.B. Colton, R.L. Dryfoos, and B.S. Kinnear. 1973. Oil and plastic contamination and fish larvae in surface waters of the northeast Atlantic. MARMAP Operational Test Survey Report: July-August 1972, January-March 1973.

Sherman, K., J.B. Colton, R.L. Dryfoos, K.D. Knapp, and B.S. Kinnear. 1974. Distribution of tar balls and nueston sampling in the Gulf Stream system, pp. 83-84. In Marine Pollution Monitoring (Petroleum). Special Publication 409. National Bureau of Standards, Gaithersburg, Md.

Shokes, R., P. Mankiewcz, R. Sims, M. Guttman, R. Jordan, J. Nemmers, and J. Payne. 1979a. Geochemical baseline study of the Texoma offshore brine disposal site: Big Hill, fall 1977-Sept. 1978. Report SAI-012-79-834-LJ.

Shokes, R., P. Mankiewicz, R. Sims, M. Guttman, R. Jordan, J. Nemmers, and J. Payne. 1979b. Geochemical baseline study of the Texoma offshore brine disposal sites: West Hackberry, fall 1977-spring 1978. Report SAI-012-79-835-LJ.

Singer, S.C., and R.F. Lee. 1977. Mixed function oxygenase activity in blue crab, Callinectes sapidus: tissue distribution and correlation with changes during molting and development. Biol. Bull. 153:377-386.

Singer, S.C., P.E. March, F. Gonsoulin, and R.F. Lee, 1979. Mixed function oxygenase activity in the blue crab, Callinectes sapidus: characterization of enzyme activity from stomach tissue. Comp. Biochem. Physiol. 65C:129.

Sleeter, T.D., B.F. Morris, and J.N. Butler. 1974. Quantitative sampling of pelagic tar in the North Atlantic. 1973. Deep Sea Res. 21:773-775.

Sleeter, T.D., B.F. Morris, and J.N. Butler. 1976. Pelagic tar in the Caribbean and Equatorial Atlantic. 1974. Deep Sea Res. 23:467-474.

Sleeter, T.D., J.N. Butler, and J.E. Barbash. 1979. Hydrocarbons in sediments from the edge of the Bermuda Platform, pp. 615-620. In Proceedings, 1979 Oil Spill Conference. American Petroleum Institute, Washington, D.C.

Smith, G.B. 1976. Pelagic tar in the Norwegian coastal current. Mar. Pollut. Bull. 7:70-72.

Smith, G.L. 1977. Determination of the leeway of oil slicks, p. 351. In D.A. Wolfe, ed. Fate and Effects of Petroleum Hydrocarbons in Marine Ecosystems and Organisms. Pergamon, New York.

Solbakken, J.E., K.H. Palmork, T. Neppelberg, and R.R. Scheline. 1980. Urinary and biliary metabolites of phenanthrene in the coalfish (Pollachius virens). Acta Pharmacol. Toxicol. 46:127.

Spaulding, M.L., P. Cornillon, and M. Reed. 1978. Modelling oil spill fates and interactions with fisheries, pp. 29-34. In D. Mackay and S. Paterson eds. Oil Spill Modelling: Proceedings of a Workshop. Publication EE-12. University of Toronto, Institute for Environmental Studies.

Sprague, J.B., J.H. Vandermeulen, and P.G. Wells eds. 1981. Oil and Dispersants in Canadian Seas--Research Appraisal and Recommendations. Environment Canada. Ottawa, Ontario.

Stainken, D. 1977. The accumulation and depuration of No. 2 fuel oil by the soft shell clam (Mya arenaria) L., pp. 313-322. In D.A. Wolfe, ed. Fate and Effects of Petroleum Hydrocarbons in Marine Organisms and Ecosystems. Pergamon, New York.

Statham, C.N., M.J. Melacon, Jr., and J.J. Lech, 1976. Bioconcentration of xenobiotics in trout bile: a proposed monitoring aid for some water-borne chemicals. Science 193:680-681.

Statham, C.N., C.R. Elcombe, S.P. Szyjka, and J.J. Lech. 1978. Effect of polycyclic aromatic hydrocarbons on aepatic microsomal enzymes and disposition of methylnaphthalene in rainbow trout in vivo. Xenobiotica 8:65-71.

Stegeman, J.J. 1977. Fate and effects of oil in marine animals. Oceanus 20:59.

Stegeman, J.J. 1978. Influence of environmental contamination on cytochrome P-450 mixed-function oxygenases in fish: implications for recovery in the Wild Harbor Marsh. J. Fish. Res. Board Can. 35:668.

Stegeman, J.J. 1979. Temperature influence on basal activity and induction of mixed-function oxygenase activity in Fundulus heteroclitus. J. Fish. Res. Board Can. 36:1400-1405.

Stegeman, J.J. 1981. Polynuclear aromatic hydrocarbons and their metabolism in the marine environment, pp. 1-60. In H.V. Gelboin and P.O.P. Ts'O eds. Polycyclic Hydrocarbons and Cancer. Vol. 3. Academic Press, New York.

Stegeman, J.J., and M. Chevion. 1980. Sex differences in cytochrome P-450 and mixed-function oxygenase activity in gonadally mature trout. Biochem. Pharmacol. 28:554-559.

Stegeman, J.J., and H.B. Kaplan. 1981. Mixed-function oxygenase activity and benzo(a)pyrene metabolism in the barnacle Babanus eburneus (Crusteacea, Cirripedia). Comp. Biochem. Physiol. 68C:55.

Stegeman, J.J., and J.H. Teal. 1973. Accumulation, release and retention of petroleum hydrocarbons by the oyster, Crassostrea virginica. Mar. Biol. 22:37-44.

Stegeman, J.J., R.L. Binder, and A. Orren. 1979. Hepatic and extrahepatic microsomal electron transport components and mixed-function oxygenases in the marine fish Stenotomus versicolor. Biochem. Pharmacol. 28:3431-3439.

Stegeman, J.J., A.V. Klotz, B.R. Woodin, and A.M. Pajor. 1981. Induction of hepatic cytochrome P-450 in fish and the indication of environmental induction in scup (Stenotomus chrysops). Aquat. Toxicol. 1:(in press).

Stegeman, J.J., T.R. Shopek, and W.G. Whilly. 1982. Bioactivation of polynuclear aromatic hydrocarbons to cytotoxic and mutagenic products by marines fish, pp. 201-211. In N. Richards and B.L. Jackson, eds. Carcinogenic Polynuclear Aromatic Hydrocarbons in the Marine Environment. Rep. EPA 600-/9-82-003. U.S. Environmental Protection Agency, Washington, D.C.

Sutcliffe, W. H., E.R. Baylor, and D.W. Menzel. 1963. Sea surface chemistry and Langmuir circulation. Deep Sea Res. 10:233-243.

Swinnerton, J.W., and R.A. Lamontagne. 1974. Oceanic distribution of low-molecular-weight hydrocarbons: baseline measurement. Environ. Sci. Technol. 8:657-663.

Tait, R.V., and R.S. De Santo. 1972. Elements of marine ecology. Springer, New York.

Teal, J.M. 1976. Hydrocarbons uptake by deep-sea benthos, pp 358-371. In Proceedings of Symposium on Sources, Effects and Sinks of Hydrocarbon in the Aquatic Environment. American University, Washington, D.C.

Teal, J.M., K. Burns, and J. Farrington. 1978. Analyses of aromatic hydrocarbons in intertidal sediments resulting from two spills of No. 2 fuel oil in Buzzards Bay, Massachusetts. J. Fish. Res. Board Can. 35:510-520.

Thomas, R.E., and S.D. Rice. 1981. Excretion of aromatic hydrocarbons and their metabolites by freshwater and saltwater Dolly Varden char, pp. 425-448. In F.J. Vernberg, F.P. Thurberg, A. Calabrese, and W.B. Vernberg, eds. Biological Monitoring of Marine Pollutants. Academic Press, New York.

Thompson, S., and G. Eglinton. 1978. Composition and sources of pollutant hydrocarbons in the Severn Estuary. Mar. Pollut. Bull. 9:133-136.

Traxler, R.W., and J.M. Bernard. 1969. The utilization of n-alkanes by Pseudomonas aeruginosa under conditions of anaerobiosis. Int. Biodeterior. Bull. 5:21-25.

Traxler, R.W., P.R. Proteau, and R.N. Traxler. 1965. Action of microorganisms on bituminous materials. I. Effect of bacteria on asphalt viscosit. Appl. Microbiol. 13:838-841.

Tripp, B.W., J.W. Farrington, and J.M. Teal. 1981. Unburned coal as a source of hydrocarbons in surface sediments. Mar. Pollut. Bull. 12:122-126.

Trudgill, P.W. 1978. Microbial degradation of alicyclic hydrocarbons, pp. 47-84. In J.R. Watkinson ed. Developments in Biodegradation of Hydrocarbons. Vol. 1. Applied Science Publishers, London.

Twardus, E.M. 1980. A Study to Evaluate the Combustibility and Other Physical and Chemical Properties of Aged Oils and Emulsions. R & D Division, Environmental Emergency Branch, Environmental Impact Control Directorate, Environmental Protection Service, Environmental Canada, Ottawa, Ontario.

Van der Linden, A.C. 1978. Degradation of oil in the marine environment, pp. 165-200. In J.R. Watkinson, ed. Developments in Biodegradation of Hydrocarbons. Vol. 1. Applied Science Publishers, London.

Vandermeulen, J.H. 1981. Scientific studies during the Kurdistan tanker incident: Proceedings of workshop. Report Bl-R-80-3. Bedford Institute of Oceanography, Dartmouth, N.S. Canada.

Vandermeulen, J.H. 1982. Some conclusions regarding long-term biological efffects of some major oil spills. Phil. Trans. R. Soc. London, Ser. B 297:335-351.

Vandermeulen, J.H., and D.C. Gordon, Jr. 1976. Reentry of 5-year-old stranded Bunker C fuel oil from a low-energy beach into the water, sediments and biota of Chedabucto Bay, Nova Scotia. J. Fish. Res. Board Can. 33:2002-2010.

Vandermeulen, J.H., B.F.N. Long, and T.P. Ahern. 1981. Bioavailability of stranded Amoco Cadiz oil as a function of environmental self-cleaning, pp. 585-598. In G. Conan, ed. Amoco Cadiz: Fates and Effects of the Oil Spill. Centre National pour l'Exploit. Oceans Brest, France.

van Dolah, R.F., V.G. Burrell, Jr., and S.B. West. 1980. The distribution of pelagic tars and plastics in the South Atlantic Bight. Mar. Pollut. Bull. 2:352-356.

Van Vleet, E.S., and J.G. Quinn. 1977. Input and fate of petroleum hydrocarbons entering the Providence River and upper Narragansett Bay from waste water effluents. Environ. Sci. and Technol. 11:1086-1092.

Van Vleet, E.S., and J.G. Quinn. 1978. Contribution of chronic petroleum inputs to Narragansett Bay and Rhode Island Sound sediments. J. Fish. Res. Board. Can. 35:536-543.

Van Vleet, E.S., W.M. Sackett, F.F. Weber, and S.B. Reinhardt. 1982a. Spatial and temporal variation of crude oil residues in the eastern Gulf of Mexico. In Advances in Organic Geochemistry, 1981. Pergamon, New York (in press).

Van Vleet, E.S., W.M. Sackett, F.F. Weber, and S.B. Reinhardt. 1982b. Input of pelagic tar into the northwest Atlantic from the Gulf loop current: chemical characterization and its relationship to Ixtoc I oil Con. Jr. Offshore Aquat. Sci. (in press).

Varanasi, U., and D.J. Gmur. 1981. Hydrocarbons and metabolites in English sole (Parophrys vetulus) exposed simultaneously to (^{3}H)benzo(a)pyrene and (^{14}C) naphthalene in oil-contaminated sediment. Aquat. Toxicol. 1:49-68.

Varanasi, U., D.J. Gmur, and P.A. Treseler. 1979. Influence of time and mode of exposure on biotransformation of naphthalene by juvenile starry flounder (Platichthys stellatus) and rock sole (Lepidopsetta bilineata). Arch. Environ. Contam. Toxicol. 8:673-692.

Varanasi, U., D.J. Gmur, and W.L. Reichert. 1981. Effect of environmental temperature on naphthalene metabolism by juvenile starry flounder (Platichthys stellatus). Arch. Environ. Contam. Toxicol. 10:203-214.

Veith, D.G., D.L. Defoe, and B.V. Bergstedt. 1979. Measuring and estimating the bioconcentration factor of chemicals in fish. J. Fish. Res. Board Can. 36:1040.

Venkatesan, M.I., S. Brenner, E. Ruth, J. Bonilla, and I.R. Kaplan. 1980. Hydrocarbons in age-dated sediment cores from two basins in the Southern California Bight. Geochim. Cosmochim. Acta 44:789-802.

Wade, T.L., and J.G. Quinn. 1975. Hydrocarbons in the Sargasso Sea surface microlayer. Mar. Pollut. Bull. 6:54-57.

Wakeham, S.G., and J.W. Farrington. Contemporary aquatic sediments, pp. 3-32. In R.A. Baker, ed. Contaminants and Sediments. Vol. 1. Ann Arbor Science Publishers. Ann Arbor, Mich.

Walker J.D., and R.R. Colwell. 1976a. Oil, chlorinated biphenyl, mercury and microrganism interactions. Environ. Sci. Technol. 10:1145-1147.

Walker, J.D., and R.R. Colwell. 1976b. Long-chain n-alkanes occurring during microbial degradation of petroleum. Can. J. Microbiol. 22:886-891.

Walker, J.D., R.R. Colwell, and L. Petrakis. 1975. Degradation of petroleum by an alga, Prototheca zopfii. Appl. Microbiol. 30:79-81.

Walker, J.D., R.R. Colwell, and L. Petrakis. 1976. Biodegradation rates of components of petroleum. Can. J. Microbiol. 22:1209-1213.

Ward, D.M., and T.D. Brock. 1978. Anaerobic metabolism of hexadecane in marine sediments. Geomicrobiol. J. 1:1-9.

Ward, D.M., R.M. Atlas, P.D. Boehm, and J.A. Calder. 1980. Microbial biodegradation and the chemical evolution of Amoco Cadiz oil pollutants. Ambio 9:277-283.

Wheeler, R.B. 1978. The fate of petroleum in the marine environment. Special Report. Exxon Production Research Company, Houston Tex.

Whittle, K.J., J. Murray, P.R. Mackie, R. Hardy, and J. Farmer. 1977. Fate of hydrocarbons in fish. Rapp. P.-v. Reun. Cons. Int. Explor. Mer 171:139-142.

Wiesenburg, D.A., G. Bodennec, and J.M. Brooks. 1981a. Volatile hydrocarbons around a production platform in the northwest Gulf of Mexico. Bull. Environ. Contam. Toxicol. 27:167-174.

Wiesenberg, D.A., J.M. Brooks, and R.A. Burke, Jr. 1981b. Gaseous hydrocarbons around an active offshore gas and oil field. Environ. Sci. Technol. 16:278-282.

Wong, C.S., D. MacDonald, and R.D. Bellegay. 1974a. Distribution of tar and other particulate pollutants along the Beaufort Sea coast, Victoria, B.C. 1974 Interim Report. Environment Canada, Ocean Chemicals Division, Ocean Aquatic Affairs, Pacific Regulations, Ottawa, Ontario.

Wong, C.S., D.R. Green, and W.J. Cretney. 1974b. Quantitative tar and plastic waste distribution in the Pacific Ocean. Nature 247:30-32.

Wong, C.S., D.R. Green, and W.J. Cretney. 1976. Distribution and source of tar on the Pacific Ocean. Mar. Pollut. Bull. 7:102-106.

Wong, W.C. 1976. Uptake and retention of Kuwait crude oil and its effect on oxygen uptake by the soft-shell clam, Mya arenaria. J. Fish. Res. Board Can. 33:2774-2780.

Wood, A.W., W. Levin, A.Y.H. Lu, H. Yagi, O. Hernandez, D.M. Jerma, and A.H. Cooney. 1976. Metabolism of benzo(a)pyrene derivatives to mutagenic products by highly purified hepatic microsomal enzymes. J. Biol. Chem. 251:4882.

Wu, J., and L.K. Wong. 1981. Microbial transformations of 7,12-dimethyl-benzo(a)anthracene. Appl. Environ. Microbiol. 41:843-845.

Yamada, K., Y. Monoda, K. Komada, S. Nakatani, and T. Akasaki. 1968. Microbial conversion of petro-sulfur compounds. I. Isolation and identification of dibenzothiophene-utilizing bacteria. Agric. Biol. Chem. 32:840-845.

Yang, W.C., and H. Wang. 1977. Modeling of oil evaporation in aqueous environment. Water Res. 11:879-887.

Zafiriou, O.C. 1973. Petroleum hydrocarbons in Narragansett Bay. II. Chemical and isotopic analysis. Estuarine Coastal Mar. Sci. 1:81-87.

Zafiriou, O.C. 1977. Marine organic photochemistry previewed. Mar. Chem. 5:497.

Zajic, J.E., B. Supplisson, and B. Volesky. 1974. Bacterial degradation and emulsification of No. 6 fuel oil. Environ. Sci. Technol. 8:664-668.

Zepp, R.G. 1978. Quantum yields for reaction of pollutants in dilute aqueous solution. Environ. Sci. Technol. 12:327-329.

Zepp, R.G., and D.M. Cline. 1977. Rates of direct photolysis in the aquatic environment. Environ. Sci. Technol. 11:359-366.

Zepp, R.G., and G.L. Baughman. 1978. Production of photochemical transformation of pollutants in the aquatic environment. In O. Hutzinger, I.H. van Lelyveld, and B.C.J. Zoeteman, eds. Aquatic Pollutants--Transformation of Biological Effects. Pergamon, New York.

Zika, R.G. 1980. Marine Organic Photochemistry, Chap. 10. In E.K. Duursma and R. Dawson eds. Marine Organic Chemistry. Elsevier, New York.

Zitko, V. 1971. Determination of residual fuel oil contamination of aquatic animals. Bull. Environ. Contam. Toxicol. 5:559-563.

ZoBell, C.E. 1969. Microbial modification of crude oil in the sea, pp. 317-326. In Proceedings, Joint Conference on Prevention and Control of Oil Spills. American Petroleum Institute, Washington, D.C.

ZoBell, C.E. 1973. Bacterial degradation of mineral oils at low temperatures, pp. 153-161. In D.G. Ahearn and S.P. Meyers, eds. The Microbial Degradation of Oil Pollutants. Publication LSU-SG-73-01. Center for Wetland Resources, Louisiana State University, Baton Rouge.

ZoBell, C.E., and J. Agosti. 1972. Bacterial oxidation of mineral oils at sub-zero Celsius. Abstracts of the 72nd Annual Meeting. Abstract Ell. American Society of Microbiology, Philadelphia, Pa., April 23-28, 1972.

Zsolnay, A. 1972. Preliminary study of the dissolved hydrocarbons and hydrocarbons on particulate material in the Gotland Deep of the Baltic. Kieler Meeresforsch. 27:129-134.

Zsolnay, A. 1973a. The relative distribution of nonaromatic hydrocarbons in the Baltic in September 1971. Mar. Chem. 1:127-136.

Zsolnay, A. 1974. Hydrocarbon content and chlorophyll correlation in the waters between Nova Scotia and the Gulf Stream, pp. 255-256. In Marine Pollution Monitoring (Petroleum). Special Publication 409. National Bureau of Standards, Gaitherburg, Md.

Zsolnay, A. 1977a. Inventory of nonvolatile fatty acids and hydrocarbons in the oceans. Mar. Chem. 5:465-475.

Zsolnay, A. 1977b. Hydrocarbon content and chlorophyll correlations in water between Nova Scotia and the Gulf Stream. Deep Sea Res. 24:199-207.

Zsolnay, A. 1979. Hydrocarbons in the Mediterranean Sea, 1974-1975. Mar. Chem. 7:343-352.

Zsolnay, A., B.F. Morris, and J.N. Butler. 1978. Relationship between aromatic hydrocarbons and pelagic tar in the Mediterranean Sea, 1974-1975. Environ. Conserv. 5:295-297.

Zürcher, F., and M. Thüer. 1978. Rapid weathering processes of fuel oil in natural waters: analyses and interpretations. Environ. Sci. Technol. 12:838-843.

5
Effects

INTRODUCTION

A vast amount of data and literature has accumulated since the 1975 NRC report. Much of this accumulation has been brought together at conferences (e.g., American Petroleum Institute, 1975, 1977, 1979, 1981; Wolfe, 1977), at specialized symposia (American Institute of Biological Sciences, 1976, 1978; Fisheries Research Board, 1978; Conan et al., 1981), and in two major reviews (Malins, 1977; Sprague et al., 1981). This has come about in response to increased funding since the early 1970s, partly due to increasing human concerns over oil in the marine environment, and partly as a natural outcome of continued spills and accidental discharges. One interesting and encouraging development has been a noticeable change in research emphasis, from descriptive to more process-oriented research, as in studies of physiological impact and ecological change (Table 5-1). During the early days of oil pollution research following the Torrey Canyon accident, most research was aimed at quantifying toxicity thresholds. At the same time there was little scientific consistency, in that researchers developed their own exposure methodology and analytical preferences. As a result, intercomparison of laboratory data was difficult.

This began to change in the mid-1970s with a redirection of research interest toward understanding the mechanisms of hydrocarbon toxicity and the sites of toxic action. This effort was paralleled by concerted efforts of various workers to standardize analytical methods, using certain reference oils set aside by the American Petroleum Institute (API). As a result, more meaning and comparability have come into the field of toxic effects of petroleum, and a type of data is being produced with which many members of the scientific community can agree upon (Rice et al., 1977). This change in emphasis in recent years represents a significant advancement since the 1975 NRC report.

The field of oil pollution impact presents unusual and major difficulties to the researcher in that at virtually every turn of study new techniques and analytical and sampling methods have to be devised. This is due to the newness of this research area, having come into its own only since 1967 with the breakup of the Torrey Canyon. Prior to that event most of the research interest with respect to petroleum concerned its physicochemical aspects, and the analytical methods and

TABLE 5-1 Emphasis of Oil-Pollution-Related Study Reports for the
Temperate and Northern Marine Environment Between 1967 and 1977

Emphasis	Pre-1974	Post-1974
1. Oil--physical-chemical changes, fate and distribution, environmental concentrations	35% (83)	33% (107)
2. Gross biological effects: mortality, toxicity	43% (100)	31% (99)
3. Physiological, developmental, and ecological change	6% (15)	22% (70)
4. Microbiology: hydrocarbon-utilizing bacteria	16% (37)	14% (44)

NOTE: Numbers in parentheses denote number of studies.

SOURCE: Environmental Protection Agency (1977).

expertise reflected these interests. However, with the Torrey Canyon a
new scientific discipline was required, including an understanding of
the behavior of oil in water, sediment, and even in tissues, and requiring analytical methods capable of resolving petrogenic compounds in
unfamiliar environmental samples. This has called for a much more
interdisciplinary approach, with constant and new exchange of expertise
and ideas among the different disciplines. Scientists in this field
are often competent in several areas, combining organic chemistry with
biochemistry or biology, and often a more than passing acquaintance
with microbiology or geology. Environmental teams of scientists have
developed, working with integrated effort.

This is not to say that the problem of understanding oil pollution
in the marine environment is now well in hand. Major inroads have been
made in analytical capabilities and in understanding petroleum effects
on two levels--physiological and ecological. However, generally there
is a good appreciation of oil effects in temperate and northern
temperate waters. At the same time the area of subcellular effects has
received less attention.

Ecological studies have been done primarily in the field, using
spills of opportunity and, more recently, large field enclosures
(mesocosms) with known dosages of oil. At spills of opportunity,
studies have been done mainly in the soft sediment areas such as salt
marshes and shallow embayments. Such areas have been documented as
sites where spilled oil will persist for long periods (years to

decades) (see Chapter 3) and have provided the basis for most of what is known about how oil affects on marine populations and communities. The mesocosm studies have provided experimental evidence of similar perturbations occurring in planktonic and benthic communities.

Investigation into the physiological (e.g., photosynthesis, respiration, growth, neurotransmission) effects of oil is largely through laboratory investigations, although some fundamental work has been done in the field. One reason for the laboratory emphasis is the need to control experimental conditions. The coverage of this level of investigation has been uneven. As in mammalian physiology, there are preferred invertebrate and algal species. Certain bivalves (Mya arenaria, Mytilus sp., Macoma spp.) and crustaceans (Cancer spp., Uca pugnax, Crangon spp., Penaeus aztecus), by virtue of their accessibility and ease of culture, are far easier to work with than benthic or pelagic organisms available only seasonally and/or by dredging or trawling. The same thing holds for the marine algae. It seems preferable, however, to attempt to understand in good detail the toxicology of petroleum in two or three well-studied representative organisms, rather than to attempt to establish simple toxic tolerance levels across all the phyla. (Although it should be noted that some of these may not be the most vulnerable.)

Study at the subcellular level has received far less attention, despite recent concerns over certain hydrocarbons interacting with cellular macromolecules such as nucleic acids.

We discuss primarily the impact of petroleum hydrocarbons, leaving the possible impact of other contaminants or compounds—either contained in oil or in some manner by-products of petroleum- or gas-related activities—to other more specialized discussions. Such contaminants would include, for example, trace metals, chemical dispersants, and drilling muds. Again, the available literature on petroleum impact alone is so vast that to include detailed discussions on these other materials, beyond merely mentioning them, would inevitably lead to an unmanageable exercise.

The approach taken in this chapter is to discuss and review the impact of petroleum on marine biota and communities, by proceeding from one level to the next—from effects on processes (cellular), through a discussion of effects on the marine foodchain (organismic), to the effects on communities (ecosystem). Inevitably this leads to some repetition or duplication, but this approach makes the most sense in unravelling and describing an extremely complex problem, involving a complex pollutant and the complexities of marine life. We hope that the index to this report will aid the reader in finding his or her way through it.

Inevitably in a task such as this, some studies and reports will not have been referenced in the writing of this chapter because of the limited space available. Throughout we tried to refer to those studies that illustrated a particular point or aspect of petroleum pollution most aptly or most concisely. In other instances we referenced those studies that would lead the interested reader in turn to other studies.

In an appendix to this chapter we have included a discussion of some well-known oil spills and oil seep problems, largely to add some real dimensions to the at times very detailed discussions in the main body of the chapter. The examples were selected for their general appropriateness and because they represented in each instance a particular spill type under certain conditions.

Finally, any report tries to be up to date in its coverage, but inevitably there has to be a cut-off date. For various reasons, editorial and technical, the review process for this chapter was very lengthy. We have tried to maintain as current a reference list as possible, but have not been able to go much beyond 1982-1983. We regret therefore having missed much excellent new literature and newer findings published in the last year.

Toxicity

In its most general sense, toxicity can be defined as the imparting of a deleterious effect, whether lethal or sublethal, to an organism, population, or community. The toxic effect can result in a permanent perturbation or change, for example, the crooked-back syndrome in larval fish, stunted growth, deformed shell formation in mollusks, and changed population patterns. However, not all effects are disruptive, and there exist adaptive mechanisms, both at the cellular level (e.g., detoxifying enzyme systems) and at the population and community levels (Capuzzo, 1981).

Toxic effects from petroleum exposure vary widely and for reasons that are not well understood. Certainly these have to do with the complexity of its chemical composition, with different products or even crude oils differing markedly in their chemical makeup. Another factor is the variability in sensitivity to oil found among marine organisms, differing not only with the species (Figure 5-1) but even for life-cycle stages (Figure 5-2). While it is generally true that the younger stages of organisms are more sensitive to petroleum hydrocarbons, there are exceptions. Unfortunately, not many studies have compared the sensitivity of organisms at various life stages under identical experimental conditions for any one species.

While the absolute toxicity of petroleum hydrocarbons appears to be greater for the higher-molecular-weight compounds (for example, 3- and 4-ring aromatics), most of the toxic effect of petroleum in water is thought to be due to the lower-molecular-weight (C_{12}-C_{24}) n-paraffin compounds and to the monoaromatic fraction, for the simple reason that these compounds are the most water-soluble (Chapter 3, Table 3-4). From examinations of the concentrations present in water-soluble fractions (WSF), it is clear that the contribution of compounds higher in molecular weight than the alkylnaphthalenes is very small and may be insignificant in terms of producing acute toxicity.

Bioassay tests have been used to a considerable extent to determine the toxicities of various crude oils and of refined products. Most of these tests have used mortality as the index of toxicity, expressed for example, as LC_{50} (the lethal concentration yielding 50% mortality over pre-determined exposure time, for example, 24, 48, or 96 hours). In practice, however, their usefulness as a research technique is

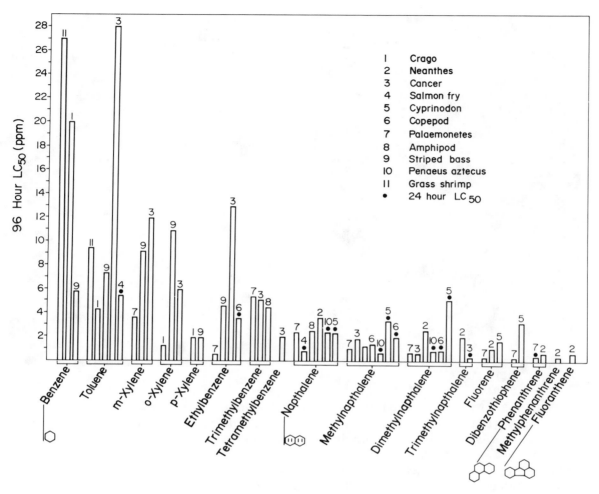

FIGURE 5-1 Acute toxicity (24- and 96-hour LC$_{50}$ static tests) of some aromatic hydrocarbons for selected marine macroinvertebrates and fish.

SOURCES: Caldwell et al. (1977), Benville and Korn (1977), Neff et al. (1976), R.E. Thomas and Rice (1979), Young (1977), Rossi and Neff (1978), Ott et al. (1978), and W.Y. Lee and Nicol (1978).

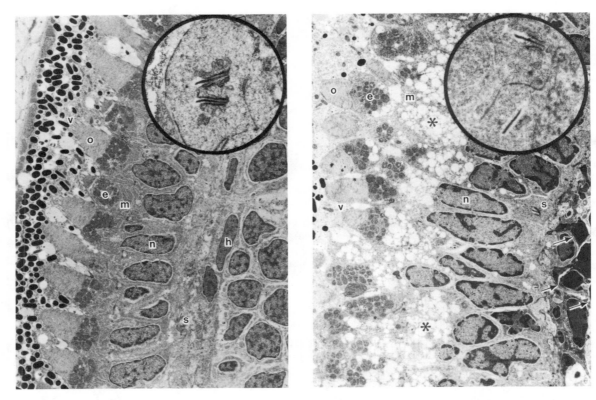

FIGURE 5-2 Effects of oiling on fine structure of surf smelt embryo
retinas. (Left) Retinal cells of an unoiled embryo (x5000). Inset is
enlargement of a synaptic junction (x18,000). (Right) Retinal cells of
an embryo exposed to 113 ppb Cook Inlet crude oil (x5000). Vesiculation
is evident in the myoid regions (asterisk) of the receptor cells. Note
also necrotic neurons (arrows). Synaptic junction (viz., inset,
x18,000) appears normal. Ellipsoid region (e) of inner segment of
receptor cell, nucleus (m) of receptor cell, outer segment (o) of
receptor cell, and synaptic (s) junctional complex are indicated on
figure. (Photo by J. Hawkes.)

limited in that they provide no data except on mortality. Toxicity
tests are subject to several variables such as complex mixture of the
oil, test parameters, and various biological factors such as age, sex,
and contamination history of the organism. For these reasons they are
somewhat imprecise measures of toxicity, and many researchers feel that
their output, the LC_{50}, has little relevance to what may happen to an
organism as a result of a spill. By its very nature the LC_{50} gives
no indication of sublethal toxic problems that the organism may be
experiencing, and gives no measure of any long term impacts that may be
occurring, measuring only the death of the organism. Instead, acute
lethal bioassays serve best as tests to compare the relative toxicities
of complex, unknown toxicants, or the comparison of relative
sensitivities of species or life stages.

Laboratory Versus Field Studies

Most studies of effects have been developed in the laboratory for the simple reason that field studies often depend on spills of opportunity, which are highly unpredictable. Also, field studies are often expensive and difficult to carry out. On the other hand, laboratory studies have frequently been criticized because experimental conditions do not simulate field conditions. Also, concentrations of oil or hydrocarbons frequently exceed those encountered in the field.

Both criticisms are probably correct. However, in recent years attempts have been made to bring laboratory conditions closer to field conditions by simulating the hydrocarbon composition and concentrations more closely through the use of flow-through systems and by careful management of the test organisms chosen. Most promising in this respect have been various studies carried out in "mesocosms," large enclosures that allow control of some environmental variables under near-open ocean conditions (for example, Marine Ecosystems Research Laboratory (MERL), CEPEX, Loch Ewe, viz., Table 5-5). In this respect, natural oil seeps also offer certain opportunities to study the impact of petroleum under more open-ocean conditions.

As for the criticism of high experimental dosages, there are situations where seemingly high concentrations of oil or hydrocarbon are warranted, for example, in the initial establishment or detection of certain toxic effects and in the analysis of metabolic pathways. High initial concentrations of toxicants are frequently necessary to establish a toxic effect that otherwise might be indistinguishable from the background "noise" or masked by other changes. Using high concentrations allows the experimenter to better define the toxic effect or response. Again, the detection and identification of primary or secondary metabolites, or of short-lived intermediates, often require unusually large doses of toxicant. In these instances the object is not so much to determine a toxic effect, as it is to better understand certain aspects of hydrocarbon metabolism for which low dosages would not elicit a measurable response. However, care has to be taken in working with high dosages and in interpreting results because of the possibility that extraordinary metabolic pathways may be expressed.

In the end, both laboratory and field study have merit. Although field studies are fraught with uncontrolled and interfering factors, they nonetheless are the ultimate testing ground. For that reason, spill sites should be visited, and revisited, whenever possible. On the other hand, laboratory studies support field studies by providing the opportunity to investigate an effect in detail and to study its underlying mechanism.

Factors Affecting Impact of Oil

When an oil spill occurs, many factors determine whether that spill will cause heavy, long lasting biological damage; comparatively little or no damage; or some intermediate degree of damage. An example of the variability that exists among the effects of oil spills on the marine biota is outlined by C.T. Mitchell et al. (1970) in their description of the widely different effects resulting from the Tampico Maru and the Santa Barbara oil spills. Nonetheless, there are some patterns beginning to emerge that are useful in identifying those physical and biological features that can influence the ultimate impact of a spill or chronic pollution (e.g., Michael et al., 1978; Gundlach and Hayes, 1978; Owens and Robilliard, 1981; Vandermeulen, 1977, 1982).

Geographic Location

In many reports, organisms from any one geographic location are apparently no different from any other location in terms of their vulnerability to petroleum hydrocarbons. While there are of course species-specific genetic differences, arctic fish or invertebrates do not appear to differ physiologically from similar organisms at lower or tropical latitudes in terms of lethal toxic concentration thresholds or toxic responses.

However, there are a number of physical features related to geographic location--mainly temperature and ice cover, together with differences in community diversity, which is latitude dependent--that will influence both impact and biological recovery. Temperature, for example, plays a significant role in the solubility of hydrocarbons in the water column and in the rate of their degradation through microbial activity.

Similarly, community diversity at different latitudes (low diversity in polar regions, high diversity in tropical environments) can lead to differences in both times and patterns of biological recovery following an oil impact.

Oil Dosage and Impact Area

If the spill occurs in a small, confined area so that the oil is unable to escape, damage will be greater, almost without exception, for a given volume and type of oil spilled than if that same volume were released in a relatively open area. For example, at the Arrow spill site in Chedabucto Bay, Nova Scotia, about 2.5 million gallons of Bunker C fuel were spilled in an embayment, whereas the Argo Merchant spilled about 7.7 million gallons of No. 6 fuel oil into the open ocean off Nantucket Island, Massachusetts. Although a considerable amount of Arrow spill eventually was swept out to sea, the confined nature of the oil during the first days resulted in nearly uniform oiling of the entire bay coastline and in considerable damage to the associated fauna and flora. However, this generalization is not inflexible. The spill of the supertanker Amoco Cadiz occurred offshore, but prevailing winds were such that the oil was kept nearshore of North Brittany for several weeks and continuously driven onto shore. Similarly, there are differences resulting from the manner of the spillage--whether low level but chronic or consisting of a sudden accidental release. The former is covered in greater detail elsewhere (for example, see Chronic Oiling section), but generally the impact from such chronic releases differs both in severity and in kind from accidental spills, where the spilled oil will eventually disappear with time due to physical-chemical processes and microbial and other biological degradation. On the other hand, in the case of chronic releases the spilled oil becomes a continuing irritant or toxicant to which the community must adjust, for example, in selection for hydrocarbon-utilizing species.

Oceanographic Conditions

Currents, sea state, coastal topography, and tidal action all combine to influence the impact of a given spill. Currents and wave action--in open water or open bays--act to break up the oil into smaller slicks, and also act to disperse some of the oil into the water column. In areas of large tidal ranges the oil can become distributed over a broad range of the intertidal zone, and can be deposited far above the high tide mark by extreme "spring" tides coinciding with high winds and strong tidal flow.

Coastal topography plays a large role in the residual impact of a spill, with low energy environments (salt marshes, lagoons, estuaries, embayments) acting as long term hydrocarbon "sinks." Impact on biota in such systems is usually long lasting.

Meteorological Conditions

Normally, storms increase wave action and wind speed and thereby aid in evaporation of the lower-molecular-weight, more volatile toxic components. On occasion, however, wave action may intensify the problems, as apparently occurred at the Florida No. 2 fuel oil spill near West Falmouth. Soon after this spill, the surf drove the oil ashore into the sediments and the surrounding marshland (Sanders, 1978). The oiled marshland and sediments then became a long term reservoir of oil with persistence in some areas to this day. Also storm-induced resuspension of subtidal sediments probably brought these sediments into contact with oil from more intertidal areas. Similar events occurred following the breakup of the Amoco Cadiz, where the winter storms drove the oil deep inland up the nearby estuarine tidal rivers (Hess, 1978).

Season

Season is particularly important in terms of the biota that might lie in the path of a spill or in the vicinity of a chronic oiling situation. For example, if a spill occurs in an area where seabirds are feeding or nesting, bird mortality might be in the thousands; at some other time of the year the mortality might be much lower. Similarly the coincidence of a spill with fish spawning events or hatching and development of larval fish migration might result in higher than normal larval mortalities.

Seasonal changes in the physical parameters of the marine environment might also influence the potential impact of a spill, as for example, seasonally timed changes in local circulation patterns that might lead to local containment of slicks.

Oil Type

Oil type determines both the short term and the long term impact. Immediate impact can be very high from such highly toxic oils as diesel and jet fuel. However, these dissipate readily and leave relatively little residue, unlike the crude and Bunker oils, which can persist in certain sediments for up to several decades. This aspect is not as simple and clear cut as it seems, however. Traces of No. 2 fuel oil, a relatively volatile product, still persist in sediments of Falmouth, Massachusetts, 13 years after the spill of the <u>Florida</u>.

Oil Metabolites and Photochemical Reaction Products

This subject deserves separate mention as it was not raised in any detail in the 1975 NRC report, but has become of interest in more recent years, following observations that some petroleum metabolites or intermediate products can be quite toxic and may even have mutagenic properties.

Oil metabolites can be formed from the parent oil by biological conversion of compounds taken up by marine biota (including bacteria), and by photochemical processes. The relatively few data available on either method come mainly from laboratory studies. The formation of compounds by photochemical processes has been addressed earlier (Chapter 4). In general, irradiated samples of petroleum or water-soluble preparations appear to be more toxic than the parent compounds. For example, Lacaze and Villedon de Naide (1976) cite field studies suggesting that the irradiated water-soluble fractions (WSF) of Kuwait crude oil were 3 times as toxic, depressing C-fixation, as nonirradiated WSF after 40- and 64-hour exposure of the alga <u>Phaeodactylum</u> <u>cornutum</u>. In similar studies, Scheier and Gominger (1976) examined the toxic effects of irradiated versus nonirradiated No. 2 fuel oil, using a Sylvania sunlamp, and compared the results with solar-irradiated WSF. They observed that (1) sunlight was nearly 10 times more effective than sunlamp exposure in raising the toxicity of the irradiated WSF, as indicated by the anthracene-dianthracene conversion ratio, and (2) both significantly increased the toxicity of the WSF due to the irradiation.

There is little or no information on the potential toxicity of biological metabolites of petroleum compounds, and any conclusion is difficult, for metabolites have been demonstrated in only a few instances (e.g., Corner and Harris, 1976; Sanborn and Malins, 1977, 1980; Varanasi and Gmur, 1980). There is no evidence to date that the bulk of the petroleum hydrocarbon metabolites formed by biological activity are any more toxic than their parent compounds. However, a small proportion of petroleum compounds do give rise to mutagenic intermediates and to metabolites capable of binding with nucleic acids (Varanasi and Gmur, 1980; Varanasi et al., 1980, 1982). This potential appears to be limited only to the polycyclic aromatic hydrocarbons. The evidence to date is sparse but does not indicate that a mutation load has been introduced generally into the marine environment by this mechanism as a direct result of petroleum spillage or chronic spillage. However, this possibility cannot be ruled out in isolated incidents.

Remedial Measures

A great deal of effort continues to be expended, on countermeasures and various cleanup and control methods. These generally fall into one of two categories--mechanical and chemical--and because of their nature they inevitably leave some traces on the landscape, be it some form of physical disruption following mechanical cleanup or the risk of chemical alteration following application of chemical methods. As these are an almost automatic response to oiling incidents, a brief discussion of their potential effects on the marine environment seems appropriate.

Mechanical Containment and Cleanup

This category includes those methods which focus on the actual removal of oil or oiled debris, as by bulldozing or hosing with water under pressure. Most of this activity involves the intertidal zone. Offshore oiling incidents rarely are suitable for mechanical cleanup except by surface skimmers or possibly the cropping of oiled kelp using mechanical aquatic weed cutters. Neither of these is very likely to have much of an adverse effect on the environment. However, the problem becomes more serious in the intertidal zone, largely due to the physical disruption of habitats.

Rocky coastlines present the least problem in terms of cleanup. Oiled rocky surfaces are cleaned most often with either flushing, steam cleaning, sand blasting, or manual scraping. None of these is likely to alter the substrate to any extent, and the main damage is the removal of fauna and flora. The biological recovery process may take several years, but nevertheless, recovery will occur. As the settling surfaces have probably not been chemically or physically altered in the cleanup process to any great extent, the only limitations to recovery are biological ones. Of course, the rerelease of the stranded and flushed oil into the water column may pose additional problems.

The problem becomes greater with the oiling of finer-grained sediments such as cobble-boulder beaches or the fine silt sediments of lagoons and marshes. Because of the penetration of oil into such sediments, removal of oiled sediments often accompanies cleanup. Excessive removal can result in the disturbance of physical and ecological equilibrium. Excessive removal of beach sediments can lead to beach retreat or backshore (cliff) erosion. This was observed following the Arrow disaster, where a 20-m landward movement of pebble-cobble beach was recorded following large scale removal of oiled cobble from the shorelines (Owens and Drapeau, 1973). Pebble-cobble beaches present a particular problem in that oil is likely to penetrate rapidly and deeply. Thus, if cleanup is advised, it necessarily involves large scale removal of the beach material with heavy machinery. Pebble, cobble, and boulder sediments usually are replaced only slowly by natural coastal processes (Owens, 1973).

Another physical consequence of excessive sediment removal is that of habitat alteration, which in turn leads to long term ecological perturbation. An extreme example is the Ile Grande salt marsh in North Brittany, France, which was cleaned following the 1978 Amoco Cadiz oiling (Hess, 1978; Long and Vandermeulen, 1979, in press; Vandermeulen et al., 1981). Large volumes of the marsh sediment were removed along with oiled vegetation. The marsh surface and stream bed were severely disturbed by movement of heavy machinery. Dredging of the adjacent marsh area for a marina aggravated the conditions. The marsh has since undergone a series of degradative stages, including deposition of sand over the finer marsh silts, undercutting of secondary and tertiary marsh tidal channels, and erosion of the marsh surfaces, all as a direct result of the increased water flow through the marsh following bulldozing. This sediment removal has resulted in net erosion of the shoreline, which ranges between 6.5 and 17 m/year in some parts of this marsh. Marsh recovery, in this extreme case, is thought to run into decades if not centuries.

Cleanup of lagoons and coastal marsh systems (salt marsh, mangrove swamp) presents problems in that both the physical (habitat) and the biological aspects of these systems are easily damaged. In such areas, natural cleaning may frequently be the most effective, appropriate, and least damaging option (e.g., Baker, 1975; Lindstedt-Siva, 1979, 1980, 1981; Tramier et al., 1981), although cosmetically it is the least appealing one. Some progress is now being made in studies of replanting marsh vegetation and mangroves in oiled and cleaned sediments, and this line of investigation appears useful for extreme cases such as the Ile Grande oiling.

A problem with mechanical cleanup is the ultimate disposal of oiled debris, including oiled seaweed, sediment, and shoreline material. These often can be in volume or mass far greater than that of the stranded oil itself.

An understanding of both the physical and the ecological environment is necessary for the development of efficient, effective, and least damaging countermeasure techniques. Considerable progress has been made in this area in recent years, mostly through combined input from ecologists, geologists, coastal sedimentologists, and physical oceanographers.

Chemical Control and Cleanup

The use of chemicals for the control and cleanup of oil spills appears to be an alternative to mechanical removal of the oil, especially in certain situations such as in offshore spills and in polar regions. However, the use of these chemicals continues to be a subject of concern and debate and has generated much research (for recent review, viz. Wells, 1984).

A range of chemical means has been tried on oil slicks, with varying results. These include sinkants, gelling agents, herders, chemical dispersants, and deemulsifiers. Of these the last two have seen the most widespread use. Sinkants were used widely during the early days of oil pollution, for example, during the Torrey Canyon spill. Herders remain largely in an experimental stage. Deemulsifiers are now used extensively in petroleum production systems but are also being studied with an eye on field application on "mousse" slicks and for transfer of collected oil to storage systems. Except for sinkants, use of these products in the field has been limited.

The use of chemical dispersants as a spill response tool remains controversial. The formulations, essentially mixtures of chemical surfactants and stabilizers in a carrier solvent, are specifically designed to reduce the interfacial tension between oil and water and thus result in a break up of the oil slick into smaller droplets. These droplets can then be distributed into the water column by the natural actions of surface and subsurface turbulence. Much of the success of conventional dispersants depends on the parallel application of physical dispersive energy, such as by the use of high speed "breaker boards," which increase surface turbulence. The more recently developed "concentrate" dispersants do not require additional dispersive energy in any but the calmest of waters. Concentrate dispersants have a "self-mix" action when applied to the oil because they contain high concentrations of surfactant molecules per unit.

Chemical dispersants first gained international prominence following their use at the Torrey Canyon spill. Much of their notoriety stems from the high toxicity of early formulations as well as improper application techniques used at the spill site. In fact, the high acute toxicity of those products was due primarily to the carrier solvents (aromatic hydrocarbons) and not the dispersant surfactants themselves. Considerable advances have been made in refining their chemical composition and design and in understanding their potential toxicity (Table 5-2). Today's formulations are developed for specific purposes and consist of compositions that retain their dispersant effectiveness but require less mixing energy and are at reduced toxicity.

While the newer formulas may be inherently less toxic, their increased effectiveness results in greater amounts of oil being put into the water column and thereby becoming available for contact with the pelagic and benthic communities (Swedmark et al., 1973; Doe and Wells, 1978). However, there exist no field data from recent years to suggest that damage from dispersants has been greater than if the oil had been left alone. Thus, no evidence of damage due to the use of dispersants has come from U.K. coastal waters, where the use of chemical dispersants is common in oil slick control. More studies are needed to look at this aspect of oil pollution. To be fair, it should be pointed out that the simultaneous monitoring of offshore pelagic and benthic communities for significant effects is extremely difficult.

One argument is that use of chemical dispersants simply represents the introduction of yet another contaminant into the marine environment.

TABLE 5-2 Acute Lethal Toxicity of Some Oil Spill Dispersants to Marine Organisms--A Selection of Current Data

Species/Stage	Dispersant	Threshold Concentrations Expressed as Four-Day LC_{50}'s, mg/L[a]
Invertebrates		
Stony coral (Madracis mirabilis)	Shell dispersant LTX[c]	162 (1 day)
Oligochaete (Marionina subterranea)	Corexit 7664	
	Finasol OSR-2	>1000
	Finasol OSR-5	
Intertidal limpet (Patella vulgata)	BP1100X	3700 (approx.)
	BP1100WD	270 (approx.)
Crustaceans		
Amphipods (Gammarus spp.)	water-based dispersants	>10000
	petroleum-based dispersants	200 ± 130
Mysids (Neomysis sp.)	water-based dispersants	>4500
	petroleum-based dispersants	~150
Amphipod (Gammarus oceanicus)	AP oil dispersant	10-100 (1.5 days)
Brown shrimp (Crangon crangon)	10 conventional dispersants	3300->10000 (2 days)
	7 concentrated dispersants (unnamed)	2800->10000 (2 days)
Grass shrimp (Palaemonetes pugio)	Corexit 7664	>10^4 (27°C), nontoxic (17°C)
	Atlantic-Pacific	1000 (27°C), 1800 (17°C)
	Gold Crew	150 (27°C), 380 (17°C)
	Nokomis-3	140 (27°C), 250 (17°C)
Fish		
Fish larvae (Pleuronectes platessa, Solea solea)	Corexit 7664	400
Gobies (Chasmichthys, Luciogobius)	Shell dispersant LT	440-480
Stickleback (Gasterosteus aculeatus)	water-based dispersants	950 ± 250
	petroleum-based dispersants	>10000
Dace (Phoxinus phoxinus)	water-based dispersants	1400 ± 200
Coho salmon (Oncorhynchus kisutch)	BP1100X	1700
Killifish (adult) (Fundulus heteroclitus	AP oil dispersant (GFC Chemical Co.)	approx. 100 (2 days), 50-100 (3 days)

[a]Unless otherwise noted.

SOURCE: From Wells (1984).

This argument is answered, in part, by the design of less toxic and more biodegradable dispersants.

From the available evidence there would seem to be little point in the application of chemical dispersants in far offshore spills except to control the onshore movement of slicks, to protect critical seabird populations and other such marine biota aggregations, and to prevent oil foulings of ships and pleasure craft transiting the area. Their use may well be more attractive in inshore and polar waters. Appropriate application of chemical dispersants in the former could avert wholesale oiling of the intertidal region, as seen with the Amoco Cadiz spill. It is difficult to assess, however, the potential impact had that volume of oil been mixed into the water column instead. Chemical dispersants are being considered seriously for use in polar spills, many of which would be expected to be inaccessible by traditional countermeasures. Their effectiveness and potential toxicity, both alone and in combination with oil, are now under study.

Indirect Effects

Not all impact, whether from a spill or from chronic oiling, is due directly to a specific toxic effect of the oil. For example (see Behavior section), mortality of bivalves during oil spillage often is due directly to suffocation or toxicity. However mortality may also result out of the habit of bivalves in oiled sediments to live closer to the surface, thereby becoming easier prey for seabirds and other predators. Following the Santa Barbara oil spill, it was observed that initial settlement by larvae of the intertidal barnacle Chthamalus fissus was heaviest on solidly oiled black surfaces (Straughan, 1976). However, subsequent survival of larvae was lowest on such surfaces, apparently because of the heat-absorbing capacity of the black substrate during exposed tidal periods.

Tarred sediments forming either layers or near-permanent pavements have been observed after the Arrow (Owens, 1976), the Metula (Gundlach, 1979), and the Amoco Cadiz (Vandermeulen et al., 1979; d'Ozouville et al., 1981) spills. Just how these oily or tarry layers affect the physical environment is not clear, but studies of an experimentally oiled sandy beach suggest that heavy concentrations of stranded oil can cause measurable changes in beach water flux and in interstitial fauna (McLachlan and Harty, 1981, 1982). Also the normal processes of sedimentation and vegetative propagation in marshes by tarry surface crusts clearly are inhibited (Levasseur and Jory, 1982). Similarly, the immobilization of beach surface substrate by heavy residual stranded tar as seen in Chedabucto Bay (Owens, 1981) alters the usual seasonal beach dynamics, with implications for the stability of the back-beach community.

EFFECTS ON BIOLOGICAL PROCESSES

Developmental Problems

Chromosomal Aberrations

Damage by oil to chromosomes or chromosome functions is only poorly understood, and research efforts to date on oil-induced aberrations in marine organisms are limited. Even so, while there exists no broad threat to chromosome function, there is indirect and direct evidence suggesting that measurable increases in the burden of chromosome mutation can occur, at least in fish, under certain conditions of petroleum contamination.

Studies of the mutagenicity of whole oils or refined products or their water-soluble products are few, even with mammalian tissues. The most recent study available, an examination of chromosome mutagenicity of heavy oil extracts, demonstrated the capacity of this material to induce chromosome aberrations in hamster cell cultures at 0.06 and 0.03 mg/mL (Matsuoka et al., 1982). Subsequent fractionation of the oil showed the mutagenic potential to be associated with a fraction containing neutral or weak basic nitrogen-containing compounds.

Most of the research to date has focused instead on individual mutagenic compounds or their mutagenic metabolites. Results from studies with mammalian and other tissues indicate that several aromatic hydrocarbons, metabolites of the polycyclic aromatic hydrocarbons, and heavy metals found in crude petroleum and in refined products can impair fidelity of DNA synthesis, increase sister-chromatid exchange and chromosome mutation, and/or cause abnormalities in chromosome number (Table 5-3). These include benzene, phenanthrene, naphthalene, chrysene, benzo(a)pyrene, 3-methylcholanthrene, benzoanthracene, 7,12-dimethylbenzo(a)anthracene, benzo(b)fluoranthrene, benzo(e)pyrene, dibenzanthracene, phenol, nickel, arsenic, vanadium, and lead. Fish and most marine invertebrates are now known to have the requisite enzymatic systems for converting promutagens, such as the polycyclic aromatic hydrocarbons, into gene-active metabolites capable of inducing gene and chromosome aberrations (e.g., Payne and May, 1979; R.F. Lee and Singer, 1980; R.F. Lee, 1981; Stegeman, 1980). Also, fish are known to be particularly susceptible to irradiation as a mutagenic agent, producing a range of responses with harmful effects on reproduction not unlike those found in mammals. One might suspect then that mutagenic agents, such as petroleum hydrocarbons, might provoke generally similar responses in fish chromosomes.

Chromosomal aberrations (chromosome breaks, chromatid breaks, gaps, and interchanges) were all observed in gill cells of a freshwater tooth carp (Notobranchius rachow) exposed for 4-6 days to a 10^{-4} mg/L concentration of benzo(a)pyrene (Hooftman and Vink, 1981). In another study, English sole (Parophrys vetulus), collected from a relatively unpolluted site in Puget Sound, when injected intraperitoneally with benzo(a)pyrene, were found to have enhanced sister-chromatid exchanges in kidney cells when compared with nontreated sole from the same area (Stromberg et al., 1981). Similarly, cultured cells of rainbow trout gonadal tissue, incubated with either 3-methylcholanthrene (1-16 µg/mL) or with benzo(a)pyrene (0.1-1 µg/mL) showed increased incidence and severity of chromosome abnormalities over control cultures (Kocan et al., 1981). Fertilized eggs of English sole exposed to 0.1-4.2 µg/L benzo(a)pyrene (B(a)P) revealed significantly increased incidences in chromosome abnormalities of their yolk-sac membranes (Hose et al., 1982). Treatment of the sperm and eggs of the purple sea urchin (Strongylocentrotus purpuratus) for 15 and 30 minutes with concentrations of B(a)P of 1.0-50 µg/L yielded embryos with significant increases in the number of mitotic abnormalities (Hose et al., in press).

Cytologic-cytogenetic analyses have been applied, in one instance, to samples taken directly from the scene of the Argo Merchant spill on the Nantucket Shoals (Longwell, 1977, 1978). Cod and pollock species eggs from the two most severely contaminated sampling stations showed markedly greater cytological deterioration than eggs from less severely oiled stations. Abnormal differentiation or dedifferentiation of the embryonic cells, grossly malformed embryos, and fouling of the outer egg membrane with tar were striking at these two stations. There was also indication of effects on the embryonic mitotic index in these samples. Cod eggs collected from all five oiled sampling stations showed greater incidence of mitotic abnormalities than similar laboratory-spawned cod eggs. Twice as many pollock eggs as cod eggs experienced mitotic difficulties, possibly because pollock eggs were higher in the water column. Many pollock eggs were contaminated with oil on the chorion.

The overall heredity risk factors to marine populations from oil are difficult to assess because of unknown natural mutation rates, the occurrence of natural mutagens, and a multiplicity of unrelated muta- genic pollutants. To this must be added varying species sensitivity and acclimation, variable cell and tissue sensitivities, and variable tissue accumulation and metabolism. The enormous number of aromatic hydrocarbons in petroleum probably act synergistically or antagonisti- cally with one another, and with other classes of marine contaminants to increase or decrease mutagenic potential (Longwell and Hughes, 1980).

Increased mutation reduces genetic fitness of individuals and populations, but demonstrations of mutagenic potential in highly sensitive laboratory tests cannot be taken as evidence of similar genetic damage affecting the population under normal in vivo exposure in nature. Enormously fecund marine species and those with par- ticularly short generation times can tolerate greater mutation frequency and gametic wastage than can less fecund ones with longer generation times. However, the effect of genetic damage on localized populations of some marine species is a possibility. The use of contaminated coastal waters by commercial fish stocks for breeding, spawning, and nursery grounds increases their genetic risk. The risk in these specialized cases is enhanced by the greater susceptibility and sensitivity of the embryonic and juvenile stages. The regularity with which these abnormalities can occur, even after relatively short exposures, suggests this as a possibility. For these reasons, further attention to research in this area seems worthwhile.

Reproduction and Development

Both laboratory and field studies have shown that a broad range of reproductive and developmental processes can be affected by hydrocarbon exposure. As successful reproduction and early development are essential to the survival of species and populations, there is much-needed research emphasis in this area, particularly in light of their sensitivity to petroleum contamination.

Most results have been obtained in laboratory experimental studies, with fewer confirming data from field situations. For the most part laboratory exposures have been with single hydrocarbon compounds or with known mixtures in either water-soluble or water-dispersed phases, over various time periods. Observed field exposures have been with dispersed crude and fuel oils over more extended durations.

TABLE 5-3 Mutagenicity of Petroleum and Its Components

Substance in test or associated with a field measurement	Human--in vitro or epidemiologic	Laboratory mammals	Marine mammals[a]	Sea-birds[a]	Fresh-water fish	Marine fish	Drosophila	Marine invertebrates[a]	Terrestrial plants	Fresh-water plants	Marine algae[a]
Benzene	1,2 +	2,3,4,5 +									
Toluene	?	1,2 +									
Xylene	-	1,2 +,- / 1,2 -									
Naphthalene									7 +		
Phenanthrene		8 +,-									
Pyrene		9 -									
Anthracene	8 +	9 -				10 -					
Chrysene	8 -	8 +									
Benzo(a)pyrene	8 +	11,12,13 +			10,14 +	15,16 +					
3-methylcholanthrene	+	11 +			10 +						
7,12-dimethylbenz(a)anthracene	8,17 +	8,11,17,18,19 +			+		20 +				
Benzoanthracene	8 +	9 +									
Benzo(b)fluoranthrene	8 +										
Benzo(e)pyrene	8 +	9 -									
Dibenzanthracene	8 +										
Extremely high number of other polycyclic aromatic hydrocarbons of petroleum[a]											
Phenol											
Nickel		21,22,23 +							1 +		

Arsenic	24,25,26,27 +	24,28,29,39 +	24,31,32 +
Vanadium			33,34 +
Lead	35 +,?	35 +	35 +
Crude petroleum		36 +	
Neutral fractions of crude petroleum (containing PAH) and their subfractions	38 –		38 –
Basic fractions of crude petroleum	38 +		38 +
Refined petroleum			
Synthetic crude			
Chemical fractions of 4 shale oils			
High boiling point coal distillate			
Aromatic hydrocarbons with marine contaminants		41,42 + 15,43 +	
Weathered or photooxidized crude[a]			
Crude with dispersants[a]			
Crude with drilling compounds[a]			
Used crankcase oil			

[a]No information.

NOTE: A plus sign indicates positive or measurable mutagenicity. A minus sign denotes absence of mutagenic potential. A question mark denotes questionable mutagenicity.

SOURCES: [1]Dean (1978), [2]Gerner-Smidt and Friedrich (1978), [3]Siou et al. (1981), [4]Lyon (1975), [5]Beljanski (1979), [6]Bos et al. (1981), [7]Avanzi (1950), [8]Rosinzky-Köcher et al. (1979), [9]Tong et al. (1981), [10]Kocan et al. (1982), [11]Hollstein et al. (1979), [12]Cole et al. (1961), [13]Pederson et al. (1978), [14]Kocan et al. (1979), [15]Hooftman (1961), [16]Stromberg et al. (1981), [17]Hose et al. (1982), [18]Hose et al. (in press), [19]Kato (1966), [20]Kato et al. (1969), [21]Kurlis et al. (1969), [22]Forbes (1980), [23]Miyski et al. (1980), [24]Strover and Loeb (1976), [25]Nishimura and Umeda (1974), [26]Leonard and Lauwerys (1980), [27]Petres and Berger (1972), [28]Nordenson et al. (1978), [29]Nakamuro and Sayoto (1981), [30]King and Lunford (1950), [31]Paton and Allison (1972), [32]Rossner et al. (1972), [33]El-Sadek (1972), [34]Nygren (1949), [35]Meisch and Benzschawel (1978), [36]U.S. EPA (1975), [37]Gerber et al. (1980), [38]Matsuoka et al. (1982), [39]Longwell (1977), [40]Payne et al. (1978), [41]Epter et al. (1978), [42]Petroy et al. (1981), [43]Kitahara et al. (1978), [44]Whon et al. (1982), [45]Prein et al. (1978), [46]Alink et al. (1980), [47]Longwell and Hughes, 1980, [48]Dixon (1982).

TABLE 5-3 (continued)

Substance in test or associated with a field measurement	Bacteria test	Sources of Material — Laboratory	Field	Oil spill	Tissue tested — In vitro	Somatic cells	Reproductive cells	Specific genetic test used in assay — Sister-chromatid exchange	Indirect chromosomal	Direct chromosomal	Dominant lethal	Sperm abnormality	Gene mutation	Cell transformation	Fidelity of DNA synthesis
Benzene	2	+,-	?		+	+,?	-	-	+,-	+,-	-				+
Toluene	-	+,-	-		-	+		-	+,-	+,-	-				
Xylene	2,6	+,-	-		-	+		-		+,-					
Naphthalene	-	-			-	+		-	+	-					
Phenanthrene	8	+				+		+	+						
Pyrene	+,-	+,-	+,-		+,-	+,-		+						-	
Anthracene	8	-			-	-		-							
Chrysene	8	-			-	-		-							
Benzo(a)pyrene	8	+,-	+,-		+	+	+	+	+	+,-	?	+	+	+	
3-methylcholanthrene	11	+,-	+,-		+	+	?,-	+	+	+	?,-	+	+	+	
7,12-dimethylbenz(a)anthracene	11	+	-		+	+	+	+	+	+	+	+	+	+	
Benzoanthracene	8	+			+	+		+							
Benzo(b)fluoranthrene	8	+			+			+							
Benzo(e)pyrene	8	+,-	+,-		+,-	+,-		+							
Dibenzanthracene	8	+			+,-	+,-		+	+						
Extremely high number of other polycyclic aromatic hydrocarbons of petroleum															
Phenol	1	+,-				+									
Nickel	11	+			+					+					+

Arsenic	24	+,−	+		+		+		+	+
Vanadium	34	−	+,−		+		+		+	
Lead	35	−	+,?		+		+		+	+
Crude petroleum	37,38,39	+,−	+	+	+	+	+	+	+	
Neutral fractions of crude petroleum (containing PAH) and their subfractions	38	+	+,−	−	−				−	
Basic fractions of crude petroleum	38	+	+	+	+		+		+	
Refined petroleum	37	−		−						
Synthetic crude	37,38	+	+							
Chemical fractions of 4 shale oils	40	+	+							
High boiling point coal distillate	40	+	+							
Aromatic hydrocarbons with marine contaminants			+	+	+	+		+		
Weathered or photooxidized crude[a]		+								
Crude with dispersants[a]										
Crude with drilling compounds[a]										
Used crankcase oil	37	+	+							

[a]No information.

NOTE: A plus sign indicates positive or measurable mutagenicity. A minus sign denotes absence of mutagenic potential. A question mark denotes questionable mutagenicity.

SOURCES: [1]Dean (1978), [2]Gerner-Smidt and Friedrich (1978), [3]Siou et al. (1978), [4]Lyon (1975), [5]Beljanski (1979), [6]Bos et al. (1981), [7]Avanzi (1950), [8]Rosinzky-Köcher et al. (1979), [9]Tong et al. (1981), [10]Kocan et al. (1982), [11]Hollstein et al. (1979), [12]Cole et al. (1961), [13]Pederson et al. (1978), [14]Kocan et al. (1979), [15]Hooftman (1961), [16]Stromberg et al. (1981), [17]Hose et al. (1982), [18]Hose et al. (in press), [19]Kato (1966), [20]Kato et al. (1969), [21]Kurlis et al. (1969), [22]Forbes (1980), [23]Miyski et al. (1977), [24]Strover and Loeb (1976), [25]Nishimura and Umeda (1974), [26]Leonard and Lauwerys (1980), [27]Petres and Berger (1972), [28]Nordenson et al. (1978), [29]Nakamuro and Sayoto (1981), [30]King and Lunford (1950), [31]Paton and Allison (1972), [32]Rossner et al. (1972), [33]El-Sadek (1972), [34]Nygren (1949), [35]Meisch and Benzschawel (1978), [36]U.S. EPA (1975), [37]Gerber et al. (1980), [38]Matsuoka et al. (1982), [39]Longwell (1977), [40]Payne et al. (1978), [41]Epter et al. (1978), [42]Petroy et al. (1981), [43]Kitahara et al. (1978), [44]Whon et al. (1982), [45]Prein et al. (1978), [46]Alink et al. (1980), [47]Longwell and Hughes, 1980, [48]Dixon (1982).

Gametogenesis Petroleum compounds and whole oils are known to interfere with sex pheromone responses of algal gametes, but the concentrations at which this occurs (0.02-1.0 mg/L) are at least 3 times above
the levels of naturally occurring pheromones (Derenbach et al., 1980;
Derenbach and Gerek, 1980). Algal (Steele, 1977) and molluskan
(Renzoni, 1975) sperm are sensitive to oil, but there are differences
in sensitivity. Thus algal sperm demonstrate sensitivity down to 0.2
μg/L oil in water, but echinoid sperm are apparently much less
sensitive (Lonning, 1977).

The development of gonadal tissue has been found to be sensitive to
petroleum hydrocarbons, but the mechanism is not understood. Thus,
corals exposed to the water-soluble fractions of crude oils created by
floating oil (nominally 3,000 mg/L) on the surface of a flowing water
system for 2 months resulted in lower numbers of female gonads per
polyp (Rinkevich and Loya, 1979); in another study, exposure to No. 2
fuel oil caused degenerated ova and abnormal gonadal development (Peters
et al., 1980a,b). Larvae can also be prematurely released from oiled
corals (Loya and Rinkevich, 1979).

Hydrocarbons may commonly be transferred from the gonads to early
developmental stages, as shown with polychaetes (Rossi and Anderson,
1977) and scaphopod mollusks (Koster and Vanden Biggelaar, 1980).
Early embryogenesis with a wide variety of species is a particularly
sensitive stage of development (Ceas, 1974; Donahue et al., 1977; Ernst
et al., 1977; Lonning, 1977; Kuhnhold, 1978; Linden, 1978; Vashchenko,
1980; among others), which has been demonstrated in detail with
echinoids and teleosts. Embryological effects among birds exposed
internally or externally to oil have been shown to occur (Coon et al.,
1979; Grau et al., 1977). The hatching success of teleost and avian
eggs exposed to hydrocarbons is markedly reduced, often but not always
at low levels (less than 1 mg/L) (J.W. Anderson et al., 1977; Ernst et
al., 1977; Sharp et al., 1979; Kuhnhold, 1978; Albers, 1977; Szaro and
Albers, 1977; White et al., 1979).

Development The general sensitivity of larvae to hydrocarbons,
especially of crustaceans and teleosts, is now well recognized, but
concentrations causing effects vary between life stages and can be
influenced markedly by physiological condition. The delayed development of some larvae, notably of decapods (Wells and Sprague, 1976;
Caldwell et al., 1977; Laughlin et al., 1978, among others), often
occurs at initial total hydrocarbon levels well below 1 mg/L.

In a series of studies, Lonning and colleagues (Lonning and
Falk-Petersen, 1982; Falk-Petersen et al., 1982; Kjorsvik et al., 1982)
have demonstrated deleterious effects on cod eggs and on sea urchin
eggs and embryos when exposed to a variety of aromatic hydrocarbons
(naphthalene, methylnaphthalenes, benzene, phenanthrene, xylene).
Short term exposure (3-6 hours) of cod eggs to water solutions of
xylene showed a range of reactions, from irregular cell cleavage (at
2-7 μg/L) to no cleavage and poor accumulation of cytoplasm (16-35
μg/L). Treatment of sea urchin embryos with benzene (10^{-2} M) or
phenanthrene (10^{-4} M) resulted in abnormal differentiation and
abnormal larvae. Methylnaphthalenes (1- and 2-methylnaphthalenes)

elicited a high incidence of skeletal abnormalities in sea urchin larvae, including aberrant skeletal forms or the formation of an extra skeleton rod.

Laboratory studies have also shown that some reproductive and developmental processes are fairly resistant to hydrocarbon exposure, but these are fewer in number relative to those processes found to be vulnerable. For example, no effects were noted among oysters chronically exposed to crude oil, even though their gonads accumulated high concentrations of total hydrocarbons (Vaughan, 1973). It seems that sperm of echinoderms are quite tolerant of oil dispersions or extracts, and fertilization among echinoderms is seldom prevented by oil exposures (Allen, 1971; DeAngelis and Giordano, 1974; Lönning, 1977; Nicol et al., 1977).

Field Studies   Some marked effects of hydrocarbons on reproduction and early development have been observed in the field. The fecundity of coral populations in chronically oiled areas of the Red Sea was sharply reduced (Rinkevich and Loya, 1977), and mussels at the West Falmouth oil spill were sterilized for at least one season (Blumer et al., 1970). Reproductive problems, including a reduced ratio of females to males, reduced juvenile settlement, and long term (>7 years) inhibition of recruitment and low population densities, have been noted for the fiddler crab Uca pugnax and have been directly related to high oil-in-sediment content in salt marshes contaminated by the Florida oil spill (Krebs and Burns, 1977).

Seabirds are very susceptible to direct oiling or oiling of their egg clutches. R.G.B. Brown (1982) states that oil may act indirectly by affecting the laying rate, the hatchability of eggs, and the growth of chicks--sublethal effects which in the long term could reduce survival of individual birds and breeding populations. However, very little direct information exists for this problem.

Apparently not all populations in the field are affected seriously by hydrocarbons, and indeed some appear to be notably tolerant even during their reproductive and early developmental phases. Thus beds of kelp (Laminaria digitata) at the Amoco Cadiz spill site in North Britanny, France, apparently were not affected by the oil, at least judging from the few studies and surveys that were carried out at that time. Reproduction in 1978 and densities in 1979 were also believed to be normal (Kaas, 1981). Meiofaunal copepods in experimental beach plots treated chronically with oil reportedly increased in reproductive activity (Feder et al., 1976). Reproduction among barnacles and mollusks near oil seeps or at spill sites appears unimpaired (Straughan, 1971, 1977a,b), and settlement of larvae is not always unaffected (Woodin et al., 1971).

Conclusion   The above work, most of it conducted in the last decade, shows that sensitivities to hydrocarbons vary between phyla and across and between developmental stages and covers processes from the maturation of gonads to the survival and settlement of larvae. This feature of variable sensitivities of life stages, organisms, and processes is particularly clear from studies of crustacea, echinoids, and teleosts and is now well recognized by aquatic toxicologists. The reported threshold concentrations for the adverse developmental effects, based on initial measured hydrocarbons by spectroscopy or chromatography, are well below 1 mg/L and even down to 1 μg/L, for acute exposures in the laboratory. In contrast, some processes appear tolerant to much higher hydrocarbon levels.

It is important to reemphasize that significant reproductive impairment in oiled field conditions has seldom been observed, although few field studies have been performed. Based on available studies at the population level, annelids, gastropods, and copepods seem to suffer no long lasting damage. Macrophytes, barnacles, and birds may sometimes be affected. However, corals, bivalves, and decapod crustacea can suffer marked and sometimes long term (years) reproductive damage at oiled sites.

There is sufficient information now available on effects at relatively low concentrations to cause concern about spills and chronic discharges of oil into protected or enclosed coastal waters. Of equal concern, perhaps, is the frequent absence of threshold data for many processes in groups of organisms when exposed to realistic concentrations of hydrocarbons.

Pathological Consequences

Laboratory studies have shown that individual aromatic hydrocarbons, whole petroleum, and fractions of petroleum induce a variety of cellular and subcellular alterations in marine teleosts and invertebrates (Malins, 1982); however, the concentrations of petroleum components used were often substantially higher than those found in petroleum-contaminated marine environments.

In teleosts the effects include abnormalities of the eyes, e.g, alterations in lens fiber cells (J.W. Hawkes, 1977) and of the chloride cells in the gills (Engelhardt et al., 1981) (Table 5-4). In invertebrates, the effects include branchial, e.g., pigment body formation, and renal lesions, e.g., necrosis (Gardner et al., 1975), and changes in the cellular and lysosomal structure of digestive cells (Lowe et al., 1981).

Embryos and larvae appear to be particularly susceptible to petroleum exposure, sometimes at lower concentrations than those inducing morphological changes in mature organisms. For example, petroleum-induced morphological changes in teleost embryos include retinal damage, e.g., necrotic neurons (J.W. Hawkes and Stehr, 1982) (Figure 5-2) and the failure of fins to differentiate from the lateral line body wall (R.L. Smith and Cameron, 1979).

Overall, the laboratory findings have not as yet provided a thorough insight into how the petroleum-induced pathological changes relate to the degree or time of exposure, or to a variety of other experimental conditions. Moreover, real difficulties exist in translating the findings into the field situation, where, for example, there are compounding problems arising from the complexity of ecosystems and the influence of other contaminants (Malins and Collier, 1981).

TABLE 5-4 Summary of Gill Morphology in Rainbow Trout (Salmo gairdneri) Following 7 Days of Oil Treatment, Comparing Different Modes of Exposure

Exposure Mode	Degree of Fusing	Epithelial Separation	Chloride-Type Cell		
			Exposure	Location	Vacuoles
Freshwater					
Control	none	none	none	basal	rare to none
Emulsion in paraffin	none	none	rare	basal	none
Emulsion in NW	some	some	extensive	length[a]	extensive
Emulsion in VEN	extensive	extensive	extensive	length	some
WSF of NW	none	none	rare	basal	few to none
WSF of VEN	none	none	little	basal	none
i.p. with NW	little	none	some	(basal)[b]	rare
Seawater					
Control	rare	none	none	basal	none
Emulsion in NW	some	none	little	(basal)	some
Emulsion in VEN	variable	none	some	(basal)	common
i.p. with NW	rare	none	none	basal	none

NOTES: NW, Norman Wells crude oil, weathered 3 days to 200 µL/L; VEN, Venezuelan crude oil, weathered 3 days to 200 µL/L; WSF, water-soluble fraction; i.p., intraperitoneal injection, 100 µL/kg fishweight per day.

[a]Chloride-type cells found along length of secondary lamellae.
[b]Chloride-type cells found in basal region and extending part way up secondary lamellae
    (from Engelhardt et al., 1981).

Contaminant-induced lesions or tissue abnormalities (e.g., tumors) also are difficult to distinguish from those caused by some other agent, such as viruses, e.g., ovarian neoplasmic responses in clams from a Maine oil spill site (Yevich and Barszcz, 1977; R.S. Brown et al., 1977).

Field studies, notably those conducted after the Amoco Cadiz spill, have shown that gross pathological conditions, e.g., fin erosion in teleosts (Haensly et al., 1981), and cellular/subcellular alterations, e.g., abnormal proliferation of chloride cells in invertebrates (Lopez et al., 1981), can be found in association with spilled petroleum. Again, it is difficult to link these pathological abnormalities directly to the spilled oil, either because of absence of proper controls or because of the presence of other contaminants or interfering conditions. Also, the data provide little understanding about how these petroleum-induced morphological changes can affect the viability of important resource species. For example, certain petroleum-related morphological changes, such as the hepatocellular lipid vacuolization in teleosts

(McCain et al., 1978) are questionable indicators of altered organism health. Further, relationships between chemically and biologically induced transformations in petroleum components in the field and deleterious effects at the cellular/subcellular level remain largely unknown (Malins et al., 1980). Also, only a limited understanding of teleost and invertebrate pathology has been attained, especially with regard to the significance and frequency of lesions associated with petroleum exposure.

Despite the various limitations, such as our inability to describe dose-response relationships with any degree of precision or, in field studies, our inability to separate effects of polycyclic aromatic hydrocarbon from those due to other pollutants, evidence is accumulating that petroleum exposure can cause gross and cellular abnormalities in marine organisms. The recognition that alterations in cellular and subcellular structures do result from petroleum exposure is an important first step in understanding relations between petroleum and effects on the biological structure of marine organisms. In this respect it is regrettable that more work has not been done on spill sites or in regions (e.g., coastal waters) receiving chronic petroleum inputs.

Growth and Metabolism

Since the early 1970s, research interests have shifted demonstrably toward studies of the mechanisms of petroleum toxicity, and toward an understanding of petroleum hydrocarbon toxicity both at the cellular and at the organismic level. This is reflected in the nature of the large amount of data that have been assembled since the National Research Council (1975) report and in the quality of current research.

Petroleum hydrocarbons are now known to affect nearly all aspects of physiology and metabolism, although the ultimate site or sites of toxic action within the cell or organism still are not understood.

Plants

The literature dealing with petroleum effects on the metabolism and physiology of marine plants is not extensive, and is restricted mainly to work with phytoplankton species. Processes studied most often are those of growth, photosynthesis, and to a lesser extent, respiration. Thus, there is little information available that might lead to general-izations about the comparative effects of petroleum hydrocarbons on a taxonomic basis (see Macrophytes section below).

Photosynthesis Phytoplankton exhibit widely varying responses to oil and to oil products, sometimes showing, for example, enhanced photo-synthesis and at other times inhibited processes. Species also differ widely in their vulnerability (Pulich et al., 1974). In general, the green algae have been found to be the most sensitive, blue-green algae next, and the diatoms most tolerant to petroleum (Winters et al., 1977). How much of this is due to experimental procedures, however, is not known. Thus, R.F. Lee et al. (1978) observed the opposite, with diatoms highly susceptible to petroleum.

In a series of experiments comparing the phytotoxicity of several No. 2 fuel oils, differences in toxicity were ascribed to the chemical composition of the water-soluble fraction (WSF) (Batterton et al., 1978a). The presence of p-toluidine in this WSF was found to be correlated with the increasing phytotoxicity observed. Indeed, low concentrations of p-toluidine (50 μg/L) would arrest growth in a blue-green alga (Batterton et al., 1978b). Curiously, in another study (Winters et al., 1977), light was found to influence the effect of a constituent of No. 2 fuel oil, perinaphthenone, on growth of two green algae. For unknown reasons a shift from white to yellow light resulted in an increased inhibitory threshold from 0.125 to 5 mg/L (Winters et al., 1977).

There is some suggestion that the inhibitory effect, at low concentrations of the toxicant, may be reversible. Cells of the unicellular alga _Monochrysis_ _lutheri_, after inhibition of photosynthesis on exposure to naphthalene, were found to recover their photosynthetic ^{14}C fixation when transferred to clean nonnaphthalene containing culture medium (Vandermeulen and Ahern, 1976).

Cellular Mechanisms Little is known of the mechanisms by which petroleum affects either photosynthesis or metabolism in marine plants. Studies with freshwater unicellular algal species suggest a correlation between toxicity and "leakage" of several ions from the exposed cells (Hutchinson et al., 1979). Other studies indicate disruption of intracellular macromolecule pools, including an alteration of the ATP/ADP balance in unicellular algae (e.g., Vandermeulen and Ahern, 1976), and of DNA and RNA synthesis in certain macroalgae (Davavin and Yerokhin, 1979). Comparatively little is known of metabolism of petroleum hydrocarbons by marine algae, such as by the mixed function oxidase system based on the cytochrome P450 complex as found widely throughout the animal phyla.

However, recent studies by Cerniglia et al. (1980a,b, 1981a, 1982) indicate that blue-greens (Cyanobacteria) and a number of green algae, as well as a red and a brown alga can metabolize naphthalene and aniline (Cerniglia et al., 1981b). Whether the process is P450 based remains to be seen.

Abnormal Growth Certain polynuclear aromatic hydrocarbons appear to be able to affect growth form in macroalgae, possibly through altering normal apical growth of the plants (Boney, 1974). Tumor-like growths found in _Porphyra_ _tenera_, in areas of industrial pollution, would seem to correlate with the occurrence of several polycyclic aromatic hydrocarbons in the sediments (Ishio et al., 1971, 1973). In this instance, however, as is often the case, it is difficult to separate the hydrocarbon effects from those of other pollutants or from other sources.

Animals

Evaluation of the effects of petroleum hydrocarbons on growth and
metabolism of animals is more extensive, with a sizable literature
dealing with feeding, respiration, growth, and enzymatic detoxification
systems.

Feeding Of the various responses that animals are capable of with
respect to toxicant exposure, feeding seems to be the first or one of
the first to be affected, generally negatively. Although feeding rate
is, of course, ultimately important to the well-being of the organism,
it is only one factor in the total energy budget. Therefore, the direct
consequences of an altered feeding rate are difficult to assess. They
are probably most serious in planktonic animals with little food
reserves and those with short life cycles or durations in the plankton
(e.g., meroplankton and microzooplankton).

Effects on feeding have been observed in most phyla, at concentrations approaching those measured initially under oil spill situations
(e.g., Berman and Heinle, 1980; Elmgren et al., 1980a). The effects are
found both in the planktonic and pelagic groups, e.g., lobster larvae
(Wells and Sprague, 1976) and adult copepods, and in benthic organisms,
e.g., lugworms (Augenfeld, 1980). The consequences of reduced feeding
rates can be further manifested in reduced production rates (Ott et
al., 1978) and in the case of benthic systems, in reduced rates of
sediment reworking, e.g., Arenicola marina (Gordon et al., 1978).

Respiration Respiration rates are very labile and can be affected in
both directions by many factors other than petroleum. Again, respiration is only one factor in an organism's total energy budget. Nonetheless, respiration is easily measured by a range of laboratory methods
and can serve as an indicator of the organism's well-being, provided
the proper controls are included. For that reason it has formed the
basis for many oil-related studies over the past 10-15 years.

Depression of respiration has been observed in a range of marine
organisms, including crustacea, mollusks, and fish (for example,
Malins, 1977; Gilfillan and Vandermeulen, 1978; Fisheries Research
Board, 1978; Thomas and Rice, 1979). Again, the results are variable,
and in some instances respiration has been found to increase in response
to certain oils. To date, there is no clear relationship between oil
type and the observed response. In general, petroleum levels of 1
mg/L, or higher are required for effects to be visible in juvenile and
adult crustacea and mollusks. In fish and planktonic crustacea the
effects are observed in the laboratory at concentrations of petroleum
less than 1 mg/L, approaching those levels observed under oil spill
conditions.

Growth Growth in animals bears directly on the capacity of the organism
to survive in its environment. Measured in a variety of ways (changes
in length or weight, scope for growth), growth has been found to be
markedly affected by exposure to petroleum in a variety of marine organisms. Reduced growth in oiled field situations has been observed,

reproducibly, in both mollusks and fish (Gilfillan et al., 1976, 1977a,b; Gilfillan and Vandermeulen, 1978; Desauney, 1981; J.W. Anderson et al., in press). Particularly susceptible are those benthic organisms, such as bivalves, inhabiting chronically oiled sediments.

The degree of reduction in growth seems to be a function of the animal's feeding mode; thus in mollusks the deposit-feeding species appear to be more affected than the filter-feeding species (Augenfeld, 1980; Roesijadi and Anderson, 1979). In fish, growth depression is most pronounced in such species as flatfish, which live in continued intimate contact with oiled bottom sediments (e.g., McCain et al., 1978). Transfer of hydrocarbons from such oiled sediments to associated benthic organisms occurs readily, as has been demonstrated under controlled simulated conditions.

Detoxification Systems Little information is available on the direct interaction of petroleum hydrocarbons with the more fundamental processes of metabolism such as enzyme activity and ion transport. One area that has received considerable attention in the past 7 years is that of detoxification mechanisms, mainly via a mixed function oxidase (MFO) system involving cytochrome P450. This enzyme system has been found in all marine animals investigated, including both vertebrates and invertebrates, and has been found to be readily inducible (i.e., will rise to elevated levels) in organisms on exposure to oil or certain other compounds. The MFO activity appears to be affected by several factors, including seasonality, sex, and maturity of the organism, as well as its pollution history (Stegeman, 1980). The latter becomes important in assessing MFO capability, particularly in coastal waters with chronic low level pollution. The inducibility of the enzyme system appears to be in response to the polycyclic aromatic hydrocarbon content of the oil, particularly the 3-to-5-ring aromatics, such as benzo(a)pyrene, chrysene, and benzanthracene (viz., Chapter 4).

Several field studies report elevated MFO activity in fish taken from contaminated areas (Burns, 1976; Bend et al., 1978, 1979; Bend, 1980; Iwaoko et al., 1977; Vandermeulen et al., 1978; Payne et al., 1978b; Spies et al., 1982). The only exception to this inducibility appears to be the bivalve mollusks, which even after several years in chronically oiled sediments did not show elevated MFO levels in their tissues (Vandermeulen and Penrose, 1978).

Because of the correlation between enhanced MFO activity and environmentally available aromatic hydrocarbons, e.g., MFO in English sole versus sediment benzo(a)pyrene levels (Dunn, 1980), there has been interest in using this system as an environmental indicator of oil pollution (Dunn and Stich, 1975, 1976; Payne, 1976). However, many factors can clearly influence the degree of inducibility of the MFO system. In addition, there are fundamental differences between the MFO system in fish and that found in invertebrates. Also, there are many other compounds besides those derived from oil that may induce MFO activity in marine organisms, including some organic compounds of biogenic origin.

<u>Metabolism</u> Various studies have shown oil impact on such metabolic functions as lysosome stability (M.N. Moore et al., 1978; M.N. Moore, 1979), taurine/glycine ratios in bivalves (Roesijadi and Anderson, 1979), oxygen to nitrogen ratios and lipid utilization in zooplankton (Vargo, 1981; Capuzzo and Lancaster, 1981), hepatic lipogenesis in certain fish (Stegeman and Sabo, 1976; Sabo and Stegeman, 1977), and reduced plasma copper (Dillon, 1981). Changes in plasma chloride have been noted in fish exposed for 6 months to floating oil in simulated environments (Payne et al., 1978). Together, these and other observations indicate the general toxicity of petroleum at the metabolic level of cell and organ function. Most of these observations come from laboratory studies, but interesting findings come from recent studies on changes in digestive enzyme levels in zooplankton from <u>Amoco</u> <u>Cadiz</u> impacted waters (Samain et al., 1979). Analyses of zooplankton taken from the English Channel during and following oiling from the tanker showed changes both in the individual levels and in the ratio of trypsin to amylase. While there is, of course, again the question of relating the observed effects to the spilled oil in the water column, this approach would appear to be a useful avenue to explore further.

Behavior

Behavior may be considered the first line of defense for dealing with an environmental perturbation (Slobodkin, 1968), and the response may be avoidance by movement or other changes that would tend to reduce exposure.

Oil, at sublethal concentrations, can significantly alter the behavior of marine organisms, including both microorganisms (bacteria, motile phytoplankton) and invertebrates, fish, and larger organisms. Changes in this behavior ultimately can reflect on such processes as feeding, reproduction, and larval settling. In microorganisms the behavior patterns are primarily those of changes in motility. In higher organisms they include more complex patterns of general activity, avoidance, burrowing, feeding, and reproduction.

Microorganisms

Sublethal concentrations of oil can change the motile behavior of unicellular organisms or alter important metabolic processes that are closely allied with the motility apparatus. Several ecological consequences have been suggested, including inhibition of bacterial-mediated nutrient regeneration and pollutant removal, disruption of intermicrobial predation and of alga-bacterial interactions, and prevention of phenomena mediated by the settling of mobile microbes on surfaces (Mitchell and Chet, 1978).

Bacteria are repelled by several known components of petroleum, including benzene, aniline, and phenol; thresholds for detection average 10^{-4} M (Young and Mitchell, 1973; Tso and Adler, 1974). Most repellents are cytotoxic at concentrations well above those which produce negative chemotaxis. Among eucaryotic cells, as in the algae, important attractants include the terpene hydrocarbon derivatives, which act as sexual pheromones for the male gametes of the marine brown algae Fucus, Sargassum, and Dictyota (Kochert, 1978; Kajiwara et al., 1980; Muller et al., 1981). More recently, Vandermeulen et al. (in press) have described altered swimming patterns in the unicellular phytoplankton Monochrysis lutheri in response to both whole oil and individual hydrocarbons, at concentrations well below those eliciting effects on photosynthesis and other physiological processes (<500 µg/L Kuwait and naphthalene).

The mechanism of inhibition or blockage is not understood. Bacteria detect chemical stimuli via specific protein chemoreceptors, some of which double as active transport enzymes for the substrates with which they combine. The resulting "signal" is transduced to the flagellar apparatus via separate membrane-bound chemotaxis proteins (MacNab, 1978). There is circumstantial evidence for the existence of highly specific membrane-bound chemoreceptors in algal gametes as well. Chemotaxis can be inhibited by blocking chemoreception, signal transduction, or the normal functioning of the flagellar apparatus. These various processes are dependent upon the normal functioning of the cell membrane, which therefore provides a highly accessible target for the action of various petroleum hydrocarbons.

Positive chemotaxis functions to maintain bacterial cells in a nutritionally favorable environment (Bell and Mitchell, 1972). Negative responses serve to remove cells from potentially toxic conditions. Prevention of normal chemotactic behavior can inhibit what is in fact an important contribution to the general homeostatic mechanism of the bacterial cell and thus adversely affect microbial activity. Interference with sexual pheromone reception in algal gametes has obvious implications for reproductive success.

Additional ecological studies are needed before we can adequately assess the effects of sublethal concentrations of petroleum hydrocarbons on native microbial populations. An oil spill is likely to result in an initial reduction or even inhibition of many aspects of native microbial activity, including chemotaxis (Bartha and Atlas, 1977). However, oil pollution creates a new set of intensely selective environmental conditions which rather quickly result in the development of a hydrocarbon-based microbial ecosystem (Barsdate et al., 1980). This may bring with it the development of certain resistances to the otherwise toxic effects of the hydrocarbons as, for example, carriage of plasmids conferring the ability to metabolize components of oil (Hada and Sizemore, 1981). Bacteria isolated in the presence of petroleum hydrocarbons have been found to exhibit normal chemotactic responses in the presence of these compounds, unlike those in non-oil-exposed cells, suggesting an underlying cellular resistance to their effects (Bitton et al., 1979). Such bacteria may be more representative of the microbial populations which develop after oil spills.

Higher Organisms

Petroleum hydrocarbons may cause large alterations in the behavior of marine invertebrates and fishes. However, critical comparison of results is difficult because of the variation in the specific behavior measured, the diversity of the species studied, and in most cases, the insufficient data on exposure conditions.

The influence of petroleum hydrocarbons on behavior of invertebrates and fish is less well understood than it is at the microbial level. While a small number of studies have shown that behavior, which is mediated by chemoreception, can be affected by oil, it has yet to be demonstrated whether effects are manifested at the sensory or at the behavioral level. A similar lack of understanding of the underlying causes of a behavioral alteration caused by oil is characteristic of most of the behavioral studies on the higher organisms published to date.

Avoidance Invertebrates and fish and possibly other marine organisms (viz., marine mammals, seabirds) can avoid polluted waters, although only a very small number of experimental studies have dealt specifically with avoidance of oiled waters. This is regrettable, especially in view of the many claims that pelagic adult fish, for example, have the ability to avoid spills. In fact, there is little evidence to support this claim. Wherever avoidance has been studied, however, it always is dependent on the species and on such factors as concentration and type of petroleum (Percy, 1977), the aquatic environment, the season (Rice, 1973), as well as on the internal state and ecological requirements of the species.

Even when an animal does avoid oil successfully, it is not necessarily the appropriate response. For example, in the process of avoiding, other critical resources may be denied, such as food or shelter. This is shown in a study of the effect of oiled sediment on the burrowing behavior of the littleneck clam Prototheca staminea. In this case, experimental clams remained closer to the surface in oiled sediments than did individuals residing in clean sediments (Pearson et al., 1981), thereby avoiding the deeper oiled sediments but at the same time becoming more susceptible to predation. As the tests included a natural predator, the Dungeness crab, predation rates were observed to increase significantly. Similar impact on oiled hard clams (Mercenaria mercenaria) has been described by Olla et al. (1983). Thus risk factors can be shifted from those associated with petroleum toxicity to those associated with habitat, e.g., predator-prey interaction.

Chemoreception Blumer (1969) was possibly the first to suggest that petroleum hydrocarbons could produce serious consequences by interfering with chemoreception. As chemical senses play a major role in mediating critical aspects in the behavior of marine organisms, including feeding, reproduction, habitat selection, and predator recognition, the implications of this phenomenon may be far reaching. Indications of this possibility have come out of the work of Kittredge et al. (1974), who described alterations of sexual/mating behavior in the shore crab

<u>Pachygrapsus</u> exposed to oil, and received confirmation from the studies on altered lobster feeding behavior in the presence of crude oil, carried out by Atema (e.g., Atema et al., 1973; Atema and Stein, 1974). To this list must be added decreased reproductive success of certain eucaryotic macroalgae through interference with pheromone chemoreception by male gametes (Derenbach and Gereck, 1980). As yet there is no certainty that these alterations were at the sensory level. The most convincing evidence thus far, without the use of neurophysiological techniques, is the observation that the chemosensory antennular response in the Dungeness crab <u>Cancer</u> <u>magister</u> is impaired by low concentrations (less than 1 ppm) of petroleum hydrocarbons (Pearson et al., 1981).

<u>Feeding</u>   Feeding and behavior associated with feeding are well known to be affected by sublethal concentrations of petroleum as low as a few µg/L. This includes reduction in the capability to respond to a food source (Jacobson and Boylan, 1973; Pearson et al., 1981), recognition of food (Atema et al., 1973), and behavior associated with location of food (Atema and Stein, 1974). Measurements of amounts of food ingested have also shown effects of petroleum, both in animals residing in the water column (Berdugo et al., 1977; Berman and Heinle, 1980) and in the sediment (Prouse and Gordon, 1976; Gordon et al., 1978).

Recurrent throughout these various studies is the observation that organisms vary widely in their response to petroleum or its components. This was shown perhaps most conclusively in the response differences to oil of two arctic marine amphipods and two isopod species (Percy, 1977). The one amphipod species, <u>Onisimus</u> <u>affinis</u>, avoided oil-contaminated sediment. In contrast, this response in a second amphipod, <u>Corophium</u> <u>clarencense</u>, and of two isopods, <u>Mesidotea</u> <u>entomon</u> and <u>M</u>. <u>sibirica</u> were either totally absent or much reduced. This illustrates the point that generalizations regarding behavioral responses to oil and regarding certain common species are not reasonable. It also underscores the risk in the use of "indicator species" or "indicator organisms" in assessing and predicting generalized effects of oiled ecosystems (Olla, 1974; Olla et al., 1980a,b).

EFFECTS ON THE MARINE FOOD WEB

Food Web Microbes

The effect of oil on marine microorganisms depends on the type and amount of oil spilled, the physical nature of the area (e.g., open ocean or estuarine marsh), nutritional status, oxygen concentration, and previous history of the impacted area with regard to hydrocarbon exposure. Such prior impact may be reflected in changes in number and types of microorganisms, as well as in their chemical composition and changes in microbial activities.

Light crude oils and refined products tend to be more toxic and to affect biological activities more than the heavy crude oils. Chronic low level spills, which allow time for selection of naturally occurring microbial degradation of oil. Both types of spills produce changes in the community structure and activities of the microbial population, the chronic one less dramatically, the catastrophic one very quickly. The net result is that those sensitive to hydrocarbons will be killed or their growth suppressed, while those with the genetic potential to utilize hydrocarbons as a carbon and energy source will grow and increase in numbers and/or biomass.

The best documented microbial response to the intrusion of oil in marine systems is the increase in proportion of oil-utilizing bacteria to the total heterotrophic bacteria and is summarized by Atlas (1981): "In unpolluted ecosystems, hydrocarbon utilizers generally constitute less than 0.1% of the microbial population; in oil polluted ecosystems, they can constitute up to 100% of the viable microorganisms. The degree of elevation above the unpolluted comparison reference sites appears to quantitatively reflect the degree or extent of exposure of that eco-system to hydrocarbon contaminants." Such increases usually refer to bacteria, but Ahearn and Meyers (1972) reported a selective effect of oil on the incidence of marine hydrocarbon degrading yeast and/or fungi. Hence, the fate of spilled petroleum and, therefore, an oil-contaminated environment, lies in the microbes' ability to use hydrocarbons as sources of carbon and energy.

Because of the problems in the enumeration and identification of marine heterotrophic bacterial populations (Staley, 1980), little information is available on the effect of oil on species diversity. Species of oil-utilizing bacteria that have been isolated varied with the geographic site of the sample and differed between pelagic and benthic samples taken from a given site (Colwell and Walker, 1977). An increase in the number of plasmid-containing (extrachromosomal genetic material that enhances biochemical activities including hydrocarbon oxidation) strains of _Vibrio_ spp. and a greater number of plasmids per strain were found in samples from an active oil field in the northwest-ern Gulf of Mexico (Hada and Sizemore, 1981). The authors suggested that oil field discharges might be responsible for this increased plasmid incidence.

Petroleum has been shown to have an effect on the uptake and mineralization of low-molecular-weight compounds such as glucose, glutamate, acetate, and glycoxolate. Griffiths et al. (1981) reported that mineralization of glutamic acid and glucose was inhibited by Cook Inlet crude oil and related products, the pelagic populations being more sensitive to the effect of oil than benthic ones. Regions with a history of chronic hydrocarbon input showed less sensitivity to the effects of oil than pristine ones.

The addition of oil to environmental samples has resulted in a decrease in microorganisms showing proteolytic, chitinolytic, and cellulytic activities (Walker and Colwell, 1975). Similarly, Griffiths et al. (1982b) found that Cook Inlet crude oil reduced the levels of cellulose and chitinase activities observed in sediments but stimulated the activities of enzymes involved in degradation of starch and algin. Nitrogen fixation (acetylene reduction) in northern marine sediments has been reported to be significantly reduced by exposure to "fresh" but not "weathered" Prudhoe Bay oil (Griffiths et al., 1982b).

Effects on Marine Plankton

Phytoplankton

There are ample observations that petroleum hydrocarbons can have immediate and marked effects, in terms of minutes or hours, on the rate of photosynthesis of natural phytoplankton assemblages (Shiels et al., 1973; Gordon and Prouse, 1973; Lacaze, 1974; Le Pemp et al., 1976; R.F. Lee and Takahasi, 1977; Brooks et al., 1977; Hsiao et al., 1978; Federle et al., 1979). Laboratory observations with unialgal cultures and pure cultures have provided a similar picture (Pulich et al., 1974; Soto et al., 1975a; Parsons et al., 1976; Vandermeulen and Ahern, 1976; Kusk, 1978; Karydis, 1979). Both laboratory and field studies clearly show that hydrocarbons can inhibit algal growth, although at the lower concentrations of oil, occasionally an enhancement is noted (e.g., Gordon and Prouse, 1973). Unfortunately little more is available to present a better understanding of both the toxic effects and the impact on natural populations under spill conditions.

Growth and Metabolism Most of the research to date into the effects of oil on phytoplankton has dealt with either the effects on culture growth or on photosynthesis. Presumably that is because of the ease with which these can be assayed, either in the laboratory or on shipboard. Far less attention has been directed to research on effects on the more fundamental aspects of metabolism or cellular fine structure. And indeed most of what is known of the latter comes from work with freshwater species.

Growth of phytoplankton is readily depressed by a wide range of petroleum hydrocarbons, including both whole oils as well as specific compounds. This includes blue-green algae, green algae, diatoms, dinoflagellates, and chrysophytes. Effects on growth vary widely, depending on the oil or compound used and on the algal species, but generally growth lags or lethality has been noted in the range of 1-10 mg/L (e.g., Mommaerts-Billiet, 1973; Pulich et al., 1974; Soto et al., 1975b; Prouse et al., 1976; Batterton et al., 1978b; Hsiao, 1978; Mahoney and Haskin, 1980). Recent work suggests that algal sensitivities may not only be species specific, but also clone specific depending on the environmental origin of the clone (Eppley and Weiler, 1979).

The growth response varies considerably and is somewhat dose respondent. Thus at low concentrations the main response consists of a lag period, after which culture growth occurs normally, growing through the usual log phase and reaching the plateau stage. Cell size under these conditions does not differ from normal control cultures. At higher concentrations the inhibition becomes increasingly marked, with little or no growth at the highest concentrations used. Interestingly, at very low concentrations (<0.1 mg/L) growth enhancement has been observed in both laboratory and field collections (e.g., Prouse et al., 1976).

Photosynthesis is equally depressed in phytoplankton exposed to petroleum. Enhancement, at very low hydrocarbon concentrations, as seen for growth has not been noted. One study has examined the potential recovery of a phytoplankton species after exposure to naphthalene (Vandermeulen and Ahern, 1976). On return to noncontaminated medium, the culture, <u>Monochrysis lutheri</u>, showed a partial recovery of photosynthesis following initial inhibition.

Jordan et al. (1978) reported that nitrogen fixation in the epiphytic blue-green alga <u>Nostic</u> sp. failed to recover for 1 year following a single dose of ASA 90 crankcase oil.

<u>Effect on Populations</u> To date, no mass toxicity to phytoplankton has been reported from the field, either from a spill or for chronic input conditions. In part this is due to the fact that very few field studies involving phytoplankton have been done during an oil spill. One data set that is available comes from a sampling program carried out fortuitously when the tanker <u>Kurdistan</u> broke up off Nova Scotia (O'Boyle, 1980). No effect of any sort was observed on phytoplankton of the Scotia Shelf waters. However, the spill occurred in a prebloom period, and although oil patches and tar were reported from the Scotia Shelf (Vandermeulen, 1980), there is little indication that the algae in fact encountered appreciable oil concentrations. Observations on phytoplankton biomass and primary productivity carried out following the <u>Tsesis</u> spill (in Sweden, 1977, No. 5 fuel oil) revealed no significant differences between noncontaminated and contaminated areas (Johansson et al., 1980). In fact, if anything, for a brief postspill period both primary productivity and cell numbers were found to be slightly higher in the contaminated areas, perhaps because of reduced grazing by zooplankton. This points out the difficulty of studying phytoplankton in isolation from the rest of the ecosystem.

Even if a large number of algal cells were affected during a spill, regeneration time of the cells (9-12 hours), together with the rapid replacement by cells from adjacent waters, probably would readily obliterate any major impact on a pelagic phytoplankton community. The situation may not be as salutory, however, in more chronically oiled waters such as inshore and coastal embayment systems, where flushing rates may be low and where concomitant oil contamination may be higher and prolonged. Under those conditions, periodic events such as plankton blooms which are critical to oceanic planktonic processes may well be affected.

Zooplankton

The impact of oil on zooplankton has been studied extensively since the 1975 NRC report, primarily because of the importance of zooplankton in marine ecosystems as secondary producers. Work has been carried out in both the laboratory and at spill sites, as well as in a number of "mesoscale" field enclosures ("mesocosms").

The responses to petroleum exposure are numerous (e.g., Kuhnhold, 1977; Corner, 1978). As individuals, most zooplankters studied to date in acute and chronic exposure experiments in the laboratory and the field appear to be highly vulnerable to dispersed and dissolved petroleum constituents, and less so to floating oils. The acute lethal toxicity of dispersions and WSF, mostly expressed as 96-hour LC_{50} using initial measured concentrations, range between 0.05 and 9.4 mg/L, with a few higher values (Table 5-5). These values, based on measurements of actual rather than nominal concentrations, are very close to the lethality thresholds predicted, 0.1-10 mg/L, of soluble hydrocarbons for fish eggs, larvae, and pelagic crustaceans, derived from an earlier intensive survey of the literature (S.F. Moore and Dwyer, 1974). It must be noted that sublethal deleterious effects can set in well before these high concentrations are achieved.

Many components of the zooplankton should, therefore, be considered sensitive to dispersed and solubilized hydrocarbons in seawater, based on these laboratory studies and on the combination of observations in field enclosures and at oil spill sites. On the positive side, concentrations of hydrocarbons in open waters may not persist long enough to always cause many of the toxic effects, either lethal or sublethal (McAuliffe, 1977; also Chapter 4 of this report).

Based on the available data on lethality of WSF (Table 5-5), there may be no marked differences in vulnerability among the planktonic ctenophores, mollusks, crustacea, and teleosts. This comparison should be considered very tentative, as the data were not collected in one laboratory or under one set of conditions.

Enclosure experiments with zooplankton (Lytle, 1975; R.F. Lee and Anderson, 1977; R.F. Lee and Takahashi, 1977; R.F. Lee et al., 1977, 1978a,b; Davies et al., 1980; Elmgren et al., 1980a; Vargo, 1981) (see also Chapter 3, Biological Methods section) point out the different capabilities within natural but confined zooplankton communities for accommodating the presence of hydrocarbons. These experiments also show the type and variability of toxic responses (lethality, lowered feeding and reproduction, community changes) that might be expected among natural zooplankton communities when continuously exposed to low levels of oil-derived hydrocarbons.

Field observations on zooplankton have now been made at several spills and chronically polluted sites. Collectively, these studies have shown that biological effects and changes have been detected (Spooner, 1978; Johansson et al., 1980), but these appear to be short lived; there are seldom significant prolonged changes in biomass or standing stocks of zooplankters in the open water near spills. Individual organisms in spills have been affected in a number of ways: direct mortality (fish eggs, copepods, mixed plankton), external contamination by oil (chorion of fish eggs, cuticles and feeding appendages of crustacea), tissue contamination by aromatic constituents, abnormal development of fish embryos, possibly temporary inhibition of feeding in copepods, and altered metabolic rates (Longwell, 1977; Gilfillan et al., 1983; Samain et al., 1980, 1981). In addition, ingested oil has been seen in copepods on some occasions. But zooplankton populations and communities experiencing spills or chronic discharges in open waters appear to recover eventually and maintain themselves, due largely to their wide distribution and rapid regeneration rates (Michael, 1977). This type of recovery may not occur in enclosed waters. In fact, Sanborn (1977) has suggested that the lowered zooplankton production in the Caspian Sea may be a reflection

TABLE 5-5 Lethality (Median Lethal Concentrations) of Physically Dispersed and Water-Soluble Fractions of Various Oils to Marine Zooplankton

Group	Dispersion				Water-Soluble Fraction		
	Crude Oil	No. 2 Fuel	No. 4 Fuel	No. 6 Fuel	Crude Oil	No. 2 Fuel	No. 6 Fuel
Protozoa (ciliates)					1.7(90h)[1]		
Ctenophora						0.59(1d)[2]	
Mollusca							
Clams							
Embryo					0.23-12(2d)[3]	0.43(2d)[3]	1.0(2d)[3]
Larvae					0.25->25(2d)[3]	1.3(2d)[3]	3.2(2d)[3]
Larvae					0.05-2.1(10d)[3]	0.53(10d)[3]	1.6(10d)[3]
Pteropods		<0.2(2d)[4]					
Crustacea							
Barnacles				5.1[8]		2.6(1-h LC_{50})[5]	
Copepods	>13,5[6], 73[7]					1.0(3d)[9]	
Amphipods larvae	0.8(2d)[10]	0.3(2d)[10]	6.2(2d)[10]			2.5(3d)[9]	
Decapods							
Shrimp larvae	>1,000[14]				0.5-8.5[11-13]		
Postlarvae		1.7[15], 9.4[14]			>19.8[14]	1.2-6.6[15,16]	1.9[14,15]
Lobators	0.86-4.9[17]						
Crab larvae	0.14(30d)[17]				1.3->10.8[11-13]		
Mysids	0.05-0.17[19]				1-12[20]	1.5[21]	0.02-0.2(6d)[22]
Teleosts							
Eggs	43[23]						
Larvae							

NOTE: Values given are for 4-day LC_{50} unless otherwise noted. All concentrations are in mg/L, total measured (extractable) hydrocarbons, calculated using initial concentrations measured by spectroscopic and chromatographic methods.

[1] Lanier and Light (1978), [2] R.F. Lee (1978), [3] Byrne and Calder (1977), [4] Winters et al. (1977), [5] Blundo (1978), [6] R.F. Lee and Singer (1980), [7] Sekerak and Foy (1978), [8] Hollister et al. (1980), [9] W.Y. Lee and Nicol (1977), [10] Linden (1976a), [11] Mecklenburg et al. (1977), [12] Brodersen et al. (1977), [13] Rice et al. (1976), [14] J.W. Anderson et al. (1974), [15] Tatem et al. (1978), [16] Neff et al. (1976), [17] Wells and Sprague (1976), [18] Cucci and Epifiano (1978), [19] Wells (1976), [20] see Table 5-7, [21] Sharp et al. (1979), [22] Kuhnhold (1978), and [23] Vuorinen and Axell (1980).

of this. Likewise, if eggs or larvae of a species are highly concentrated in the area of a large spill or chronic discharge, individual and population responses may be greater, and recruitment may become affected (Food and Agricultural Organization, 1977). Such an area is exemplified by the under-ice environment of the poles, where the under-ice communities appear very vulnerable to spilled oils (see Polar Environments section).

The main routes of zooplankton contamination by oil are direct uptake from the water, uptake from food, and direct ingestion of oil particles. Metabolic capabilities to metabolize and detoxify hydrocarbons appear to vary somewhat among zooplankton. Thus in scyphozoans and ctenophores they are discharged chemically unchanged, while in crustaceans and ichthyoplankton they are discharged as metabolites. Depuration of the zooplankton tissues is often incomplete, even after 1-month exposure to clean seawater, as in some copepods, euphausiids, and shrimp larvae (R.F. Lee, 1975; Lee and Anderson, 1977; Lee and Takahashi, 1977).

Adverse sublethal effects have been observed among oil-exposed zooplankton at concentrations often well below 1 mg/L (total measured hydrocarbons). The exposure times involved varied from several hours (behavior, physiological perturbations), to days (developmental abnormalities), to weeks (growth, developmental, and reproductive problems). Sublethal responses of many important groups of zooplankters have not yet been adequately studied, or given any attention at all, with the bulk of the studies focused on a few groups.

The high sensitivities of certain developmental stages, especially before and during fertilization and during early embryonic development, hatching, and larval phases, suggest that substantial damage may occur to localized populations of zooplankton during oil spills. Indeed, some acute damage has been observed at spills. Particularly vulnerable would be those stages that float or swim weakly near or on the surface. Less vulnerable would be the more advanced stages and adult zooplankton capable of diel migration. Much of the susceptibility of zooplankton populations will depend on the persistence of oil in the water column after a spill or under chronic discharge conditions. For example, the susceptibility of local populations of surface-dwelling zooplankton to significant sublethal damage would be high if total hydrocarbon concentrations greater than 0.05-0.3 mg/L persist for days or a few weeks. Lethal effects would be expected if concentrations between 0.5 and 1.0 mg/L persist. However, such concentrations rarely persist for longer than a few days following a spill, and only in limited areas (see Chapter 4 and Appendix A).

The recovery of an oil-impacted zooplankton community is probably fairly rapid, mainly due to recruitment from other areas and to intrinsic biological characteristics of wide distribution, large numbers, short generation times, and high fecundity. Little is known of the physiological recovery of oiled zooplankton, but this may be of little importance in open water in view of the other recovery factors. In more enclosed waters, however, where recruitment from outside becomes less important, these intrinsic factors may well become limiting to recovery. Chronic pollution may also pose long term problems.

TABLE 5-6 Experimental Conditions of Three Field Enclosure Systems or Mesocosms

Condition	MERL, Narragansett Bay[a]	Loch Ewe, Scotland[b]	CEPEX, Saanich Inlet, B.C., Canada[c]
Volume of water column	13.1 m^3	304 m^3	60 m^3 (?)
Number of columns	14	3	
Mixing	yes	no	no
Flow-through/ turnover time	30 days	no	no
Oil	No. 2 fuel	North Sea (Forties Field)	Prudhoe Bay crude No. 2 fuel
Concentration	Mean=181 ppb and 93 ppb in two experiments	100 ppb	10, 20, and 40 ppb initial concentration
	Oil-water dispersion 2/week for 163-122 days	Several additions over 2 weeks to reach initial concentration; declined to 25 ppb over 5 weeks	Single additions
Total added	414 and 153 mL	30 g	
Major effect	Reduced zooplankton at 190 ppb; Reduced benthic fauna at 90 and 190 ppb[d]	Loss of calanoid copepods due to increase in predators[e]	Changes in biomass, and zooplankton and phytoplankton communities[f]

[a]Elmgren et al. (1980).
[b]Davies et al. (1980).
[c]R.F. Lee et al. (1977, 1978a,b).
[d]Oviatt et al. (1982).
[e]Giesy (1980).
[f]R.F. Lee and Takahashi (1977).

Studies in Field Enclosures (Mesocosms)

Difficulties in studying the effects of oil on biota in the water column, in open water, and the problems in overcoming extrapolation from the laboratory results to the field, have led to the development of several "mesocosm" experiments. These have been in the form either of large tanks situated on land and inoculated with a volume of nearby bay water (e.g., Gordon et al., 1976; MERL, Narragansett Bay) or have consisted of large, suspended bag systems placed in the field (Loch Ewe, Scotland; CEPEX, Saanich Inlet, B.C., Canada) (Table 5-6).

These enclosure experiments allow simultaneous testing of the effects of relatively low levels of oil on both phytoplankton and zooplankton and their interactions. They also permit simulation of both the effects of transitorily elevated concentrations in the water after an oil spill by using single oil additions and the effects of chronic low level additions to mimic the input from refineries, terminal activities, sewage effluents, and drainage in the coastal zone. The advantage of the mesocosm approach is that this scale lies midway between the laboratory setup and the open field. Mesocosms allow one to run controls and to have known concentrations of pollutants and yet still retain many or most of the species interactions, animal-sediment interactions, etc., which occur in natural systems.

The strength of the mesocosm studies has been the ability to work at low concentrations (less than 100 µg/L) and to use environmentally realistic marine population assemblages. With these various schemes, detrimental effects were found at approximately 90 µg/L in the MERL studies, the lowest concentration tested there (Elmgren and Frithsen, 1982), and at 20 µg/L in the CEPEX experiments (R.F. Lee and Takahashi, 1977). The effects on the enclosed plankton and microbial communities differed from experiment to experiment, but in each instance there were profound changes in the balance between species in the zooplankton and phytoplankton communities and in the biomass of the compartments. These changes included shifts in dominant species (e.g., a sharp decline in the abundance of the dominant diatom Ceratulina bergonii, CEPEX), radical changes in phytoplankton species composition and marked increase in biomass, possibly due to reduced grazing pressure by zooplankton and benthic filter feeders (MERL), decimation of a calanoid copepod population and increased numbers of predators (siphonophores and ctenophores) (Loche Ewe) (Giesy, 1980). The Loch Ewe experiment also resulted in adverse developmental effects in copepod eggs and nauplii larvae (Figure 5-3).

Clearly, the results of these three mesocosm experiments were not identical, and replication of experiments in any one mesocosm showed low precision. The results probably reflect the starting components, conditions of the various systems, and the feeding links between the systems' components. At the same time these large scale studies demonstrated that communities can experience dramatic changes and shifts due to low level oiling and that a wide range of changes can be expected at surprisingly low concentrations of oil in water. The experiments also demonstrated the usefulness of this approach and provided a much-needed intermediate step between the laboratory and studies of spills of opportunity.

Macrophytes--Intertidal and Subtidal

Little is known about oil effects on macrophytes, including the intertidal species and the larger offshore species. What there is has come largely from observations made during spills, with very little direct follow-up work and virtually no laboratory studies.

The habitats involved here include both high energy rocky shore-lines (e.g., Fucus spp.) and low energy mud flats and salt marshes (e.g., Spartina spp.) as well as the nearshore subtidal environment (Laminaria spp.). Of these, the plants growing in the intertidal environment are most vulnerable to oiling and suffer the more severe impact during a spill. Thus, during the Arrow spill in Chedabucto Bay,

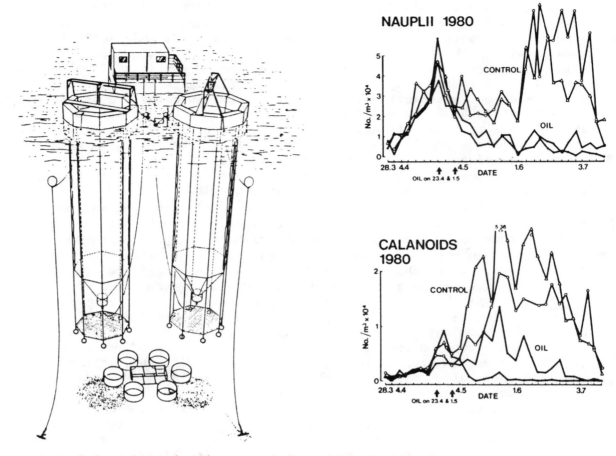

FIGURE 5-3 Schematic diagram of the offshore "mesocosm" study at Loch Ewe, Scotland, and changes in zooplankton populations observed with the introduction of petroleum.

SOURCE: Adapted from Davies et al. (1981).

Nova Scotia, masses of oil and algae, mostly <u>Fucus</u> <u>serratus</u>, washed ashore (M.L.H. Thomas, 1973), presumably having been broken away from their rocky substrate by the weight of the oil clinging to the fronds. Similar losses of intertidal algal cover have been described for the <u>Amoco</u> <u>Cadiz</u> spill where much of the intertidal zone was covered by newly arrived oil slicks for periods up to 2 or 3 weeks (e.g., Hess, 1978). However, recovery of intertidal algae appears to occur quite readily. Thus, investigations of the impact of <u>Amoco</u> <u>Cadiz</u> oil on <u>Fucus</u> species from nine oiled sites along the north Brittany coast showed growth rates to fall within the normal expected ranges (Topinka and Tucker, 1981). Some decrease in population density was observed in the red alga <u>Chondrus</u> <u>crispus</u>, economically important for its carrageenan content, at two heavily oiled sites fully a year after the spill (Kaas, 1981). However, such drops in density are not that unusual, and the intrinsic growth of the species was not affected.

Impact on more subtidal or submerged flora, while they are perhaps less vulnerable to oiling due to the depth of the water column, nonetheless does occur. For example, a chronic oil situation, resulting from a grounded troop ship, the General M.C. Meigs on the northwest coast of Washington, provided observations on oil damage to a wide range of marine flora (R.C. Clark and Finley, 1973). Damage included loss of fronds in the subtidal Laminaria andersonii and bleaching of tissues, sometimes complete, in several red algae and in the false eel grass Phylospadix sp.

These observations have led to some concern for potential oil impact on sea grass communities, especially where they represent either a commercial/economic significance (as in Southern California or North Brittany) or where they provide the basis for a valuable and complex invertebrate and vertebrate community, as in Florida (e.g., Zieman, 1982). Data on oiling impact of such submerged sea grass communities are scarce. Where it has been observed and evaluated, the impact has been varied, ranging from little (Amoco Cadiz, Den Hartog and Jacobs, 1980; Topinka and Tucker, 1981; Maurin, 1981) to severe with high mortalities (Zoe colocotronis, Nadeau and Bergquist, 1977). It would seem that the impact depends in part on the depth of the water column, the type of oil released, and the local mixing conditions.

One seemingly positive consequence of oiling is the often-observed proliferation of certain algal species, such as Enteromorpha sp., Ulva sp., and some Porphyra spp., following a spill (e.g., J.E. Smith, 1968). However, such proliferation is invariably a direct result of the elimination, by the oil, of the naturally occurring grazers-limpets and other intertidal herbivores (e.g., North et al., 1964; Southward, 1982). With the elimination of these consumers other algae proliferated, and by virtue of their spreading green cover gave the appearance of rapid recovery by the oiled vegetation.

In fact, long term recovery of oiled macrophytes is variable, depending in part on their intrinsic tolerance. Both laboratory tests and field observations suggest that the various species of bladder wrack (Fucales) are resistant to moderate levels of petroleum and to brief exposures (Ganning and Billing, 1974; Ravanko, 1972; Notini, 1978; Percy, 1981). There has been one suggestion that this resistance may be due to the inability of the oil to cling to the mucilagenous coating of the plant wall (Nelson-Smith, 1973). Another possibility is that the wracks are not rooted in sediments, which absorb oil readily, but live attached to a rocky substrate above the oiled sediments and are therefore subject to oil concentrations much lower than found in the water (Vandermeulen and Gordon, 1976).

While generally speaking the reestablishment of a floral community may be quite rapid, especially following cleaning, imbalances in the recovering community can persist for several years. Final return to a stable floral community similar to nonoiled communities probably requires a decade or more (M.L.H. Thomas, 1978; Southward and Southward, 1978). In one instance, the Arrow spill, one species of bladder wrack, Fucus spiralis, that had disappeared during the spill in 1970 had not yet reappeared at the time of a 1976 follow-up survey (M.L.H. Thomas, 1978).

One interesting feature of oil impact and biological recovery is delay in the decline of certain species following the initial oiling incident. Thus some species do not necessarily exhibit maximum impact until a year or two after the spill, e.g., Spartina sp. (M.L.H. Thomas, 1973). Too few data are available to date to determine if this is a significant feature of macrophyte recovery following oiling, but if so, this could have some significance for assessing spill impact.

One area that has been totally neglected is that of spill impact on tropical or warm-water microphytes and macrophytes. Of particular concern in this respect are the giant algal flats that constitute important components of tropical trophic systems and barrier reef systems. These are highly vulnerable to oiling because of their shallowness. Except for a single follow-up study (Lopez, 1978), nothing is known of their recovery potential following oiling.

Benthic and Intertidal Invertebrates

Considerable work has been done studying the effects of oil on the macroinvertebrates, mainly the intertidal species, from both physiological and population viewpoints. It is in this group of organisms also that most advances have been made in determining much more precisely the relationships between oil-hydrocarbon concentration, exposure time, and toxic responses, mainly because of the growing realization that a description of effects without mention of these relationships is of little value. The intertidal species have been the subjects of choice: they are obvious victims of oil spills, several are of economic significance, and several serve well as experimental subjects in laboratory studies. Hence, a good deal of information has been assembled on bivalves (Mya arenaria, Mytilus edulis, Crassostrea sp., Littorina sp.), crustaceans (crabs, lobster, amphipods, and isopods), and annelids (Arenicola spp.). Results of these studies can be found in several reviews and sources published over the last 5 years (Wolfe, 1977; Malins, 1977; Fisheries Research Board, 1978; Neff, 1979; Sprague et al., 1981; Sampson et al., 1980; Connell and Miller, 1980).

Generalizations about the vulnerability of the benthic and intertidal invertebrates are difficult because there exists a great deal of variation among the genera and species, as well as among the various life cycle stages of any one species (Figure 5-4). However, intertidal invertebrates, although highly vulnerable to oiling, may exhibit some slightly greater tolerance to petroleum hydrocarbons than do offshore benthic or pelagic species. Also, juvenile and molting stages tend to be more sensitive than the mature adult stages. However, these two statements must be made with considerable care, and each instance should be interpreted in its own context.

Vulnerability

Most susceptible are those species inhabiting the intertidal zone, especially those found in lagoons, embayments, estuaries, marshes, and

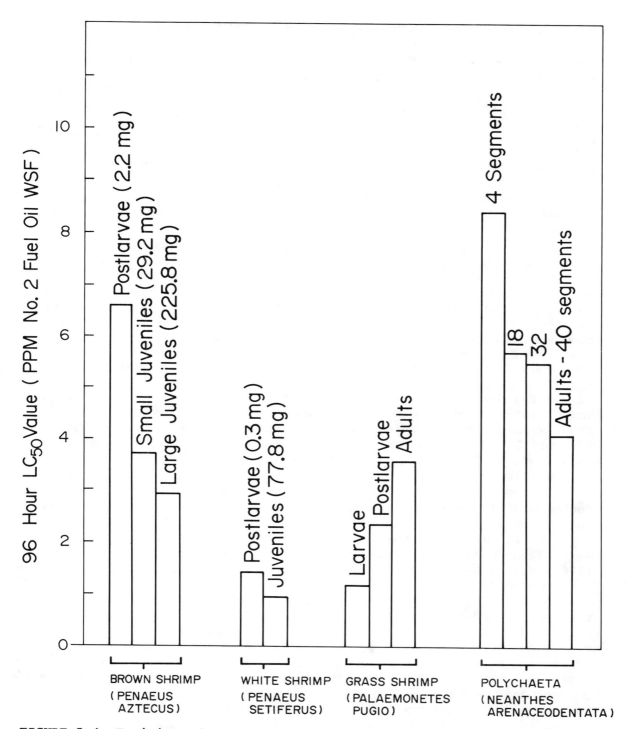

FIGURE 5-4 Toxicity of No. 2 fuel oil to life-cycle stages of selected marine shrimp and polychaetes. Life-cycle stages are indicated by size or segment number.

SOURCE: Adapted from Neff et al. (1976) and Rossi and Anderson (1976).

tidal flats. This risk derives from two factors: high oil concentrations and shallow depth of the water column. Oil coming ashore during spills becomes highly concentrated in a very narrow band along the shoreline, while the shallowness results in high concentrations throughout the water column down to the sediments. These factors combine to raise the hydrocarbon concentrations in the intertidal zone beyond the tolerance thresholds of most organisms. These same low energy areas are also prime targets for chronic pollution, either from catastrophic spills or from chronic input, for the fine sediments characterizing these environments tend to act as sorbents or "sinks" for the spilled petroleum components (e.g., Blumer et al., 1973; Sanders et al., 1980; Vandermeulen and Gordon, 1976).

Besides the physiologically toxic effect of oil, there is also a physical hazard of spilled oil, and in this respect, intertidal organisms are especially vulnerable. For example, large numbers of intertidal invertebrates were killed during the Arrow Bunker C spill simply by smothering (M.L.H. Thomas, 1973). Sessile species such as barnacles are easy prey to smothering by heavier oils (Straughan, 1972), and the more mobile invertebrates become immobilized and thus more susceptible to either the toxic effects of the hydrocarbons or fall to predators, as was also observed during the Arrow spill (e.g., M.L.H. Thomas, 1973). Following a Bunker C spill in Puerto Rico, oiled tree crabs were found stuck to the mangrove branches by dried oil (Gundlach et al., 1979). Intertidal species also are exposed to volatile components of spilled petroleum. To what extent this happens in the field is not known, but laboratory studies have shown that such volatiles also can readily narcotize Littorina sp. (Straughan, 1979), rendering them susceptible to being washed off the rocky intertidal, either being washed out to sea or becoming easy prey.

This is not to say that benthic organisms living subtidally are safe from oil impact, protected by an overlying water column. With vertical mixing of hydrocarbons into and throughout the water column, the impact can be felt subtidally, as was learned during the 1978 Amoco Cadiz spill, in which numerous subtidal razor clams and heart urchins were killed during the first weeks following the spill (e.g., Hess, 1978) (Figure 5-5). In a 1-year follow-up study, Cabioch and colleagues have monitored the disappearance of a benthic amphipod species offshore from North Brittany, correlating with increasing hydrocarbon levels in the benthic sediments down to 30 m (Cabioch et al., 1980; Beslier et al., 1980). Also Addy et al. (1978) have demonstrated that chronic oil pollution in relatively deep, offshore waters can have deleterious effects on benthic invertebrates in a limited area around an open-bottom oil storage structure. In contrast, benthos at the Arrow were contaminated but apparently unaffected (Anon, 1970).

Speculation that arctic or polar species may be more sensitive to oiling than more temperate species has been made in the literature. There is, in fact, no evidence to date that intrinsic differences in oil sensitivity exist between polar and temperate species. About the only generalization that can be made is that polar organisms exhibit the same variability in sensitivity to petroleum hydrocarbons, over much the same concentration range, as is found at other latitudes

FIGURE 5-5 Subtidal invertebrates such as sea urchins and razor clams proved to be highly vulnerable to the oil spilled from the Amoco Cadiz tanker, off the coast of France, 1978. Shown here are numbers of dead urchins and clams typically found in 1 m^2 areas during the days following the spill.

SOURCE: Hess (1978).

(Percy, 1981; see also Polar Environments section). However, there is one physical factor, that of reduced evaporation at the lower polar temperatures and therefore slower loss of the toxic lower-molecular-weight compounds, that may make a difference regarding the potentially acute impact of oil spilled in the Arctic or Antarctic. A similar effect presumably occurs in temperate regions during the winter.

Lethal and Sublethal Effects

Sublethal effects are seen in all phyla examined, ranging from alterations in respiration, growth, reproduction, and behavior to the more specific processes of calcification, molting, ion transport, and enzyme function. Much of the work so far has come from laboratory studies, which enable precise control of various experimental conditions. When the 1975 NRC report was written, the concentrations of hydrocarbons used were often unrealistically high. This excess is now being corrected, and more recently, attempts are made generally to use realistic concentrations in such experimental studies. There are very few studies of metabolic or physiological perturbations, due to oiling, that have been made at actual oil spills or at oil pollution sites (e.g., Fisheries Research Board, 1978; Centre National pour l'Exploitation des Oceans, 1981). This omission is unfortunate in that such studies can suggest new approaches or new areas of research. In an effort to close this gap in our knowledge, a number of simulated field experiments have been designed and carried out during the last few years (e.g., Roesijadi et al., 1978; Roesijadi and Anderson, 1979; J.W. Anderson et al., in press; Bieri and Stamoudis, 1977; R.F. Lee and Ryan, 1983).

The broad range of effects that oil or its components elicit can be seen from studies done with one typical benthic invertebrate, the soft-shell clam Mya arenaria. They include mortality due to smothering (M.L.H. Thomas, 1973, 1977); long term mortality (M.L.H. Thomas, 1977); altered population composition (Gilfillan and Vandermeulen, 1978); altered metabolic rates and feeding rates (J.W. Anderson et al., 1974; Avolizi and Nuwayhid, 1974; Stainken, 1978); reduced filtration and C-assimilation rate (Gilfillan et al., 1976, 1977b; Gilfillan and Vandermeulen, 1978), reduced survival, condition index, and levels of tissue amino acids (Roesijadi and Anderson, 1979); changes in tissue structure (Stainken, 1976); restriction in growth (Gilfillan and Vandermeulen, 1978; M.L.H. Thomas, 1978); and alteration in shell formation (Gilfillan and Vandermeulen, 1978). These, and other effects, not yet looked for in the soft-shell clam, are found as well in other invertebrates exposed to oil or to oil fractions. Included are effects both at the subcellular level, e.g., genetic alteration (Cole, 1978) as well as long term perturbations at the population and community levels (e.g., Elmgren et al., 1981).

An apparent enigma observed at both spill sites and in experimental studies is the puzzling survival of some benthic macroinvertebrates in highly oiled sediments. For example, several studies have noted the lack of significant mortality during exposures to concentrations of oil in sediments in excess of 1,000 μg/g (Wells and Sprague, 1976; J.W. Anderson et al., 1977; Roesijadi et al., 1978; Vandermeulen and Gordon, 1976; Gordon et al., 1978). Krebs and Burns (1977) remarked on the adverse responses of Uca sp. apparently surviving in sediments containing 1,000-7,000 μg/g petroleum hydrocarbons. Similarly, Shaw and Cheek (1976) made measurements on the clam Macoma balthica from sediments containing 640-3,890 μg/g (dry weight) hydrocarbons.

While this phenomenon is by no means fully understood, there are several aspects for which some answers now exist. First, the very toxic lower-molecular-weight components (see Toxicity section) rarely persist in highly oiled sediments in the field. Second, the feeding habits of certain of these invertebrates may preclude uptake directly from contaminated sediments because they feed instead on the lesser contaminated waters above the sediments. Third, adult stages generally show a relative resistance to the hydrocarbon toxic effects. Another possibility is that of influx of mature organisms into these oiled areas from nonoiled "seed populations." In this context, J.W. Anderson et al. (1978) describe near-normal recruitment of a range of macroinvertebrates (polychaetes, bivalves, gastropods, crustaceans) into experimentally oiled sediment plots set out in trays in the field. On the other hand, judging from field studies of biological recovery, larval recruitment under such high oil concentrations appears to be very sensitive, frequently with poor survival (e.g., Gilfillan and Vandermeulen, 1978; M.L.H. Thomas, 1978; Sanders, 1978).

Effects at the Population Level

There is ample evidence that oil and oil spills can perturb entire invertebrate populations. As alluded to above, Mya arenaria populations in an oiled lagoon of Chedabucto Bay, 6 years after the Arrow spill, still experienced recruitment problems (M.L.H. Thomas, 1978; Gilfillan and Vandermeulen, 1978). In another study, isozyme patterns in a population of snails, Urosalpinx cinera, 6 years after the Wild Harbor marsh oiling, showed a persistently greater variation than that found in nonoiled populations (Cole, 1978), including a persistent imbalance in genetic patterns. Long term recruitment inhibition and low population densities in the salt marsh crab Uca pugnax were found 8 years after oiling (Krebs and Burns, 1977). Similar patterns of long term impact on benthic invertebrates are beginning to unfold following the more recent spills, e.g., Amoco Cadiz (Cabioch et al., 1980; Beslier et al., 1980; Glemarec and Hussenot, 1982) and Tsesis (Elmgren et al., 1981).

Fish

Fish can be affected directly by petroleum, either by ingestion of oil or oiled prey, through uptake of dissolved petroleum compounds through the gills and other body epithelia, through effects on fish eggs and larval survival, or through changes in the ecosystem which support fish. In addition, the commercial fisheries as well as the sports and subsistence fisheries can be affected through contamination of gear, closure of fishing seasons, or buyer resistance to tainted or suspect products.

The potential impact of spilled oil on fish populations has stimulated many studies to determine the lethal and sublethal effects of oil on fish. Most of the studies to date have been done in the laboratory, mainly because both the test organism and the test conditions can be controlled. Field studies are far fewer. As a result, the effects of oil on fish in the field are only poorly understood.

Natural stresses may predispose fish to increased sensitivity to hydrocarbon toxicity. These stresses may include changes in salinity, temperature, and food abundance, as well as competitive or potentiating stress from other pollutants. For example, Moles et al. (1979) found Prudhoe Bay crude oil WSF as well as benzene, although at admittedly high concentrations (>1 mg/L), were twice as toxic to several salmonid species in seawater as in fresh water, this at the stage of their life cycle when they would normally migrate to seawater. Levitan and Taylor (1979) found that euryhaline killifish acclimated at extremes of salinity were less tolerant to naphthalene than fish exposed in dilute seawater. Rainbow trout also, when acclimated to seawater, showed greater susceptibility to crude oil-in-water dispersions (Englehardt et al., 1981).

Temperature as a natural variable becomes important in relation to hydrocarbon pollution, affecting its toxicity by affecting the persistence of hydrocarbons in water and by imposing physiological stress on the fish at both ends of the temperature range. The interaction of these stresses can be complex and severe. For example, in colder waters, hydrocarbons will tend to persist longer in the water column, while at the same time, hydrocarbon metabolism and clearance rates in the fish may become reduced. This would explain the increased sensitivity of pink salmon exposed to toluene at low temperatures (Korn et al., 1979). Salmon exposed to toluene at low temperatures had greater stress responses than salmon exposed at high temperatures (R.E. Thomas and Rice, 1979). Also, the tissue burden at lower temperatures may persist longer than at the higher temperatures.

It would appear that fish, in comparison to invertebrates, might be more sensitive to short term, acute exposures, requiring relatively shorter exposure periods to absorb lethal quantities of hydrocarbons. The sensitivity of fish to oil WSF varies widely. Pelagic species seem to be more tolerant than benthic species, while fish inhabiting intertidal areas appear to be the most tolerant (Rice et al., 1979). Of these, the benthic species are of special concern because of their lifelong association with benthic sediments, which are known to become hydrocarbon traps in the event of oil spillage.

Life-Cycle Stages

The most vulnerable portion of a fish life cycle to environmental/ pollution stress probably occurs during the development of the germ-line primordia, early embryo (pregastrulation), and especially the larval transition to exogenous food sources (Rosenthal and Alderdice, 1976) (Tables 5-7 and 5-8). And indeed, from the various data available on the vulnerability of gametes, developing eggs, larvae, juveniles, and adult fish (e.g., Kuhnhold et al., 1978; Struhsaker et al., 1974), some stages of eggs and most stages of larvae probably are more vulnerable to oil than are juveniles and adults. There are, however, exceptions and nuances to the generalization that all eggs are sensitive. For example, Moles et al. (1979) found salmon eggs (postgastrula) to be extremely tolerant to benzene and oil WSF. Also, fish have a variety of reproductive strategies that cause further variations and differences; egg incubation times vary from days to a year, and yolk content varies tremendously.

Damage to an embryo may not become apparent until after hatching of the eggs into the larvae. For example, Kuhnhold (1978) found that the effect of oil on the eggs was far worse if one used larval deformities as an index instead of egg survival. Many of the larvae, after hatching, was found to be deformed and incapable of swimming. These abnormalities would not be included or recognized if only the survival

TABLE 5-7 Effects of Oil on Eggs and Embryos of Marine Fish

Type of Response	Threshold Concentrations for Response (mg/L)[1]	Reference
Lethal		
Increased mortality	5.1(TH) (1-d LC_{50})	
	1.3-2.2(TH), 0.28-0.47(TN)	15
	<0.68(TH)	2
	0.1->44(TH) - 2(4-d LC_{50})	9
	0.02-0.2(TH)(6-d LC_{50})	
	0.01-0.5(TH)	17
	1.5(TH), 0.4(TN) (4-d LC_{50})	21
		17,18
Sublethal		
Fertilization decrease	>14.1(TH)	14
Uptake, accumulation of hydrocarbons		
Contamination chorion	<0.25(TH)	17
Uptake, metabolism, and retention	0.01-2.1(B)	5
Physiology		
Heartbeat changes	0.01-0.1(TH)	10
	<3.1-11.9(TH)	14
	5.0-6.0(TH), 1-1.2(TN)	1
	<2.1(TH), 0.54(TN)	21
Respiration affected	<2.1(B)	4
	2.4-8.0(TH)	3
Behavior		
Embryonic movements	<3.1-11 9(TH)	14
Development		
Pathological tissues	1.1-2.2(TH), 0.6-1.1(TN)	6
Morphological anomalies	<0.1(TH)	11
Rate changes	0.0001(B(a)P)	29
Growth		
Length, height changes	<1.7(TH), 0.38(TN)	16
Tissue	2.1(B)	4
Hatching		
Occurrence affected	<1.9(TH)	26
	2.2-4.4(TH), 1.1-1.3(TN)	6
Premature or delayed	<0.1(TH)	11
	<3.1-11.9(TH)	14
Success decreases	<2.1(TH), 0.54(TN)	21
	<0.8(B)	24
	0.01-0.1(TH)	11
	5-6(TH), 1.0-1.2(TN)	1
	0.0001(B(a)P)	29
Duration changes	<1.3(TH), 0.28(TN)	15
	>0.68(TH)	22

NOTE: TH, total hydrocarbons; TN, total naphthalenes; B, benzene B(a)P, Benzo(a)pyrene.

REFERENCES: 1. J.W. Anderson et al. (1977). 2. Cameron and Smith (1980). 3. Davenport et al. (1979). 4. Eldridge et al. (1977). 5. Eldridge et al. (1978). 6. Ernst et al. (1977). 9. Kuhnhold (1974). 10. Kuhnhold (1978). 11. Kuhnhold et al. (1978). 12. Linden (1975). 14. Linden (1978). 15. Linden et al. (1979). 16. Linden et al. (1980). 17. Longwell (1977). 18. Longwell (1978). 21. Sharp et al. (1979). 22. R.L. Smith and Cameron (1979). 24. Struhsaker (1977). 26. Vuorinen and Axell (1980). 27. Whipple et al. (1981). 29. Hose et al. (1982). Also see Johnson et al. (1979), Kuhnhold (1972), Linden (1976), Lonning (1977b), Mironov (1967), Stoss and Haines (1979), Struhsaker (1977), Struhsaker et al. (1974).

TABLE 5-8 Summary of Effects of Oil on Marine Fish Larvae

Type of Response	Threshold Concentration for Response (mg/L)	Reference
Lethal		
Increased mortality	43(TH)(4-d LC$_{50}$)	Voorinen and Axell (1980)
	<0.8(B)	Struhsaker (1977)
	1.5(TH), 0.4(TN) 4dLC$_{50}$	Sharp et al. (1979)
Sublethal		
Uptake, accumulation of hydrocarbons	0.01-2.1(B)	Eldridge et al. (1978)
Physiology		
Respiration changes	2.4-8.0(TH)	Davenport et al. (1979)
	<0.85(TH), 0.21(TN)	Sharp et al. (1979)
Development		
Delays	<0.68(TH)	R.L. Smith and Cameron (1979)
Tissue pathologies	<0.1-2.2(TH)	Vuorinen and Axell (1980)
Morphological anomalies	<3.1-11.9(TH)	Linden (1978)
	<0.68(TH)	R.L. Smith and Cameron (1979)
	<8.0(TH)	Davenport et al. (1979)
	<3.0(TH)	
Eye pigmentation		
Irregular cytolysis		
Growth		
Length decreases	<5.4-5.8(TH)	Linden (1978)
	0.1-2.2(TH)	Vuorinen and Axell (1980)
Length increases	<0.68	R.L. Smith and Cameron (1979)
	<2.1(B)	Eldridge et al. (1977)

NOTE: Threshold concentrations are based on direct hydrocarbon measurements. TH, total hydrocarbons; B, Benzene; TN, total naphthalenes.

of the eggs were considered. Linden (1978) also noted that many oil-exposed Baltic herring eggs that had appeared to develop normally ultimately failed to hatch.

Typical fish larval responses to toxic concentrations of petroleum include a brief increase in activity, followed by reduced activity, sporadic twitching, narcosis, and ultimately, death (Struhsaker et al., 1974; Linden, 1975). Morphological and physiological effects include deformed spinal column, tissue destruction (particularly of the fish fins), and reduced growth (Linden, 1975, 1976). They seem generally to become more sensitive to oil as their yolk sac becomes used up (Kuhnhold, 1972; Rice et al., 1975).

Sublethal Effects

Although fish can accumulate hydrocarbons from contaminated food, there is no evidence of food web magnification in fish. Fish have the capability to metabolize hydrocarbons and can excrete both metabolites and parent hydrocarbons from the gills and the liver. There is some evidence that metabolites can persist in the tissues longer than parent hydrocarbons but their toxicity is not known (see Chapter 4, Invertebrates and Vertebrates section and Rates of Petroleum Biodegradation in the Marine Environment section). There are indications, however, that some metabolites or intermediates may be toxic or even mutagenic.

Oil effects in fish can occur in many ways: histological damage, physiological and metabolic perturbations, and altered reproductive potential. Histological studies have documented damage to liver (McCain et al., 1978; Sabo and Stegeman, 1977), gill (Ernst et al., 1977; Englehart et al., 1981), gut (J.W. Hawkes et al., 1980), vertebrae (Linden et al., 1980), eye lens (J.W. Hawkes, 1977), stomach (Wang and Nicol, 1977), brain (DiMichelle and Taylor, 1978; Cameron and Smith, 1980), and olfactory organs (Gardner, 1975).

Physiological changes include increased heart beat (Wang and Nicol, 1977; J.W. Anderson et al., 1977; Linden, 1978), increased coughing (Rice et al., 1977; Barnett and Toews, 1978), ionic and osmotic imbalances (McKeown and March, 1978; Englehardt et al., 1981) changes in respiration (Barnett and Toews, 1978; J.W. Anderson et al., 1974; R.E. Thomas and Rice, 1979), changes in blood parameters (Fletcher et al., 1979; P. Thomas et al., 1980), decreased energy reserves (Stegeman and Sabo, 1976; Sabo and Stegeman, 1977; J.W. Hawkes, 1977; Kovaleva, 1979), and changes in gill enzymes (Wong and Englehardt, 1982).

Long Term Effects

Many sublethal effects of oil on fish are symptomatic of stress and may be transient. Others may persist longer but may be only slightly debilitating. However, because all repair or recovery requires some energy, these sublethal effects can ultimately lead to increased vulnerability to disease or to decreased growth and reproductive success, even though the individual may continue to live for some time.

Several studies have indicated a correlation between hydrocarbon stress and increased vulnerability to disease. For example, hydrocarbon exposure has been found to be associated with increased fin erosion or fin rot (Minchew and Yarbrough, 1977; Giles et al., 1978), reduction in external bacterial flora (Giles et al., 1978), and reduction in the rate of tissue repair or regeneration (Fingerman, 1980). There also appears to be some relationship, albeit poorly understood, between hydrocarbon exposure and parasitism. For example, juvenile coho infested with parasites were more sensitive to oil, toluene, and naphthalene than were uninfested fish (Moles, 1980).

Decreases in growth after hydrocarbon exposure have been observed in several studies: herring embryos (Linden, 1978; R.L. Smith and Cameron, 1979), killifish larvae (Sharp et al., 1979; Linden et al., 1980), cutthroat trout juveniles (D.F. Woodward et al., 1981), pink salmon alevins (Rice et al., 1975), English sole (McCain et al., 1978), and coho salmon (Moles et al., 1981). Several of these studies were of considerable length, and involved flow-through systems where exposures may be comparable to field exposure conditions. Growth decreases may be a highly significant factor with respect to the fisheries, for the entry of fish into the fishery may be delayed.

Low levels of oil contamination (less than 1 ppm) can also affect the ability of individuals to reproduce, either by causing malformation of gonads or gametes or by simply decreasing the energy that the fish has left to invest in growth. Such concentrations are found in the field under spill conditions, and damage to gonadal tissues has been noted at very low hydrocarbon concentrations. For example, starry flounder, exposed to 50-100 µg/L monoaromatic hydrocarbons, showed changes in gonadal maturation (Whipple et al., 1978). Also Kuhnhold et al. (1978) have demonstrated that exposure of adult fish during gonad maturation to 10-µg/L WSF can result in reduced survival of the larvae hatched from the eggs laid by these fish.

Vulnerability and Avoidance

It is difficult to state definitely whether fish will avoid hydrocarbons in the field. Observations are few and circumstantial. When Atlantic silverside were exposed to crude oil, they lost their normal schooling behavior, possibly because the olfactory organs and lateral lines of the fish were blocked or damaged (Gardner, 1975). On the other hand, the appeal of easy prey may attract fish into an oiled area. Thus other factors in the fish life habit may override avoidance of oiled waters. For example, in a study of adult salmon returning to a home stream, about 50% of the salmon avoided a contaminated fish ladder containing 3.2 mg/L of monoaromatic hydrocarbons, indicating that a considerable proportion of the salmon could avoid much lower concentrations. But it was also clear that an equally large proportion would pass through even higher concentrations that approached acutely toxic levels (Weber et al., 1981).

In contrast, pelagic eggs and larvae at or near the surface clearly are unable to avoid contaminated surface waters and are more vulnerable to oil pollution.

Effect on Fish Stocks

Oil has interfered with or affected the fishery in several ways. Some fish have become tainted through ingestion of oil or through contact with oil or oiled gear, and are therefore unmarketable (e.g., Vandermeulen and Scarratt, 1979). Oil spills have also interfered more indirectly through fouling of gear and closing of harbors and fishing seasons. In many instances, fish even slightly suspect of tainting or thought to originate from oiled waters, have been found to be unmarketable. However, a direct impact on fishery stocks has not been observed, nor has it been looked for directly in most cases, although close inspection is made of fish catch statistics.

The problem of tainting has been reviewed by Howgate et al. (1977), Stansby (1978), and Whittle (1978). Tainting, or the presence of off flavors in fish meat, is a troublesome problem in that often the source of the off flavors is not known, and off flavors can be due to a number of causes. To make matters worse, not infrequently can different sources give rise to similar off flavors. For example, a kerosene-like flavor in a catch of tainted mullet could be linked to the presence of crude-oil-derived hydrocarbons later found in the fish tissues (Shipton et al., 1970). However, in another apparently similar case, the kerosene-like flavor in a different batch of tainted mullets arose from thermal decomposition of naturally occurring components in fish (Vale et al., 1970).

Tainting has occurred in nearly all types of oil spillage, including diesel fuel in brown trout (Mackie et al., 1972) and unidentified petroleum refiner effluent (Nitta et al., 1965), and have been frequently reported for spill accidents such as the Juliana (Motohiro and Inoue, 1973), the Torrey Canyon, the Ekofisk blowout (Mackie et al., 1978), and the Amoco Cadiz (Chasse, 1978). In two instances the tainting compounds may have entered the fish from contaminated sediments instead of through the water column (Connell, 1971, 1974; Nitta et al., 1965).

Invertebrate tissues also are liable to tainting, including lobsters (Wilder, 1970; Scarratt, 1980), mussels (Nelson-Smith, 1970; Brunies, 1971), and clams (A.L. Hawkes, 1961). Oysters from beds near drilling operations offshore from Louisiana were found to contain oily taints, correlating with the proximity to the drilling operations (Mackin and Sparks, 1962; St. Amant, 1958; Menzel, 1947, 1948).

The precise route of entry of the tainting substances is uncertain but probably includes both the respiratory process and uptake through tainted food.

Little is known of the actual substances responsible for the tainting of fish meat. No unequivocal identifications of substances in crude oil and distillates exist which exactly match the taint profile in contaminated fish products (Whittle, 1978), but it is thought that the more polar components (aromatic hydrocarbons, substituted benzenes and naphthalenes, naphthenic acids, organosulphur compounds, and olefins) are involved. Representatives of these compounds have all been identified in contaminated marine products (Ogata and Miyake, 1973; Lake and Hershner, 1977; Ogata et al., 1977; Shipton et al., 1970).

It seems that under natural conditions, tainting is unlikely to occur at the sort of petroleum hydrocarbon concentrations found in the water column after a spill, barring a major catastrophe such as the Amoco Cadiz. It is likely, however, in more restricted spill incidents, as in bays, or under chronic oiling conditions.

There are two major factors due to pollution that severely complicate the detection of impacted fishery stocks. First, there are extremely wide variations in the recruitment to fisheries that are caused by both natural environmental factors and by current overfishing (e.g., Ware, 1982). Predictions of fish abundance often deviate from the observed for no apparent cause, and our present capability to assess a standing fish stock is very limited, due mainly to sampling methodology (e.g., Sinclair, 1982). Moreover, natural environmental factors, often poorly understood, can cause fluctuations in year-class abundance which bear little relation to the size of the parent stock. It would therefore be very difficult to quantitatively assess the impact of an oil spill on populations that already possess broad and unpredictable changes in year-class size.

The second factor is that massive fish kills during oil spills probably have not occurred. Some mortalities have been observed at a number of spills, but generally only in limited areas, and then not in large amounts. Fish have the ability to move away from an impacted area, either laterally or by moving to a greater depth (whether in fact this occurs is still not known). If any large mortalities do occur, they probably occur in the egg and larval stages found in the surface waters. Being more sensitive than the adult stages, eggs and larvae may have been killed in large numbers during spills. However, such kills are extremely difficult to document, simply because these fragile life stages are difficult to sample, since the dead fall out of the water column and decompose within hours, becoming unrecognizable tissue debris. In this respect the available evidence from the Argo Merchant spill, although open to criticism for lack of statistical corroboration, suggests a kill of cod and pollock.

Marine Mammals

Information on oil impact for marine mammals is still limited. There are a number of documented or inferred oil spill fouling incidents but only few definitive experimental studies. Despite this paucity of information, because of their life habits, the potential exists for many marine mammal species to come into contact with spilled petroleum in the seas. As a large proportion of the world's marine mammals spends part or all of its life in ice-infested waters, this risk seems to be especially great in the polar regions. Adding to this is the grooming habit of several species that adds petroleum toxicity through ingestion of oil. Recent reviews, including indicated research needs, have been completed by Geraci and St. Aubin (1980), Smiley (1982), and Englehardt (1983).

Vulnerability

The dependence of pelagic seals and whales on air and the amphibious habit of polar bear (Ursus maritimus) enhance the possibility of unavoidable contact with spilled oil on the sea surface. This is readily visualized for ice-covered waters where the diving mammal can surface only in restricted areas of open water.

Ice will modify the pattern of spreading and disappearance of an oil spill (Glaeser, 1971; Snow and Scott, 1975), tending to move or concentrate oil in leads and breathing holes. In addition, wind may "herd" oil between moving ice floes or up against ice edges (Ayers et al., 1974). In this way, preferred travel routes of many of the marine mammals could become highly contaminated with oil. For example, the preferred travel route for narwhal (Monodon monocerus) is the ice edge. The situation for bowhead whales (Balaena mysticetus) may be particularly serious because this endangered species uses ice leads extensively during its arctic migration (Braham et al., 1980). The entire population of bowhead whales (Bering, Chukchi, Beaufort stocks), comprising only some 3,000-4,000 animals, travels from the Bering Strait to Point Barrow, Alaska, through a rather well-defined ice-lead system.

Ringed seals (Phoca hispida) are another good example of a specialized ice dependency and potential oiling vulnerability. Birth and postnatal care occur in this species in subnivean (under snow) birth lairs, with access through the ice surface (T.G. Smith and Stirling, 1975), and it has been suggested that ringed seals use under-ice air pockets on extended subice travels (Bertram, 1940; Milne, 1974). In both instances, coating and inhalation toxicity may be a problem, for oil would concentrate in these areas. Also, the colonial breeding habit of most seals and of certain other marine mammals creates a particular vulnerability of these animals to oil spills within the breeding period.

Oiling Incidents

Documented cases of oil pollution incidents involving marine mammals are few, even if one includes news media accounts. In most instances these are not case histories but are obituary accounts that only implicate but do not define oil as the cause of death. One of the most detailed of these occurred during the 1969 blowout in the Santa Barbara Channel in southern California. Nelson-Smith (1970) causally related this event to the deaths of gray whales (Eschrichtius gibbosus), a dolphin, northern fur seals (Callorhinus ursinus), California sea lions (Zalophas californianus), and northern elephant seals (Mirounga angustirostris). However, the mortality link was subsequently interpreted as tenuous (Simpson and Gilmartin, 1970; Brownell and Le Boeuf, 1971; Le Boeuf, 1971).

A number of other incidents have been associated with deaths of marine mammals. These have included gray seals (Halichoerus grypus) on the coast of Wales (J.E. Davis and Anderson, 1976) and seal mortalities during the Torrey Canyon (Spooner 1967) and Arrow spills (Anon, 1970). Harp seals (Phoca groenlandica) were found dead and coated with Bunker C oil after a spill in the Gulf of Saint Lawrence, Canada (Warner, 1969). Gray and harbor seal (Phoca vitulina) mortalities were associated temporally with coating by oil from the Kurdistan Bunker C spill (Parsons et al., 1980) (Figure 5-6). Other reports describe oil fouling in various species without being identified with a spill incident or necessarily with mortality. Oil-fouled individuals have been reported in the Arctic (Hess and Trobaugh, 1970; Morris, 1970; Muller-Willie, 1974), in the Antarctic (Lillie, 1954), and in European waters (J.L. Davis, 1949; Spooner, 1967; Van Haaften, 1973).

Oil Adherence to Body Surfaces

The extent of oil adhesion to the skin or pelage of marine mammals depends at least on the following: texture of the exposed body surface,

FIGURE 5-6 (Top) Oiled and (bottom) unoiled harbor seal pups seen on Sable Island, Nova Scotia, Canada, in June 1979. Causes of oiling was unknown. Note the clean ring, result of tearing, around the eyes of both animals. (Photo by J. Parsons.)

SOURCE: Parsons et al. (1980).

frequency and duration of exposure, and characteristics of the oil. A heavy, viscous oil would seem more likely to adhere to skin or pelage, and marine mammals possessing well-developed pelage would be expected to have oil adhere readily.

This assertion is supported by laboratory studies involving ringed seals (T.G. Smith and Geraci, 1975; Geraci and Smith, 1976), sea otters (Williams, 1978), and polar bears (Oritsland et al., 1981; Engelhardt, 1981). Similar evidence comes from oil-fouled animals such as harp seals (Warner, 1969), gray seal pups (J.E. Davis and Anderson, 1976), and elephant seal pups (Le Boeuf, 1971) taken from spilled oil areas. The persistence of adherent oil varied among the species examined. Ringed seal pelage cleaned itself from complete coating by 1 day in seawater. Captive sea otters and polar bears groomed their oil-fouled hair readily, although this resulted in oil ingestion and subsequent illness or death. In an examination of 58 free-ranging elephant seal pups, in which initially more than 75% were coated with spilled oil, all but one were clean when examined 1 month later (Le Boeuf, 1971).

Marine mammals with a poorly developed or no pelage (e.g., walrus, sirenians, cetaceans) would seem to be less likely to have oil adhere to them. Thus there have been no substantiated reports of oil fouling of cetacean skin, although whether due to the inability of oil to adhere to their generally smooth skin or due to avoidance by the mammals of oiled areas is not known. The skin of most cetaceans is generally quite smooth over the entire body surface, although there are exceptions. These include the gray whale, with large numbers of attached barnacles, right whales with their prominent rostral callosities (Slijper, 1979), and the bowhead whale with dozens to hundreds of eroded areas, particularly involving the skin of the head (Albert, 1981). Such roughened body areas would seem to increase the likelihood of oil adherence.

Spilled oil may also be expected to interfere with baleen functioning, for the inner aspect of the baleen plates presents a very roughened surface. This is supported from laboratory studies (Braithwaite, 1981) which demonstrated that the filtering efficiency of bowhead whale baleen was reduced by about 10% when coated with Prudhoe Bay crude oil and by 85% when coated with oil with a higher wax content. Geraci and St. Aubin (1982) reported transient changes in water flow through oil-exposed baleen from fin (Balaenoptera physalus) and gray whales, and suggested that the temporary inhibition of baleen function would not be important in the long term feeding strategies of baleen whales. The consequences of a possibly increased ingestion of oil from this coated filtering apparatus cannot be evaluated at this time.

Avoidance of Oiled Waters

Little is known of the capability or the willingness of marine mammals to detect or avoid oil-contaminated waters. Multiple observations of oil-coated seals suggest that these species do not actively avoid oil. Limited observations on captive sea otters (Williams, 1978) and seals (Spooner, 1967) also showed that they did not necessarily avoid oil-covered water. In the Canadian north, free-ranging polar bears have repeatedly been reported to consume small quantities of petroleum oils from dump sites of Arctic camps.

A recently released report indicates that a broad range of cetaceans do not actively avoid oil. Observations in an oil slick resulting from the Regal Sword spill (Goodale et al., 1981) of Bunker C and No. 2 fuel oil showed cetaceans swimming and feeding in both oil-covered and oil-free waters. Both surface and below-surface feeding was observed in humpback whales (Megaptera novaeangliae), fin whales and in white-sided dolphins (Lagenorhynchus acutus). Another species, probably right whales (Eubalaena glacialis), was also observed in the oil slick. Their behavior was apparently normal when observed in the oil slick. Part of the gray whale migration along the California coast goes through the oil seep waters off Coal Oil Point, California, apparently in clear disregard of the contaminated waters (Geraci and St. Aubin, 1982). Yet in an experimental setting, captive bottlenosed dolphins (Tursiops truncatus) were able to detect and chose to avoid oil on the water surface both visually and apparently by echolocation (Geraci et al. 1983).

Although these observations do not answer the question of detection ability, they at least suggest that under normal conditions, species observed may not actively avoid oil-covered waters.

Thermoregulation and Metabolism

A survey study of stranded cetaceans has shown the presence of petroleum hydrocarbons, particularly in blubber tissue (Geraci and St. Aubin, 1982).

Oil uptake and excretion in marine mammals have been examined experimentally in ringed seals (Engelhardt et al., 1977; Engelhardt, 1978) and in polar bears (Engelhardt, 1981; Oritsland et al., 1981). Major sites of accumulation of petroleum residues in the seals were in the blubber and liver, with excretion of petroleum residues and of metabolites via the urine and bile. In the polar bears, ingestion of oil was brought on primarily by grooming of the oiled pelage. There was evidence of some long term contamination of internal tissues, as shown by continued hydrocarbon concentrations in the serum some weeks after initial oiling. This may have been due to continued uptake of oil through grooming, or through transfer of contaminated fatty tissue from fat stores to excretory routes. In the polar bears, highest hydrocarbon levels were found in the kidney, brain, and bone marrow.

Effects on thermoregulation and metabolic stress have been studied in some detail in the sea otter (Kooyman et al., 1977; Costa and Kooyman, 1980). Coating of the fur resulted in major conductance changes, increasing with the degree of oil coverage. In comparison, isolated skins from seals and sea lions showed less or little conductance change when oiled. A consequence of the increased conductance is a greater heat flow across the body surface. Loss of heat from the body core must be compensated for by an increase in metabolism. Although core temperatures did not change, the oiled sea otters showed a 5°-10°C decrease in subcutaneous temperatures below the selectively oiled areas when they were in the water. The lack of core temperature change was attributed to a near doubling of body metabolism. Increases in metabolism also showed a dose response in relation both to the amount of oil applied and to the amount of body surface covered. Failure of an increase in metabolism to compensate for heat loss may lead to hypothermia, thought to be the cause of death in the case of one oiled sea otter (Williams, 1978).

Studies with polar bears suggest similar implications from skin oil fouling. Oil coating of isolated polar bear fur led to as much as a tripling in the conductance of heat across the skin (Hurst et al., 1982; Oritsland et al., 1981). This was increased even further in the presence of wind or by increasing the viscosity of the test oil. The consequences of exposing bears experimentally to oil were changes in temperature and metabolism. Subcutaneous temperatures increased as a result of reactive vasodilation in the skin, leading to a major heat loss promoted by the increased conductance. Resting deep body temperatures decreased slightly but metabolism nearly doubled. Such thermal and metabolic responses in polar bears can eventually lead to hypothermia, particularly in the presence of wind or in instances of poor health.

Unlike the case in sea otters and polar bears, oil-coated ringed seals and late stage harp seal pups showed no core temperature response to crude oil (T.G. Smith and Geraci, 1975). In this case the thick insulative blubber layer was thought to serve as an adequate thermal barrier. Newborn phocid seals, having little or no blubber (Engelhardt and Ferguson, 1980) may be more seriously affected, as they rely on fur (lanugo) and metabolic activity for their thermal balance (Oritsland and Ronald, 1973; Blix et al., 1979).

Pathological Consequences

Marine mammals may be expected to show clinical and toxicological responses to petroleum hydrocarbons similar to those in other mammals. Aberrations in hematological parameters, in diagnostic enzymes, and in tissue structure have been recorded, although not consistently.

No indications of clinical abnormalities were found in the oiled sea otters (Costa and Kooyman, 1980; Williams, 1978). Ringed seals showed only a limited toxic response (T.G. Smith and Geraci, 1975; Geraci and Smith, 1976). Twenty-four hours of oil exposure resulted in complete coating of the body, including the eyes, and caused eye damage such as conjunctivitis and corneal lesions. The eye involvement subsided, however, after removal from the oil and was considered reparable. Oil-induced eye damage has also been suggested by Nelson-Smith (1970) and Morris (1970). However, eye damage is by no means uncommon in normal seal populations (King, 1964; Ridgway, 1972) and any circumstantial link to oil spill events should be treated with caution.

Although petroleum exposure in seals, either by immersion or by ingestion, has led to petroleum uptake and to its accumulation in tissues (Engelhardt et al., 1977; Engelhardt, 1978), there was no evidence in either ringed or harp seals of major haematological, plasma chemical, or histological/clinical changes which could be associated with the oiling (Geraci and Smith, 1976). Similarly, a study of the effect of hydraulic oil (Caldwell and Caldwell, cited by Geraci and St. Aubin, 1982) found no clinical pathology. From this it would seem that at least phocid seals are not likely to show a lasting toxic effect from short term exposures to crude oil, even at high doses. The same conclusion cannot be drawn, however, for long term exposures. Aside from the clinical damage, long exposures may lead to adrenal steroid exhaustion. In ringed seals for example, oil ingestion was found to lead to a greatly elevated plasma cortisol level, while cortisol breakdown rates were nearly doubled (Engelhardt, 1982).

The situation is different in the case of oiling of polar bears. Exposure, and presumably continued exposure as well, to unknown quantities of oil by short term immersion and ingestion led to severe clinical pathological abnormalities (Oritsland et al., 1981; Engelhardt, 1981). Predominant were extensive anemia, caused by peripheral hemolysis and erythropoietic dysfunction, and renal abnormalities associated with a buildup of nitrogenous metabolites and hydromineral imbalance. Adrenocortical, pulmonary, and skin changes were also evident. Renal failure was proposed as the ultimate cause of death of two bears (Juck, in Oritsland et al., 1981).

Summary and Conclusion

Although information of oil effects on marine mammals is scanty at best, the sum of the results indicates that both habit and habitat make the various species highly vulnerable to oil pollution at sea. Experimental studies have shown that seals, sea otters, whales, and polar bears differ in their sensitivities to oil exposure. Fur-insulated marine mammals respond to oil contact by a compromised ability to thermoregulate. Continued contact may result in skin and eye lesions. Both seals and polar bears can absorb oil readily, distributing it through the body tissues, including the fatty reserves. Seals showed endocrine stress responses but few other tissue problems, while polar bears were severely affected in blood and renal functions. Cetaceans were little or only transiently affected by oil exposure. Odontocetes appear to be able to detect oil under captive circumstances, but do not necessarily avoid slicks at sea.

Birds

Birds are probably the most conspicuous casualties of oil pollution in the sea. Death of seabirds from oil pollution receives great publicity and in several countries attracts a public reaction such as the death of few other animals does. In addition, probably because of its visual impact, the death of birds from oiling evokes an emotional reaction stronger than would their death from other pollutants. These reactions are not susceptible to scientific assessment, but their strength cannot be ignored.

Much of the information about the effects of oil on seabirds has been reviewed fairly thoroughly in recent years (e.g., Bourne, 1976;

Group of Experts on the Scientific Aspects of Marine Pollution, 1977; Holmes and Cronshaw, 1977; Royal Society for the Protection of Birds, 1979; R.G.B. Brown, 1982). Many of the data relate to west European waters, and it is largely to this area that we must turn to assess the short term and long term effects of oil pollution on seabirds. Fortunately, due to recent data gathering in western Atlantic waters, the interpretations can be applied to the North American waters with some degree of confidence.

Effect of Oil on Individual Birds

The direct effect of oil on a bird is to clog the fine structure of its feathers, which is responsible for maintaining water-repellance and heat insulation (Holmes and Cronshaw, 1977). The plummage absorbs water as the bird sinks and drowns. Unfortunately, it is not known to what extent this effect occurs at sea or just how many birds are lost in this way.

The loss of thermal insulation produces a more immediate response. It results in greatly increased metabolic activity to maintain body temperature (Hartung, 1967). As a result, fat and muscular energy reserves are rapidly exhausted, leading to mortality (Croxall, 1977). For these reasons it may be anticipated that birds are more likely to succumb from oiling in colder climates than in warmer climates (R.G.B. Brown, 1982) or after prolonged stormy weather when feeding has been limited or energy reserves are low.

Birds also ingest oil, probably mainly from preening their oiled plumage. Autopsies of oiled seabirds have revealed, in addition to wasting of fat and muscle tissues, abnormal conditions in the lungs, adrenals, kidneys, liver, nasal salt gland, and gastrointestinal tract, and a reduction in white blood cell count (Croxall, 1977; R.G.B. Brown, 1982). Few analyses are available from oiled birds collected live, but one has shown elevated levels of the mixed function oxidase system (MFO) in the presence of hydrocarbon contamination of liver, kidney, and muscle tissues (Vandermeulen et al., 1978). However, whether pathological conditions are related to petroleum hydrocarbons or to generalized stress is uncertain. Nor is there evidence to suggest that any of the observed abnormalities were a primary cause of death, which in the majority of cases, is likely to have been drowning or hypothermia.

A variety of physiological changes have been recorded in experimental studies involving ingested oil. Osmoregulator and hormone changes have been found (Holmes, 1975; Peakall et al., 1981), including retardation of weight gain of young birds (Miller et al., 1978), induction of hepatic enzymes (Gorsline et al., 1981), and generalized pathological effects (Holmes et al., 1978). However, other workers have reported conflicting findings (i.e., Gorman and Simms, 1978; McEwan and Whitehead, 1978). In some cases these differences can be ascribed to different oils (Peakall et al., 1982), and differences in species and dosing regimes are also important. Relatively small amounts of ingested oil can cause a temporary depression of egg laying and reduce the hatching success of those eggs that are laid (Ainley et al., 1981). The importance of all of these biochemical and physiological change in the wild are unknown.

Ingested crude oils may interfere with water and sodium ion transfer in the intestine and with excretion of salt by the nasal gland in some species, and may retard the growth of young birds, but the evidence is conflicting (R.G.B. Brown, 1982). Better evidence exists to show that relatively small amounts of ingested oil can cause a temporary depression of egg laying and can reduce the hatching success of those eggs that are laid (e.g., Ainley et al., 1981). Small quantities of oil applied to the surface of the egg are known to destroy the embryo at certain stages of development in the laboratory (e.g., R.G.B. Brown, 1982) and the field (Birkhead et al., 1973). There is no evidence, however, to suggest that this happens in practice on a widespread scale.

Seabird Mortalities Due to Oiling

An accurate estimate of the number of seabird casualties from oil pollution is not now and may never be possible. The only firm figures available are from counts of oiled birds found on shore surveys, but these are subject to severe limitations imposed by the intensity of the search, accessibility of the shore, reporting efficiency by cleanup crews, etc., and there is often doubt about the proportion of corpses found on beaches that were in fact oiled after death.

A very large and probably significant source of error is the unknown proportion of oiled birds that die at sea but which never reach the coast, thereby escaping the shore surveys. The evidence suggests that, in fact, 30% or fewer of bird corpses drift ashore (Hope-Jones et al., 1978). Estimates of actual losses in major incidents are therefore usually little more than informed guesses. They do probably indicate the orders of magnitude, however, whether hundreds, thousands, or exceptionally, tens of thousands (Bourne, 1976; Holmes and Cronshaw, 1977; see also Table 5-9).

There is even less certainty about the number of casualties from oil pollution in regions such as the North Sea where frequent small oil slicks are formed as a consequence of the discharge of oily bilge and ballast water from ships. Pollution from this source approaches the chronic and is suspected of accounting, in aggregate, for at least as many seabird deaths as those resulting from more spectacular incidents (Croxall, 1977). The estimate by Tanis and Morzer-Bruyns (1968) that 150,000-450,000 seabirds annually are killed by oil pollution in the North Sea and North Atlantic has a slender factual base but may well indicate the order of magnitude of losses.

In view of these various uncertainties and in the differences in oil spills (oil type, weather, etc.) and available bird communities, it is therefore not surprising that there is little relationship between the size of an oil spillage and the number of seabird deaths (Table 5-9). One of the largest kills of seabirds by oil on record was in the Skaggerrak, Denmark, in January 1981, when some 30,000 oiled birds appeared on neighboring beaches (Mead and Baillie, 1981); this was

TABLE 5-9 Relationship Between Amount of Oil Spilled in an Incident and Number of Dead Birds Found

Incident	Tons Spilled	Dead Birds Found	Dead Birds per Kilometer
Waddensee, Netherlands, Feb. 1969[a]	under 1,000	14,564	40
Poole, U.K. Jan. 1961[a]	300	487	?
Seestern, Medway, U.K., Sept. 1966[a]	1,700	2,772	?
Torrey Canyon, English Channel April 1967[a]	119,328	7,815	?
Tank Duchess, Tay, U.K., Feb.-March 1968[a]	87	1,368	?
Loch Indaal, U.K., Oct. 1969[a]	115	449	25
Hamilton Trader, Irish Sea, May 1969[a]	700	4,092	?
San Francisco Jan. 1971	2,700	7,380	?
Arrow, Cape Breton, Feb.-March 1970[b]	10,400	567	26
Irving Whale, SE Newfoundland, Feb. 1970[b]	under 30	625	47
Kurdistan, Cape Breton, March-April 1979[b]	7,900	1,697	26
Amoco Cadiz, Brittany, March 1978[c]	200,000	4,572	?

NOTE: The actual numbers killed, allowing for extrapolation for known length of contaminated shoreline, and for the sinking of dead birds before they can come ashore, would be at least an order of magnitude greater in each of these cases. "Dead birds per kilometer" is a rather more meaningful figure than "dead birds found" for comparisons between spills.

[a]Bourne (1976, Table V).
[b]Brown and Johnson (1980, Table 3).
[c]Hope-Jones et al. (1978).

caused by relatively small amounts of oil discharged by perhaps two ships (Royal Commission on Environmental Pollution, 1981).

Heavy casualties are sustained when floating oil encounters concentrations of seabirds on the water. The risk is greater in heavily traveled sea lanes and near oil industry operations but is critically influenced by seasonal and climatic factors.

Species Vulnerability

The species most commonly oiled are well known and well documented (Bourne, 1976). For obvious reasons the most susceptible birds are those which are gregarious, spend most of their time on the water, and dive rather than fly up when disturbed. Auks, especially murres (Uria sp.) and dovekies (Alle alle), form large concentrations on the water both at their breeding colonies in the summer and in their wintering areas, and they have suffered very heavy casualties from oil pollution. Their counterparts in the southern hemisphere are the penguins, and of these, the jackass penguin (Spheniscus demersus) suffers regular losses from oil pollution, being close enough to shipping routes along the coast of South Africa.

Diving ducks such as scoters (Melanitta spp.), oldsquaw (Clangula hyemalis), scaup and canvasback (Aythya spp.), and mergansers (Mergus spp.) suffer heavy casualties when concentrated on their winter feeding grounds. Common eiders (Somateria mollissima) appear to be vulnerable at most times of the year. Grebes (Podiceps spp.) and loons (Gavia spp.) tend to concentrate in coastal waters in winter, when they may be oiled; the numbers affected are rarely large, but in view of their small world populations, even small losses may be significant. Phalaropes (Phalaropus spp.) congregate in winter at sea at narrow convergence fronts (R.G.B. Brown, 1982), and they would be vulnerable there to offshore spills, though such mortality has not as yet been reported.

Many waterfowl and shorebirds flock on salt marshes and mud flats, and would be vulnerable to the destruction of their feeding habitat by oil spills. This would be especially significant for species such as the greater snow goose (Anser caerulescens atlanticus) which are highly concentrated on a very small stretch of shoreline in the Saint Lawrence estuary and Chesapeake Bay, and the semipalmated sandpiper Ereunetes pusillus in the Bay of Fundy.

Unlike the foregoing, which are all from temperate/cold regions, seabirds of tropical and subtropical waters are less vulnerable because the species occurring there usually do not feed by pursuing their prey under water (e.g., Ashmole, 1971). The one exception is the cormorants (Phalacrocorax spp.), which do feed under water and occur in tropical seas; in one incident, several hundred of these birds were killed by a leakage of fish oil off Namibia (Berry, 1976).

Impact on Seabird Populations

Despite these various concerns and considering the large losses of seabirds from oil pollution, there may not be a material impact on the total population of a given species. This apparently is true of the auks of the British Isles and probably also for the diving ducks. In contrast, the northern (Arctic) auk populations are already in serious decline from other forms of human interference, and increased exposure to oil pollution is likely to affect them more seriously.

The reproductive strategy adopted by auks and most other primary seabirds (as opposed to diving ducks) is a low reproductive turnover combined with great longevity and a long adolescent period (e.g., Ashmole, 1971; R.B. Clark, 1969). This allows them to cope with erratic natural catastrophes such as fluctuations in their food supply. However, such a strategy makes the species slow to recover from additional and persistent man-induced mortalities due to excessive hunting, drowning in fishing nets, and oil. For these reasons most auk populations in the Atlantic have declined sharply in the course of this century (e.g. Cramp et al., 1974; Norderhaug et al., 1977). The surprising exception to this is in the British Isles, where the most recent census figures show that 25% of the colonies of common murres (Uria aalge) and razorbills (Alca torda) have maintained stable numbers over the last decade, and 50% have actually increased. Colonies of Atlantic puffins (Fratercula arctica) in western and northern Scotland seem to be following the same pattern (Harris and Murray, 1981). Despite the chronic oil pollution in the North Sea and adjacent waters, which is estimated to kill several hundred thousand seabirds a year, many of them auks (Tanis and Morzer-Bruyns, 1968), it indicates that the pollution has not been severe enough to check the regeneration of the populations following protection or perhaps climatic amelioration (e.g., Burton, 1981).

On the other hand, auk populations elsewhere in the Atlantic (with the exception of dovekies) continue to decline. These declines are probably mainly due to overhunting and drowning in fishing nets at present, though Arctic auks have suffered mortality from oil spills on their winter quarters off Scandinavia and eastern Canada (e.g., R.G.B. Brown and Johnson, 1980). The risks there will undoubtedly increase with the development of offshore oil production off Alaska, northern Canada, and Newfoundland, along with associated shipping transport.

There are reasons to believe that these northern seabirds will be more seriously affected by oil spills than auks and other vunerable species in the more temperate waters. Oil spilled in these cold waters may remain unweathered for a longer period than in tropical waters due to slow or no evaporation of the more volatile fractions. Oil frozen into the sea ice and released during thaw may further prolong the effects of a spill. Added to this is the fact that the Arctic auks tend to breed in few very large colonies of ca. 100,000-plus birds, rather than in small to medium colonies as occurs further south (e.g., Cramp et al., 1974; R.G.B. Brown et al., 1975). These facts taken together suggest a real potential for oil on the water close to an Arctic auk colony to cause greater mortality to the northern populations.

The diving ducks, the other principal seabird group vulnerable to oil because of their diving habit, are probably less sensitive to oiling because of a different breeding strategy. They have a high reproductive potential so that adult losses can be replaced rapidly. In recent years there have been large losses of common eiders in a number of oil pollution incidents, but the numbers in the breeding colonies appear to recover within a year or two (e.g., Leppakoski, 1973). There may, however, be short term economic losses in areas such as Greenland, where these birds are a staple food for hunting communities (Salomonsen, 1967).

Remedial and Preventive Measures

There is no doubt that the death of birds from oil pollution is regrettable. On this account there are continuing efforts to devise methods of reducing the number of casualties through rescuing and cleaning those birds that are oiled (Research Unit on the Rehabilitation of Oiled Seabirds, 1972a,b). While from a humanitarian point of view the cleaning and rehabilitation of oiled birds are commendable, from a conservation point of view the practice seems to have little value (R.B. Clark, 1978). The number of birds saved and returned to the wild by this practice would be entirely too few to make any difference to breeding colonies.

A more positive conservation measure is the restocking of depleted colonies, as has been practiced with apparent success with Atlantic puffins on the coast of Maine (Kress, 1980). However, the cost, difficulty, and protracted nature of such a program suggests that, although it is probably practical, the restocking of colonies can make only a very local contribution to seabird conservation. For polar colonies the logistics would become awesome.

The very limited scope for remedial measures has led conservation bodies to look instead to the prevention of seabird losses from oil pollution. This includes chemical dispersion of the slicks, scaring birds from the path of threatening oil slicks, and the protection of sensitive colonies by the use of booms (Koski and Richardson, 1976; Royal Society for the Protection of Birds, 1979). However, experience suggests that these also have only very limited applicability and in the end will add very little to the saving of birds from oil pollution. Regardless of safeguards and remedial measures, a small amount of oil in the wrong place and at the wrong time can kill a disproportionately large number of birds.

EFFECTS ON COMMUNITIES AND ECOSYSTEMS

The earlier sections on biological effects have been concerned principally with the impact of oil spills and chronic pollution on processes and on individual species and populations of similar life habit. Beyond these specific effects there is the further and more difficult question of how perturbations of individual populations affect other members of the biological association and the overall balance of the ecosystem. These ecosystem effects most commonly operate by altering either food web interrelations or competition for available space, although some other effects have been noted. The problems are complex, and few areas have been studied in sufficient detail or for a long enough time to provide satisfactory information about the whole ecosystem. Conclusions are meager, and many problems deserve further attention.

Wetlands, Saltmarshes, and the Intertidal Zone

The margins of the sea are particularly susceptible to the impact of oil pollution. They are subject to heavy oiling when a large spill drifts ashore, with a fraction of the oil becoming sequestered in sediments and persisting in some cases for years. This is in marked contrast to conditions in the open sea, where currents and diffusion usually rapidly reduce the concentration of petroleum, making it less toxic and most likely more amenable to degradation processes.

The immediate effects of heavy oiling of the shore zone are obvious. There is widespread death of plants and animals due to smothering and toxic effects. In the longer term the effects are more variable and subtle, and a first step in their evaluation is a knowledge of the behavior and persistence of the petroleum. Southward (1982), Vandermeulen (1982), and Teal and Howarth (1984) reviewed this matter as part of a general account of long term effects observed after some major oil spills.

As has been discussed in Chapter 4, oil behaves differently in the different coastal environments, dependent very much on the porosity of the sediments and the wave-erosion activity acting on them. In high energy environments, mainly rocky shores, the stranded oil coats the rocks and gradually hardens by weathering into a tough, tarry "skin." The oil is gradually removed by wave erosion, although the rate of removal declines as weathering progresses. As much as 50% is lost within a year and a half to two years, although pools of oil are likely to collect in hollows among the rocks, protected by a skin of weathered oil, and may remain essentially unchanged for a long time.

On cobble and sandy beaches the oil can sink more deeply into the sediments and can remain longer than on bare rocks. Wave erosion becomes less effective, and microbial degradation assumes a more important role. However, as the oil is mobile in these porous systems, some of it is gradually returned to the overlying water, where it is more subject to dissipation but may also have toxic effects on the organisms. A general hypothesis has emerged which appears to be applicable to the low energy systems: sandy beaches (Long et al., 1981a; Gundlach et al., 1982), marshes (Vandermeulen, 1980), and tidal rivers (Vandermeulen et al., 1981). Tidal pumping is the active factor causing penetration into the sediments. Sediment grain size controls the rate of penetration (Owens and Robilliard, 1981). In muddy sediments, penetration is minimal, and only the upper few centimeters are affected. However, because these are low energy environments with little physical weathering, stranded oil can persist here for up to decades, frequently becoming bound up in the soft organic sediments.

All these variations are important in shaping ecosystem responses to oil spills. The macrophytes in wetlands--mainly mangrove swamps in

TABLE 5-10 Summary of Known Long Term Biological Effects of Some Major
Marine Oil Spills

Spill	Effect	Reference
Torrey Canyon[a] 1967	Continued community perturbations during recovery, 1967-1978.	Southward and Southward (1978)
Searsport 1971	C-flux abnormalities in Mya, 1976. Suppressed recovery of Mya population, 1976.	Gilfillan et al. (1976) Mayo et al. (1978)
Arrow 1970	Intertidal species abnormalities, 1976. Population abnormalities in Mya, 1976. C-flux depression in Mya, 1976. Abnormal shell formation in Mya, 1976.	M.L.H. Thomas (1978) Gilfillan and Vandermeulen (1978)
Florida[a] 1969	Community abnormalities. Changes in genetic structure of Urosalpinx populations, 1976. Mussel sterility, 1970. Long term inhibition of recruitment and low population densities in Uca pugnax. Behavioral disorders.	Sanders (1978) Sanders et al. (1980) Cole (1978) Blumer et al. (1970) Krebs and Burns (1977) Michael et al. (1978)
Metula 1974	Slow marsh recovery. Alteration of total microbial ecology.	Gundlach et al. (1982) Colwell et al. (1978)
Bouchard 1974	Impaired reseeding and rhizome growth in salt marsh vegetation, reduced interstitial fauna, increased marsh erosion.	Hampson and Moull (1978)
Argo Merchant 1974	No known long term effects.	

the tropics and salt marshes in high latitudes--are broadly susceptible
to a variety of hydrocarbons (Baker, 1979, 1981; Bender et al., 1980;
Getter et al., 1981; Golley et al., 1962; Hampson and Moul, 1978). The
effect of either physical or chemical stress from an oil spill in these
systems almost invariably is a severe reduction in population and growth
rate, amounting in some cases to complete obliteration (Baker, 1979;
Hershner, 1977; Hershner and Moore, 1979; Hershner and Lake, 1980; Chan,
1977). However, in the absence of continued stress there is likely to

TABLE 5-10 (continued)

Spill	Effect	Reference
Tsesis 1977	Reproductive effects leading to reduced Pontoporeia population, 1980.	Elmgren et al. (1980, 1981)
	Persistent disturbed community composition, 1980.	Elmgren et al. (1980b, 1981)
Amoco Cadiz 1978[b]	Benthic offshore sublittoral community perturbations, 1979.	Cabioch et al. (1980)
	Elevated mortalities in plaice, 1980	Friha and Conan (1981)
	Persistence of benthic infauna perturbations.	Glemarec et al. (1980), LeMoal (1981), Glemarec and Hussenot (1982)

[a]Dispersants used.
[b]Dispersants used only offshore.

SOURCE: Adapted from Vandermeulen (1982).

be some degree of recovery within a time span of one generation. This varies from 1 year for some marsh grass species to a decade for mangroves. Probably recovery is facilitated in part because the oil does not penetrate deeply enough into the mud to kill all of the extensive system of underground rhizomes. In this report the marsh grass (Zostera sp.), growing in oiled sediments, has been found to have a higher level of contamination than, for example, the rockweed Fucus (Vandermeulen and Gordon, 1976), which was growing in a contaminated shore zone but, of course, was attached to rocks above the oiled sediments rather than being in direct contact with them. On the other hand, the persisting hydrocarbon load in such oiled marshes is now known to have long term impact on growth characteristics of such marsh vegetation as Juncus maritimus, a species common to both shores of the North Atlantic.

Studies in an Amoco Cadiz oiled marsh have shown changes in reproductive capacity, growth abnormalities, abnormalities in seed formation, and reduction in plant spikes in samples collected in August 1981, 3 years after the spill (Levasseur and Jory, 1982). These kinds of impact, if found in a dominant member of a community, must inevitably lead to continued imbalances in recovery patterns at the community level. These may be exacerbated with persistence of oil in the associated sediments. On the other hand, vegetative recovery is possible, although of the order of decades, once the source of the contaminant is reduced and eliminated, and natural weathering processes reduce the more toxic components of residual oil (e.g., Dicks and Iball, 1981).

Petroleum hydrocarbons can apparently serve as a supplemental energy source for microbial populations in wetlands systems (L.R. Brown and Tischer, 1969; Herwig, 1978; Kator and Herwig, 1977), and there is at least a temporary increase in the number of bacteria and yeasts that are capable of degrading oil. Nevertheless, in the few cases that have been followed for as much as 10 years, considerable quantities of petroleum compounds can remain in the soft sediments.

Rapid decimation of the animal population is the immediate effect of a major spill. The recovery period is slow. Table 5-10 summarizes toxicity effects that have been observed during the recovery period. There are abnormalities in the development and recruitment of individual species and, in some cases, large scale community perturbations. In general, species diversity and total abundance are reduced, but resistant and opportunistic species often increase dramatically, temporarily making use of vacated space and then crashing from overproduction. Aside from individual species differences in tolerance, the life habit can be important in survival. For example, epibenthic filter feeders would seem to be subjected to less direct contamination than sediment feeders.

This recovery process has been examined in detail for intertidal infauna by Glemarec and colleagues following the Amoco Cadiz spill in north Brittany (Aelion and LeMoal, 1981; Glemarec, 1981; LeMoal and Quillien-Monot, 1981; Bodin and Boucher, 1981; Glemarec and Hussenot, 1981, 1982; LeMoal, 1981, 1982). Based on a scheme of species gradients developed for domestic sewage-polluted marine coastal wetlands (Glemarec and Hily, 1981), they describe progressive temporal fluctuations within the infaunal communities of two coastal inlets. After the complete destruction of the original communities, recolonization following the Amoco Cadiz spill went through a series of transitory faunal species, before proceeding through intermediate phases to the development of stable population characteristics of the unpolluted communities of the area (Figure 5-7). Glemarec and colleagues were able to recognize particular groups of infaunal species (polychaetes, bivalves) according to the extent of their sensitivity to petroleum. Appearance and disappearance of the various transitory or intermediate species were closely tied to the persistence of oil in the sediments, in turn related to the intensity of hydrodynamic ("weathering") processes in these coastal systems.

Despite a growing amount of descriptive information about ecosystem perturbations, there are big gaps in the data and uncertainties about interpretation. Toxicity is generally blamed for the overall decrease in the animal population, but primary productivity is reduced too, and its role is undefined. The fluctuations in benthic populations probably affect the fishes that normally feed upon them, but little is known about that. Also, the effects of effluents from oiled wetlands on adjacent intertidal and subtidal waters have not been studied.

Coastal and Offshore Waters and Sediments

Broadly defined, "coastal" and "offshore" are any areas seaward of the low tide level and include bays, open coastal waters, and the deep ocean. Oil spills in the open ocean do not appear to have as severe an effect on the biota as oil in coastal water or in the shore zone. The latter also, of course, are subject to serious effects from chronic pollution.

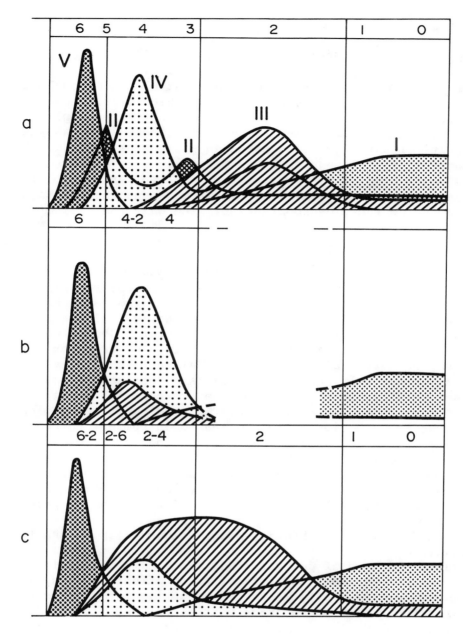

FIGURE 5-7 (a) Schema of ecological succession of benthic infauna species after oiling of intertidal sediments (stage 6). (b) Establishment of opportunistic species (groups IV and V) after stabilization of pollution. (c) Regression of opportunistic species and reestablishment of more vulnerable species (group III). Group I. Sensitive species, dominant under normal conditions (e.g., Bathyporeia spp., Ampelisca spp.). Group II. Tolerant species, normally present in small numbers (e.g., Nephtys, Glycera, and Platynereis spp.). Group III. Species vulnerable to hydrocarbon pollution which reestablish rapidly on cleanup (e.g., Phyllodoce, Spio, and Nereis spp.). Group IV. Opportunistic species, fewer in number, but which occur in high densities in maximally polluted environments (e.g., Capitella capitata, Scolelepis fuliginosa, oligochaetes).

SOURCE: Adapted from Glemarec and Hussenot (1982).

Coastal and Offshore Waters

The effects of oil in offshore waters have proven difficult to study and have in fact received little investigation.

Thus far no evidence has been found that plankton populations are significantly altered by oil spills on the high seas. However, in near-shore waters, impact has been documented. For example, after the Tsesis spill in the Baltic Sea there was a decrease in zooplankton in the vicinity of the wreck (Johansson et al., 1980). The quantity of phytoplankton, consisting mostly of microflagellates, increased briefly. There was no increase in the rate of production per unit of biomass, and Johansson therefore concluded that the change was due to a decrease in the amount consumed by zooplankton. The only analyses of oil in the water column gave values of 50-60 µg/L at a depth of 0.5-1 m at a distance of 5 km from the wreck after 2-5 days of weathering of the slick.

Quite similar results were obtained in long term oiling experiments in the 13 m^3 MERL mesocosms (Elmgren et al., 1980a; Elmgren and Frithsen, 1982; Oviatt et al., 1982; see also Studies in Field Enclosures section). The concentration of No. 2 fuel oil ranged from 60 to 350 µg/L, averaging 180 µg. This average probably was equivalent to about 75 µg/L of total aromatics (Gearing et al., 1979). Phytoplankton biomass declined initially but later increased to several times the values in control tanks. Species composition in the oiled tanks differed radically from the controls, shifting toward excessive dominance of nanoplankton. At the same time zooplankton declined, while bacterioplankton increased in number, thereby paralleling the field observations made following the Tsesis spill.

Although the Tsesis is the only spill where effects of this sort have been noted, and indeed one of the few where such effects have been looked for, chronic pollution in some bays and estuaries may be sufficient to produce similar perturbations of the plankton part of the ecosystem. Values as high as 74 µg/L have been measured in inshore waters around Britain, although typical values offshore are 1-3 µg/L (Law, 1981).

In fact, the type of perturbation that has been described may be a common result of various kinds of pollution effects. Thus, in Long Island Sound (Riley, 1956), where the pollution load includes sewage and minor amounts of oil and trace metals, primary production is maintained at a high level, with a much larger proportion of nanoplankton than in open coastal waters. The copepod population is small, relatively speaking; the cause might be either sublethal toxic effects or inefficient feeding on small phytoplankton species. Poor utilization of phytoplankton results in an increase in the quantity of nonliving particulate organic matter and in bacterioplankton. In short, both in Long Island Sound and in oil-polluted environments, there is a tendency to shift from the typical grazing pattern of coastal waters toward a detritus food web. The most obvious causal factor is competitive exclusion of some of the larger phytoplankton species by hardy and opportunistic nanoplankton; however, the microzooplankton, usually the main predators on nanoplankton, did not increase correspondingly either in the MERL tanks or in the aftermath of the Tsesis spill. As they usually respond quickly to an abundance of food, a toxic effect on these organisms was suspected (e.g., Oviatt et al., 1982).

Experiments indicate that fish eggs and larvae can be affected by exposure to petroleum hydrocarbons in water at levels similar to those found in the more polluted marine areas discussed above (Kuhnhold et al., 1978; Rice et al., 1979). This has been a matter of concern, particularly when oil operations are conducted in an area that supports an important commercial fishery. Particularly vulnerable in this respect are both that fraction of the plankton population, the neuston, which normally lives in the immediate surface layer, and the eggs and larvae of some of the commercially important fishes and invertebrates spending part of their life cycle in or near the surface layer. There are a few reports in the literature of deleterious effects, but new and better devices need to be developed for sampling under slicks before this problem can be evaluated properly.

However, eggs and larvae are subjected also to many natural hazards, and survival of most of these species depends upon the production of enormous numbers of eggs. During the feeble planktonic stage, storm-generated currents can carry them off into deep-water areas that offer no support for developing juveniles. Along the continental slope of the eastern seaboard and the southern edge of Georges Bank, warm-core eddies from the Gulf Stream can have the same effect. These natural catastrophic events generally involve larger water masses than are affected by the largest oil spills observed to date. In addition, eggs and larvae are fed upon by a variety of predators throughout the developmental period. The number surviving long enough to enter the commercial fishery is literally one in a million for cod and perhaps as much as 5×10^{-4} for herring and mackerel. Because of the variety of hazards, some sudden and catastrophic and others requiring long and careful observation, attempts to predict year-class fluctuations have not been very successful. Little connection has been demonstrated between the success of egg and larval stages and the success of the year class when it eventually enters the fishery (Hennemuth et al., 1980). Thus, observed mortality arising from an oil spill does not necessarily imply that recruitment into the commercial stock will be reduced, and to prove that there was a significant effect because of oiling, among all the other variables, would be virtually impossible. On the other hand, this cannot be taken to mean that such an impact does not exist.

Benthic Subtidal Environments

As has been described earlier (see Chapter 4), oil spilled onto the surface of the water column can be transferred to bottom sediments in a variety of ways: sorption on clay particles and subsequent sinking, sinking of dead organisms, uptake and packaging as fecal pellets by zooplankton, and direct mixing to the bottom in shallow water. Thus even the deepest ocean bottoms are potentially liable to pollution, although shallow bottoms are matters of more practical concern because of possible pollution effects on fisheries. Deposit feeders, living in intimate contact with pore waters, can be expected to show a rapid uptake from sediments, while filter feeders and carnivores are more likely to derive oil from food or from water overlying the sediments (J.W. Anderson et al., 1978, 1979).

The MERL experiments discussed earlier provided detailed information on the effects of oil on a benthic system. At the end of 20 weeks there were 109 μg/g of oil in the upper 2 cm of sediments in the bottom of the tanks. During the experimental period the macrofauna decreased drastically. There was an average of 325 individuals with a total weight of 845 g in the control tanks and 95 animals weighing 78 g in oiled tanks. There was also a significant decrease in metazoan meio-fauna populations (especially harpacticoids and ostracods, with a smaller reduction in nematodes), although benthic diatoms, ciliates, and foraminifera increased significantly. These increases were believed to be due to decreased bioperturbation, predation, and grazing as a result of reduction in the quantity of larger animals. The observed changes could not have been predicted on the basis of knowledge of the effects of oil on the individual populations involved (Grassle et al., 1981; Elmgren et al., 1980a). As for ecosystem interactions, so many possibilities exist that observed results are seldom intuitively obvious. The MERL tanks, despite certain limitations and in part because of them, have proven very useful in describing these inter-actions.

An unexpected ecosystem effect was found in studies of the Tsesis spill. The reproduction of herring was significantly reduced in the oiled area, and Nellbring et al. (1980) reported that it was not due to a direct effect of oil on the eggs, but rather was a consequence of a decrease in amphipod populations which ordinarily graze fungi growing on the fish eggs, thus preventing fungal damage.

Significant oiling of the sediments frequently has a deleterious effect of one sort or another on ground fish populations. For example, following the Amoco Cadiz spill in North Brittany, populations of plaice (Pleuronectes platessa) and sole (Solea vulgaris) from the oiled Aber Benoit river system showed a number of degradative features including reduction in growth and increased incidence of fin and tail rot (Conan and Friha, 1981). These and other toxic defects are now well known (R.D. Anderson and Anderson, 1976; Fletcher et al., 1981; Jackson et al., 1981; McCain et al., 1978). Although some fishes may avoid contaminated areas, this is not a universal type of behavior, as illustrated by the results of Weber et al. (1979) with juvenile English sole. In fact, flounder may be a particularly susceptible group. A majority of the species live in shallow, inshore waters, exhibit little or no migratory behavior, and spend a considerable amount of time lying on the bottom or even partially buried in the sediments.

In the final analysis, the benthic communities appear to be most vulnerable to oiling, although different spill conditions will lead to different degrees of impact. This confirms the conclusion of the 1975 NRC report, which was based on a few isolated studies. Two examples come from the Amoco Cadiz and the Tsesis spills which point out very remarkably the sorts of impacts that may occur, albeit unexpectedly.

In the case of the Amoco Cadiz, studies by Cabioch and coworkers (Cabioch et al., 1980; Beslier et al., 1980) have demonstrated the destruction of an entire population of the amphipod Ampelisca. Of particular significance in this case is that repopulation is expected to be difficult if not unlikely because of the absence of other such populations upstream on the southern side of the English Channel. Studies following the Tsesis spill (e.g., Elmgren et al., 1980b, 1981, 1983) show rather lucidly the results that may arise from the different sensitivities of marine organisms to oiling. Three years after the accident, the population of amphipods, Pontoporeia, remained depressed below prespill levels, partly due to heavy mortality of the amphipods by oiling, and partly because the expected repopulation from other nonoiled areas had not occurred. At the same time, the bivalve Macoma balthica population persevered and increased, apparently because it is less sensitive to spilled oil (No. 5 fuel oil plus some Bunker fuel), and possibly because of heavy recruitment during the time when Pontoporeia was eliminated. Further imbalance has been added to the community by Macoma's long life span, relative to the amphipod. Current studies indicate that full recovery is likely to require more than 5 years and possibly up to a decade (Elmgren et al., 1983).

Chronic Oiling

While the discussion has dealt primarily with spills, because they have provided the main source of information, chronic pollution in fact is a more serious problem statistically (see Chapter 2), and deserves more attention. It results from the continuous long term release of low concentrations of petroleum hydrocarbons as may occur from refinery effluent and general petroleum activities. Bays and estuaries are particularly subject to damage. As for coastal waters and offshore fishing grounds, few adequate investigations have taken place. Those that have been published show that such regions are not immune to the dangers of chronic pollution.

The development of the Ekofisk oil field provided opportunity to monitor environmental response in an area of potential long-term oiling. Observations were made in the vicinity of Platform B and a nearby storage tank. The initial survey (Dicks, 1975) was conducted in 1973 at about the time production began, and further studies were made in 1975 and 1977 (Addy et al., 1978). During this period there was a decrease in total abundance and number of species of benthic fauna and an increase in the hydrocarbon content of the sediments in the immediate vicinity of the operation, but the effect was limited to a radius of a few kilometers. Thus the impact was minor and may have been due to the open-bottom design of the storage tank (a situation since corrected) rather than due to the platform operation.

A different kind of long term oiling potential is found in the waters off Louisiana and eastern Texas, following initial drilling there in the 1930s. Production in the area became significant, and by the 1970s, navigational charts showed more than 2,600 platforms extending from east of the mouth of the Mississippi River to a point beyond the Louisiana-Texas border. Because of the scale of drilling and the controversy surrounding some of the environmental studies, it is worthwhile to consider this situation in some detail.

In the early 1970s a major effort was launched to determine whether environmental quality had been affected by drilling and production operations, using sites in the estuarine waters of Timbalier Bay and on the adjacent continental shelf off central Louisiana. This program, known as Offshore Ecology Investigation by the Gulf Universities Research Consortium (OEI/GURC), produced a series of reports by 23 principal investigators from 14 research institutes and universities and was funded by oil companies and oil-related industries. The Project Planning Council released a consensus report (Morgan et al., 1974), which stated that conditions near the platforms were not very different from control stations and concluded that little if any damage had been done.

The original reports were microfilmed, and some were later published in the Rice University Studies, together with a general review by the editors. In the meantime the conclusions of the consensus report were challenged on the basis of data from the original report (Sanders, 1981). Major points in the critique were that the so-called control stations were too contaminated to serve as adequate controls and that the collection and analysis of bottom fauna had been improperly conducted.

However, this controversy is put into a larger context by another survey conducted in 1978-1979 by the Southwest Research Institute under the auspices of the U.S. Bureau of Land Management. The series of reports coming from this survey was summarized by Bedinger (1981) and revealed a situation much more complex than was apparent in the earlier investigation.

A portion of the outflow from the Mississippi River moves westward as a coastwise density current, and particularly during periods of peak runoff it distributes a heavy sediment load and a variety of contaminants including trace metals and hydrocarbons. In addition, vertical stability associated with the reduction of salinity in the surface layer leads to depletion of oxygen in the bottom waters. The net result was that large areas of "dead bottom" were found, extending as far as 300 km west of the main Mississippi tributaries. However, this was a varying situation in which storm mixing tended to restore healthy bottom conditions and permitted repopulation by benthic fauna. Bedinger concluded "that the entire Louisiana OCS is experiencing chronic contamination from the Mississippi River and probably production activities, but that the periodic flooding of the river and irregular but relatively frequent tropical cyclones cause such serious effects that they mask any platform related effects." The quotation refers to the area in general; biotic effects and increased hydrocarbon content in the sediments were noted in the immediate vicinity of some of the platforms.

Another Gulf Coast survey has been described in a book of composite authorship edited by Middleditch (1981). It was conducted in the Buccaneer Gas and Oil Field, about 50 km south of Galveston, Texas. Environmental investigations of possible oil impact from this operation did not begin until some 15 years after start-up of the field activities. It is therefore possible that the studies missed the period of biotic transition that might have occurred with an environmental perturbation. The Louisiana coastal current continues through this area most of the year, with maximum reduction in salinity 1-2 months after the peak outflow from the Mississippi, but there is no evidence that significant hydrocarbon contamination is carried that far from its source. Currents commonly are of the order of 10-30 cm/s and occasionally are much stronger during storms and are able to resuspend and disperse soft sediments as well as to remove dissolved materials. Because of this rapid dispersion, hydrocarbon contamination appeared to be negligible except in the biofouling community on the platform.

The Buccaneer and Ekofisk surveys suggest that thus far oil production has posed only localized environmental hazard of a chronic nature in open waters. However, there is concern that chronic releases in some coastal and continental shelf areas when coupled with restrictive circulation of water or mesoscale gyres could result in adverse effects over a period of years. While this concern appears to us justifiable, the current data related to this concern are not conclusive. Whatever the expected activity, baseline surveys of any inshore area prior to petroleum operations should include a thorough study of flushing rates and existing pollution levels.

Another example of chronic oil impact concerns a saltmarsh near Southampton, U.K., which was continuously receiving refinery and petrochemical effluents between the years 1953 and 1971. Studies at the time indicated considerable ongoing damage to marsh biota, especially because of repeated light oiling of the vegetation, with considerable concern for long-term recolonization (Baker, 1970, 1971). One of the remarkable consequences accompanying this regression in marsh quality was the progressive die-back of Spartina between 1950 and 1970 to the point where much of the impacted marsh was denuded of this grass, and consisted largely of areas of bare mud.

With the early 1970s an active effluent improvement program, on the part of the refinery/petrochemical industry involved, led to a reduction in effluent hydrocarbons, while at the same time an ecological monitoring program documented a steady recolonization of the mud areas, both by marsh grasses and by infauna (Dicks and Iball, 1981). The recovery showed a number of interesting features: (1) Regrowth by the dominant marsh plants, Salicornia and Spartina, paralleled the decrease in effluent hydrocarbon concentration, following the effluent remodeling program, and could be directly related to this. (2) This vegetative recovery did lag behind the effluent hydrocarbon reduction by about two years. (3) Vegetative recovery consisted of a graded process, with Salicornia the primary recolonizer and Spartina the secondary recolonizer, while in particularly polluted mud areas near the effluent outfalls a similar graded progression of algae were documented, coinciding with distance from the outfalls. (4) Vegetative recovery did proceed despite long-term hydrocarbon pollution of the marsh sediments, apparently correlating with considerable weathering of the aliphatic fraction.

These observations indicate that saltmarshes, despite their obvious vulnerability and sensitivity, can recover from oiling impact. One must be careful, however, in extrapolating these results and this recovery to situations involving more massive oiling, such as in the Ile Grande Marsh following the Amoco Cadiz, where much of the marsh was inundated by a thick layer of whole crude oil and mousse.

A different kind of chronic pollution is found in areas subject to repeated oil spillage, such as the English Channel, scene of several large spills beginning with the Torrey Canyon. In these cases the environmental impact of one spill may overlap on the impact of a previous spill, with the result potentially exceeding that of either of the individual spills. Little is known of either the cumulative impact or of the long term biological recovery after such repeated spillage, partly because of the lack of sufficient follow-up studies. Thus there are inherent difficulties in distinguishing ecological perturbations due to a previous spill from those resulting from a subsequent accident.

The ecological survey method, using benthic infauna species, developed by Glemarec and coworkers (viz., Wetlands, Marshes, and the Intertidal Zone section, Figure 5-7) seems one useful approach. In a comparison of benthic infauna from six North Brittany sites (three known oiled by the 1978 Amoco Cadiz, and three by the 1980 Tanio), their data suggest that one site, the bay of Ste. Anne (Tregastel), was already in ecological imbalance at the time of the Tanio oiling, possibly as a result of the Amoco Cadiz oiling 2 years earlier (LeMoal, 1982).

Effects on Communities and Ecosystems

Despite a growing amount of descriptive information about ecosystem perturbations resulting from oil pollution, there are big gaps in the data and uncertainties about interpretation. Where there have been heavy spills, and severe damage still exists after some years, further monitoring is needed to examine the whole history of the recovery process.

In the cases that have now reached a late stage of recovery, investigation is hampered by lack of sufficient knowledge about normal, unstressed ecosystems. There are virtually no surveys that have been continued long enough to provide a standard for comparing long term variability in stressed and unstressed environments. In Table 5-10, continued community perturbations have been noted in the vicinity of the Torrey Canyon disaster despite the fact that the beaches were thoroughly cleaned with dispersants (Southward and Southward, 1978). M.L.H. Thomas (1978) has also found long term effects in rocky intertidal areas where self-cleaning was rapid after the Arrow grounding. These are presumed to be relaxation responses from the original perturbation. However, there are perturbations in natural systems, too (Jones, 1982; Ware, 1964; Bowman, 1978) and we do not know the extent of their effect on the ecosystem as a whole. Year-class fluctuations in some of the commercial fishery stocks are well known, and there are similar variations in recruitment of some of the long-lived benthic invertebrates.

SPECIAL PROBLEM AREAS

Tropical Regions

Coral Communities

At the writing of the 1975 NRC report, little was known of the potential impact of oil on coral reefs or on associated organisms, and indeed the general consensus at that time was that "there appears to be no conclusive evidence that oil floating above reef corals damages them" (Johannes, 1975). Studies carried out since that time, principally those of Loya, Rinkevich, and coworkers, have established very clearly the vulnerability and sensitivity of the hermatypic or reef-building corals and their communities to oil and oil components. In addition, their importance, both as major marine ecosystems and in terms of human use, ranks them near the top in terms of spill concerns.

Data on the fate and effects of petroleum hydrocarbons on coral systems are still limited, despite the vigorous efforts of a few workers (for recent reviews, e.g., Loya and Rinkevich, 1980; Ray, 1981; Knap et al., 1983; Vandermeulen and Gilfillan, 1984). Unfortunately, the variety of field and laboratory methodologies used, the various exposure conditions, and the differing hydrocarbon preparations and concentrations all add to the difficulty in assessing the problem. In fact, in several studies, reliable estimates of oil composition or concentration during the experiment are not available. Fortunately, more information is now beginning to become available, although much more is needed to assess properly both the vulnerability and sensitivity of reef systems, and their recoverability following a spill (Figure 5-8).

A wide range of responses to oil has been observed for hermatypic corals, including decreases in reproduction and colonization capacity, and effects on feeding and behavior (Table 5-11). Oiling can cause problems, particularly with the reproductive process, which in turn has consequences for colonization and recolonization. A comparison, for example, of populations of the reef coral _Stylophora pistillata_, from chronically oiled and clean reefs and from laboratory populations, has shown a higher mortality rate in the oiled colonies. The oiled reefs also exhibited smaller numbers of breeding colonies, a decrease in the average number of ovaria per polyp, smaller numbers of planula larvae produced per coral head (fecundity was 4 times higher in the clean reef), and a lower settlement rate of planulae on artificial substrates (Rinkevich and Loya, 1977). Subsequent long term laboratory studies (2-6 months), using an aquarium-type flow-through system with surface oiling to simulate the field oiling conditions, showed a significant decrease in the number of female gonads per polyp in 75% of the polluted _Stylophora_ colonies as compared to nonoiled control specimens (Rinkevich and Loya, 1979). Similar results, due to No. 2 fuel oil, were described for the Bahamian coral _Manicina areolata_ by Peters et al. (1981). In this case, degeneration of the coral's ova and lack of gonadal development were noted, under both static and flow-through bioassay procedures.

FIGURE 5-8 Oil pollution at coral nature reserve of Eilat, Red Sea.
The oil is coming in to shore as broad, 3-m-wide slicks and can be seen
along the shore up to the high water mark. (Photo by Y. Loya.)

Premature expulsion of the coral planula larvae appears to be a
common feature of oil pollution of reef systems examined to date.
Larval extrusion due to sublethal concentrations of a crude oil, after
72-hour exposure, were reported for the soft coral Heteroxenia
fuscescens (Cohen et al., 1977). Loya and Rinkevich (1979) described
immediate mouth opening in the coral Stylophora pistillata in response
to sublethal concentrations of Iranian crude oil, followed by premature
extrusion of planulae larvae. Under natural conditions, S. pistillata
releases its planula only during the night. In the presence of
water-soluble fractions of this crude oil the shedding was immediate,
regardless of time of the day, decreasing their chances of survival and
settling.
Detrimental effects on the feeding response and tactile stimuli
have been reported by J.B. Lewis (1971) for four Caribbean corals. The
effects included a marked decrease in the number of tentacles expanded
during feeding, and increased extrusion of septal filaments. Reimer
(1975a) has described abnormal feeding reactions in four scleractinian
corals and in one zoanthid coral, Palythoa sp., due to oil floating
over the coral (Reimer, 1975b). Other effects, such as thinning of the
coenosarc (Peters et al., 1980a,b) and changes in the pulsation rate of
the polyps of Heteroxenia fuscescens in response to Iranian crude
(Cohen et al., 1977) have also been reported. Most corals produce
excessive mucus on exposure to oil. This probably is a generalized
protective mechanism whereby the coral cleans itself of irritating
foreign material. In fact, abundant mucus producers such as Fungia and

TABLE 5-11 Summary of Effects of Oil Spills on Coral Reefs

Spill	Amount Spilled	Reported Effects	Reference
World War II, unknown, Japtan Island (Enewetak)		Many porous rocks and boulders near the remains of the ship still heavily tarred in 1974	Johannes (1975)
WW II, several tankers, Gulf of Mexico and Caribbean Sea			Dennis (1959), U.S. Coast Guard (1959, 1969)
WW II, Dry Tortugas		Young mangroves of 4-5 years were killed	Odum and Johannes (1975)
1967, Argea Prima Puerto Rico	10,000 tons crude	Mortalities of adult and juvenile lobster, crabs, sea urchins, sea stars, sea cucumbers, gastropoda, octopus, and fish; Thalassia beds degenerated, rocky areas denuded of algae; extensive damage to mangrove-swamp habitat	Diaz-Piferrer (1964)
1966, British Crown, Persian Gulf	25,000 tons Qater crude		Beynon (1971), Nelson-Smith (1973)
1967, R.C. Staner, Wake Island, Pacific Ocean	22,000 tons gasoline, jet fuel, turbine fuel, diesel oil, Bunker C	About 2,500 kg reef fish killed and stranded; large mortalities snails and sea urchins	Gooding (1971)
1968, General Colocotronis Installation, Eleuthera-Bahamas	4,500 tons crude		Spooner and Spooner (1968)
1968, Ocean Eagle, San Juan, Puerto Rico		Many mortalities among intertidal organisms due to oil and emulsifier, including fish, mollusks, and algae; recovery good	Cerame-Vives (1968)
1968, Witwater, Galeta Island, Canal Zone	20,000 barrels diesel oil, Bunker C	Harmful effects to meiofauna, mangroves, fiddler crabs, elimination of algae; reef corals least affected	Rutzler and Sterrer (1970)
1970 Ennerdale, Seychelles	20,000 tons fuel		Nelson-Smith (1973)
1970 Pipeline break, Tarut Bay, Saudi Arabia	100,000 barrels Arabian light crude	Mortalities among crabs, bivalves, gastropods, fish; mangrove trees less affected; no detrimental effects on corals and associated fauna; good subsequent recovery	Spooner (1970)
1970, Oceanic Grandeur, Torres Strait, N. Great Barrier Reef	1,100 tons crude	Heavy mortalities of pearl oysters	Smithsonian Institute (1970a)

TABLE 5-11 (continued)

Spill	Amount Spilled	Reported Effects	Reference
1970, unknown tanker near Pennekamp coral reef, Florida Keys	Unknown; slick 75 mi long, 0.5 mi wide	No apparent effects; no apparent stranding of oil	Smithsonian Institute (1970b)
1971 MV Solar Trader, West Fayu, Caroline I	520 tons fuel and lubricating oils	Numerous dead lobsters and clams; survey 8 months afterward reported large algal growth on corals in the area	Smithsonian Institute (1970c)
1974 Sygma, Stockton Bight, east coast of Australia	400 tons heavy fuel	13 km beaches affected; no damage reported to marine life	Hughes (1974)
1975 MV Lindenbank, Fanning Atoll, Pacific Ocean	10,000 tons copra palm oil, coconut oil, cocoa beans	Mortalities of fish, crustacea, and mollusks; afterward extensive growth of Enteromorpha and Ulva; reportedly complete recovery of original coralline algal community after 11 months	D.J. Russell and Carlson (1978)
1969-1979, two oil terminals, Eilat, Red Sea, Israel	Many small scale oil spills, various tankers (crude oil)	Decrease in coral and fish diversity; lack of colonization by hermatypic corals in reef areas chronically polluted by oil; damage to reproductive system of corals	Fishelson (1973, 1977), Loya (1975, 1976), Rinkevich and Loya (1977, 1979), Loya and Rinkevich (1979)

SOURCE: After Loya and Rinkevich (1980).

Symphyllia showed good survival when directly contaminated with oil (Johannes et al., 1972). There appears to be some relation between mucus production and associated bacterial growth (Mitchell and Chet, 1975), but this is by no means clearly understood.

Some observations suggest that corals may ingest oil, but nothing further is known of the fate of such ingested oil, or indeed of any metabolism that might be involved (Reimer, 1975b; Cohen et al., 1977). Certain species would seem to have a greater affinity for oiling, possibly because of their morphology and surface texture. For example, J.B. Lewis (1971) noted that the branching species (e.g., Acropora) had a greater affinity for oil than the encrusting species (e.g., Agaricia).

Summary Recent quantitative field studies have considerably advanced our knowledge of both coral reef ecology and on the potential effects of petroleum in these ecosystems. However, there are many gaps (see also Chapter 3, Biological Methods section). One item missing from most field studies involving oil is reliable information on the actual concentrations or composition of petroleum hydrocarbons found in the water, particularly during exposure times. Clearly this information is needed for critical comparison of data. Even more fundamental is the need for a better understanding of the reproductive physiology and metabolism of reef corals, which for many genera is totally lacking. Without this basic information and understanding the marine toxicologist is working in the dark.

At present the potential hazard of petroleum to reef corals and reef communities can best be estimated from the sensitivity of the larval stages. Despite their awesome mass and apparent robustness, reef corals are very sensitive organisms, forming the matrix for an extensive and highly intricate food web. In many parts of the southern hemisphere the coral reef is also the basis for human economy. Considerations of evidence to date rank this marine tropical system high in terms of potential impact of oil and the need for further research (e.g., Gundlach et al., 1979).

Mangroves

Most of the information on oil impact on mangrove systems comes from studies done on spills of opportunity (see Table 5-12). Unfortunately, in many cases the studies were done only sporadically and shortly after their occurrence, but without subsequent follow-up work. In others, several years had elapsed before work was begun, in the absence of data on the on-scene spill impact data. To make matters more difficult, very little of the available information is in the primary scientific literature, and has to be sought from conference reports, workshop proceedings, etc.

Nonetheless, the survey of work assembled for this report shows that the mangrove system represents a unique problem regarding oiling and oil impact, exceeding that of the salt marsh of the temperate zone.

Mangroves are essential to the tropical marine environment, representing a major component in the productivity of tropical coastal systems. They are found in most tropical areas of the world, occurring along an estimated 75% of the coastlines between 25°N and 25°S latitude. In Australia they extend to 39°S, and in Japan and Bermuda to 32°N. In the United States, mangroves are represented by four species: the red mangrove (Rhizophora mangle), black mangrove (Avicennia germinans), white mangrove (Laguncularia racemosa Gaertn), and Bruguiera gymnorrhiza. The later is found only in Hawaii as an introduced species.

Mangrove systems, in terms of the marine environment, provide two essential functions. They act to protect coastal systems against storm and current erosion through trapping and stabilizing sediments and debris (e.g., Socffin, 1970; Carlton, 1974; Teas et al., 1975). They also provide food and shelter for a large number of invertebrate and vertebrate species through a complex detrital food web (e.g., Blasco, 1982).

Many commercially important species depend on mangroves for part of their life cycles, including the spiny lobster, snapper, drum, sea trout, crabs, shrimp, mullet, and menhaden. Based on analyses of juvenile fish species present in mangroves, Carter et al. (1973) concluded that the most important nursery grounds in the marine environment were the mangrove-fringed bays. A similar relationship between

TABLE 5-12 Comparisons of the Impact of Some Oil Spills on Mangroves

Spill	Amount Spilled	Mangrove Species Affected	Reported Impact	Reference
1962 Argea Prima, Guanica, Puerto Rico	10,000 tons crude	unidentified	Virtual destruction of habitat (reportedly rediscovered in mid-1970; Teas, personal communication)	Diaz-Piferrer (1964)
1968 Witwater, Galeta Island, Panama	20,000 bbls diesel oil and Bunker C	Rhizophora mangle Avicennia sp.	Death of young mangroves, loss of sessile animals and algae on prop roots (loss visible 66 months after spill)	Rutzler and Sterrer (1970), Birkeland et al. (1976)
1970 Pipeline break Tarut Bay, Saudi Arabia	light crude	A. marina	Defoliation, but many trees survived	Spooner (1970)
1976 St. Peter, Colombia/Ecuador	243,442 bbls cargo; unknown quantity spilled	Rhizophora sp., Avicennia sp.	No noticeable long term effects; temporary decline in fishery and clam harvesting	Jernelov et al. (1976), Jernelov and Linden (1980), Hayes (1977)
1975 Garbis, Florida Keys	1,500-3,000 bbls crude oil	R. mangle, A. germinans (nitida)	Death of young red mangrove seedlings and some dwarf black mangroves	Chan (1976, 1977), VAST/TRC (1975)
1973 Zoe Colocotroni, Cabo Rojo, Puerto Rico	37,000 bbls Venezuelan crude	R. mangle and A. germinans	Death of adult trees (red and black) over 1.0-2.7 hectare area within 3 years	Tosteson et al. (1977), Nadeau and Bergquist (1977), Page et al. (1979), Gilfillan et al. (1981)

Incident	Amount/Type of Oil	Species	Effects	References
1975 Showa Maru, Indonesia	54,000 bbls Arabian Light, Berri and Murban crude	Sonneratia sp., Rhizophora sp.	Unquantified number of dead trees of both species; greatest impact in sheltered bays; low numbers of crabs and snails associated with oiled sediments afterward	Baker et al. (1981), Baker (1981)
1976 Pipeline rupture, Corpus Christi, Texas	377 bbls crude	A. germinans	Mangroves burned to remove oiled uncleaned trees; recovered after minor defoliation	Holt et al. (1978)
1977 unidentified vessel, Guayanillu Bay, Puerto Rico	1,000 bbls Venezuelan crude	R. mangle	Damage to mangrove root community; trees survived	Lopez (1978)
1971 Santa Augusta, St. Croix, U.S. Virgin I.	12.5 million liters crude	R. mangle	5 hectares completely destroyed; little or no recolonization after 7 years	R.R. Lewis (1979a,b), R.R. Lewis and Haines (1980)
1980 Funiwa 5, Offshore oil well, Nigeria	8.4 million gallons, crude	R. racemosa, A. africana, L. aguncularia racemosa	Under study	Baker (1981), OSIR (1980a,b), Teas and Gilfillen (in preparation)
1978 Pack Slip barge, Puerto Rico	440–460,000 gallons Bunker C	R. mangle	significantly affected mangrove, crab, snail, and epiphyte population	Gundlach et al. (1979), Robinson (1979), Getter et al. (1981)
1978 Howard Star, Tampa, Florida	40,000 gallons 20% diesel and 80% Bunker C	R. mangle, A. germinans, L. racemosa	Mortalities in all three species of mangroves, in mollusks and polychaetes; root abnormalities	Gundlach et al. (1979), R.C. Lewis (1980a,b), Getter et al. (1980, 1981), Snedaker et al. (1981)

SOURCE: Adapted from Lewis (1980c) and Baker et al. (1981).

mangroves and prawn and finfish catches has been argued for portions of the Indo-Pacific. In Florida the mangrove habitat is considered a nursery ground for such commercially valuable crustaceans and fish as the pink shrimp (Penaeus duorarum), blue crab (Callinectes sapidus), mullet (Mugil cephalus), gray snapper (Lutjanus griseus), red drum (Sciaenops ocellata), and sea trout (Cynoscion nubulosus) (Odum and Heald, 1972). In many tropical regions where they occur, mangroves also represent an important source of wood and other products and are a significant feature and staple of local culture and commerce (Teas, 1979; Blasco, 1982).

The vulnerability of the mangrove system to oil residues resides in two features, both unique to this ecosystem--the aerial root system and the permeability of the mangrove swamp to tidal water and therefore to oil (Figure 5-9). The roots of mangroves are highly adapted to anaerobic soils or muds, emerging above the surface as aerial prop roots (red mangrove) or pneumatophores (e.g., black mangrove). The surfaces of these structures are marked by numerous small pores termed lenticels, through which oxygen passes into the air passages within the root system. While it is a remarkable adaptation to an otherwise anaerobic environment, this aerial root system is also the Achilles heel of the mangrove in the event of oiling, for the aerial roots are highly susceptible to oiling, with clogging of the lenticels and inner air passages, eventually choking off the respiratory system (e.g., Bacon, 1970; Odum and Johannes, 1975).

The second problem is that of permeability of this coastal system. Mangroves are the tropical equivalent of the more temperate salt marshes (Teas, 1979) and share many of the physical features that make salt marshes highly sensitive to oiling--low wave energy, large numbers of small channels, fine sediments. Their highly organic detritus-derived sediments make them especially susceptible to oil entrapment.

Most of what is known about oil impact on mangroves comes from opportunistic investigations at the time of an oiling incident (for reviews viz. Baker, 1982; Vandermeulen and Gilfillan, 1984). Few of these observations have been reinvestigated or confirmed in a rigorous way, either in the field or in the laboratory. Mangrove seedlings show considerable sensitivity to oil. Oiled propagules show reduced rooting rates in comparison to nonoiled controls. Young trees generally appear to be more sensitive than the mature trees.

A common feature of oiling of a mangrove forest is leaf loss, and severely impacted areas have seen complete defoliation, frequently accompanied by either root or leaf exposure to the oil. Partial defoliation usually occurs in any spill impact area. Recovery from leaf loss is a varying phenomenon. Totally defoliated trees have not been known to recover. In partially defoliated trees, recovery was found to be initiated within 4 months. In others, recovery was still occurring only slowly 7 years after the spill incident. The leaves appear to be particularly sensitive to direct oiling, possibly because the trichomes on the lower surface of the leaves are damaged, resulting in disruption of the plant's ability to regulate water loss through its stomates. The wilting and desiccation often seen in oiled trees supports this suggestion.

FIGURE 5-9 Effects of oil on a mangrove forest. (Top) Aerial view
(lighter regions along river drainage system show river banks through
leafless trees) and (bottom) close-up view. (Photo by E. Gilfillan.)

Recovery has been seen in most spill areas, although the few rates that have been estimated vary in their magnitude. Microbial degradation of petroleum in these tropical sediments appears to be more rapid than in temperate climates (Gilfillan et al., 1981), but very little else is known of the capacity of the associated microbial community to degrade the spilled and residual oil. Oil, once buried within the anaerobic sediments below the surface zone of biological activity, may be effectively isolated from such activity. The level of activity of macroorganisms in these sediments, for example, burrowing crabs, would have considerable influence on the long term fate of the impacted area. Similarly, estimates of biological recovery are uncertain. Totally devastated areas have appeared to be on climax cycles of approximately 20 years. However, functional recovery probably occurs over a shorter period. Canopy recovery occurs within 6-7 years, and deposition of the rootmat detritus cover is well established after 10 years (H.J. Teas, unpublished data, 1983). Some efforts at artificially reseeding damaged areas have been successful, and more work in this direction is continuing.

Summary Mangrove regions represent a large unknown for the tropical areas of the world's ocean in terms of petroleum impact. Both the mangroves and the dependent communities appear vulnerable to oiling, with mangroves in particular highly sensitive to oil impact. Further research in all aspects is recommended.

Polar Environments

Dunbar's (1968) timely warning that oil represents a serious threat to the arctic marine ecosystem still appears justified today. The rapid pace of petroleum activities in the Arctic during the past decade has aroused intense public concern about the effects of spilled oil, and the potential for oil spillage will increase even more as proposed production and transportation networks spread more widely across the Arctic. Unfortunately, the nature and magnitude of the threat have been difficult to assess in specific terms, and predicted effects range from the virtually negligible to the cataclysmic. During the past decade, however, sufficient basic and applied research has been undertaken (viz., Table 5-13; also Arctic Marine Oil Spill Program, 1978-1982) to at least place the problem in a more realistic perspective and to identify areas of particular concern. Considerable efforts are being expended and some progress seems to have been made on developing spill countermeasures.

In contrast, petroleum development in the Antarctic is not imminent, and as a result, little attention has been paid to the effects of oil on the fauna and flora of the southern polar ocean. The many pronounced physical and biological differences between the two polar marine areas (e.g., Knox and Lowry, 1977; George, 1977) will make it difficult to utilize the information gathered in the Arctic for anticipating any ecological impact of oil in the Antarctic.

TABLE 5-13 Summary of Studies on the Sublethal Effects of Petroleum Hydrocarbons on Polar Marine Organisms

Species	Location	Sublethal Effect	Reference
Microbiota Bacteria	Beaufort Sea coast	Heavy oils more resistant to biodegradation than lighter oils, light volatiles toxic to bacteria at low temperatures, biodegradation enhanced by nutrients	Atlas (1974)
Bacteria	Point Barrow	Increase in oil-degrading microorganisms after contamination, biodegradation slow at low temperature	Atlas (1977) Atlas et al. (1976, 1978)
Bacteria	Point Barrow, Arctic Ocean	Alkane metabolism by surface microflora widespread in Arctic Ocean	Arhelger et al. (1977)
Bacteria	Beaufort Sea	Ability of mixed cultures to degrade oil at 0°C in laboratory	Bunch and Harland (1976)
Phytoplankton	Beaufort Sea	Growth of diatoms and green flagellates inhibited by exposure to high concentrations	Hsiao (1978)
Phytoplankton	Beaufort Sea	Increasing inhibition of primary production with increasing oil concentration	Hsiao et al. (1978)
Phytoplankton (ice-edge stations)	Kachemak Bay, Alaska	Naphthalene metabolism, changed growth characteristics sensitivity to crude oil	Cerniglia et al. (1982), Van Baaken and Gibson (1982)

TABLE 5-13 (continued)

Species	Location	Sublethal Effect	Reference
Phytoplankton	Cape Parry	Slight enhancement of primary production in water under oiled ice even though light reduced 50%	Adams (1975)
Phytoplankton	Cape Parry	Growth and photosynthesis inhibited at high concentrations	NORCOR (1975)
Macroflora			
Laminaria saccharina Phyllophora truncata	Beaufort Sea	Primary production inhibited	Hsiao et al. (1978)
Marsh vegetation	Cape Parry	Chlorophyll content and photosynthetic capacity reduced	NORCOR (1975)
Coelenterates			
Halitholus cirratus	Beaufort Sea	Locomotory activity impaired	Percy and Mullin (1977)
Cyanea capillata unidentified medusae	Cape Parry	Locomotory activity impaired	NORCOR (1975)
Ctenophores			
Pleurobrachia pileus	Cape Parry	Oil exposure, tolerated behavior unaffected	NORCOR (1975)
P. pileus	Fletcher's ice island	Benzo(a)pyrene taken up but not metabolized	R.F. Lee (1975)

TABLE 5-13 (continued)

Species	Location	Sublethal Effect	Reference
Polychaetes:			
Pectinaria hyperborea Nephythys longosetosa Spio sp.	Point Barrow	Attraction to oil-contaminated sediments	Atlas et al. (1978); Busdosh (1978)
Acanthostephieia behrengiensis	Point Barrow	No recolonization of oil-contaminated sediments	Atlas et al. (1978); Busdosh (1978)
Nereis vexillosa Harmothoe imbricata	Alaska	Sensitivity	Rice et al. (1979)
Mollusks			
Macoma balthica	Gulf of Alaska	Forced to surface when exposed to dissolved or sediment-absorbed oil	Taylor and Karinen (1977)
Mytilus edulis	Gulf of Alaska	Exposure to hydrocarbons resulted in lower rate of byssal thread formation	Rice et al. (1978)
Wide range of mollusks	Alaska	Sensitivity	Rice et al. (1979)
Echinoderms			
Strongylocentrotus droebachiensis Opiopholis aculeata	Spitzbergen	Uptake of lower weight aromatic hydrocarbons after spill	Carstens and Sendstad (1979)
Cucumaria vega S. droebachiensis Leptasterias hexactis Eupentacta quinquesimita	Alaska	Sensitivity	Rice et al. (1979)
Crustaceans **Copepods**			
Calanus finmarchicus C. glacialis	Spitzbergen	Uptake of lower weight aromatic hydrocarbons after spill	Carstens and Sendstad (1979)
C. hyperboreus	Fletcher's ice island	Uptake and depuration of benzo(a)pyrene	R.F. Lee (1975)
	Davis Strait	Depression of feeding by seep oil	Gilfillan et al. (1984)

TABLE 5-13 (continued)

Species	Location	Sublethal Effect	Reference
Isopods			
<u>Mesidotea entomon</u>	Beaufort Sea	Inhibition of growth and molting only at high concentrations	Percy (1977b)
<u>M. entomon</u> <u>M. sibirica</u>	Beaufort Sea	Neutral response to presence of oil, oil-tainted food, and contaminated sediments	Percy (1976, 1977a)
<u>M. entomon</u>	Point Barrow	Readily recolonized contaminated sediment	Atlas et al. (1978)
<u>M. entomon</u>	Baltic Sea	Abortions in gravid females exposed to phenol and 4-chlorophenol, exposed animals more aggressive	Oksama and Kristoffersson (1979)
Amphipods			
<u>Boeckosimus affinis</u> <u>Gammarus zaddachi</u>	Point Barrow	Food search success reduced, recovery occurs; no effect on respiration; reduced burrowing in contaminated sediments; reduced locomotory activity	Busdosh (1978)
Unidentified amphipods	Cape Parry	Reduced activity following prolonged exposure	NORCOR (1975)
<u>B. affinis</u>	Beaufort Sea	Reduction in respiration at low oil concentrations but increase at high concentrations	Percy (1977b)
<u>B. affinis</u> <u>Gammarus oceanicus</u>	Beaufort Sea	Avoid oil, oil-tainted food, and contaminated sediments	Percy (1976)
<u>B. affinis</u>	Beaufort Sea	Reduction in locomotory activity	Percy and Mullin (1977)

TABLE 5-13 (continued)

Species	Location	Sublethal Effect	Reference
Corophium clarencense	Beaufort Sea	Neutral behavioral response to contaminated sediments	Percy (1977a)
Pontoporeia femorata Aceroides latipes Melita formosa Gammaracanthus loricatus Monoculodes sp.	Point Barrow	Little or no recolonization of contaminated sediments within 60 days	Atlas et al. (1978)
Range of crustacea	Alaska	Sensitivity	Rice et al. (1979)

Teleosts

Species	Location	Sublethal Effect	Reference
Myoxocephalus verrucosus	Bering Sea	Exposure to napthalene, rapid uptake, loss of equilibrium, cessation of feeding, no effect on ionic and osmotic regulation, decrease in hematocrit, reduced respiration, no effect on protein synthesis by liver	DeVries (1979)
Range of fish (including salmon herring, flounder)	Alaska	Sensitivity	Rice et al. (1979) Moles and Rice (1983)
Oncorhynchus kisutch (Coho salmon)	Alaska	Reduced growth in salmon fry	Moles et al. (1981)

Mammals

Species	Location	Sublethal Effect	Reference
Phoca hispida	Cape Parry, NWT, Canada	Seals immersed in floating oil; immediate signs of stress; severe eye irritation, recovery occurs; no evidence of haematologic, biochemical, or physiological disturbances; hydrocarbons in urine and bile	T.G. Smith and Geraci (1975), Geraci and Smith (1976, 1977)
Callorhinus ursinus Erignathus barbatus Phoca groenlandica Ieptonychotes weddelli Odobenus rosmarus	Various areas	Small amounts of crude oil have great effect on thermal conductance of fur-bearing pelts, but little effect on non-fur-bearing pelts	Kooyman et al. (1976)

TABLE 5-13 (continued)

Species	Location	Sublethal Effect	Reference
C. ursinus	Bering Sea	Light oiling of pelt surface, increased metabolic rate when animal immersed in cold water, effect lasts at least 2 weeks	Kooyman et al. (1976)
Thalarctos maritimus	Churchill, Manitoba, Canada	Ingested oil affected the hematopoietic system culminating in renal failure, oil on the skin and fur increased heat loss and reduced insulation drastically	Oritsland et al. (1981)

Polar Studies

Much of the useful information about effects of oil on temperate environments has come from studies on oil released into the sea accidentally, experimentally, or naturally (as from seeps). In polar seas, little of ecological significance has been learned from the many minor spills there (Keevil and Ramseier, 1975; E.L. Lewis, 1979). This is due to a combination of logistic difficulties, interference from cleanup activities, and lack of prespill baseline information. Thus, there has been a continuing interest in the use of small-scale experimental spills. These have been particularly useful in increasing our understanding of the behavior and fate of oil spilled on or under ice (Glaeser and Vance, 1971; McMinn, 1972; McMinn and Golden, 1973; Adams, 1975; NORCOR, 1975; Martin, 1979; Arctic, 1978) but have rarely been effectively utilized for the examination of biological effects (Busdosh and Atlas, 1977; Atlas et al., 1978). The Baffin Island oil spill (BIOS) program now nearing completion in the Canadian Arctic involves the first moderate-sized experimental spill devoted in part to an integrated and multidisciplinary study of biological effects (Blackall and Sergy, 1981, 1983).

Although oil seeps occur in many areas of the Arctic, the total quantity of oil released is probably low, and elevated hydrocarbon concentrations in the water column are detectable only in their immediate vicinity (Levy and Ehrhardt, 1981). Little attempt has been made to examine the effects of these seeps on the surrounding biota, except for one study now in press (Gilfillan et al., 1983; see also Effects of Natural Seeps section). With these few local exceptions, petroleum hydrocarbon concentrations in seawater and biota are generally low (<50 μg/L) in both the Arctic (Shaw and Cheek, 1976; C.S. Wong et al., 1976; R.C. Clark and Finley, 1982) and the Antarctic (Swinnerton and Lamontagne, 1974).

Vulnerability

Features unique to polar waters and relevant to oil pollution include ice leads, polynyas (open-water areas), and the subice and ice-edge habitat. These are heavily utilized at certain times of the year by many species of birds (Vermeer and Anweiler, 1975; Cross, 1980; Bradstreet and Cross, 1980) and by marine mammals (Stirling et al., 1975, 1977; Fraker et al., 1978; Stirling, 1980). Leads and polynyas are generally well defined (e.g., Stirling and Cleator, 1981) and appear to be essential for the feeding, migration, and reproduction of many species (e.g., R.G.B. Brown and Nettleship, 1981). These habitats are also especially susceptible to oil contamination. Thus their potential for severe oil impact is considerable. This may be of particular significance in the case of the bowhead whale which, by virtue of its low population numbers and its annual migration through the lead systems in the Bering and Beaufort seas, could be very much threatened by spilled oil.

The ice-water interface presents special problems in the event of a submarine blowout. Associated with this interface in the spring is a well-developed under-ice community based on a highly productive algal layer, often extending several centimeters into the ice (Meguro et al., 1967; Horner, 1977). While there remains much uncertainty about the trophic relationships and about the quantitative contribution of subice primary and secondary production to the annual values for the marine ecosystem as a whole, there is general agreement that timing rather than magnitude is of critical importance to certain species. It supplements the abbreviated season of water column production at a time when many species are releasing young.

The intertidal zone, often subjected to severe oil impacts in lower latitudes, tends to be extremely depauperate in the Arctic, a consequence of low air temperature, freshwater runoff, and severe ice abrasion. The ecological impact of oil in this zone is likely to be minimal.

On the other hand, the shallow coastal lagoons found in many parts of the Arctic present greater problems in terms of oil impact. Although their total areal extent is small, they are heavily utilized by bird (Johnson and Richardson, 1981, 1982; Richardson and Johnson, 1981) and fish populations at certain times of the year (Craig and Haldorson, 1980).

Physiological Sensitivity

Data on sensitivities of polar organisms to oil are too sparse and fragmentary to permit many useful generalizations. There is, however, little evidence that polar species are sensitive to petroleum hydrocarbons than are comparable species from more temperate areas. In some instances the lower temperatures may enhance the apparent toxicity of oil by altering the physicochemical characteristics of its components or by prolonging the residence time of toxic fractions in the water column (Rice et al., 1977). However, we cannot be too dogmatic. Thus

Van Baalen and Gibson (1982) and Cerniglia et al. (1982), in studying the capacity of Bering Sea ice algae to metabolize petroleum hydrocarbons (Figure 5-10), observed that their particular cold-adapted diatoms proved highly sensitive to Cook Inlet and Prudhoe Bay crude oils. This vulnerability together with the very slow growth rates at near 0°C, led these authors to suggest that such arctic diatoms will generally prove more sensitive to any accidental crude oil spills in or around the ice edge than might be expected from toxicity data alone.

A number of sublethal effects, similar to those occurring in temperate species, have been reported for polar organisms (Table 5-13). It is likely that certain types of physiological disturbances, particularly those interfering with carbon flux (Gilfillan, 1975) and energy storage (W.Y. Lee et al., 1981), could have more serious ecological consequences for polar species, many of which experience extreme seasonal oscillations in food supply. Any interference with the very abbreviated period of summer production may be critical for overwintering (Littlepage, 1964) and reproductive success (Dunbar, 1968).

Habitat and Biological Recovery

There is little doubt that the rate of recovery following a severe spill will be much slower than at lower latitudes. Oil cleanup in ice-infested waters presents many problems (Barber, 1970; Milne and Smiley, 1976, 1978; Wadhams, 1980). The low temperature and presence of ice will probably result in much of the oil remaining unweathered for extended periods (Mackay, 1977; NORCOR, 1975), especially when coupled with less rapid microbial degradation than in temperate zones (Atlas and Bartha, 1972; Atlas, 1977; Bunch and Harland, 1976; Arhelger et al., 1977). Although psychrophlic bacteria can grow rapidly (Morita, 1975), biomass can be reduced at polar temperatures (Bunch and Harland, 1976; Colwell et al., 1978).

Biological recovery will also be delayed because of the reduced fecundity, dispersal, and growth rates of many polar species (Clarke, 1979; Dunbar, 1968).

Natural self-cleaning of oiled intertidal sediments, however, can be very rapid, judging from preliminary results obtained at the BIOS study site (Owens et al., 1983). This was found to depend greatly on the degree of shoreline exposure and the amount of fetch offshore. On the other hand, in very low energy areas, persistence of stranded oil was estimated at longer than 10 years, similar to that recorded at lower latitudes.

Summary

Information on oil impact on polar environments is still fragmentary, with large knowledge gaps making spill impact assessment more guesswork than sound appraisal. Underlying much of the uncertainty is the absence of data about the basic biology of many important polar marine species.

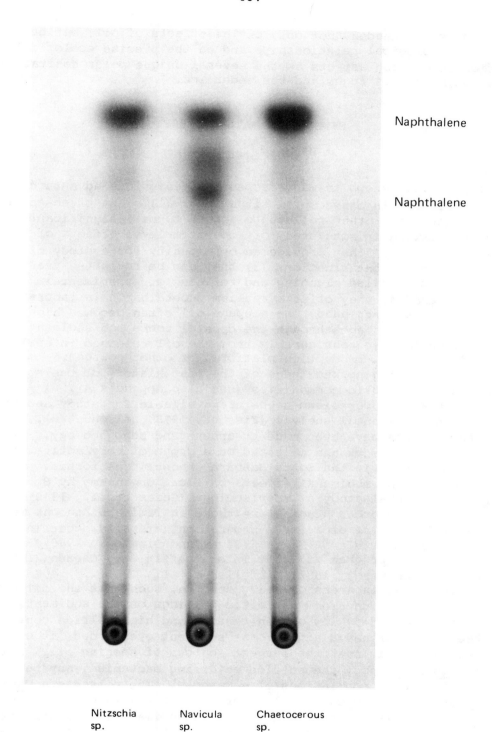

Naphthalene

Naphthalene

Nitzschia
sp.

Navicula
sp.

Chaetocerous
sp.

FIGURE 5-10 Radioautogram of metabolites formed from (1-^{14}C) naphthalene by ice-edge diatoms isolated from the Bering sea.

SOURCE: Adapted from Van Baalen and Gibson (1982).

Thus studies are needed, not only on the effects of oil, but perhaps more so on ecological relationships and on the precise ecological significance of such aspects as the several unique polar habitats--leads, polynyas, ice edge, and the under ice.

EFFECTS OF NATURAL SEEPS

Seep Studies

Natural petroleum seeps in many respects offer unique advantages not readily available in other spill situations: (1) seeps often occur in coastal areas where other pollutants appear to be insignificant, therefore allowing in situ study of ecosystem-level effects of petroleum alone; (2) they provide an opportunity for a study of really long term chronic contamination; (3) they can be revisited year after year, allowing detailed planning and refining of experimental approaches; and (4) they offer a realism unobtainable in laboratory studies. On the other hand, the seepage oil often becomes highly weathered in its passage through the crustal rocks and sediments and, therefore, can differ considerably from the oil released in spills, whether from tanker, production platform, or other sources. Also, there is often no knowledge of the past history of the seep area, of its biota and biotic communities (see also Appendix A).

While submarine petroleum seeps are available for study and some are found on continental shelves (Fischer, 1978; Nelson et al., 1978), only a few efforts have been made to understand seep ecology. Of these, most of the work has centered on a group of very active seeps near Coal Oil Point in the Santa Barbara Channel, California. They include research on sublethal effects on local organisms by Straughan (1976) and a general study of seep ecology (Spies et al., 1980). Elsewhere the only other studies are those in Tamiahua Lagoon, Mexico (Giammona, 1980), part of a large program on naturally occurring hydrocarbons in the Gulf of Mexico (Geyer and Giammona, 1980), and an exploratory study program in Scott Inlet, Baffin Bay, Canada (Levy, 1981; Levy and Ehrhardt, 1981; Gilfillan et al., 1983).

In California, in areas of heavy seepage, such that the oil oozes out and forms tar mounds and asphaltic coatings on the sediment, this is often accompanied by low oxygen tension and high sulfide content. Where the sea floor has a significant sediment overburden, it often is completely anaerobic near the largest sources of seeping oil. White mats of Beggiatoa spp., the sulfide oxidizing bacterium, may be found together with large populations of nematodes. The polychaete Capitella capitata also occurs occasionally, but few other infaunal invertebrates appear able to exist in such heavily oiled sediments. The starfish Patiria miniata, Astropectin spp., and the surf perch Pheneroden furcata are sometimes found in these isolated areas of heavy oil accumulation (Spies and Davis, 1979).

No direct measurements on hydrocarbon utilization by microbes in such sediments have been made, but it is thought to be vigorous. In addition to the visible mats of Beggiatoa, ATP measurements indicate that the microbial biomass is 2-3 times higher than in background sediments and there is an abundance of isotopically light H_2S (Spies et al., 1980), indicating a microbial origin. Also the oil droplets and streamers emerging from the sediment have a low n-alkane content, characteristic of microbially degraded oils. Thus the oil entering the sedimentary environment is already partially degraded, and a complex microbial community of sulfur bacteria, sulfide oxidizers, and hydrocarbon metabolizers is associated with it. These are the kinds of changes that generally accompany large sources of organic matter in the sediments (Stanley et al., 1978) and probably represent an intensification of processes that normally occur in coastal sediments (cf. Novitsky and Kepkay, 1981). Interestingly, despite the loss of many of the lighter fractions and the considerable weathering that this particular seep oil has undergone in its passage through the sediments, in one case at least, its toxicity, as measured by the diminution of growth in starfish embryos in laboratory bioassays, was comparable to or greater than several crude oils (Spies and Davis, 1982).

Biodegradation in buried reservoir source rocks may be responsible for the heavy, naphthenic nature of shallow oil deposits that give rise to seeps (Phillipi, 1977). Evidence from seeps in a quarried Gulf Coast salt dome also links H_2S production with biodegradation processes (Sassen, 1980).

Low oxygen tension and high sulfide content presumably are not the only reasons for the impoverished macrofauna. Concentrations of dissolved hydrocarbons in interstitial water (sediment-bound water) as high as 1,300 µg/L have been determined in some seep areas. One such sample was composed almost exclusively of relatively toxic monoaromatic and diaromatic compounds (Stuermer et al., 1982). This concentration and composition are probably sufficient to inhibit colonization by most infaunal species.

Areas with more moderate seepage show correspondingly less impact, and fewer differences from control area. A detailed 24-month study of benthic macrofauna in the Coal Oil Point area showed that the benthic population, despite some differences, was generally representative of the shallow water extension of the Nothria/Tellina community described by Jones (1969) for large portions of the southern California shelf. In such areas at least a few drops of oil were found in every sediment core taken, and interstitial water samples contained 45-117 µg/L of dissolved hydrocarbons. Nearby control areas contained no free oil, although there was some weathered oil in the sediments, and the pore water contained 0.2-5 µg/L of hydrocarbons. The species in the seep and in the comparison stations were both representative of this same community, but some deposit feeders, especially oligochaetes, were more common in the seep, and phoxocephalid amphipods were less common. In general, the populations in the seep area were significantly greater and tended to fluctuate more rapidly. However, the community structure generally was nearly constant and identical in the two areas (Spies and Davis, 1979; P.H. Davis and Spies, 1980).

Closer examination of the incorporation of bacterial carbon into some benthic polychaetes suggests that microbial activity, at least in areas of moderate seepage, and the microbial use of seep hydrocarbons

may play a dominant role in the establishment of benthic macrocom-
munities. For example, tissues of maldanid polychaetes from such a
seep area exhibited a shift toward isotopically light bacterial carbon
(Spies et al., 1980). It has now been confirmed, through the use of
naturally occurring abundances of ^{14}C, that petroleum carbon contri-
butes some 15% more carbon to this species than the nearby comparison
area (Spies and DesMarais, 1983). In this same study it was estimated
that the petroleum-carbon source for the polychaetes had a δ^{13}C of
-35.9%, suggesting it is mostly the lighter hydrocarbons that contribute
to the food webs near the seep. Thus it would seem that the seepage
oil and the accompanying H_2S can make a contribution of energy and
substrates to the benthic food web.

The Tamiahua Lagoon (Mexico) and Scott Inlet (Baffin Island, Canada)
are areas of relatively low seepage and have not yet provided much
further insight into the ecological effects in such environments. In
the case of the Tamiahua Lagoon, the harsh physical conditions related
to large sections of the tidal flat lying continuously exposed during
the dry season make it difficult to attribute any apparent impact on
the benthic community to that of the seep petroleum alone. Thus large
differences in the two unnamed amphipod species may have been due to
oiling; however, differences in the distribution of a dominant plant,
Ruppia maritime, may also have caused this. While there are obvious
anomalies in the benthic populations during the course of a year,
studies have shown (Giammona, 1980) that the cause of the distribution
and abundances of organisms in this area is not understood.

The near-Arctic seep at Scott Inlet did not come under study until
recently, primarily with efforts to identify the type and source of the
seep oil (e.g., Levy, 1981). Subsequently, observations have been made
on hydrocarbon-utilizing bacteria (Westlake and Vandermeulen, unpublished results) and on effects of zooplankton from the area (Gilfillan
et al., 1984). The seep source lies in a deep-water inlet on the east
coast of Baffin Island (MacLean et al., 1981) and is in an area of high
water turnover due to the southerly currents in Davis Strait. The seep
itself has been difficult to define accurately (Loncarevic and Falconer,
1977; Levy and MacLean, 1981). Large areas of oil sheen are frequently
visible, but because of the ice and general weather conditions, continuous observations are nearly impossible. Work on microbes from both
water column and bottom sediment samples have shown little or no
evidence of enhancement of hydrocarbon-utilizing microbial species
(Westlake and Vandermeulen, unpublished observations), possibly due to
both the low levels of hydrocarbons found in the water (<88 µg/L)
and sediments (<41 µg/g) (Levy, 1981) and due to the rapid turnover
of the water mass through the area. On the other hand, studies of
feeding and respiration of zooplankton have shown a decrease in feeding
in copepods taken from the seep waters in direct relationship to the
hydrocarbon content in the water column. This may be due to narcotization of the zooplankton as they are swept through the seep area.

Sublethal Effects and Adaptation

Most of the work on possible sublethal effects was done by Straughan (1976). In addition to examining a large number of invertebrates for malformations, there were comparisons of growth and reproduction in Coal Oil Point organisms with those at several control sites in the Channel Islands well offshore. Coal Oil Point abalone, mussels, and nonstalked barnacles appeared to grow and reproduce as might be expected in unpolluted environments. However, variations were observed at the different control sites, so that there was difficulty in discriminating between possible effects of petroleum and from other environmental factors. A comparison of oiled and unoiled goose barnacles, Pollicipes polymerus, at Coal Oil Point showed that differences in brooding rates between oiled and unoiled animals correlated with oiling, apparently due to thermal effects associated with enhanced heat absorption by tar on the carapaces of the oiled animals.

Spies et al. (1980) reported that the starfish Patriria miniata at Coal Oil Point had consistently lower gonadal indices and a shorter breeding season than a population at Naples Reef, about 8 km away. However, no relationship has been found between the gonadal index and hydrocarbon levels in the tissues, so that again there is a possibility of other causitive factors.

Malformations have not been reported for animals from Coal Oil Point. Thousands of benthic organisms have been collected by Straughan (1976) and P.H. Davis and Spies (1980), and no gross malformations were seen, but neither study included detailed studies of tissues for evidence of pathological conditions. Straughan and Lawrence (1975) examined bryozoans for ovicell hyperplasia, which was reported from areas exposed to coal-tar derivatives and petroleum (Powell et al., 1970). Again the results were negative, although the account did not state the number of colonies that were examined.

Whether seep organisms are adapted to petroleum is still an unanswered question. Laboratory experiments have given mixed but generally negative results. In tests with sea urchin embryos (Straughan, 1976), starfish embryos (Spies and Davis, 1982), and several adult invertebrates (the mollusks Acanthia punctata and Tegula funebralis, the isopod Excirolana linguifrons, the sea anemone Anthopleura elegantissima), seep populations have not proved to be more resistant than organisms from nonoiled areas. On the other hand, a study of southern California mussels showed that a Coal Oil Point population was better able to withstand the rigors of an oil-covered aquarium (Straughan, 1976) than were nonseep mussels. However, it is not clear that there has been much opportunity for the sort of genetic selection through series of generations that might lead to significant adaptation in the organisms examined. Most of these species that have been studied have pelagic young, and inevitably there has been scattering and intermixture of inshore and offshore populations over the years with each breeding period.

A different line of evidence developed to assess possible adaptation by seep organisms to the oil was an examination of the levels of hydrocarbon-metabolizing enzymes (see Chapter 3, Part B for a general review of this subject). Hepatic aryl hydrocarbon hydroxylase (AHH) activity in the ambiotocid fish _Phenerodon furcatus_ taken from a seep was significantly higher than in specimens from a nearby comparison area (Spies et al., 1980). Similarly, the flatfishes _Citharictys sordidus_ and _C. stigmaeus_ had elevated AHH levels, and in related work, AHH activity was induced in both species by giving them food contaminated with oil (Spies et al., 1982).

The general interpretation of these limited experiments is difficult because of the variety of organisms tested and because the dosage was much larger than the concentrations that seep populations would normally experience. The high doses were applied in order to get a measurable toxic effect in a short period of time. However, the fact that the results show little evidence of superior resistance to acute toxicity in seep organisms does not necessarily have a bearing on the degree of resistance to sublethal doses. Thus it is possible that these "ppm" doses overwhelmed adaptive mechanism(s) that might be effective at more environmentally realistic "ppb" and lower concentrations.

HUMAN HEALTH

During the last decade there have been various suggestions and warnings that petroleum hydrocarbons entering the marine environment may constitute a health hazard to humans. These concerns have been amplified as a result of several major maritime transportation and production accidents over the last 5 years. Contact with the human population following such accidents may be acute or chronic. Cleanup crews, members of the investigating scientific community, and coastal residents may be subjected to acute exposures following a local spill. In such cases, human uptake may be by inhalation, skin contact, or even by ingestion of the petroleum or refined products. Exposure may also occur by direct contact with hydrocarbon components in the pelagic tars that are washed ashore frequently, such as the tar balls now found commonly on many of the world's beaches. Chronic nonoccupational exposure may occur through the accumulation and transfer of potentially harmful hydrocarbons from contaminated seafoods (Figure 5-11).

The effects of the majority of petroleum components on human health are unknown. Studies with human subjects in the main are lacking. Most of the available information on the toxicity of natural crude oils and compounds pertains primarily to animals and not to humans. However, in most instances the animal studies serve as models for the human situation, and the data obtained have implications for human responses.

Toxicity of Hydrocarbons to Humans--Acute Exposures

Because of the concerns surrounding petroleum pollution to humans, some realistic and others more emotional, it is worthwhile to briefly review what is known of petroleum hydrocarbon toxicity generally. The direct ingestion of a variety of petroleum distillates will cause a number of toxic symptoms. The organ systems affected include the lung, gastro-

FIGURE 5-11   One potential route of oil contact with man is through eating of oiled seafood, as by this North Brittany coastal resident digging for clams in sediments oiled by the Amoco Cadiz tanker spill in 1978. (Photo by J.H. Vandermeulen.)

intestinal tract, liver, kidney, central nervous system (CNS), and the hematopoietic system (Zieserl, 1979; Vaziri et al., 1980; Poklis and Burkett, 1977). In such cases, CNS symptoms are generalized weakness, lethargy, dizziness, convulsions, and coma. A greater risk of acute CNS toxicity is associated with those refined petroleum products having high concentrations of the more volatile aromatic hydrocarbons. Symptomatic involvement of the respiratory tract is the most common complication. If sufficient material is ingested, death will result presumably from aspiration of ingested materials into the lungs.

Knowledge of human response to acute exposure comes primarily from studies with benzene-containing solvents and gasoline. Thirty-minute exposures to the solvents "Stoddard" and "70," containing 22% and 74% alkylbenzene, respectively, resulted in eye, nose, and throat irritations. However, there were no observable differences in eye blink, swallowing, or respiratory rates (American Petroleum Institute, 1976). Other inhalation studies using rats, dogs, and the sensory response of human subjects with Stoddard solvent suggested a hygienic exposure level of 1.2 mg/L air for humans (Carpenter et al., 1975b).

A few studies have been reported for the acute toxicity of gasoline to humans (Poklis and Burkett, 1977). Thirty- to sixty-minute exposures to 500-1,000 ppm of gasoline vapor produced eye, nose, and throat irritation and dizziness. Exposures to higher concentrations for the same periods of time resulted in varying degrees of nausea, headache, numbness, and anaesthesia. At 10,000 ppm, deep anaesthesia was achieved in 4-10 minutes for all subjects. The acute toxicities of a series of other petroleum hydrocarbon products have also been reported from inhalation studies with both human subjects and with experimental animals. Recommended human hygienic standards from these studies are summarized in Table 5-14.

From a human health point of view, the volatile aromatic hydrocarbon benzene occupies a unique position in that it is one of the few hydrocarbon compounds that have been established as human carcinogens. Normally the benzene content in fuels ranges from as low as 0.1% in some crude oils (H.M. Smith, 1968) to a high of 16% in some refined products, although it apparently constitutes 30-40% of oil discharged in formation waters in Alaska. Acute toxicity to benzene can be induced very rapidly via inhalation exposure. The most prominent effect is CNS stimulation, followed by depression and respiratory failure (Leong, 1977). Subacute and chronic exposures as low as 44 ppm can lead to a sequence of hematopoietic tissue changes. Lymphoid tissues and the myeloid bone marrow itself, are affected, resulting in anemia and leukopenia. The adverse effect of prolonged exposure to benzene on myeloid and lymphoid tissues also can result in a deterioration of the immunological defense mechanisms. There is also clinical evidence that hyperplastic bone marrow leukemia is associated with benzene exposure. A recent clinical study of tank-cleaning personnel chronically exposed to petroleum vapors suggested an exposure-connected relationship between vapors and chromosomal aberrations in bone marrow cells (Hogstedt et al., 1981). For these reasons, benzene occupational exposure limits have been set at approximately 10 ppm.

TABLE 5-14 Suggested Inhalation Exposure Standards in Humans for a
Variety of Petroleum Hydrocarbons

Petroleum Vapor	Suggested Hygiene Standard for Humans[a] (mg/L)	Reference
Mixed xylenes	0.46 (110 ppm)[b]	Carpenter et al. (1975c)
Rubber solvent	1.7 (430 ppm)	Carpenter et al. (1975b)
Varnish maker's and painter's naphtha	2.0 (430 ppm)	Carpenter et al. (1975a)
Toluene concentrate	1.9 (480 pm)	Carpenter et al. (1976d)
Deodorized kerosene	0.1 (14 ppm)	Carpenter et al. (1976b)
"60 Solvent"	0.44 (90 ppm)	Carpenter et al. (1975d)
"80 Thinner"	0.45 (100 ppm)	Carpenter et al. (1976a)
"40 Thinner"	0.15 (25 ppm)	Carpenter et al. (1976c)
"High aromatic solvent"	0.15 (26 ppm)	Carpenter et al. (1977a)
"High naphthenic solvent"	2.1 (380 ppm)	Carpenter et al. (1977b)

[a]Based on sensory response in humans.
[b]For conversion to ppm, see U.S. Department of Health, Education and
Welfare (1973).

Direct Skin Contact

Acute dermal tests have been done primarily with laboratory animals.
These usually produce slight irritation but do not necessarily result
in systemic toxicity. However, subacute dermal testing using 8 mL/kg
of No. 6 fuel oil (Bunker C) produced dermal irritation as well as
other dose-related responses. Histopathologic observations confirmed
both dermal and hepatic tissue toxicity (Beck and Hepler, 1980a). In

another study, diesel fuel was found to be extremely irritating to the skin of rabbits when it was allowed to stay in contact with the skin for 24 hours (Beck and Hepler, 1980b). Subacute dermal testing with 4 and 8 mL/kg dosage levels produced treatment-related responses. Skin in the test area became necrotic, and weight loss occurred. There was also a 67% mortality rate in the 8-mL/kg dose group. Liver and kidney involvement was also observed. A similar set of toxicity tests has been conducted for motor oil, where 24-hour exposures produced only slight primary skin irritation (Beck and Hepler, 1980c). Acute and subacute dermal toxicity tests showed skin irritation and dermal corrosion, respectively. However, no signs of systemic toxicity were reported.

Pregnancy and Development

A series of studies has been carried out to determine the influence of particular products on pregnant laboratory animals and their offspring. These included kerosene (106 and 364 ppm), diesel fuel (101 and 401 ppm), and n-hexane (93 and 408 ppm) airborne concentrations, applied on days 6-15 of gestation (Mecler and Beliles, 1979a,b; Ament et al., 1979). Only decreases in food consumption, at the higher exposure levels of diesel fuel were observed. There were no indications of compound-induced abnormal growths, variation in sex ratios, embryo toxicity, or inhibition of fetal growth and development.

Several biochemical and pathological alterations are known to occur within hours or days following exposure to certain specific PAHs (Zedek, 1980). Huggins et al. (1961) observed a delay in growth for several days, and the development of leukopenia after the second day, following a single oral dose of 7,12-dimethylbenzo(a)anthracene in rats. The inhibition of DNA synthesis and mitoses in proliferating hepatic cells of rats has been reported (Marquardt and Philips, 1970). In this case, however, the impact lasted only 24 hours and was reversible. Hematopoietic functions in bone, spleen, and thymus, as well as the cell renewal in gonads and the intestinal lining, are found to be especially sensitive to acute exposure. PAHs can also be toxic to the developing rat fetus (Currie et al., 1970).

Human Exposure During Spills

Field observations are few and mostly qualitative and anecdotal. Nonetheless there are several reports of human response to acute exposure during spills. Symptoms characteristic of acute toxicity to petroleum vapors were reported following the Amoco Cadiz oil spill along the coast of France (Menez et al., 1978). An estimated 40,000 metric tons of light hydrocarbons may have been released into the atmosphere of the coastal area during the spill, creating a potential health hazard to the inhabitants, as well as the personnel in spill cleanup activities. Exposure to the workers was increased by mists and aerosols resulting from the high-pressure projection of water and steam during various phases of the cleanup operation. Such activities contributed to increased dermal contact and even ingestion of small amounts of petroleum by the cleanup personnel.

Among the symptoms reported by workers and coastal inhabitants, as well as by some of the scientific groups studying aspects of the spill, were headaches, dizziness, nausea, sensation of inebriation, vomiting, and abdominal pains. Workers coming in direct contact with the oil also reported skin irritations and erythema on the hands and limbs. Biochemical tests of blood samples taken from cleanup personnel who had worked at least 15 days in the vicinity of maximum impact, and from inhabitants of a local community, revealed, however, no significant changes in blood chemistry or enzymatic activity.

Atmospheric samples taken adjacent to the Amoco Cadiz cleanup activities revealed the presence of many volatile aromatic and aliphatic hydrocarbons (Dowty et al., 1981). However, the concentrations of benzene, toluene, and alkylbenzenes were substantially lower than those measured for a nearby urban center. The reverse was true for naphthalenes, which were substantially higher in the spill-impacted areas.

Carcinogenic Potential in Humans

It has been recognized for many years that specific hydrocarbon constituents commonly found in natural crudes, refined products, and other related fossil fuel sources can result in the induction of cancer in humans and animals. Thus the introduction of large quantities of these materials, whether accidentally from spills, land runoff, or natural sources, into the marine environment becomes of concern, particularly where such materials may become incorporated into common seafoods. The ingestion of hydrocarbon-contaminated seafoods over a long period of time then may conceivably result in humans obtaining a carcinogenic insult.

While the potential exists for ingestion of these carcinogenic components from the marine environment, the actual likelihood of their inducing cancer in humans is from small to negligible.

Carcinogens in Oil

Crude oils, several of the refined products, and such products as coal tar, crude shale oil, and furnace tar have long been known to have carcinogenic or mutagenic potential (e.g., Bonser, 1932; Leitch, 1922; Passey, 1922; Cook et al., 1933; Twort and Twort, 1931; Woodhouse and Irwin, 1950; Bingham et al., 1965). For an extensive review of the literature see Bingham et al. (1979). This carcinogenic potential is invariably associated with the PAH fraction, the distillate fractions of crude oils with boiling point above 350°C. The PAH content of oils varies quite widely, both in amount and composition of PAHs. Content varies from a low of 0.2-7.4% in a range of crude oils (Gilchrist et al., 1972) up to 7.7-14.6% in three synthetic crude oil samples (Woodward et al., 1976).

TABLE 5-15 Benzo(a)pyrene Content of Selected Petroleum Products

Product	B(a)P Content (ppm)	Reference
Crude oil		
Libya	1.32	Graf and Winter (1968)
Venezuela	1.66	Graf and Winter (1968)
Persian Gulf	0.4	Graf and Winter (1968)
Wilmington	2.5-2.7	Tomkins et al. (1980)
Gasoline	0.21-0.48	Walcave et al. (1971)
No. 2 fuel oil	0.6	Pancirov and Brown (1975, 1977)
Bunker C	44	Pancirov and Brown (1975, 1977)
Asphalt	27	Walcave et al. (1971)
Shale-derived oil	4	Weaver and Gibson (1979)

There are 13 PAHs listed as carcinogenic by the International Agency for Research on Cancer (1973), including the commonly detected compounds 7,12-dimethylbenzo(a)anthracene, benzo(a)pyrene (B(a)P), dibenz(ah)anthracene, and 3-methylcholanthrene, as well as the B(a)P metabolite 7-hydroxymethyldimethylbenzo(a)pyrene.

B(a)P and related isomers plus the other carcinogenic hydrocarbons, despite their notoriety, are often present in crude oils in extremely small quantities relative to other PAHs (Table 5-15). However, the used or spent petroleum products appear to be enriched in B(a)P and other PAH content. For example, Graf and Winter (1968) observed that the B(a)P content increased by as much as 200-fold in used oils.

A potential future source of PAHs are the synthetic oils derived from oil shale and coal. A number of studies have found that overall mutagenicity of these products is greater than that observed for natural crude sources (Guerin et al., 1981). They are also more active skin carcinogens (Holland et al., 1979). While presently these products are transported primarily on land, accidental discharges relating to bulk transportation and handling may allow synthetic fuel products to find their way into the marine environment. Besides the potential mutagenic/carcinogenic risk from the synthetic parent products, there is also the possible increase in their mutagenicity after use, as has been found in the case of used motor oils (Payne et al., 1978). Judging from the range of PAHs and derivatives identified so far in these products (Weaver and Gibson, 1979; Robbins, 1980; Ho et al., 1981; Guerin et al., 1981), the future use and discharge of these synthetic products should be monitored with care.

PAHs in the Marine Environment

PAHs enter into the marine environment from a wide range of sources, including surface runoff from land, atmospheric fallout and rainout, as well as from the more traditional sources of spills and discharges of petroleum. A major contributor of PAHs is the direct combustion of fossil fuels, i.e., gasoline- and diesel-powered vehicles, electrical and heat-generation operations, catalytic cracking of crude oils in refining and related industrial processes, and refuse burning. Most significant in the formation of PAHs are those processes that utilize high temperature pyrolysis of organic material. Thus, it is estimated that from forest and agricultural fires alone, circa 420 t (metric tons, or tonnes) enter the atmosphere annually (Suess, 1976).

Clearly, the direct contribution to the marine PAH load from petroleum spills and discharges is only one factor. Neff (1979), using a series of assumptions, estimated that the total PAH input into the global aquatic environment would be 230,040 t/yr. Of that, some 170,000 t/yr were estimated to be derived from petroleum spillage. However, when viewed instead from the point of view of B(a)P input, an estimated 697 t B(a)P/yr enter the aquatic environment, of which only 20-30 t are due to petroleum spillage. This matches closely the calculations of about 10-20 tons of B(a)P entering the oceans as a direct result of natural crude and oil petroleum product discharges (Sullivan, 1974; quoted in O'Conner et al., 1981).

Risk to Humans

The main concern regarding the risk to humans is the known carcinogenicity of several of the oil components. Because of their lipophilic nature, hydrocarbons will accumulate in seafoods and can potentially be passed on to man. The existence of PAHs in marine animals has been known for some time, and considerable documentation has been assembled over the past 5-10 years (Zechmeister and Koe, 1952; Group of Experts on the Scientific Aspects of Marine Pollution, 1977; Neff, 1979). Certain marine organisms are also known to accumulate PAHs from the environment very readily, without losing them rapidly over time either by metabolism or simple depuration. Mollusks, for example, are known to have low or little enzymatic activity to degrade PAHs. However, they readily accumulate these compounds, and one can find a strong relationship between the level of PAH in the tissues of mussels and their proximity to anthropogenic hydrocarbon inputs (Dunn and Stich, 1976; Dunn and Young, 1976; Farrington et al., 1983).

These PAHs in seafood are not the only source of potentially carcinogenic material to humans. Other sources include the full range of foods, from plant products to meats (Table 5-16), Over 100 PAHs have been reported in environmental and food sources. But of these, only 11 have been shown to be carcinogenic to test animals.

In addition to exposures from other foodstuffs, humans are exposed to PAHs from a variety of other sources (Table 5-17). While the nature of personal habits, such as cigarette smoking, will vary exposure to

TABLE 5-16 PAH Levels in Foodstuffs (μg/kg Wet Weight)

Foodstuffs	PAH Benzo(a)pyrene	Benzo(a)anthracene
Meats		
Fresh	-	-
Cooked	0.17-4.2	0.2-1.1
Charcoal broiled	2.6-11.2	1.4-31
Smoked	0.02-14.6	up to 12
Fish		
Cooked	0.9	up to 2.9
Smoked	0.3-60	0.02-2.8
Grains and cereal products		
Grains	0.2-4.1	0.4-6.8
Flour and bread	1.1-4.1	0.4-6.8
Baker's dry yeast	1.8-40.4	2.9-93.5
Fruits and vegetables		
Soybean	3.1-12.8	-
Salad	2.8	4.6-15.4
Spinach	7.4	16.1
Kale	12.6-48.1	43.6-230
Apples	0.1-0.5	-

SOURCE: Group of Experts on Scientific Aspects of Marine Pollution (1977).

PAHs and to B(a)P considerably, nonetheless, uptake from seafood appears not to be unusually high. There is no epidemiologic evidence for human cancer from intake of PAH-contaminated food (Lo and Sandi, 1978). The preparation of seafood by smoking or grilling may be a factor in some isolated cases. However, for the continental United States and for other countries with similar food habits, combinations of source foods with high PAH content may be a factor along with occupational and other forms of exposure. There may be exceptions in those isolated areas were certain seafoods, especially if smoked, constitute a major portion of the staple diet.

The limited information available on the metabolism of PAHs by humans suggests that the majority of these compounds are rapidly absorbed and excreted and do not tend to accumulate in humans. Small quantities of PAHs can accumulate in body fat, adrenals, and ovaries up to 8 days, but those that do remain are apparently rapidly metabolized. For example, in one study with laboratory rats 70-80% of the B(a)P injected was metabolized to binary and secondary oxidation products within 6 hours (Falk et al., 1962). There does exist some concern over the fate of these secondary products of metabolites, because several of them may possess either mutagenic and/or carcinogenic activity, while

TABLE 5-17 Estimated Human Exposure to Benzo(a)pyrene (B(a)P) Through Respiratory and Gastrointestinal Intake

Source	Daily Consumption	Estimated Annual Intake of B(a)P (μg)
Respiratory intake[a]		
Air		0.05-500
Cigarette smoking	20 cigarettes	15-900
Gastrointestinal intake		
Drinking water	2.5 L	6-70
Food Normal diet		250-500
Smoked food diet	1.5 kg[b]	550-3000
Potential seafood contribution	100 g[d]	36.5-1825
Contaminated seafood burden[e]	24-48 g	263-920

[a]Respiratory intake is assumed to be 5,000 m^3/person/yr.
[b]For 0.5 μg B(a)P/Kg of food.
[c]Assumed to be contaminated with 1-5 μg/kg B(a)P.
[d]Assumed B(a)P levels to be from 1 to 50 μg/kg.
[e]Group of Experts on the Scientific Aspects of Marine Pollution (1977).

SOURCE: Connell and Miller (1980).

others have been shown to bind positively to nucleic acids (e.g., Varanasi and Gmur, 1980; Varanasi et al., 1980, 1982).

Conclusion

There are recognized human biochemical and physiological responses associated with acute exposure to natural crudes or their refined products. In general these responses appear to be transient and short lived unless the exposure levels are unusually high. Prolonged subacute exposures, however, can result in tissue damage. In situations where the potential for human contact exists, efforts should be made to limit exposure through the use of respiratory and protective equipment.

The quantities of PAHs found in seafoods are for the most part equal to or less than the levels reported in other food sources. The major exceptions to this are seafoods harvested in the vicinity of municipal outfalls, creosote pilings, or local petroleum hydrocarbon sources. However, in most instances the natural intake, depuration, and metabolic processes are such that the PAH tissue levels will drop rapidly if the source is removed.

Thus at present there is no demonstrated relationship that chronic exposures through eating petroleum-derived PAH-contaminated seafood are related to the incidence of cancer or other diseases in humans (King, 1977; Cowell, 1976; Group of Experts on the Scientific Aspects of Marine Pollution, 1977).

Exceptions to these conclusions may arise in localized areas, as in the case of isolated fishing villages where seafood constitutes a major portion of the annual diet. No data are available, however, for these cases.

SUMMARY

Petroleum in the marine environment can elicit a broad range of toxic responses, at low concentrations (less than 1 mg/L), to many marine organisms, both plant and animal.

However, in sediments and in the water column there is no compelling evidence to date indicating permanent damage to the world's ocean resources or even to a particular part of it. Nor is there yet evidence of increased pathological abnormalities in marine biota, due to petroleum hydrocarbons alone.

Prior to the 1973 workshop, which formed the basis for the 1975 NRC report, much of the focus was on establishing toxicity and lethality thresholds or on the assessment of hydrocarbon concentrations in environmental samples. Since that time, research activity has broadened to include work on the site of action of toxicity, regarding petroleum toxicity as a dynamic process affecting the living organism at various levels: enzymatic, metabolic, ultrastructural, molecular. This has become evident from the breadth of studies brought together for this chapter, ranging from effects studied at the ecosystem level down to the effects observed and measured at the chromosomal level. There is also increased ability to predict, in the event of oil spills and of chronic oiling, the vulnerable and sensitive components of ecosystems, and to identify those parts of the ecosystem where petroleum hydrocarbons are likely to persist.

One can see an increased direction of the research effort, with more integration between different disciplines as well as between different research teams. Research needs are being identified more clearly, and after considerable critical discussion with peers. New research areas are springing up, and the research effort is using more novel and critical approaches and technology. The last few years have seen an increased recognition of the importance of intercalibration, both of methods and of intercomparison of experimental results. There has also been a better understanding of the variability that exists in different organisms or different life-cycle stages. The past 8 years or so of effects-related research can be marked by two main advances-- the increasingly strong shift toward understanding the physiological impact of petroleum toxicity, and the recognition that both organisms and life-cycle stages vary widely in their sensitivity and responses. As a result there is an encouraging trend toward establishing the study of petroleum pollution on a solid basis of interrelated chemical and biological data.

Evidence of this change in petroleum-pollution-related research is the increasing quality of published scientific papers that have come out of this work, and their appearance in the ranks of first-rate scientific literature. Mention has been made of the particular problems in the study of oil in the marine environment--the extreme complexity of petroleum and its refined derivatives, the state of analytical capability which is still far behind the requirements, and the complexity of the marine environment itself which is only poorly understood, as well as the lack of knowledge of other toxic contaminant effects in the sea that could have served as models.

Impact of Petroleum

Without reiterating the individual findings and conclusions set out in the preceding pages, it is nevertheless useful to focus on some specific observations in order to develop an understanding of how petroleum does affect the marine environment.

First, no one marine organism has been found capable of actively excluding petroleum hydrocarbons from its tissues, be it plant or animal. In fact, all marine biota appear to be readily permeable to hydrocarbons, and readily accumulate them from their environment either directly from the water column or from pore water, or through their food. Also petroleum hydrocarbons can affect and can cause changes in a broad range of organisms and at all levels--cellular, organismic, and community. However, much of this information has come from laboratory studies, and just how these data apply to the field, in either a spill or with chronic input, is the subject of much discussion and much needed research. Part of the problem lies with the range and variation of laboratory studies, and the difficulties encountered in comparing different laboratory data. Part of the problem also lies in the lack of corroborative field data. There are difficulties in applying results of the necessarily simple laboratory systems to the complex, real ecosystems. Nonetheless, there are enough pieces of quantitative information available now that some broad outlines begin to emerge. For example, there is such a wide range of data available for petroleum impact on fish that we can at least suggest the potential impact of a spill or of chronic input of petroleum on a stock or a region (e.g., Longhurst, 1982). Thus, we have considerable information on the effects of petroleum at very low levels (less than 100 µg/L), i.e., there is the knowledge that oil exposure can enhance susceptibility to disease, that there exists a differential sensitivity of the various life-cycle stages and a greater susceptibility of larval stages, that there can occur genetic effects (although not documented in all of its forms, yet indicative of a problem area), and that there is a wide range of deleterious effects on metabolism. Together these observations frame the understanding of a potential and real impact on fish stock, both of lethality in the short term and deleterious and debilitating impact in the long term, under certain environmental conditions. Our present inability to measure several of these effects in the field in a fish population does not contradict the possibility of an impact.

A similar conception of toxic impact due to petroleum can be sketched out for other marine biota, although for some the information is as yet slim. Relatively little attention has been paid, for example, to marine flora, either phytoplankton and macroalgae. The information for macroalgae especially is far from complete. Surprising also is that the data base for larval and juvenile fish is not nearly as plentiful and complete as it is for adult forms. Probably best understood are the macroinvertebrates, especially the intertidal forms of commercial interest (bivalves, crustaceans).

Marine organisms differ widely in their sensitivity to petroleum. For example, arctic amphipods are highly susceptible to relatively low concentrations of crude oil, but isopods from the same sediments appear to be totally unaffected. Similar differences in susceptibility can be found throughout the phyla. Nonetheless there are some common features beginning to emerge from the very large volume of data assembled since 1975. There are suggestions that intertidal biota may be somewhat more resistant to petroleum than are the offshore benthic or pelagic forms. Whether this is related to some intrinsic higher tolerance, having to do with their highly variable intertidal habitat, is not known. Another generality is that larval fish are more sensitive than are juveniles and adults. Even certain stages in the development of fish eggs apparently are more resistant. However, again one encounters great variability between growth stages and between species, and while it is tempting to draw a sensitivity curve for fish life-cycle stages, in fact, this still cannot be done with certainty. Where this is needed, as for the commercially important species, the sensitivities will have to be determined separately for each species.

Impact on Processes and Organisms

Studies of oil impact on development at present seem to raise more questions than answers. The developmental process appears to be particularly sensitive to petroleum, and even relatively low concentrations of petroleum (less than 1 mg/L; see Table 5-18) can result in measurable abnormalities, including spinal abnormalities in larval fish, symptomatic not only of petroleum but also of other contaminant pollution. The interaction between petroleum hydrocarbons and the chromosomal/genetic fraction of the cell is a relatively new area of research. We know that polynuclear hydrocarbons can bind with nucleic acids, and in other ways can perturb the normal meiotic and mitotic processes, resulting in subsequent abnormalities in development.

TABLE 5-18 Effects of Low Concentrations (Less than 1 ppm) of Petroleum Hydrocarbons on Marine Organisms in Laboratory and Mesocosm Studies

Reference	Organism	Exposure Period	Type of Hydrocarbon	Lowest Concentration Tested	Concentration Having Effect	Effect
Mironov (1972)	phytoplankton, various spp.	hours	oil and oil products	10 ppb	10-100 ppb	growth inhibitions
Pulich et al. (1974)	phytoplankton, Thalassiosira pseudonana	hours	No. 2 fuel oil	40 ppb	40 ppb	growth inhibitions
R.F. Lee and Takahashi (1977)	phytoplankton, various spp.	19 days	No. 2 fuel oil	10 ppb	20 ppb	microflagellates replaced diatoms
Mironov (1968)	plaice eggs	days	oil and oil products	10 ppb	10 ppb	40% mortality
Linden (1976)	amphipod, Gammarus obsoletus	20-23 days	Venezuelan crude oil	1 ppm added 3.0-0.4 ppm	0.3-0.4 ppm	reduced reproduction
Jacobson and Boylan (1973)	gastropod, Nassarius obsoletus	minutes	Kerosene extract	4 ppb	4 ppb	interference with chemoreception
Stegeman (1976)	fish	up to one month	No. 2 fuel oil	125-100 ppb	125-200 ppb	altered metabolic systems
Stegeman and Teal (1973)	oyster, Crassostrea virginica	50 days	No. 2 fuel oil	106 ppb	106 ppb	20-fold increase in hydrocarbon content
Corner et al. (1976)	copepod, Calanus helgolandicus	hours	naphthalene	0.1 ppb	0.1 ppb	storage of naphthalene
Nelson-Smith (1973)	oyster, Crassostrea virginica	months	unreported oil constituents	10 ppb	10 ppb	tainted flesh
Oviatt et al. (1982)	zooplankton, benthic fauna	1 year	No. 2 fuel oil	90 ppb	90 ppb	depression
Vandermeulen et al. (1983)	phytoplankton, Pavlova lutheri	<1/2 h	naphthalene	100 ppb	100 ppb	interference with mobility
			diesel fuel	10 ppb	20 ppb	interference with mobility

SOURCE: Adapted from Hall et al. (1978).

Certain of the petroleum hydrocarbons have a histopathologic and/or mutagenic potential, although the research effort in these areas is still limited to only a few workers. Tumor formation in a range of organisms has been linked to oil exposure on several occasions, although the relationship has never been firmly established for field-collected samples. Similarly, such pathologic features as fin rot in groundfish appear to be more than casual correlates of oiling, being reported for some oil spills or chronic pollution sites. The mutagenic potential of petroleum stands on somewhat firmer grounds, albeit mainly from laboratory studies, being associated definitely with its polycyclic aromatic hydrocarbon fraction. To date, however, this mutagenic potential has been demonstrated principally with certain microbes and is only now being extended to higher marine organisms. As most marine invertebrates and vertebrates alike possess the metabolic conversion mechanism (cytochrome P450 dependent) capable of producing the genotoxic intermediates from these PAHs, these intermediates are probably produced following oil spillage. Whether persistent mutagenic intermediates occur, in invertebrates or vertebrates, is totally unknown. However, any serious potential mutagenic threat to marine biota will probably come, not directly from oil spillage into the oceans, but more so from chronic inputs such as combustion products, street runoff, and used lubricating oils entering marine coastal waters.

Of all the processes examined, the perturbation of normal behavior at very low concentrations of petroleum (as low as 10 µg/L) suggests a particular concern. The continuance of normal behavior underlies and is absolutely critical to larval settling, feeding, reproduction, substrate recognition, and homing. In this context a change in or cessation of feeding is one of the first indications of oil pollution in many test animals. Yet most of the data available are largely anecdotal, and at least for the higher organisms, the effect of petroleum on behavior is poorly understood.

The last 5 years have seen a dramatic increase in the number of studies investigating the effects on various metabolic and physiological processes and have demonstrated a broad range of effects on those fundamental biological processes that govern normal growth and development. However, as yet no site of action has been determined for even a single petroleum hydrocarbon, with only gross perturbations noted. While such processes as respiration, photosynthesis, ATP production, carbon assimilation, and lipid formation are known to be affected by single hydrocarbons (e.g., naphthalene), the ultimate site or sites of action have not yet been determined. In fact, there is still uncertainty over the mechanism of uptake of hydrocarbons, whether active or passive, and whether it involves the lipid component of the cell membrane or enters the cell in some other manner. The lipid portion of cell membranes is thought by many investigators to aid in lipophilic hydrocarbon transport through the membrane into the cell, based mainly on some early studies on hydrocarbon solubility in artificial lipid bilayers.

Impact on Communities

Damage from oil can be extensive and catastrophic, affecting several hundred kilometers of shoreline as in the case of the supertanker Amoco Cadiz breakup. Entire communities have been impacted or even eliminated. However, with time such communities do recover. The recovery time varies, depending on the degree of oiling, the physical conditions of the ecosystem, and the nature of the community.

Recovery to something approaching prespill conditions begins within a matter of months, and general prespill appearances will return within a year or two. However, there will be local pockets or "hot spots" of particularly heavy oiling and residue where impact may persist for 15 years or longer. Also, certain community perturbations may continue for a decade or longer before return to a stable situation. In Chedabucto Bay, for example, the main portion of the oiled shorelines had returned to near-normal conditions within 2-3 years. At the time of this writing there is no visual evidence of any impact over 99% of the coastline, even though trace levels of petroleum hydrocarbon persist in many of the coastal sediments.

The recovery process is a function of the degree of self-cleaning due to wave action and the proportion of soft-sediment low-energy systems (lagoons, estuaries, marshes) in the oiled areas. Thus an oiled rocky coast will be mostly self-cleaned within several months to a year, whereas an oiled lagoon or salt marsh can retain stranded oil residue for several years. For these reasons the time period of recovery differs for different oiled environments. The course of recovery, i.e., the pattern of biological recovery, differs also, depending in part of the biomass composition at the time of the spill, the climate, and other factors. Thus the survival of a dominant predator can influence the recovery process significantly over several years. Again, elimination of a key species, without a ready nearby available source for reintroduction, can result in a long-term altered community.

The matter of "recovery to prespill conditions" has been the subject of some discussion, particularly as it relates directly to spill impact assessment. The notion of prespill conditions, of course, implies return to the ecosystem function and structure that existed prior to the spill. In reality, that is neither likely nor possible, for ecosystems and communities are dynamic assemblages, forever undergoing change and cycles of composition. A coastal community or benthic assemblage is never static, and what may have been its composition in one year becomes a different composition 5 years hence. Therefore, the best one can hope for is a return to the sort of community composition, in terms of biomass and species diversity and their cycles, characteristic of that particular environment. Recovery can thus be reasonably addressed only by comparison with what would have occurred in an undisturbed but otherwise similar ecosystem in the same time period. Depending on the degree of impact, the recovery process will proceed through a series of fluctuations, eventually to return to some stability.

Polar and Tropical Regions

Special attention was given, in discussions and the workshop leading up to this report, to tropical and polar regions, areas largely ignored in the 1975 NRC report.

Since 1975 considerable attention has been diverted to problems of oil pollution in the arctic regions, largely as a direct response to increasing oil exploration activities in the Arctic. Much of this work originated with the Beaufort Sea studies of the early 1970s and has since been expanded by Canadian, Norwegian, and U.S. researchers.

The Arctic presents special problems because of nearly year-round ice cover and inaccessibility. These are compounded by the large gaps in the data base on arctic biology, there existing only slim understanding of biological events during the brief summer open-water season and virtually no understanding of winter events. The Arctic possesses unique features such as marine mammals, under-ice algal/crustacean communities, and seabird nesting areas. On the other hand, the threat of ice and ice scouring, the apparently slower degradation of stranded oil by arctic hydrocarbon-utilizing microbes, and inaccessibility place this region near the top of environmental concern.

Some encouragement may be gained from the knowledge that the arctic marine environment is not nearly as fragile and pristine as it was once popularly thought to be. The northern marine biota are no more fragile than their more temperate counterparts. Natural oil seeps form part of the arctic substrate, and petroleum hydrocarbons have probably been part of the Arctic seas for unknown eons just as they have been for the rest of the world's oceans. However, those hydrocarbons probably have existed at very low concentrations, largely beyond the present detection limits, and the potential impact of a major oil spill on an arctic ecosystem can presently not be estimated with confidence. The arctic (polar) ecosystem experiences special problems because of the sharply reduced ice-free season, and the particular conditions imposed on polar biota critically in tune with this climate.

The tropics with their coastal mangrove swamps and coral reefs pose an entirely different kind of problem. Our present knowledge suggests that these systems may be highly vulnerable to oil. They are common coastal systems and highly porous in terms of oil penetration. Damage is not limited to the local biological component of these systems but has further ramifications in that these systems also support local fisheries and in much of the tropics form part of the human economic resources (mangrove timber, atoll human settlements).

Unfortunately, little research has been carried out on the effects of oil on mangroves or reef corals, or on their associated biota, and the amount of work is significantly less than that done in the temperate zones (North America, Europe, USSR). This is especially paradoxical in view of the vast tonnage of crude oil and refined products that annually are shipped through these environments. Certainly oil spills are not a rarity in the tropics.

Impact on Human Health

Concern for human health centers mainly on the ingestion of either toxic or mutagenic components of oil, either directly or via tainted seafood. Inhalation of these components is less of a concern, and direct contact with oil as on contaminated holiday resort beaches in some parts of the world is viewed more as a socioeconomic concern.

There is no evidence to date of a deleterious impact of petroleum via the marine environment on human health, although admittedly there exist few direct data. However, examination of circumstantial evidence from industrial hydrocarbon inhalation studies and from hydrocarbon levels found in marine biota suggests that any contact with either toxic or mutagenic oil components via the seas is far smaller than that affecting human health from other sources (atmospheric input, smoked foods, etc.). One contributing factor is probably the absence of food web magnification of hydrocarbons through the food chain.

But there exists the possibility that under certain conditions, as in chronically oiled areas or in situations where man is highly dependent on a certain seafood, man may well come into contact with fractions of oil that are deleterious to health, in concentrations higher than those encountered by the average world resident.

Petroleum and Other Chemical Contaminants

Many research workers have expressed concerns over the potential synergistic effects of petroleum, acting in concert with other contaminants, as in waters adjacent to highly industrialized areas (Puget Sound, North Sea, New York Bight). The phenomenon is poorly understood, but there are strong indications that the presence of one contaminant can enhance the toxic effect of a second. Whether this is related directly to some sort of molecular potentiation or whether it is simply a matter of lowering resistance to contaminants is as yet not known. Whatever the mechanism, the net effect is that while the potential damage of one compound may be minimal, the potential can become intensified in the presence of a second compound.

Several workers feel strongly that this aspect of oil pollution deserves more emphasis and that the potential impact through synergistic processes is much more of a problem than that of oil alone.

CONCLUSION

A great deal of productive and relevant work has been done since the 1975 NRC report. One outcome has been a shift from concerns over the cataclysmic tanker spill to more long term chronic input of petroleum into the marine environment. There is no evidence to date, using present-day assessment techniques, that tanker spills have unalterably changed the world's oceans or marine resources. Studies from such spill sites have shown that oiled environments do recover with time. However, it has become equally evident that petroleum can affect local environments, where under certain conditions, oil may persist for several decades. In this respect, we find that in contrast to offshore situations where impact may be minimal and transient (although information is scanty), there are greater immediate concerns for coastal waters (for example, biologically productive estuaries) receiving on the average a greater proportion of discharged petroleum.

Much more impressive, and fundamentally more disquieting, is the broad range of biological processes that can be affected negatively by petroleum hydrocarbons, particularly changes that can be elicited by some hydrocarbons in the genetic framework of marine biota. Petroleum hydrocarbons have demonstrably deleterious effects in laboratory and mesocosm experiments at concentrations as low as a few parts per billion, and such concentrations are now found over wide areas in the coastal oceans. Unfortunately we do not yet know the extent of any deterioration or damage, if any, from these low concentrations that may have occurred from oil hydrocarbon pollution in nature. This is because of the difficulty in separating the effects of oil from other kinds of pollution, from overfishing, or from natural changes and perturbations. In this respect, petroleum may well presage potential hazards due to other, more persistent and less biodegradable chemical contaminants entering our world's oceans, particularly along industrialized coastlines.

RESEARCH RECOMMENDATIONS

To give further depth to our understanding of the effects of petroleum in the marine environment, more work in certain areas is strongly recommended. Some of these areas are already subject to research, while others are relatively new.

1. Mutagenicity/Tumorigenicity. While recognized as a problem area, insufficient study has been directed toward solving this aspect of petroleum effects. Information is scant on invertebrate and vertebrate marine animals and marine plants.

2. Interference of Behavior by Petroleum. The perturbation of normal behavior at very low concentrations of petroleum (less than 0.1 μg/L) is a matter of particular concern. Change in or cessation of feeding is one of the first indications of oil pollution in many test animals. Yet most of the available data are anecdotal, and at least for the higher organisms, effects on behavior are poorly understood.

3. Mechanisms of Toxicity. The present focus on research into perturbations of physiological processes should be encouraged. While such processes as respiration, photosynthesis, ATP production, carbon assimilation, and lipid formation are known to be affected by single hydrocarbons (e.g., naphthalene), the ultimate site or sites of action have not yet been determined.

4. Polar and Tropical Environments. The polar environment presents special problems because of its nearly year-round ice cover and inaccessibility. These are compounded by large gaps in the data base on polar biology. Data are scanty even for the brief summer season, and

there is virtually no understanding of winter events. The potential impact of a major oil spill on an arctic ecosystem cannot now be estimated with confidence. With respect to tropical regions, there is also only a minimal amount of information on the effects of oil on mangroves, coral reefs, and their associated biota.

5. Synergistic Toxicity. Interaction of various petroleum compounds is poorly understood, as is their interaction with other contaminants. Further work along these lines is badly needed, for chronic pollution of inshore waters commonly involves a number of contaminants.

6. Ecosystem Effects. Population changes caused by an oil spill or chronic pollution inevitably alter food-web relations and interspecific competition in the ecosystem as a whole. Each oil spill is different, and the effects are sometimes quite unexpected. Continued study of the history of recovery from oil spills is essential in order to assess their biological and economic significance.

7. Chronic Pollution. Much effort has been directed toward understanding the impact of petroleum hydrocarbons from episodic events such as tanker spills, pipeline breaks, or leaks from shore-based facilities. Much less has been done on incidents involving chronic pollution, where the concentration of petroleum is frequently low but released continuously over extensive periods of time.

We encourage research toward a better understanding of the impact of chronic and accidental oil input at the ecosystem level. We recognize that this requires a much better understanding of both the natural processes occurring in ecosystems and the interaction of oil with other anthropogenic influences.

8. Pollution Indices. Further attention must be directed toward developing capability to assess petroleum impact at sea, in the pelagic environment, involving especially zooplankton and larval fish.

The single most significant difficulty is transferring information obtained from laboratory studies to predicting and/or evaluating potential impact of petroleum on living marine resources in the field, especially in the case of spill impact on such commercially important stocks as fish and shellfish.

REFERENCES

Adams, W.A. 1975. Light intensity and primary productivity under sea ice containing oil. Beaufort Sea Project Technical Report 29. Department of Environment, Victoria, B.C. 156 pp.

Addy, J.M., D. Levell, and J.P. Hartley. 1978. Biological monitoring of sediments in Ekofisk oil field, pp. 515-539. In Proceedings, Conference on Assessment of Ecological Impacts of Oil Spills. American Institute of Biological Sciences, Arlington, Va.

Aelion, M., and Y. LeMoal. 1981. Impact écologique de la marée noire du Tanio sur les plages de Tregastel (Bretagne nord-occidentale). Rapport de contrat CNEXO. 80 65 96. Institut d'Etudes Marines, Univ. de Bretagne Occidentale, Brest, France. 30 pp.

Ahearn, D.G., and S.P. Meyers. 1972. The role of fungi in the decomposition of hydrocarbons in the marine environment, pp. 12-18. In A.H. Walters and E.H. Hueck-vander Plas, eds. Biodeterioration of Materials, Vol. 2. Applied Science Publishers, London.

Ainley, D.G., C.R. Grau, T.E. Roudybush, S.H. Morrell, and J.M. Utts. 1981. Petroleum ingestion reduces reproduction in Cassin's Auklets. Mar. Pollut. Bull. 12.

Albers, P.H. 1977. Effects of external applications of fuel oil on hatchability of mallard eggs, pp. 158-163. In D.A. Wolfe, ed. Fate and Effects of Petroleum Hydrocarbons in Marine Organisms and Ecosystems. Pergamon, New York.

Albert, T.F. 1981. Some thoughts regarding the effect of oil contamination on bowhead whales, _Balaena mysticetus_, pp. 945-953. In T.F. Albert, ed. Tissue, Structural Studies and Other Investigations on the Biology of Endangered Whales in the Beaufort Sea. Appendix 9. Bureau of Land Management, Washington, D.C.

Alink, G.M., E.M.H. Frederik-Wolters, M.A. vander Gaag, J.F.J. van der Kerkhoff, and C.L.M. Poels. 1980. Induction of sister-chromatid exchanges in fish exposed to Rhine water. Mut. Res. 78:369-374.

Allen, H. 1971. Effects of petroleum fractions on the early development of a sea urchin. Mar. Pollut. Bull. 2:138-140.

Ament, M.L., F.J. Meder, and R.P. Beliles. 1979. Teratology study in rats. n-Hexane. Medical Research Publication 27-32177. American Petroleum Institute, Washington, D.C.

American Institute of Biological Sciences. 1978. Proceedings, Conference on Assessment of Ecological Impacts of Oil Spills. NTIS AD-A072 859. National Technical Information Service, Springfield, Va. 945 pp.

American Petroleum Institute. 1975. Proceedings, 1975 Conference on Prevention and Control of Oil Pollution. Washington, D.C.

American Petroleum Institute. 1976. Detectability and irritability of hydrocarbons in human subjects. _Rep._ U-15-14-PS-5. Washington, D.C.

American Petroleum Institute. 1977. Proceedings, 1977 Oil Spill Conference. (Prevention, Behavior, Control, Cleanup). Publication American Petroleum Institute 4284. Washington, D.C. 640 pp.

American Petroleum Institute. 1979. Proceedings, 1979 Oil Spill Conference. (Prevention, Behavior, Control, Cleanup). API Publication 4308. Washington, D.C. 728 pp.

American Petroleum Institute. 1981. Proceedings, 1981 Oil Spill Conference. Publication 4334. Washington D.C. 742 pp.

Aminot, A., and R. Kerouel. 1978. Premiers resultats sur 1-hydrologie, l'oxygene dissous et les pigments photosynthetiques en Manch Occidentale apres l'echouage de l'_Amoco Cadiz_, pp. 51-68. In Conan et al., eds. _Amoco Cadiz_: Consequences d'une pollution accidentelle par les hydroocarbures. CNEXO, Paris.

Anderson, J.W. 1979. An assessment of knowledge concerning the fate and effects of petroleum hydrocarbons in the marine environment, pp. 3-22. In W.B. Vernberg, F.J. Vernberg, A. Calabrese, and F.P. Thurberg, eds. Marine Pollution: Functional Responses. Academic Press, New York.

Anderson, J.W., J.M. Neff, B.A. Cox, H.E. Tatem, and G.H. Hightower. 1974a. Characteristics of dispersions and water-soluble extracts of crude and refined oil and their toxicity to estuarine crustaceans and fish. Mar. Biol. 27:75-88.

Anderson, J.W., J.M. Neff, B.A. Cox, H.E. Tatem, and G.M. Hightower 1974b. The effects of oil on estuarine animals: toxicity, uptake and depuration, respiration, pp. 285-310. In F.J. Vernberg and W.B. Vernberg, eds. Pollution and Physiology of Marine Organisms. Academic Press, New York.

Anderson, J.W., D.B. Dixit, G.S. Ward, and R.S. Foster. 1977. Effects of petroleum hydrocarbons on the rate of heart beat and hatching success of estuarine fish embryos, pp. 241-258. In F.J. Vernberg, A. Calabrese, F.P. Thurberg, and W.B. Vernburg, eds. Physiological Responses of Marine Biota to Pollutants. Academic Press, New York.

Anderson, J.W., R.G. Riley, and R.M. Bean, 1978a. Recruitment of benthic animals as a function of petroleum hydrocarbon concentrations in the sediment. J. Fish. Res. Board Can. 35:679-680.

Anderson, J.W., G. Roesijadi, and E.A. Crecelius. 1978b. Bioavailability of hydrocarbons and heavy metals to marine detritivores from oil-impacted sediments, pp. 130-148. In D.A. Wolfe, ed. Marine Biological Effects of OCS Petroleum Development. National Oceanic and Atmospheric Administration, Bethesda, Md.

Anderson, J.W., J.W. Blaylock, and S.L. Kiesser, 1979. Comparative uptake of naphthalenes from water and oiled sediment by benthic amphipods, pp. 579-584. In Proceedings, 1979 Oil Spill Conference. API Publication 4308. American Petroleum Institute, Washington, D.C.

Anderson, J.W., R.G. Riley, S.L. Kiesser, B.L. Thomas, and G.W. Fellingham. 1983. In press. Natural weathering of oil in marine sediments: tissue contamination and growth of the clam Protothaca staminea. Can. J. Fish. Aquat. Sci. Special Suppl.

Anderson, R.D., and J.W. Anderson. 1976. Effects of salinity and selected petroleum hydrocarbons on the osmotic and chloride regulation of the American oyster, Crassostrea virginica. Physiol. Zool. 48:420-429.

Anon. 1970. Report of the Task Force--Operation Oil (Cleanup of the Arrow Oil Spill in Chedabucto Bay) Vols. I, II, III, IV. Canadian Minister of Transport.

Arctic. 1978. Proceedings of the workshop on ecological effects of hydrocarbon spills in Alaska. Arctic 31(3):155-411.

Arctic Marine Oil Spill Program. 1978-1982. Technical Seminars. 1st, 1978, 259 pp.; 2nd, 1979, 384 pp.; 3rd, 1980, 580 pp.; 4th, 1981, 741 pp.; 5th, 1982, 618 pp. Environmental Protection Service, Environmental Emergency Branch, Ottawa, Ontario.

Arhelger, S.D., B.R. Robertson, and D.K. Button. 1977. Arctic hydrocarbon biodegradation, pp. 270-275. In D.A. Wolfe, ed. Fate and Effects of Petroleum Hydrocarbons in Marine Organisms and Ecosystems. Pergamon, New York.

Ashmole, N.P. 1971. Seabird ecology and the marine environment, pp. 224-286. In D.S. Farner and J.R. King, eds. Avian Biology, Vol. 1. Academic Press, New York.

Atema, J., and L.S. Stein. 1974. Effects of crude oil on the feeding behaviour of the lobster Homarus americanus. Environ. Pollut. 6:77-86.

Atema, J., S. Jacobson, J. Todd, and D. Boylan. 1973. The importance of chemical signals in stimulating behavior of marine organisms: effects of altered environmental chemistry on animal communication, pp. 177-197. In G.E. Glass ed. Bioassay Techniques and Environmental Chemistry. Ann Arbor Science Publishers, Ann Arbor, Mich.

Atlas, R.M. 1974. Fate and effects of oil pollutants in extremely cold marine environments. NTIS AD/A-003 554. National Technical Information Service, Springfield, Va.

Atlas, R.M. 1977. Studies on petroleum biodegradation in the arctic, pp. 261-269. In D.A. Wolfe, ed. Fate and Effects of Petroleum Hydrocarbons in Marine Organisms and Ecosystems. Pergamon, New York.

Atlas, R.M. 1981. Microbial degradation of petroleum hydrocarbons: an environmental perspective. Microbiol. Rev. 45:180-209.

Atlas, R.M., and R. Bartha. 1972. Biodegradation of petroleum in seawater at low temperatures. Can. J. Microbiol. 18:1851-1855.

Atlas, R.M., E.A. Schofield, F.A. Morelli, and R.E. Cameron. 1976. Effects of petroleum pollutants on arctic microbial populations. Environ. Pollut. 10(1):35-43.

Atlas, R.M., A. Horowitz, and M. Busdosh. 1978. Prudhoe crude oil in arctic marine ice, water and sediment ecosystems: degradation and interactions with microbial and benthic communities. J. Fish. Res. Board Can. 35:585-590.

Atlas, R.M., G.E. Roubal, A. Bronner, and T.R. Haines. 1982. Biodegradation of hydrocarbons in mousse from the Ixtoc I well blowout. American Chemical Society, Washington, D.C. (in press).

Atwood, D.K., convener. 1980. Proceedings Symposium on Preliminary Scientific Results From the Researcher/Pierce Cruise to the Ixtoc I Blowout. U.S. Department of Commerce, National Oceanic and Atmospheric Administration, Boulder, Colo. 591 pp.

Augenfeld, J.M. 1980. Effects of Prudhoe Bay crude oil contamination on sediment working rates of Abarenicola pacifica. Mar. Environ. Res. 3:307-313.

Avanzi, M.G. 1950. Frequenz e tipi di aberrazioni chromosomiche indotte da alcuni derivati dell' a-naftalene. Caryologia 3:165-180.

Avolizi, R.J., and M. Nuwayhid. 1974. Effects of crude oil and dispersants on bivalves. Mar. Pollut. Bull. 5:149.

Ayers, R.C., Jr., H.O. Johns, and J.L. Glaeser. 1974. Oil spills in the Arctic Ocean: extent of spreading and possibility of large scale thermal effects. Science 186:842-845.

Bacon, P.R. 1970. The Ecology of Caroni Swamp. Special Publication. Central Statistical Office, Trinidad. 68 pp.

Baker, J.M. 1970. Studies on saltmarsh communities - refinery effluent, pp. 33-43. In E.B. Cowell, ed. The Ecological Effects of Oil Pollution on Littoral Communities. Elsevier Press.

Baker, J.M. 1971. The Effects of Oil Pollution and Cleaning on the Ecology of Saltmarshes. Ph.D. dissertation. University Colleges of South Wales, Swansea, U.K.

Baker, J.M. 1975. Effects on shore life and amenities, pp. 85-91. In H.A. Cole, ed. Petroleum and the Continental Shelf of North-west Europe. Vol. 2. John Wiley and Sons, New York.

Baker, J.M. 1979. Responses of salt marsh vegetation to oil spills and refinery effluents, pp. 529-542. In R.C. Jeffries and A.J. Davy, eds. Ecological Processes in Coastal Environments, The First European Ecological Symposium and the 19th Symposium of the British Ecological Society, Norwich, September 1977. Blackwell Scientific Publications, Oxford.

Baker, J.M. 1981. The investigation of oil industry influences on tropical marine ecosystems. Mar. Pollut. Bull. 12:6-10.

Baker, J.M. 1982. Mangrove swamps and the oil industry. Oil Petrochem. Pollut. 1(1):5-22.

Baker, J.M., I.M. Suryowinoto, P. Brooks, and S. Rowland. 1981. Tropical marine ecosystems and the oil industry; with a description of a post-oil spill survey in Indonesia mangroves, pp. 679-703. In Petromar 80. Petroleum and the Marine Environment, Eurocean. Graham & Trotman Ltd., London.

Barber, F.G. 1970. Oil spills in ice: some cleanup options. Arctic 23:285-286.

Barnett, J., and D. Toews. 1978. The effect of crude oil and the dispersant, Oilsperse 43, on respiration and coughing rates in Atlantic salmon (Salmo salar). Can. J. Zool. 56:307-310.

Barsdate, R.J., M.C. Miller, V. Alexander, J.R. Vestal, and J.E. Hobbie. 1980. Oil spill effects, pp. 388-406. In J.E. Hobbie ed. Limnology of Tundra Ponds. Dowden, Hutchinson and Ross, New York.

Bartha, R., and R.M. Atlas. 1977. The microbiology of aquatic oil spills. Adv. Appl. Microbiol. 22:225-266.

Batterton, J., K. Winters, and C. Van Baalen. 1978a. Sensitivity of three microalgae to crude oils and fuel oils. Mar. Environ. Res. 1:31-41.

Batterton, J., K. Winters, and C. Van Baalen. 1978b. Anilines: selective toxicity to blue-green algae. Science 199:1068-1070.

Beck, L.S., and D.I. Hepler. 1980a. Acute toxicity tests of #6 heavy fuel oil (11.7 API/2.7S) (API-78-6). Medical Research Publication 27-32817. American Petroleum Institute, Washington D.C.

Beck, L.S., and D.I. Hepler (Elars Bio-research Lab.). 1980b. Acute toxicity tests of API-79-6 diesel fuel. Medical Research Publication 27-32817. American Petroleum Institute, Washington, D.C.

Beck, L.S., and D.I. Hepler. 1980c. Acute toxicity tests of API 78-1 new composite motor oil. Medical Research Publication 27-32131. American Petroleum Institute, Washington, D.C.

Bedinger, C.A. 1981. Ecological investigations of petroleum production platforms in the central Gulf of Mexico. Vol. 3. Executive Summary. S.W. Research Institute, San Antonio, Tex. 29 pp.

Beljanski, M. 1979. Oncotest: a DNA assay system for the screening of carcinogenic substances. IRCS Med. Sci. 7:476.

Bell, W., and R. Mitchell. 1972. Chemotactic and growth responses of marine bacteria to algal extracellular products. Biol. Bull. 143:265-277.

Bend, J.R. 1980. Induction of drug-metabolizing enzymes by polycyclic aromatic hydrocarbons: mechanisms and some implications in environmental health research, p. 83. In Environmental Chemicals, Enzymatic Mechanisms and Human Disease. Ciba Foundation Symposium 76. Excerpta Medica, Amsterdam.

Bend, J.R., G.L. Foureman, and M.O. James. 1978. Partially induced hepatic mixed-function oxidase systems in individual members of certain marine species from coastal Maine and Florida, pp. 483-486. In O. Hutzinger et al., eds. Aquatic Pollutants: Transformation and Biological Effects. Pergamon, Oxford.

Bend, J.R., L.M. Ball, T.H. Elmamlouk, M.O. James, and R.M. Philpot. 1979. Microsomal mixed function oxidation in untreated and polycyclic aromatic hydrocarbon treated marine fish, pp. 297-318. In M.A.Q. Kahn et al., eds. Pesticide and Xenobiotic Metabolism in Aquatic Organisms. Symposium Series 99. American Chemical Society, Washington, D.C.

Bender, M.E., E.A. Shearls, L. Murray, and R.J. Huggett. 1980. Ecological effects of experimental oil spills in eastern coastal plain estuaries. Environ. Int. 3:121-133.

Benville, P.E., Jr., and S. Korn. 1977. The acute toxicity of six monocyclic aromatic crude oil components to striped bass (Marone saxatilis) and bay shrimp (Crago franciscorum). Calif. Fish Game 63:204-209.

Berdugo, V., R.P. Harris, and S.C. O'Hara. 1977. The effect of petroleum hydrocarbons on reproduction of an estuarine planktonic copepod in laboratory cultures. Mar. Pollut. Bull. 8:138-143.

Berman, M.S., and D.R. Heinle. 1980. Modification of the feeding behavior of marine copepods by sub-lethal concentrations of water-accommodated fuel oil. Mar. Biol. 56:59-64.

Berry, H.H. 1976. Mass mortality of Cape Cormorants, caused by fish oil, in the Walvis Bay region of South West Africa. Madoqua 9(4):57-62.

Bertram, G.C.L. 1940. The biology of the Weddell and Crabeater seals. Br. Mus. (Nat. Hist.) Br. Graham Land Exped. 1934-1937 Sci. Rep. 1:10.

Beslier, A., J.L. Birrien, L. Cabioch, C. Larsonneur, and L. LeBorgne, 1980. La pollution des Baies de Morlaix et de Lannion par les hydrocarbures de l'Amoco Cadiz: repartition sur les fonds et evolution. Helgolander Meeresunters. 33:209-224.

Beynon, L.R. 1971. Dealing with major oil spills at sea, pp. 187-193. In P. Hepple, ed., Water Pollution by Oil. Institute of Petroleum, London.

Bieri, R.H., and V.C. Stamoudis. 1977. The fate of petroleum hydrocarbons from a No. 2 fuel oil spill in a seminatural estuarine environment, pp. 332-344. In D.A. Wolfe, ed., Fate and Effects of Petroleum Hydrocarbons in Marine Organisms and Ecosystems. Pergamon, New York.

Bingham, E., A.W. Horton, and R. Tye. 1965. The carcinogenic potency of certain oils. Arch. Environ. Health 10:449-451.

Bingham, E., R.P. Trosset, and D. Warshansky. 1979. Carcinogenic potential of petroleum hydrocarbons. J. Environ. Pathol. Toxicol. 3:483-563.

Birkeland, C., A.A. Reimer, and J.R. Young. 1976. Survey of marine communities in Panama and experiments with oil. Ecological Research Series PB 253 409. EPA-600/3-76-028. Environmental Protection Agency, Washington, D.C. 176 pp.

Birkhead, T.R., C. Lloyd, and P. Corkhill. 1973. Oiled seabirds successfully cleaning their plumage. Br. Birds 66:535-537.

Bitton, G., D.A. Chuchran, I. Chet, and R. Mitchell. 1979. Resistance of bacterial chemotaxis to blockage in petroleum waters. Mar. Pollut. Bull. 10:48-49.

Blackall, D.J., and G.J. Sergy. 1981. The BIOS Project-Frontier oil spill countermeasure research, pp. 167-172. In Proceedings, 1981 Oil Spill Conference. American Petroleum Institute, Washington, D.C.

Blackall, D.J., and G.J. Sergy. 1983. The BIOS Project--An Update, pp. 451-456. In Proceedings, 1983 Oil Spill Conference. American Petroleum Institute, Washington, D.C.

Blasco, F. 1982. Ecosystemes mangroves: fonctionnement, utilite, evolution. Oceanol. Acta. Proc's Int'l. Symp. Coastal Lagoons. SCOR/IABO/UNESCO, Boreaux, France 8-14 Sept. 1981. pp. 225-230.

Blix, A.S., H.J. Grav, and K. Ronald. 1979. Some aspects of temperature regulation in newborn harp seal pups. Am. J. Physiol. 236:R188-R197.

Blumer, M. 1969. Oil pollution of the ocean, pp. 5-13. In D.P. Hoult ed. Oil on the Sea. Plenum, New York.

Blumer, M., G. Souza, and J. Sass. 1970. Hydrocarbon pollution of edible shellfish by an oil spill (Buzzards Bay, Mass.). Mar. Biol. 5:195-202.

Blumer, M., M. Ehrhardt, and J.H. Jones. 1973. The environmental fate of stranded crude oil. Deep Sea Res. 20:239-259.

Blundo, R. 1978. The toxic effects of the water-soluble fraction of No. 2 fuel oil and of three aromatic hydrocarbons on the behavior and survival of barnacle larvae. Contrib. Mar. Sci. 21:35-37.

Bodin, Ph., and D. Boucher. 1981. Evolution temporelle du meiobenthos et du microphytobenthos sur quelques plages touchees par la maree noire de l'Amoco Cadiz, pp. 327-346. In Amoco Cadiz: Fate and Effects of the Oil Spill. CNEXO, Paris.

Boehm, P., A.D. Wait, D.L. Fiest, and D. Pilson. 1982. Chemical assessment-hydrocarbon analyses, Section 2. Ixtoc Oil Spill Assessment Final Report. Contract AA851-CTO-71. Bureau of Land Management, U.S. Department of the Interior, Washington, D.C.

Boney, A.P. 1974. Aromatic hydrocarbons and the growth of marine algae. Mar. Pollut. Bull. 5:185-186.

Bonser, G.M. 1932. Tumors of the skin produced by blast furnace tar. Lancet 1:775-778.

Bos, R.P., R.M.E. Brouns, R. van Doorn, J.L.G. Theuws, and P. Th. Henderson. 1981. Non-mutagenicity of toluene, o-, m- and p-xylene, o-methylbenzylalcohol and o-methylbenzylsulphate in the Ames assay. Mut. Res. 88:273-279.

Bourne, W.R.P. 1976. Seabirds and pollution, pp. 403-502. In R. Johnston ed. Marine Pollution. Academic Press, New York.

Bowman, R.S. 1978. Dounreay oil spill: Major implications of a minor incident. Mar. Pollut. Bull. 9:269-273.

Bradstreet, M.S.W., and W.E. Cross. 1980. Studies near the Pond Inlet ice edge: trophic relationships. Unpublished report for Petro-Canada Ltd., Calgary. LGL Ltd., Toronto, Ontario. 41 pp.

Braham, H., M. Fraker, and B. Krogman. 1980. Spring migration of the western arctic population of bowhead whales. Mar. Fish. Rev. 42:36-46.

Braithwaite, L. 1981. The effects of oil on the feeding mechanism of the bowhead whale. Bureau of Land Management, Washington, D.C. (in press).

Brodersen, C.C., S.D. Rice, J.W. Short, T.A. Mecklenburg, and J.F. Karinen. 1977. Sensitivity of larval and adult Alaskan shrimp and crabs to acute exposures of the water-soluble fraction of Cook Inlet crude oil, pp. 575-578. In Proceedings, 1977 Oil Spill Conference. API Publication 4284. American Petroleum Institute, Washington, D.C.

Brooks, J.M., G.A. Fryxell, D.F. Reid, and W.M. Sackett. 1977. Gulf underwater flare experiment (GUFEX): effects of hydrocarbons on phytoplankton, pp. 45-75. In C.S. Giam, ed. Pollutant Effects on Marine Organisms. Lexington Books, Lexington, Mass.

Brooks, J.M., B.B. Bernard, T.C. Saner, Jr., and H.A. Reheim. 1978. Environmental aspects of a well blowout in the Gulf of Mexico. Environ. Sci. Technol. 12:695-703.

Brown, L.R., and R.G. Tischer. 1969. The decomposition of petroleum products in our natural waters. Water Resources Research Institute, Mississippi State University, State College, Miss.

Brown, R.G.B. 1982. Birds, oil and the Canadian environment, pp. 105-112. In J.B. Sprague, J.H. Vandermeulen, and P.G. Wells., eds. Oil and Dispersants in Canadian Seas--Research Appraisal and Recommendations. Economic and Technical Report EPS-3-EC-82-2. Environment Canada, Environmental Protection Service.

Brown, R.G.B., and B.C. Johnson. 1980. The effects of Kurdistan oil on seabirds. In J.H. Vandermeulen ed. Report Series BI-R-80-3. Scientific Studies During the Kurdistan Tanker Incident. Bedford Institute of Oceanography, Dartmouth, N.S.

Brown, R.G.B., and D.N. Nettleship. 1981. The biological significance of polynyas to arctic colonial seabirds, pp. 59-65. In I. Stirling and H. Cleator eds. Polynyas in the Canadian Arctic. Occasional Paper 45. 73 pp. Canadian Wildlife Service Ottawa, Ontario.

Brown, R.G.B., D.N. Nettleship, P. Germain, C.E. Tull, and T. Davis. 1975. Atlas of Eastern Canadian Seabirds. Canada Wildlife Service, Ottawa, Ontario. 220 pp.

Brown, R.S., R.E. Wolke, S.B. Saila, and C.W. Brown. 1977. Prevalence of neoplasia in ten New England populations of soft-shell clam Mya arenaria. Ann. N.Y. Acad. Sci. 298:522-534.

Brownell, R., and B. Le Boeuf. 1971. California sea lion mortality; natural or artifact, p. 287. In D. Straughan, ed. Biological and Oceanographic Survey of the Santa Barbara Channel Oil Spill, 1969-1970, Vol. 1. Sea Grant Publication 2. Allan Hancock Foundation. University of Southern California, Los Angeles.

Brunies, A. 1971. Taint of mineral oil in mussels. Arch. Lebensmittelhug. March 1971:63-64.

Bunch, J.B., and R.C. Harland. 1976. Biodegradation of crude petroleum by the indigenous microbial flora of the Beaufort Sea. Beaufort Sea Project Technical Report 10. Department of the Environment, Victoria, B.C. 52 pp.

Burns, K.A. 1976a. Hydrocarbon metabolism in intertidal fiddler crab, Uca pugnax. Mar. Biol. 36:5-11.

Burns, K.A. 1976b. Microsomal mixed function oxidases in an estuarine fish, Fundulus heteroclitus, and their induction as a result of environmental contamination. Comp. Biochem. Physiol. 53B:443-446.

Burns, K.A., and J.M. Teal. 1979. The West Falmouth oil spill; hydrocarbons in the saltmarsh ecosystem. Estuarine Coastal Mar. Sci. 8:349-360.

Burton, J. 1981. Half north, half south: Britain's rich wildlife. New Sci. 90:16-18.

Busdosh, M., 1978. The effects of Prudhoe crude oil fractions on the arctic amphipods Boeckosimus affinis and Gammarus zaddachi. Ph.D. dissertation, University of Louisville, Louisville, Ky. 111 pp.

Busdosh, M., and R.M. Atlas. 1977. Toxicity of oil slicks to arctic amphipods. Arctic 30:85-92.

Byrne, C.J., and J.A. Calder. 1977. Effect of the water-soluble fractions of crude, refined and waste oils on the embryonic and larval stages of the Quahog clam Mercenaria sp. Mar. Biol. 40:225-231.

Cabioch, L., J.C. Dauvin, J. Mora Bermudez, and C. Rodriguez Babio, 1980. Effets de la maree noire de l'Amoco Cadiz sur le benthos sublittoral du nord de la Bretagne. Helgolander Meeresunters. 33:192-208.

Calder, J.A., and P.D. Boehm. 1981. One-year study of weathering processes acting on the Amoco Cadiz oilspills. In Amoco Cadiz: Fate and Effects of the Oil Spill. CNEXO Publication, Paris.

Caldwell, R.S., E.M. Calderone, and M.H. Mallon. 1977. Effects of a seawater-soluble fraction of Cook Inlet crude oil and its major aromatic components on larval stages of the Dungeness crab, Cancer magister Dana, pp. 210-220. In D.A. Wolfe, ed. Fate and Effects of Petroleum Hydrocarbons in Marine Organisms and Ecosystems. Pergamon, New York.

Cameron, J.A., and R.L. Smith. 1980. Ultrastructural effects of crude oil on early life stages of Pacific herring. Trans. Am. Fish. Soc. 109:224-228.

Capuzzo, J.M. 1981. Predicting pollution effects in the marine environment. Oceanus 24:25-33.

Capuzzo, J.M., and B.A. Lancaster. 1981. The physiological effects of South Louisiana crude oil on larvae of the American lobster (Homarus americanus), pp. 405-423. In F.J. Vernberg, A. Calabrese, F.P. Thurberg, and W.B. Vernberg, eds. Biological Monitoring of Marine Pollutants. Academic Press, New York.

Carlton, J.M. 1974. Land building and stabilization by mangroves. Environ. Conserv. 1:285-294.

Carpenter, C.R., E.R. Kinkead, D.L. Geary, L.J. Sullivan, and J.M. King. 1975a. Petroleum hydrocarbon toxicity studies. II. Animal and human response to vapors of varnish markers and painters naphtha. Toxicol. Appl. Pharmacol. 32(2)63-281.

Carpenter, C.R., E.R. Kinkead, D.L. Geary, L.J. Sullivan, and J.M. King. 1975b. Petroleum hydrocarbon toxicity studies. IV. Animal and human response to vapors of rubber solvent. Toxicol. Appl. Pharmacol. 33:526-542.

Carpenter, C.R., E.R. Kinkead, D.L. Geary, L.J. Sullivan, and J.M. King. 1975c. Petroleum hydrocarbon toxicity studies. V. Animal and human response to vapors of mixed xylenes. Toxicol. Appl. Pharmacol. 33:543-558.

Carpenter, C.R., E.R. Kinkead, D.L. Geary, L.J. Sullivan, and J.M. King. 1975d. Petroleum hydrocarbon toxicity studies. VI. Animal and human response to vapors of "60 solvent". Toxicol. Appl. Pharmacol. 35:374-394.

Carpenter, C.R., E.R. Kinkead, D.L. Geary, L.J. Sullivan, and J.M. King. 1976a. Petroleum hydrocarbon toxicity studies. IX. Animal and human response to vapors of "80 thinner". Toxicol. Appl. Pharmacol. 36:409-425.

Carpenter, C.R., E.R. Kinkead, D.L. Geary, L.J. Sullivan, and J.M. King. 1976b. Petroleum hydrocarbon toxicity studies. XI. Animal and human response to vapors of deodorized kerosene. Toxicol. Appl. Pharmacol. 36:443-456.

Carpenter, C.R., E.R. Kinkead, D.L. Geary, L.J. Sullivan, and J.M. King. 1976c. Petroleum hydrocarbon toxicity studies. XII. Animal and human response to vapors of "40 thinner". Toxicol. Appl. Pharmacol. 36:457-472.

Carpenter, C.R., E.R. Kinkead, D.L. Geary, L.J. Sullivan, and J.M. King. 1976d. Petroleum hydrocarbon toxicity studies. XIII. Animal and human response to vapors of toluene concentrate. Toxicol. Appl. Pharmacol. 36:473-490.

Carpenter, C.R., E.R. Kinkead, D.L. Geary, L.J. Sullivan, and J.M. King. 1977a. Petroleum hydrocarbon toxicity studies. XIV. Animal and human response to vapors of "high aromatic solvent". Toxicol. Appl. Pharmacol. 41:235-249.

Carpenter, C.R., E.R. Kinkead, D.L. Geary, L.J. Sullivan, and J.M. King. 1977b. Petroleum hydrocarbon toxicity studies. XV. Animal and human response to vapors of "high naphthenic solvent". Toxicol. Appl. Pharmacol. 41:251-260.

Carstens, T., and E. Sendstad. 1979. Oil spill on the shore of an ice-covered fjord in Spitzbergen. Proceedings, 5th International Conference on Port and Ocean Engineering under Arctic Conditions. Norwegian Inst. of Technol. 2:1227-1242.

Carter, M.R., L.A. Burus, T.R. Cavinder, K.R. Dugger, P.F. Fore, D.B. Hicks, H.L. Revells, and T.W. Schmidt. 1973. Ecosystem analysis of the Big Cypress swamp and estuaries. U.S. Environmental Protection Agency, Region IV, Atlanta, Ga.

Ceas, M.P. 1974. Effects of 3-4-benzopyrene on sea urchin egg development. Acta Embryol. Exp. 1974:267-272.

Centre National pour l'exploitation des Oceans. 1981. Amoco Cadiz: Consequences d'une pollution accidentelle par les hydrocarbures. Actes du Colloque International Center Océanologique de Bretagne, p. 881. CNEXO, Paris.

Cerame-Vivas, M.T. 1968. The wreck of the Ocean Eagle. Sea Frontiers 15:222-231.

Cerniglia, C.E., C. Van Baalen, and D.T. Gibson. 1980a. Metabolism of naphthalene by the Cyanobacterium, Oscillatoria sp., strain JCM. J. Gen. Microbiol. 116:485-494.

Cerniglia, C.E., C. Van Baalen, and D.T. Gibson. 1980b. Oxidation of biphenyl by the Cyanobacterium, Oscillatoria sp., strain JCM. Arch. Microbiol. 125:203-207.

Cerniglia, C.E., D.T. Gibson, and C. Van Baalen. 1981a. Oxidation of naphthalene by Cyanobacteria and microalgae. J. Gen. Microbiol. 116:495-500.

Cerniglia, C.E., J.P. Freeman, and C. Van Baalen. 1981b. Biotransformation and toxicity of aniline and aniline derivatives in Cyanobacteria. Arch. Microbiol. 130:272-275.

Cerniglia, C.E., D.T. Gibson, and C. Van Baalen. 1982. Naphthalene metabolism by diatoms isolated from the Kachemak Bay region of Alaska. J. Gen. Microbiol. 128:987-990.

Chan, E.I. 1976. Oil pollution and tropical littorial communities: biological effects of the 1975 Florida Keys oil spill, pp. 539-542. In Proceedings, 1977 Oil Spill Conference. API Publication 4284. American Petroleum Institute, Washington, D.C.,

Chan, E.I. 1977. Oil pollution and tropical littoral communities: biological effects of the 1975 Florida Keys oil spill, pp. 539-542. In Proceedings, 1977 Oil Spill Conference. API Publication 4284. American Petroleum Institute, Washington, D.C.

Chasse, C. 1978. The ecological impact on and near shores by the Amoco Cadiz oil spill. Mar. Pollut. Bull. 9:298-301.

Clark, R.B. 1969. Oil pollution and the conservation of seabirds. Proc. Int. Conf. Oil Pollut. Sea, Rome 1968:76-112.

Clark, R.B. 1978. Oiled seabird rescue and conservation. J. Fish. Res. Board Can. 35:675-678.

Clark, R.C., Jr., and J.S. Finley. 1973. Paraffin in hydrocarbon patterns in petroleum-polluted mussels. Mar. Pollut. Bull. 4:172-178.

Clark, R.C., Jr., and J.S. Finley, 1982. Occurrence and impact of petroleum on arctic environments, pp. 295-341. In L. Rey, ed. The Arctic Ocean: The Hydrographic Environment and the Fate of Pollutants. Macmillan, London.

Clarke, A. 1979. Living in cold water: K-strategies in antarctic benthos. Mar. Biol. 55:111-119.

Cohen, Y., A. Nissenbaum, and R. Eisher. 1977. Effects of Iranian crude oil on the Red Sea octocoral Heteroxenia fuscescens. Environ. Pollut. 12:173-185.

Cole, T.J. 1978. Preliminary ecological-genetic comparison between uperturbed and oil-impacted Urosalpinx cinerea (Prosobranchia: Gastropoda) populations, Nobska Point (Woods Hole) and Wild Harbour (West Falmouth), Massachusetts. J. Fish. Res. Board Can. 35:624-629

Cole, T.J., N. Taylor, J. Cole, and C.F. Arlett. 1981. Short-term tests for transplacentally active carcinogens. I. Micronucleus formation in fetal and maternal erythroblasts. Mut. Res. 80:141-157.

Colwell, R.R., and J.D. Walker. 1977. Ecological aspects of microbial degradation of petroleum in the marine environment. CRC. Crit. Rev. Microbiol. 5:423-445.

Colwell, R.R., A.L. Mills, J.D. Walker, P. Garcia-Tello, and P.V. Campo. 1978. Microbial ecology studies of the Metula spill in the Straits of Magellan. J. Fish. Res. Board Can. 35(5):573-580.

Conan, G., and M. Friha. 1981. Effets des pollutions par les hydrocarbures du petrolier Amoco Cadiz sur la croissance des soles et des plies dans l'estuaire de l'Aber Benoit, pp. 749-773. In G. Conan et al., eds. Amoco Cadiz: Consequences d'une pollution accidentelle par les hydrocarbures. CNEXO, Paris.

Conan, G., L. d'Ozouville, and M. Marchand, eds. 1978. Amoco Cadiz: Preliminary Observations of the Oilspill Impact on the Marine Environment. Actes de Colloques 6. CNEXO, Paris.

Conan, G., L. Laubier, M. Marchand, and L. d'Ozouville. eds. 1981. Amoco Cadiz: consequences d'une pollution accidentelle par les hydrocarbures. CNEXO, Paris. 881 pp.

Connell, D.W. 1971. Kerosene-like tainting in Australian mullet. Mar. Pollut. Bull. 2:188-190.

Connell, D.W. 1974. A kerosene-like taint in the sea mullet, Mugil cephalus (Linnaeus). I. Composition and environmental occurrence of the tainting substance. Aust. J. Mar. Freshwater Res. 25:7-24.

Connell, D.W., and G.J. Miller. 1980. Petroleum hydrocarbons in aquatic ecosystems--behavior and effects of sublethal concentrations. Part I. Crit. Rev. Environ. Control 11:37-104.

Conover, R.J. 1971. Some relations between zooplankton and Bunker C oil in Chedabucto Bay following the wreck of the tanker Arrow. J. Fish. Res. Board Can. 28:1327-1330.

Cook, J.W., C.L. Hewett, and I. Hieger. 1933. The isolation of a cancer-producing hydrocarbon from coal tar. Parts I, II, and III. J. Chem. Soc.:397-420.

Coon, N.C., P.H. Albers, and R.C. Szaro. 1979. No. 2 fuel oil decreases embryonic survival of great black-backed gulls. Bull. Environ. Contam. Toxicol. 21:152-156.

Corner, E.D.S. 1978. Pollution studies with marine plankton. 1. Petroleum hydrocarbons and related compounds. Adv. Mar. Biol. 15:289-380.

Corner, E.D.S., and R.P. Harris. 1976. Hydrocarbons in marine zooplankton and fish, pp. 71-205. In A.P.M. Lockwood, ed. Effects of Pollutants on Aquatic Organisms. Cambridge University Press, New York.

Corner, E.D.S., R.P. Harris, C.C. Kilvington, and S.C.M. O'Hara. 1976. Petroleum compounds in the marine food web; short term experiments on the fate of naphthalene in Calanus. J. Mar. Biol. Assoc. U.K. 56:121-133.

Cornillon, P. 1978. Oil droplet measurements made in the wake of the Argo Merchant, pp. 43-47. In M.P. Wilson, J.G. Quinn, and K. Sherman, eds. In the Wake of the Argo Merchant. Center Ocean Management Studies, University of Rhode Island, Kingston.

Costa, D.P., and G.L. Kooyman. 1980. Effects of oil contamination in the sea otter, Enhydra lutris. Outer Continental Shelf Environmental Assessment Program, National Oceanic and Atmospheric Administration, Alaska.

Cowell, E.P. 1976. Oil pollution of the sea, pp. 389-391. In R. Johnston, ed. Marine Pollution. Academic Press, New York.

Craig, P.C., and L. Haldorson. 1980. Environmental assessment of the Alaskan continental shelf. Part 4, Fish, pp. 384-677. Report for U.S. Department of Commerce and U.S. Department of the Interior, Washington, D.C. LGL Ltd., Toronto, Ontario.

Cramp, S., W.R.P. Bourne, and D. Saunders. 1974. The Seabirds of Britain and Ireland. Collins, London. 287 pp.

Cross, W.E. 1980. Studies near the Pond Inlet ice edge: underice biota at the ice edge and in adjacent fast ice areas during May-July. 1979. Unpublished Report for Petro-Canada Ltd., Calgary. LGL Ltd., Toronto, Ontario. 65 pp.

Croxall, J.P. 1977. The effects of oil on seabirds. Rapp. P-v. Reun. Cons. Int. Explor. Mer 171:191-195.

Cucci, T.L., and C.E. Epifiano. 1979. Long-term effects of water-soluble fractions of Kuwait crude oil on the larval and juvenile development of the mud crab Eurypanopeus depressus. Mar. Biol. 55:215-220.

Currie, A.R., C.C. Bird, A.M. Crawford, and P. Sims. 1970. Embryopataic effects of 7,12-dimethylbenz(a)anthracene and its hydroeylmethyl derivatives in the Sprague-Dawley rat. Nature 226:911-914.

Davavin, I.A., and V.Y. Yerokhin. 1979. Changes in biochemistry of some aquatic organisms of the Barents Sea coast in experimental oil intoxication. Hydrobiol. J. 14:63-66.

Davies, J.M., I.E. Baird, L.C. Massie, S.J. Hay, and A.P. Ward. 1980. Effects of oil derived hydrocarbons on the pelagic food web from observations in an enclosed ecosystem and a consideration of their implications for monitoring. Rapp. P.-v. Reun. Cons. Int. Explor. Mer 179:201-211.

Davies, J.M., R. Hardy, and A.D. McIntyre. 1981. Environmental effects of North Sea oil operations. Mar. Pollut. Bull. 12(12):412-416.

Davis, J.E., and S.S. Anderson. 1976. Effects of oil pollution on breeding grey seals. Mar. Pollut. Bull. 7:115-118.

Davis, J.L. 1949. Observations on the grey seal (Halichoerus grypus) at Ramsey Island, Pembrokeshire. Proc. Zool. Soc. London 119:673-692.

Davis, P.H., and R.B. Spies. 1980. Infaunal benthos of a natural petroleum seep: study of community structure. Mar. Biol. 59:31-41.

Dean, B.J. 1978. Genetic toxicology of benzene, toluene, xylenes and phenols. Mut. Res. 47:75-97.

DeAngelis, E., and G.G. Giordano. 1974. Sea urchin egg development under the action of benzo(a)pyrene and 7,12-dimethylbenz(a)anthracene. Cancer Res. 34:1275-1280.

Den Hartog, C., and R.P.W.M. Jacobs. 1980. Effects of the Amoco Cadiz oil spill on an eel grass community at Roscoff (France) with special reference to the mobile benthic fauna. Helgolander Meeresunters. 33:182-191.

Dennis, J.V. 1959. Oil Pollution Survey of the U.S. Atlantic Coast. American Petroleum Institute, Washington, D.C.

Department of Environment. 1977. Selected bibliography of fate and effects of oil pollution relevant to the Canadian marine environment. Report EPS-3-EC-77-23. Environmental Protection Service, Victoria, B.C.

Derenbach, J.B., and M.V. Gereck. 1980. Interference of petroleum hydrocarbons with the sex pheromone reaction of Fucus vesiculosus (L.). J. Exp. Mar. Biol. Ecol. 44:61-65.

Derenbach, J.B., W. Boland, E. Folster, and D.G. Muller. 1980. Interference tests with the pheromone system of the brown alga Cutleria multifida. Mar. Ecol. Program. Ser. 3:357-361.

Desaunay, Y. 1981. Evolution des stocks des poissons plats dans la zone contamin e par l'Amoco Cadiz, pp. 727-736. In Amoco Cadiz: Fates and Effects of the Oil Spill. CNEXO, Paris.

DeVries, A.L. 1979. The effect of naphthalene on synthesis of peptide antifreeze in the Bering Sea sculpin Myoxocephalus verrucosus, pp. 53-68. In W.B. Vernberg et al., eds. Marine Pollution--Functional Responses. Academic Press, New York.

Diaz-Piferrer, M. 1964. The effects of an oil spill on the shore of Guanica, Puerto Rico (abstract). Deep Sea Res. 11(5):855-856.

Dicks, B. 1975. Offshore biological monitoring, pp. 325-440. In J.M. Baker, ed. Marine Ecology and Oil Pollution. Applied Science Publishers, Barking, Essex.

Dicks, B., and K. Iball. 1981. Ten years of saltmarsh monitoring--the case history of a Southampton water saltmarsh and a changing refinery effluent discharge, pp. 361-374. In Proceedings, 1981 Oil Spill Conference. API Publication 4334. American Petroleum Institute, Washington, D.C.

Dillon, T.M. 1981. Effects of dimethylnaphthalene and fluctuating temperatures on estuarine shrimp, pp. 79-85. In Proceedings, 1981 Oil Spill Conference. API Publication 4334. American Petroleum Institute, Washington, D.C.

DiMichelle, L., and M.H. Taylor. 1978. Histopathological and physiological responses of Fundulus heteroclitus to naphthalene exposure. J. Fish. Res. Board Can. 35:1060-1066.

Dixon, D.R. 1982. Aneuploidy in mussel embryos (Mytilus edulis L.) originating from a polluted dock. Mar. Biol. Lett. 3:155-161.

Doe, K.G., and P.G. Wells. 1978. Acute aquatic toxicity and dispersing effectiveness of oil spill dispersants: Results of Canadian oil dispersant testing program (1973-1977), pp. 50-65. In L.T. McCarthy, Jr., G.P. Lindblom, and H.F. Walter, eds. Chemical Dispersants for the Control of Oil Spills. ASTM Special Technical Publication 659. American Society for Testing and Materials, Philadelphia, Pa.

Donahue, W.H., R.T. Wang, M. Welch, and J.A.C. Nicol. 1977. Effects of water-soluble components of petroleum oils and aromatic hydrocarbons on barnacle larvae. Environ. Pollut. 13:187-202.

Dowty, B.J., J.W. Brown, F.N Stone, J. Lake, and J.L. Laseter. 1981. GC-MS analysis of volatile organics from atmospheres impacted by the Amoco Cadiz oil spill, pp. 13-22. In Amoco Cadiz: Fate and Effects of the Oil Spill. CNEXO, Paris.

Dronenburg, R.B., G.M. Carroll, D.J. Rugh, and W.M. Marquette. 1983. Report of the 1982 spring bowhead whale census and harvest monitoring including 1981 fall harvest results. Rept. Internatl. Whaling Commission 33:525:537.

Dunbar, M.J. 1968. Ecological Development of Polar Regions. Prentice-Hall, Englewood Cliffs, N.J. 199 pp.

Dunbar, M.J. 1977. Are arctic ecosystems really as fragile as everyone assumes? Sci. Forum 10:26-29.

Dunn, B.P. 1980. Benzo(a)pyrene in the marine environment: analytical techniques and results. In A. Bjorseth, ed. Polynuclear Aromatic Hydrocarbons: Fourth International Symposium on Analysis, Chemistry and Biology. 367 pp.

Dunn, B.P., and H.F. Stich. 1975. The use of mussels in estimating benzo(a)pyrene contamination of the marine environment. Proc. Soc. Exper. Biol. Med. 150(1):49-51.

Dunn, B.P., and H.F. Stich. 1976a. Monitoring procedures for chemical carcinogens in coastal waters. J. Fish. Res. Board Canada 33(9):2040-2046.

Dunn, B.P., and H.F. Stich. 1976b. Release of carcinogenic benzo(a)pyrene from environmentally contaminated mussels. Bull. Environ. Contam. Toxicol. 14:398-401.

Dunn, B.P., and D.R. Young. 1976. Baseline levels of benzo(a)pyrene in southern California mussels. Mar. Pollut. Bull. 7:231-234.

Eldridge, M.B., T. Echeverria, and J.A. Whipple. 1977. Energetics of Pacific herring (Clupea harengus pallasi) embryos and larvae exposed to low concentrations of benzene, a monoaromatic component of crude oil. Trans. Am. Fish. Soc. 106:452-461.

Eldridge, M.B., T. Echeverria, and S. Korn. 1978. Fate of ^{14}C-benzene in Pacific herring (Clupea harengus pallasi) eggs and larvae. J. Fish Res. Board Can. 35:861-865.

Elmgren, R., and J.B. Frithsen. 1982. The use of experimental ecosystems for evaluating the environmental impact of pollutants: a comparison of an oil spill in the Baltic Sea and two long-term, low-level oil addition experiments in mesocosms, 153-165. In G.D. Grice and M. Reeve, eds. Marine Ecosystems: Biological and Chemical Research in Experimental Ecosystems. Springer-Verlag, New York.

Elmgren, R., G.A. Vargo, G.A. Grassle, J.F. Grassle, J.P. Heinle, P.R. Langlois, and S.L. Vargo. 1980a. Trophic interactions in experimental marine ecosystems, perturbed by oil, pp. 779-800. In J.P. Giesy, ed. Microcosms in Ecological Research. DOE Symposium Series 781101. National Technical Information Service, Springfield, Va.

Elmgren, R., S. Hansson, U. Larsson, and B. Sundelin. 1980b. Impact of oil on deep soft bottoms, pp. 97-126. In J.J. Kineman, R. Elmgren, and S. Hansson, eds. The Tsesis oil spill. National Oceanic and Atmospheric Administration, Boulder, Colo.

Elmgren, R., S. Hansson, U. Larsson, and B. Sundelin 1981. The Tsesis oil spill; acute and long-term impact on the benthic ecosystem.

Elmgren, R., S. Hansson, U. Larsson, B. Sundelin, and P.D. Boehm. 1983. The _Tsesis_ oil spill: Acute and long-term impact on the benthos. Mar. Biol. 73:51-65

El-Sadek, L.M. 1972. Mitotic inhibition and chromosomal aberrations induced by some arylarsonic acids and its compounds in root-tips of maize. Egypt. J. Genet. Cytol. 1:218-224.

Engelhardt, F.R. 1978. Petroleum hydrocarbons in arctic ringed seals, _Phoca hispida_, following experimental oil exposure, pp. 614-628. In Proceedings, Conference on Assessment of Ecological Impacts of Oil Spills. American Institute of Biological Sciences, Arlington, Va.

Engelhardt, F.R. 1981. Oil pollution in polar bears: exposure and clinical effects, pp. 139-179. In Proceedings, Fourth Arctic Marine Oil Spill. Program Technical Seminar. Edmonton, Alberta.

Engelhardt, F.R. 1982. Hydrocarbon metabolism and cortisol balance in oil-exposed ringed seal, _Phoca hispida_. Comp. Biochem. Physiol. 72C:133-136.

Engelhardt, F.R. 1983. Petroleum effects on marine mammals. Aquat. Toxicol. 1:175-186.

Engelhardt, F.R., and J.M. Ferguson. 1980. Adaptive hormone changes in harp seals, _Phoca groenlandica_, and gray seals, _Halichoerus grypus_, during the postnatal period. Gen. Comp. Endocrinol. 40:434-445.

Engelhardt, F.R., J.R. Geraci, and T.G. Smith. 1977. Uptake and clearance of petroleum hydrocarbons in the ringed seal, _Phoca hispida_. J. Fish. Res. Board Can. 34:1143-1147.

Engelhardt, F.R., M.P. Wong, and M.E. Duey. 1981. Hydro-mineral balance and gill morphology in rainbow trout, _Salmo gairdneri_, acclimated to fresh and sea water, as affected by petroleum exposure. Aquat. Toxicol. 1:175-186.

Environmental Protection Service. 1977. Selected Bibliography of Fate and Effects of Oil Pollution Relevant to the Canadian Marine Environment. EPS-3-EC-77-23. Department of Environment, Victoria, B.C.

Epler, J., J. Young, A. Hardigree, T. Rao, M. Guerin, I. Rubin, C. Ho, and B. Clark. 1978. Analytical and biological analysis of test materials from the synthetic fuel technologies. I. Mutagenicity of crude oil determined by the _Salmonella typhimurium_/microsomal activation. Mut. Res. 57:265-276.

Eppley, R.W., and C.S. Weiler. 1979. The dominance of nanoplankton as an indicator of marine pollution: a critique. Oceanol. Acta 2:241-245.

Ernst, V.V., J.M. Neff, and J.W. Anderson. 1977. The effects of the water-soluble fractions of No. 2 fuel oil on the early development of the estuarine fish, _Fundulus grandis_, Baird and Girard. Environ. Pollut. 14:25-35.

Falk, H.L., P. Kotin, S.K. Lee, and A.J. Nathan. 1962. Intermediary metabolism of benzo(a)pyrene in the rat. J. Nat. Cancer Inst. 28:699.

Falk-Petersen, I.-B., L.J. Saethre, and S. Lonning. 1982. Toxic effects of naphthalene and methylnaphthalenes on marine plankton organisms. Sarsia 67:171-178.

Farrington, J.W., E.D. Goldberg, R.W. Risebrough, J.H. Martin, and V.T. Bowen. 1983. U.S. "Mussel Watch" 1976-1978: An overview of the trace-metal, DDE, PCB, hydrocarbon, and artificial radionuclide data. Environ. Sci. Technol. 17:490-496.

Feder, H.M., L.M. Cheek, P. Flanagan, S.C. Jewitt, M.H. Johnston, A.S. Naidu, and S.A. Norell. 1976. The sediment environment of Port Valdez, Alaska: the effect of oil on this ecosystem. U.S. Environmental Protection Agency Report No. EPA-600/3-76-086. Corvallis Environmental Research Lab., Corvallis, Oreg. 282 pp.

Federle, T.W., J.R. Vestal, G.R. Hater, and M.C. Miller. 1979. Effects of Prudhoe Bay crude on primary production and zooplankton in arctic tundra thaw ponds. Mar. Environ. Res. 2:3-18.

Fingerman, S.W. 1980. Differences in the effects of fuel oil, an oil dispersant, and three polychlorinated byphenyls on fin regeneration in the Gulf Coast killifish, Fundulus grandis. Bull. Environ. Contam. Toxicol. 25:134-240.

Fischer, P.J. 1978. Natural gas and oil seeps, Santa Barbara Basin, California, pp. 1-62. In California Offshore Gas, Oil and Tar Seeps. California State Lands Commission, Sacramento, Calif.

Fishelson, L. 1973. Ecology of coral reefs in the Gulf of Aquaba (Red Sea) influenced by pollution. Oecologia 12:55-67.

Fishelson, L. 1977. Stability and instability of marine ecosystems, illustrated by examples from the Red Sea. Helgolander Wiss. Meeresunters. 30:18-29.

Fisheries Research Board. 1978. Recovery potential oiled marine northern environments. J. Fish. Res. Board Can. 35(5):499-795.

Fletcher, G.L., J.W. Kiceniuk, M.J. King, and J.F. Payne. 1979. Reduction of blood plasma copper concentrations in a marine fish following a six-month exposure to crude oil. Bull. Environ. Contam. Toxicol. 22:548-551.

Fletcher, G.L., J.W. Kiceniuk, and U.P. Williams 1981. Effects of oiled sediments on mortality, feeding and growth of winter flounder Pseudopleuronectes americanus. Mar. Ecol. Program Ser. 4:91-96.

Food and Agricultural Organization of the United Nations. 1977. Impact of oil on the marine environment. Reports and Studies No. 6. IMCO/FAO/UNESCO/WMO/WHO/IAEA/UN Joint Group of Experts on Scientific Aspects of Marine Pollution (GESAMP). United Nations, Rome.

Forbes, C. 1980. Sex-linked lethal mutations induced in Drosophila melanogaster by 7,12-dimethylbenz(a)anthracene. Mut. Res. 79:231-237.

Fraker, M.A., D.E. Sergeant, and W. Hoek 1978. Bowhead and white whales in the Southern Beaufort Sea. Beaufort Sea Project Technical Report No. 4. Department of the Environment, Victoria, B.C.

Friha, M., and G. Conan. 1981. Long-term impact of hydrocarbon pollution from the Amoco Cadiz on the mortality of plaice (Pleuronectes platessa) in the Aber Benoit estuary, p. 55. ICES Contribution C.M. 1981/E. International Council for the Exploration of the Seas, Charlottenlund, Denmark.

Galt, J.A. 1981. Transport, distribution, and physical characteristics of the oil. I. Offshore movement and distribution pp. 13-39. In C.

Hooper, ed. The _Ixtoc_ I Oil Spill: The Federal Scientific Response." U.S. Department of Commerce, National Oceanic and Atmospheric Administration, Office of Marine Pollution Assessment, Washington, D.C.

Ganning, B., and U. Billing 1974. Effects on community metabolism of oil and chemically dispersed oil on Baltic bladder wrack, _Fucus vesiculosis_, pp. 53-61. In L.R. Beynon and E.B. Cowell, eds. Ecological Aspects of Toxicity Testing of Oils and Dispersants. Proceedings, Workshop at the Institute of Petroleum, London. John Wiley & Sons, New York.

Gardner, G.R. 1975. Chemically induced lesions in estuarine or marine teleosts, pp. 657-693. In W.E. Ribelin and G. Migaki, eds. The Pathology of Fishes. University Wisconsin Press, Madison.

Gardner, G.R., P.P. Yevich, and P.F. Rogerson. 1975. Morphological anomalies in adult oyster, scallop, and Atlantic silversides exposed to waste motor oil, pp. 473-477. In Proceedings, 1977 Conference on the Prevention and Control of Oil Pollution. American Petroleum Institute, Washington, D.C.

Gearing, J.N., P.J. Gearing, T. Wade, J.G. Quinn, H.B. McCarty, J. Farrington, and R.F. Lee. 1979. The rates of transport and fates of petroleum hydrocarbons in a controlled marine ecosystem, and a note on analytical variability, pp. 555-564. In Proceedings, 1979 Oil Spill Conference. API Publication 4308. American Petroleum Institute, Washington, D.C.

George, R.Y. 1977. Dissimilar and similar trends in antarctic and arctic marine benthos, pp. 391-408. In M.J. Dunbar, ed. Polar Oceans. Arctic Institute of North America. Calgary, Alberta.

Geraci, J.R., and T.G. Smith. 1976. Direct and indirect effects of oil on ringed seals (_Phoca hispida_) of the Beaufort Sea. J. Fish. Res. Board Can. 33:1976-1984.

Geraci, J.R., and T.G. Smith. 1977. Consequences of oil fouling on marine mammals, pp. 399-410. In D.C. Malins, ed. Effects of Petroleum on Arctic and Subarctic Marine Environments and Organisms. Vol. 2, Biological Effects. Academic Press, New York.

Geraci, J.R., and D.J. St. Aubin. 1980. Offshore petroleum resource development and marine mammals: a review and research recommendations. Mar. Fish. Res. 42:1-12.

Geraci, J.R., and D.J. St. Aubin. 1982. Study of the effects of oil on cetaceans. Contract AA-551-CT9-29. U.S. Department of the Interior, Bureau of Land Management, Washington, D.C. 274 pp.

Geraci, J.R., D.J. St. Aubin, and R.J. Reisman. 1983. Bottlenose dolphins, _Tursiops truncatus_, can detect oil. Can. J. Fish. Aquat. Sci. 40(9).

Gerber, G.B., A. Leonard, and P. Jacquet. 1980. Toxicity, mutagenicity, and teratogenicity of lead. Mut. Res. 76:115-141.

Gerner-Smidt, P., and V. Friedrich. 1978. The mutagenic effect of benzene, toluene and xylene studied by the SCE technique. Mut. Res. 58:313-316.

Getter, C.D., S.C. Snedaker, and M.S Brown. 1980. Assessment of biological damages at _Howard Star_ oilspill site, Hillsborough Bay and Tampa Bay, Florida. Florida Department of Natural Resouces, Tallahassee. 65 pp.

Getter, C.D., G.I. Scott, and J. Michel. 1981. The effects of oil spills on mangrove forests: a comparison of five oil spill sites in the Gulf of Mexico and the Caribbean Sea, pp. 535-540. In Proceedings, 1981 Oil Spill Conference. API Publication 4334. American Petroleum Institute, Washington, D.C.

Geyer, R.A., and C.P. Giammona 1980. Naturally occurring hydrocarbons in the Gulf of Mexico and Carribean Sea, pp. 37-106. In R. Geyer ed. Marine Environmental Pollution. I. Hydrocarbons. Elsevier, New York.

Giammona, C.P. 1980. Biota near natural marine hydrocarbon seeps in the western Gulf of Mexico, pp. 207-228. In R. Geyer, ed. Marine Environmental Pollution. I. Hydrocarbons. Elsevier, New York.

Giesy, J.P., Jr., ed. 1980. Microcosms in Ecological Research. U.S. TIC Conference 781101.

Gilchrist, C.A., A. Lvnes, G. Steel, and B.T. Whitham 1972. The determination of polycyclic aromatic hydrocarbons in mineral oils by thin-layer chromatography and mass spectrometry. Analyst 97:880-888.

Giles, R.C., L.R. Brown, and C.D. Minchew 1978. Bacteriological aspects of fin erosion in mullet exposed to crude oil. J. Fish. Biol. 13:113-117.

Gilfillan, E.S. 1975. Decrease of net carbon flux in two species of mussels caused by extracts of crude oil. Mar. Biol. 29:53-57.

Gilfillan, E.S., and J.H. Vandermeulen. 1978. Alterations in growth and physiology of soft shell clams, *Mya arenaria*, chronically oiled with Bunker C from Chedabucto Bay, Nova Scotia, 1970-76. J. Fish. Res. Board Can. 35:630-636.

Gilfillan, E.S., D. Mayo, S. Hanson, D. Donovan, and L.C. Jiang 1976. Reduction in carbon flux in *Mya arenaria* caused by a spill of No. 6 fuel oil. Mar. Biol. 37:115-123.

Gilfillan, E.S., S.A. Hanson, D.S. Page, D. Mayo, S. Cooley, J. Chalfant, A. West, and J.C. Harshbarger. 1977a. A chemical, biological and histopathological examination of the Searsport oil spill site. Final Report Contract 906439. To State of Maine Department of Environmental Protection, Augusta. 192 pp.

Gilfillan, E.S., D.W. Mayo, D.S. Page, D. Donovan, and S.A. Hanson 1977b. Effects of varying concentrations of hydrocarbons in sediments on carbon flux in *Mya arenaria*, pp. 299-314. In F.S. Vernburg, ed. Physiological Responses of Marine Biota to Pollutants. Academic Press, New York.

Gilfillan, E.S., D.S. Page, R.P. Gerber, S. Hanson, J. Cooley, and J. Hothan. 1981. Fate of the *Zoe Colocotronis* oil spill and its effects on infaunal communities associated with mangroves, pp. 353-360. In Proceedings, 1981 Oil Spill Conference. API Publication 4334. American Petroleum Institute, Washington, D.C.

Gilfillan, E.S., J.H. Vandermeulen, and S. Hanson. 1984. Feeding, respiration and N-excretion of adult *Calanus hyperboreus* from Baffin Bay including from oil-seep contaminated waters. Arctic (submitted).

Glaeser, J.L. 1971. A discussion of future oil spill problems in the
Arctic, pp. 479-484. In Proceedings, Joint Conference on Prevention
and Control of Oil Spills. American Petroleum Institute,
Washington, D.C.

Glaeser, J.L., and G.P. Vance 1971. A Study of the Behavior of Oil
Spills in the Arctic. U.S. Coast Guard, Applied Technology
Division. NTIS AD717-142. National Technical Information Service,
Springfield, Va. 55 pp.

Glemarec, M. 1981. Note de synthèse, pp. 293-302. In Amoco Cadiz: Fate
and Effects Of the Oil Spill. CNEXO, Paris.

Glemarec, M., and C. Hily 1981. Perturbations apportees a la
macrofaune benthique de la baie de Concarneau par les effluents
urbains et portuaires. Acta Oecologica Decologia App. 2:139-150.

Glemarec, M., and E. Hussenot 1981. Definition d'une succession
ecologique en milieu meuble anormalement enrichi en matières
organiques à la suite de la catastrophe de l'Amoco Cadiz, pp.
499-512. In Amoco Cadiz: Fates and Effects of the Oil Spill. CNEXO,
Paris.

Glemarec, M. and E. Hussenot 1982. A three-year ecological survey in
Benoit and Wrachabers following the Amoco Cadiz oil spill.
Netherlands J. Sea Res. 16:483-490.

Glemarec, M., C. Hily, E. Hussenot, C. LeGall, and Y. Le Moal. 1980.
Recherches sur les indicateurs biologiques en milieu sedimentaire
marin. Colloque: Recherches sur les indicateurs biologiques.
Association française des Ingenieurs Ecologues, Grenoble, XI-80.

Golley, F.B., H.T. Odum, and R.F. Wilson. 1962. The structure and
metabolism of a Puerto Rican red mangrove forest in May. Ecology
43:9-19.

Goodale, D.R., M.A.M. Hyman, and H.E. Winn. 1981. Cetacean responses
in association with Regal Sword oil spill, Chapter XI, pp. XI-1 to
XI-15. In R.K. Edel, M.A. Hyman, and M.F. Tyrell, eds. A
Characterization of Marine Mammals and Turtles in the Mid and North
Atlantic Areas of the U.S. Outer Continental Shelf. Cetacean and
Turtle Assessment Program Annual Report 1979. University of Rhode
Island.

Gooding, R.M. 1971. Oil pollution on Wake Island from the tanker R.C.
Stoner. Spec. Sci. Rep. U.S. Fish Wildlife Serv. (Fish.) 636:1-10.

Gordon, D.C., and P. Michalik. 1971. Concentration of Bunker C fuel
oils in waters of Chedabucto Bay, April 1971. J. Fish. Res. Board
Can. 28:1912-1914.

Gordon, D.C., Jr., and N.J. Prouse. 1973. The effects of three oils on
marine phytoplankton photosynthesis. Mar. Biol. 22:329-333.

Gordon, D.C., Jr., P.D. Keizer, W.R. Hardstaff, and D.G. Aldous. 1976.
Fates of crude oil spilled on seawater contained in outdoor tanks.
Environ. Sci. Technol. 10(6):580-585.

Gordon, D.C., Jr., J. Dale, and P.D. Keizer 1978. Importance of
sediment working by the deposit-feeding polychaete Arenicola marina
on the weathering rate of sediment-bound oil. J. Fish. Res. Board
Can. 35:591-603.

Gorman, M.L., and C.E. Simms. 1978. Lack of effect of ingested Forties
Field crude oil on avian growth. Mar. Pollut. Bull. 9(10):273-276.

Gorsline, J., W.N. Homes, and J. Cronshaw. 1981. The effects of ingested petroleum on the napthalene metabolizing properties of liver tissue in sea water adapted mallard ducks (Anas platyrhynchos). Environ. Res. 24(2):377-390.

Graf, W., and C. Winter 1968. 3,4-benzpyrene. Erdol. Arch. Hyg. Bakteriol. 152:289-293.

Grassle, J.F., R. Elmgren, and J.P. Grassle 1981. Response of benthic communities in MERL experimental ecosystems to low level, chronic additions of No. 2 fuel oil. Mar. Environ. Res. 4:279-297.

Grau, C.R., T. Roudybush, J. Dobbs, and J. Wathen. 1977. Altered yolk structure and reduced hatchability of eggs from birds fed single doses of petroleum oils. Science 195:779-781.

Griffiths, R.P., T.M. McNamara, B.A. Caldwell, and R.Y. Morita. 1981. Field observations on the acute effect of crude oil on glucose and glutamate uptake in samples collected from arctic and subarctic waters. Appl. Environ. Microbiol. 41:1400-1406.

Griffiths, R.P., B.A. Caldwell, W.A. Broich, and R.Y. Morita. 1982a. Long-term effects of crude oil on microbial processes in subarctic marine sediments. Studies on sediments amended with organic nutrients. Mar. Pollut. Bull. 13:273-278.

Griffiths, R.P., B.A. Caldwell, W.A. Broich, and R.Y. Morita. 1982b. The long-term effects of crude oil on microbial processes in subarctic marine sediments. Estuarine Coastal Shelf Sci. 15:183-198.

Grose, P.L., and J.S. Mattson. 1977. The Argo Merchant oil spill: a preliminary scientific report. NOAA Special Report. NTIS PB-267-505. National Technical Information Service, Springfield, Va. 133 pp.

Group of Experts on the Scientific Aspects of Marine Pollution. 1977. Impact of oil on the marine environment. Reports and Studies No. 6. Food and Agricultural Organization of the United Nations, Rome. 250 pp.

Guerin, M.R., I.B. Rubin, T.K. Rao, B.R. Clark, and J.L. Epler. 1981. Distribution of mutagenic activity in petroleum and petroleum substitutes. Fuel 60:282-288.

Gundlach, E.R., and K.J. Finkelstein. 1982. Transport, distribution and physical characteristics of the oil. II. Nearshore movement and distribution, pp. 41-73. In C. Hooper, ed. The Ixtoc I Oil Spill: The Federal Scientific Response". U.S. Department of Commerce, National Oceanic and Atmospheric Administration, Office of Marine Pollution Assessment, Washington, D.C.,

Gundlach, E.R., and M.O. Hayes. 1978. Vulnerability of coastal environments to oil spill impacts. Mar. Technol. Soc. J. 12:18-27.

Gundlach, E.R., J. Michel, G.I. Scott, M.O. Hayes, C.D. Getter, and W.P. Davis. 1979. Ecological assessment of the Peck Slip (19 December 1978) oil spill in eastern Puerto Rico, pp. 303-317. In Proceedings, Ecological Damage Assessment Conference. Society of Petroleum Industry Biologists, Arlington, Va.

Gundlach, E.R., S. Berne, L. d'Ozouville, and J.A. Topinka. 1981. Shoreline oil two years after Amoco Cadiz: new complications from Tanio, pp. 525-540. In Proceedings, 1981 Oil Spill Conference. API Publication 4334. American Petroleum Institute, Washington, D.C.

Gundlach, E.R., D.D. Domeracki, and L.C. Thebeau. 1982. Persistence of _Metula_ oil in the Strait of Magellan six and one-half years after the incident. Oil Petrochem. Pollut. (in press).

Hada, H.S., and R.S. Sizemore. 1981. Incidence of plasmids in marine _Vibrio_ spp. isolated from an oil field in the northwestern Gulf of Mexico. Appl. Environ. Microbiol. 41:199-202.

Haegh, T., and L.I. Rossemyr. 1980. A comparison of weathering processes of oil from the _Bravo_ and the _Ixtoc_ blowouts, pp. 237-244. In Proceedings, 12th Annual Offshore Technology Conference, Houston, Tex. Paper OTC 3702.

Haensly, W.E., J.M. Neff, J.R. Sharp, A.C. Morris, M.F. Bedgood, and P.D. Boehm. 1981. Histophathology of _Pleuronectes platessa_ from Aber Wrach'h and Aber Benoit, Brittany, France: long-term effects of the _Amoco Cadiz_ crude oil spill (unpublished manuscript.)

Hall, Ch.A.S., R. Howarth, B. Moore, and Ch.J. Vorosarty. 1978. Environmental impacts of industrial energy systems in the coastal zone. Ann. Rev. Energy 3:395-475.

Hampson, G.R., and E.T. Moull. 1978. No. 2 fuel oil spill in Bourne, Massachusetts. Immediate assessment of the effects on marine invertebrates and a 3-year study of growth and recovery of a salt marsh. J. Fish. Res. Board Can. 35(5):731-744.

Hampson, G.R., and H.L. Sanders. 1969. Local oil spill. Oceanus 15:8-11.

Harris, M.P., and S. Murray 1981. Monitoring of puffin numbers at Scottish colonies. Bird Study 28:15-20.

Hartung, R. 1967. Energy metabolism in oil-covered ducks. J. Wildlife Manage. 31:798-804.

Hawkes, A.L. 1961. A review of the nature and extent of damage caused by oil pollution at sea. Trans. N. Am. Wildlife Natur. Resour. Conf. 26:343-355.

Hawkes, J.W. 1977. The effects of petroleum hydrocarbon exposure on the structure of fish tissues, pp. 115-128. In D.A. Wolfe, ed. Fate and Effects of Petroleum Hydrocarbons in Marine Organisms and Ecosystems. Pergamon, New York.

Hawkes, J.W., and C.M. Stehr 1981. Cytopathology of the brain and retina of embryonic surf smelt (_Hypomesus pretiosus_) exposed to crude oil. Environ. Res. (in press).

Hawkes, J.W., E.H. Gruger, Jr., and O.P. Olson. 1980. Effects of petroleum hydrocarbons and chlorinated biphenyls on the morphology of the intestine of chinook salmon (_Oncorhynchus tshawytscha_) Environ. Res. 23:149-161.

Hayes, T.M. 1977. Sinking of tanker St. Peter off Columbia, pp. 289-291. In Proceedings, 1977 Oil Spill Conference. API Publication 4284. American Petroleum Institute, Washington, D.C.

Hennemuth, R.C., J.E. Palmer, and B.E. Brown 1980. A statistical description of recruitment in eighteen selected fish stocks. J. Northwest Atlantic Fish. Sci. 1:101-111.

Hershner, C.H., Jr. 1977. Effects of petroleum hydrocarbons on salt marsh communities. Ph.D. thesis, University of Virginia, Charlottesville. 172 pp.

Hershner, C., and J. Lake. 1980. Effects of chronic oil pollution on a salt marsh grass community. Mar. Biol. 56:163-173.

Hershner, C., and K. Moore. 1979. Effects of a Chesapeake Bay oil spill on salt marshes of the lower bay, pp. 529-534. In Proceedings, 1979 Oil Spill Conference. API Publication 4308. American Petroleum Institute, Washington, D.C.

Herwig, R.P. 1978. Bacterial response to crude oil spillage in a salt marsh. M.Sc. thesis. College of William and Mary, Williamsburg, Va.

Hess, R., and L. Trobaugh. 1970. Kodiak Islands oil pollution. Event No. 26-70, pp. 150-153. Annual Report Smithsonian Institute, Center Shortlived Phenomena, Washington, D.C.

Hess, W.N., ed. 1978. The Amoco Cadiz oil spill; A preliminary scientific report. NOAA/EPA Special Report. NTIS PB-285-805. National Technical Information Service, Springfield, Va. 355 pp.

Ho, C.H., B.R. Clark, M.R. Guerin, B.D. Barkenbus, T.K. Row, and J.L. Ebler. 1981. Analytical and biological analysis of test materials from the snythetic fuel technologies. IV. Studies of chemical-mutagenic activity relationships of aromatic nitrogen compounds relevant to synfuels. Mut. Res.

Hoffman, E.J., and J.G. Quinn. 1979. Gas chromatographic analyses of Argo Merchant oil and sediment hydrocarbons at the wreck site. Mar. Pollut. Bull. 10:20-24.

Hoffman, E., and J.G. Quinn. 1980. The Argo Merchant oil spill and the sediments of Nantucket Shoals: research, litigation and legislation, pp. 185-218. In R.A. Baker, ed. Contaminants and Sediments, Vol. I. Ann Arbor Science Publishers, Ann Arbor, Mich.

Hogstedt, B., B. Gullberg, E. Mark-Vendel, F. Mitelman, and S. Skerfving. 1981. Micronuclei and chromosome aberrations in bone marrow cells and lymphocytes of humans exposed mainly to petroleum vapors. Hereditas 94(2):179-187.

Holland, J.M., R.O. Rahn, L.H. Smith, B.R. Clark, S.S. Chang, and T.J. Stephens. 1979. Skin carcinogenicity of synthetic and natural petroleums. J. Occup. Med. 21(9):614-618.

Hollister, T.A., G.S. Ward, and P.R. Parrish. 1980. Acute toxicity of a No. 6 fuel oil to marine organisms. Bull. Environ. Contam. Toxicol. 24:656-661.

Hollstein, J., J. McCann, F.A. Angelosanto, and W.W. Nichols. 1979. Short-term tests for carcinogens and mutagens. Mut. Res. 65:133-226.

Holmes, W.N. 1975. Hormones and osmoregulation in marine birds. Gen. Comp. Endocrinol. 25:249-258.

Holmes, W.N., and J. Cronshaw. 1977. Biological effects of petroleum on marine birds, pp. 359-398. In D.C. Malins, ed. effects of Petroleum on Arctic and Subarctic Marine Environments and Organisms Vol. 2, Biological Effects. Academic Press, New York.

Holmes, W.N., J. Cronshaw, and J. Gorsline. 1978. Some effects of ingested petroleum on seawater-adapted ducks (Anas platyrhynchos). Environ. Res. 17(2):177-190.

Holt, S., S. Rabalais, N. Rabalais, S. Cornelius, and J.S. Holland. 1978. Effects of an oil spill on salt marshes at Harbor Island, Texas. 1. Biology, pp. 345-352. In Conference on Assessment of Ecological Impacts of Oil Spills. American Institute of Biological Sciences, Arlington, Va.

Hooftman, R.N. 1981. The induction of chromosomal aberrations in Notobranchius rachow (Pisces: Cyprinodonditae) after treatment with ethylmethane sulhponate or benzo(a)pyrene. Mut. Res. 91:352-374.

Hooftman, R.N., and G.J. Vink. 1981. Cytogenetic effects on the eastern mud-minnow, Umbra pygmaea, exposed to ethyl methanesulphonate, benzo(a)pyrene and river water. Ecotoxicol. Environ. Safety (in press).

Hope-Jones, P., J.Y. Monnat, C.J. Cadbury, and T.J. Stowe. 1978. Birds oiled during the Amoco Cadiz incident--an interim report. Mar. Pollut. Bull. 9:307-310.

Horner, R.A. 1977. History and recent advances in the study of ice biota, pp. 269-283. In M.J. Dunbar, ed. Polar Oceans. Arctic Institute of North America, Calgary, Alberta.

Hose, J.E., J.B. Hannah, D. DiJulio, M.L Landolt, B.S. Miller, W.T. Iwaoka, and S.P. Felton. 1982. Effects of benzo(a)pyrene on early development of flatfish. Arch. Environ. Contam. Toxicol. 11:161-171.

Hose, J.E., H.W. Puffer, P.S. Oshida, and S.M. Bay. In press. Development and cytogenetic abnormalities induced in the purple sea urchin by benzo(a)pyrene. Arch. Environ. Contam. Toxicol.

Howgate, P., P.R. Mackie, K.J. Whittle, J. Farmer, A.D. McIntyre and A. Eleftheriou. 1977. Petroleum tainting in fish. Rapp. P.-v. Reun. Cons. Int. Explor. Mer 171:143-146.

Hsiao, S.I.C. 1978. Effects of crude oils on the growth of arctic marine phytoplankton. Environ. Pollut. 17:93-107.

Hsiao, S.I.C., D.W. Kittle, and M.G. Fox. 1978. Effects of crude oil and the oil dispersant Corexit on primary production of arctic marine phytoplankton and seaweed. Environ. Pollut. 15:209-221.

Huggins, C., L.C. Grand, and F.P. Brulantes. 1961. Mammary cancer induced by a single feeding of polynuclear hydrocarbons, and its suppression. Nature 189:204-207.

Hughes, D. 1974. Sygna slick. Mar. Pollut. Bull. 5:99.

Hurst, R.J., N.A. Oritsland, and P.D. Watts. 1982. Metabolic and temperature responses of polar bears to crude oil, pp. 263-280. In P.J. Rand, ed. Land and Water Issues in Resource Development. Ann Arbor Science Publishers, Ann Arbor, Mich.

Hutchinson, T.C., J.A. Hellebust, D. Mackay, D. Tamm, and P.B. Kauss. 1979. Relation of hydrocarbon solubility to toxicity and cellular membrane effects, pp. 541-547. In Proceedings, 1979 Oil Spill Conference. API Publication 4308. American Petroleum Institute, Washington, D.C.

Hutzinger, O., I.H. van Lelyveld, and B.C.J. Zoeteman, eds. Aquatic Pollutants: Transformation and Biological Effects. Proceedings, 2nd International Symposium on Aquatic Pollutants, pp. 483-486. Pergamon, New York.

International Agency for Research on Cancer. 1973. Monographs on the Evaluation of Carcinogenic Risk of the Chemical to Man, p. 271. Certain Polycyclic Aromatic Hydrocarbons and Heterocyclic Compounds, Vol. 3. World Health Organization, Lyon.

Ishio, S., T. Yaio, and H. Nakagana. 1971. Algal cancer and causal substances in wastes from the coal chemical industry. In S.H. Jenkins, ed. Proceedings, 5th International Conference, San

Francisco and Hawaii. Vol. II. Adv. Water Pollut. Res. 2(III):18/1--18/8.

Ishio, S., T. Yaio, and H. Nakagana. 1973. Cancerous disease of *Porphyra tenera* and its causes, p. 373. In K. Nisizawa ed. Proceedings, 7th International Seaweek Symposium. John Wiley & Sons, New York.

Iwaoko, W.T., M.L. Landolt, K.B. Pierson, S.P. Felton, and A. Abolins. 1977. Studies on aryl hydrocarbon hydroxylase, polycyclic hydrocarbon content, and epidermal tumors of flatfish, pp. 85-93. In Symposium on Pathobiology of Environmental Pollution: Animals as Monitors of Environmental Pollutants. National Academy of Sciences, Washington, D.C.

Jackson, L., T. Bidleman, and W. Vernberg. 1981. Influence of reproductive activity on toxicity of petroleum hydrocarbons to ghost crabs. Mar. Pollut. Bull. 12:63-65.

Jacobson, S.M., and D.B. Boylan. 1973. Effect of seawater soluble fraction of kerosene on chemotaxis in a marine snail, *Nassarius obsoletus*. Nature 241:213-215.

Jernelov, A., and O. Linden. 1980. The effects of oil pollution on mangroves and fisheries in Ecuador, Columbia. Program of the Second International Symposia on Biology and Management of Mangroves and Tropical Shallow Water Communities, Papua, New Guinea. 31 pp.

Jernelov, A., O. Linden, and J. Rosenblum. 1976. The St. Peter Oil Spill--An Ecological and Socio-Economic Study of Effects. Publication B334. Swedish Water and Air Pollution Research Institute, Stockholm. 34 pp.

Johannes, R.E. 1975. Pollution and degradation of coral reef communities, pp. 13-51. In E.G. Ferguson Wood and R.E. Johannes, eds. Tropical Marine Pollution. Elsevier Oceanography Series 12. Elsevier, New York.

Johannes, R.E., J. Maragos, and S.L. Coles. 1972. Oil damaged corals exposed to air. Mar. Pollut. Bull. 3(2):29-30.

Johansson, S., U. Larsson, and P. Boehm. 1980. The *Tsesis* oil spill impact on the pelagic ecosystem. Mar. Pollut. Bull. 11:284-293.

Johnson, A.G., T.D. Williams, J.F. Messinger III, and C.R. Arnold. 1979. Larval spotted sea trout *Cynoscion nebulosus*: a bioassay subject for the marine subtropics. Contrib. Mar. Sci. 22:57-62.

Johnson, S.R., and W.J. Richardson. 1981. Beaufort Sea barrier island-lagoon ecological process studies: Final report. Simpson Lagoon. Part 3, Birds, pp. 109-383. In Environmental Assessment of the Alaskan Continental Shelf. Final Rep. Biol. Stud., Vol. 7. NOAA, Boulder, Colo.

Johnson, S.R., and W.J. Richardson. 1982. Waterbird migration near the Yukon and Alaskan coast of the Beaufort Sea. II. Moult migration of seaducks in summer. Arctic 35(2):291-301.

Jones, G.F. 1969. The benthic macrofauna of the mainland shelf of Southern California. Allan Hancock Monograph. Mar. Biol. 4:1-219.

Jones, R. 1982. Population fluctuations and recruitment in marine populations. Philos. Trans. Roy. Soc. London, Ser. B 297:353-368.

Jordan, D.C., J.P. McNicol, and R.M. Marshall. 1978. Biological nitrogen fixation in the terrestrial environment of a high arctic ecosystem (Truelove Lowland).

Kaas, R. 1981. Evolution des peuplements algaux exploitables depuis le naufrage de l'Amoco Cadiz, pp. 687-702. In Amoco Cadiz: Fate and Effects of the Oil Spill. CNEXO, Paris.

Kajiwara, T., K. Kodana, and A. Hatanaka. 1980. Male attracting substance in a marine brown alga Sargassum horneri. Naturwiss. 67:612-613.

Karydis, M. 1979. Short term effects of hydrocarbons on the photosynthesis and respiration of some phytoplankton species. Bot. Mar. 22:281-285.

Kato, R. 1968. Chromosome breakage induced by a carcinogenic hydrocarbon in Chinese hamster cells and human leukocytes in vitro. Hereditas 59:120-141.

Kato, R., M. Bruze, and Y. Tegner. 1969. Chromosome breakage induced in vivo by a carcinogenic hydrocarbon in bone marrow cells of the Chinese hamster. Hereditas 61:1-8.

Kator, H., and R. Herwig. 1977. Microbial responses after two experimental oil spills in an eastern coastal plain estuarine ecosystem, pp. 517-522. In Proceedings, 1977 Oil Spill Conference. API Publicaton 4284. American Petroleum Institute, Washington, D.C.

Keevil, B.E., and R.O. Ramseier. 1975. Behavior of oil spilled under floating ice, pp. 497-501. In Proceedings, 1975 Conference on Prevention and Control of Oil Pollution. American Petroleum Institute, Washington, D.C.

Keizer, P.D., T.P. Ahern, J. Dale, and J.H. Vandermeulen. 1978. Residues of Bunker C oil in Chedabucto Bay, Nova Scotia, 6 years after the Arrow spill. J. Fish Res. Board Can. 35:528-535.

King, H., and R.J. Lunford. 1950. The relation between the constitution of arsenicals and their action on cell division. J. Chem. Soc. 8:2086-2088.

King, J.E. 1964. Seals of the world, pp. 124-125. In Trustees of the British Museum (Nat. Hist.). London.

King, P.J. 1977. An assessment of the potential carcinogenic hazard of petroleum hydrocarbons in the marine environment. Rapp. P.-v. Reun. Cons. Int. Explor. Mer 171:202-211.

Kitahara, Y., H. Okuda, K. Shudo, T. Okamoto, M. Nagao, Y. Seino, and T. Sugimura. 1978. Synthesis and mutagenicity of 10-azabenzo(a)pyrene, 4,5-oxide and other pentacyclic aza-arene oxides. Chem. Pharmacol. Bull. 26:1950-1953.

Kittredge, J.S., F.T. Takahashi, and F.O. Sarinana. 1974. Bioassays indicative of some sublethal effects of oil pollution, pp. 891-897. In Proceedings, Marine Technology Society, Washington, D.C.

Kjorsvik, E., L.J. Saethre, and S. Lonning. 1982. Effects of short-term exposure to xylenes on the early cleavage stages of cod eggs (Gadus morhua L.). Sarsia 67:299-308.

Knap, A.H., T.D. Sleeter, R.E. Dodge, S.C. Wyers, H.R. Frith, and S.R. Smith. 1983. The effects of oil spills and dispersant use on corals. Oil Petrochem. Pollut. 1(3):157-169.

Knox, G.A., and J.K. Lowry. 1977. A comparison between benthos of the southern ocean and north polar ocean with special reference to the amphipods and polychaeta, pp. 423-462. In M.J. Dunbar, ed. Polar Oceans. Arctic Institute of North America, Calgary, Alberta.

Kocan, R.M., M.L. Landolt, and K.M. Sabo. 1979. In vitro toxicity of eight mutagens/carcinogens for three fishcell lines. Bull. Environ. Contam. Toxicol. 23:269-274.

Kocan, R.M., M.L. Landolt, J. Bond, and E.P. Benditt. 1981. In vitro effects of some mutagens/carcinogens on cultured fish cells. Arch. Environ. Contam. Toxicol. 10:663-671.

Kocan, R.M., M.L. Landolt, and K.M. Sabo. 1982. Anaphase aberrations: a measure of genotoxicity in mutagen-treated fish cells. Environ. Mut. 4:181-187.

Kochert, G. 1978. Sexual pheromones in algae and fungi. Ann. Rev. Plant Physiol. 29:471-486.

Kooyman, G.L., R.L. Gentry, and W.B. McAlister. 1976. Physiological impact of oil on pinnipeds, pp. 3-26. In Environmental Assessment of the Alaskan Continental Shelf, Vol. 3. Environmental Research Labs., Boulder, Colo.

Kooyman, G.L., R.W. Davis, and M.A. Castellini. 1977. Thermal conductance of immersed pinniped and sea otter pelts before and after oiling with Prudhoe Bay crude, pp. 151-157. In D.A. Wolfe ed. Fate and Effects of Petroleum Hydrocarbons in Marine Organisms and Ecosystems. Pergamon, New York.

Korn, S., D.A. Moles, and S.D. Rice. 1979. Effects of temperature on the median tolerance limit of pink salmon and shrimp exposed to toluene, naphthalene, and Cook Inlet crude oil. Bull. Environ. Contam. Toxicol. 21:521-525.

Koski, W.R., and W.J. Richardson. 1976. Review of waterbird deterrent and dispersal systems for oil spills. PACE Report 76-6. Petroleum Association for Conservation of the Canandian Environment, Ottawa, Ontario. 122 pp.

Koster, A.S., and J.A.M. Vanden Biggelaar. 1980. Abnormal development of Dentalium due to the Amoco Cadiz oil spill. Mar. Pollut. Bull. 11(6):166-169.

Kovaleva, G.I. 1979. Effect of dissolved petroleum products on carbohydrate metabolism of the liver of two Black Sea species of fishes. Water Toxicol. UDC 597.05-11:578:085.2:615.9:48-52.

Krebs, C.T., and K.A. Burns. 1977. Long term effects of an oil spill on populations of the salt marsh crab Uca pugnax. Science 197(4302):484-487.

Kress, S.W. 1980. Egg Rock update. Newsletter of the Fratercula Fund. National Audubon Society, 1980 Report.

Kuhnhold, W.W. 1972. The influence of crude oils on fish fry, pp. 315-318. In M. Ruivo, ed. Marine Pollution and Sea Life. Food and Agricultural Organization of the United Nations, Fishing News Ltd., Surrey, England.

Kuhnhold, W.W. 1974. Investigations on the toxicity of seawater-extracts of three crude oils on eggs of cod (Gadus morhua L.). Ber. Deut. Wiss. Komm. Meeresforsch. 23:165-180.

Kuhnhold, W.W. 1977. The effect of mineral oils on the development of eggs and larvae of marine species. A review and comparison of experimental data in regard to possible damage at sea. Rapp. P.-v. Reun. Cons. Int. Explo. Mer 171:175-183.

Kuhnhold, W.W. 1978. Effects of the water soluble fraction of a Venezuelan heavy fuel oil (No. 6) on cod eggs and larvae, pp. 126-130. In M.P. Wilson, J.G. Quinn, and K. Sherman, eds. In the Wake of the Argo Merchant. Center for Ocean Management Studies, University of Rhode Island, Kingston.

Kuhnhold, W.W., D. Everich, J.J. Stegeman, J. Lake, and R.E. Wolke. 1978. Effects of low levels of hydrocarbons on embryonic, larval and adult winter flounder, Pseudopleuronectes americanus, pp. 677-711. In Proceedings, Conference on Assessment of Ecological Impacts of Oil Spills. NTIS AD-A072 859. National Technical Information Service, Springfield, Va.

Kurita, Y., T. Sugiyama, and Y. Nishizuka. 1969. Chromosome aberrations induced in rat bone marrow cells by 7,12-dimethyl benz(a)anthracene. J. Natl. Cancer Inst. 43:635-641.

Kusk, K.O. 1978. Effects of crude oil and aromatic hydrocarbons on the photosynthesis of the diatom Nitzschia palea. Physiol. Plant. 43:1-6.

Lacaze, J.C. 1974. Ecotoxicology of crude oils and the use of experimental marine ecosystems. Mar. Pollut. Bull. 5:153-156.

Lacaze, J.C., and O. Villedon de Naide. 1976. Influence of illumination on phytotoxicity of crude oil. Mar. Pollut. Bull. 7:73-76.

Lake, J.L., and C. Hershner. 1977. Petroleum sulphur-containing compounds and aromatic hydrocarbons in the marine molluscs Modiolus demissus and Crassostrea virginica, pp. 627-632. In Proceedings, 1977 Oil Spill Conference. API Publication 4284. American Petroleum Institute, Washington, D.C.

Lanier, J.J., and M. Light. 1978. Ciliates as bioindicators of oil pollution. Proceedings, Conference on Assessment of Ecological Impacts of Oil Spills, pp. 651-676. NTIS AD-A072 859. National Technical Information Service, Springfield, Va.

Laseter, J.L., G.C. Lawler, E.B. Overton, J.R. Patel, J.P. Holmes, M.I. Shields, and M. Maberry. 1981. Characterization of aliphatic and aromatic hydrocarbons in flat and Japanese type oysters and adjacent sediments collected from l'Aber Wrac'h following the Amoco Cadiz oil spill, pp. 633-644. In G. Conan et al., eds. Amoco Cadiz: consequences d'une pollution accidentelle par les hydrocarbures. CNEXO, Paris.

Laughlin, R.B., Jr., L.G.L. Young, and J.M. Neff. 1978. A long-term study of the effects of water-soluble fractions of No. 2 fuel oil on the survival, development rate and growth of the mud crab Rhithropanopeus harrisii. Mar. Biol. 47:87-95.

Law, R.J. 1981. Hydrocarbon concentrations in water and sediments from marine waters, determined by fluorescence spectroscopy. Mar. Pollut. Bull. 12:153-157.

Lawler, G.C., J.P. Holmes, D.M. Adamkiewica, M.I. Shields, J.-Y. Monnat, and J.L. Laseter. 1981. Characterization of petroleum hydrocarbons in tissues of birds killed in the Amoco Cadiz oil spill. In G. Conan et al., eds. Amoco Cadiz: consequences d'une pollution accidentelle par les hydrocarbures. CNEXO, Paris.

Le Boeuf, B. 1971. Oil contamination and elephant seal mortality: a "negative" finding, pp. 277-281. In D. Straugham ed. Biological and Oceanographic Survey of the Santa Barbara Channel Oil Spill, 1969-1970, Vol. 1. Alan Hancock Foundation, Sea Grant Publication 2. University of Southern California, Los Angeles.

Lee, R.F. 1975. Fate of petroleum hydrocarbons in marine zooplankton, pp. 549-553. In Proceedings, 1975 Conference on the Prevention and Control of Oil Pollution. American Petroleum Institute, Washington, D.C.

Lee, R.F. 1981. Mixed function oxygenases (MFO) in marine invertebrates. Mar. Biol. Lett. 2:87-105. 859.

Lee, R.F., and J.W. Anderson. 1977. Fate and effect of naphthalenes in controlled ecosystem enclosures. Bull. Mar. Sci. 27:127-134.

Lee, R.F., and C. Ryan. 1983. Microbial and photochemical degradation of polycyclic aromatic hydrocarbons in estuarine waters and sediments. Can. J. Fish. Aquat. Sci. 40(Suppl. 2).

Lee, R.F., and S.C. Singer. 1980. Detoxifying enzyme system in marine polychaetes: increases in activity after exposure to aromatic hydrocarbons. Rapp. P.-v. Reun. Cons. Int. Explor. Mer 179:29-32.

Lee, R.F., and M. Takahashi. 1977. The fate and effect of petroleum in controlled ecosystem enclosures. Rapp. P.-v. Reun. Cons. Int. Explor. Mer 171:150-156.

Lee, R.F., M. Takahashi, J.H. Beers, W.H. Thomas, D.R.L. Seibert, P. Koeller, and D.R. Green. 1977. Controlled ecosystems: their use in the study of the effects of petroleum hydrocarbons on plankton, pp. 323-342. In F.J. Vernberg, A. Calabrese, F.P. Thurberg, and W.B. Vernburg, eds. Physiological Responses of Marine Biota to Pollutants. Academic Press, New York.

Lee, R.F., W.S. Gardner, J.W. Anderson, J.W. Blaylock, and J. Barwell-Clarke. 1978a. Fate of polycyclic aromatic hydrocarbons in controlled ecosystem enclosures. Environ. Sci. Technol. 12:323-338.

Lee, R.F., M. Takahashi, and J. Beers. 1978b. Short term effects of oil on plankton in controlled ecosystems, pp. 634-650. In Proceedings, Conference on Assessment of Ecological Impacts of Oil Spills, NTIS AD-A072. National Technical Information Service, Springfield, Va.

Lee, W.Y., and J.A.C. Nicol. 1977. The effects of the water soluble fractions of No. 2 fuel oil on the survival and behavior of coastal and oceanic zooplankton. Environ. Pollut. 12:279-292.

Lee, W.Y., and J.A.C. Nicol. 1978. Individual and combined toxicity of some petroleum hydrocarbons to the marine amphipod, Elasmopus pectenicrus. Mar. Biol. 48:215-222.

Lee, W.Y., S.A. Macko, and J.A.C. Nicol. 1981. Changes in nesting behavior and lipid content of a marine amphipod (Amphithoe valida) to the toxicity of a No. 2 fuel oil. Water Air Soil Pollut. 15:185-195.

Leighton, F.A., D.B. Peakall, and R.G. Butler. 1983. Heinz-body hemolytic anemia from the ingestion of crude oil: A primary toxic effect in marine birds. Science 220:871-873.

Leitch, A. 1922. Paraffin cancer and its experimental production.
Br. Med. J. ii:1104-1106.

LeMoal, Y. 1981. Ecologie dynamique des plages touchees par la maree
noire de l'Amoco Cadiz. These de 3eme cycle. Universite de Bretagne
Occidentale, Brest, France. 130 pp.

LeMoal, Y. 1982. Ecologie dynamique de plages sableuses perturbes
initialement par les marees noires de l'Amoco Cadiz et du Tanio.
Rapport de Contrat CNEXO-MECV 81/6596. Institut d'Etudes
Marines--Universite de Bretagne Occidentale, Brest, France. 84 pp.

LeMoal Y. and M. Quillien-Monot. 1981. Etude des populations de la
macrofaune et de leurs juveniles sur les plages des Abers Benoit et
Wrac'h, pp. 311-326. In Amoco Cadiz: Fate and Effects of the Oil
Spill. CNEXO, Paris.

Leonard, A., and R.R. Lauwerys. 1980. Carcinogenicity, teratogenicity
and mutagenicity of arsenic. Mut. Res. 75:49-62.

Leong, B.K. 1977. Experimental benzene intoxication, pp. 45-61. In S.
Laskin and B. Goldstein, eds. Benzene Toxicity - A Critical
Evaluation. J. Toxicol Environ. Health, Suppl. 2.

Le Pemp, X., J.C. LaCaze, and O. Villedon de Naide. 1976. Toxicite
d'un petrole brut vis-a-vis de la production primaire du
phytoplacton de l'estuaire de la Rance: influence de la
temperature et du temps de contact. Bull. Mus. Natl. d'Hist. Natur.
Ecol. Gen. 32:107-110.

Leppakoski, E. 1973. Effects of an oil spill in the Northern Baltic.
Mar. Pollut. Bull. 4:93-94.

Levasseur, J.E., and M.-L. Jory. 1982. Retablissement naturel d'une
vegetation de marais maritimes alteree par les hydrocarbures de
l'Amoco Cadiz: modalites et tendances. In Ecological Study of the
Amoco Cadiz Oil Spill, pp. 329-362. NOAA-CNEXO Joint Scientific
Communication. CNEXO, Paris

Levitan, W.M., and M.H. Taylor. 1979. Physiology of
salinity-dependent naphthalene toxicity in Fundulus heteroclitus.
J. Fish Res. Board Can. 36:615-620.

Levy, E.M. 1971. The presence of petroleum residue off the East Coast
of Nova Scota, in the Gulf of St. Lawrence and the St. Lawrence
River. Water Res. 5:723-733.

Levy, E.M. 1979. Further chemical evidence for natural seepage on the
Baffin Island shelf. Current Research, Part B. Geol. Surv. Can.
Pap. 79-1B:379-383.

Levy, E.M. 1981. Background levels of petroleum residues in Baffin Bay
and the eastern canadian Arctic: role of natural seepage, pp.
346-362. In Petromar 80. Petroleum in the Marine Environment,
Eurocean, Graham & Trotman Ltd. London

Levy, E.M., and M. Ehrhardt 1981. Natural seepage of petroleum at
Buchan Gulf, Baffin Island. Mar. Chem. 10:355-364.

Levy, E.M., and B. MacLean. 1981. Natural hydrocarbon seepage at Scott
Inlet and Buchan Gulf, Baffin Island Shelf: 1980 update. Sci. and
Tech. Notes Pap. 81-1A. In Current Research Part A. Geol. Surv. of
Can.

Lewbel, G.S., R.L. Howard, and S.W. Anderson. 1982. Biological
assessment, Section 4. In Ixtoc Oil Spill Assessment Final Report.

Bureau of Land Management, U.S. Department of the Interior, Washington, D.C.

Lewis, E.L. 1979. Some possible effects of arctic industrial development on the marine environment, pp. 369-392. In Proceedings, Port and Ocean Engineering Under Arctic Conditions Conference. Norwegian Institute of Technology.

Lewis, J.B. 1971. Effects of crude oil and an oil spill dispersant on reef corals. Mar. Pollut. Bull. 2:59-62.

Lewis, R.R. 1979a. Oil and mangrove forests: the aftermath of the Howard Star oil spill. Fla. Sci. 42(Suppl.):26.

Lewis, R.R. 1979b. Large scale mangrove restoration on St. Croix, U.S. Virgin Islands, pp. 231-242. In D.P. Cole, ed. Proceedings, Sixth Annual Conference on Restoration and Creation of Wetlands. Hillsborough Community College, Tampa, Fla.

Lewis, R.R. 1980a. Oil and mangrove forests: observed impacts 12 months after the Howard Star oil spill. Fla. Sci. 43(Suppl.):23.

Lewis R.R. 1980b. Impact of oil spills on mangrove forests. Second International Symposia on the Biology and Management of Mangroves and Tropical Shallow Water Communities, 20 July-2 Aug. 1980, Port Moresby, Madang. Papau, New Guinea. 36 pp.

Lewis, R.R., and K.C. Haines. 1980. Large scale mangrove restoration on St. Croix, U.S. Virgin Islands. II. Second Year. In Proceedings, Seventh Annual Conference on Restoration and Creation of Wetlands. Hillsborough Community College, Tampa, Fla. 294 pp.

Lillie, H. 1954. Comments and discussion, pp. 31-33. In Proceedings, International Conference on Oil Pollution of the Sea. London.

Linden, O. 1975. Acute effects of oil and oil/dispersant mixtures on larvae of Baltic herring. Ambio 4:130-133.

Linden, O. 1976a. Effects of oil on the amphipod Gammarus oceanicus. Environ. Pollut. 10:239-250.

Linden, O. 1976b. Effects of oil on the reproduction of the amphipod Gammarus oceanicus. L. Ambio 5(1):36-37.

Linden, O. 1976c. The influence of crude oil and mixtures of crude oil/dispersants on the ontogenic development of the Baltic herring, Clupea harengus membras L. Ambio 5(3):136-140.

Linden, O. 1978. Biological effects of oil on early development of the Baltic herring Clupea harengus membras. Mar. Biol. 45(3):273-283.

Linden, O., J.R. Sharp, R. Laughlin, Jr., and J.M. Neff. 1979. Interactive effects of salinity, temperature and chronic exposure to oil on the survival and development rate of embryos of the estuarine killifish Fundulus heteroclitus. Mar. Biol. 51:101-109.

Linden, O., R. Laughlin, Jr., J.R. Sharp, and J.M. Neff. 1980. The combined effect of salinity, temperature and oil on the growth pattern of embryos of the killifish, Fundulus heteroclitus Walbaum. Mar. Environ. Res. 3:129-144.

Lindstedt-Siva, J. 1979. Ecological impacts of oil spill cleanup: Are they significant?, pp. 521-523. In Proceedings, 1979 Oil Spill Conference. API Publication 4308. American Petroleum Institute, Washington, D.C.

Lindstedt-Siva, J. 1980. Minimizing the ecological impacts of oil spills. Environ. Internatl. 3:185-188.

Lindstedt-Siva, J. 1981. Preparing for oil spills: Planning for optimum use of limited resources, pp. 103-112. In Petromar 80. Petroleum and the Marine Environment. Eurocean, Graham & Trotman Ltd., London.

Lissauer, I., and P. Welsh. 1978. Can oil spill movement be predicted?, pp. 22-27. In M.P. Wilson, J.G. Quinn, and K. Sherman, eds. In the Wake of the Argo Merchant. Center for Ocean Management Studies, University of Rhode Island, Kingston.

Littlepage, J.L. 1964. Seasonal variation in lipid content of two antarctic marine crustacea, pp. 463-470. In R. Carrick, M. Holdgate, and J. Prévost, eds. Biologie Antarctique. Hermann, Paris.

Lo, M.-T., and E. Sandi. 1978. Polycyclic aromatic hydrocarbons (polynuclears) in foods, pp. 35-65. In F.A. Gunther and J.D. Gunther, eds. Residue Reviews, Vol. 69. Springer-Verlag, New York.

Loncarevic, B.D., and R.K. Falconer. 1977. An oil slick off Baffin Island. Report of Activities. Geol. Serv. Can. Res. Pap. 77-1A:523-524.

Long, B.F.N., and J.H. Vandermeulen. 1979. Impact of cleanup efforts on an oiled saltmarsh in North Brittany. Spill Technol. Newslett. 4(4):218-229.

Long, B.F.N., and J.H. Vandermeulen. 1983. Geomorphological impact of clean-up of an oiled salt-marsh (Ile Grande). In Proceedings, 1983 Oil Spill Conference. American Petroleum Institute, Washington, D.C. In press.

Long, B.F.N., J.H. Vandermeulen, and T.P. Ahern. 1981a. The evolution of stranded oil within sandy beaches, pp. 519-524. In Proceedings, 1981 Oil Spill Conference. API Publication 4334. American Petroleum Institute, Washington, D.C.

Long, B.F.N., J.H. Vandermeulen, and L. D'Ozouville. 1981b. Geomorphological alteration of a heavily oiled salt marsh (Ile Grande, France) as a result of massive cleanup, pp. 347-352. In Proceedings, 1981 Oil Spill Conference. EPA/API/USCG.

Longhurst, A., ed. 1982. Consultation on the consequences of offshore oil production on offshore fish stocks and fishing operations. Can. Tech. Rep. Fish. Aquat. Sci. 1096:95.

Longwell, A.C. 1977. A genetic look at fish eggs and oil. Oceanus 20:45.

Longwell, A.C. 1978. Field and laboratory measurements of stress responses at the chromosome and cell levels in planktonic fish eggs and the oil problem, pp. 116-125. In M.P. Wilson, J.G. Quinn, and K. Sherman, eds. In the Wake of the Argo Merchant. Center for Ocean Management, University of Rhode Island, Kingston.

Longwell, A.C., and J.B. Hughes. 1980. Cytologic, cytogenic, and developmental state of Atlantic mackerel eggs from sea surface water of the New York Bight, and prospects for biological effects monitoring with ichthyoplankton. Rapp. P.-v. Reun. Cons. Int. Explor. Mer 179:275-291.

Lonning, S. 1977a. The sea urcchin as a test object in oil pollution studies. Rapp. P.-v. Reun. Cons. Int. Explor. Mer 171:186.

Lonning, S. 1977b. The effects of crude Ekofisk oil and oil products on marine fish larvae. Astarte 10:37-47.

Lonning, S., and I.-B. Falk-Petersen. 1982. Electron microscopial studies of the effects of aromatic hydrocarbons on sea urchin embryos. Sarsia 67:149-155.

Lopez, E., J. Leloup-Hatey, A. Hardy, F. Lallier, E. Martelly, P.-D. Oudot, and Y.A. Fontaine. 1981. Modifications histopathologiques et stress chez des anguilles soumises a une exposition prolongée aux hydrocarbures, pp. 645-653. In G. Conan et al., eds. Amoco Cadiz: Consequences d'une pollution accidentelle par les hydrocarbures. CNEXO, Paris.

Lopez, J.M. 1978. Ecological consequences of petroleum spillage in the coastal waters of Puerto Rico, pp. 895-908. In Proceedings, Conference on Assessment of Ecological Impacts of Oil Spills. NTIS AD-A072 859. National Technical Information Service, Springfield, Va.

Lowe, D.M., M.N. Moore, and K.R. Clarke. 1981. Effects of oil on digestive cells in mussels: quantitative alterations in cellular and lysosomal structure. Aquat. Toxicol. 1:213-226.

Loya, Y. 1975. Possible effects of water pollution on the community structure of Red Sea corals. Mar. Biol. 29:177-185.

Loya, Y. 1976. Recolonization of Red Sea corals affected by natural catastrophes and man-made perturbations. Ecology 57(2):278-289.

Loya, Y., and B. Rinkevich. 1979. Abortion effect in corals induced by oil pollution. Mar. Ecol. Program Ser. 1(1): 77-80.

Loya, Y., and B. Rinkevich. 1980. Effects of oil pollution on coral reef communities. Mar. Ecol. Program Ser. 3(2):167-180.

Lyon, J.P. 1975. Mutagenicity studies with benzene. Ph.D. thesis. University of California.

Lytle, J.S. 1975. Fate and effects of crude oil on an estuarine pond, pp. 595-600. In Proceedings, 1975 Conference on Prevention and Control of Oil Pollution. American Petroleum Institute, Washington, D.C.

MacDonald, B.A., and M.L.H. Thomas. 1982. Growth reduction in the soft-shell clam Mya arenaria from a heavily oiled lagoon in Chedabucto Bay, Nova Scotia. Mar. Environ. Res. 6:145-156.

Mackay, D. 1977. Oil in the Beaufort and Mediterranean seas. Arctic 30(2):93-100.

Mackie, P.R., A.S. McGill, and R. Hardy. 1972. Diesel oil contamination of brown trout (Salmo trutta L.). Environ. Pollut. 3:9-16.

Mackie, P.R., R. Hardy, and K.J. Whittle. 1978. Preliminary assessment of the presence of oil in the ecosystem at Ekofisk after the blowout, April 22-30, 1977. J. Fish. Res. Board Can. 35:544-551.

Mackin, J.G., and A.K. Sparks. 1962. A study of the effects on oysters of crude oil loss from a wild well. Publ. Inst. Mar. Sci. Univ. Tex. 7:230-261.

MacLean, B., R.K.H. Falconer, and E.M. Levy. 1981. Geological, geophysical, and chemical evidence for natural seepage of petroleum off the northeast coast of Baffin Island. Can. Pet. Geol. 29(1):75-95.

MacLeod, W.D., Jr., L.C. Thomas, M.Y. Uyeda, and R.G. Jenkins. 1978. Evidence of Argo Merchant cargo oil in marine biota by glass capillary GC analysis, pp. 137-151. In Proceedings, Conference on Assessment of Ecological Impacts of Oil Spills. NTIS AD-A072-859. National Technical Information Service, Springfield, Va.

MacNab, R.M. 1978. Bacterial motility and chemotaxis: the molecular biology of a behavioral system. CRC Crit. Rev. Biochem. 5:291-341.

Mahoney, B.M., and H.H. Haskin. 1980. The effects of petroleum hydrocarbons on the growth of phytoplankton recognized as food forms for the eastern oyster, Crassostrea virginica Gmelin. Environ. Pollut. 22:123-132.

Malins, D.C., eds. 1977. Effects of petroleum on Arctic and Subarctic Marine Environments and Organisms. Vol. I, Nature and Fate of Petroleum. 321 pp. Vol. II, Biological Effects. 500 pp. Academic Press, New York.

Malins, D.C. 1982. Alterations in the cellular and subcellular structure of marine telosts and invertebrates exposed to petroleum in the laboratory and field: a critical review. Can. J. Fish. Aquat. Sci. 39:877-889.

Malins, D.C., and T.K. Collier. 1981. Xenobiotic interacions in aquatic organisms: Effects on biological systems. Aquat. Toxicol. 1:257-268.

Malins, D.C., M.N. Krahn, D.W. Brown, W.D. MacLeod, Jr., and T.K. Collier. 1980. Analysis for petroleum products in marine environments. Helgolander Meeresunters. 33(1-4):257-271.

Marchand, M., and M.-P. Caprais. 1979. Suivi chimique de la pollution de l'Amoco Cadiz dans l'eau de mer et les sediments marins en Manche occidentale, Mars 1978-Mars 1979. Rapport Interne Centre Oceanologique de Bretagne. CNEXO. Paris. 103 pp.

Marchand, M., and M.-P. Caprais. 1981. Suivi de la pollution de l'Amoco Cadiz dans l'eau de mer et les sediments marins, pp. 23-54. In Amoco Cadiz: Fate and Effects of the Oil Spill. CNEXO, Paris.

Marquardt, H., and F.S. Philips. 1970. The effects of 7,12-dimethylbenz(a)anthracene on the synthesis of nucleic acids in rapidly dividing hepatic cells in rats. Cancer Res. 30:2000-2006.

Martin, S. 1979. A field study of brine drainage and oil entrapment in first year sea ice. J. Glaciol. 22(88):473-502.

Matsuoka, A., K. Shudo, Y. Saito, T. Sofuni, and M. Ishidate, Jr. 1982. Clastogenic potential of heavy oil extracts and some aza-arenes in Chinese hamster cells in culture. Mut. Res. 102:275-283.

Maurin, Cl. 1981. Impact sur les resources expoitables. Note de synthese. Consequences de l'accident de l'Amoco Cadiz sur des resources marines vivantes exploitables, pp. 667-686. In G. Conan et al., eds. Amoco Cadiz: consequences d'une pollution accidentelle par les hydrocarbures. CNEXO, Paris.

Mayo, D.W., D.S. Page, J. Cooley, E. Sorenson, F. Bradley, E.S. Gilfillan, and S.A. Hanson. 1978. Weathering characteristics of petroleum hydrocarbons deposited in fine clay marine sediments, Searsport, Maine. J. Fish. Res. Board Can. 35(5):552-562.

McAuliffe, C.D. 1977. Dispersal and alteration of oil discharged on a water surface, pp. 19-35. In D.A. Wolfe, ed. Fate and Effects of Petroleum Hydrocarbons in Marine Organisms and Ecosystems. Pergamon, New York.

McAuliffe, C.D., B.L. Steeleman, W.R. Leek, D.E. Fitzgerald, J.P. Ray, and C.D. Barker. 1981a. The 1979 Southern California dispersant treated research oil spills, pp. 269-282. In Proceedings, 1981 Oil Spill Conference. API Publication 4334. American Petroleum Institute, Washington, D.C.

McAuliffe, C.D., G.P. Canevari, T.D. Searl, and J.C. Johnson. 1981b. The dispersion and weathering of chemically treated crude oils on the sea surface. In Petromar 80. Petroleum in the Marine Environment. Monaco. Graham & Trotman Ltd., London.

McCain, B.B., H.O. Hodgins, W.D. Gronlund, J.W. Hawkes, D.W. Brown, M.S. Myers, and J.H. Vandermeulen. 1978. Bioavailability of crude oil from experimentally oiled sediments to English sole (*Parophrys vetulus*), and pathological consequences. J. Fish. Res. Board Can. 35:657-664.

McEwan, E.H., and P.M. Whitehead. 1978. Influence of weathered crude oil on liver enzyme metabolism of testosterone in gulls. Can. J. Zool. 56:1922-1924.

McKeown, B.A., and G.L. Marsh. 1978. The acute effect of Bunker C oil and an oil dispersant on: serum glucose, serum sodium and gill morphology in both freshwater and seawater acclimated rainbow trout, *Salmo gairdneri*. Water Res. 12:157-163.

McLachlan, A., and B. Harty. 1981. Effects of oil on water filtration by exposed sandy beaches. Mar. Pollut. Bull. 12(11):374-378.

McLachlan, A., and B. Harty. 1982. Effects of crude oil on the supralittoral meiofauna of a sandy beach. Mar. Environ. Res. 7:71-79.

McMinn, T.J. 1972. Crude oil behavior on arctic winter ice. Report USCG-734108. U.S. Coast Guard, Washington, D.C. 76 pp.

McMinn, T.J., and P. Golden. 1973. Behavioral characteristics and cleanup techniques of North Slope crude oil in an arctic winter environment, pp. 263-276. In Proceedings, Joint Conference on Prevention and Control of Oil Spills. American Petroleum Institute, Washington, D.C.

Mead, C., and S. Baillie. 1981. Seabirds and oil: the worst winter. Nature 292:10-11.

Mecklenburg, T.A., S.D. Rice, and J.F. Karinen. 1977. Molting and survival of king crab (*Paralithodes camtschatica*) and coonstripe shrimp (*Pandalus hypsinotus*) larvae exposed to Cook Inlet crude oil water-soluble fractions, pp. 221-228. In D.A. Wolfe, ed. Fate and Effects of Petroleum Hydrocarbons in Marine Organisms and Ecosystems. Pergamon, New York.

Mecler, F.J., and R.P. Beliles. 1979a. Teratology study in rats. Kerosene. API Medical Research Publication 27-32175. American Petroleum Institute, Washington, D.C.

Mecler, F.J., and R.P. Beliles. 1979b. Teratology study in rats. Diesel Fuel. API Medical Research Publication 27-32174. American Petroleum Institute, Washington, D.C.

Meguro, H., K. Ito, and H. Fukushima. 1967. Ice flora (bottom type): a mechanism of primary production in polar seas and the growth of diatoms in sea ice. Arctic 20(2):114-133.

Meisch, H.U., and H. Benzschawel. 1978. The role of vanadium in green plants. III. Influence on cell division of _Chlorella_. Arch. Microbiol. 116:91-95.

Menez, J.R., F. Berthou, D. Picart, and C. Riche. 1978. Impacts of the oil spill (_Amoco Cadiz_) on human biology. Penn. Ar. Bed 94:367-378.

Menzel, R.W. 1947. Observations and conclusions on the oily tasting oysters near the tank battery of the Texas Co. in Crooked Bayou at Bay Ste. Elaine. Project 9. Texas A&M Research Foundation. 6 pp.

Menzel, R.W. 1948. Report on two cases of "oily tasting" oysters at Bay Ste. Elaine oilfield. Project 9. Texas A&M Research Foundation. 9 pp.

Michael, A.D. 1977. The effects of petroleum hydrocarbons on marine populations and communities, pp. 129-137. In D.A. Wolfe, ed. Fate and Effects of Petroleum Hydrocarbons in Marine Organisms and Ecosystems. Pergamon, New York.

Michael, J., M.O. Hayes, and P.J. Brown. 1978. Application of an oil spill index vulnerability index to the shoreline of lower Cook Inlet, Alaska. Environ. Geol. 2(2):107-117.

Middleditch, B.S, ed. 1981. Environmental effects of offshore oil production. The Buccaneer Gas and Oil Field Study. Plenum Press, New York. 446 pp.

Miller, D.S., D.B. Peakall, and W.B. Kinter. 1978. Ingestion of crude oil: sublethal effects in Herring Gull chicks. Science 199:315-317.

Milne, A.R. 1974. Use of artificial subice pockets by wild ringed seals (_Phoca hispida_). Can. J. Zool. 52:1092-1093.

Milne, A.R., and B.D. Smiley. 1976. Offshore drilling for oil in the Beaufort Sea: a preliminary environmental assessment. Beaufort Sea Project Technical Report 39. Department of Environment, Victoria, B.C. 43 pp.

Milne, A.R., and B.D. Smiley. 1978. Offshore drilling in Lancaster Sound: possible environmental hazards. Institute of Ocean Sciences, Department of Fisheries and Environment, Sydney, B.C. 95 pp.

Minchew, C.D., and J.D. Yarbrough. 1977. The occurrence of fin rot in mullet (_Mugil cephalus_) associated with crude oil contamination of an estuarine pond-ecosystem. J. Fish. Biol. 10:319-323.

Mironov, O.G. 1967. Effects of small concentrations of petroleum and petroleum products on the development of eggs of Black Sea flatfish. Vop. Ikhtiol. 7(3):577-580.

Mironov, O.G. 1968. Hyrocarbon pollution of the sea and its influence on marine organisms. Helgolander Wiss. Meeresunters. 17:335-339.

Mironov, O.G. 1972. Effect of oil pollution on flora and fauna of the Black Sea, pp. 222-224. In M. Ruivo, ed. Marine Pollution and Sea Life. Fishing News (Books) Ltd., London.

Mitchell, C.T., E.A. Anderson, L.J. Jones, and W.J. North. 1970. What oil does to ecology. J. Water Pollut. Control Fed. 42(5, part 1):812-818.

Mitchell, R., and I. Chet. 1975. Bacterial attack of corals in polluted seawater. Microbial. Ecol. 2:227-233.

Mitchell, R., and I. Chet. 1978. Indirect ecological effects of pollution, pp. 177-199. In R. Mitchell, ed. Water Pollution Microbiology, Vol. 2. John Wiley & Sons, New York.

Miyaki, M., I. Murata, M. Osabe, and T. Ono. 1977. Effect of metal cations on misincorporation by E. coli DNA polymerase. Biochem. Biophys. Res. Commun. 77:854-860.

Moles, A. 1980. Sensitivity of parasitized coho salmon fry to crude oil, toluene, and naphthalene. Trans. Am. Fish. Soc. 109:293-297.

Moles, A., and S.D. Rice. 1983. Effects of crude oil and naphthalene on growth, caloric content and fat content on pink salmon juveniles in seawater. Trans. Am. Fish. Soc. 112:205-211.

Moles, A., S.D. Rice, and S. Korn. 1979. Sensitivity of Alaskan freshwater and anadromous fishes to Prudhoe Bay crude oil and benzene. Trans. Am. Fish. Soc. 108:408-414.

Moles, A., S. Bates, S.D. Rice, and S. Korn. 1981. Reduced growth of coho salmon fry exposed to two petroleum components, toluene and naphthalene, in fresh water. Trans. Am. Fish. Soc. 110:430-436.

Mommaerts-Billiet, F. 1973. Growth and toxicity tests on the marine nanoplanktonic alga Platymonas tetrathele G.S. West in the presence of crude oil and emulsifiers. Environ. Pollut. 4:261-282.

Moore, M.N. 1979. Cellular responses to polycyclic aromatic hydrocarbons and phenobarbitol in Mytilus edulis. Mar. Environ. Res. 2:255-263.

Moore, M.N., D.M. Lowe, and P.E.M. Fieth. 1978. Lysosomal responses to experimentally injected anthracene in the digestive cells of Mytilus edulis. Mar. Biol. 48:297-302.

Moore, S.F., and Dwyer, R.L. 1974. Effects of oil on marine organisms: a critical assessment of published data. Water Res. 8:819-827.

Morgan, J.P., R.J. Menzies, S.Z. El-Sayed, and C.H. Oppenheimer. 1974. The offshore ecology investigation, final project planning council consensus report. Gulf Universities Research Consortium Report 138. Houston, Tex.

Morita, R.Y. 1975. Psychrophilic bacteria. Bacteriol. Rev. 39:144-167.

Morris, R. 1970. Alaska Peninsula oil spill. Event No. 36-70, pp. 154-157. Annual Report. Smithsonian Institute, Center for Short-lived Phenomena, Washington, D.C.

Motohiro, T., and H. Inoue. 1973. n-Paraffins in polluted fish by crude oil from Juliana wreck. Bull. Fac. Fish. Hokkaido Univ. 23:204-208.

Muller, D.G., G. Gassman, W. Boland, F. Marner, and L. Jaenicke. 1981. Dictyota dichotoma (Phaeophyceae): Identification of the sperm attractant. Science 212:1040-1041.

Muller-Willie, L. 1974. How effective is oil pollution legislation in arctic waters? Musk-Ox 14:56-57.

Murphy, L.S., and R.R.L. Guillard. 1976. Biochemical taxonomy of marine phytoplankton by electrophoresis of enzymes. I. The centric diatoms Thalassiosira pseudonana and T. fluviatilis. J. Phycol. 12:9-13.

Nadeau, R.J., and E.T. Bergquist. 1977. Effects of the March 18, 1973, oil spill near Cabo Rojo, Puerto Rico on tropical marine communities, pp. 535-538. In Proceedings, 1977 Oil Spill Conference. API Publication 4284. American Petroleum Institute, Washington, D.C.

Nakamuro, K., and Y. Sayoto. 1981. Comparative studies of chromosomal aberrations induced by trivalent and pentavalent arsenic. Mut. Res. 88:73-80.

National Oceanic and Atmospheric Administration. 1983. Assessing the Social Cost of Oil Spills: The Amoco Cadiz Case Study. Office of Ocean Resources Coordination and Assessment, U.S. Department of Commerce, Washington, D.C. 144 pp.

National Research Council. 1975. Petroleum in the Marine Environment. National Academy of Sciences, Washington, D.C. 107 pp.

Neff, J.M. 1979. Polycyclic Aromatic Hydrocarbons in the Aquatic Environment. Sources, Fates and Biological Effects. Applied Science Publishers, London.

Neff, J.M., J.W. Anderson, B.A. Cox, R.B. Laughlin, Jr., S.S. Rossi, and H.E. Tatem. 1976. Effects of petroleum on survival, respiration and growth of marine animals, pp. 515-539. In Sources, Effects and Sinks of Hydrocarbons in the Aquatic Environment. American Institute of Biological Sciences, Arlington, Va.

Nellbring, S., S. Hansson, G. Aneer, and L. Weshn. 1980. Impact of oil on local fish fauna, pp. 193-201. In Kineman, Elmgren, and Hansson, eds. The (Tsesis) Oil Spill. U.S. Department of Commerce, National Oceanic and Atmospheric Administration, Rockville, Md.

Nelson, H., K.A. Kvenvolden, and E.C. Clukeyo. 1978. Thermogenic gases in near surface sediments of Norton Sound, Alaska. OTC 3354. Offshore Technol. Conf. 6:2623-2633.

Nelson-Smith, A. 1970. The problem of oil pollution of the sea. Adv. Mar. Biol. 8:215-306.

Nelson-Smith, A. 1973. Oil Pollution and Marine Ecology. Plenum, New York.

Nicol, J.A.C., W.H. Donahue, R.T. Wang, and K. Winters. 1977. Chemical composition and effects of water extracts of petroleum on eggs of the sand dollar Melitta quininquiesperforata. Mar. Biol. 40:309-316.

Nishimura, M., and M. Umeda. 1979. Induction of chromosomal aberrations in cultured mammalian cells by nickel compounds. Mut. Res. 68:337-349.

Nitta, T., K. Arakawa, K. Okubo, T. Okubo, and K. Tabata. 1965. Studies on the problems of offensive odor in fish caused by wastes from petroleum industries. Bull. Tokai Region Fish. Res. Lab. 43:23-37.

NORCOR. 1975. The interaction of crude oil with arctic sea ice. Beaufort Sea Project Technical Report 27. Department of the Environment, Victoria, B.C. 145 pp.

Nordenson, I., G. Beckman, L. Beckman, and S. Nordström. 1978. Occupational and environmental risks in and around a smelter in northern Sweden. II. Chromosomal aberrations in workers exposed to arsenic. Hereditas 88:47-50.

Norderhaug, M., E. Brun, and G.U. Mollen. 1977. Barentshavets sjofuglressurser. Meddelelser 104:1-119.

North, W.J., M. Neushul, and K. Clendenning. 1964. Successive biological changes observed in a marine cove exposed to a large spillage of mineral oil, p. 335. In Symposium, Pollution of Marine Microorganisms by Products of Petroleum, Monaco.

Notini, M. 1978. Long-term effects of an oil spill on Fucus macrofauna in a small Baltic Bay. J. Fish. Res. Board Can. 35(5):745-753.

Novitsky, J.A., and P.G. Kepkay. 1981. Patterns of microbial heterotrophy through changing environments in a marine sediment. Mar. Ecol. Programs 4:1-7.

Nygren, A. 1949. Cytological studies of the effects of 2,4-D, MCPA, and 2,4,5-T on Allium cepa. Ann. R. Agric. Coll. (Sweden) 16:723-728.

O'Boyle, R.N. 1980. Distribution of oil, chlorophyll, and larval fish on the Scotian Shelf during April and May 1979 following the Kurdistan spill, pp. 167-192. In J.H. Vandermeulen, ed. Scientific Studies During the Kurdistan Tanker Incident: Proceedings of a Workshop. Report Series BI-R-80-3. pp. 167-192. Bedford Institute of Oceanography. Dartmouth, N.S.

O'Connor, J.M., J.B. Klotz, and Th. J. Kneip. 1981. Sources, sinks, and distribution of organic contaminants in the New York Bight ecosystem, pp. 631-653. In G.F. Mayer, ed. Ecological Stress and the New York Bight: Science and Management. Estuarine Research Foundation, Columbia, S.C.

Odum, W.E., and E.J. Heald. 1972. Trophic analyses of an estuarine mangrove community. Bull. Mar. Sci. 22:671-738.

Odum, W.E., and R.E. Johannes. 1975. The response of mangroves to man-induced environmental stress, pp. 52-62. In E.J.F. Wood and R.E. Johannes, eds. Tropical Marine Pollution. Elsevier Oceanography Series 12. Elsevier, New York.

Ogata, M., and Y. Miyake. 1973. Identification of substances in petroleum causing objectionable odour in fish. Water Res. 7:1493-1504.

Ogata, M., Y. Miyake, S. Kina, K. Matsunga, and M. Imanaka. 1977. Transfer to fish of petroleum paraffins and organic sulphur compunds. Water Res. 11:333-338.

Oil Spill Intelligence Report. 1980. Special Report: Ixtoc I. Vol. III, No. 1. 36 pp.

Oil Spill Intelligence Report. 1980a. Nigeria plans clean-up inspection following well blow-out. 7 March 1980. Cahners Publishing, Boston.

Oil Spill Intelligence Report. 1980b. Texaco supplies food to Nigerian villages. 18 April 1980. Cahners Publishing, Boston.

Oksama, M., and R. Kristoffersson. 1979. The toxicity of phenol to Phoxinus phoxinis, Gammarus duebeni and Mesidotea entomon in brackish water. Ann. Zool. Fenn. 16(3):209-216.

Olla, B.L. 1974. Behavioral measures of environmental stress, pp. 1-31. In Proceedings of a Workshop on Marine Bioassays. Marine Technological Society, Washington, D.C.

Olla, B.L., W.H. Pearson, and A.L. Studholme. 1980a. Applicability of behavioral measures in environmental stress assessment. Rapp. P.-v. Reun. Cons. Int. Explor. Mer 179:162-173.

Olla, B.L., J. Atema, R. Forward, J. Kittredge, R.J. Livingston, D.W. McLeese, D.C. Miller, W.B. Vernberg, P.G. Wells, and K. Wilson. 1980b. The role of behavior in marine pollution monitoring: behavior panel report. Rapp. P.-v. Reun. Cons. Int. Explor. Mer 179:174-181.

Oritsland, N.A., and K. Ronald. 1973. Effects of solar radiation and windchill on skin temperature of the harp seal, Pagophilus groenlandicus (Erxleben, 1777). Comp. Biochem. Physiol. 44:519-525.

Oritsland, N.A., F.R. Engelhardt, F.A. Juck, R. Hurst, and P.O. Watts. 1981. Effect of crude oil on polar bears. Department of Indian Affairs and Northern Development Canada Publication QS-8283-020-EE-AI. Canada Catalog No. R71-19/24-1981E. 268 pp.

O'Sullivan, A.J. 1978. The Amoco Cadiz oil spill. Mar. Pollut. Bull. 9(5):123-128.

Ott, F.S., R.P. Harris, and S.C.M. O'Hara. 1978. Acute and sublethal toxicity of naphthalene and three methylated derivatives to the estuarine copepod Eurytemora affinis. Mar. Environ. Res. 1:49-58.

Oviatt, C., J. Frithsen, J. Gearing, and P. Gearing. 1982. Low chronic additions of No. 2 fuel oil: chemical behavior, biological impact and recovery in a simulated estuarine environment. Mar. Ecol. Program Ser. 9:121-136.

Owens, E.H. 1973. The cleaning of gravel beaches polluted by oil. Proceedings, 13th International Coastal Engineering Conference, pp. 2549-2556. American Society of Civil Engineers, New York.

Owens, E.H., and G. Drapeau. 1973. Changes in beach profiles at Chedabucto Bay, N.S., following large-scale removal of sediments. Can. J. Earth Sci. 10(8):1226-1232.

Owens, E.H., and G.A. Robilliard. 1981. Shoreline sensitivity and oil spills--A re-evaluation for the 1980's. Mar. Pollut. Bull. 12(3):75-78.

Owens, E.H., J.R. Harper, C.R. Foget, and W. Robson. 1983. Shoreline experiments and the persistence of oil on arctic beaches, pp. 261-268. In Proceedings, 1983 Oil Spill Conference. API Publication 4356. American Petroleum Institute, Washington, D.C.

Page, D.S., D.W. Mayo, J.F. Cooley, E. Sorenson, E.S. Gilfillan, and S. Hanson. 1979. Hydrocarbon distribution and weathering characteristics at a tropical oil spill site, pp. 709-712. In Proceedings, 1979 Oil Spill Conference. API Publication 4308. American Petroleum Institute, Washington, D.C.

Pancirov, R.J., and R.A. Brown. 1975. Analytical methods for polynuclear aromatic hydrocarbons in crude oils, heating oils, and marine tissues, pp. 103-113. In Proceedings, 1975 Conference on Prevention and Control of Oil Pollution. American Petroleum Institute, Washington, D.C.

Pancirov, R.J., and R.A. Brown. 1977. Polynuclear aromatic hydrocarbons in marine tissues. Environ. Sci. Technol. 11:989-992.

Parsons, J., J. Spry, and T. Austin. 1980. Preliminary observations on the effect of Bunker C fuel oil on seals on the Scotian shelf, pp. 193-202. In J.H. Vandermeulen, ed. Scientific Studies During the Kurdistan Tanker Incident: Proceedings of a Workshop. Report Series BI-R-80-3. Bedford Institute of Oceanography, Dartmouth, N.S.

Parsons, K.R., W.K.W. Li, and R. Waters. 1976. Some preliminary observations on the enhancement of phytoplankton growth by low levels of mineral hydrocarbons. Hydrobiology 51:85.

Passey, R.D. 1922. Experimental soot cancer. Br. Med. J. ii:1112-1114.

Paton, G.R., and A.C. Allison. 1972. Chromosome damage in human cell cultures induced by metal salts. Mut. Res. 16:332-336.

Patton, J.S., M.W. Rigler, P.D. Boehm, and D.L. Fiest. 1981. Ixtoc I oil spill: flaking of surface mousse in the Gulf of Mexico. Nature 290:235-238.

Payne, J.F. 1976. Field evaluation of benzo(a)pyrene hydroxylase induction as a monitor for marine petroleum pollution. Science 191(4320):945-946.

Payne, J.F., and N. May. 1979. Further studies on the effect of petroleum hydrocarbons on mixed function oxidases in marine organisms, pp. 339-347. In M.A.Q. Kahn et al., ed. Pesticide and Xenobiotic Metabolism in Aquatic Organisms. ACS Symposium Series 99. American Chemical Society, Washington, D.C.

Payne, J.F., I. Martins, and A. Rahimtula. 1978a. Crankcase oils: are they a major mutagenic burden in the aquatic environment? Science 200(4339):329-330.

Payne, J.F., J.W. Kiceniuk, W.R. Squires, and G.L. Fletcher. 1978b. Pathological changes in a marine fish after a 6-month exposure to petroleum. J. Fish. Res. Board Can. 35:665-667.

Peakall, D.B., J. Tremblay, W.B. Kinter, and D.S. Miller. 1981. Endocrine dysfunction in seabirds caused by ingested oil. Environ. Res. 24:6-14.

Peakall, D.B., D.J. Hallett, J.R. Bend, G.L. Foureman, and D.S. Miller. 1982. Toxicity of Prudhoe crude oil and its aromatic fractions to nestling Herring Gulls. Environ. Res. 27:206-215.

Pearson, W.H., D.L. Woodruff, P.C. Sugarman, and B.L. Olla. 1981a. Effects of oiled sediment on predation on the littleneck clam, Protothaca staminea, by the Dungeness crab, Cancer magister. Estuarine Coastal Shelf Sci. 13:445-454.

Pearson, W.H., P.C. Sugarman, D.L. Woodruff, and B.L. Olla. 1981b. Impairment of the chemosensory antennular flicking response in the Dungeness crab, Cancer magister, by petroleum hydrocarbons. Fish. Bull. U.S. 79:641-647.

Pederson, R.A., A.I. Spindle, and S. Takehisa. 1978. Inhibition of mouse embryo development in vitro by benzo(a)pyrene, pp. 152-161. In D.D. Mahlum, M.R. Sikov, P.L. Hackett, and F.D. Andrew, eds. Developmental Toxicology of Energy-Related Pollutants. Technical Information Service, U.S. Department of Energy, Washington, D.C.

Pelroy, R.A., D.S. Sklarew, and S.P. Downey. 1981. Comparison of the mutagenicities of fossil fuels. Mut. Res. 90:233-245.

Percy, J.A. 1976. Responses of arctic marine crustaceans to crude oil and oil-tainted food. Environ. Pollut. 10(2):155-162.

Percy, J.A. 1977a. Response of arctic marine benthic crustaceans to sediments contamined with crude oil. Environ. Pollut. 13(1):1-10.

Percy, J.A. 1977b. Effects of dispersed crude oil upon the respiratory metabolism of an arctic marine amphipod, Onisimus (Boeckosimus) affinis, pp. 192-200. In D.A. Wolfe, ed. Fate and Effects of Petroleum Hydrocarbons in Marine Organisms and Ecosystems. Pergamon, New York.

Percy, J.A. 1981. Benthic and intertidal organisms, pp. 87-104. In J.B. Sprague, J.H. Vandermeulen, and P.G. Wells, eds. Oil and Dispersants in Canadian Seas: Research Appraisal and Recommendations. Economic and Technical Review Report. EPS-3-EC-82-2. Environment Canada, Ottawa, Ontario.

Percy, J.A., and T.C. Mullin. 1977. Effects of crude oil on the locomotory activity of arctic marine invertebrates. Mar. Pollut. Bull. 8(6):35-40.

Peters, E.C., P.A. Meyers, P.P. Yerich, and N.J. Blake. 1981. Bioaccumulation and histopathological effects of oil on a stony coral. Mar. Pollut. Bull. 12(10):333-339.

Petres, J., and A. Berger. 1972. Zum Einfluss anorganischer Arsens auf die DNA Synthese menschlicher Lymphocyten in vitro. Arch. Dermatol. Forsch. 242:343-352.

Pfaender, F.K., E.N. Buckley, and R. Ferguson. 1980. Response of the pelagic microbial community to oil from the Ixtoc I blowout. I. In situ studies, pp. 545-562. In D.K. Atwood, convener. Proceedings, Conference on the Preliminary Scientific Results from the Researcher/Pierce Cruise to the Ixtoc I Blowout. U.S. Department of Commerce, National Oceanic and Atmospheric Administration, Boulder, Colo.

Philippi, G.T. 1977. On the depth, time and mechanism of origin of the heavy to medium-gravity naphthenic crude oils. Geochim. Cosmochim. Acta 41:33-52.

Poklis, A., and C.D. Burkett. 1977. Gasoline sniffing: a review. Clinical Toxicol. II(1):35-41.

Powell, N.A., C.S. Sayce, and D.F. Tufts. 1970. Hyperplasia in an estuarine bryzoan attributable to coal tar derivatives. J. Fish. Res. Board Can. 27:2095-2096.

Powers, K.D., and W.T. Rumage. 1978. Effect of the Argo Merchant oilspill on bird populations off the New England coast, 15 December 1976-January 1977, pp. 142-148. In M.P. Wilson, J.G. Quinn, and K. Sherman, eds. In the Wake of the Argo Merchant. Center for Ocean Management Studies, University of Rhode Island, Kingston.

Pratt, S.D. 1978. Interactions between petroleum and benthic fauna at the Argo Merchant spill site, pp. 131-136. In M.P. Wilson, J.G. Quinn, and K. Sherman, eds. In the Wake of the Argo Merchant. Center for Ocean Management Studies, University of Rhode Island, Kingston.

Prein, A.E., G.M. Thie, G.M. Alink, C.L.M. Poels, and J.H. Koeman. 1978. Cytogenetic changes in fish exposed to water of the river Rhine. Sci. Total Environ. 9:287-291.

Prouse, N.J., and D.C. Gordon, Jr. 1976. Interactions between the deposit feeding polychaete Arenicola marina and oiled sediment, pp. 408-422. In Sources, Effects and Sinks of Hydrocarbons in the Aquatic Environment. American Institute of Biological Sciences, Arlington, Va.

Prouse, N.J., D.C. Gordon, Jr., and P.D. Keizer. 1976. Effects of low concentrations of oil accommodated in sea water on the growth of unialgal marine phytoplankton cultures. J. Fish. Res. Board Can. 33:810-818.

Pulich, W.M., Jr., K. Winters, and C. Van Baalen. 1974. The effects of a No. 2 fuel oil and two crude oils on the growth and photosynthesis of microalgae. Mar. Biol. 28:87-94.

Ravanko, O. 1972. The Palva oil tanker disaster in the Finnish southwestern archipelago. V. The littoral and aquatic flora of the polluted area. Aqua Fenn. 1972:142-144.

Ray, J.P. 1981. The effect of petroleum hydrocarbons on corals, pp. 705-726. In Petromar-80. Graham and Trotman, Ltd. London.

Reimer, A.A. 1975a. Effects of crude oil on corals. Mar. Pollut. Bull. 6:39-43.

Reimer, A.A. 1975b. Effects of crude oil on the feeding behavior of the zoanthid Palythoa variabilis. Environ. Physiol. Biochem. 5:258-266.

Renzoni, A. 1975. Toxicity of three oils to bivalve gametes and larvae. Mar. Pollut. Bull. 6(8):125-128.

Research Unit on the Rehabilitation of Oiled Seabirds. 1972a. Second Annual Report of the Advisory Committee on Oil Pollution of the Sea. 32 pp. Zoology Department, University of Newcastle upon Tyne.

Research Unit on the Rehabilitation of Oiled Seabirds. 1972b. Recommended treatment of oiled seabirds. University of Newcastle upon Tyne. 10 pp.

Rice, S.D. 1973. Toxicity and avoidance tests with Prudhoe Bay crude oil and pink salmon fry, pp. 667-670. In Proceedings, Joint Conference on Prevention and Control of Oil Spills. American Petroleum Institute, Washington, D.C.

Rice, S.D., D.A. Moles, and J.W. Short. 1975. The effect of Prudhoe Bay crude oil on survival and growth of eggs, alevins, and fry of pink salmon, Oncorhynchus gorbuscha, pp. 503-507. In Proceedings, 1975 Conference on Prevention and Control of Oil Pollution. American Petrleum Institute, Washington, D.C.

Rice, S.D., J.W. Short, C.C. Brodesen, T.A. Mecklenburg, D.A. Moles, C.J. Misch, D.L. Cheatham, and J.F. Karinen. 1976. Acute toxicity and uptake-depuration studies with Cook Inlet crude oil, Prudhoe Bay crude oil, No. 2 fuel oil, and several subarctic marine organisms. Processed Report. Northwest Fisheries Center, Auke Bay Lab., NMFS, NOAA, Auke Bay, Alaska. 90 pp.

Rice, S.D., R.E. Thomas, and J.W. Short. 1977a. Effect of petroleum hydrocarbons on breathing and coughing rates and hydrocarbon uptake-depuration in pink salmon fry, pp. 259-277. In F.J. Vernberg, A. Calabrese, F.P. Thurberg, and W.B. Vernberg, eds. Proceedings, Physiological Responses of Marine Biota to Pollutants. Academic Press, New York.

Rice, S.D., J.W. Short, and J.F. Karinen. 1977b. Comparative oil toxicity and comparative animal sensitivity, pp. 78-94. In D.A. Wolfe, ed. Fate and Effects of Petroleum Hydrocarbons in Marine Organisms and Ecosystems. Pergamon, New York.

Rice, S.D., J.F. Karinen, and S. Korn. 1978. Acute and chronic toxicity, uptake and depuration, and sublethal response of Alaskan marine organisms to petroleum hydrocarbons, pp. 11-24. In D.A. Wolfe, ed. Marine Biological Effects of OCS Petroleum Development. NOAA Technical Memorandum ERL-OCSEAP-1. NTIS PB-288 935. National Technical Information Service, Springfield, Va.

Rice, S.D., A. Moles, T.L. Taylor, and J.F. Karinen. 1979. Sensitivity of 39 Alaskan marine species to Cook Inlet crude oil and No. 2 fuel oil. In Proceedings, 1979 Oil Spill Conference. API Publication 4308. American Petroleum Institute, Washington, D.C.

Richardson, W.J., and S.R. Johnson. 1981. Waterbird migration near the Yukon and Alaskan coast of the Beaufort Sea. I. Timing routes and numbers in spring. Arctic 34(2):108-121.

Ridgway, R.H. 1972. Homeostasis in the aquatic environment. pp. 59-747. In S.H. Ridgway, ed. Mammals of the Sea-Biology and Medicine. Charles C. Thomas, Publisher. Springfield, Ill.

Riley, G.A. 1956. Review of the oceanography of Long Island Sound. Deep Sea Res. 3(Suppl.)224-238.

Rinkevich, B., and Y. Loya. 1977. Harmful effects of chronic oil pollution on a Red Sea scleractinian coral population, pp. 585-591. In D.L. Taylor, ed. Proceedings, Third International Coral Reef Symposium. II. Geology Rosenstiel School of Marine and Atmospheric Science, University of Miami, Miami, Fla.

Rinkevich, B., and Y. Loya. 1979. Laboratory experiments on the effects of crude oil on the Red Sea coral Stylophora pistillata. Mar. Pollut. Bull. 10(11):328-330.

Robbins, W.K. 1980. Analysis of shale oils and downstream products. American Petroleum Report Sp S-5-ERE (85808).

Robinson, J.H., ed. 1979. The Peck Slip oil spill, a preliminary scientific report. NOAA Special Report. Office of Marine Pollution Assessment, U.S. Department of Commerce, Boulder, Colo. 190 pp.

Roesijadi, G. and J.W. Anderson. 1979. Condition index and free amino acid content of Macoma inquinata exposed to oil-contaminated marine sediments, pp. 69-83. In W.B. Vernberg, A. Calabrese, F. Thurberg, and F.J. Vernberg, eds. Marine Pollution: Functional Responses. Academic Press, New York.

Roesijadi, G., J.W. Anderson, and J.W. Baylock. 1978. Uptake of hydrocarbons from marine sediments contaminated with Prudhoe Bay crude oil: influence of feeding type of test species and availability of polycyclic aromatic hydrocarbons. J. Fish Res. Board Can. 35:608-614.

Rosenthal, H., and D.F. Alderdice. 1976. Sublethal effects of environmental stressors, natural and pollutional, on marine fish eggs and larvae. J. Fish. Res. Board Can. 33:2047-2065.

Ross, S.L., C.W. Ross, F. Lepine, and R.K. Langtry. 1979. Ixtoc-I oil blowout. Spill Technol. Newslett. (Environ. Can.) July-August 1979:245-256.

Ross, S.L., C.W. Ross, F. Lepine, and R.K. Langtry. 1980. Ixtoc-I oil blowout, pp. 25-40. In D.K. Atwood, convener. Proceedings, Symposium on Preliminary Scientific Results From the Researcher/Pierce Cruise to the Ixtoc I Blowout. U.S. Department of Commerce, National Oceanic and Atmospheric Administration, Boulder, Colo.

Rossi, S.S., and J.W. Anderson. 1976. Toxicity of water-soluble fractions of No. 2 fuel oil and South Louisiana crude oil to selected stages in the life history of the polychaete Neanthes arenaceodentata. Bull. Environ. Contam. Toxicol. 16:18-24.

Rossi, S.S., and J.W. Anderson. 1977. Accumulation and release of fuel-oil-derived diaromatic hydrocarbons by the polychaete, Neanthes arenaceodentata. Mar. Biol. 39:51-55.

Rossi, S.S., and J.M. Neff. 1978. Toxicity of polynuclear aromatic hydrocarbons to the marine polychaete, Neanthes arenaceodentata. Mar. Pollut. Bull. 9:220-223.

Rossner, P., J. Cinatl, and Vl. Bencko. 1972. The effect of sodium arsenate on cell cultures. Cesk. Hygiene 17:58-63.

Roszinsky-Kocher, G., A. Basler, and G. Rohrborn. 1979. Mutagenicity of polycyclic hydrocarbons. V. Induction of sister-chromatid exchanges in vivo. Mut. Res. 66:65-67.

Royal Commission on Environmental Pollution. 1981. Eighth report, Marine oil pollution. HMSO, London.

Royal Society for the Protection of Birds. 1979. Marine Oil Pollution and Birds. Sandy, U.K. 126 pp.

Russell, D.J., and B.A. Carlson. 1978. Edible-oil pollution on Fanning Island. Pacific Sci. 32:1-15.

Rutzler, K., and W. Sterrer. 1970. Oil pollution: damage observed in tropical communities along the Atlantic seaboard of Panama. Biol. Sci. 20:222-224.

Sabo, D.J., and J.J. Stegeman. 1977. Some metabolic effects of petroleum hydrocarbons in marine fish, pp. 279-287. In F.J. Vernberg, A. Calabrese, F.P. Thurberg, and W.B. Vernberg, eds. Proceedings, Physiological Responses of Marine Biota to Pollutants. Academic Press, New York.

Salomonsen, F. 1967. Fuglene pa Gronland. Kobenhaven. 343 pp.

Samain, J.F., J. Moal, J.Y. Daniel, and J. Boucher. 1979. Ecophysiological effects of oil spills from Amoco Cadiz on pelagic communities--Preliminary results, pp. 175-186. In Proceedings, 1979 Oil Spill Conference. API Publicaton 4308. American Petroleum Institute. Washington, D.C.

Samain, J.F., J. Moal, A. Coum, J.R. Le Coz, and J.Y. Daniel. 1980. Effects of the Amoco Cadiz oil spill on zooplankton: A new possiblity of ecophysiological survey. Helgolander Meeresunters. 33:225-235.

Samain, J.F., J. Moal, J.R. Le Coz, J.Y. Daniel, and A. Coum. 1981. Impact de l'Amoco Cadiz sur l'ecophysiologie du zooplancton: une nouvelle possibilite de surveillance ecologique. In G. Conan et al., eds. Amoco Cadiz: consequences d'une pollution accidentelle par les hydrocarbures. CNEXO, Paris.

Sampson, A.L., J.H. Vandermeulen, P.G. Wells, and C. Moyse. 1980. A selected bibliography on the fate and effects of oil pollution relevant to the Canadian marine environment. 2nd edition. Economic and Technical Review Report EPS-3-EC-80-5. Environment Canada, Environmental Protection Service, Ottawa, Ontario.

Sanborn, H.R. 1977. Effects of petroleum on ecosystems, Chap. 6, pp. 337-357. In D.C. Malins, ed. Effects of Petroleum on Arctic and Subarctic Marine Environments and Organisms. Vol. II, Biological Effects. Academic Press, New York.

Sanborn, H.R., and D.C. Malins. 1977. Toxicity and metabolism of naphthalene: a study with marine larval invertebrates. Proc. Soc. Exp. Biol. Med. 154:151-155.

Sanborn, H.R., and D.C. Malins. 1980. The disposition of aromatic hydrocarbons in adult spot shrimp (Pandalus platyceros) and the formation of metabolites of naphthalene in adult and larval spot shrimp. Xenobiotica 10(3):193-200.

Sanders, H.L. 1978. Florida oil spill impact on the Buzzards Bay benthic fauna: West Falmouth. J. Fish. Res. Board Can. 35(5):717-730.

Sanders, H.L. 1981. Safety and offshore oil. In Background papers of the Committee on Assessment of Safety of OCS Activities. National Academy of Sciences, Washington, D.C.

Sanders, H.L., J.F. Grassle, and G.R. Hampson. 1972. The West Falmouth oil spill. I. Biology. Technical Report WHOI-72-20. Woods Hole Oceanographic Institute. Woods Hole, Mass. 23 pp.

Sanders, H.L., J.F. Grassle, G.R. Hansson, L.S. Morse, S. Price-Gartner, and C.C. Jones. 1980. Anatomy of an oil spill: long-term effects from the grounding of the barge Florida off West Falmouth, Massachusetts. J. Mar. Res. 38:265-380.

Sassen, R. 1980. Biodegradation of crude oil and mineral deposition in a shallow Gulf Coast salt dome. Org. Geochem. 2:153-166.

Scarratt, D.J. 1980. Taste panel assessments and hydrocarbon concentrations in lobsters, clams and mussels following the wreck of the Kurdistan, pp. 212-227. In J.H. Vandermeulen, ed. Scientific Studies During the Kurdistan Tanker Incident: Proceedings of a Workshop. Report Series BI-R-80-3. Bedford Institute of Oceanography, Dartmouth, N.S.

Scheier, A., and D. Gominger. 1976. A preliminary study of the toxic effects of irradiated vs. non-irradiated water soluble fractions of a No. 2 fuel oil. Bull. Environ. Contam. Toxicol. 16:595-603.

Scoffin, T.P. 1970. The trapping and binding of subtidal carbonate sediment by marine vegetation in Bimini Lagoon, Bahamas. J. Sedimentol. Pet. 40:249-273.

Sekerak, A., and M. Foy. 1978. Acute lethal toxicity of Corexit 9527/ Prudhoe Bay crude oil mixtures to selected Arctic invertebrates. Spill Technol. Newslett. 3(2):37-41.

Seneca, E.D., and S.W. Broom. 1982. Restoration of marsh vegetation impacted by the Amoco Cadiz oil spill and subsequent cleanup operations at Ile Grande, France, pp. 363-420. In Ecological Study of the Amoco Cadiz Oil Spill. NOAA-CNEXO Joint Scientific Commission.

Sharp, J.R., K.W. Fucik, and J.M. Neff. 1979. Physiological bases of differential sensitivity of fish embryonic stages to oil pollution, pp. 85-108. In W.B. Vernberg, A. Calabrese, F.P. Thurberg, and F.J. Vernberg, eds. Marine Pollution: Functional Responses. Academic Press, New York.

Shaw, D.G., and L.M. Cheek. 1976. Hydrocarbon studies in the benthic environment at Prudhoe Bay, pp. 425-431. In D.W. Hood, and D.C. Burrell, eds. Assessment of the Arctic Marine Environment: Selected Topics. Occasional Publication 4. Institute of Marine Sciences, University of Alaska, Fairbanks.

Sherman, K., and D. Busch. 1978. The _Argo Merchant_ oil spill and the fisheries, pp. 149-165. In M.P. Wilson, J.G. Quinn, and K. Sherman, eds. In the Wake of the _Argo Merchant_. Center for Ocean Management Studies, University of Rhode Island, Kingston.

Shiels, W.E., J.J. Goering, and D.W. Hood. 1973. Crude oil phytotoxicity studies, pp. 413-446. In D.W. Hood, W.E. Shiels, and E.J. Kelly, eds. Environmental Studies of Port Valdez. Occasional Publication 3. Institute of Marine Sciences, University of Alaska, Fairbanks.

Shipton, J., J.H. Last, K.E. Murray, and G.L. Vale. 1970. Studies on a kerosene-like taint in mullet (_Mugil cephalus_). II. Chemical nature of the volatile constituents. J. Sci. Food Agric. 21:433-436.

Simpson, J.G., and W.G. Gilmartin. 1970. An investigation of elephant seal and sea lion mortality on San Miguel Island. Bioscience 20:289.

Sinclair, M. 1982. From probabilities concerning the accidental release of oil and its physiological consequences, what kind of observational programs would be required to detect the biotic effects. In A. Longhurst, ed. Consultation on the Consequences of Offshore Oil Production on Offshore Fish Stocks and Fishing Operations. Can. Tech. Rep. Fish. Aquat. Sci. 1096:54-63.

Siou, G., L. Conan, and M. el Haitem. 1981. Evaluation of the clastogenic action of benzene by oral administration with 2 cytogenetic techniques in mouse and Chinese hamster. Mut. Res. 90:273-278.

Sirover, M.A., and L.A. Loeb. 1976. Metal-induced infidelity during DNA synthesis. Proc. Natl. Acad. Sci. 73:2331-2335.

Slijper, E.J. 1979. Whales. Cornell University Press. Ithaca, N.Y. 511 pp.

Slobodkin, L.B. 1968. Toward a predictive theory of evolution, pp. 187-205. In R.C. Lewontin, ed. Population Biology and Evolution. Syracuse University Press, Syracuse, N.Y.

Smiley, B.D. 1982. Effects of oil on marine mammals, pp. 113-122. In J.B. Sprague, J.H. Vandermeulen, and P.G. Wells, ed. Oil and Dispersants in Canadian Seas: Research Appraisal and Recommendations. Economic and Technology Review Report EPS-3-EC-82-2. Environment Canada, Ottawa, Ontario.

Smith, H.M. 1968. Qualitative and quantitative aspects of crude oil composition, p. 136. Bureau of Mines Bulletin 642. U.S. Department of the Interior, Washington, D.C.

Smith, J.E. 1968. _Torrey Canyon_ pollution and marine life. A Report by the Plymouth Laboratory of Marine Biological Associations of the United Kingdom. Columbia University Press, New York.

Smith, R.L., and J.A. Cameron. 1979. Effect of water soluble fraction of Prudhoe Bay crude oil on embryonic development of Pacific herring. Trans. Am. Fish. Soc. 108:70-75.

Smith, T.G., and J.R. Geraci. 1975. The effect of contact and ingestion of crude oil on ringed seals of the Beaufort Sea. Beaufort Sea Project Technical Report 5. Department of the Environment, Victoria, B.C. 67 pp.

Smith, T.G., and I. Stirling. 1975. The breeding habitat of the ringed seal (Phoca hispida). The birth lair and associated structure. Can. J. Zool. 53:1297-1305.

Smithsonian Institute. 1970a. S.I. Event No. 24-70. Center for Short-lived Phenomena, Washington, D.C.

Smithsonian Institute 1970b. S.I. Event No. 186-70. Center for Short-lived Phenomena, Washington, D.C.

Smithsonian Institute. 1972. S.I. Event No. 1-72. Center for Short-lived Phenomena, Washington, D.C.

Snedaker, S.C., J.A. Jimenez, and M.S. Brown. 1981. Anomalous aerial roots in Avicennia germinans (L.) in Florida and Costa Rica. Bull. Mar. Sci. 31(2):467-470.

Snow, N.B., and B.F. Scott. 1975. The effect and fate of crude oil spilt on two arctic lakes, pp. 527-534. In Proceedings, Joint Conference on Prevention and Control of Oil Spills. American Petroleum Institute, Washington, D.C.

Soto, C., J.A. Hellebust, and T.C. Hutchinson 1975a. Effect of naphthalene and aqueous crude oil extracts on the green flagellate Chlamydomonas angulosa. II. Phytosynthesis and the uptake and release of naphthalene. Can. J. Bot. 53:118-126.

Soto, C., Hellebust, J.A., Hutchinson, T.C. and Sawa, T. 1975b. Effect of naphthalene and aqueous crude oil extracts on the green flagellate Chlamydomonas angulosa. I. Growth. Can. J. Bot. 53:109-117.

Southward, A.J. 1982. An ecologist's view of the implications of the observed physiological and biochemical effects of petroleum compounds on marine organisms and ecosystems. Phil. Trans. R. Soc. London, Ser. B 297:241-255.

Southward, A.J., and E.C. Southward. 1978. Recolonization of rocky shores in Cornwall after use of toxic dispersants to clean up the Torrey Canyon spill. J. Fish. Res. Board Can. 35(5):682-706.

Souza, G. 1970. Report of the Shellfish Warden, pp. 161-165. In Annual Report of the Finances of the Town of Falmouth for the year ending December 31, 1970.

Spaulding, M.L. 1978. Surface and subsurface spill trajectory forecasting: application to the Argo Merchant, pp. 37-42. In M.P. Wilson, J.G. Quinn, and K. Sherman, eds. In the Wake of the Argo Merchant. Center for Ocean Management Studies, University of Rhode Island, Kingston.

Spies, R.B., and P.H. Davis. 1979. The infaunal benthos of a natural oil seep in the Santa Barbara Channel. Mar. Biol. 50:227-238.

Spies, R.B., and P.H. Davis. 1981. Toxicity of Santa Barbara seep oil to starfish embryos. III. Influence of parental exposure and the effects of other crude oils. Mar. Environ. Res. In press.

Spies, R.B., and P.H. Davis. 1982. Toxicity of Santa Barbara seep oil to starfish embryos. 3. Influence of parental exposure and the effects of other crude oils. Mar. Environ. Res. 6:3-11.

Spies, R.B., and D.J. DesMarais. 1983. A natural isotope study of trophic enrichment of marine benthic communities by petroleum seapage. Mar. Biol. (in press).

Spies, R.B., P.H. Davis, and D.H. Stuermer. 1980. Ecology of a submarine petroleum seep off the California Coast, pp. 208-263. In R. Geyer, ed. Environmental Pollution. I. Hydrocarbons. Elsevier, New York.

Spies, R.B., J. Felton, and L. Dillard. 1982. Hepatic mixed-function oxidases in California flatfishes are increased in contaminated environments and by oil and PCB ingestion. Mar. Biol. 79:117-127.

Spooner, M.F. 1967. Biological effects of the Torrey Canyon disaster. J. Devon Trust Nat. Conserv. Suppl. 1967:12-19.

Spooner, M.F. 1970. Oil spill in Tarut Bay, Saudi Arabia. Mar Pollut. Bull. 1:166-167.

Spooner, M.F. 1978. Editorial introduction. Amoco Cadiz oil spill. Mar. Pollut. Bull. 9:281-284.

Spooner, M.F., and G.M. Spooner. 1968. The problem of oil spills at sea, illustrated by the strandings of the general Colocotronis on Eleuthera, Bahamas, March 7, 1968. Marine Biological Association of the U.K., Plymouth. 21 pp.

Sprague, J.B., J.H. Vandermeulen, and P.G. Wells. 1981. Oil and dispersants in Canadian Seas: research appraisal and recommendations. Economic and Technical Review Report EPS-3-EC-82-2. Environment Canada, Ottawa, Ontario. 185 pp.

Stainken, D.M. 1976. A descriptive evaluation of the effects of No. 2 fuel oil and the tissues of the soft shell clam, Mya arenaria L. Bull. Environ. Contam. Toxicol. 16:730-738.

Stainken, D.M. 1978. Effects of uptake and discharge of petroleum hydrocarbons on the respiration of the soft-shell clam, Mya arenaria. J. Fish. Res. Board. Can. 35:637-642.

Staley, J.T. 1980. Diversity of aquatic heterotrophic bacterial communities, pp. 321-322. In D. Schlessinger, ed. Microbiology 1980. American Society for Microbiology, Washington, D.C.

St. Amant, L.S. 1958. Investigation of oily taste in oysters caused by oil drilling operations, pp. 75-77. 7th Biennial Report. Louisiana Wildlife Fisheries Commission.

Stanley, S.O., T.H. Pearson, and C.M. Brown. 1978. Marine microbial ecosystems and the degradation of organic pollutants, pp. 60-79. In K. Chater and H. Sommerville, eds. The Oil Industry and Microbial Ecosystems. Heyden and Son, London.

Stansby, M.E. 1978. Flavors in fish from petroleum pickup. Mar. Fish. Rev. 40(1):13-17.

Steele, R.L. 1977. Effects of certain petroleum products on reproduction and growth of zygotes and juvenile stages of the alga Fucus edentatus de la Pyl (Phaeophyceae:Fucales), pp. 138-142. In D.A. Wolfe, ed. Fate and Effects of Petroleum Hydrocarbons in Marine Organisms and Ecosystems. Pergamon, New York.

Stegeman, J.J. 1976. Aspects of the effects of petroleum hydrocarbons on intermediary metabolism and xenobiotic metabolism in marine fish, pp. 424-436. In Sources, Effects and Sinks of Hydrocarbons in the Aquatic Environment. American Institute of Biological Sciences, Arlington, Va.

Stegeman, J.J. 1978. Influence of environmental contamination on cytochrome P-450 mixed-function oxygenases in fish: implications for recovery in the Wild Harbor marsh. J. Fish. Res. Board Can. 35:668-674.

Stegeman, J.J. 1980. Cytochrome P-450 and benzo(a)pyrene metabolism in cardiac tissue of the marine fish Stenotomus versicolor. Pharmacologist 20:248.

Stegeman, J.J., and D.J. Sabo. 1976. Aspects of the effects of petroleum hydrocarbons on intermediary metabolism and xenobiotic metabolism in marine fish, pp. 423-431. In Sources, Effects and Sinks of Hydrocarbons in the Aquatic Environment. American Institute of Biological Sciences, Washington, D.C.

Stegeman, J.J., and J.M Teal. 1973. Accumulation, release and retention of petroleum hydrocarbons by the oyster Crassostrea virginica. Mar. Biol. 22(1):37-44.

Stirling, I. 1980. The biological importance of polynyas in the Canadian arctic. Arctic 33(2):303-315.

Stirling, I., and H. Cleator. 1981. Polynyas in the Canadian Arctic. Occasional paper 45. Canadian Wildlife Service, Ottawa, Ontario. 73pp.

Stirling, I., D. Andriashek, P. Latour, and W. Calvert. 1975. The distribution and abundance of polar bears in the eastern Beaufort Sea. Beaufort Sea Project Technical Report 2. Department of the Environment, Victoria, B.C. 59 pp.

Stirling, I., R. Archibald, and D. Demaster. 1977. Distribution and abundances of seals in the eastern Beaufort Sea. J. Fish. Res. Board Can. 34:976-988.

Stoss, F.W., and T.A. Haines. 1979. The effects of toluene on embryos and fry of the Japanese medaka Oryzias latipes with a proposal for rapid determination of maximum acceptable toxicant concentration. Environ. Pollut. 20:139-148.

Straughan, D. 1971. Breeding and larval settlement of certain intertidal invertebrates in the Santa Barbara Channel following pollution by oil, pp. 223-244. In D. Straughan, ed. Biological and Oceanographical Survey of the Santa Barbara Channel Oil Spill 1969-1970, Vol. 1. Sea Grant Publication 2. Allan Hancock Foundation, University of Southern California, Los Angeles.

Straughan, D. 1972. Factors causing environmental changes after an oil spill. J. Pet. Technol. Offshore Issue 1972:250-254.

Straughan, D. 1976. Temperature effects of crude oil in the upper intertidal zone. EPA-600/2-76-127. Industrial Environmental Research Lab., Office R&D, U.S. Environmental Protection Agency.

Straughan, D. 1977a. The sublethal effects of natural chronic exposure to petroleum on marine invertebrates, pp. 563-568. In Proceedings, Joint Conference on Prevention and Control of Oil Spills, API Publication 4284. American Petroleum Institute, Washington, D.C.

Straughan, D. 1977b. Effects of natural chronic exposure to petroleum hydrocarbons on size and reproduction in Mytilus californianus Conrad, pp. 289-298. In F.J. Vernberg, A. Calabrese, F.P. Thurberg, and W.B. Vernberg, eds. Physiological Responses of Marine Biota to Pollutants. Academic Press, New York.

Straughan, D. 1979. Variability in chemical exposure of marine organisms to petroleum, pp. 467-475. In Symposium on Chemistry and Economics of Ocean Resources. Publication of Chemical Marketing and Economics Division. American Chemical Society, Staten Island, N.Y.

Straughan, D., and D.M. Lawrence. 1975. Investigation of ovicell hyperlasia in bryzoans chronically exposed to natural oil seepage. Water Air Soil Pollut. 5:39-45.

Stromberg, P.T., M.L. Landolt, and R.M. Kocan. 1981. Alterations in the frequency of sister chromatid exchanges in flatfish from Puget Sound, Washington, following experimental and natural exposure to mutagenic chemicals. NOAA Technical Memorandum OMPA-10. National Oceanic and Atmospheric Administration, Rockville, Md. 43 pp.

Struhsaker, J.W. 1977. Effects of benzene (a toxic component of petroleum) on spawning Pacific herring, Clupea harengus pallasi. Fish. Bull. (U.S.) 75:43-49.

Struhsaker, J.W., M.B. Eldridge, and T. Echeverria. 1974. Effects of benzene (a water-soluble component of crude oil) on eggs and larvae of Pacific herring and northern anchovy, pp. 253-284. In F.J. Vernberg and W.B. Vernberg, eds. Pollution and Physiology of Marine Organisms. Academic Press, New York.

Stuermer, D.H., R.B. Spies, P.H. Davis, D.J. Ng, C.J. Morris, and S. Neal. 1982. The hydrocarbons in the Isla Vista marine seep environment. Mar. Chem. 11:413-426.

Suess, M.J. 1976. The environmental load and cycle of polycyclic aromatic hydrocarbons. Sci. Total Environ. 6:239-250.

Swedmark, M., A. Granmo, and S. Kollberg. 1973. Effects of oil dispersants and oil emulsions on marine animals. Water Res. 7:1649-1672.

Swinnerton, J.W., and R.A. Lamontagne. 1974. Oceanic distribution of low molecular weight hydrocarbons. Baseline measurements. Environ. Sci. Technol. 8:657-663.

Szaro, R.C., and P.H. Albers. 1977. Effects of external applications of No. 2 fuel oil on common eider eggs, pp. 164-167. In D.A. Wolfe, ed. Fate and Effects of Petroleum Hydrocarbons in Marine Organisms and Ecosystems. Pergamon, New York.

Tanis, J.J.C., and M.F. Morzer-Bruyns. 1968. The impact of oil pollution on seabirds in Europe. Proc. Int. Conf. Oil Pollut. Sea, Rome 1967:67-74.

Tatem, H.E., B.A. Cox, and J.W. Anderson. 1978. The toxicity of oils and petroleum hydrocarbons to estuarine crustaceans. Estuarine Coastal Mar. Sci. 6:365-373.

Taylor, T.L., and J.F. Karinen. 1977. Response of the clam Macoma balthica (L.) exposed to Prudhoe Bay crude oil as unmixed oil-water soluble fraction and oil-contaminated sediment in the laboratory, pp. 229-237. In D.A. Wolfe, ed. Fate and Effects of Petroleum Hydrocarbons in Marine Organisms and Ecosystems. Pergamon, New York.

Teal, J.M., and R.W. Howarth. 1984. Oil spill studies: a review of ecological effects. Environ. Manage. In press.

Teal, J.M., K. Burns, and J. Farrington. 1978. Analyses of aromatic hydrocarbons in intertidal sediments resulting from two spills of

No. 2 fuel oil in Buzzards Bay, Massachusetts. J. Fish Res. Board Can. 35(5):510-520.

Teas, H.J. 1979. Silviculture with saline water, pp. 117-161. In The Biosaline Concept. A. Hollaender, ed. Plenum, New York.

Teas, H.J., W. Jurgens, and M.C. Kimbell. 1975. Proceedings of the Second Annual Conference on Restoration of Coastal Vegetation in Florida. (Hillsborough) Community College, Tampa, Fla.

Thebeau, L.C., J.W. Tunnell, Jr., Q.R. Dokken, and M.E. Kindinger. 1981. Effects of the Ixtoc I oil spill on the intertidal and subtidal infaunal populations along lower Texas coast barrier island beaches, pp. 467-475. In Proceedings, 1981 Oil Spill Conference. API Publication 4334. American Petroleum Institute, Washington, D.C.

Thomas, M.L.H. 1973. Effects of Bunker C oil on intertidal and lagoonal biota in Chedabucto Bay, Nova Scotia. J. Fish. Res. Board Can. 30(1):83-90.

Thomas, M.L.H. 1977. Long term biological effects of Bunker C oil in the intertidal zone, pp. 238-245. In D.A. Wolfe, ed. Fate and Effects of Petroleum Hydrocarbons in Marine Organisms and Ecosystems. Pergamon, New York.

Thomas, M.L.H. 1978. Comparison of oiled and unoiled intertidal communities in Chedabucto Bay, Nova Scotia. J. Fish Res. Board Can. 35:707-716.

Thomas, P., B.R. Woodin, and J.M. Neff. 1980. Biochemical responses of the striped mullet Mugil cephalus to oil exposure. I. Acute responses, interrenal activations and secondary stress responses. Mar. Biol. 59:141-149.

Thomas, R.E., and S.D. Rice. 1979. The effect of exposure temperatures on oxygen consumption and opercular breathing rates of pink salmon fry exposed to toluene, naphthalene, and water-soluble fractions of Cook Inlet crude oil and No. 2 fuel oil, pp. 39-52. In W.B. Vernberg, A. Calabrese, F.P. Thurberg, and F.J. Vernberg, eds. Marine Pollution: Functional Responses. Academic Press, New York.

Thornton, D.E., and P.J. Blackall. 1980. Experimental oil spills: The Baffin Island oil spill in particular, pp. 486-494. In Proceedings, 3rd Arctic Marine Oilspill Program. Technical Seminar. Research and Development Division, Environmental Protection Service, Ottawa, Ontario.

Tilseth, S., T.S. Solberg, and K. Westrheim. 1984. Sublethal effects of the water-soluble fraction of Ekofisk crude oil on the early larval stages of cod (Gadus morhua L.). Mar. Environ. Res. 11:1-16.

Tomkins, B.A., H. Kubota, W.H. Griest, J.E. Caton, B.R. Clark, and M.R. Guerin. 1980. Determination of benzo(a)pyrene in petroleum substitutes. Anal. Chem. 52(8):1331-1334.

Tong, C., S. Ved Brat, and G.M. Williams. 1981. Sister-chromatid exchange induction by polycyclic aromatic hydrocarbons in an intact cell system of adult rat-liver epithelia cells. Mut. Res. 91:467-473.

Topinka, J.A., and L.R. Tucker. 1981. Long-term oil contamination of fucoid marcoalgae following the Amoco Cadiz oil spill, pp. 393-404. In Conan et al., eds. Amoco Cadiz: consequences d'une pollution accidentelle par les hydrocarbures. CNEXO, Paris.

Tosterson, T.R., et al. 1977. Bahai Sucia: a re-evaluation of the biota affected by petrochemical contamination in March, 1973. University of Puerto Rico, Department of Marine Science, C.A.A.M. 138 pp. + appendices.

Tramier, B., G.H.R. Aston, M. Durrieu, A. Lepain, J.A.C.M. van Oudenhoven, N. Robinson, K.W. Sedlacek, and P. Sibra. 1981. A field guide to coastal oil spill control and cleanup techniques. Report 9/81. CONCAWE, Den Haag.

Tso, W.-W., and J. Adler. 1974. Negative chemotaxis in Escherichia coli. J. Bacteriol. 118:560-576.

Twort, C.A., and J.M. Twort. 1931. The carcinogenic potency of mineral oils. J. Ind. Hygiene 13:204-206.

U.S. Coast Guard. 1959. Efforts to reduce oil pollution. Proc. Merchant Mar. Counc. 16:199-203.

U.S. Coast Guard. 1969. Sunken tanker project report. Washington, D.C.

U.S. Department of Health, Education and Welfare. 1973. The Industrial Environment--Its Evaluation and Control. Public Health Service, Center for Disease Control, National Institute for Occupational Safety and Health. U.S. Government Printing Office, Washington, D.C.

U.S. Environmental Protection Agency. 1975. Supplement to development document, hazardous substances regulations, Federal Water Pollution Control Act, as amended 1972. U.S. EPA Office of Water Planning and Standards Report EPA-440/9-75-009. National Technical Information Service, U.S. Department of Commerce, Springfield, Va.

Vale, G.L., G.S. Sidhu, W.A. Montgomery, and A.R. Johnson. 1970. Studies on a kerosene-like taint in mullet (Mugil cephalus). I. General nature of taint. J. Sci. Food. Agric. 21:429-432.

Van Baalen, C., and D.T. Gibson. 1982. Biodegradation of aromatic compounds by high latitude phytoplankton. Final Report RD/MPF24-Effects-675. U.S. Department of Commerce, NOAA/OMPA.

Vandermeulen, J.H. 1977. The Chedabucto Bay spill--Arrow, 1970. Oceanus 20(4):31-39.

Vandermeulen, J.H. 1980a. Chemical dispersion of oil in coastal low-energy systems: saltmarshes and tidal rivers. In D. Mackay, P.G. Wells, and S. Patterson, eds. Chemical Dispersion of Oil Spills. Publication No. EE-17. Institute of Environmental Studies, University of Toronto.

Vandermeulen, J.H., ed. 1980b. Scientific studies during the Kurdistan Tanker Incident. Proceedings of a workshop. Report Series BI-R-80-3. Bedford Institute of Oceanography, Dartmouth, N.S. 227 pp.

Vandermeulen, J.H. 1981. Contamination des organismes marins par les hydrocarbures. Note de synthese, pp. 563-572. In Amoco Cadiz: Fates and Effects of the Oil Spill. CNEXO, Paris.

Vandermeulen, J.H. 1982. Some conclusions regarding long-term biological effects of some major oil spills. Philos. Trans. R. Soc. London, Ser. B 297:335-351.

Vandermeulen, J.H., and T.P. Ahern. 1976. Effect of petroleum hydrocarbons on algal physiology: review and progress report, pp. 107-126. In A.P.M. Lockwood, ed. Effects of Pollutants on Aquatic Organisms. Cambridge University Press, New York.

Vandermeulen, J.H., and E.S. Gilfillan. 1984. Petroleum pollution, corals and mangroves. Amer. Technol. Soc. J. In press.

Vandermeulen, J.H., and D.C. Gordon, Jr. 1976. Re-entry of five year old standard Bunker C fuel oil from a low-energy beach into the water, sediments, and biota of Chedabucto Bay, Nova Scotia. J. Fish. Res. Board Can. 33(9):2002-2010.

Vandermeulen, J.H., and W.R. Penrose. 1978. Absence of aryl hydrocarbon hydroxylase (AHH) in three marine bivalves. J. Fish. Res. Board Can. 35:643-647.

Vandermeulen, J.H., and D.J. Scarratt. 1979. Impact of oil spills on living natural resources and resource-based industry, pp. 91-96. In Evaluation of Recent Data Relative to Potential Oil Spills in the Passamaquoddy Area. Fisheries and Marine Service Technical Report 901. Fisheries and Environment Canada, Ottawa, Ontario.

Vandermeulen, J.H., D.E. Buckley, E.M. Levy, B.F. Long, P. McLaren, and P.G. Wells. 1978. Immediate impact of Amoco Cadiz environmental oiling: oil behavior and burial, and biological aspects, pp. 159-173. In G. Conan, L. D'Ozouville, and M. Marchand, eds. Amoco Cadiz: Preliminary Observations of the Oil Spill Impact on the Marine Environment. Actes de Colloques 6. CNEXO, Paris.

Vandermeulen, J.H., B.F.N. Long, and T.P. Ahern. 1981. Bioavailability of stranded Amoco Cadiz oil as a function of environmental self-cleaning April 1978 - January 1979, pp. 585-598. In Amoco Cadiz: Fates and Effects of the Oil Spill. CNEXO, Paris.

Vandermeulen, J.H., Wm. Silvert, and A. Foda. In press. Sublethal hydrocarbon phytotoxicity in the marine unicellular alga Pavlova lutheri Droop. Aquat. Toxicol.

Van Haaften, J.L. 1973. Die Bewirtschaftung von Seehunden in den Niederlanden. Beitr. Jagdund Wildforsch. 8:345-349.

Varanasi, U., and D.J. Gmur. 1980. Metabolic activation and covalent binding of benzo(a)pyrene to deoxyribonucleic acid catalyzed by liver enzymes of marine fish. Biochem. Pharmacol. 29:753-761.

Varanasi, U., D.J. Gmur, and M.M. Krahn. 1980. Metabolism and subsequent binding of benzo(a)pyrene to DNA in pleuronectic and salmonid fish, pp. 455-470. In Polynuclear Aromatic Hydrocarbons: Fourth International Symposium on Analysis, Chemistry, and Biology. Battelle Press, Columbus, Ohio.

Varanasi, U., J.E. Stein, M. Nishimoto, and T. Hom. 1982. Benzo(a)pyrene metabolites in liver, muscle, gonads and bile, pp. 1221-1234. In M.W. Cooke and A.J. Dennis, eds. Polynuclear Aromatic Hydrocarbons: 7th International Symposium on Formation, Metabolism and Measurement. Battelle Press, Columbus, Ohio.

Vargo, S.L. 1981. The effects of chronic low concentrations of No. 2 fuel oil on the physiology of a temperate estuarine zooplankton community in the MERL microcosms, pp. 295-322. In F.J. Vernberg, A. Calabrese, F.P. Thurberg, and W.B. Vernberg, eds. Biological Monitoring of Marine Pollutants. Academic Press, New York.

Vashchenko, M.A. 1980. Effects of oil pollution on the development of sex cells in sea urchins. Helgolander Meeresunters. 33:297-300.

VAST/TRC. 1976. Oil spill - Bahia Sucia, Puerto Rico, 18 March 1973: environmental effects. EPA Contract 68-10-0542.

Vaughan, B.E. 1973. Effects of oil and chronically dispersed oil on selected marine biota - a laboratory study. API Publication 4191. American Petroleum Institute, Washington, D.C. 105 pp.

Vaziri, N.D., P.J. Smith, and A. Wilson. 1980. Toxicity with intravenous injection of naphtha in man. Clinical Toxicol. 16(3):335-343.

Vermeer, K., and G.G. Anweiler. 1975. Oil threat to aquatic birds along the Yukon coast, Alaska. Wilson Bull. 87:467-480.

von Borstel, R.C., and M.L. Rekeymer. 1959. Radiation-induced and genetically contrived dominant lethality in Habrobracon and Drosophila. Genetics 44:1053-1074.

Vuorinen, P., and M.B. Axell. 1980. Effects of the water-soluble fraction of crude oil on herring eggs and pike fry. ICES C.M. 1980/E:30. 10 pp. International Council for the Exploration of the Seas, Charlottenlund, Denmark.

Wadhams, P. 1980. Oil and ice in the Beaufort Sea--the physical effects of a hypothetical blowout, pp. 231-250. In Petromar 80: Petroleum in the Marine Environment. Association Europeenne Oceanique, Monaco.

Walcave, L., H. Garcia, R. Feldman, W. Lijinsky, and P. Shubik. 1971. Skin tumorigenesis in mice by petroleum asphalts and coal tar pitches of known polynuclear aromatic hydrocarbon content. Toxicol. Appl. Pharmacol. 18:41-52.

Walker, J.D., and R.R. Colwell. 1975. Some effects of petroleum on estuarine and marine microorganisms. Can. J. Microbiol. 21: 305-313.

Walton, D.G., W.R. Penrose, and J.M. Greene. 1978. The petroleum-inducible mixed-function oxidase of cunner (Tautogolabrus adspersus Walbaum 1972): some characteristics relevant to hydrocarbon monitoring. J. Fish. Res. Board Can. 35(12):1547-1552.

Wang, R.T., and J.A.C. Nicol. 1977. Effects of fuel oil on sea catfish: feeding activity and cardiac responses. Bull. Environ. Contam. Toxicol. 18:170-176.

Ware, D.M. 1982. Recruitment variability--can we detect the effect of oil pollution? In A Longhurst, ed. Consultation on the Consequences of Offshore Oil Production on Offshore Fish Stocks and Fishing Operations. Can. Technol. Rep. Fish. Aquat. Sci. 64-73.

Warner, R.F. 1969. Experimental effects of oil pollution in Canada. An evaluation of problems and research needs. Miscellaneous Report 645. Canadian Wildlife Service, Ottawa, Ontario.

Weaver, N.K., and R.L. Gibson. 1979. The U.S. oil shale industry: a health perspective. Am. Ind. Hygiene J. 40:460-467.

Weber, D.D., W. Gronlund, T. Scherman, and D. Brown. 1979. Non-avoidance of oil-contaminated sediment by juvenile English sole (Parophrys vetulus). In A. Calabrese and F.P. Thurburg, eds. Proceedings, Symposium on Pollution and Physiology of Marine Organisms.

Weber, D.D., D.J. Maynard, W.D. Gronlund, and V. Konchin. 1981. Avoidance reactions of migrating adult salmon to petroleum hydrocarbons. Can. J. Fish. Aquat. Sci. 38. In press.

Wells, P.G. 1976. Effects of Venezuelan crude oil on young stages of the American lobster, <u>Homarus</u> <u>americanus</u>. Ph.D. thesis. University of Guelph, Guelph, Ontario.

Wells, P.G. 1982. Zooplankton, Chap. 6, pp. 65-80. In J.B. Sprague, J.H. Vandermeulen, and P.G. Wells, eds. Oil and Dispersants in Canadian Seas - Research Appraisal and Recommendations. Economic and Technical Review Report EPS-3-EC-82-2. Environment Canada, Ottawa, Ontario.

Wells, P.G. 1984. The toxicity of oil spill dispersants to marine organisms: A current perspective. In Th.E. Allen, ed. Oil Spill Chemical Dispersants: Research, Experience, and Recommendations, STP 840. American Society for Testing and Materials, Philadelphia.

Wells, P.G., and J.B. Sprague. 1976. Effects of crude oil on American lobster (<u>Homarus</u> <u>americanus</u>) larvae in the laboratory. J. Fish. Res. Board Can. 33(7):1604-1614.

Whipple, J.A., T.G. Yocom, D.R. Smart, and M.H. Cohen. 1978. Effects of chronic concentrations of petroleum hydrocarbons on gonadal maturation in starry flounder (<u>Platichthys</u> <u>stellatus</u> [Pallas]), pp. 757-806. In Proceedings of Conference on Assessment of Ecological Impacts of Oil Spills. American Institute of Biological Sciences, Arlington, Va.

Whipple, J.A., M.Bv. Eldridge, and P. Benville, Jr. 1981. An ecological perspective of the effects of monocyclic aromatic hydrocarbons on fishes, pp. 483-551. In F.J. Vernberg, A. Calabrese, F.P. Thurberg, and W.B. Vernberg, eds. Biological Monitoring of Marine Pollutants. Academic Press, New York

White, D.H., K.A. King, and N.C. Coon. 1979. Effects of No. 2 fuel oil on hatchability of marine and estuarine bird eggs. Bull. Environ. Contam. Toxicol. 21(1/2):7-10.

Whittle, K.J. 1978. Tainting in marine fish and shellfish with reference to the Mediterranean Sea, pp. 89-108. In Data Profiles for Chemicals for the Evaluation of their Hazards to the Environment of the Mediterranean Sea. Vol. II. IRPTC Data Profile Series, Number One. International Register of Potentially Toxic Chemicals, UNEP, Geneva.

Whong, W.-Z., W.G. Sorenson, J.A. Elliott, J. Stewart, J. Simpson, L. Piacitelli, M. McCawley, and T. Ong. 1982. Mutagenicity of oil-shale ash. Mut. Res. 103:5-12.

Wilder, D.G. 1970. The tainting of lobster meat by Bunker C oil alone or in combination with the dispersant Corexit. J. Fish. Res. Board Can. Misc. Rep. 1087:25.

Williams, T.D. 1978. Chemical immobilization, baseline parameters and oil contamination in the sea otter. Report No. MMC-77/06. Marine Mammal Commission, Washington, D.C.

Wilson, M.P., J.G. Quinn, and K. Sherman, eds. 1978. In the Wake of the <u>Argo</u> <u>Merchant</u>. Center for Ocean Management Studies, University of Rhode Island, Kingston. 181 pp.

Winters, K., J.C. Batterton, and C. Van Baalen. 1977. Phenalen-1: occurrence in a fuel oil and toxicity to microalgae. Environ. Sci. Technol. 11:270-272.

Wolfe, D.A., ed. 1977. Fate and Effects of Petroleum Hydrocarbons in Marine Organisms and Ecosystems. Pergamon, New York. 478 pp.

Wong, C.S., W.J. Cretney, P. Christensen, and R.W. MacDonald. 1976. Hydrocarbon levels in the marine environment of the southern Beaufort Sea. Beaufort Sea Project Technical Report 38. Department of the Environment, Victoria, B.C. 113 pp.

Wong, M.P., and F.R. Engelhardt. 1982. Effects of petroleum hydrocarbons on branchial adenosine triphosphatases (ATPases) in fresh and salt water acclimated rainbow trout (Salmo gairdneri), pp. 281-296. In P.J. Rand, ed. Land and Water Issues in Resource Development. Ann Arbor Science Publishers, Ann Arbor, Mich.

Woodhouse, D.L., and J.O. Irwin. 1950. The carcinogenic activity of some petroleum fractions and extracts: comparative results in test on mice reported after an interval of eighteen months. J. Hygiene 47:121-134.

Woodin, S.A., C.F. Nyblade, and F.S. Chia. 1971. Oil spill on Guemes Island, Washington. Preliminary Report, Friday Harbor Laboratories, University of Washington, Seattle. 8 pp.

Woodward, D.F., P.M. Mehrle, Jr., and W.L. Mauck. 1981. Accumulation and sublethal effects of a Wyoming crude oil in cutthroat trout. Trans. Am. Fish. Soc. 110:437-445.

Woodward, P.W., G.P. Sturm, Jr., J.W. Vogh, S.A. Holmes, and J.E. Dooley. 1976. Compositional analyses of synthoic from West Virginia. Cal. BERC/RI 76/2 January. Bartlesville Energy Research Center, Bartlesville, Okla.

Yevich, P.P., and C.A. Barszcz. 1977. Neoplasia in soft-shell clams (Mya arenaria) collected from oil-impacted sites. Ann. NY Acad. Sci. 298:409-426.

Young, G.P. 1977. Effects of naphthalene and phenanthrene on the grass shrimp Palaemonetes pugio (Holthius). Master's thesis. The Gradute College, Texas A&M University, College Station. 67 pp.

Young, L.Y, and R. Mitchell. 1973. Negative chemotaxis of marine bacteria to toxic chemicals. Appl. Microbiol. 25:972-975.

Zechmeister, L., and B.K. Koe. 1952. The isolation of carcinogenic and other polycyclic aromatic hydrocarbons from barnacles. Arch. Biochem. Biophys. 35:1-11.

Zedeck, M.S. 1980. Polycyclic aromatic hydrocarbons--A Review. J. Environ. Pathol. Toxicol. 3:537-567.

Zieman, J.C. 1982. The ecology of the seagrasses of South Florida: A community profile. Report No. FWS/OBS-82/25. U.S. Fish and Wildlife Services, Office of Biological Services, Washington, D.C. 158 pp.

Zieserl, E. 1979. Hydrocarbon ingestion and poisoning. Compr. Ther. 5(6):35-42.

APPENDIX A

Impact of Some Major Spills
(Spill Case Histories)

INTRODUCTION

Both acute spills and oil seeps have provided opportunities for field
study of the impact of oil on the surrounding marine ecosystems.
Popularly, they are thought to resemble each other rather closely, and
indeed it is often assumed that the natural oil seep provides the sort
of controlled experimental situation for field studies that cannot be
found with the unpredictable acute tanker spills.

Scientific studies of tanker spills present several problems for
the serious scientist--awesome difficulties in field sampling, and
readiness of personnel and equipment. Spills are not anticipated, and
in the past, personnel and equipment have seldom been readily available.
Also, most spills occur in areas that have not been studied previously,
and adequate controls are rare. Spills frequently occur in weather
conditions that make sampling difficult or impossible. These problems
are compounded in offshore spills, where sampling becomes much more
difficult, background data are less available, and the expense of large
ship operations is difficult to finance on short notice.

Despite these difficulties, it is encouraging that acute spills
have not only continued to receive scientific attention since the
writing of the last NRC report, but also that field studies have
increased in number and in scope and have yielded some valuable data.
The long term follow-up studies have provided further understanding,
both of the vulnerability of the various ecosystems and of the
biological recovery processes.

Natural oil seeps, on the other hand, have received much less
attention, despite their seemingly obvious availability as a natural
experimental spill situation. Geographically, these sites are not
easily accessible, generally located in unsettled offshore locations,
and removed from marine laboratories. The major exception to this, the
seep off southern California, is located near a heavily populated area
with several major marine laboratories, and indeed has been the subject
of a number of studies. More recently, work has been initiated on two
seeps off the east Baffin Island coast in the Canadian Arctic. In this
case the seeps lie in an area of considerable geological interest, and
on the cruise track of annual Arctic research cruises out of eastern
Canada.

However, seeps differ in several aspects from tanker spills and from well blowouts as to the chemical composition of the oil fraction accomodated in the water and the generally lower rate of release and local concentration. Seeps therefore are not truly analogous to controlled tanker spills, but they do provide an opportunity to obtain parallel observations, particularly on the weathering processes of oil and on the impact of such aged oil on surrounding biota under chronic conditions.

Only a few major oil spills have been the object of detailed scientific study, but among these are two of the largest that have ever occurred--the Ixtoc I blowout of June 9, 1979, in the Gulf of Campeche, Mexico, and the breakup of the tanker Amoco Cadiz off the coast of Brittany, France, on March 16, 1978. In addition to these giants there are a number of smaller spills that have received considerable scientific attention--the Florida spill in Buzzards Bay, Massachusetts (1969), the Arrow spill in Chedabucto Bay, Nova Scotia (1970), the Metula spill in the Strait of Magellan (1974), the Argo Merchant breakup off the east coast of the United States (1976), the Tsesis in the Baltic south of Stockholm (1977), and the Kurdistan in Cabot-Strait off Nova Scotia (1979). The first spill to receive serious scientific attention, the Torrey Canyon off the south coast of England (1967), remains of interest to this date. However, the heavy use of dispersants and other chemical and physical treatment agents places this spill in a separate category, and many of the impacts observed were due largely to the awesome cleanup efforts used and not to the spilled oil. A considerable amount of information on both the behavior and the fate of the spilled oil in the marine environment and on the impact on the living resources has been derived through these studies. However, all these spills differ both in the fate of the oil and in their biological impacts, with each spill presenting yet another set of conditions and facets of petroleum efforts in the oceans. The examples described below demonstrate this variability and dissimilarity, but at the same time they indicate some common features.

AN INSHORE SPILL: THE BARGE FLORIDA

Two spills in relatively protected waters were intensively studied for many years: the Florida barge spill of 1969 (Sanders et al., 1972, 1980) in the West Falmouth area of Buzzards Bay, Massachusetts, and the Arrow spill of 1970 in Chedabucto Bay, Nova Scotia.

The barge Florida grounded on rocks off West Falmouth Harbor, Buzzards Bay, Massachusetts, and lost 630 tons of No. 2 fuel oil. A storm the following day drove the oil ashore, mixing it into water and sediments. There were immediate kills of small fishes, benthic invertebrates, and marsh organisms. Some dispersants were used, and booms were deployed in an attempt to keep the oil out of West Falmouth and Wild Harbor. Visible oil never appeared in West Falmouth, but the booms were unable to keep oil out of Wild Harbor. The extent of the immediate kill was documented by timely sampling before the dead

organisms decomposed. Oil and its effects persisted for at least 10
years after the spill (Sanders et al., 1980).

Oil Fate

The slick was washed ashore and mixed into the water rapidly. There
were no special studies of any effects that the intact slick might have
had in the short period before its breakup. No concentration measurements were made on oil in the water column.

In the sediments the spilled oil fared rather differently, with
recognizable components of the spilled No. 2 fuel oil persisting for at
least 8 years. Sediments from the Wild Harbor station yielded 1-3 mg
aromatic fraction per gram of dry weight until at least July 1976,
compared with 0.02-0.04 mg/g for control stations (Teal et al., 1978).

Impact on Biota

Fish

An immediate fish kill was documented at the time of the spill, with
fish washed ashore in windrows (Hampson and Sanders, 1969). Measurements on the mixed function oxygenase (MFO) system in Fundulus
heteroclitus from Wild Harbor revealed induction, i.e., enhanced levels
of the activity of this hydrocarbon-metabolizing enzyme system, in
comparison with similar fish taken from control stations (Burns,
1976). Four years after the spill, the fish showed a reduction of body
burden of hydrocarbons to near background levels, presumed to be the
result of the MFO enzyme activity (Burns and Teal, 1979). High MFO
levels continued to be measured 8 years after the spill in F.
heteroclitus from this area, correlating with the persistence of oil in
the sediments (Stegeman, 1978).

Benthos

The most persistent impact was in the benthic macrofaunal communities
(Sanders, 1978). Within 48 hours after the arrival of the oil from the
Florida, there was nearly total eradication of the macrobenthos at the
most heavily oiled sites, with oil concentrations exceeding 133 µg/g
wet weight (ca. 400 µg/g dry weight). At sites with intermediate oil
levels (9-100 µg/g), there were intermediate reductions as compared
with control sites. Soft-bodied animals killed by the oil disappeared
within 1 week. The deaths would not have been detectable if sampling
had been initiated later than a few days after the spill (Sanders et
al., 1980). Ampeliscid amphipods were particularly vulnerable to oil,
in part because of the habit of these organisms to move into contaminated sediments. These declines in the macrobenthos continued until
the oil had decreased sufficiently in concentration and toxic components
to permit their survival (Sanders et al., 1972).

Opportunist species typified by <u>Capitella</u> in the inshore stations increased greatly in abundance, monopolizing the otherwise defaunated sediments for the first 11 months following the spill. At that time this opportunistic species peaked in population, and subsequently "crashed." A similar course was observed with another opportunistic species <u>Mediomastis</u> in sites further offshore, i.e., steady increase in population numbers while overwhelming the defaunated area, followed by a rapid crash, or drop in numbers. It differed from <u>Capitella</u> only in that it peaked somewhat later and its opportunistic period occurred for several months later.

In general the changes in fauna matched the extent of pollution by the No. 2 fuel oil, both in intensity and duration (Sanders et al., 1980). Faunal changes included decreases in diversity, in density, and in numbers of species. The recovery pattern of the impacted communities also showed abnormalities that could be linked to the extent of oiling. For example, at the minimally oiled sites the recovery process was rapid with short term effects. Recovery was essentially complete within a year after the spill. However, at the intermediate and heavily oiled sites the recovery process was markedly different. At the intermediate polluted sites the recovery process was dominated by the initial defaunation and subsequent high postlarval settlment (half million/cm^{-2}), leading to high numbers of individuals (high richness) but low numbers of species (low evenness). Here a more normal recovery pattern was not evident until 3 years after the spill. At the most heavily oiled sites a "normal" recovery pattern was not evident for the 52 months of study following the spill (Sanders, 1978; Sanders et al., 1980).

There were no detailed meiofaunal studies, although in some of the initial samples taken in the most heavily oiled stations, field notes showed the presence of large numbers of nematodes.

Intertidal Communities

Marsh grass (<u>Spartina alterniflora</u>) was completely killed off on the most heavily oiled parts of the intertidal area (oil concentrations over 2,000 µg/g). By 1981 recovery was not yet complete, although most areas seemed normal in appearance at first visual inspection.

Bivalves were particularly susceptible to the oil and its effects. Approximately 77 bushels of soft-shell clams (<u>Mya arenaria</u>) and 11,200 bushels of seed clams were reported killed in Wild Harbor (Souza, 1970).

Fiddler crabs (<u>Uca pugnax</u>) were reduced in density in the oiled marsh, which, as in the case with the benthic amphipods, acted as a lethal trap for these territorial organisms. Behavioral changes caused by the oil included slowing of movements and digging of burrows, the latter being shallower than normal. Newly settled animals appeared to be more susceptible to the oil than the adults, and their settling success was sharply reduced (Krebs and Burns, 1977). Recovery was found to be highly correlated with the loss of the naphthalene fraction of the oil trapped in the sediments. However, recovery of the fiddler crab population was not complete in 1977, 7 years after the spill

(Krebs and Burns, 1977). Induction of MFO system activity was detected in tissues of the fiddler crabs, but it would seem that the MFO levels were insufficient to deal with the body burden of hydrocarbons within the lifetime of these organisms (Burns, 1976).

Changes in populations, similar to those described for the inter-tidal areas, were found in the soft-bottom intertidal areas below the salt marsh (Sanders et al., 1980) (Figure A-1).

AN OPEN BAY SPILL: THE ARROW

The Arrow spill (Anon, 1970) occurred on February 4, 1970, in Chedabucto Bay, Nova Scotia, when the tanker ran aground on Cerberus Rock on her way into the off-loading facilities in the Strait of Canso. She was carrying 15,000 tons of Bunker C fuel oil, of which about two-thirds were released into the waters of the bay (Anon, 1970).

Although Chedabucto Bay in some ways represents a relatively sheltered environment (its northern half consists of numerous small lagoons and shallow embayments), the entrance to the bay opens directly onto the Atlantic Ocean, and at the time of the accident the prevailing winds caused high sea state conditions within the bay. As a result, oil driven by wind (Figure A-2) and wave action coated over 300 km of the bay's shorelines (Figure A-3), before the remainder of the oil was swept out of the bay and into the Atlantic. Eventually oil from the Arrow was traced as far south as Halifax, N.S., and Bermuda.

Oil Fate

In May 1970, 3 months after the spill, levels as high as 100 µg/L were found in the water column (Levy, 1971); but by April 1971, concentrations had dropped to background levels, ca. 1 µg/L (Gordon and Michalik, 1971).

Oiling along the southern and western shores of the bay resulted in a mixture of oil with sand, gravel, and rocks to yield a resistant pavement of tar along much of the coastline. By 1976 such surface oiling was sharply reduced, either by wave erosion or by burial, and could be found visually primarily in a few "hotspots" (Janvrin Lagoon, Inhabitants Bay, Black Duck Cove) (Vandermeulen, 1977) (Figure A-4). However, a parallel chemical analysis of subsurface sediments indicated high concentrations of Arrow oil persisting below the surface within the beaches (1,280 µg Bunker C per gram of sediment at 7-11 cm, compared with 106 at the surface and 27 µg/g at 12-15 cm), represent-ing a potential long term source of reentry of spilled oil (Figure A-5) (Keizer et al., 1978). But by this time, the origins of these hydro-carbons could no longer be unequivocally traced to the Arrow because of weathering and contamination from subsequent spills (Keizer et al., 1978).

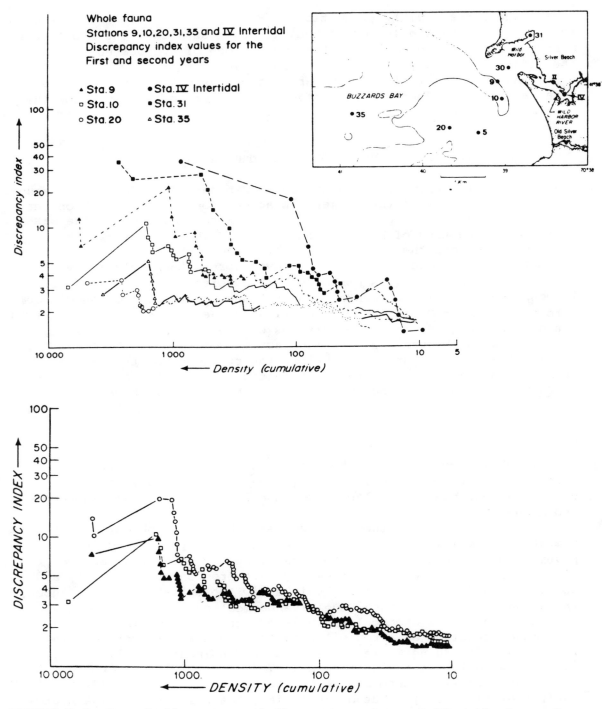

FIGURE A-1 Faunal discrepancy indices at various stations in Buzzards Bay, Massachusetts, following the _Florida_ spill. (Top) Indices obtained by comparing the composition of the fauna in the first with the second year of study. (Bottom) Indices at station 10 (viz., inset) for the first and second, first and third, and second and third year.

SOURCE: Adapted from Sanders (1978).

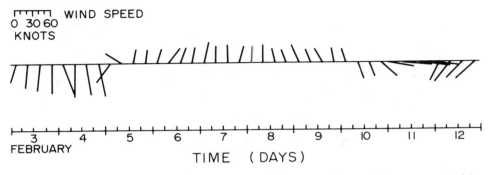

FIGURE A-2 Mean wind vectors over Chedabucto Bay, February 3-12, 1970. Shown are the 6-hourly mean vectors of the reduced geostrophic winds.

SOURCE: Based on data from Anon (1970).

Impact on Biota

Most of the work on organisms was done on littoral communities, with very few observations made in the water column. Thus, for example, no data of impact on fish are available.

Conover (1971) found incorporation of oil droplets by copepods in the bay water column. Most apparently passed through the animals without modification, although no detailed uptake or tissue hydrocarbon studies were done at the time. The oil droplets, many of the general size range of the food of the copepods, were apparently filtered from the water column by the animals. Eventually a considerable portion of the oil droplets became associated with the fecal pellets. As much as 10% of the oil in the water was associated with the copepods, and up to 7% was found in the fecal pellets, sugggesting that this route may be an important sedimentation route for spilled oil. There appeared to be no obvious effect of the oil on the copepods, although no data are available on this.

Benthos

Most studies on benthic organisms were carried out in the rocky and sedimentary intertidal areas (Thomas, 1973, 1977, 1978), although in follow-up studies, attention was focused on the low energy, silt-dominated lagoons. The oil was concentrated primarily in the upper two-thirds of the intertidal zone. Studies showed the oil to be most persistent when stranded along the mean high tide line, where in sheltered lagoons it was still present visually 10 years after the event. The rockweed Fucus vesiculosus was reduced in vertical distribution for about 5 years. Fucus spiralis, which is confined generally to the region up to the high tide line, was killed off completely and had not reappeared in the oiled region by 1976, 6 years after the spill. In sheltered areas the marsh grass Spartina

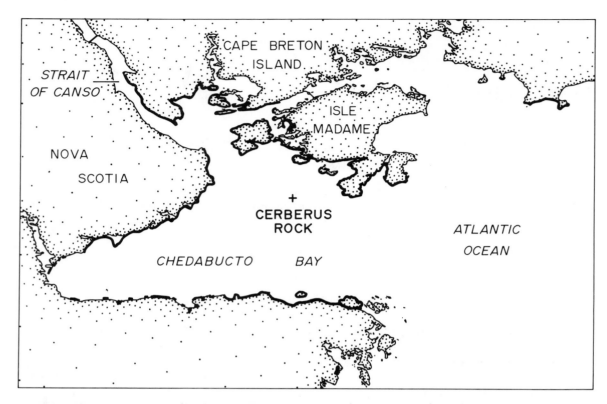

FIGURE A-3 Geographical extent of shoreline contamination in Chedabucto Bay, February 1970, immediately following Arrow breakup.

SOURCE: From Anon (1970).

alterniflora population declined steadily after the spill, with few surviving plants remaining 1 year later, However, it recovered 2 years later, by 1973. Rocky shore animals including barnacles and periwinkles did not change in abundance or in distribution except where their habitat had been altered by changes in the rockweed, demonstrating the significance of community associations (Thomas, 1978). Larvae of the common barnacle Balanus balanoides apparently settled and grew normally, even during the spill year, 1970 (Thomas, 1977). In contrast, follow-up studies suggested changes in bivalve larval recruitment 6 years afterward (Gilfillan and Vandermeulen, 1978).

A detailed follow-up study was done in 1976, 6 years after the spill, when sediments from oiled sites still contained 10-25,000 µg/g of oil (measured with fluorescence). Species diversity (Shannon-Weiner index) was lower at oiled than at unoiled control sites. Macrofaunal biomass was ca. 1,400 wet g/m^2 at oiled sites, versus approximately 4,400 wet g/m^2 at control stations. Oil concentrations in living clams in 1976 averaged between 150 and 350 µg/g, compared with 650 µg/g in recently dead bivalves in 1970. Periwinkles also were found to be contaminated with oil, but the average level of contamination was only 12-18 µg/g. The marsh grass S. alterniflora from six oiled

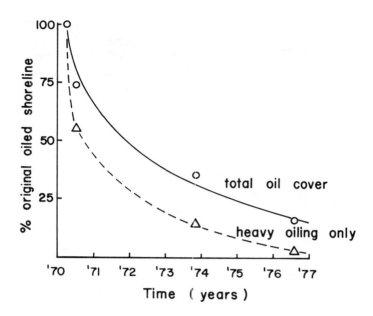

FIGURE A-4 Erosion pattern of stranded Arrow Bunker C fuel oil on
Chedabucto Bay shorelines, 1970-1976. Curves based on shoreline
surveys and visual inspection of residual stranded oil.

SOURCE: Vandermeulen (1977).

sites still showed surprisingly high contamination of about 15,000
µg/g, compared with less than 70 in control (Thomas, 1978).

Six and seven years after the spill, populations of soft-shelled
clams from oiled sites were still stressed (Gilfillan and Vandermeulen,
1978). Fewer mature adults were found at oiled stations. Individuals
showed lower shell growth, lower assimilation rates, and lags of 1-2
years in tissue growth. These observations were confirmed in parallel
studies by Thomas (1978), and reduced weight of body and shell
persisted through 1979 (MacDonald and Thomas, 1982).

On the other hand, populations of the lugworm Arenicola were more
abundant in oiled sediments in 1976 than anywhere else in Nova Scotia.
They did exhibit elevated hydrocarbon concentrations, suggesting that
they are relatively resistant to oil pollution (Gordon et al., 1978).

AN OPEN OCEAN SPILL WITH OFFSHORE WINDS: THE ARGO MERCHANT

The Argo Merchant spill (Grose and Mattson, 1977; Wilson et al., 1978),
in several ways, represents the opposite to the Arrow spill. Both
spills occurred in winter, on the northeast coast of North America,
with the same oil cargo (Bunker C fuel oil). However, whereas the
Arrow broke up in a large embayment, with initially onshore winds, the
Argo Merchant ran aground and broke up in open waters, with prevailing
offshore winds for most of the spill period. In the end, much of the

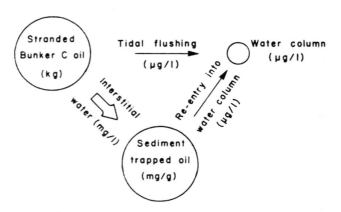

FIGURE A-5 Summary of stranded Bunker C fuel oil reentry pattern into marine environment by oil stranded on low energy gravel-cobble beach.

SOURCE: Vandermeulen and Gordon (1976).

Arrow's cargo became stranded on adjacent coastlines, while the oil from the Argo Merchant disappeared from view in the Atlantic Ocean.

The Argo Merchant ran aground on Nantucket Shoals, off Massachusetts, on December 15, 1976 (Figure A-6), and over the next month spilled almost her entire cargo (29,000 tons) of No. 6 fuel oil (Grose and Mattson, 1977). The cargo also contained about 20% of its volume in cutting stock (equivalent to No. 2 fuel oil) for thinning purposes. Storms broke up the vessel after grounding, and attempts to pump the oil into another vessel failed. Burning the oil was tried without success. No dispersants were used.

Oil escaped from the wreck for 1 month after grounding, but surprisingly little oil was found later in its immediate vicinity. In February 1977, significant contamination was found near the wreck, extending at least down to 8- to 13-cm depth, but by July 1977 no evident cargo oil remained. It is speculated that the bow forced oil into the sand, or that sand was forced against the hull by currents and carried the oil away from the wreck along the bottom. Most of the oil that appeared on the surface was formed into large floating "pancakes" and disappeared into the ocean to the east (Figure A-6). Parts of the cutting stock dissolved and could be detected under the slick at concentrations up to 250 µg/L.

It was the occasion for one of the most elaborate slick monitoring efforts up to that time (e.g., Grose and Mattson, 1977; Spaulding, 1978), but because of the bad weather, relatively few samples of water, sediment, or biota were obtained (Grose and Mattson, 1977; Wilson et al., 1978). Despite the relatively high potential toxicity of the cutting stock in the cargo, there was little evidence of impact on the marine fauna or phytoplankton. The accident occurred at the time when the fewest potential effects on pelagic organisms would be expected: a period of low productivity in the water column, with few fish eggs and larvae present. The spill did provide, however, the first indications

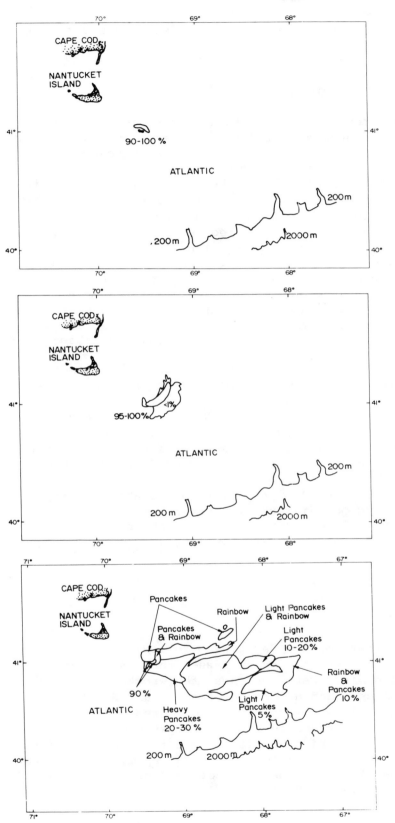

FIGURE A-6 Horizontal dispersion of oil spilled from the _Argo Merchant_, December 17, 20, and 23, 1977. SOURCE: Lissauer and Welsh (1978).

of potential impact of oil on the development of fish eggs under field spill conditions.

Oil Fate

The resulting slick soon broke up into a mixture of thick pancakes surrounded by sheen. The pancakes varied in thickness, averaging about 1 cm with smooth surface. Underside examination of such pancakes by divers showed them to be similar in appearance.

Dispersion of the oil as droplets into the water column was studied on three cruises. Droplets of pure oil larger than 100 µm in diameter were found in five out of forty-two samples. Samples of water taken from 6 m above the bottom contained oil-coated sediment particles, ranging in diameter from 100 to 500 µm (Cornillon, 1978).

In the water column, concentrations of oil ranged up to 340 µg/L, and in several stations were uniform down to 20 m. Concentrations, however, varied greatly. Thus at one station the surface sample concentration was below detectability (20 µg/L), while at three others the concentration of the 3-m sample was below detectability, although the surface samples were 90-170 µg/L. One sample, taken from 79 m, contained 170 µg/L of oil in water.

Winds, during the spill period, were offshore from Massachusetts, and as a result no oil from the Argo Merchant ever reached the shore-line, and no coastal impact was incurred.

Hydrocarbon contamination of the bottom sediments seems to have been restricted to an area immediately around the wreck, and apparently was short lived. Sediment analyses 1 and 2 months after the incident showed oiling, in the form of small tar particles or droplets, in a 10-15 km^2 area around the site.

Concentrations varied between 0.1 and 327 µg hydrocarbons/g dry weight sediment (Hoffman and Quinn, 1980). However, by July 1977 (7 months after the accident) only one station showed any trace of petroleum hydrocarbons, and that at a very low level.

Impact on Biota

Birds

Bird observations showed that for a 9-day period following the wreck, about 1,120 birds from 13 different species passed through the area, mostly (92%) gulls. Oiled birds were seen near the wreck, as well as in the vicinity of the slicks, and ashore on Cape Cod and Nantucket. Over half of the herring gulls and about 41% of the Great Black-Backed gulls were oiled. Oiled birds found on the beaches were mostly alcids (murres, razorbills). Although bird abundances and total mortalities are difficult to evaluate, it was concluded that the spill probably had little effect on the coastal and marine bird populations off the New England coast (Powers and Rumage, 1978).

Plankton

Most of the focus was on copepods and fish eggs and larvae from the impacted area. Copepods at most stations within 10 km of the wreck were found to have oil either on their mandibles or in their intestines. There was no clear correlation, however, between the occurrence of oil in the copepods and the presence of detectable oil concentrations in the water column.

On the other hand, a high correlation was found between the oiling of copepods and the occurrence of dead fish eggs (Longwell, 1978). Both cod and pollock eggs were found contaminated with oil adhering to their surfaces. Examination of fish eggs and embryos indicated that about 20% of the collected cod eggs and up to 46% of the pollock eggs sampled were either moribund or dead, in comparison to 4% mortality in laboratory-spawned material. Unfortunately the weather conditions prevented collecting of more samples to improve the statistical evaluation.

Sand launce larvae were low in numbers under the slick (Sherman and Busch, 1978), but their abundance was not correlated with other biological effects of oil. Fish gut analyses showed Argo Merchant type alkanes in only two cod and one flounder out of 37 sampled. The profile of aromatic hydrocarbons matched in only one case (MacLeod et al., 1978).

Benthos

Oil was found in an interstitial harpacticoid copepod and in a polychaete species, and on the appendages of a burrowing amphipod taken from samples near the wreck. A few other organisms were contaminated with tar, but there was no evidence to suggest whether it came from the Argo Merchant. It was concluded that generally there was no indication of significant impact of the Argo Merchant on the benthos (Pratt, 1978).

A NEAR-SHORE SPILL WITH ONSHORE WIND: THE AMOCO CADIZ

The Amoco Cadiz spill (Hess, 1978; Conan et al., 1978, 1981; Gundlach et al. 1983) is of interest for several reasons, aside from the obvious one that it is the largest single spill to date originating from a tanker accident. First, the spill occurred in an area of high water movement (the English Channel). However, because of the prevailing winds, the slicks did not drift off into the nearby Atlantic Ocean but remained in the spill area for up to 4 weeks following the accident. As a result there was an almost continual oiling of the adjacent coastlines. Second, the coastline affected includes a wide range of coastal systems, from high energy rocky coast to low energy soft sediment systems such as sandy pocket beaches, salt marshes, and estuarine tidal rivers. All these were at one time or another heavily oiled during the month following the breakup of the tanker. Finally, the area represents a significant economic resource, not only for

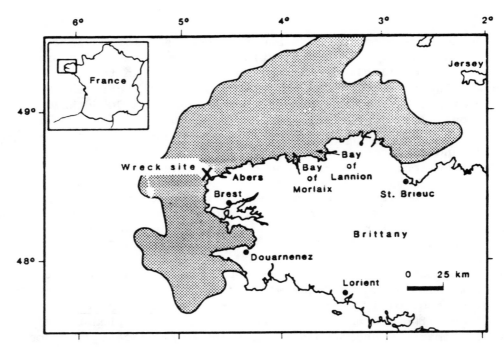

FIGURE A-7 Horizontal extent of oil movement from the _Amoco Cadiz_ spill as determined by chemical analyses on water samples and from visual observations by overflights and scientific cruises.

SOURCE: Gundlach et al. (1983).

Brittany in terms of the valuable oyster-culture industry in the two tidal rivers of the area, but also in terms of the valuable tourist industry, attracted to the long sandy beaches.

The tanker suffered steering failure 13 km north of the Ile d'Ouessant, in the English Channel west of the Brittany coast (e.g., Hess, 1978). Her cargo was 120,000 tons of light Iranian crude oil and 100,000 tons of light Arabian crude, as well as the remainder of her Bunker fuel. The tanker grounded at high tide on rocks north of Portsall, within sight of the shore. As the tide ebbed, the ship broke in two, releasing 50,000 tons of oil from 4 of her 12 cargo tanks during the following day. This oil was driven ashore by a strong northwest wind (O'Sullivan, 1978) (Figure A-7).

As the tanker continued to break up and release oil, the wind drove the oil further into the English Channel and further east along the coast, Almost 200 km from the wreck by March 25. Winds then shifted north and east, which drove the oil into formerly protected areas. In all, the spill resulted in oiling of up to ca. 300 km of Brittany coastline.

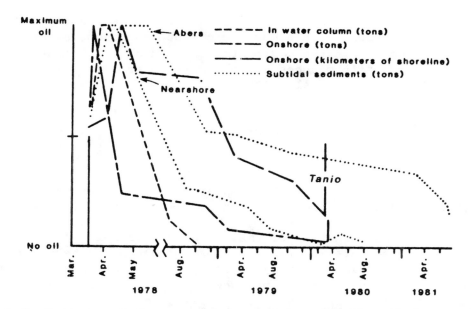

FIGURE A-8 Pattern of the persistence of Amoco Cadiz oil in coastal waters and sediments of Brittany, France, April 1978 to April 1981.

SOURCE: Gundlach et al. (1983).

Oil Fate

The oil released from the Amoco Cadiz was converted to a reddish-brown water-in-oil emulsion (mousse) by tide and wave-induced mixing with water entering the ruptured tanks. Emulsification was very rapid, and no authentic cargo oil sample could be obtained. Adjacent to the wreck, mousse contained 40-60% water, whereas mousse collected from the beach contained up to 75% water.

Concentrations of oil in the water column varied rather widely, but for several weeks immediately following the tanker breakup, much of the water column along the Brittany north coast was contaminated with oil (3-20 µg/L offshore, 2-200 µg/L nearshore, and 30-500 µg/L in estuaries) (e.g., Marchand and Caprais, 1979, 1981). Concentrations in the English Channel itself decreased to near-background levels during the following to 2-3 months. Concentrations in the estuaries remained elevated for some time thereafter.

Changes in the spilled oil have been extensively analyzed and documented (e.g., Gundlach et al., 1983). Evaporation of the more volatile components is thought to have carried from 20% to 40% of the spilled oil from the sea surface into the atmosphere (Figure A-8). The Amoco Cadiz apparently differed from other similar spills in that the amount of the cargo that eventually became entrapped or incorporated into the water column was greater than seen elsewhere, probably because of the higher wave energy and the vertical mixing of the water column typifying that part of the English Channel. This greater incorporation of the oil into the water column is also probably the reason for the unexpectedly high mortality of subtidal organisms.

A considerable portion of the oil that did come ashore, and that was not removed manually in a massive cleanup campaign, eventually became either buried in the sediments (Long et al., 1981) or entrapped in the low energy salt marshes and estuaries (see below) (also, Boehm et al., 1982; Gundlach et al., 1983).

Impact on Biota

Birds

Birds were migrating through the area at the time of the spill, and over 4,500 oiled birds were recovered (Hope-Jones et al., 1978), 3,200 dead on the beaches (Hess, 1978). Of the 33 species found dead, most were alcids and cormorants. Of these, the former belong to three species considered rare in France. Cause of death in these birds is not known with certainty. Analyses of specimens representing four species oiled during the incident showed Amoco Cadiz oil in only one, the common shag (Phalacrocorax aristotelis), indicating that the others may have died from physical effects of the oil on the birds (Lawler et al., 1981). Measurements of aryl hydrocarbon hydroxylase (AHH) levels in tissues from several oiled birds (three species) showed sixfold higher levels of AHH in birds collected from a bird-cleaning station near Brest in comparison with birds from the same species collected offshore from Nova Scotia (Vandermeulen, 1978).

Plankton

Decreases in biomass of phytoplankton, as measured by chlorophyll-a content, were observed for several weeks in the immediate vicinity of the wreck and in the highly contaminated tidal rivers, the Aber Benoit and Aber Wrach. In contrast, at a further distance from the wreck the phytoplankton production was elevated, perhaps stimulated as a result of either low levels of petroleum hydrocarbons in the water column or a result of nutrient release from oiled dead organisms (Aminot and Kerouel, 1978).

Zooplankton experienced high mortalities, but not until about 20 days after the wreck, The Aber Benoit river contained large amounts of zooplankton debris, coinciding with high oil levels, and the surviving copepod Temora longicornis showed depressed levels of digestive enzymes. Seventy days later the biomass showed no signs of recovery (Samain et al., 1979). The effect on the digestive enzymes in these pelagic zoo-plankton persisted into 1979 in the areas that continued to exhibit elevated levels of oil in the water (Marchand and Caprais, 1981; Samain et al., 1981).

Fish

There was some mortality of fish, generally within 10 km of the wreck, and consisting mainly of rockfish, gobies, and one gadid species.

Mortalities among commercially important species were insignificant (Hess, 1978).

Growth of flatfish, plaice, and sole in the oiled rivers was gradually reduced in the year of the spill (Conan and Friha, 1981). The effect was greatest among the younger sole and in the adult plaice. Thus young sole grew at only 30% of their normal rate. Also, up to 80% of the examined flatfish showed fin rot, up to 9 months after the spill. This decreased to about 10% over the following 11 months (Conan and Friha, 1981). One study, involving eels from a harbor that was very heavily oiled at the time of the spill, showed a variety of physiological and histopathological abnormalities in gills, ovaries, and kidneys. Although it is difficult to relate these abnormalities directly to the oiling from the Amoco Cadiz, 2-8 months earlier, such changes are characteristic of many vertebrates subjected to long term stress (Lopez et al., 1981).

Benthos

Oil reached the bottom over a large area, including both the bottom sediments of the tidal rivers and a large portion of the western English Channel. In places it penetrated to a depth of 7 cm, with highest concentrations found in the muddy sediments. In one area of the English Channel, ampeliscid amphipods, comprising ca. 40% of the bottom biomass in fine sand sediments, were totally eliminated. The absence of upstream "seed" populations lessens the chances of their replacement (Cabioch et al., 1980). Inshore there was massive mortality of some species such as heart urchins, razor clams, and the amphipod Bathyporeia (Hess, 1978). Other species, such as the clam Tellina teniu and the polychaete Owenia survived. However, T. fabula gradually disappeared in the 2 years after the wreck.

In the tidal rivers there was a strong correlation between the benthic populations and sediment-hydrocarbon concentrations. At less than 50 μg/g oil, there was a reduction in total numbers of animals, but with no evidence of a change in population structure. At higher concentrations of oil, 100-1,000 μg/g, one observed the appearance of certain opportunistic polychaete species such as Cirratulids and Spionids. At the highest levels of pollution, over 10,000 μg/g, only the Cirratulids and Capitellids were present (Glemarec and Hussenot, 1981).

This pattern of community response to the presence of oil in the benthic sediments is similar to that observed near sources of heavy industrial and urban pollution and has also been observed at other spills (see Sanders, 1978; Sanders et al., 1980). These various observations suggest that in some parts of the Brittany coastline the oil contamination in the sediments has approached the limits of biological response and recovery.

Amoco Cadiz oil has been found in the tissues of a broad range of benthic organisms, although both the concentrations and the rates of depuration, i.e., disappearance from the tissues, vary widely. Thus, although both Japanese and flat oysters showed contamination up to 3 months after the spill, the flat oysters were almost clean of oil 9 months later. The Japanese oysters, however, remained heavily contaminated at that time (Laseter et al., 1981).

Intertidal Communities

The most exposed sites, the rocky shores, were protected from oil by reflected waves. _Fucus_ sp. generally suffered little damage from the oil in either appearance or growth, except where they were damaged by cleanup operations (Topinka and Tucker, 1981). There was some local mortality, but on the whole these rockweeds fared well. On sheltered rocky shores _Ascophyllum_ was killed and replaced by _Fucus_ as long as there were _Fucus_ plants in the immediate vicinity to provide the sporelings. A variety of intertidal rocky shore animals were killed, especially limpets and periwinkles.

Sandy beaches retained oil as buried layers for several years after the spill, for the oil came ashore during the transition in beach profile from the erosional slope to the beach depositional period. Beaches were littered with dead animals immediately after the spill, including both intertidal and subtidal species.

Exposed intertidal mudflats had almost their entire fauna killed by oil from the water column (Chasse, 1978). In more sheltered areas the lugworm _Arenicola_ was commonly found alive after the spill (Gundlach et al., 1981).

Marshes and other low energy environments were severely affected where oil came ashore, with complete kill of higher plants and animals. There was no recovery in 2 years at the most heavily oiled sites (Gundlach et al., 1981). The edges of marshes and rivers were found to be especially retentive of oil, with concentrations 5-10 times higher than found in oiled mudflats (Vandermeulen, 1981). Part of the Ile Grande salt marsh was subjected to a massive cleanup effort involving both manual and mechanized labor to remove oiled debris, resulting in increased drainage and erosional drainage velocities through the marsh. Heavy traffic through the marsh primary drainage channel further weakened the channel's bottom sediment structure, enhancing the erosion of the fine sediments. The combined results were so large a change of the system that several decades plus active conservation measures may be required for return to the prespill conditions by natural depositional processes (Long and Vandermeulen, 1979, 1983).

An active recolonization program is under way in the Ile Grande marsh, involving transplanting of marsh vegetation from other nonoiled marshes (Levasseur and Jory, 1982) and from nurseries established in other coastal areas (Seneca and Broome, 1982). The approach appears to be successful, especially with applications of fertilizers, but full recovery is estimated to take several years, both because of slow growth and because of the sensitivity of some of the plants toward the oil persisting in much of the sediments.

AN UNDERWATER BLOWOUT: THE IXTOC I

The Ixtoc I (Atwood, 1980; Brooks et al., 1978) produced the largest man-made oil spill in history, but unlike a tanker accident, the blowout released an estimated half million tons of oil into an open ocean/continental shelf environment, over a period of 9 months. There had been no studies of tropical oil spills in any great detail before 1979, and the little work that had been done was directed toward the impacts on the intertidal zone. The little work done previously on the fate of oil in the open ocean had been related largely to the occurrence of pelagic tar. The long release period of Ixtoc I, however, allowed scientists to prepare for more detailed physical and chemical studies than have generally been possible in the case of the tanker spills. In the end, the blowout studies greatly augmented the understanding of the fate of oil on the open ocean. Less was learned, from this instance, about the impact of the spilled oil on living resources of the environment.

Oil Fate

The oil emerged in a plume, together with natural gas, from the top of the blowout preventer some 38 m beneath the surface and 14 m above the sea floor, in the Bay of Campeche in Mexican waters (Figure A-9). The turbulent flow mixed the oil and water to form an emulsion of about 70% water suspended in oil, the familiar "chocolate mousse." At the surface a gas fire burned continually, consuming some of the lighter-molecular-weight compounds (less than C_{10}). The upwelling gas formed a relatively stable burning core of fire (Figure A-10) with an approximate diameter of 25-30 m and occasional bursts up to 80-m diameter. Outside the burning area, the upwelling oil plume formed an oil slick of less than 1-mm thickness in a circular area of 300 m. Evaporation further reduced many of the low ends (National Oceanic and Atmospheric Administration, 1983). Within this area and up to 3,000 m downstream, the water contained a milky suspension of fine oil-in-water emulsion which rose to the surface during some of the scientific observations (Haegh and Rossemyr, 1980).

The combination of water circulation and wind resulted in a general movement of large patches of oil slick and mousse, of several to tens of meters across, toward the west-northwest along the coast of Mexico toward the coast of Texas during July and August 1977 (Figure A-9). Although physical oceanographic knowledge of the area was not as extensive as might be desirable for the best models of oil slick movement, the forecast estimates of general direction and rate of movement of the patches of mousse and slicks were reasonably accurate as a result of the use of combinations of satellite imagery, aircraft overflights, ship observations, and previous data on average currents and meterological conditions (Galt, 1981). However, in September 1979, several severe tropical storms traversed the Gulf of Mexico, and the resulting temporary changes in winds and surface currents resulted in the transport of oil slicks and mousse to the north and northeast and

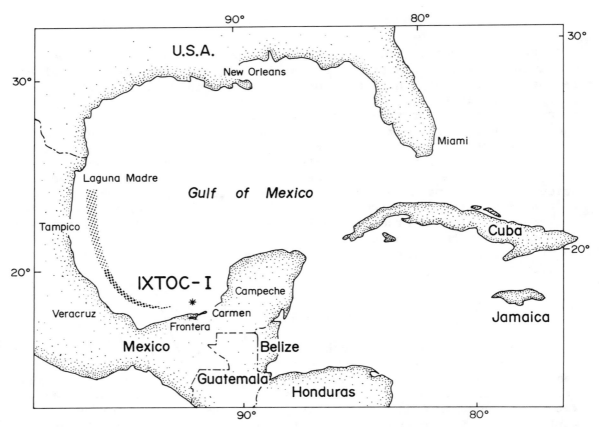

FIGURE A-9   Location of the Ixtoc I blowout and the initial path of the slick movements.

SOURCE: Adapted from Ross et al. (1980).

then to the south (Atwood et al., 1980). Then the usual seasonal flow conditions of late fall to early winter returned, with the oil moving south and southwest followed again in January with flow to the northwest (Galt, 1981).

The well was finally capped on 23 March, 1980. Estimates of the volume of spilled oil vary considerably, from 454,000 to over 1.4 million tons due to uncertainties of estimating flow from the fractured well over such a long period of time. Estimates of the amount burned vary from 30% (Ross et al., 1979) to as much as 50% (Haegh and Rossemyr, 1980). Less than 10% were recovered. Large amounts of dispersants were used on slicks approaching within 25 miles (30 km) of the Mexican coastline. None was used north of 20°N.

An intensive sampling and analysis effort near the well site in September 1979 (Atwood, 1980) showed that relatively high concentrations of gas, volatile liquid hydrocarbons, and high-molecular-weight compounds were transported subsurface away from the well site for distances up to 20-30 km. The oil slick on the surface changed appearance with increasing distance from the well. Concentration of

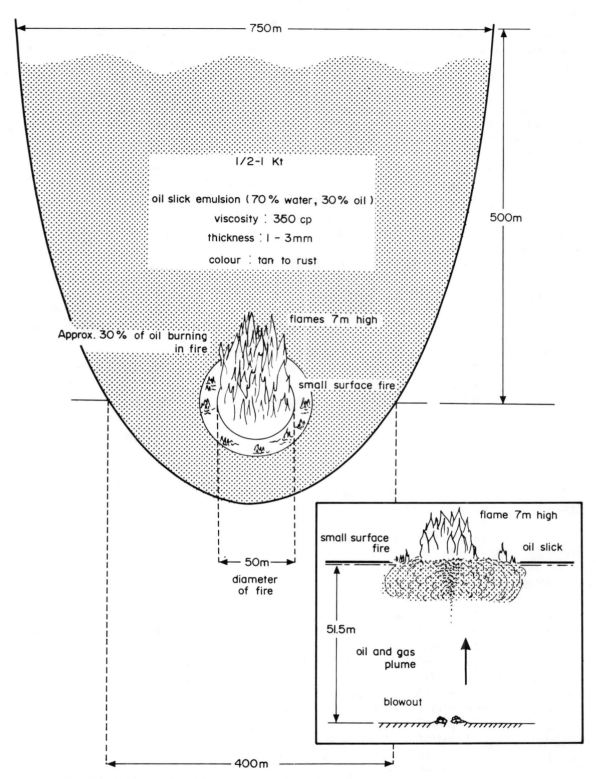

FIGURE A-10 Schematic diagram of the <u>Ixtoc</u> I blowout.

SOURCE: Ross et al. (1980).

oil slicks by Langmuir circulation cells and evaporative weathering resulted in the formation of windrows of mousse several meters wide and hundreds of meters long and globs of mousse a few centimeters in diameter. Patches of mousse up to 20-30 m in length and width and up to a meter thick were observed outside the main slick. The mechanism of formation for these large patches is not known, but microbial and photochemical processes may be involved in addition to surface currents and evaporation. Similar patches or rafts of mousse were observed off the coast of Texas in association with slicks, and some of these came ashore in Texas. Chemical analysis revealed that there were still significant amounts of substituted naphthalenes, phenanthrenes, and dibenzothiophenes remaining in samples of mousse off the coast of Texas (Patton et al., 1981). Thus, formation of mousse seems to provide a mechanism for delivery of toxic aromatic oil components to areas long distances from the site of a spill.

Analyses for hydrocarbons of sediments near the well site showed that no more than 0.5-3% of the spilled oil eventually became associated with the bottom sediments in the blowout area. However, there were considerable difficulties with quantitative sampling of the surface "floc" (fine bottom sediments) at the sediment-water interface (Boehm et al., 1982).

Between 6 August and 13 September 1979 approximately 3,000-4,000 metric tons of oil were deposited onshore in Texas, mainly on the beaches. On 13 September a tropical depression passed through the area and some of the oil was removed from the beaches to the near-shore zone (Gundlach and Finkelstein, 1982).

Impact on Biota

Most of the information on the effects of this blowout on biota comes from U.S. studies on the northern side of the Gulf. At the present time there is very little information about the effects nearer the spill site, in Mexican waters and coastal systems.

Microbes

Research on microbial degradation of the spilled oil near the well site was instructive. Hydrocarbon-utilizing bacteria populations were several orders of magnitude higher in samples associated with mousse compared to mousse-free water. Within 25 km of the well site, in the slick and mousse area, the normal microbial populations shifted to a community capable of degrading hydrocarbons, but nutrients were limiting in these Gulf of Mexico waters (Atlas et al., 1982; Pfaender et al., 1980). Thus, little microbial degradation of the oil was noted. This result is in agreement with chemical analyses data (Boehm et al., 1982). In summary, the major processes acting on the Ixtoc oil, removing it from the environment, appeared to be physical-chemical weathering.

Seabirds and Turtles

No data are available on the impact of the spilled oil on seabird communities and populations of the bay. With respect to sea turtle populations in the area, again no data are available. Because there were concerns that the Atlantic Ridley turtle hatchlings would be affected when they went to the sea after hatching in sandy beaches on Padre Island, some 9,000 hatchlings were retained ashore until late July in a preventive measure. No mortality was observed after the hatchlings left the beach, but this would be very difficult to verify. Whatever the impact, even if this year-class is low in numbers when it returns after 8 years, one can do little more than speculate about the possible link to the Ixtoc I blowout (Oil Spill Intelligence Report, 1980).

Plankton

No data are available on the biological effects in the water column.

Fish and Shrimp

Data on fish and shrimp are sketchy and not always credible. There are reports of changes in fish catch from Mexican fishermen, believed to be a result of fish avoidance of the traditional fishery areas due to the presence of the oil (Oil Spill Intelligence Report, 1980). Other than this, there are no published data on effects on fish or fish larvae.

Faint petroleum odors were detected in some shrimp samples, suggesting contamination (Woods and Hannah, 1981). Shrimp samples were taken for analysis of petroleum hydrocarbons, but the results are not yet available.

Benthos

Little, if any, Ixtoc oil could be detected in surface sediments on the South Texas outer continental shelf by grab sample techniques. However, sorbent pad samples detected Ixtoc oil associated with a mobile sedimentary material nepheloid in the so-called nepheloid layer, consisting of fine suspended particles in the water overlying the bottom (Boehm et al., 1982).

A total of 72 grab samples were taken in 1980 from the South Texas shelf area were sorted for benthic ecology studies, and compared to earlier data from 1976 and 1977. There is no doubt that there were major decreases in numbers of taxa and numbers of individuals at the 12 stations sampled (Lewbel et al., 1982). However, no quantitative cause and effect relationship with the Ixtoc oil spill can be established because of the gaps in sampling between 1977 and 1979 and the lack of sufficient data on life histories and normal cycles of abundance for benthic infauna of the area.

Intertidal and Subtidal Communities

The macroinfauna, along the Texas coast, dominated by polychaetes and haustoriid amphipods, showed decreases in population density but not a parallel decrease in numbers of species. The changes may have been due to the Ixtoc I oil, but hurricanes, seasonal changes, and cleanup techniques (dispersants) may also have been responsible (Thebeau et al., 1981). However, it should be noted that other studies in the Gulf region do not show any effect of storms on these organisms.

AN ONSHORE TROPICAL SPILL: THE ZOE COLOCOTRONI

The Zoe Colocotroni (Nadeau and Bergquist, 1977; Tosterson, 1977) spill represents one of the few tropical spills that has received any scientific attention. As a result, little is known about the potential impact of oil on tropical ecosystems, especially the common mangrove and coral reef communities, which constitute a large part of the tropical marine coastline.

The tanker Zoe Colocotroni ran aground off LaParguera, Puerto Rico, on 18 March 1973. In order to free the vessel, about 5,000 tons of crude oil were pumped overboard. An estimated 60% of this crude oil was subsequently swept into Bahia Sucia, off the extreme southwestern tip of Puerto Rico. There the oil impacted the sea grass beds, mangrove communities, and lagoons.

Oil Fate

Four years later (1977) much of the oil had disappeared, but some still remained on the west side of Bahia Sucia. Analyses showed this to be highly weathered (Page et al., 1979). Follow-up work, in 1978 and 1979, showed that Zoe Colocotroni oil, apparently unaltered, still persisted in some areas. Tarry residues were found in the bottom ooze of largely shallow salt lagoons. Droplets of tar were readily dislodged when the soft bottom sediments were disturbed. Once on the surface of the lagoon water these tar droplets slowly formed sheens of oil. Aside from such tarry deposits, however, it appears that most of the stranded Zoe Colocotroni oil had undergone extensive weathering. Analyses of oil sediments from all impacted environments showed extensive degradation of the lower-molecular-weight hydrocarbons, presumably because of high microbial activity. Leaching out of soluble alkanes and of the smaller aromatic components by tidal waters may also be a factor in these environments (Gilfillan, personal communication).

Impact on Biota

An initial assessment of the impact of the Zoe Colocotroni on biological communities of the area showed large numbers of dead sea cucumbers, conchs, prawns, sea urchins, and polychaete annelids washed

573

ashore (Nadeau and Bergquist, 1977). Dead and dying organisms were also found in offshore sea grass beds (Thalassia). The sea grass beds themselves also suffered from contact with the oil entrained into the water column by the action of the surf. Leaves turned brown and black, and a considerable amount of Thalassia died and was removed by wave action. The oil also had an acute effect on the mangrove communities, with the red mangrove most severely affected, together with the fauna living in the mangrove prop root environments.

Subsequent surveys of the impacted area have shown marked changes in the affected faunal and floral communities. By 1976 about 1 ha of red mangroves had become defoliated and eventually died, presumably through suffocation of the specialized aerial prop roots by oil. Apparently, however, degradation of spilled, stranded oil in tropical environments occurs at a greater rate than in more temperate climates, as suggested by chemical analyses of oil from the Bahia Sucia sediments. A similar rate of biological recovery appears to be occurring, except for the impacted red mangrove communities in which, at the time of the most recent survey (1979), the faunal composition was still marked by the presence of opportunistic polychaete species (Gilfillan et al., 1981). In general, changes in faunal composition appeared to be related to the degree of weathering of oil, which in turn is related to water movement over the sediments.

SUMMARY AND DISCUSSION

Clearly, these spills differed markedly in their extent and impact, much being dependent on meteorological conditions operating at the time of the spill. Thus, for example, the Arrow's spilled oil eventually washed out into the open Atlantic, but not before all of the Chedabucto Bay coastline had been heavily oiled under the influence of prevailing easterlies and southeasterlies. The coastline of Massachusetts escaped this fate 6 years later when the Argo Merchant ran aground on the Nantucket Shoal, largely because of offshore winds operating for most of the postspill period. On the other hand, Amoco Cadiz oil remained near the Brittany coast for several weeks under the action of the shifting northeasterly and northwesterly winds. The Ixtoc I blowout presented problems of a prolonged underwater oil spill source into a large coastal circulation system. The Zoe Colocotroni spill in general parallels the Florida spill in its impact on benthic communities and in long term oiling of soft lagoonal sediments. Its differences lie primarily in the higher temperatures found in the tropical environment, which appear to hasten the breakdown of spilled oil. On the other hand, the Zoe Colocotroni presented two new features of oil spills: oiling of sea grass beds and impact on mangrove communities.

Biological impacts varied as did the oil fates, from the long term problems still encountered in benthic communities at the Florida site to the virtual absence of observed effects on biota from the Ixtoc I, although the latter in large part reflects the absence of biological studies. Low energy coastal environments appear to be particularly vulnerable to the effects of oil and to oil entrapment, as was seen at the Florida, Arrow, Zoe Colocotroni, and Amoco Cadiz spills. They also demonstrate the persistence of both the oil and its effects, especially in association with soft sediments (see also Teal and Howarth, 1983), in some instances for over a decade (Arrow, Florida). However, as noted elsewhere in this report, oiled environments do clean themselves, and there appear to exist in all cases briefly examined here, mechanisms of oil degradation and biological recovery with the potential for eventual complete recovery.

If one were to rank the various factors that can influence the potential impact and persistence of oil in these various spills, then clearly, biological impact is linked closely to the extent and duration of oiling (e.g., Florida, Amoco Cadiz). In this respect, the low energy lagoonal environments appear to be most susceptible for long term impact. Another, generally unexplored, factor seems to be the rate of release of spilled oil, with the rate of oil spillage being related to spill impact. For example, the Amoco Cadiz spilled approximately 100,000 tons over 1-2 days, while the Ixtoc I blowout spilled about the same amount in about 2 months. The Santa Barbara Channel seep, on the other hand, released about that amount over the course of a century or more. While there are, of course, many differences between these three spills, one would expect qualitative differences in their impact, just because of the different release rates.

Together these five examples of tanker spills have introduced about 270,000 tons of crude, Bunker C, and No. 2 fuel oils into the world's oceans, plus at least twice that amount from the Ixtoc I blowout (estimated variously at between 454,000 and 1,400,000 tons). Remnants of this oil can still be found in five out of the six (the Argo Merchant's cargo disappeared from view totally). Remnants of their effects can also still be measured, in terms of numbers of biota and in depression of certain metabolic parameters, at the sites of the Florida, Arrow, Zoe Colocotroni, and the Amoco Cadiz, and perhaps at the Ixtoc I. Of these, the impact on seabird populations at the Amoco Cadiz site has undeniably been the most dramatic. A disastrous effect on the bird population had been feared in the case of the Amoco Cadiz, but it did not materialize, suggesting that the survival potential of seabirds, at least for the eastern North Atlantic populations, is quite good (see also Chapter 5, Impact on Seabird Populations section). In none of these examples has a widespread, immediate impact on fish populations been observed, nor generally on the pelagic plankton communities.

By far the most persistent impact is found instead in the intertidal and subtidal benthic communities, where long term perturbations can be found several years after the spill. Where the oil has become stranded on coastlines, such impact can extend to the ecological balance and stability of the coastlines as well as to economic resources, as in the case of the impacted oyster mariculture of northern Brittany (Amoco Cadiz, Maurin, 1981).

One major feature of oil spills, which has not received much attention and remains an enigma, is the fate of nonstranded oil. Of the volume of oil spilled by the tankers discussed here, about 165,000 tons (60%) did not come ashore but is largely unaccounted for, having

TABLE A-1 Estimates of the Distribution of Oil Spilled from the Tankers
Arrow, Zoe Colocotroni, and the Amoco Cadiz

Tanker	Distribution	Tons	Percentage
Arrow[a]	In tanker prior to casualty	15,000	
	Removed by pumping	5,432	
	Remaining in hull	168	
	Ashore in Chedabucto Bay	1,895	
	At the surface within Chedabucto Bay	21	
	At depth within Chedabucto Bay	?	
	Evaporated	?	
	Swept out into the Atlantic	8,421	
Zoe Colocotroni[b]	Pumped overboard	5,000	
	Ashore in Bahia Sucia	3,000	60
	At sea/evaporated	2,000	40
Amoco Cadiz[c]	Total spilled	223,000	
	Subtidal sediments	18,000	8
	Onshore	62,000	28
	Water column	30,000	13.5
	Biodegraded	10,000	4.5
	Evaporated	67,000	30
	Unaccounted for	46,000	20.5

[a]Anon (1970).
[b]Nadeau and Bergquist (1977).
[c]Gundlach et al. (1983).

either evaporated into the atmosphere or dispersed or dissolved into
the water column. In addition, most of the spilled oil from the Ixtoc
I blowout remains either at sea or in the atmosphere. Little can be
said about this significant portion of the spilled tonnage, for there
exist at present no data to enable us to assess either its fate or its
degradation rate in the open ocean.

The absence of a spill budget (mass balance) is probably the single
largest gap in the data to date. Estimates made at the time of the
Arrow spill were at best crude (Table A-1), but they have not been
refined significantly, and no estimates exist for the Florida, Argo
Merchant or Ixtoc I. For the Zoe Colocotroni, only the simplest
estimate exists for oil lost into the water column, 40% (Nadeau and
Bergquist, 1977). The best attempt at an oil budget available is
probably that calculated for the Amoco Cadiz (Table A-1). However,
even these figures, based in part on chemical analyses and in part on
theoretical extrapolations, remain estimates at best.

As a large portion of the oil spilled to date, whether from tankers or from other discharges, appears never to have reached shorelines (where amounts and effects can be assessed to some extent), the fate, and ultimate impact of the oil presents a large set of questions to be answered.

REFERENCES

Aminot, A., and R. Kerouel. 1978. Premiers resultats sur l-hydrologie, l'oxygene dissous et les pigments photosynthetiques en Manch Occidentale apres l'echouage de l'Amoco Cadiz, pp. 51-68. In Conan et al., eds. Amoco Cadiz: Consequences d'une pollution accidentelle par les hydrocarbures. CNEXO, Paris.

Anon. 1970. Report of the Task Force--Operation Oil (Cleanup of the Arrow oil spill in Chedabucto Bay). Vols. I, II, III, IV. Canadian Minister of Transport.

Atlas, R.M., G.E. Roubal, A. Bronner, and T.R. Haines. 1982. Biodegradation of hydrocarbons in mousse from the Ixtoc I well blowout. American Chemical Society, Washington, D.C. (in press).

Atwood, D.K., convener. 1980. Proceedings, Symposium on Preliminary Scientific Results From the Researcher/Pierce Cruise to the Ixtoc I Blowout. U.S. Department of Commerce, National Oceanic and Atmospheric Administration, Boulder, Colo. 591 pp.

Atwood, D.K., J.A. Benjamin, and J.W. Farrington. 1980. The mission of the September 1979 Researcher/Pierce Ixtoc I cruise and the physical situation encountered. In Proceedings of a Symposium on Preliminary Scientific Results From the Researcher/Pierce Ixtoc I Cruise. U.S. Department of Commerce, National Oceanic and Atmospheric Administration, Office of Marine Pollution Assessment, Washington, D.C.

Boehm, P., A.D. Wait, D.L. Fiest, and D. Pilson. 1982. Chemical assessment-hydrocarbon analyses, Section 2. Ixtoc Oil Spill Assessment Final Report. Contract AA851-CTO-71. Bureau of Land Management, U.S. Department of Interior, Washington, D.C.

Brooks, J.M., B.B. Bernard, T.C. Saner, Jr., and H.A. Reheim. 1978. Environmental aspects of a well blowout in the Gulf of Mexico. Environ. Sci. Technol. 12:695-703.

Burns, K.A. 1976b. Microsomal mixed function oxidases in an estuarine fish, Fundulus heteroclitus, and their induction as a result of environmental contamination. Comp. Biochem. Physiol. 53B:443-446.

Burns, K.A., and J.M. Teal. 1979. The West Falmouth oil spill; hydrocarbons in the saltmarsh ecosystem. Estuarine Coastal Mar. Sci. 8:349-360.

Cabioch, L., J.C. Dauvin, J. Mora Bermudez, and C. Rodriguez Babio. 1980. Effets de la maree noire de l'Amoco Cadiz sur le benthos sublittoral du nord de la Bretagne. Helgolander Meeresunters. 33:192-208.

Chasse, C. 1978. The ecological impact on and near shores by the Amoco Cadiz oil spill. Mar. Pollut. Bull. 9:298-301.

Conan, G., and M. Friha. 1981. Effets des pollutions par les hydrocarbures du petrolier Amoco Cadiz sur la croissance des soles et des plies dans l'estuaire de l'Aber Benoit, pp.749-773. In G. Conan et al., eds. Amoco Cadiz: Consequences d'une pollution accidentelle par les hydrocarbures. CNEXO, Paris.

Conan, G., L. d'Ozouville, and M. Marchand, eds. 1978. Amoco Cadiz: Preliminary Observations of the Oilspill Impact on the Marine Environment. Actes de Colloques 6. CNEXO, Paris.

Conan, G., L. Laubier, M. Marchand, and L. d'Ozouville, eds. 1981. Amoco Cadiz: consequences d'une pollution accidentelle par les hydrocarbures. CNEXO, Paris. 881 pp.

Conover, R.J. 1971. Some relations between zooplankton and Bunker C oil in Chedabucto Bay following the wreck of the tanker Arrow. J. Fish. Res. Board Can. 28:1327-1330.

Cornillon, P. 1978. Oil droplet measurements made in the wake of the Argo Merchant, pp. 43-47. In M.P. Wilson, J.G. Quinn, and K. Sherman, eds. In the Wake of the Argo Merchant. Center Ocean Management Studies, University of Rhode Island, Kingston.

Galt, J.A. 1981. Transport, distribution, and physical characteristics of the oil. Part I. Offshore movement and distribution, pp. 13-39. In C. Hooper, ed. The Ixtoc I Oil Spill: The Federal Scientific Response. U.S. Department of Commerce, National Oceanic and Atmospheric Administration, Office of Marine Pollution Assessment, Washington, D.C.

Gilfillan, E.S., and J.H. Vandermeulen. 1978. Alterations in growth and physiology of soft shell clams, Mya arenaria, chronically oiled with Bunker C from Chedabucto Bay, Nova Scotia, 1970-76. J. Fish. Res. Board Can. 35:630-636.

Gilfillan, E.S., D.S. Page, R.P. Gerber, S. Hanson, J. Cooley, and J. Hothan. 1981. Fate of the Zoe Colocotroni oil spill and its effects on infaunal communities associated with mangroves, pp. 353-360. In Proceedings, 1981 Oil Spill Conference. API Publication 4334. American Petroleum Institute, Washington, D.C.

Glemarec, M., and E. Hussenot. 1981. Definition d'une succession ecologique en milieu meuble anormalement enrichi en matières organiques à la suite de la catastrophe de l'Amoco Cadiz, pp. 499-512. In Amoco Cadiz: Fate and Effects of the Oil Spill. CNEXO, Paris.

Gordon, D.C., and P. Michalik. 1971. Concentration of Bunker C fuel oils in waters of Chedabucto Bay, April 1971. J. Fish. Res. Board Can. 28:1912-1914.

Grose, P.L., and J.S. Mattson. 1977. The Argo Merchant oil spill: a preliminary scientific report. NOAA special Report. NTIS PB-267-505. National Technical Information Service, Springfield, Va. 133 pp.

Gundlach, E.R., and K.J. Finkelstein. 1982. Transport, distribution and physical characteristics of the oil. II. Nearshore movement and distribution, pp. 41-73. In C. Hooper, ed. The Ixtoc I Oil Spill: The Federal Scientific Response. U.S. Department of Commerce, National Oceanic and Atmospheric Administration, Office of Marine Pollution Assessment, Washington, D.C.

Gundlach, E.R., S. Berne, L. d'Ozouville, and J.A. Topinka. 1981. Shoreline oil two years after Amoco Cadiz: new complications from Tanio, pp. 525-540. In Proceedings, 1981 Oil Spill Conference. API Publication 4334. American Petroleum Institute, Washington, D.C.

Gundlach, E.R., D.D. Domeracki, and L.C. Thebeau. 1983. Persistence of Metula oil in the Strait of Magellan six and one-half years after the incident. Oil Petrochem. Pollut. (in press).

Haegh, T., and L.I. Rossemyr. 1980. A comparison of weathering processes of oil from the Bravo and the Ixtoc blowouts, pp. 237-244. In Proceedings, 12th Annual Offshore Technology Conference, Houston, Tex. Paper OTC 3702.

Hampson, G.R., and H.L. Sanders. 1969. Local oil spill. Oceanus 15:8-11.

Hess, W.N., ed. 1978. The Amoco Cadiz oil spill; A preliminary scientific report. NOAA/EPA Special Report. NTIS PB-285-805. National Technical Information Service, Springfield, Va. 355 pp.

Hoffman, E., and J.G. Quinn. 1980. The Argo Merchant oil spill and the sediments of Nantucket Shoals: research, litigation and legislation, pp. 185-218. In R.A. Baker, ed. Contaminants and Sediments, Vol. I. Ann Arbor Science Publishers, Ann Arbor, Mich.

Keizer, P.D., T.P. Ahern, J. Dale, and J.H. Vandermeulen. 1978. Residues of Bunker C oil in Chedabucto Bay, Nova Scotia, 6 years after the Arrow Spill. J. Fish Res. Board Can. 35:528-535.

Krebs, C.T., and Burns, K.A. 1977. Long term effects of an oil spill on populations of the salt marsh crab Uca pugnax. Science 197(4302):484-487.

Laseter, J.L., G.C. Lawler, E.B. Overton, J.R. Patel, J.P. Holmes, M.I. Shields, and M. Maberry. 1981. Characterization of aliphatic and aromatic hydrocarbons in flat and Japanese type oysters and adjacent sediments collected from l'Aber Wrac'h following the Amoco Cadiz oil spill, pp. 633-644. G. Conran et al., eds. In Amoco Cadiz: consequences d'une pollutin accidentelle par les hydrocarbures. CNEXO, Paris.

Lawler, G.C., J.P. Holmes, D.M. Adamkiewica, M.I. Shields, J.-Y. Monnat, and J.L. Laseter. 1981. Characterization of petroleum hydrocarbons in tissues of birds killed in the Amoco Cadiz oil spill. In Conan et al., eds. Amoco Cadiz: Consequences d'une pollution accidentelle par les hydrocarbures. CNEXO, Paris.

Levasseur, J.E., and M.-L. Jory. 1982. Retablissement naturel d'une vegetation de marais maritimes alteree par les hydrocarbures de l'Amoco Cadiz: modalites et tendances. In Ecological Study of the Amoco Cadiz Oil Spill, pp. 329-362. NOAA-CNEXO Joint Scientific Communication. CNEXO, Paris.

Levy, E.M. 1971. The presence of petroleum residue off the East Coast of Nova Scota, in the Gulf of St. Lawrence and the St. Lawrence River. Water Res. 5:723-733.

Lewbel, G.S., R.L. Howard, and S.W. Anderson. 1982. Biological assessment, Section 4. In Ixtoc Oil Spill Assessment Final Report. Contract AA851-CTO-71. Bureau of Land Management, U.S. Department of the Interior, Washington, D.C.

Lissauer, I., and P. Welsh. 1978. Can oil spill movement be predicted?, pp. 22-27. In M.P. Wilson, J.G. Quinn, and K. Sherman, eds. In the

Wake of the <u>Argo</u> <u>Merchant</u>. Center for Ocean Management Studies, University of Rhode Island, Kingston.

Long, B.F.N., and J.H. Vandermeulen. 1979. Impact of cleanup efforts on an oiled saltmarsh in North Brittany. Spill Technol. Newslett. 4(4):218-229.

Long, B.F.N., and J.H. Vandermeulen. 1983. Geomorphological impact of clean-up of an oiled salt-marsh (Ile Grande). In Proceedings, 1983 Oil Spill Conference. API Publication. American Petroleum Institute, Washington, D.C. In press.

Long, B.F.N., J.H. Vandermeulen, and T.P. Ahern. 1981. The evolution of stranded oil within sandy beaches. In Proceedings, 1983 Oil Spill Conference. American Petroleum Institute, Washington, D.C. Pp. 519-524.

Longwell, A.C. 1978. Field and laboratory measurements of stress responses at the chromosome and cell levels in planktonic fish eggs and the oil problem, pp. 116-125. In M.P. Wilson, J.G. Quinn, and K. Sherman, eds. In the Wake of the <u>Argo</u> <u>Merchant</u>. Center for Ocean Management, University of Rhode Island, Kingston.

Lopez, E., J. Leloup-Hatey, A. Hardy, F. Lallier, E. Martelly, P.D. Oudot, and Y.A. Fontaine. 1981. Modifications histopathologiques et stress chez des anguilles soumises a une exposition prolongée aux hydrocarbures, pp. 645-653. In Conan et al., eds. <u>Amoco</u> <u>Cadiz</u>: Consequences d'une pollution accidentelle par les hydrocarbures. CNEXO, Paris.

MacDonald, B.A., and M.L.H. Thomas. 1982. Growth reduction in the soft-shell clam <u>Mya</u> <u>arenaria</u> from a heavily oiled lagoon in Chedabucto Bay, Nova Scotia. Mar. Environ. Res. 6:145-156.

Marchand, M., and M.-P. Caprais. 1979. Suivi chimique de la pollution de l'<u>Amoco</u> <u>Cadiz</u> dans l'eau de mer et les sediments marins en Manche occidentale, Mars 1978-Mars 1979. Rapport Interne Centre Oceanologique de Bretagne. CNEXO, Paris. 103 pp.

Marchand, M., and M.-P. Caprais. 1981. Suivi de la pollution de l'<u>Amoco</u> <u>Cadiz</u> dans l'eau de mer et les sediments marins, pp. 23-54. In <u>Amoco</u> <u>Cadiz</u>: Fates and Effects of the Oil Spill. CNEXO, Paris.

National Oceanic and Atmospheric Administration. Proceedings of a Symposium on Preliminary Results From the September 1979 <u>Researcher/Pierce</u> <u>Ixtoc</u> I Cruise. U.S. Department of Commerce, Office of Marine Pollution Assessment, Washington, D.C.

National Oceanic and Atmospheric Administration. 1983. Assessing the Social Cost of Oil Spills: The <u>Amoco</u> <u>Cadiz</u> Case Study. Office of Ocean Resources Coordination and Assessment, U.S. Department of Commerce, Washington, D.C. 144 pp.

Oil Spill Intelligence Report. 1980. Special Report: <u>Ixtoc</u> I. Vol. III, No. 1. Oil Spill Intelligence Report. 36 pp.

O'Sullivan, A.J. 1978. The <u>Amoco</u> <u>Cadiz</u> oil spill. Mar. Pollut. Bull. 9(5):123-128.

Patton, J.S., M.W. Rigler, P.D. Boehm, and D.L. Fiest. 1981. <u>Ixtoc</u> I oil spill: flaking of surface mousse in the Gulf of Mexico. Nature 290:235-238.

Pfaender, F.K., E.N. Buckley, and R. Ferguson. 1980. Response of the pelagic microbial community to oil from the <u>Ixtoc</u> I blowout. I. In situ studies, pp. 545-562. In D.K. Atwood, convenor. Proceedings, Conference on the Preliminary Scientific Results From the Researcher/Pierce Cruise to the Ixtoc I Blowout. U.S. Department of Commerce, National Oceanic and Atmospheric Administration, Boulder, Colo.

Powers, K.D., and W.T. Rumage. 1978. Effect of the Argo Merchant oilspill on bird populations off the New England coast, 15 December 1976-January 1977, pp. 142-148. In M.P. Wilson, J.G. Quinn, and K. Sherman, eds. In the Wake of the Argo Merchant. Center for Ocean Management Studies, University of Rhode Island, Kingston.

Pratt, S.D. 1978. Interactions between petroleum and benthic fauna at the Argo Merchant spill site. In M.P. Wilson, J.G. Quinn, and K. Sherman, eds. In the Wake of the Argo Merchant. Center for Ocean Management Studies, University of Rhode Island, Kingston.

Ross, S.L., C.W. Ross, F. Lepine, and R.K. Langtry. 1979. Ixtoc I oil blowout. Spill Technol. Newslett. (Environ. Can.) July-August 1979:245-256.

Ross, S.L., C.W. Ross, F. Lepine, and R.K. Langtry. 1980. Ixtoc I oil blowout, pp. 25-40. In D.K. Atwood, convenor. Proceedings, Conference on the Preliminary Scientific Results From the Researcher/Pierce Cruise to the Ixtoc I Blowout. U.S. Department of Commerce, National Oceanic and Atmospheric Administration, Boulder, Colo.

Samain, J.F., J. Moal, J.Y. Daniel, and J. Boucher. 1979. Ecophysiological effects of oil spills from Amoco Cadiz on pelagic communities--preliminary results. In Proceedings, 1979 Oil Spill Conference. API Publication 4308. American Petroleum Institute, Washington, D.C.

Samain, J.-F., J. Moal, J.-R. Le Coz, J.-Y. Daniel, and A. Coum. 1981. Impact de l'Amoco Cadiz sur l-ecophysiologie du zooplancton: une nouvelle possibilite de surveillance ecologique. In Conan et al., et al., eds. Amoco Cadiz: Consequences d'une pollution accidentelle par les hydrocarbures. CNEXO, Paris.

Sanders, H.L. 1978. Florida oil spill impact on the Buzzards Bay benthic fauna: West Falmouth. J. Fish. Res. Board Can. 35(5):717-730.

Sanders, H.L., J.F. Grassle, and G.R. Hampson. 1972. The West Falmouth oil spill. I. Biology. Technical Report WHOI-72-20. Woods Hole Oceanographic Institute, Woods Hole, Mass. 23 pp.

Sanders, H.L., J.F. Grassle, G.R. Hansson, L.S. Morse, S. Price-Gartner, and C.C. Jones. 1980. Anatomy of an oil spill: long-term effects from the grounding of the barge Florida off West Falmouth, Massachusetts. J. Mar. Res. 38:265-380.

Seneca, E.D., and S.W. Broom. 1982. Restoration of marsh vegetation impacted by the Amoco Cadiz oil spill and subsequent cleanup operations at Ile Grande, France, pp. 363-420. In Ecological Study of the Amoco Cadiz Oil Spill. NOAA-CNEXO Joint Scientific Commission.

Sherman, K., and D. Busch. 1978. The Argo Merchant oil spill and the fisheries, pp. 149-165. In M.P. Wilson, J.G. Quinn, and K. Sherman, eds. In the Wake of the Argo Merchant. Center for Ocean Management Studies, University of Rhode Island, Kingston.

Souza, G. 1970. Report of the Shellfish Warden, pp. 161-165. In Annual Report of the Finances of the Town of Falmouth for the year ending December 31, 1970.

Spaulding, M.L. 1978. Surface and subsurface spill trajectory forecasting: application to the Argo Merchant, pp. 37-42. In M.P. Wilson, J.G. Quinn, and K. Sherman, eds. In the Wake of the Argo Merchant. Center for Ocean Management Studies, University of Rhode Island, Kingston.

Stegeman, J.J. 1978. Influence of environmental contamination on cytochrome P-450 mixed-function oxygenases in fish: implications for recovery in the Wild Harbor marsh. J. Fish. Res. Board Canada 35:668-674.

Teal, J.M., and R.W. Howarth. 1983. Oil spill studies: a review of ecological effects. Environ. Manage. In press.

Teal, J.M., K. Burns, and J. Farrington. 1978. Analyses of aromatic hydrocarbons in intertidal sediments resulting from two spills of No. 2 fuel oil in Buzzards Bay, Massachusetts. J. Fish Res. Board Can. 35(5):510-520.

Thebeau, L.C., J.W. Tunnell, Jr., Q.R. Dokken, and M.E. Kindinger. 1981. Effects of the Ixtoc I oil spill on the intertidal and subtidal infaunal populations along lower Texas coast barrier island beaches, pp. 467-475. In Proceedings, 1981 Oil Spill Conference. API Publication 4334. American Petroleum Institute, Washington, D.C.

Thomas, M.L.H. 1973. Effects of Bunker C oil on intertidal and lagoonal biota in Chedabucto Bay, Nova Scotia. J. Fish. Res. Board Can. 30(1):83-90.

Thomas, M.L.H. 1977. Long term biological effects of Bunker C oil in the intertidal zone, pp. 238-245. In D.A. Wolfe, ed. Fate and Effects of Petroleum Hydrocarbons in Marine Organisms and Ecosystems. Pergamon, New York.

Thomas, M.L.H. 1978. Comparison of oiled and unoiled intertidal communities in Chedabucto Bay, Nova Scotia. J. Fish Res. Board Can. 35:707-716.

Topinka, J.A., and L.R. Tucker. 1981. Long-term oil contamination of fucoid marcoalgae following the Amoco Cadiz oil spill, pp. 393-404. In Conan et al., eds. Amoco Cadiz: Consequences d'une pollution accidentelle par les hydrocarbures. CNEXO, Paris.

Tosterson, T.R., et al. 1977. Bahai Sucia: a re-evaluation of the biota affected by petrochemical contamination in March, 1973. University of Puerto Rico, Department of Marine Science, C.A.A.M. 138 pp. + appendices.

Vandermeulen, J.H. 1977. The Chedabucto Bay spill--Arrow, 1970. Oceanus 20(4):31-39.

Vandermeulen, J.H. 1981. Contamination des organismes marins par les hydrocarbures. Note de syntheses. In Amoco Cadiz: Fates and Effects of the Oil Spill, pp. 563-572. CNEXO, Paris.

Vandermeulen, J.H., and D.C. Gordon, Jr. 1976. Re-entry of five year old standard Bunker C fuel oil from a low-energy beach into the water, sediments, and biota of Chedabucto Bay, Nova Scotia. J. Fish Res. Board Can. 33(9):2002-2010.

Woods, E.G., and R.P. Hannah. 1981. Ixtoc I oil spill--The damage assessment program and ecological impact, pp. 439-443. In Proceedings, 1981 Oil Spill Conference. Publication 4334. American Petroleum Institute, Washington, D.C.

APPENDIX B
Abbreviated Terms

AOSS	Airborne Oil Surveillance System (USCG)
API	American Petroleum Institute
ASTM	American Society of Testing Materials
ATP	adenosine triphosphate
BaP	Benzo(a)pyrene
bbl	barrels
bbl/d	barrels per day
BLM	Bureau of Land Management
^{14}C	radioactive carbon
CBT	clean ballast tanks
COW	crude oil washing
DCMU	3-(3,4-dichlorophyll) 1,1-dimethyl urea
DNA	deoxyribonucleic acid
EPA	Environmental Protection Agency
FAME	fatty acid methyl esters
GC	gas chromatography
GC/MS	gas chromatography/mass spectrometry
g/d, g/cap/d	grams per day, grams per capita per day
g, kg, mg, µg, ng	gram, kilogram, milligram, microgram, nanogram
gt	gross tonnage
ha	hectare ($10,000$ m^2)
HMWH	high-molecular-weight hydrocarbons
HP	horse power
HPLC	high pressure liquid chromatography
ICES	International Council for the Exploration of the Seas
IFP	French Institute of Petroleum (Institut Francais du Petrol)
IGOSS	Integrated Global Ocean Station Systems
IMO	International Maritime Organization
INTERTANKO	International Association of Independent Tanker Owners
IOC/WMO	International Oceanographic Commission/World Meteorological Organization
IR	infrared

ITOPF	International Tanker Owners Pollution Federation Ltd.
JASIN	Joint Air/Sea Interaction project
L, mL, μL, nL	liter, milliliter, microliter, nanoliter
LC	liquid chromatography
LC_{50}	median lethal concentration
LMWH	low-molecular-weight hydrocarbons
LOT	load-on-top
m, km, mm, μm	meter, kilometer, millimeter, micrometer (micron)
MARPOL 73/78	International Convention for the Prevention of Pollution from Ships, 1973 as modified by the protocol of 1978 relating thereto
MERL	Marine Ecosystems Research Laboratory (University of Rhode Island)
MESA	Marine Ecosystems Analysis project
MFO	microsomal cytochrome P450 mixed-function oxidase
mg/L	milligrams per liter
MO	monooxygenases
MPN	most probable number
mt	million metric* tons
mta	million metric* tons per annum
NAS	National Academy of Sciences
NASA	National Aeronautics and Space Administration
NBS	National Bureau of Standards
NOAA	National Oceanic and Atmospheric Administration
NPC	National Petroleum Council
NSO	nitrogen, sulfur, and oxygen components
OCS	outer continental shelf
OILPOL 54/69	International Convention for the Prevention of Pollution of the Sea by Oil 1954, as amended in 1969
O:N ratio	the ratio of oxygen consumed to nitrogen excreted
PAH	polynuclear aromatic hydrocarbons
PDB	Peedee Belemnite, geological formation from which rock reference standard for C stable isotope was obtained
PHC	petroleum hydrocarbons
PTM	petroleum transforming microorganisms
RNA	ribonucleic acid
SBT	segregated ballast tanks
SCCWRP	Southern California Coastal Water Research Project
SCEP	Study of Critical Environmental Problems
SGO	silica-gel-oil medium
SLAR	Side-Looking Airborne Radar
SRM	standard reference mixture
TC	total cargo carrying capacity
THC	total hydrocarbons
UCM	unresolved complex mixture
USCG	United States Coast Guard
USGS	United States Geological Survey

UV, UV/F	ultraviolet, ultraviolet fluorescence
VLCC	very large crude carriers
VLH	volatile liquid hydrocarbons
WHOI	Woods Hole Oceanographic Institution
WSF	water-soluble fraction

*Metric understood when abbreviated.

Public Meeting Participants, November 13, 1980, Washington, D.C.

John R. Botzum, Ocean Science News
Eric B. Cowell, American Petroleum Institute
*Rita R. Colwell, University of Maryland
*John W. Farrington, Woods Hole Oceanographic Institution
Jack R. Gould, American Petroleum Institute
James C. Greene, National Resources and the Environment Subcommittee
Gerald Haff, Outboard Marine
Thomas Hruby, Massachusetts Audubon Society
*C. Bruce Koons, Exxon Production Research Company
Albert H. Lasday, Texaco, Inc.
Debbie Lipman, Natural Resources Defense Council
Clayton D. McAuliffe, Chevron Oil Field Research Company
Norman Mead, Ocean Resources Coordination Assessment Committee
Elizabeth Mullally, American Institute of Shipping
Harold Muth, The American Waterways Operators, Inc.
Alvin M. Natkin, Exxon Corporation
*Gordon A. Riley, Bedford Institute of Oceanography, Canada (Retired)
*William M. Sackett, University of South Florida
Donald H. Seaman, U.S. Department of Commerce
Robert Smith, U.S. Coast Guard
James Thornton, Natural Resources Defense Council
*John H. Vandermeulen, Bedford Institute of Oceanography, Canada
**Richard C. Vetter, National Research Council
**Jan Vorhees, University of South Florida
Susan Walton, Bio Science

---

*Steering committee members.
**Staff.

Index